Inspection Authorization Test Prep

Eighth Edition

INSPECTION IA AUTHORIZATION

A comprehensive study tool to prepare for the FAA Inspection Authorization Knowledge Exam

Based on the original text by **Dale Crane** Edited by **Terry Michmerhuizen**

READER TIP:
The FAA Knowledge Exam can change throughout the year. Stay current with test changes; sign up for ASA's free email update service at **asa2fly.com/testupdate**

AVIATION SUPPLIES & ACADEMICS, INC.
NEWCASTLE, WASHINGTON

Inspection Authorization Test Prep
Eighth Edition
By Dale Crane
Edited by Terry Michmerhuizen

Aviation Supplies & Academics, Inc.
7005 132nd Place SE
Newcastle, Washington 98059
asa@asa2fly.com | 425.235.1500 | asa2fly.com

Eighth Edition published 2018.

ASA has online resources for this IA Test Prep: go to the ASA Product Updates pages to check for the latest question updates: **asa2fly.com/testupdate**.

See also the Reader Resources section at **asa2fly.com/reader/ia** for additional resources and information applicable to the IA certificate (see explanation on Page vii).

ASA-IA-8
ISBN 978-1-61954-823-7

Printed in the United States of America

2026 2025 2024 2023 9 8 7 6 5 4 3

Library of Congress Cataloging-in-Publication Data:

Crane, Dale.
IA test prep : test preparation and study references for the FAA Inspection Authorization Knowledge Test / by Dale Crane.
p. cm. — (A fast-track series guide)
Includes bibliographical references.
1. Airplanes—Inspection—Examinations—Study guides.
2. Airplanes—Maintenance and repair—Examinations—Study guides.
3. United States. Federal Aviation Administration—Examinations—Study guides. I. Title. II. Series.
TL671.7.C73 1997
629.134'6'076—dc21 96-45676
CIP

[03]

Stay informed with ASA online resources.

Website

asa2fly.com

Updates

asa2fly.com/testupdate

Follow, like, share.

- instagram.com/asa2fly
- facebook.com/asa2fly
- youtube.com/asa2fly
- linkedin.com/company/asa2fly

Contents

Continued

Chapter 7 **Type Certificate Data Sheets, Aircraft Specifications and Listings**

Appendix

Preface

How to Use This IA Test Prep

This IA Test Prep has been prepared to provide you with the information you will need to pass the IA Knowledge Test and help you become familiar with the privileges and limitations of this, the highest level of maintenance airman certification.

The knowledge test for IA is different from other FAA certification tests in that you are furnished with a more extensive supplement with which to take the test—the latest revision of the *Computer Testing Supplement for Inspection Authorization* (CT-8080-8). This lengthy supplement contains excerpts from the Federal Regulations, Advisory Circulars, Type Certificate Data Sheets, charts and figures from AC 43.13-1B and AC 43.13-2A, and examples of FAA forms. However, there are questions on the IA Knowledge Test regarding the core knowledge the FAA expects of an airframe and powerplant mechanic that are not covered by the reference material included in the latest revision of CT-8080-8. As it is explained in the FAA's *IA Knowledge Test Guide* (FAA-G-8082-11):

"The inspection authorization knowledge test has been considered by some as an open book test because of the use of reference material during the test. To view the test in this manner is a misconception. There has always been a core knowledge requirement for which no reference material was provided. Therefore, it should be noted that, during the tests, there are subject areas for which reference material is not included in the test supplement. These areas will draw on skills acquired as an airframe and powerplant mechanic and which are necessary to properly inspect work performed by others."

Therefore the IA Knowledge Test also differs from the other FAA tests in that it remains a "closed test," which means the exact database of questions is not available to the public. The sample questions included in this book have been derived based on history and experience with the IA testing process, and the Learning Statement Codes (LSC) from both the latest revision of CT-8080-8 and the airframe and powerplant mechanics LSC listing. For this reason, it is recommended that in addition to studying this *Inspection Authorization Test Prep*, you also study the *General Test Guide* (ASA-AMG), the *Airframe Test Guide* (ASA-AMA), and the *Powerplant Test Guide* (ASA-AMP).

The Reader Resources section at **asa2fly.com/reader/ia** provides additional helpful resources, such as links to copies of pertinent FAA Advisory Circulars, and most importantly, a PDF of the most recent test supplement (the FAA-CT-8080-8). To become familiar with the contents of this FAA test supplement, review this downloadable PDF. If you know ahead of time how the supplement is organized and how to access it while answering questions, this will help you prepare to take the actual exam.

The Federal Regulations that should be studied for the IA knowledge test have been reprinted by ASA and are available in one volume, ***FAR-AMT: Federal Aviation Regulations for Aviation Maintenance Technicians***.

The Advisory Circulars that contain information required for the IA knowledge test are reprinted either in ASA's FAR-AMT book, or in this IA Test Prep. An exception to this is that AC 43.13-1B *Acceptable Methods, Techniques, and Practices—Aircraft Inspection and Repair* and AC 43.13-2B *Acceptable Methods, Techniques, and Practices—Aircraft Alterations* have been reprinted and bound into a single volume as **AC 43.13-1B/2B *Acceptable Methods, Techniques, and Practices—Aircraft Inspection, Repair, and Alterations***, reprinted by ASA and sold separately.

Continued

The proven effective ASA "Fast-Track" format is used for this test guide and the questions and their answer alternatives are similar to those in the FAA Knowledge Test. Examine the question and the alternatives carefully, then select the alternative that is the best answer for the question. Read the explanation directly below the alternatives to verify your answer. At the bottom of the page in smaller type are the question number, the chosen answer alternative, the LSC, and the actual reference from which the question is derived. There is also a complete answer key in the Appendix, beginning on Page A-1, that shows the question number, chosen answer alternative, LSC, and the reference source from which the answer was derived.

Dale Crane

Terry Michmerhuizen

Note: Although the IA Knowledge Exam is based upon FAA documentation (i.e. regulations, Orders, and Advisory Circulars), the actual FAA documents change more frequently than the agency updates the CT-8080-8 Testing Supplement. The answers to the FAA test questions are based upon the information in the CT-8080-8 Testing Supplement. Additionally, the AC 43.13-2B was released in March 2008, but the CT-8080-8D continues to reference the AC 43.13-2A.

Chapter 1
Overview of Inspection Authorization

Introduction

The questions in this manual are typical of those asked on an IA Knowledge Test, and therefore their primary purpose is to help you become familiar with the reference materials. However, ASA's *Inspection Authorization Test Prep* is not merely an aid to passing the FAA test, but has been prepared to help you understand the materials used by an IA in his/her daily conduct of business.

What You Should Know About IA Certification

Maintenance Airmen

The regulations regarding certification of maintenance airmen are included in Title 14 of the Code of Federal Regulations (14 CFR) Part 65, *Certification: Airmen Other Than Flight Crewmembers*, §65.91. This regulation identifies three categories of maintenance airmen: mechanic, inspector, and repairman.

Mechanic is the basic certification, and there are two ratings available for it: Airframe and Powerplant.

The Inspection Authorization is available to the holder of a Mechanic certificate with both Airframe and Powerplant ratings who meets certain additional experience and knowledge requirements.

Repairman certification is issued to persons who have specialized experience and who work at a specific job in an FAA-certificated facility, such as a repair station or an air carrier. There is another category of Repairman certification that allows the builder of an amateur-built aircraft to perform condition inspections on the aircraft he or she has built.

Basic Privileges of an IA

With the exception of aircraft maintained on a Continuous Airworthiness Program under 14 CFR Part 121 (*Operating Requirements: Domestic, Flag, and Supplemental Operations*), an IA may inspect and approve for return to service any aircraft or related part or appliance after a major repair or major alteration. Also the holder of an IA may perform an annual inspection and may supervise or perform a progressive inspection.

Eligibility Requirements for an IA

Eligibility is established at the local FAA Flight Standards District Office (FSDO) prior to taking the Inspection Authorization Knowledge Test.

You are eligible for the Inspection Authorization Knowledge Test if you meet the requirements of 14 CFR Part 65, §65.91(c).

§65.91 Inspection Authorization

(c) To be eligible for an inspection authorization, an applicant must—

(1) Hold a currently effective mechanic certificate with both an airframe rating and a powerplant rating, each of which is currently effective and has been in effect for a total of at least 3 years;

(2) Have been actively engaged, for at least the two-year period before the date he applies, in maintaining aircraft certificated and maintained in accordance with this chapter;

(3) Have a fixed base of operations at which he may be located in person or by telephone during a normal working week, but it need not be the place where he will exercise his inspection authority;

(4) Have available to him the equipment, facilities, and inspection data necessary to properly inspect airframes, powerplants, propellers, or any related part or appliance; and

(5) Pass a written test on his ability to inspect according to safety standards for returning aircraft to service after major repairs and major alterations and annual and progressive inspection performed under Part 43 of this chapter.

Duration of an Inspection Authorization

Each IA expires on March 31 of each odd-numbered year. However, the holder may exercise the privileges of that authorization only while he holds a currently effective mechanic certificate with both a currently effective airframe and powerplant rating.

An IA ceases to be effective whenever any of the following occurs:

- The authorization is surrendered, suspended, or revoked.
- The holder no longer has a fixed base of operation.
- The holder no longer has the equipment, facilities, and inspection data required for the issuance of the authorization.

Renewal of an Inspection Authorization

To be eligible for renewal of an inspection authorization for a two-year period, an applicant must present evidence at renewal, during the month of March in odd-numbered years, at an FAA FSDO or International Field Office that the applicant still meets the requirements of §65.91(c)(1) through (4) for each year they have held the IA certificate. The applicant must show that during the current period the inspection authorization has been held, the applicant has —

- Performed at least one annual inspection for each 90 days the applicant has held the current authority; or
- Performed inspections of at least two major repairs or major alterations for each 90 days the applicant has held the current authority; or
- Performed or supervised and approved at least one progressive inspection in accordance with standards prescribed by the Administrator; or
- Attended and successfully completed a refresher course, acceptable to the Administrator of not less than 8 hours of instruction during each 12-month period; or
- Passed an oral test by an FAA inspector to determine that the applicant's knowledge of applicable regulations and standards is current.

The holder of an inspection authorization that has been in effect for less than 90 days before the expiration date need not comply with these requirements.

Change of Fixed Base of Operation

If the holder of an IA changes his fixed base of operation, he may not exercise the privileges of the authorization until he has notified, in writing, the FAA FSDO or International Field Office for the area in which the new base is located, of the change.

What You Should Know About the IA Knowledge Test

The Knowledge Test for Inspection Authorization is different from any of the other FAA certification test in that you must get permission to take the test by having a personal interview with an Aviation Safety Inspector (ASI) in your local FSDO.

Steps For Taking the Inspection Authorization Knowledge Test

We appreciate feedback from individuals who have taken their Inspection Authorization test so we may continually make improvements to this publication.

1. Contact your local FSDO to make an appointment to interview with an ASI (airworthiness) to determine your eligibility to take the test.
2. When the ASI is satisfied that you have met all of the requirements for IA, furnish positive proof of identification and complete FAA Form 8610-1, *Mechanic's Application for Inspection Authorization.*
3. Register with the computer testing designee at the test center indicated by the ASI to schedule a test and make financial arrangements for test payment.
4. You will not need to take (nor will you be allowed to carry in) any of your IA reference material to the test center; however, you will need proper identification.
5. Before you take the actual test, you will have the option to take a sample test. Since there is no time limit on the sample test, be sure to work through it completely. It will not only help you become familiar with the computer testing, but will also provide valuable information concerning charts and graphs referenced on the test and included in the *Computer Testing Supplement for Inspection Authorization.* Finally, it will help you understand how to "flag" questions you want to research and return to later. This is an important feature that prevents you from getting bogged down on a particular question, and instead allows you to keep up your momentum. The actual test is time-limited; however, you should have sufficient time to complete and review your test.
6. Make a chart of your progress as you go through the test. This chart has four columns with the first labeled "***Question Number***" and runs 1–50. The second is labeled "***Finished.***" The third is labeled "***Review In.***" The last column is labeled "***Calculation Required.***" The object is to help you keep track of what you have completed and which questions need more attention. If you run through the actual test using this method without stopping to research anything, you may find you have a large portion of the test completed with a high degree of confidence. For questions that you know you must research such as ADs or TCDS data, put that reference information in the third column and come back to it later. Sometimes, more than one question will direct you to the same reference material. This way you minimize lost time in redundant searches. Finally, you should use any remaining time for doing the computation questions, such as weight and balance, and rivet-spacing.
7. Upon completion of the test, you will receive your Airman Test Report with the testing center's embossed seal, which reflects your score. This test report lists the learning statement codes (LSC) for questions answered incorrectly. Study the LSC subjects to increase your knowledge of the subject matter.
8. You will be given 10 minutes to review any questions you missed (without the answer choices or your selected answer). This is helpful for determining where future study and learning can be focused.
9. The minimum passing score is 70; however, if you fail the test you must wait 90 days before you are allowed to retest. Because the 8610-1 form is only good for a period of 30 days, you will have to complete a new form and have your local FSDO again approve you for testing. You must also pay the testing center for this second test.
10. After passing the test, present your Airman Test Report to an ASI at the FSDO where you interviewed. It is best to return to the original interviewer if possible; however, any available ASI can complete the authorization process. At that time, the ASI will again review your application and discuss any questions you may have. When the ASI is satisfied that you have met all of the requirements, your IA certificate will be issued.

Description of the IA Knowledge Test

The test contains 50 objective multiple-choice type questions, each of which can be answered by the selection of a single response. Each test question is independent of any other questions; therefore, a correct response to one does not depend upon, or influence the correct response to another.

The maximum time allowed for the test is 3 hours. This time is based on previous experience and is considered more than adequate if you are properly prepared.

At the test center, you will be provided with the latest revision of CT-8080-8. This supplement is the ***only*** reference you may use and contains excerpts from the applicable parts of the Federal Regulations (14 CFR), representative Airworthiness Directives, charts and diagrams from pertinent Advisory Circulars, and examples of Type Certificate Data Sheets and Specifications and pertinent FAA forms. Before you start the test, take a few minutes to look through the supplement to familiarize yourself with its contents.

Carefully read the information and instructions given with the tests, as well as the introductory statements in each test item.

When taking a test, keep the following points in mind:

- Answer each question in accordance with the latest regulations and procedures, unless the data provided in the computer testing supplement differs.
- Read each question carefully before looking at the possible answer choices. You should clearly understand the problem before attempting to solve it.
- After formulating an answer, determine which of the alternatives most closely corresponds with that answer. The answer chosen should resolve the problem *completely*.
- From the answers given, it may appear that there is more than one possible answer; however, only one answer is correct and complete. The other answers are either incomplete, or they reflect popular misconceptions.
- If a certain question is difficult for you, it is best to mark it for review and proceed to the other questions. After you answer the less difficult questions, return to those which you marked for review and answer them. The review-marking procedure will be explained to you prior to starting the test. When you have finished taking the test, make sure an answer has been recorded for each question — the computer will alert you to all unanswered questions. This procedure will enable you to use the available time to the maximum advantage.
- When solving a calculation problem, select the answer closest to your solution. The problem has been checked with various types of calculators; therefore, if you have solved it correctly, your answer will be closer to the correct answer than any of the other choices.

Note: Sometimes a test will have more than 50 questions. This occurs when the FAA includes additional new "sample" questions for determining user understanding and validating properly-worded questions. Usually there are no more than five of these. Do not assume that the last five questions are the additional sample questions. Instead, they are randomly placed throughout the test, so you must answer ***all*** questions to the best of your ability. These additional questions will not count towards your final score, but if you leave any blank they will be counted against you.

Test Aids You May Use

The IA Knowledge Test requires you to analyze all of the variables needed to solve the problems. When solving problems involving mathematical calculation you are tested on concepts rather than rote calculation ability. This allows you to use certain calculators, computers, or similar devices designed for aviation-related activities provided they are used within these guidelines.

- Applicants may use test aids, such as scales, straightedges, protractors, plotters, navigation computers, log sheets, and all models of aviation-oriented calculating devices that are directly related to the test. In addition, applicants may use any test materials provided with the test.
- Manufacturer's permanently inscribed instructions on the front and back of these test aids such as formulas, conversions, regulations, signals, weather data, holding pattern diagrams, frequencies, weight and balance formulas, and air traffic control procedures are permissible.
- The test proctor may provide calculating devices to applicants and deny them use of their personal calculating devices if the applicant's device does not have a screen that indicates all memory has been erased. The test proctor must be able to determine the calculating device's erasure capability. You are not allowed to use calculating devices incorporating permanent or continuous-type memory circuits without erasure capability.
- Magnetic cards, magnetic tapes, modules, computer chips, or any other device upon which prewritten programs or information related to the test can be stored and retrieved are not allowed. Printouts of data will be surrendered at the completion of the test if the calculating device used incorporates this design feature.
- The use of any booklet or manual containing instructions related to the use of the applicant's calculating device is not permitted.
- Dictionaries are not allowed in the testing area.
- The test proctor makes the final determination relating to test materials and personal possessions that the applicant may take into the testing area.

Cheating or Other Unauthorized Conduct

Computer testing centers follow strict security procedures to avoid test compromise. These procedures are established by the FAA and are covered in FAA Order 8080.6, *Conduct of Airman Knowledge Tests*. The FAA has directed all testing centers to terminate a test at any time a test proctor suspects a cheating incident has occurred. An FAA investigation will then follow. If the investigation determines that cheating or other unauthorized conduct has occurred, any airman certificate that you hold may be revoked, and you may not be allowed to take a test for one year.

Retesting Procedures

If you fail the IA Knowledge Test, you may not apply for retesting until 90 days after the date that you failed the test. Any attempt to retest prior to the 90-day waiting period is contrary to 14 CFR Part 65, and could result in revocation of any airman certificates that you hold.

Preparation for the IA Knowledge Test

Aviation Supplies & Academics has a comprehensive array of books to prepare you for the IA test:

ASA-FAR-AMT *Federal Aviation Regulations for Aviation Maintenance Technicians*

This volume contains reprints of pertinent parts of 14 CFR and ACs that apply to aviation maintenance.

ASA-IA *Inspection Authorization Test Prep*

Contains explanations of the documents used in the IA Knowledge Test with example questions similar to those that will be on the test.

AC 43.13-1B/2B *Acceptable Methods, Techniques, and Practices— Aircraft Inspection, Repair, and Alterations*

This single volume contains reprints of both of these essential Advisory Circulars. The procedures and techniques described are *acceptable* for inspections, repairs, and alterations but may not necessarily be used as *approved* data unless specifically approved by an FAA Aviation Safety Inspector.

FAA-H-8083-1 *Aircraft Weight and Balance Handbook*

Provides information on determining the empty weight and EWCG of an aircraft, and information on loading and operating an aircraft to keep the weight and CG within allowable limits.

ASA-DAT *Dictionary of Aeronautical Terms*

A comprehensive dictionary of aeronautical terms and abbreviations.

ASA-MHB *Aviation Mechanics Handbook*

A handy toolbox-sized reference manual of charts, tables, diagrams, formulas, and other information useful to the aircraft mechanic.

ASA-AMG *General Test Guide*

ASA-AMA *Airframe Test Guide*

ASA-AMP *Powerplant Test Guide*

These three volumes contain answers, and explanations for all the questions that may be asked on the mechanic knowledge tests. They are a good source of review for the basic core knowledge questions that may be asked on the IA test.

CT-8080-8 *Computer Testing Supplement for Inspection Authorization*

This large loose-leaf notebook is the same as that furnished for use during the IA Knowledge Test, and contains the necessary excerpts and figures for the test questions. All of the pertinent information, tables, charts, and figures in this expensive test supplement are included in the other materials listed here. *Note:* Read the instructions on Pages ii and vii regarding how to download a PDF version of this supplement.

Visit **www.asa2fly.com/reader/ia** to access documents important to your IA test preparation.

Studying for the IA Knowledge Test

The computer-based IA test is straightforward, but you should prepare for it to the best of your ability. Here are some specific suggestions for studying for this test.

- Study all of the regulations and technical data listed in the FAA Learning Statement Codes subject listing (*see* next page).
- Learn to use the indexes in the publications efficiently, especially those for the Type Certificate Data Sheets and Specifications.
- Learn to identify the revision dates and change numbers for all FAA publications.
- Study 14 CFR Part 43 and its Appendixes, for detailed information regarding major repairs, major alterations, and annual inspections.
- Learn the use of graphs and tables in AC 43.13-1B, *Acceptable Methods, Techniques and Practices—Aircraft Inspection and Repair*, and in AC 43.13-2B *Acceptable Methods, Techniques and Practices—Aircraft Alterations.**
- Practice researching ADs, Type Certificate Data Sheets, and Specification Sheets on different makes and models of aircraft, engines, and propellers.
- Practice filling out FAA Form 337, *Major Repair and Alteration (Airframe, Powerplant, Propeller, or Appliance).* Guidance is provided in AC 43.9-1F, *Instructions for Completion of FAA Form 337.*
- Practice filling out maintenance and inspection record entries in accordance with 14 CFR §43.11.
- Practice making changes to an aircraft weight and balance report by simulating installation or removal of equipment, then computing the forward, aft, and empty-weight center of gravity (CG).
- Practice the use of the CT-8080-8 supplement prior to taking the actual test.*

* See the Reader Resources page on the ASA website for free downloadable PDFs of the AC 43.13-1B/2B, and the current CT-8080-8: **www.asa2fly.com/reader/ia**

Learning Statement Codes (LSC)

When you take the applicable airman knowledge test required for an airman pilot certificate or rating, you will receive an Airman Knowledge Test Report. The test report will list "learning statement codes" (LSC) for questions you answered incorrectly. Match the code given on your test report to the ones in the list of official FAA Learning Statement Codes (shown below). The Airman Knowledge Test Report must be presented to the examiner conducting the practical test. This examiner may evaluate the noted areas of deficiency.

The expression "learning statement," as used in airman testing, refers to measurable statements of knowledge that a student should be able to demonstrate following a certain segment of training. In order that each learning statement may be read and understood as a complete sentence, precede each LSC with the words: "Upon the successful completion of training the student should be able to..." — then complete the phrase with the subject indicated by the LSC given in your knowledge test results.

FAA Learning Statement Codes are prefixed with a letter-identifier (for example, IAR031). For the purposes of reference within this IA Test Prep, the letter prefix is omitted; therefore throughout the book in the reference lines, LSCs are referred to by their number-identifiers only, in parantheses.

The FAA appreciates testing experience feedback. You can contact the branch responsible for the FAA Knowledge Exams directly at:

Federal Aviation Administration
AFS-630, Airman Test Standards Branch
P.O. Box 25082
Oklahoma City, OK 73125
Email: AFS630comments@faa.gov

LSC	Subject area
IAR001	Calculate alteration specification
IAR002	Calculate center of gravity
IAR003	Calculate electrical load
IAR004	Calculate proof loading
IAR005	Calculate repair specific
IAR006	Calculate sheet metal repair
IAR007	Calculate temperature conversion
IAR008	Calculate weight and balance — adjust weight / fuel
IAR009	Determine alteration parameters
IAR010	Determine alteration requirements
IAR011	Determine correct data
IAR012	Determine data application
IAR013	Determine design specific
IAR014	Determine fabrication specification
IAR015	Determine process specific
IAR016	Determine regulatory requirement
IAR017	Determine regulatory requirements
IAR018	Determine repair parameters
IAR019	Determine repair requirements
IAR020	Interpret data
IAR021	Interpret regulations
IAR022	Recall alteration /design fundamentals
IAR023	Recall engine repair fundamentals
IAR024	Recall fundamental inspection principles — airframe / engine
IAR025	Recall MEL requirements
IAR026	Recall principles of corrosion control
IAR027	Recall principles of sheet metal forming
IAR028	Recall principles of system fundamentals
IAR029	Recall principles of weight and balance
IAR030	Recall regulatory requirements
IAR031	Recall regulatory specific
IAR032	Recall repair fundamentals

Chapter 2
The Function of an IA

Introduction

An Inspection Authorization accords two additional privileges to an FAA-certificated mechanic with an Airframe and Powerplant rating. These are:

1. The authorization to inspect and approve for return to service of an aircraft or related part or appliance after a major repair or a major alteration.
2. The authorization to perform an annual inspection or to perform or supervise a progressive inspection.

This chapter considers four items of concern to an IA:

1. The approval of major repairs and major alterations
2. Annual and progressive inspections
3. Maintenance records
4. The relationship between an IA and the aircraft owner

Approving Major Repairs and Major Alterations

A primary responsibility of the holder of an IA is to determine airworthiness by inspecting repairs or alterations for conformity to approved data, and ensuring that the aircraft is in a condition for safe operation. During inspection of major repairs or major alterations, the holder of an IA must also determine that these are compatible with previous repairs and alterations made to the aircraft.

14 CFR §43.11 requires the inspector to include in his or her maintenance record entry, either the following statement or a similarly worded one, upon completion of determining airworthiness***: "I certify that this aircraft has been inspected in accordance with _________ inspection and was determined to be in an airworthy condition."*** Therefore it is important to understand the definition for the term "airworthy." Although the term is fundamental to all maintenance and inspection activities, only recently did the FAA specifically define it in regulatory material (September 16, 2005). *See* 14 CFR Part 3 on the next page.

Prior to this regulatory reference, the definition was inferred by reading the requirements found in 14 CFR §21.183 necessary to obtain an Airworthiness Certificate, and the related comment on the Standard Airworthiness Certificate (FAA Form 8100-2), or referred to in nonregulatory guidance information found in FAA Order 8130.2. These references and now the formal definition in 14 CFR Part 3 state that the definition of airworthiness is based upon two specific issues:

1. **Conformity** to type design data (or approved alterations).
2. **Condition** for safe operation.

It is the aviation inspector's job to determine that these two criteria are met before signing off the required maintenance record entry required in §43.11.

It may help in understanding this definition if the IA considers that "conformity" to type design is really an *objective* determination. In other words, do all the numbers match? Are the model numbers, part numbers, and serial numbers when applicable, consistent with approved design and alteration data?

The second part — the condition for safe operation — can be considered much more *subjective* to the technician. This is where the years of personal experience, lunchroom conversations, team meetings, networking at professional tradeshows and seminars, and discussions with coworkers come into

Continued on Page 2–5

14 CFR Part 3—General Requirements

Authority: 49 U.S.C. 106(g), 40113, 44701, and 44704.
Source: 70 FR 54832, Sept. 16, 2005, unless otherwise noted.

§3.1 Applicability.

(a) This part applies to any person who makes a record regarding:

(1) A type-certificated product, or

(2) A product, part, appliance or material that may be used on a type-certificated product.

(b) Section 3.5(b) does not apply to records made under part 43 of this chapter.

§3.5 Statements about products, parts, appliances and materials.

(a) Definitions. The following terms will have the stated meanings when used in this section:

Airworthy means the aircraft conforms to its type design and is in a condition for safe operation.

Product means an aircraft, aircraft engine, or aircraft propeller.

Record means any writing, drawing, map, recording, tape, film, photograph or other documentary material by which information is preserved or conveyed in any format, including, but not limited to, paper, microfilm, identification plates, stamped marks, bar codes or electronic format, and can either be separate from, attached to or inscribed on any product, part, appliance or material.

(b) Prohibition against fraudulent and intentionally false statements. When conveying information related to an advertisement or sales transaction, no person may make or cause to be made:

(1) Any fraudulent or intentionally false statement in any record about the airworthiness of a type-certificated product, or the acceptability of any product, part, appliance, or material for installation on a type-certificated product.

(2) Any fraudulent or intentionally false reproduction or alteration of any record about the airworthiness of any type-certificated product, or the acceptability of any product, part, appliance, or material for installation on a type-certificated product.

(c) Prohibition against intentionally misleading statements.

(1) When conveying information related to an advertisement or sales transaction, no person may make, or cause to be made, a material representation that a type-certificated product is airworthy, or that a product, part, appliance, or material is acceptable for installation on a type-certificated product in any record if that representation is likely to mislead a consumer acting reasonably under the circumstances.

(2) When conveying information related to an advertisement or sales transaction, no person may make, or cause to be made, through the omission of material information, a representation that a type-certificated product is airworthy, or that a product, part, appliance, or material is acceptable for installation on a type-certificated product in any record if that representation is likely to mislead a consumer acting reasonably under the circumstances.

(d) The provisions of §3.5(b) and §3.5(c) shall not apply if a person can show that the product is airworthy or that the product, part, appliance or material is acceptable for installation on a type-certificated product.

the picture. Most differences of opinion regarding airworthiness will occur here, but good and honest dialogue with an open mind are valuable tools for the technician. The IA must learn to balance both a willingness to learn from others, and also to defend one's own professional opinion — with the final goal of determining whether an aircraft is airworthy or not.

The holder of an IA must personally perform the inspection. The Code of Federal Regulations (CFRs) does not provide for delegation of this responsibility. Approving major repairs and major alterations is a serious responsibility. The approval action should consist of a detailed investigation to establish at least that:

- All replacement parts installed conform to approved design and/or have traceability to the original equipment manufacturer (OEM), or to the appropriate Parts Manufacturing Approval (PMA) or Technical Standard Order (TSO).
- As installed, the installation conforms to approved data that is applicable to the installation.
- Workmanship meets the requirements of 14 CFR §43.13, which specifies that the aircraft or product is equal to its original or properly altered condition.
- The data used is appropriate to the aircraft certification rule (e.g., CAR 3 or 14 CFR Part 23).
- Work is complete and compatible with other structures or systems.

Approved Data

The holder of an IA ***cannot*** approve the ***data*** for major repairs or major alterations. He or she may, however, inspect to see that alterations conform to data ***previously approved by the Administrator*** (14 CFR §65.95). This means the holder of an IA ensures that approved data is available and is used as the basis for the approval. This availability determination should be made prior to beginning the repair or alteration. If data is unavailable, or if the holder of an IA is unsure of the acceptability of the available data, the local Aviation Safety Inspector (ASI) should be consulted. The ASI may, as the circumstances warrant, be able to:

- establish an acceptable basis for approval;
- approve the data; or
- recommend application for a supplemental type certificate.

Often, major repairs are performed that are eventually covered by fabric, metal skin, or another structure. When this situation exists, the holder of an IA should have a clear understanding with the mechanic performing the repair that a precover inspection is necessary. The inspection should ensure that the repair was made in accordance with acceptable methods, techniques, and practices prescribed by 14 CFR Part 43 and that the structure to be covered is free from defects, corrosion, or wood rot, and is protected from the elements. In addition, the holder of an IA should inspect other affected areas for hidden damage, if the aircraft has been involved in an accident or incident. An entry is required to be made in the maintenance record, and FAA Form 337 *Major Repair and Alteration* must be completed.

Minor deviation from approved data is permissible if the change is one that could be approved as a minor alteration when considered by itself. Be sure to list the deviations on FAA Form 337 and make an entry in the maintenance record when completing the aircraft records. When in doubt, contact the local ASI who may decide the change is not minor and would need specific approval or an amendment of the original approval.

Approved data to be used for major repairs and major alterations may be one or more of the following.

- Type Certificate Data Sheets
- Aircraft Specifications
- Supplemental Type Certificates (STCs) (Note: Persons interested in using an STC must have approval in accordance with 49 U.S.C., §44704, from the holder of the STC prior to its use.)
- Airworthiness Directives (ADs)
- Designated Engineering Representative (DER) Approved Data With FAA Form 8110-3, *Statement of Compliance*
- Organizational Designation Authorization (ODA) Approved Data
- Appliance Manufacturer's Manuals (excluding installation instructions)

AC 43.13-1, *Acceptable Methods, Techniques, and Practices (Aircraft Inspection and Repair),* may be used directly as approved data (for repairs only) without further approval only when there is no manufacturer's repair or maintenance instructions that address the repair, and the user has determined that it is:

- appropriate to the product being repaired;
- directly applicable to the repair being made; and
- not contrary to manufacturer's data.

This data may also be used as a basis to gain FAA data approval for major repairs.

FAA field approval (occurs when FAA enters a statement and signs block 3 of FAA Form 337) issued for duplication of identical aircraft may be used as approved data only when the identical alteration is performed on an aircraft of identical make, model, and series by the original modifier. FAA Form 337s approved in 1955 or earlier may be used as approved data.

Inspecting Repairs or Alterations

Inspecting repairs or alterations consists of these basic operations:

- Determine that the repair or alteration data has FAA approval.
- Inspect the configuration of the repair or alteration for conformity to the approved data and the performance standards of 14 CFR Part 43. At the same time, the aircraft should still comply with applicable airworthiness requirements, and the repair or alteration be compatible with all other installations.
- All operating limitations affected by an alteration should be appropriately revised. Sometimes limitations are in the form of flight manual supplements, instrument range markings, placards, or combinations of these. See the local ASI for limitations on changes that can be made.
- Determine that aircraft record entries have been made and the weight and balance data and equipment list have been revised, when appropriate. There should be a statement on FAA Form 337 to the effect that the weight and balance data and equipment list have been revised. When an alteration results in a change in the center-of-gravity (CG) position, the affected CG limit should be investigated under adverse loading conditions unless the new CG falls within an approved empty CG range. For instance, if the CG has shifted aft, the loading conditions should be computed to see that the aircraft does not exceed the aft CG limit. It is the pilot's responsibility to have the aircraft correctly loaded. However, when approving an alteration, it is the IA's responsibility to see that weight and balance data have been revised. The aircraft record entries may refer to FAA Form 337 for details, such as: "Installed STOL kit in accordance with STC SA 940 CE drawing number 5084 dated April 24, 1996. *See* FAA Form 337, this date, for details."
- Indicate approval in block 7 of FAA Form 337, and return both copies to the person who performed the work, for disposition in accordance with 14 CFR Part 43, Appendix B.

Annual, 100-Hour, and Progressive Inspections

The procedures and scope for annual and 100-hour inspections are set forth in 14 CFR Part 43 Appendix D, and 14 CFR §43.15. These should be followed in detail. The scope and detail for a progressive inspection is established by the owner or operator in accordance with 14 CFR §91.409(d). There are additional requirements for annual and progressive inspections listed in 14 CFR §43.15.

The scope and detail of 100-hour and annual inspections are the same.

Record entries are very important as they are the only evidence an aircraft owner has to show compliance with the inspection requirements of 14 CFR §91.409. The following reminders should help in determining that the aircraft complies with all airworthiness requirements (refer to 14 CFR §43.15(a)):

Inspection for Configuration ("conformity to type design")

The aircraft should conform to the Aircraft Specification or Type Certificate Data Sheet, any changes by Supplemental Type Certificates, and/or its properly altered condition. When the aircraft does not conform, use the procedures for "unairworthy" items listed in 14 CFR §43.11(a)(5).

1. Alterations to the product may have changed some of the operating limitations.
2. Unrecorded alterations or repairs may have been made in the past and warrant one of the following:
   a. Contact owner for pertinent information.
   b. If approved data is available, conduct inspection and personally approve for return to service by completing FAA Form 337.
   c. Contact local ASI for assistance.

3. The Aircraft Specification or Type Certificate Data Sheet indicates when a flight manual is required. It also identifies the limitations that must be displayed in the form of markings and placards.
4. Unlike the Aircraft Specifications, Type Certificate Data Sheets do not contain a list of equipment approved for a particular aircraft. The list of required and optional equipment can be found in the equipment list furnished by the manufacturer of the aircraft. Sometimes a later issue of the list is needed to cover recently approved items. Serial number eligibility should always be considered.

Inspection for Condition ("condition for safe operation")

The holder of an IA may use the checklist in 14 CFR Part 43, Appendix D, the manufacturer's inspection sheets, or a checklist designed by the holder of an IA, that includes the scope and detail of the items listed in Appendix D, to check the condition of the entire aircraft. This includes checks of the various systems listed in 14 CFR §43.15.

1. Routine servicing is ***not*** a part of the annual inspection. The inspection itself is essentially a visual evaluation of the condition of the aircraft and its components and certain operational checks. The manufacturer may recommend certain services to be performed at various operating intervals. These can often be done conveniently during an annual inspection, and in fact should be done, but are not considered to be a part of the inspection itself.
2. It is very important that the holder of an IA be familiar with the manufacturer's service manuals, bulletins, and letters for the product being inspected. Use these publications to avoid overlooking problem areas.
3. AC 43-16, *Aviation Maintenance Alerts*, is also an important source of service experience. The articles for the alerts are taken from selected service difficulties reported to the FAA on FAA Form 8010-4, *Malfunction or Defect Reports*. Monthly copies of the alerts are provided on the Internet at the following address:

   www.faa.gov

   Comments may be sent by letter, with name and address typed or legibly printed to:

   Federal Aviation Administration
   Designee Standardization Branch, AFS-640
   P.O. Box 25082,
   Oklahoma City, OK 73125
4. When the holder of an IA approves an aircraft for return to service, he or she will be held responsible for the condition of the aircraft *as of the time of approval.*

Minimum Equipment List (MEL)

The minimum equipment list (MEL) is intended to permit operations with certain inoperative items of equipment for the minimum period of time necessary until repairs can be accomplished. It is important that repairs are accomplished at the earliest opportunity in order to return the aircraft to its design level of safety and reliability.

1. When inspecting aircraft operating with an MEL, the holder of an IA should review the document where inoperative items are recorded, (aircraft maintenance record, logbook, discrepancy record, etc.) to determine the state of airworthiness with regard to those recorded discrepancies. Inspections of aircraft with approved MELs will be in accordance with the 14 CFR Part under which the MEL was issued.
2. Those MELs specifying repair intervals through the use of A, B, C, D codes require repairs of deferred items at or prior to the repair times established by the letter designated category. In such instances, some items previously deferred may not be eligible for continued deference at the

inspection or may require additional maintenance. Where repair intervals are not specified by codes in the MEL, all MEL-authorized inoperative instruments and/or equipment should be repaired or inspected and deferred before approval for return to service.

3. Aircraft established on a progressive inspection program require that all MEL-authorized inoperative items be repaired or inspected and deferred at each inspection whether or not the item is encompassed in that particular segment.
4. When inspecting aircraft operating without an MEL, 14 CFR §91.213(d) allows certain aircraft not having an approved MEL to be flown with inoperative instruments and/or equipment. These aircraft may be presented for annual or progressive inspection with such items previously deferred or may have inoperative instruments and equipment deferred during an inspection. In either case, the holder of an IA is required by 14 CFR §43.13(b) to determine that:
    a. The deferrals are eligible within the guidelines of that rule.
    b. All conditions for deferral are met, including proper recordation in accordance with 14 CFR §§43.9 and 43.11; and
    c. Deferral of any item or combination of items will not affect the intended function of any other operable instruments and/or equipment, or in any manner constitute a hazard to the aircraft. When these requirements are met, such an aircraft is considered to be in a properly altered condition with regard to those deferred items.

Airworthiness Directives (ADs)

The holder of an IA is required by 14 CFR §43.13 to determine that all applicable airworthiness directives (ADs) for aircraft, powerplants, propellers, instruments, and appliances have been accomplished.

1. If the maintenance records indicate compliance with an AD, the holder of an IA should make a reasonable attempt to verify the compliance. It is not uncommon for a component to have compliance with an AD accomplished and properly recorded then later be replaced by another component on which the AD has not been accomplished. The holder of an IA is not expected to disassemble major components such as cylinders, crankcases, etc., if adequate records of compliance exist.
2. When the maintenance records ***do not*** contain indications of AD compliance, the holder of an IA should:
    a. make the AD an item on a discrepancy list provided to the owner, in accordance with 14 CFR §43.11(b);
    b. with the owner's concurrence, do whatever disassembly is required to determine the status of compliance; or
    c. obtain concurrence of the owner to comply with the AD.
3. Often, an AD calls for an inspection at one time with a modification or inspection required at a later date. It is very important to identify, in the maintenance record entry, the portion of the AD complied with and the exact method of compliance.
4. 14 CFR §91.417(a)(2)(v) requires each registered owner or operator to keep a record of the current status of applicable ADs. This status includes, for each, the method of compliance, AD number, and revision date. If the AD involves recurring action, the time and date should be recorded when the next action is required. As a vital part of the services performed, the holder of an IA may wish to provide the owner with information he/she is expected to keep.

Continued

5. The owner should also be informed of any subsequent requirements of an AD or whether a reinspection is required at operating intervals other than at annual inspections. Often, the subsequent requirements are at 100-hour intervals and will need to be done whether or not the aircraft is required to have 100-hour inspections. Where a progressive inspection is involved, the approved program should state how and when the AD review will be accomplished. However, as a mechanic or IA, you should be aware of an AD that is pending or due, and even though is not in the area you are inspecting, it is good customer relations to inform the owner or pilot of the situation.

Malfunction or Defect Reports

All malfunctions or defects that come to the attention of the holder of an IA should be reported on FAA Form 8010-4. Copies of the self-addressed form are available at all FSDOs, easy to fill out and require no postage. Prompt reporting will contribute much toward improving air safety by helping correct unsafe conditions. In addition to the old style "postcards," it is now possible to submit these M & D reports electronically via the web. The FAA still accepts both. Some technicians will find it easier to keep the cards in their toolbox.

Paperwork Review

The owner or operator is responsible for maintaining the equipment list, CG and weight distribution computations, and loading schedules, if necessary.

1. The holder of an IA is required by 14 CFR §43.13 to determine that the required placards and documents set forth in the aircraft specification or type certificate data sheet are available and current. The aircraft should be reported as being in an unairworthy condition if these placards and documents are not available. Missing, incorrect, or improperly located placards are regarded as an unairworthy item, and the owner or operator should be informed that, under the requirements of 14 CFR §91.9, the aircraft may not be operated until they are correctly installed.
2. The holder of an IA should refer to the registration and airworthiness certificates for the owner's name and address; the aircraft make, model, registration, and serial numbers needed for recording purposes. Be sure not to use manufacturers' trade names as they do not always coincide with the actual model designation (Cessna Skylane is 182, Piper Seneca III is PA 34 220T, etc.). If registration and airworthiness certificates are not available, the aircraft does not need to be reported in unairworthy condition; however, the owner or operator should be informed that the documents required by 14 CFR §91.203 (a) and (b), should be in the aircraft and the airworthiness certificate displayed, *when the aircraft is operated.*
3. Other documents often needed but not a part of the airworthiness requirement might be a state registration, and if the aircraft is equipped with a transceiver, a Federal Communications Commission radio license. The owner or operator is responsible for maintaining these documents. However, the IA holder will be performing an appreciated service by informing the operator of any deficiencies in the display and carriage of these documents.
4. On aircraft for which no approved flight manual is required, the operating limitations prescribed during original certification, and as required by 14 CFR §91.9, must be carried in or be affixed to the aircraft. Range markings on the instruments, placards, and listings are required to be worded and located as specified in the Type Certificate Data Sheet.

Aircraft Markings

Required aircraft identification markings are discussed in 14 CFR Part 45. It is the owner's or operator's responsibility to have the nationality and registration markings properly displayed on the aircraft (14 CFR §91.9(c)). The holder of an IA can, and should, offer advisory service to owners and operators in regard to any deficiencies in markings; however, such deficiencies are not cause to report an aircraft in "unairworthy" condition.

Aircraft with Discrepancies or Unairworthy Conditions

If the aircraft is not approved for return to service after a required inspection, use the procedures specified in 14 CFR §43.11. This will permit an owner to assume responsibility for having the discrepancies corrected prior to operating the aircraft.

1. The discrepancies can be cleared by a person who is authorized by 14 CFR Part 43 to do the work. Preventive maintenance items could be cleared by a pilot who owns or operates the aircraft, provided the aircraft is not used under 14 CFR Parts 121, 129, or 135; except that approval may be granted to allow a pilot operating a rotorcraft in a remote area under 14 CFR Part 135 to perform preventive maintenance.
2. The owner may want the aircraft flown to another location to have repairs completed, in which case the owner should be advised that the issuance of FAA Form 8130-7, *Special Flight Permit*, is required. This form is commonly called a ferry permit and is detailed in 14 CFR §21.197. The certificate may be obtained in person or by fax at the local FSDO or from a Designated Airworthiness Representative.
3. If the aircraft is found to be in an unairworthy condition, an entry will be made in the maintenance records that the inspection was completed and a list of unairworthy items was provided to the owner. When all unairworthy items are corrected by a person authorized to perform maintenance and that person makes an entry in the maintenance record for the correction of those items, the aircraft is approved for return to service.

Incomplete Inspection

If an annual inspection is not completed, the IA holder should:

- Indicate any discrepancies found in the aircraft records.
- *Not* indicate that an annual inspection was conducted.
- Indicate the extent of the inspection and all work accomplished in the aircraft records.

Maintenance Records

The holder of an IA and other maintenance personnel or agencies are required to record maintenance, inspections, or alterations performed or approved in accordance with the requirements of 14 CFR §§43.9 and 43.11. The owner or operator is required by 14 CFR §91.417 to keep maintenance records. The holder of an IA is also required to indicate the total aircraft time in service when a required inspection is done.

Significance of Maintenance Record Entries

Responsibility for maintenance work performed rests with the person whose signature and certificate number are entered on the appropriate maintenance record and/or forms. The responsibility for annual and progressive inspections and approval for return to service of major repairs or major alterations is assumed by the holder of an IA whose signature and certificate number appear on the appropriate maintenance records.

Completion of FAA Form 337

FAA Form 337, *Major Repair and Alteration (Airframe, Powerplant, Propeller, or Appliance)*, serves two purposes. One is to provide owners and operators a record of major repairs and major alterations indicating details and approval. The other purpose is to provide the FAA with a copy for the aircraft records.

1. The person who performed or supervised the major repair or major alteration prepares the original FAA Form 337 (two copies, or three if extended-range fuel tanks are installed in the passenger or baggage compartment). The holder of an IA then further processes the forms when they are presented for approval.
2. Instructions for the completion of FAA Form 337 appear in AC 43.9-1F, *Instructions for Completion of FAA Form 337, Major Repair and Alteration (Airframe, Powerplant, Propeller, or Appliance).*
3. Disposition of FAA Form 337.
   a. The IA holder who has found a major alteration or a major repair to be in conformity with FAA-approved data should review the FAA Form 337 for completeness and accuracy, and complete item 7.
   b. The person performing a major repair or major alteration shall in accordance with Part 43:
      (1) Give a signed copy of FAA Form 337 to the aircraft owner.
      (2) Make the proper entry in the maintenance records.
      (3) Forward the duplicate copy to the local FAA Aircraft Registration branch in Oklahoma City within 48 hours.
   c. The holder of an IA should ensure that the duplicate copy is an exact and legible reproduction of the original. The signatures should not be carbon copies but original signatures written in ink.
   d. If the FAA Form 337 is completed for extended-range fuel tanks installed within the passenger compartment or a baggage compartment, the person who performs the work and the person authorized to approve the work by 14 CFR §43.7 shall execute an FAA Form 337 in at least triplicate, as required by 14 CFR Part 43, Appendix B. One (1) copy of the FAA Form 337 shall be placed on board the aircraft as specified in 14 CFR §91.417. The remaining forms shall be distributed as previously noted.
   e. If FAA Form 337 has been completed for engines, propellers, spare parts or components, both copies of the form, with the approval portion completed, should be attached to the part or component until it is installed on an aircraft.
      (1) The mechanic who makes the installation will, in accordance with 14 CFR §43.9(a)(4), complete both copies of FAA Form 337 by filling in blocks 1 and 2 and sign for the installation in the aircraft records, making reference to the FAA Form 337 in the record entry.
      (2) He or she will give a copy to the owner and forward a copy to the FAA Aircraft Registration branch in Oklahoma City.

Weight and Balance

Although weight and balance data are no longer required to be entered on FAA Form 337, it is still imperative that weight and balance checks and computations be made very carefully. Since practically every aircraft manufacturer uses a different method of weight and balance control, it would be impossible to provide a universally adaptable method. When revising weight and balance data, these general guidelines should be followed.

1. The weight and balance data should be kept together in the aircraft records.
2. When making revisions, use a permanent, easily-identified method, with full-size sheets of paper large enough to contain complete computations and minimize the possibility of them becoming detached or lost.
3. Each page should be identified with the aircraft by make, model, serial number, and registration number.
4. The pages should be signed and dated by the person making the revision.
5. The nature of the weight change should be described.
6. The old weight and balance data should be marked "superseded" and dated.
7. A new page should show the date of the old figures it supersedes.
8. Appropriate fore and/or aft extreme loading conditions should be investigated and the computations shown.
9. Example loading computations may be helpful.
10. On large aircraft, be careful to distinguish between empty weight and operating weights that may include items, such as commissary supplies, spare parts, lavatory water, etc.
11. On small aircraft, it is often convenient to post a placard in the aircraft indicating the empty weight, useful load, and empty CG, along with example loadings or general instructions, to cover the most likely loading conditions.

Aircraft Owner/IA Relationships

Be sure to come to a mutual agreement with the aircraft owner concerning exactly what work is to be performed. Misunderstandings usually result from a lack of clear communication. Attention to the following details will usually avoid the ill will a later disagreement may generate.

1. Itemize the work to be done so the owner will have a clear understanding of the work order.
2. Establish a firm understanding about the cost, or range of cost, anticipated for the job.
3. If an annual inspection is involved, indicate that certain maintenance is required to perform the inspection, such as:
    a. Removing cowling and fairing, opening inspection plates, etc.
    b. Cleaning the aircraft and engine.
    c. Disassembling wheels and other components to determine their condition.
4. Advise the owner that an annual inspection involves determination of compliance with aircraft specifications and airworthiness directives (ADs).

5. Agree whether routine servicing is to be included as part of the inspection or if it is to be performed separately. Such servicing is not a part of the inspection, but may be conveniently done while conducting the inspection. Such items might be:
   a. Cleaning spark plugs.
   b. Servicing landing gear shock struts.
   c. Changing oil.
   d. Making minor adjustments.
   e. Servicing brakes.
   f. Dressing nicked propeller blades.
   g. Lubricating where necessary.
   h. Stop-drilling small cracks and minor patching of cowling and baffles.
6. The owner should be made aware that the annual or progressive inspection does not include correction of discrepancies or unairworthy items, and that such maintenance will be additional to the inspection. Maintenance and repairs may be accomplished simultaneously with the inspection by a person authorized to perform maintenance if agreed on by the owner and holder of the IA. This method would result in an aircraft that is approved for return to service upon the completion of the inspection.

   A written list of discrepancies and unairworthy items not repaired concurrently with the inspection must be made and given to the owner. Record uncorrected discrepancies and unairworthy items in the maintenance records. The owner must make arrangement for correction or deferral of items on the list of discrepancies and unairworthy items with a person authorized to perform maintenance prior to returning the aircraft to service.

   The holder of the IA ensures that any item permitted to be inoperative by an MEL or under 14 CFR §91.213(d)(2) are properly placarded and any maintenance for deferral has been carried out. Any deferred items are to be included on the list of discrepancies and unairworthy items. The owner should be informed that the aircraft should not be operated until the discrepancies and unairworthy items are corrected or are appropriately deferred.
7. Establish a reasonable time frame to accomplish the inspection.
8. Request the owner to supply the complete aircraft records (airframe, engines, and propellers) for study, review, and entries. Point out that this is necessary to properly conduct an annual inspection.
9. Complete the inspection as soon as practicable. Often, an aircraft will sit around the shops waiting for parts, even though the inspection has actually been completed. In this cases, it is advisable to officially report the aircraft unairworthy. (Refer to 14 CFR §43.11(a)(5).) When the parts arrive, the repairs can be completed and the aircraft approved for return to service in the usual manner by the person who makes the repairs. The time lapse may represent several weeks, or even months, and things can deteriorate on the aircraft. Also, there is the chance that an AD involving some part of the aircraft may have been issued in the interim. In these cases, it might be unwise to complete the repairs originally intended and sign off the aircraft as "airworthy" without doing another complete inspection.
10. Complete the aircraft record entries as required by 14 CFR §§43.9 and 43.11 and provide sufficient information for the owner to comply with 14 CFR §91.417(a)(2)(i). Make adequate descriptions of repairs or alterations if accomplished along with the inspection.

11. Record compliance with all ADs actually accomplished. Provide sufficient information for the owner to comply with 14 CFR §91.417(a)(2)(v). A general statement, such as "All ADs complied with" is *not* an adequate entry and should be avoided. Many owners keep a separate record of AD compliance in the back of the logbook or in a section specifically provided for this record. This is a good place to identify the ADs of a recurring nature and show when the next compliance is required.
12. When approving repairs and alterations, the holder of an IA should be available as work progresses on major jobs. In this way, affected areas and structures can be seen more readily than after completion of the entire job. In many cases, the workmanship can be inspected and improved easier during the process of the job rather than having to redo it later.
13. Remind the owners or operators that they are responsible for operational requirements, such as:
    a. VOR equipment checked in accordance with 14 CFR §91.171.
    b. Altimeter and altitude reporting equipment test and inspections in accordance with 14 CFR §91.411*.
    c. ATC transponder test and inspection in accordance with 14 CFR §91.413*.

    *These tests and inspections are not part of the annual inspection.

Chapter 3
Title 14 of the Code of Federal Regulations (14 CFR)

Introduction

The documents in Title 14 of the Code of Federal Regulations (14 CFR), formerly called the Federal Aviation Regulations, are the actual legal documents that govern civil aviation operations. Throughout this book the term "14 CFR" is directly interchangeable with the former "FAR."

It is the responsibility of an IA to have current copies of all the applicable parts of 14 CFR that pertain to aviation maintenance. IAs should refer to **rgl.faa.gov** to determine the currency of any document they are using.

This chapter contains typical questions taken from applicable Federal Regulations.

Note: On October 16, 2009, the FAA released major amendments to a number of regulations—14 CFR Parts 1, 21, 43 and 45 were affected by these revisions. Although the *effective* date for these in the Federal Register (FR) was April 14, 2010, only a portion of the changes became effective on that date. The remaining amendments have compliance dates of 18 months after the FR publication date. Currently it is not known when the FAA will update the IA test database and computer testing supplement accordingly with this amendment; therefore the proactive IA applicant will be prepared to answer questions that deal with the regulations as they currently exist as well as those affected by the new amendments. For further information, see the "Reader Resources" page on the ASA website (details are contained in a downloadable PDF of this October 2009 FR).

14 CFR Part 1
Definitions and Abbreviations

This part of 14 CFR contains a number of definitions and abbreviations that pertain to aviation operation. These are the legal definitions that take precedence over all others.

1. An alteration made in accordance with an aircraft specification would be

A—a minor alteration.
B—a major alteration.
C—accomplished with acceptable data.

A major alteration is an alteration not listed in the aircraft, aircraft engine, or propeller specifications—

(1) That might appreciably affect weight, balance, structural strength, performance, powerplant operation, flight characteristics, or other qualities affecting airworthiness; or

(2) That is not done according to accepted practices or cannot be done by elementary operations.

A minor alteration is an alteration other than a major alteration. Therefore, an alteration made in accordance with an aircraft specification is a minor alteration.

2. An alteration was done that normally would be considered a major alteration. How could this be signed off as a minor alteration?

A—If you, as the inspecting IA, determine that the repair will not appreciably change the aircraft's flight characteristics.
B—If the alteration is listed in the appropriate aircraft specification.
C—If the alteration is done in accordance with a Supplemental Type Certificate.

A major alteration is an alteration not listed in the aircraft, aircraft engine, or propeller specifications—

(1) That might appreciably affect weight, balance, structural strength, performance, powerplant operation, flight characteristics, or other qualities affecting airworthiness; or

(2) That is not done according to accepted practices or cannot be done by elementary operations.

A minor alteration is an alteration other than a major alteration. Therefore, an alteration made in accordance with an aircraft specification is a minor alteration.

Answers *All answer references in this chapter are to Title 14 of the Code of Federal Regulations (14 CFR)*

1 [A] (022) 14 CFR Part 1
2 [B] (022) 14 CFR Part 1

3. As the holder of an inspection authorization, you are performing an annual inspection on a fabric-covered airplane. During the review of the aircraft records you notice that the cover was recently replaced with a synthetic cover. The aircraft documents indicate only grade-A cotton material was certificated for the aircraft. The change in covering material would constitute a

A— minor repair.
B— major repair.
C— major alteration.

A major alteration is an alteration not listed in the aircraft, aircraft engine, or propeller specifications that might appreciably affect weight, balance, structural strength, performance, powerplant operation, flight characteristics, or other qualities affecting airworthiness; or that is not done according to accepted practices or cannot be done by elementary operations.

The material used for covering can affect structural strength, and this alteration was not listed in the specifications. Therefore, this is a major alteration.

3a. What would ***not*** be considered maintenance?

A— Inspection.
B— Preservation.
C— Preventive maintenance.

Maintenance means inspection, overhaul, repair, preservation, and the replacement of parts, but excludes preventive maintenance.

4. "Time in Service," where turbo-propeller powered airplanes are concerned, is defined in the regulation as

A— engine startup to engine shutdown.
B— airborne to touchdown.
C— the time the aircraft first begins to move until it comes to a final stop after the flight (chock to chock).

Time in service, with respect to maintenance time records, means the time from the moment the aircraft leaves the surface of the earth until it touches down at the next point of landing.

5. What is meant by the abbreviation TCAS?

A— Total Corrected Air Speed.
B— Terrain Clearance Activation System.
C— Traffic Alert and Collision Avoidance System.

The abbreviation and symbols section of 14 CFR Part 1 defines TCAS as Traffic Alert and Collision Avoidance System.

6. What is meant by the abbreviation V_2?

A— Stalling speed.
B— Takeoff safety speed.
C— Speed for best rate of climb.

The abbreviation and symbols section of 14 CFR Part 1 defines V_2 as takeoff safety speed.

6a. What is meant when the FAA uses the word "shall" in a regulation?

A— It is "advisable" to accomplish the action.
B— It is "permissible" to accomplish the action.
C— It is required that you accomplish the action.

When the word "shall" is used, it is in the imperative sense. It is required you accomplish the action.

6b. A mechanic replaces the engine on a Cessna 182 with another engine, which is listed on the aircraft TCDS. The mechanic should complete which of the following?

A— A logbook entry as this is only a minor alteration.
B— A 337 form, and get IA approval for this major alteration.
C— A 337 form, and get local FSDO "field approval."

The definition of "major alteration" is "...an alteration not listed in the aircraft...specifications." Since the engine was listed, it is only a minor alteration and nothing more than the logbook entry is required.

Answers Note: *All Learning Statement Codes in parentheses are preceded by "IAR"— see explanation on Page 1–10.*

3	[C]	(022)	14 CFR Part 1	3a	[C]	(030)	14 CFR Part 1	4	[B]	(030)	14 CFR Part 1
5	[C]	(031)	14 CFR Part 1	6	[B]	(031)	14 CFR Part 1	6a	[C]	(031)	14 CFR Part 1
								6b	[A]	(031)	14 CFR Part 1

6c. The term "consensus standard"

A— means everybody agrees with the FAA's interpretation.
B— is not an FAA-recognized term.
C— is an industry-developed standard that applies to LSA certification.

Consensus standard means, for the purpose of certificating light-sport aircraft, an industry-developed consensus standard that applies to aircraft design, production, and airworthiness. It includes, but is not limited to, standards for aircraft design and performance, required equipment, manufacturer quality assurance systems, production acceptance test procedures, operating instructions, maintenance and inspection procedures, identification and recording of major repairs and major alterations, and continued airworthiness.

14 CFR Part 21

Certification Procedures for Products, Articles and Parts

This part of 14 CFR prescribes the procedural requirements for the issue of type certificates and changes to those certificates: the issue of airworthiness certificates and the issue of export airworthiness approvals.

7. "Certification Procedures for Products, Articles and Parts" are found in which Part of Title 14 of the CFR?

A— Part 23.
B— Part 21.
C— Part 121.

14 CFR Part 21 is entitled "Certification Procedures for Products, Articles and Parts."

8. Aircraft certificated after what date were required to have an approved flight manual?

A— July 18, 1978.
B— November 1, 1981.
C— March 1, 1979.

With each airplane or rotorcraft that was not type-certificated with an Airplane or Rotorcraft Flight Manual and that has had no flight time prior to March 1, 1979, the holder of a type certificate (including a supplemental type certificate) or the licensee of a type certificate shall make available to the owner at the time of delivery of the aircraft a current approved Airplane or Rotorcraft Flight Manual.

9. If a person alters an aircraft by installing special wing tips, he or she may get approval to duplicate this alteration on other aircraft by the issuance of a

A— Provisional Type Certificate.
B— Supplemental Type Certificate.
C— Production Certificate.

Any person who alters a product by introducing a major change in type design, not great enough to require a new application for a new application for a type certificate under §21.19, shall apply to the Administrator for a supplemental type certificate, except that the holder of a type certificate for the product may apply for an amendment of the original type certificate.

10. An aircraft with a standard airworthiness certificate could be in which of the following categories?

A— Utility, acrobatic, and commuter.
B— Primary, restricted, and experimental.
C— Normal, provisional and Transport

Standard airworthiness certificates are airworthiness certificates issued for aircraft that are type-certificated in the normal, utility, acrobatic, commuter, or transport category and for unmanned free balloons and for aircraft designated by the Administrator as special classes of aircraft.

11. An aircraft with a special airworthiness certificate would be in which of the following categories?

A— Utility, acrobatic, and commuter.
B— Primary, restricted, and experimental.
C— Normal, provisional, and transport.

Special airworthiness certificates are primary, restricted, limited, and provisional airworthiness certificates, special flight permits, and experimental certificates.

Answers

6c	[C]	(031)	14 CFR Part 1	7	[B]	(030)	14 CFR Part 21	8	[C]	(030)	14 CFR Part 21
9	[B]	(022)	14 CFR Part 21	10	[A]	(030)	14 CFR Part 21	11	[B]	(030)	14 CFR Part 21

12. Which section of 14 CFR provides for the fabrication of aircraft replacement and modification parts (PMA)?

A— 14 CFR §21.303.
B— 14 CFR, section 23, Appendix B.
C— 14 CFR, section 45.21.

Part 21 is titled "Certification Procedures for Products and Articles" and contains regulations for certifying products and articles ranging from small PMA replacement parts to complete aircraft products and parts.

13. Before a new type of oil filter for a type certificated aircraft engine, manufactured by a person other than the original engine manufacturer, can be sold for installation on that engine, the manufacturer of the filter must be issued a

A— Parts Manufacturer Approval.
B— Supplemental Type Certificate.
C— Production Certificate.

Except as provided in paragraph (b) of this section no person may produce a modification or replacement part for sale for installation on a type certificated product unless it is produced pursuant to a Parts Manufacturer Approval issued under this subpart.

14. An export certificate of airworthiness may be issued for

A— a new or used aircraft if it meets the requirements of 14 CFR Part 21, Subpart E.
B— an engine, propeller or article, only if it meets the definition of being airworthy.
C— an engine or propeller that is not airworthy, if the importing country accepts the deviation from approved design or condition for safe operation on a form acceptable to the FAA.

Part 21 is titled "Certification Procedures for Products and Articles" and contains regulations for certifying products and articles ranging from small PMA pieces to complete aircraft, including export certificates. Normally the exported product or article must meet the requirements of being airworthy. However, the country seeking the import may waive that requirement if they specify the acceptance of the deviation on a form and in a manner acceptable to the FAA.

15. An altimeter that has been approved for installation on a civil aircraft may be manufactured under a

A— Supplemental Type Certificate.
B— Technical Standard Order.
C— Production Certificate.

A Technical Standard Order (TSO) is issued by the Administrator and is a minimum performance standard for specified articles (for the purpose of this subpart, articles means materials, parts processes, or appliances) used on civil aircraft.

Answers

12 [A] (030) 14 CFR Part 21
13 [A] (016) 14 CFR Part 21
14 [C] (016) 14 CFR Part 21
15 [B] (016) 14 CFR Part 21

14 CFR Part 23
Airworthiness Standards: Normal Category Airplanes

This part of 14 CFR prescribes airworthiness standards for the issue of type certificates, and changes to those certificates for airplanes in the normal, utility, acrobatic, and commuter categories.

16. Which regulation prescribes airworthiness standards for airplanes in the normal category, having a passenger seating configuration of 19 or less and a maximum certified takeoff weight of 19,000 pounds or less?

A— 14 CFR Part 21.
B— 14 CFR Part 23.
C— 14 CFR Part 25.

The FAA has recently amended its airworthiness standards for normal, utility, acrobatic, and commuter category airplanes by replacing current prescriptive design requirements with performance-based airworthiness standards (the effective date was August 30, 2017). These standards also replace the current weight and propulsion divisions in small airplane regulations with performance- and risk-based divisions for airplanes with a maximum seating capacity of 19 passengers or less and a maximum takeoff weight of 19,000 pounds or less. 14 CFR §23.2000 prescribes airworthiness standards for airplanes in the normal category. 14 CFR §23.2005 (a) limits passenger configuration and maximum certified takeoff weight.

17. When defining the weight limits and center of gravity for an airplane certificated under Part 23,

A— the applicant must determine those limits that provide for safe operation of the airplane.
B— the condition of the airplane at the time of determining the empty weight and center of gravity as defined by the FAA.
C— the FAA specifies only the critical combinations of weight and center of gravity.

The FAA has recently amended its airworthiness standards for normal, utility, acrobatic, and commuter category airplanes by replacing current prescriptive design requirements with performance-based airworthiness standards (the effective date was August 30, 2017). Therefore the burden is now on the applicant to establish the appropriate weight and balance procedures. The applicant must determine limits for weights and centers of gravity that provide for safe operation of the airplane.

18. When weighing an aircraft certified under 14 CFR Part 23,

A— the aircraft must be drained of all fluids.
B— the applicant must determine a well-defined and easily repeatable condition of the airplane.
C— the FAA determines the limits for empty weight and center of gravity.

The FAA has recently amended its airworthiness standards for normal, utility, acrobatic, and commuter category airplanes by replacing current prescriptive design requirements with performance-based airworthiness standards (the effective date was August 30, 2017). Therefore, the burden is now on the applicant to establish the appropriate weight and balance procedures. The condition of the airplane at the time of determining its empty weight and center of gravity must be well defined and easily repeatable, and must now be defined by the applicant.

19. The trim system for a Level 4 airplane must

A— maintain lateral and directional control during cruise.
B— maintain longitudinal control during cruise.
C— maintain lateral and directional control during normal operations.

The range of airplanes certified under Part 23 is diverse in terms of capability and complexity. The Level 4 airplane (which is one having a maximum seating configuration of 10-19 passengers) must maintain lateral and trim directional control without further force in normal operations.

20. The structure of an airplane certificated under 14 CFR Part 23 must have a factor of safety of

A— 2.3.
B— 1.5.
C— 4.0.

Unless otherwise provided, a factor of safety of 1.5 must be used.

Answers

16 [B] (030) 14 CFR Part 23
17 [A] (008) 14 CFR §23.2100(a)
18 [B] (008) 14 CFR §23.2100(c)
19 [C] (008) 14 CFR §23.2140(a)
20 [B] (030) 14 CFR Part 23

21. Regarding materials and processes in a Part 23 airplane,

A— the FAA determines all material design characteristics.
B— the applicant must determine the suitability and durability of the materials used.
C— the airplane must be capable of continued safe flight following a sudden release of cabin pressure.

The range of airplanes certified under Part 23 is diverse in terms of capability and complexity, varying from Level 1 (seating only one passenger) to Level 4 (seating up to 19 passengers). The FAA has recently replaced current prescriptive design requirements with performance-based airworthiness standards, placing much of the responsibility for the design of the aircraft on the applicant. Per 14 CFR §23.2260(a), the applicant must determine the suitability and durability of the materials used.

22. The fuel system of a Part 23 airplane must

A— provide the fuel necessary to each powerplant and the APU functions properly in all likely operating conditions.
B— provide fuel for at least one hour of operation at maximum continuous power or thrust.
C— be designed and arranged to provide for ignition of the fuel.

It is critical for safe flight for the fuel system to provide fuel as controlled by the pilot, to the engine(s). The range of airplanes certified under Part 23 is diverse in terms of capability and complexity, varying from Level 1 (seating only one passenger) to Level 4 (seating up to 19 passengers). Per 14 CFR §23.2430 (a)(3), each fuel system must provide the fuel necessary to each powerplant and the APU functions properly in all likely operating conditions.

23. Regarding the distribution of electrical power in a Part 23 airplane,

A— there must be enough capacity so that if the primary electrical source fails, the system will be capable of supplying power for essential loads.
B— the function at the airplane level is not adversely affected during the time it is exposed to lightning.
C— the function at the airplane level is not adversely affected during the time it is exposed to the HIRF environment.

It is critical for safe flight that the electrical system be capable of meeting "essential loads" required by the aircraft to continue flying. The range of airplanes certified under Part 23 is diverse in terms of capability and complexity, varying from Level 1 (seating only one passenger) to Level 4 (seating up to 19 passengers). Per 14 CFR §23.2525(c), the power generation, storage, and distribution for any system must be designed and installed to have enough capacity, if the primary source fails, to supply essential loads.

24. Regarding the installation of flight and navigation systems in a Part 23 airplane, the information must

A— not include any limitations.
B— be presented in a manner that the crewmember can monitor the parameters and determine trends.
C— inhibit the primary display of flight parameters.

During each phase of flight the aircraft instruments must provide the information necessary regarding flight, navigation, and powerplant operation. Per 14 CFR §23.2615(a), installed systems must provide the flight crewmember a presentation in such a way that the crewmember can monitor the parameters and determine trends.

25. Regarding instrument markings and placards in a Part 23 airplane,

A— the FAA will define the required placards and markings as necessary.
B— the applicant must include instrument marking and placard information in the Airplane Maintenance Manual.
C— each airplane must display in a conspicuous manner any placard or instrument marking necessary for operation.

It is critical for safe flight that the markings and placards determined to be necessary for operation and included in the Airplane Flight Manual be conspicuously displayed. Per 14 CFR §23.2610(a), each airplane must display in a conspicuous manner any placard or instrument marking necessary for operation.

Answers

21 [B] (030) 14 CFR Part 23
24 [B] (013) 14 CFR Part 23
22 [A] (030) 14 CFR Part 23
25 [C] (017) 14 CFR Part 23
23 [A] (013) 14 CFR Part 23

26. If the airplane has a flight data recorder installed, it must be

A— visible to crew at all segments of the flight.
B— installed so there is an aural or visual means for preflight checking of the recorder for proper recording of data in the storage medium.
C— painted either red or black.

The range of airplanes certified under Part 23 is diverse in terms of capability and complexity, varying from Level 1 (seating only one passenger) to Level 4 (seating up to 19 passengers). This requirement for a flight data recorder is one of only two paragraphs retained without change from the previous Part 23. 14 CFR §91.609 states "No holder of an air carrier operating certificate... unless that aircraft complies with any applicable flight data recorder and cockpit voice recorder requirements." Per 14 CFR §23.1459(a), each flight recorder required by the operating rules of Part 91 must be installed so that there is an aural or visual means for preflight checking of the recorder for proper recording of data in the storage medium.

27. Regarding the airplane fuel storage system, it must

A— provide fuel for at least one hour of operation at maximum continuous power or thrust.
B— provide the flight crew with a means to determine the total unusable fuel available.
C— be designed to prevent improper filling or recharging.

The range of airplanes certified under Part 23 is diverse in terms of capability and complexity, varying from Level 1 (seating only one passenger) to Level 4 (seating up to 19 passengers). The fuel for these aircraft varies from Avgas (100LL) to Turbine required (Jet A). These fuels are not compatible nor are the engines operable if they are improperly fueled. Per 14 CFR §23.2430(c), each fuel storage refilling or recharging system must be designed to prevent improper filling or recharging.

28. If the maintenance manual contains required maintenance, it will be listed in the Airworthiness Limitations Section. This section must include

A— items having time change intervals.
B— airframe specified inspection intervals and related procedures.
C— both A and B.

This statement is found in the G23.4 "Airworthiness Limitations" section of Appendix A to 14 CFR Part 23: "This section must set forth each mandatory replacement time, structural inspection interval, and related structural inspection procedure required for type certification."

28a. Aircraft certified under 14 CFR Part 23 must have "Airworthiness Limitations" documentation, including:

A— Special inspection techniques.
B— Structural inspection intervals.
C— Special tool catalog.

This statement is found in the G23.4 "Airworthiness Limitations" section of Appendix A to 14 CFR Part 23: "This section must set forth each mandatory replacement time, structural inspection interval, and related structural inspection procedure required for type certification."

28b. Aircraft certified under 14 CFR Part 23 must have "Airworthiness Limitations" documentation, including:

A— Service information.
B— Structural inspection intervals.
C— Maintenance Manual.

This statement is found in the G23.4 "Airworthiness Limitations" section of Appendix A to 14 CFR Part 23: "This section must set forth each mandatory replacement time, structural inspection interval, and related structural inspection procedure required for type certification."

28c. Within 14 CFR Part 23, certain required documents are outlined. Which of the following is correct?

A— The "Instruction for Continued Airworthiness" and "Airworthiness Limitation Section" are contained in the Manufacturer's Maintenance Manual.
B— The "Airworthiness Limitation Section" contains the "Instruction for Continued Airworthiness" and the "Manufacturer's Maintenance Manual."
C— The "Airworthiness Limitation Section" and the "Manufacturer's Maintenance Manual" are contained in the "Instruction for Continued Airworthiness."

The ICA must contain the "Airplane Maintenance Manual" and a section titled "Airworthiness Limitations Section" that is separated and clearly distinguishable from the rest of the document.

Answers

26	[B]	(017)	14 CFR Part 23
27	[C]	(024)	14 CFR Part 23
28	[C]	(017)	14 CFR Part 23
28a	[B]	(017)	14 CFR Part 23
28b	[B]	(017)	14 CFR Part 23
28c	[C]	(031)	14 CFR Part 23

28d. In addition to other requirements set forth in the maintenance manual of a 14 CFR part 23 airplane, commuter category additionally specifies which of the following maintenance instructions?

A— Towing instructions.
B— Servicing points.
C— Special repair methods.

For commuter category airplanes, the following information must be furnished:

1. *Electrical loads*
2. *Balancing methods*
3. *Identification of structure*
4. *Special repair methods*

29. When you are performing an annual inspection on an aircraft that has instructions for continued airworthiness, which of the following would be considered approved data?

A— Structural repair.
B— Maintenance manual.
C— Airworthiness limitations section.

The following statement is found in the G23.4 "Airworthiness Limitations" section of Appendix A to 14 CFR Part 23: "The Instructions for Continued Airworthiness must contain a section titled Airworthiness Limitations that is segregated and clearly distinguishable from the rest of the document. This section must set forth each mandatory replacement time, structural inspection interval, and related structural inspection procedure required for type certification." This section must contain a legible statement in a prominent location that reads, "The airworthiness limitations section is FAA-approved and specifies maintenance required under §43.16 and §91.403 of the Federal Aviation Regulations unless an alternate program has been FAA-approved."

14 CFR Part 27
Airworthiness Standards: Normal Category Rotorcraft

This part prescribes airworthiness standards for the issue of type certificates, and changes to those certificates, for normal category rotorcraft with maximum weights of 7,000 pounds and nine or less passenger seats.

30. Part 27 of 14 CFR prescribes airworthiness standards for which of the following?

A— Normal category rotorcraft.
B— Transport category rotorcraft.
C— Both A and B.

14 CFR Part 27 prescribes airworthiness standards for normal category rotorcraft.

30a. 14 CFR Part 27 prescribes airworthiness standards for which of the following?

A— Small rotorcraft.
B— Large rotorcraft.
C— Both A and B.

14 CFR Part 27 prescribes airworthiness standards for rotorcraft with a maximum weight of 7,000 pounds or less and nine or less passengers.

31. What are the designed limit maneuvering load factors required for a helicopter certificated under 14 CFR Part 27?

A— + 4.0 to -1.6.
B— + 3.8 to -1.5.
C— + 3.5 to -1.0.

The rotorcraft must be designed for a limit maneuvering load factor ranging from a positive limit of 3.5 to a negative limit of -1.0.

32. The materials used in the seats of a helicopter must be rated at least

A— flame resistant.
B— fireproof.
C— fire resistant.

For each compartment used by the crew or passengers the materials must be at least flame resistant.

Answers

28d [C] (031) 14 CFR Part 23
29 [C] (024) 14 CFR Part 23
30 [A] (031) 14 CFR Part 27
30a [A] (017) 14 CFR Part 27
31 [C] (031) 14 CFR Part 27
32 [A] (031) 14 CFR Part 27

33. The reciprocating engine for a helicopter must be approved under 14 CFR Part

A— 27.
B— 33.
C— 29.

According to 14 CFR §27.903, reciprocating engines for use in helicopters must be qualified in accordance with §33.49(d) or be otherwise approved for the intended usage.

34. What is the maximum deviation error allowed in a magnetic direction indicator installed in a helicopter?

A— 5 degrees.
B— 10 degrees.
C— 15 degrees.

According to 14 CFR §27.1327, each magnetic direction indicator must be installed so that its accuracy is not excessively affected by the rotorcraft's vibration or magnetic fields; and the compensated installation may not have a deviation, in level flight, greater than 10 degrees on any heading.

35. Where are the requirements found for the preparation of Instructions for Continued Airworthiness for a rotorcraft?

A— In Appendix A of Part 27.
B— In Appendix B of Part 21.
C— In Appendix B of Part 91.

The applicant must prepare Instructions for Continued Airworthiness in accordance with Appendix A of Part 27 that are acceptable to the Administrator.

14 CFR Part 43
Maintenance, Preventive Maintenance, Rebuilding, and Alteration

This part of 14 CFR prescribes rules governing the maintenance, preventive maintenance, rebuilding and alteration of any aircraft having a U.S. airworthiness certificate, foreign-registered civil aircraft used in common carriage or carriage of mail under the provisions of Part 121 or 135 of this chapter and airframe, aircraft engines, propellers, appliances and component parts of such aircraft. This part does not apply to any aircraft for which an experimental airworthiness certificate has been issued, unless a different kind of airworthiness certificate had previously been issued for that aircraft.

36. Certain systems of rotorcraft are specified by 14 CFR Part 43 to be inspected in accordance with the maintenance manual or Instructions for Continued Airworthiness as provided by the manufacturer concerned. Of the following, which is such a system?

A— Engine.
B— Flight controls.
C— Drive shaft.

Each person performing an inspection required by Part 91 on a rotorcraft shall inspect the following systems in accordance with the maintenance manual or Instructions for Continued Airworthiness of the manufacturer concerned:

- *The drive shafts or similar systems*
- *The main rotor transmission gear box for obvious defects*
- *The main rotor and center section (or the equivalent area)*
- *The auxiliary rotor on helicopters*

Answers

33 [B] (031) 14 CFR Part 27
34 [B] (030) 14 CFR Part 27
35 [A] (031) 14 CFR Part 27
36 [C] (017) 14 CFR Part 43

37. As an IA, you are performing an annual inspection on a rotorcraft in accordance with 14 CFR Part 43, Appendix D. Would a maintenance manual, or a section thereof, also be required to perform the inspection?

A— They are required when part of an approved inspection program.
B— They are required if made mandatory by the type certificate data sheet.
C— Maintenance manuals are optional if an equivalent procedure is used.

Each person performing an inspection required by Part 91 (this includes inspections conducted in accordance with 14 CFR Part 43, Appendix D) on a rotorcraft shall inspect the following systems in accordance with the maintenance manual or Instructions for Continued Airworthiness of the manufacturer concerned:

- *The drive shafts or similar systems*
- *The main rotor transmission gear box for obvious defects*
- *The main rotor and center section (or the equivalent area)*
- *The auxiliary rotor on helicopters*

Many helicopter TCDS refer to the manufacturer maintenance manuals as required information.

38. Each person performing an annual inspection must use a checklist while performing the inspection. The checklist must include the scope and detail of the items in 14 CFR Part 43, Appendix D. In addition to the items in the appendix, additional specific items are required by 14 CFR Part 43 if the aircraft is a

A— balloon.
B— multiengine airplane.
C— rotorcraft.

Each person performing an inspection required by Part 91 on a rotorcraft shall inspect the following systems in accordance with the maintenance manual or Instructions for Continued Airworthiness of the manufacturer concerned:

- *The drive shafts or similar systems*
- *The main rotor transmission gear box for obvious defects*
- *The main rotor and center section (or the equivalent area)*
- *The auxiliary rotor on helicopters*

39. Following an annual inspection, if the aircraft is found unairworthy, regulations require a list of discrepancies and unairworthy items be given to whom?

A— The owner or operator.
B— The owner or operator and the Flight Standards District Office (FSDO).
C— The owner or operator, the FSDO, and a copy for your file.

If the person performing any inspection required by 14 CFR Part 91, Part 125 or §135.411(a)(1) finds that the aircraft is unairworthy or does not meet the applicable type certificate data, airworthiness directives, or other approved data upon which its airworthiness depends, that person must give the owner or lessee a signed and dated list of those discrepancies.

40. Which Canadian maintenance entity may perform an annual inspection on a U.S.-registered aircraft?

A— Aviation maintenance engineer.
B— Approved maintenance organization.
C— Neither A nor B.

A person holding a valid Canadian Department of Transport license (Aviation Maintenance Engineer) or an (Approved Maintenance Organization) (AMO) may perform maintenance, preventive maintenance, and alterations on a U.S.-registered aircraft located in Canada (including any inspection required by 14 CFR §91.409, except an annual inspection).

41. Which Canadian maintenance entity may approve major repairs and major alterations on U.S.-registered aircraft?

A— Aviation maintenance engineer.
B— Approved maintenance organization.
C— Neither A nor B.

An Approved Maintenance Organization (AMO) whose system of quality control for the maintenance, preventive maintenance, alteration, and inspection of aeronautical products has been approved by the Canadian Department of Transport, or an authorized employee performing work for such an AMO, may approve (certify) a major repair or major alteration performed under this section if the work was performed in accordance with technical data approved by the Administrator.

Answers

37 [B] (017) 14 CFR Part 43
38 [C] (017) 14 CFR Part 43
39 [A] (017) 14 CFR Part 43
40 [C] (021) 14 CFR Part 43
41 [B] (021) 14 CFR Part 43

42. Which Canadian maintenance entity may approve data for major repairs to U.S.-registered aircraft?

A— Aviation maintenance engineer.
B— Approved maintenance organization.
C— Neither A nor B.

An Approved Maintenance Organization (AMO) whose system of quality control for the maintenance, preventive maintenance, alteration, and inspection of aeronautical products has been approved by the Canadian Department of Transport, or an authorized employee performing work for such an AMO, may approve (certify) a major repair or major alteration performed under this section if the work was performed in accordance with technical data approved by the Administrator.

Note: *the question does not ask who can do the work, but rather who may approve the data. As the reference subparagraph states, "...performed in accordance with technical data approved by the FAA."*

43. The overhaul of a system or component of an aircraft on a progressive inspection program is considered to be what type of inspection?

A— Detailed.
B— Routine.
C— Component segment.

Detailed inspections consist of a thorough examination of the appliances, the aircraft, and its components and systems, with such disassembly as is necessary. For the purposes of this subparagraph, the overhaul of a component or system is considered to be a detailed inspection.

44. In accordance with 14 CFR Part 43, a component part that has been disassembled, cleaned, inspected, repaired as necessary, reassembled, and tested to the same tolerances and limits as a new item is classified as

A— refurbished.
B— remanufactured.
C— rebuilt.

No person may describe, in any required maintenance entry or form, an aircraft, airframe, aircraft engine, propeller, appliance, or component part as being rebuilt unless it has been disassembled, cleaned, inspected, repaired as necessary, reassembled, and tested to the same tolerances and limits as a new item, using either new parts or used parts that either conform to new part tolerances and limits or to approved oversized or undersized dimensions.

45. In accordance with 14 CFR Part 43, an aircraft engine that has been disassembled, cleaned, inspected, repaired as necessary and reassembled, then tested in accordance with approved standards and technical data is classified as

A— overhauled.
B— remanufactured.
C— rebuilt.

No person may describe in any required maintenance entry or form an aircraft, airframe, aircraft engine, propeller, appliance, or component part as being overhauled unless using methods, techniques, and practices acceptable to the Administrator, it has been disassembled, cleaned, inspected, repaired as necessary, and reassembled; and has been tested in accordance with approved standards and technical data, or in accordance with current standards and technical data acceptable to the Administrator, which have been developed and documented by the holder of the type certificate, supplemental type certificate, or a material, part, process, or appliance approval under §21.305.

46. What Part of the Federal Aviation Regulations contains the standards of performance under which the holder of an inspection authorization performs his/her activities?

A— 14 CFR Part 43.
B— 14 CFR Part 61.
C— 14 CFR Part 91.

The standards of performance under which an IA performs are delineated in 14 CFR §43.13, "Performance rules (general)" and §43.15, "Additional performance rules for inspections."

Answers

42	[C]	(021)	14 CFR Part 43
43	[A]	(031)	14 CFR Part 43
44	[C]	(031)	14 CFR Part 43
45	[A]	(031)	14 CFR Part 43
46	[A]	(031)	14 CFR Part 43

47. When performing an annual inspection of a helicopter main rotor transmission gearbox, the inspection will be accomplished in accordance with what data?

A— 14 CFR Part 43, Appendix B.
B— Maintenance manual or instructions for continued airworthiness.
C— Manufacturer's service bulletins.

Each person performing an inspection required by Part 91 on a rotorcraft shall inspect the following systems in accordance with the maintenance manual or Instructions for Continued Airworthiness of the manufacturer concerned:

- *The drive shafts or similar systems*
- *The main rotor transmission gear box for obvious defects*
- *The main rotor and center section (or the equivalent area)*
- *The auxiliary rotor on helicopters*

48. As specified by 14 CFR Part 43, which of the following inspections is NOT part of a progressive inspection program?

A— Initial.
B— Routine.
C— Segment.

Each person performing a progressive inspection shall, at the start of a progressive inspection system, inspect the aircraft completely. After this initial inspection, routine and detailed inspections must be conducted as prescribed in the progressive inspection schedule.

49. During the annual inspection, you note a minor repair was made to an airframe component by a repair station. The repair was recorded on a work order containing reference to acceptable data and the date of completion. What other information must be on the work order?

A— Signature and certificate number of person approving the repair.
B— A maintenance release signed by an authorized representative of the repair station.
C— A yellow tag signed by an authorized representative of the repair station.

Except as provided in 14 CFR §43.9(b) and (c), each person who maintains, performs preventive maintenance, rebuilds, or alters an aircraft, airframe, aircraft engine, propeller, appliance, or component part shall make an entry in the maintenance record of that equipment containing the following information:

1. *A description (or reference to data acceptable to the Administrator) of work performed.*
2. *The date of completion of the work performed.*
3. *The name of the person performing the work if other than the person specified in paragraph (a)(4) of this section.*
4. *If the work performed on the aircraft, airframe, aircraft engine, propeller, appliance, or component part has been performed satisfactorily, the signature, certificate number, and kind of certificate held by the person approving the work.*

50. What Part of Title 14 of the CFR referred to in §65.95(a)(1) contains the standards of performance under which a holder of an inspection authorization performs his or her activity?

A—Part 43.
B—Part 73.
C—Part 135.

14 CFR §43.13 specifies the general performance rules, and §43.15 gives additional performance rules for inspections.

51. Within Title 14 of the Code of Federal Regulations, the term "remanufactured"

A— means an engine has been overhauled to serviceable limits.
B— means an engine has been overhauled to standards that allow zero time to be established.
C—has no specific meaning.

The categories of maintenance in 14 CFR Part 43 and Part 91 include overhaul and rebuilding, but there is no reference to remanufacturing.

Answers

47 [B] (021) 14 CFR Part 43
48 [C] (021) 14 CFR Part 43
49 [A] (017) 14 CFR Part 43
50 [A] (017) 14 CFR Part 43
51 [C] (031) 14 CFR Part 43 and 91

52. Certain maintenance records are required to be retained until the work is repeated, superseded by other work or for 1 year. An exception to this rule occurs with maintenance records concerning what maintenance?

A— Engine remanufacture records.
B— 100-hour inspection.
C— Altimeter test and inspection.

14 CFR §91.417(b) lists certain maintenance records that must be retained for one year or until the work is repeated or superseded by other work. There are exceptions to this specified in §91.417(a) and AC 43-9C (Maintenance Records). Records for inspection of altimeter systems and altitude reporting equipment must be retained for 24 months or until the work is repeated or superseded.

53. 14 CFR Part 43 requires persons performing maintenance to make entries in the Maintenance records. If the logbook entry is made per the requirements of §91.417, does this entry also meet the requirements of 14 CFR Part 43?

A— No, it would not be sufficient.
B— Yes, if it is for an inspection only.
C— Yes, it would be sufficient for maintenance and inspections.

14 CFR §91.417(a) requires that the maintenance record include the date of completion of the work and the signature and certificate number of the person approving the aircraft for return to service.

14 CFR §43.9(a)(4) requires that in addition to the name and certificate number that the kind of certificate held by the person be specified.

54. 14 CFR Part 43 requires persons performing maintenance to make entries in the Maintenance records. If the logbook entry is made per the requirements of 14 CFR Part 43, does this entry also meet the requirements of 14 CFR §91.417?

A— No, it would not be sufficient.
B— Yes, if it is for an inspection only.
C— Yes, it would be sufficient for maintenance and inspections.

14 CFR §91.417(a) requires that the maintenance record include the date of completion of the work and the signature and certificate number of the person approving the aircraft for return to service.

14 CFR §43.9(a)(4) requires that in addition to the name and certificate number that the kind of certificate held by the person be specified.

55. During the review of an aircraft's records while performing an annual inspection, you discover that the altimeter has not been inspected and tested within the preceding 24 calendar months. If the aircraft meets all other airworthiness requirements, as the inspection authorization, you should

A— list the altimeter as a deficiency and report the aircraft as unairworthy.
B— approve the aircraft for return to service after placarding the instrument "for day VFR usage only."
C— approve the aircraft for return to service and advise the owner or operator that the aircraft should not be used under instrument flight rules.

No person may operate an airplane, or helicopter, in controlled airspace under IFR unless within the preceding 24 calendar months, each static pressure system, each altimeter instrument, and each automatic pressure altitude reporting system has been tested and inspected and found to comply with Appendix E of 14 CFR Part 43.

The aircraft can be approved for return to service, but the owner or operator should be advised that the aircraft should not be used for IFR flight in controlled airspace.

56. According to Title 14 of the CFR, an engine may be granted zero time by the manufacturer, if it has been

A— newly overhauled.
B— rebuilt.
C— remanufactured.

The owner or operator may use a new maintenance record, without previous operating history, for an aircraft engine rebuilt by the manufacturer or by an agency approved by the manufacturer.

Answers

52 [C] (021) 14 CFR Part 91
53 [A] (021) 14 CFR Part 43 and 91
54 [C] (021) 14 CFR Part 43 and 91
55 [C] (017) 14 CFR Part 91
56 [B] (031) 14 CFR Part 91

57. You are performing an annual inspection on a small twin-engine airplane. The engine has been "zero-timed" by the manufacturer. What would you expect to find in the maintenance records?

A— A new engine logbook.
B— An appropriate entry by the manufacturer.
C— Both A and B.

Each manufacturer or agency that grants zero time to an engine rebuilt by it shall enter in the new record a signed statement of the date the engine was rebuilt.

58. You are performing an annual inspection on a small twin-engine airplane. The engine has been "zero-timed" by the manufacturer. What would you expect to find in the maintenance records?

A— A new engine logbook.
B— A Form 337, work order, and a new engine logbook.
C— An 8130-3 approval tag, maintenance release and a new engine logbook.

The owner or operator may use a new maintenance record, without previous operating history, for an aircraft engine rebuilt by the manufacturer or by an agency approved by the manufacturer. Neither a Form 337 nor an 8130-3 tag would be used.

59. You are performing an annual inspection on an aircraft where the engine has been "zero-timed" by the manufacturer. What does this indicate?

A— That the parts have been inspected and found to meet the manufacturer's new specifications (minimum and maximum dimensions).
B— That the parts used are new parts.
C— That the parts used meet the serviceable limits as listed in the manufacturer's overhaul manual.

An engine that has been "zero timed" by the manufacturer has been rebuilt. A rebuilt engine is a used engine that has been completely disassembled, inspected, repaired as necessary, reassembled, tested, and approved in the same manner and to the same tolerances and limits as a new engine with either new or used parts. However, all parts used must conform to the production drawing tolerances and limits for new parts or be of approved oversized or undersized dimensions for a new engine.

60. When a normally-aspirated engine having spur-type reduction gearing is overhauled, an FAA Form 337 is

A— completed and retained until the engine is again overhauled or removed from the airframe and replaced with a different engine.
B— not required to be completed.
C— completed and retained until the work is repeated or superseded by other work or for 1 year after the work is performed.

The overhaul of a normally-aspirated (not equipped with an integral supercharger) reciprocating engine having a spur-type propeller reduction gearing is considered to be a minor repair, and as such, it does not require the completion of an FAA Form 337.

61. Airframe alterations of the type found in 14 CFR Part 43, Appendix A, are major alterations except when listed in

A— aircraft specifications issued by FAA.
B— supplemental type certificates.
C— manufacturers' maintenance manuals.

Alterations of the following parts and alterations of the following types, when not listed in the aircraft specifications issued by the FAA, are airframe major alterations.

62. A duplicate copy of FAA Form 337 shall be forwarded to the local FAA Aircraft Registration Branch within how many hours after the work is inspected?

A— 24.
B— 48.
C— 72.

Each person performing a major repair or major alteration shall execute FAA Form 337 at least in duplicate. Give a signed copy of that form to the aircraft owner, and forward a copy of that form to the FAA Aircraft Registration branch in Oklahoma City within 48 hours after the aircraft, airframe, aircraft engine, propeller, or appliance is approved for return to service.

Answers

57 [C] (021) 14 CFR Part 91
58 [A] (017) 14 CFR Part 91
59 [A] (021) 14 CFR Part 91
60 [B] (010) 14 CFR Part 43
61 [A] (010) 14 CFR Part 43
62 [B] (010) 14 CFR Part 43

63. During the inspection of a complete electrical system installation, you determine the field coils of the generator installed had been rewound. For this work to be acceptable, what kind of maintenance record would you expect to find?

A— FAA Form 337.
B— Repair station maintenance release.
C— Either A or B.

Rewinding generator field coils is an appliance major repair. This may be recorded either with an FAA Form 337 or with a repair station maintenance release.

64. For which of the following would you expect to find a maintenance release?

A— Straightening of steel blades.
B— Repair of a propeller governor.
C— Either A or B.

Both straightening steel propeller blades and repairing propeller governors are major repairs. They may be recorded with a repair station maintenance release.

64a. Which of the following would be considered a propeller major repair?

A— Straightening of steel blades.
B— Repair of a propeller governor.
C— Both A and B.

14 CFR Part 43, Appendix A provides specific examples of major repairs and alterations to airframes, powerplants, propellers and appliances. Straightening of blades and repair of governors are both listed in the propeller section under major repairs.

64b. Which of the following would be considered a propeller major repair?

A— Straightening of aluminum blades.
B— Repair of a propeller governor.
C— Both A and B.

14 CFR Part 43, Appendix A provides specific examples of major repairs and alterations to airframes, powerplants, propellers and appliances. Straightening of blades and repair of governors are both listed in the propeller section under major repairs.

64c. During the records review at annual inspection, for which of the following would you expect to find a repair station maintenance release?

1. Straightening of an aluminum propeller blade.
2. Repair of a propeller governor.

A— 1.
B— 2.
C— Both 1 and 2.

14 CFR 43, Appendix A provides specific examples of major repairs and alterations to airframes, powerplants, propellers and appliances. Straightening of blades and repair of governors are both listed in the propeller section under major repairs.

65. If a repair station does a major repair and uses a maintenance release and an FAA Form 337, which of the following would require a statement for return to service?

A— Maintenance release.
B— Form 337.
C— Both A and B.

For major repairs made in accordance with a manual or specifications acceptable to the administrator, a certificated repair station may in place of the Form 337 use the customer's work order and give the aircraft owner a maintenance release that includes a statement such as "The aircraft, airframe, aircraft engine, propeller, or appliance identified above was repaired and inspected in accordance with current Regulations of the Federal Aviation Administration and is approved for return to service. Pertinent details of the repair are on file at this repair station under Order No. ________." This must be dated and signed by an authorized representative of the repair station.

Answers

63	[C]	(017)	14 CFR Part 43
64	[C]	(017)	14 CFR Part 43
64a	[C]	(021)	14 CFR Part 43
64b	[C]	(021)	14 CFR Part 43
64c	[C]	(032)	14 CFR Part 43
65	[A]	(019)	14 CFR Part 43

66. If a repair station does a major repair and uses a maintenance release and an FAA Form 337, which of the following would require a description of the work performed?

A— Maintenance release.
B— Form 337.
C— Both A and B.

For major repairs made in accordance with a manual or specifications acceptable to the administrator, a certificated repair station may use a Form 337. The back of the form (Item 8) is used to describe the work performed.

67. A major repair has been completed and you as the inspecting IA have approved the aircraft for return to service. How will you distribute the copies of FAA Form 337?

A— Both copies are returned to the person who performed the work.
B— One copy is placed in the aircraft records and one copy is sent to the FAA District Office within 48 hours.
C— One copy is given to the aircraft owner, one is sent to the FAA District Office within 48 hours and one copy is placed in your records.

An IA who has found a major alteration or major repair to be in conformity with FAA-approved data, should review the Form 337 for completeness and accuracy, then complete and sign item 7 (Approval for return to service) on both the original and the duplicate copies. Both copies are returned to the person who performed the work. This person will give a signed copy of the Form 337 to the aircraft owner and forward a copy to the FAA Registration Branch in Oklahoma City, OK within 48 hours after the aircraft is approved for return to service. The basic question here is not where the forms eventually go, but what responsibility the IA has in that distribution process. The reference given is very clear that this is the responsibility of the person doing the work.

68. 14 CFR Part 43 contains several appendixes that deal with maintenance, inspection, repair, alteration, and recordation. Which appendix specifies various alterations, the performance of which is a major alteration?

A— Appendix A.
B— Appendix B.
C— Appendix C.

14 CFR Part 43, Appendix A lists a number of alterations that are identified as major alterations unless they are listed in the aircraft specifications.

69. 14 CFR Part 43 contains several appendixes that deal with maintenance, inspection, repair, alteration, and recordation. Which appendix provides information on the recording of major repairs and major alterations?

A— Appendix A.
B— Appendix B.
C— Appendix C.

14 CFR Part 43, Appendix B describes the procedures for recording major repairs and major alterations.

70. During an annual inspection, in accordance with 14 CFR Part 43, Appendix D, the altimeter instrument must be inspected for

A— marking.
B— calibration.
C— both of the above.

Each person performing an annual or 100-hour inspection shall inspect (where applicable) the following components of the cabin and cockpit group: "Instruments—for poor condition, mounting, marking, and (where practicable) improper operation."

71. The maintenance release used by a repair station to approve a major repair for return to service must be signed by whom?

A— An inspection authorization.
B— The A&P mechanic performing the work.
C— An authorized representative of the repair station.

For major repairs made in accordance with a manual or specifications acceptable to the Administrator, a certificated repair station may give the aircraft owner a maintenance release signed by an authorized representative of the repair station.

Answers

66	[B]	(019)	14 CFR Part 43	67	[A]	(019)	14 CFR Part 43	68	[A]	(031)	14 CFR Part 43
69	[B]	(031)	14 CFR Part 43	70	[A]	(031)	14 CFR Part 43	71	[C]	(031)	14 CFR Part 43

72. Who has the primary responsibility for determining if a repair or alteration is major or minor?

A— The FAA inspector in whose area of responsibility the work is accomplished.
B— The person making the repair or alteration.
C— The FAA regional engineer responsible for the area where the work is performed.

The person making the repair or alteration must determine whether it is major or minor by referring to 14 CFR Part 43 Appendix A. This determines the approval needed for the data used.

73. While reviewing the records of an aircraft during an annual inspection, a major repair by a certificated repair station was noted. The repair was recorded on a work order containing a maintenance release from the repair station. However, no FAA Form 337 could be found in the records. What action should you take as the holder of an inspection authorization?

A— Contact the repair station and have them send you an FAA Form 337 for the repair.
B— Inform the aircraft owner of the oversight and allow he or she to make arrangement for the form.
C— No further action is necessary.

A major repair made by a certificated repair station may be recorded with an FAA Form 337 or by a maintenance release and a signed copy of the work order.

74. Which of the following would NOT be a propeller major repair?

A— Replacement of tip fabric.
B— Refinishing the blades.
C— Shortening of blades on a wood propeller.

Refinishing the blades of a propeller is not considered to be a propeller major repair.

75. You have been asked to inspect and approve for return to service the relocation of a battery. The specific installation will follow the general guidelines of a Supplemental Type Certificate, with the exception that the battery is located 6 inches aft of the location shown on the Supplemental Type Certificate. The aft center-of-gravity limit is exceeded by 0.1 inch. Concerning the installation, which of the following is true?

A— The deviation to the Supplemental Type Certificate is in itself a major alteration and will require additional approval.
B— The installation could be approved for return-to-service if an inspection authorization inspects and approves the deviation.
C— The exceeding of the aft center-of-gravity by 0.1 inch is shown to be negligible by AC 43.13-1B and may be approved as a minor alteration.

The deviation from the STC constitutes a major alteration. Since this alteration has not been done in accordance with the STC, it is not done according to approved data, and it has moved the CG out of its approved limits. Additional approval is required.

76. A repair station may use a maintenance release in place of an FAA Form 337 for which of the following work?

A— Major repairs.
B— Major alterations.
C— Either A or B.

For major repairs made in accordance with a manual or specifications acceptable to the Administrator, a certificated repair station may, in place of FAA Form 337, use the customer's work order upon which the repair is recorded. The aircraft owner must be given a maintenance release signed by an authorized representative of the repair station.

Answers

72 [B] (031) 14 CFR Part 43
73 [C] (017) 14 CFR Part 43
74 [B] (019) 14 CFR Part 43
75 [A] (017) 14 CFR Part 43
76 [A] (019) 14 CFR Part 43

77. Following the installation of extended range fuel tanks in the baggage compartment of a Cessna model 310, you, as the approving authorized inspector, make the required record entries in the aircraft's records and complete the FAA Form 337. What is the distribution of the form?

A— One copy is given to the aircraft owner and one is sent to the FAA Aircraft Registration Branch in Oklahoma City, OKwithin 48 hours.
B— One copy is sent to the FAA Aircraft Registration Branch in Oklahoma City, OK within 48 hours, one copy is placed in the aircraft and one is given to the owner.
C— One copy is placed in the aircraft, one copy is sent to the FAA Aircraft Registration Branch in Oklahoma City, OK within 48 hours, and one copy is placed in your records.

When extended range fuel tanks are installed in a passenger compartment or a baggage compartment, the person who performs the work and the person authorized to approve the work shall execute an FAA Form 337 in at least triplicate. One copy shall be placed aboard the aircraft as specified in §91.417. One copy is given to the owner, and the other copy is sent to the FAA Aircraft Registration Branch in Oklahoma City, OK within 48 hours after the aircraft is approved for return to service.

78. An aircraft has a cracked aileron control bracket. To restore the aircraft to an airworthy condition, a certificated airframe mechanic reinforces the bracket. This procedure would constitute a

A— major repair.
B— major alteration.
C— minor repair.

Repairs to the following parts of an airframe and repairs of the following types, involving the strengthening, reinforcing, splicing, and manufacturing of primary structural members or their replacement, when replacement is by fabrication such as riveting or welding, are airframe major repairs.

79. When a repair station makes a major repair to an appliance, the repair may be recorded on a tag attached to the product or on the customer's work order. In either case, the document used must include what information?

A— Name of part.
B— Date of manufacture.
C— Both A and B.

The Authorized Release Certificate or the maintenance release must include the name of the part, including the manufacturer's name, name of the part, model, and serial number (if any).

80. When a repair station makes a major repair to an appliance, the repair may be recorded on a tag attached to the product or on the customer's work order. In either case, the document used must include what information?

A— Name of part.
B— Manufacturer's name.
C— Both A and B.

The Authorized Release Certificate or the maintenance release must include the name of the part, including the manufacturer's name, name of the part, model, and serial number (if any).

81. When a repair station makes a major repair to an appliance, the repair may be recorded on a tag attached to the product or on the customer's work order. In either case, the document used must include what information?

A— Manufacturer's name.
B— Date of manufacture.
C— Both A and B.

The Authorized Release Certificate or the maintenance release must include the name of the part, including the manufacturer's name, name of the part, model, and serial number (if any).

Answers

77 [B] (017) 14 CFR Part 43
78 [A] (019) 14 CFR Part 43
79 [A] (031) 14 CFR Part 43
80 [C] (031) 14 CFR Part 43
81 [A] (031) 14 CFR Part 43

82. When a repair station makes a major repair to an appliance, the repair may be recorded on a tag attached to the product or on the customer's work order. In either case, the document used must include what information?

A— Model.
B— Manufacturer's name.
C— Both A and B.

The Authorized Release Certificate or the customer's work order for a major repair to an appliance must include the manufacturer's name, name of the part, model, and serial number (if any).

82a. When performing a static pressure system test, as required by 14 CFR 91.411, the technician must

A— determine that leakage is within established and allowable limits.
B— ensure the system is free from entrapped moisture and restrictions.
C— Both A and B.

Each person performing the altimeter system tests and inspections required by 14 CFR §91.411 shall comply with the following:

(a) Static pressure system:

(1) Ensure freedom from entrapped moisture and restrictions.

(2) Determine that leakage is within the tolerances established in 14 CFR §23.1325 or §25.1325, whichever is applicable.

82b. Which of the following readings is not an acceptable altitude readout for a barometric pressure reading of 25.842?

A— 4,025 feet.
B— 4,046 feet.
C— 3,965 feet.

For an altitude of 4,000 feet, the equivalent pressure is 25.842. There is a tolerance of ±35 feet. Therefore, the acceptable altitude reading ranges from 3,965 (minimum) to 4,035 (maximum).

82c. When conducting a case leak test at an indicated altitude of 18,000 feet, the leakage of the altimeter case shall not change the altimeter reading by more than

A— ±50 feet.
B— ±100 feet.
C— Neither A nor B.

For a case leak: the leakage of the altimeter case, when the pressure within it corresponds to an altitude of 18,000 feet, shall not change the altimeter reading by more than the tolerance shown in Table II during an interval of 1 minute. The tolerance at 18,000 feet is ±100 feet.

82d. Upon successful completion of an altimeter test, what information is to be recorded on the altimeter?

A— The date of the inspection and the minimum altitude to which it was tested.
B— The date of the inspection and the certificate number of the repair station performing the test.
C— The date of the inspection and the maximum altitude to which it was tested.

The person performing the altimeter tests shall record on the altimeter the date and the maximum altitude to which the altimeter has been tested and the person approving the aircraft for return to service shall enter that data in the airplane log or other permanent record.

Answers

82 [C] (031) 14 CFR Part 43
82a [C] (031) 14 CFR Part 43
82b [B] (031) 14 CFR Part 43
82c [B] (031) 14 CFR Part 43
82d [C] (031) 14 CFR Part 43

14 CFR Part 45
Identification and Registration Marking

This part prescribes the requirements for identification of aircraft, and identification of aircraft engines and propellers that are manufactured under the terms of a type or production certificate; identification of certain replacement and modified parts produced for installation on type certificated products; and nationality and registration marking of U.S.-registered aircraft.

83. In conjunction with your performance of an annual inspection on an aircraft, you determine that the aircraft cannot be approved for return to service due to the fact that one of the following items is missing. Which item is it?

A— Airworthiness certificate.
B— Aircraft registration.
C— Aircraft data plate.

Aircraft covered under 14 CFR §21.182 must be identified. The aircraft identification plate must be secured to the aircraft fuselage exterior so that it is legible to a person on the ground, and must be either adjacent to and aft of the rear-most entrance door, or on the fuselage surface near the tail surfaces.

83a. When conducting the ATC transponder test, the radio reply frequency for a Class 1A transponder is

A— 1090 ±3 MHz.
B— 1090 ±1 MHz.
C— Neither A nor B.

For Classes 1A, 2A, 3A and 4 Mode S transponders, interrogate the transponders and verify that the reply frequency is 1090 ±1 MHz.

83b. When conducting the RF peak output power test, the minimum RF peak output for Class 1B and 2B transponders is

A— 70 watts.
B— 125 watts.
C— 500 watts.

Verify the transponder RF output power is within specifications for the class of transponder. For Class 1B and 2B ATCRBS transponders, verify the minimum RF peak output power is at least 18.5 dbw (70 watts) and for classes 1B, 2B, and 3B Mode S transponders, verify the minimum output power is at least 18.5 dbw (70 watts).

84. Parts Manufacturer Approval (PMA) parts not having mandatory replacement times, inspection intervals, or related procedure specified in the airworthiness limitations section of a manufacturer's maintenance manual, or instructions for continued airworthiness, are NOT required to be marked with what information?

A— The letters "FAA PMA."
B— The trademark.
C— The serial number.

Unless the part is too small to be marked in this way, all replacement or modification parts manufactured under a Parts Manufacturer Approval shall be permanently and legibly marked with the letters "FAA-PMA" and the name, trademark, or symbol of the holder of the PMA.

85. Parts Manufacturer Approval (PMA) parts are required to be marked with certain information among which is the name and model designation of each type certificated product on which the part is eligible for installation. When that information is so extensive it is impractical to mark on the part, a tag attached to the part or the part's container the information may be

A— included on an FAA Form 337 to accompany the part and be completed at the time of installation.
B— included in a readily available catalog.
C— included in the original type certificate data sheet.

If the part is too small or is otherwise impractical to mark with all of the required information, a tag attached to the part or its container must include the information that could not be marked on the part. If this marking is too extensive to place on a tag, the tag may refer to a specific, readily available manual, or catalog for part eligibility information.

Answers

83 [C] (017) 14 CFR Part 45
84 [C] (017) 14 CFR Part 45
83a [B] (017) 14 CFR Part 43
85 [B] (017) 14 CFR Part 45
83b [A] (017) 14 CFR Part 43

86. Aircraft, aircraft engines, propellers, and propeller blades and hubs are required to be marked with certain identification data. An aircraft engine requires which of the following?

(1) Builder's name.
(2) Model designation.
(3) Builder's serial number.
(4) Type certificate number, if any.
(5) Production certificate number, if any.
(6) The established rating for aircraft engines.

A— 1, 3, 4, and 6.
B— 1, 2, 3, and 6.
C— All of the above.

The identification required on an aircraft engine by §45.11(a) and (b) shall include the following information:

1. *Builder's name*
2. *Model designation*
3. *Builder's serial number*
4. *Type certificate number, if any*
5. *Production certificate number, if any*
6. *For aircraft engines, the established rating*

87. Aircraft, aircraft engines, propellers, and propeller blades and hubs are required to be marked with certain identification data. A fixed-pitch propeller requires which of the following?

(1) Builder's name.
(2) Model designation.
(3) Builder's serial number.
(4) Type certificate number, if any.
(5) Production certificate number, if any.
(6) Part number (or equivalent).

A— 1, 3, 4, and 6.
B— 1, 2, 3, and 6.
C— 1, 2, 3, 4, and 5.

The identification required on a propeller by §45.11(a) and (b) shall include the following information:

1. *Builder's name*
2. *Model designation*
3. *Builder's serial number*
4. *Type certificate number, if any*
5. *Production certificate number, if any*

88. Aeronautical parts, for which a replacement time is specified in the airworthiness limitations section of the manufacturer's maintenance manual, shall be marked with which of the following information?

(1) Builder's name.
(2) Model designation.
(3) Serial number (or equivalent).
(4) Type certificate number, if any.
(5) Production certificate number, if any.
(6) Part number (or equivalent).

A— 3 and 6.
B— 1, 2, and 3.
C— All of the above.

Each person who produces a part for which a replacement time, inspection interval, or related procedure is specified in the Airworthiness Limitations section of a manufacturer's maintenance manual or Instructions for Continued Airworthiness shall permanently and legibly mark that component with a part number (or equivalent) and a serial number (or equivalent).

89. A small U.S.-registered aircraft built at least 30 years ago may display registration marks of what size?

A— 2 inch.
B— 12 inch.
C— Either A or B.

A small U.S.-registered aircraft built at least 30 years ago or a U.S.-registered aircraft for which an experimental certificate has been issued under §§21.191(d) or 21.191(g) for operation as an exhibition aircraft or as an amateur-built aircraft which has the same external configuration as an aircraft built at least 30 years ago may be operated without displaying marks in accordance with §§45.21 and 45.23 through 45.33 (12-inch numbers) if it displays in accordance with §45.21(c) marks at least 2 inches high on each side of the fuselage or vertical tail surface consisting of the Roman capital letter "N" followed by the U.S. registration number of the aircraft.

Answers

86 [C] (017) 14 CFR Part 45
87 [C] (017) 14 CFR Part 45
88 [A] (017) 14 CFR Part 45
89 [C] (017) 14 CFR Part 45

90. What are the height requirements for the registration marks of a 1975 Cessna model 177RG repainted in 1994?

A— 2 inches.
B— 3 inches.
C— 12 inches.

The nationality and registration marks must be of equal height, and on fixed-wing aircraft must be at least 12 inches high in most cases. However, aircraft built at least 30 years ago can operate with 2-inch registration markings.

91. What regulation in 14 CFR sets forth the required identification markings on U.S.-registered civil aircraft?

A— 14 CFR Part 43.
B— 14 CFR Part 45.
C— 14 CFR Part 47.

The required identification markings for U.S.-registered civil aircraft are found in 14 CFR Part 45, Subpart C.

91a. What size registration marks would an Experimental "exhibition" aircraft registered in 1994 have?

A— 2-inch.
B— 3-inch.
C— 12-inch.

A small U.S.-registered aircraft built at least 30 years ago, ***or*** *a U.S.-registered aircraft for which an experimental certificate has been issued under §§21.191(d) or 21.191(g) …may be operated without displaying the marks in accordance with §45.21 and §45.23 through §45.33* ***if*** *(1) it displays marks at least 2 inches high in accordance with §45.21(c).*

91b. An amateur-built aircraft recently registered in the U.S. has a maximum speed of 180 mph. What size nationality and registration markings are required?

A— 2-inch.
B— 3-inch.
C— 12-inch.

The required identification markings for U.S.-registered civil aircraft are found in 14 CFR Part 45, Subpart C; 3-inch marking is acceptable when the aircraft maximum cruising speed does not exceed 180 knots CAS.

91c. Aircraft registration and markings (N Numbers) on fixed wing aircraft must be 12 inches high except

A— exhibition aircraft.
B— corporate aircraft.
C— commercial aircraft.

With the exception of exhibition aircraft, the nationality and registration marks must be equal height, and on fixed-wing aircraft, must be at least 12 inches high.

91d. Two-inch letters for registration marking are

A— never allowed.
B— allowed only on spherical balloons and powered parachutes.
C— allowed on aircraft manufactured before January 1, 1983.

Nationality and registration marks must be at least 12 inches high in most cases. However, aircraft built at least 30 years ago can operate with 2-inch registration markings.

91e. The width of the letter or numeral in the registration number must be

A— twice the height dimension.
B— two thirds of the height.
C— one half the height dimension (except for the number 1).

Nationality and registration marks must be two-thirds as wide as they are high, except the number "1" which must be one-sixth as wide as it is high, and the letter "M" and "W" which may be as wide as they are high.

Answers

90 [A] (017) 14 CFR Part 45
91 [B] (017) 14 CFR Part 45
91a [A] (017) 14 CFR Part 45
91b [B] (017) 14 CFR Part 45
91c [A] (017) 14 CFR Part 45
91d [C] (017) 14 CFR Part 45
91e [B] (017) 14 CFR Part 45

14 CFR Part 65
Certification: Airmen Other Than Flight Crewmembers

This part of 14 CFR prescribes the requirements for issuing the following certificates and associated ratings and the general operating rules for the holders of those certificates and ratings:

1. Air-traffic control-tower operators,
2. Aircraft dispatchers,
3. Mechanics,
4. Repairmen, and
5. Parachute riggers.

92. A person holding an inspection authorization (IA) may perform an annual inspection and approve aircraft for return to service. The IA may also supervise the performance of which of the following?

A— Progressive inspection.
B— Continuous airworthiness maintenance program.
C— Both A and B.

The holder of an Inspection Authorization may inspect and approve for return to service any aircraft or related part or appliance (except any aircraft maintained in accordance with a continuous airworthiness program under 14 CFR Part 121) after a major repair or major alteration to it in accordance with Part 43, if the work was done in accordance with technical data approved by the Administrator; and perform an annual, or perform or supervise a progressive inspection according to §§43.13 and 43.15.

93. Which of the following is NOT a privilege of the holder of an inspection authorization?

A— Performing annual inspections and approving for return to service.
B— Approving major repairs and alterations for return to service when the work is done in accordance with data approved by the Administrator.
C— Inspecting and approving an aircraft for return to service when the aircraft is on a continuous airworthiness program.

The holder of an Inspection Authorization may inspect and approve for return to service any aircraft or related part or appliance (except any aircraft maintained in accordance with a continuous airworthiness program under 14 CFR Part 121) after a major repair or major alteration to it in accordance with Part 43, if the work was done in accordance with technical data approved by the Administrator.

94. Select the statement which identifies one of the privileges of a mechanic who holds an inspection authorization.

A— Inspect and approve for return to service any aircraft or related part or appliance after a major repair or major alteration, if the work was done in accordance with technical data approved by the Administrator.
B— Perform or supervise an annual inspection.
C— Inspect and approve for return to service a major repair to an aircraft maintained in accordance with an Approved Aircraft Inspection Program.

The holder of an Inspection Authorization may inspect and approve for return to service any aircraft or related part or appliance (except any aircraft maintained in accordance with a continuous airworthiness program under 14 CFR Part 121) after a major repair or major alteration to it in accordance with Part 43, if the work was done in accordance with technical data approved by the Administrator.

Answers

92 [A] (031) 14 CFR Part 65
93 [C] (031) 14 CFR Part 65
94 [A] (031) 14 CFR Part 65

95. Which of the following is the holder of an inspection authorization privileged to perform?

A— An annual inspection, or perform or supervise a progressive inspection according to 14 CFR Sections 43.13 and 43.15.
B— Return to service large aircraft maintained in accordance with a continuous airworthiness program under 14 CFR Part 121 after they have undergone major alterations.
C— Perform or supervise an annual inspection of larger aircraft.

The holder of an inspection authorization may inspect and approve for return to service any aircraft or related part or appliance (except any aircraft maintained in accordance with a continuous airworthiness program under 14 CFR Part 121) after a major repair or major alteration to it in accordance with Part 43, if the work was done in accordance with technical data approved by the Administrator; and perform an annual, or perform or supervise a progressive inspection according to §§43.13 and 43.15.

96. The inspection authorization will cease to be effective when the holder no longer has

A— available facilities.
B— a current subscription to technical standards orders.
C— a business location where he or she will exercise his or her inspection authority.

An inspection authorization ceases to be effective whenever any of the following occurs: the authorization is surrendered, suspended, or revoked, the holder no longer has a fixed base of operation, the holder no longer has the equipment, facilities, and inspection data required by §65.91(c)(3) and (4) for issuance of his authorization.

97. Before exercising the privileges of the authorization after changing his or her fixed base of operation, the holder of an inspection authorization must

A— notify in writing the FAA district office in the area in which the new base is located of the change.
B— not exercise the authority until both the FAA district offices, in the old and new districts, are notified in writing of the change.
C— notify the FAA district office where the original authorization was held that he or she intends to relocate and the date of the intended move.

If the holder of an inspection authorization changes his fixed base of operation he may not exercise the privileges of the authorization until he has notified the FAA Flight Standards District Office or International Field Office for the area in which the new base is located, in writing, of the change.

98. The holder of an inspection authorization may NOT approve a major repair for return to service unless

A— he or she has performed or supervised the performance of a similar repair for which he or she intends to inspect and return to service.
B— the repair was done in accordance with technical data approved by the Administrator.
C— the product was repaired or altered by a certificated mechanic holding both airframe and powerplant ratings.

The holder of an inspection authorization may inspect and approve for return to service any aircraft or related part or appliance (except any aircraft maintained in accordance with a continuous airworthiness program under 14 CFR Part 121) after a major repair or major alteration to it in accordance with Part 43, if the work was done in accordance with technical data approved by the Administrator.

99. What Part of 14 CFR referred to in 14 CFR Section 65.95(a)(1) contains the standards of performance under which a holder of an inspection authorization performs his or her activity?

A— 14 CFR Part 43.
B— 14 CFR Part 61.
C— 14 CFR Part 135.

The holder of an inspection authorization may inspect and approve for return to service any aircraft or related part or appliance (except any aircraft maintained in accordance with a continuous airworthiness program under 14 CFR Part 121) after a major repair or major alteration to it in accordance with Part 43, if the work was done in accordance with technical data approved by the Administrator; and perform an annual, or perform or supervise a progressive inspection according to §§43.13 and 43.15.

Answers

95 [A] (031) 14 CFR Part 65
96 [A] (031) 14 CFR Part 65
97 [A] (031) 14 CFR Part 65
98 [B] (031) 14 CFR Part 65
99 [A] (031) 14 CFR Part 65

100. Under the privileges of an inspection authorization, the holder may approve for return to service any

A— aircraft after a major repair, when the repair is accomplished using manufacturer's data.
B— major repair to an aircraft, when the technical data has been approved by the person holding the inspection authorization and the holder of the inspection authorization is familiar with the type of aircraft being repaired.
C— aircraft major alteration (with the exception of aircraft maintained under 14 CFR Part 121) when the work was accomplished in accordance with technical data approved by the Administrator.

The holder of an inspection authorization may inspect and approve for return to service any aircraft or related part or appliance (except any aircraft maintained in accordance with a continuous airworthiness program under 14 CFR Part 121) after a major repair or major alteration to it in accordance with Part 43, if the work was done in accordance with technical data approved by the Administrator.

101. Select the items that are within the privileges and limitations of the inspection authorization.

(1) Perform a progressive inspection.
(2) Supervise a progressive inspection.
(3) Perform an annual inspection.
(4) Supervise an annual inspection.

A— 1, 3, and 4.
B— 1, 2, and 3.
C— 1, 2, and 4.

The holder of an inspection authorization may inspect and approve for return to service any aircraft or related part or appliance (except any aircraft maintained in accordance with a continuous airworthiness program under 14 CFR Part 121) after a major repair or major alteration to it in accordance with Part 43, if the work was done in accordance with technical data approved by the Administrator; and perform an annual, or perform or supervise a progressive inspection according to §§43.13 and 43.15.

102. Of the following certificated operators' aircraft, which may the holder of an inspection authorization NOT approve for return to service after a major repair or alteration?

A— 14 CFR Part 125 operator.
B— 14 CFR Part 121 operator.
C— 14 CFR Part 133 operator.

The holder of an inspection authorization may inspect and approve for return to service any aircraft or related part or appliance (except any aircraft maintained in accordance with a continuous airworthiness program under 14 CFR Part 121) after a major repair or major alteration to it in accordance with Part 43, if the work was done in accordance with technical data approved by the Administrator.

103. When may the holder of an inspection authorization approve for return to service a major repair that is recorded on an FAA Form 337?

A— Any time a repair is accomplished by a properly certificated and rated mechanic.
B— When the repair has been described in detail on the reverse side of the FAA Form 337.
C— Whenever a major repair is accomplished or supervised by an appropriately certificated and rated mechanic, and the data is approved by the Administrator.

The holder of an inspection authorization may inspect and approve for return to service any aircraft or related part or appliance (except any aircraft maintained in accordance with a continuous airworthiness program under 14 CFR Part 121) after a major repair or major alteration to it in accordance with Part 43, if the work was done in accordance with technical data approved by the Administrator.

Answers

100 [C] (031) 14 CFR Part 65
101 [B] (031) 14 CFR Part 65
102 [B] (031) 14 CFR Part 65
103 [C] (031) 14 CFR Part 65

104. A certificated mechanic holding an inspection authorization is NOT permitted to

A— approve data used for the repair or alteration of aircraft.
B— perform conformity inspections and approve for return to service.
C— perform annual inspections and approve for return to service.

The holder of an inspection authorization may inspect and approve for return to service any aircraft or related part or appliance (except any aircraft maintained in accordance with a continuous airworthiness program under 14 CFR Part 121) after a major repair or major alteration to it in accordance with Part 43, if the work was done in accordance with technical data approved by the Administrator.

105. (1) Each inspection authorization expires annually on the anniversary date of issuance.

(2) Each inspection authorization expires on March 1 of each year.

Regarding the above statements, which one of the following is true?

A— Only 1 is true.
B— Only 2 is true.
C— Neither 1 nor 2 is true.

Each inspection authorization expires on March 31 of every odd numbered year.

106. Inspection authorizations expire on

A— December 31 of each year.
B— March 31 of every odd numbered year.
C— the last day of the month, 1 year from date of issue.

Each inspection authorization expires on March 31 of every odd numbered year.

107. The holder of an inspection authorization must

A— have a fixed base of operations at which he/she can be contacted in person or by telephone during a normal working week.
B— present evidence satisfactory to the Administrator that he/she has performed at least 4 annual inspections during the last 2 years.
C— have some means of being contacted at the place where he or she will exercise the inspection authorization.

To be eligible for an inspection authorization, an applicant must have a fixed base of operations at which he/she may be located in person or by telephone during a normal working week, but it need not be the place where he/she will exercise his/her inspection authority.

108. What are the minimum activity requirements for renewal of an inspection authorization that has been in effect for a 1 year period?

A— Four annual inspections and eight inspections of major repairs or major alterations; or four progressive inspections.
B— Four annual inspections; or eight inspections of major repairs or major alterations; or one progressive inspection.
C— One progressive inspection and eight inspections of major repairs or major alterations; or one annual inspection.

To be eligible for renewal of an inspection authorization for a 1-year period, an applicant must present evidence at renewal, during the month of March in odd-numbered years, at an FAA Flight Standards District Office or an International Field Office that the applicant still meets the requirements of §65.91(c)(1) through (4) for each year they have held the IA certificate. The applicant must show that during the current period the inspection authorization has been held, the applicant:

1. *has performed at least one annual inspection for each 90 days that the applicant held the current authority; or*
2. *has performed inspections of at least two major repairs or major alterations for each 90 days that the applicant held the current authority; or*
3. *has performed or supervised and approved at least one progressive inspection in accordance with the standards prescribed by the Administrator.*

Answers

104 [A] (031) 14 CFR Part 65
105 [C] (031) 14 CFR Part 65
106 [B] (031) 14 CFR Part 65
107 [A] (031) 14 CFR Part 65
108 [B] (031) 14 CFR Part 65

109. What are the minimum requirements for renewal of an inspection authorization that has been in effect for 1 year?

A— Three annual inspections.
B— Two progressive inspections.
C— Eight inspections of major repairs or major alterations.

To be eligible for renewal of an inspection authorization for a 1-year period, an applicant must present evidence at renewal, during the month of March in odd-numbered years, at an FAA Flight Standards District Office or an International Field Office that the applicant still meets the requirements of §65.91(c)(1) through (4) for each year they have held the IA certificate. The applicant must show that during the current period the inspection authorization has been held, the applicant:

1. *has performed at least one annual inspection for each 90 days that the applicant held the current authority; or*
2. *has performed inspections of at least two major repairs or major alterations for each 90 days that the applicant held the current authority; or*
3. *has performed or supervised and approved at least one progressive inspection in accordance with the standards prescribed by the Administrator.*

110. An inspection authorization may be renewed if the holder has performed or supervised, during the year of authorization,

A— an annual inspection and four major repairs.
B— a progressive inspection.
C— six major repairs or major alterations within the last 6 months.

To be eligible for renewal of an inspection authorization for a 1-year period, an applicant must present evidence at renewal, during the month of March in odd-numbered years, at an FAA Flight Standards District Office or an International Field Office that the applicant still meets the requirements of §65.91(c)(1) through (4) for each year they have held the IA certificate. The applicant must show that during the current period the inspection authorization has been held, the applicant:

1. *has performed at least one annual inspection for each 90 days that the applicant held the current authority; or*
2. *has performed inspections of at least two major repairs or major alterations for each 90 days that the applicant held the current authority; or*
3. *has performed or supervised and approved at least one progressive inspection in accordance with the standards prescribed by the Administrator.*

111. The holder of an inspection authorization that has been in effect for less than 90 days needs to have performed

A— at least one annual inspection.
B— at least two major repairs or major alterations.
C— neither of the above.

The holder of an inspection authorization that has been in effect for less than 90 days before the expiration date need not comply with paragraphs (a)(1) through (5) of this section.

112. The holder of an inspection authorization has changed his fixed base of operation. Who must be notified of the change before he may exercise the privileges of his authorization?

A— FAA Flight Standards District Office or International Field Office for the area in which the old base is located.
B— FAA Regional Office for the area in which the new base is located.
C— FAA Flight Standards District Office or International Field Office for the area in which the new base is located.

If the holder of an inspection authorization changes his fixed base of operation, he may not exercise the privileges of the authorization until he has notified the FAA Flight Standards District Office or International Field Office for the area in which the new base is located, in writing, of the change.

Answers

109 [C] (031) 14 CFR Part 65
110 [B] (031) 14 CFR Part 65
111 [C] (031) 14 CFR Part 65
112 [C] (031) 14 CFR Part 65

113. The holder of an inspection authorization who changes his/her fixed base of operation would, before exercising the privileges of the authorization, do which of the following?

A— Notify the FAA Flight Standards District Office, in the area in which the new base is located, in writing, of the change.
B— Not exercise the authority until such time that both the FAA Flight Standards District Offices, in the old and new districts, are notified in writing of the change.
C— Continue to operate using the old authorization until March 31 when the new authorization will be renewed.

If the holder of an inspection authorization changes his fixed base of operation, he may not exercise the privileges of the authorization until he has notified the FAA Flight Standards District Office or International Field Office for the area in which the new base is located, in writing, of the change.

114. What must the holder of inspection authorization do before exercising his/her privileges after changing his fixed base of operation?

A— Apply, in writing, to the FAA Flight Standards District Office or FAA International Field Office in which the new fixed base of operations is located.
B— Request a renewal of his/her inspection authorization from the FAA Flight Standards District Office or FAA International Field Office for the area in which the new base is located.
C— Notify in writing the FAA Flight Standards District Office or FAA International Field Office for the area in which the new base is located.

If the holder of an inspection authorization changes his fixed base of operation, he may not exercise the privileges of the authorization until he has notified the FAA Flight Standards District Office or International Field Office for the area in which the new base is located, in writing, of the change.

115. Title 14 of the Code of Federal Regulations (CFR) Section 65.95 prescribes the

A— privileges and limitations of an airframe mechanic.
B— requirements for renewal of the inspection authorization.
C— privileges and limitations of an inspection authorization.

14 CFR §65.95 contains the privileges and limitations for an inspection authorization.

116. Under the privileges and limitations of the Inspection Authorization you could

A— overhaul and approve for return to service a float-type carburetor.
B— exchange engines of different makes, per the type certificate data sheet, and approve for return to service.
C— approve of return-to-service a major repair made in accordance with an Airworthiness Directive.

The holder of an inspection authorization may inspect and approve for return to service an aircraft or related part or appliance after a major repair or major alteration to it in accordance with Part 43 if the work was done in accordance with technical data approved by the Administrator.

The item in alternative A is a minor repair and in alternative B is a minor alteration. Neither of these requires an IA.

117. An inspection authorization ceases to be effective when the holder

A— fails to meet the renewal activity requirements of 14 CFR Section 65.93 within a 90-day interval.
B— no longer has a fixed base of operation.
C— violates Title 14 of the CFR.

An inspection authorization ceases to be effective whenever any of the following occurs: the authorization is surrendered, suspended, or revoked, the holder no longer has a fixed base of operation, the holder no longer has the equipment, facilities, and inspection data required by §65.91(c)(3) and (4) for issuance of his authorization.

118. Title 14 of the CFR requires the inspection of all civil aircraft at specific intervals to determine the overall condition. An amateur-built aircraft with an experimental airworthiness certificate requires what type of inspection?

A— Annual inspection.
B— Condition inspection.
C— Approved aircraft Inspection program.

The holder of a repairman certificate (experimental aircraft builder) may perform condition inspections on the aircraft constructed by the holder in accordance with the operating limitations of that aircraft.

Answers

113	[A]	(031)	14 CFR Part 65	114	[C]	(031)	14 CFR Part 65	115	[C]	(031)	14 CFR Part 65
116	[C]	(031)	14 CFR Part 65	117	[B]	(031)	14 CFR Part 65	118	[B]	(030)	14 CFR Part 65

14 CFR Part 91
General Operating and Flight Rules

This part of 14 CFR prescribes rules governing the operation of aircraft (other than moored balloons, kites, unmanned rockets, and unmanned free balloons, which are governed by Part 101, and ultralight vehicles operated in accordance with Part 103) within the United States.

119. Which of the following is required to be installed on an airplane manufactured after July 18, 1978?

A— Shoulder harness for each front seat.
B— An operable emergency locator beacon.
C— An equipment list showing the minimum equipment required for aircraft operations under specified conditions.

For small civil airplanes manufactured after July 18, 1978, an approved shoulder harness for each front seat.

120. When would data in aircraft maintenance manuals be considered approved data?

A— Aircraft maintenance manuals from the original manufacturer are always considered approved data.
B— A manual or parts thereof become approved if the manufacturer assigns it's approval in the manual.
C— When a manual includes an airworthiness limitations section, that section is considered approved data.

No person may operate an aircraft for which a manufacturer's maintenance manual or instructions for continued airworthiness has been issued unless the mandatory replacement times, inspection intervals, and related procedures specified in that section or alternative inspection intervals and related procedures set forth in an operations specification approved by the Administrator under 14 CFR Parts 121 or 135 or in accordance with an inspection program approved under §91.409(e) have been complied with.

121. When, if ever, is the manufacturer's maintenance manual, or a section thereof, considered mandatory during an annual inspection?

A— They are required when part of an approved inspection program.
B— They are required if made mandatory by the type certificate data sheet.
C— Maintenance manuals are optional if an equivalent procedure is used.

No person may operate an aircraft for which a manufacturer's maintenance manual or instructions for continued airworthiness has been issued unless the mandatory replacement times, inspection intervals, and related procedures specified in that section or alternative inspection intervals and related procedures set forth in an operations specification approved by the Administrator under 14 CFR Part 121 or 135 or in accordance with an inspection program approved under §91.409(e) have been complied with.

122. Following a major repair that might appreciably change the flight characteristics of an aircraft, other than the maintenance record entry required by 14 CFR Section 43.9, what must be entered in the aircraft records?

A— The supplemental type certificate number.
B— Record of flight test.
C— FAA Form 337.

No person may carry any person (other than crewmembers) in an aircraft that has been maintained, rebuilt, or altered in a manner that may have appreciably changed its flight characteristics or substantially affected its operation in flight until an appropriately rated pilot with at least a private pilot certificate flies the aircraft, makes an operational check of the maintenance performed or alteration made, and logs the flight in the aircraft records.

123. A 14 CFR Part 91 operator of a turbine-powered rotorcraft may elect to use which of the following inspection provisions?

A— An annual inspection.
B— A current inspection program recommended by the manufacturer.
C— Either A or B.

14 CFR §91.409 states that no person may operate a large airplane, turbojet multi-engine airplane, turbopropeller-powered multi-engine airplane, or turbine-powered rotorcraft unless...it is inspected in accordance with an inspection program selected under the provisions of §91.409(f), except that the owner or operator

Continued

Answers

119 [A] (031) 14 CFR Part 91
120 [C] (031) 14 CFR Part 91
121 [A] (031) 14 CFR Part 91
122 [B] (017) 14 CFR Part 91
123 [C] (017) 14 CFR Part 91

of a turbine-powered rotorcraft may elect to use the inspections of §91.409(a), (b), (c) or (d) in lieu of an inspection option of §91.409(f).

Paragraph (a) specifies an annual inspection and paragraph (f)(3) specifies an inspection program recommended by the manufacturer.

123a. A Cessna 421B operated as a 14 CFR Part 91 airplane would have inspections conducted in accordance with

A— 14 CFR Part 43, Appendix D.
B— 14 CFR Part 91, §91.409 (Inspections).
C— 14 CFR Part 43, §43.16 (Airworthiness Limitations).

Use the TCDS for the Cessna 421B aircraft (A7CE) to determine that this is a CAR 3 aircraft (maximum certificated takeoff weight is either 7,250 lbs or 7,450 lbs, depending on specific serial number). The aircraft is not turbine powered, but rather has two Continental GTSIO-520-H powerplants. Depending on the specific aircraft logbooks, any one of the answers could be acceptable. However, the best answer is the one referring to 91.409, the "umbrella regulation" under which Part 43 rules apply.

124. When may an authorized inspector (IA) perform an annual inspection on an aircraft?

A— 12 calendar months from the day the last annual inspection was performed.
B— The IA may perform an annual inspection anytime the owner requests it.
C— The IA may only perform an annual inspection only during the month the inspection is due.

No person may operate an aircraft unless within the preceding 12 calendar months, it has had an annual inspection and has been approved for return to service by a person authorized by §43.7. There is no regulation that would prevent additional annual inspections being performed within the 12 calendar month period.

125. If an aircraft that has a separate record for airframe and engine receives a replacement engine between annual inspections, what action would be required?

A— An inspection as required by the manufacturer and a maintenance record entry.
B— A 100-hour inspection of the engine followed by a flight test.
C— Only the engine portion of the annual inspection with an entry in the engine's record.

The entire aircraft (this includes the engine) must have an annual inspection each 12 calendar months. If a replacement engine was installed between annual inspections it must be given the engine portion of the annual inspection and this information entered into the engine records.

126. A 14 CFR Part 91 operator of a turbine-powered rotorcraft may elect to use which of the following inspection provision?

A— An annual inspection.
B— An inspection program established by the operator and approved by the Administrator.
C— Either A or B.

14 CFR §91.409(a)(1) allows the rotorcraft to be given an annual inspection. 14 CFR §91.409(c)(4) allows turbine-powered rotorcraft to be inspected according to an inspection program established by the operator and approved by the Administrator under the provisions of §91.409(f).

127. As the holder of an inspection authorization, you are approached by the owner of an aircraft that has recently been imported from a foreign country. The new owner wants to know when the aircraft will require an annual inspection. The aircraft's records indicate that the aircraft has been maintained under a French inspection program and that the last complete inspection under that program was four months earlier. The record further indicates that the aircraft entered the U.S. two months ago and was issued a U.S. standard airworthiness certificate at that time. When will the annual inspection be due?

A— In 8 calendar months.
B— Immediately.
C— In 10 calendar months.

14 CFR §91.409(a)(1), (2) states that no person may operate an aircraft unless, within the preceding 12 calendar months it has had an annual inspection in accordance with Part 43 and has been approved for return to service by a person authorized by §43.7, or has had an inspection for the issuance of an airworthiness certificate in accordance with 14 CFR Part 21.

The imported aircraft received a standard airworthiness certificate when it entered the U.S. two months ago and therefore it must have its next annual inspection in 10 calendar months.

Answers

123a [B] (021) 14 CFR Part 43 and 91
124 [B] (031) 14 CFR Part 91
125 [C] (017) 14 CFR Part 91
126 [C] (017) 14 CFR Part 91
127 [C] (017) 14 CFR Part 91

128. Title 14 of the Code of Federal Regulations allow aircraft established on certain inspection programs to exceed an inspection interval by 10 hours while en route to a place where the inspection can be done. Which of the following may exceed the inspection interval by 10 hours?

A— An aircraft using a 100-hour inspection program.
B— An aircraft using a progressive inspection program.
C— Both A and B.

14 CFR §91.409 allows aircraft operating on either a 100-hour or progressive inspection program to exceed the 100-hour inspection interval by not more than 10 hours while en route to a place where the inspection can be done.

129. Which one of the following aircraft could require an annual inspection?

A— A small single-engine turboprop with a restricted airworthiness certificate.
B— A rotorcraft with a provisional airworthiness certificate.
C— A small multiengine turboprop aircraft with a standard airworthiness certificate.

Exceptions to aircraft that are required to have an annual inspection are aircraft having a provisional airworthiness certificate and turbopropeller-powered multi-engine airplanes.

130. Each registered owner or operator of an aircraft using a progressive inspection must designate a person to supervise or conduct the inspection. Which of the following could perform those duties?

A— A and P mechanic.
B— Certificated airframe repair station.
C— Both A and B.

Each registered owner or operator of an aircraft desiring to use a progressive inspection program must submit a written request to the FAA Flight Standards district office having jurisdiction over the area in which the applicant is located, and shall provide a certificated mechanic holding an inspection authorization, a certificated airframe repair station, or the manufacturer of the aircraft to supervise or conduct the progressive inspection.

131. Each registered owner or operator of an aircraft using a progressive inspection must designate a person to supervise or conduct the inspection. Which of the following could perform those duties?

A— The aircraft manufacturer.
B— Certificated airframe repair station.
C— Both A and B.

Each registered owner or operator of an aircraft desiring to use a progressive inspection program must submit a written request to the FAA Flight Standards district office having jurisdiction over the area in which the applicant is located, and shall provide a certificated mechanic holding an inspection authorization, a certificated airframe repair station, or the manufacturer of the aircraft to supervise or conduct the progressive inspection.

132. The holder of an inspection authorization may supervise a progressive inspection on a turbine-powered rotorcraft weighing

A— 7,000 lbs or less.
B— 7,001 lbs or more.
C— both A and B.

The owner or operator of a turbine-powered rotorcraft may elect to use the inspection provisions of §91.409(d), a progressive inspection. No weight limitation is specified.

132a. The holder of an inspection authorization may supervise a progressive inspection on a turbine-powered rotorcraft weighing

A— 12,500 lbs or less.
B— over 12,500 lbs.
C— both A and B.

The owner or operator of a turbine-powered rotorcraft may elect to use the inspection provisions of §91.409(d), a progressive inspection. No weight limitation is specified.

Answers

128 [C] (017) 14 CFR Part 91
129 [A] (017) 14 CFR Part 91
130 [B] (017) 14 CFR Part 91
131 [C] (017) 14 CFR Part 91
132 [C] (031) 14 CFR Part 91
132a [C] (031) 14 CFR Part 91

133. Each registered owner or operator of a large airplane, turbojet multiengine airplane, turbopropeller powered multiengine airplane and turbine-powered rotorcraft operated under 14 CFR Part 91, must elect one of the inspection programs of 14 CFR Section 91.409(e). However, one of these operators may additionally select an annual inspection. Which operator is it?

A— Large airplane.
B— Multiengine turbopropeller airplane.
C— Turbine rotorcraft.

14 CFR §91.409(e) excuses turbine-powered rotorcraft from the inspection requirements of other large aircraft (12,500 pounds or more) and may be inspected by an annual or progressive inspection.

The owner or operator of a turbine-powered rotorcraft may elect to use the inspection provisions of §91.409(d), a progressive inspection. No weight limitation is specified. Section 91.409(e) lists numerous different types of aircraft and specifies that they must be inspected in accordance with one of the types of inspection programs stated in §91.409(f).

A 12-seat multi-engine turbine aircraft operating under 14 CFR Part 135 (Commuter and On-Demand Operations) could be maintained under an approved inspection program by the operator (§135.411(a)).

134. Which one of the following aircraft could be maintained under an approved inspection program by the operator?

A— 30 seat large airplane operating under Part 25.
B— 12 seat multi-engine turbine airplane under commuter.
C— Neither A or B.

14 CFR §91.409(e) lists numerous different types of aircraft and specifies that they must be inspected in accordance with one of the types of inspection programs stated in §91.409(f), but excludes large airplanes operated under 14 CFR Part 125 (which specifies airplanes having a seating capacity of 20 or more passengers, and not operating under 121 or 135).

A 12-seat multi-engine turbine aircraft operating under 14 CFR Part 135 (Commuter and On-Demand Operations) could be maintained under an approved inspection program by the operator.

134a. An Approved Aircraft Inspection Program could apply to which of the following?

1. Large 30-passenger turbine powered helicopter.
2. A 12-passenger commuter category multi-engine turboprop aircraft.

A— Both 1 and 2.
B— Only 1.
C— Only 2.

135. When is an aircraft equipped for "Day VFR Only" operation, required to have it's altimeter tested and inspected in accordance with 14 CFR Part 43, Appendix E? Compliance with Appendix E is

A— not required.
B— required each 12 calendar months.
C— required each 24 calendar months.

14 CFR §91.411 requires that airplanes and helicopters operated in controlled airspace under IFR have their altimeter and altitude reporting equipment tested in accordance with 14 CFR Part 43, Appendix E. There is no such requirement for aircraft operated day VFR only.

136. How long must an FAA Form 337 for the overhaul of a reciprocating engine having a planetary propeller reduction gearing be retained?

A— It is retained in the engine records until the engine is removed from the airframe and replaced with a different engine or for 24 calendar months whichever occurs first.
B— It is retained as a permanent document in the engine records or until the engine is zero timed.
C— It is retained until the work is repeated or superseded by other work or for 1 year after the work is performed.

Records of the maintenance, preventive maintenance, and alteration and records of the 100-hour, annual, progressive, and other required or approved inspections as appropriate for each aircraft (including the airframe) and each engine, propeller, rotor, and appliance of an aircraft must be retained until the work is repeated or superseded by other work or for 1 year after the work is performed.

Answers

133 [C] (031) 14 CFR Part 91
134 [B] (031) 14 CFR Part 91 and 135
134a [A] (031) 14 CFR Part 91, 125, and 135
135 [A] (031) 14 CFR Part 91
136 [C] (031) 14 CFR Part 91

137. How long must an FAA Form 337 for the overhaul of a reciprocating engine having an integral supercharger be retained?

A— It is retained until the engine is removed from the airframe and replaced with a different engine.
B— It is retained until the work is repeated or superseded by other work or for 1 year after the work is performed.
C— It is retained as a permanent document in the engine records or until the engine is zero timed.

Records of the maintenance, preventive maintenance, and alteration and records of the 100-hour, annual, progressive, and other required or approved inspections as appropriate for each aircraft (including the airframe) and each engine, propeller, rotor, and appliance of an aircraft must be retained until the work is repeated or superseded by other work or for 1 year after the work is performed.

138. While reviewing an aircraft's records during an annual inspection, you notice three different FAA Forms 337 for the overhaul of the aircraft's engine. How long must these major repair records be retained?

These records shall be retained

A— and transferred with the engine.
B— for a period of 2 years after the work was performed.
C— for a period of 1 year from the date of completion.

Records of the maintenance, preventive maintenance, and alteration and records of the 100-hour, annual, progressive, and other required or approved inspections as appropriate for each aircraft (including the airframe) and each engine, propeller, rotor, and appliance of an aircraft must be retained until the work is repeated or superseded by other work or for 1 year after the work is performed.

139. Federal Aviation Regulations require certain information to be included in an aircraft's maintenance records for airworthiness directive compliance. Which of the following information would be part of a required record entry for the airworthiness directive?

A— Amendment number.
B— Serial number of aircraft.
C— Revision date.

The information regarding ADs that are required in the aircraft maintenance records is: The current status of applicable airworthiness directives (AD) including, for each the method of compliance, the AD number, and revision date. If the AD involves recurring action, the time and date when the next action is required.

140. The regulations require each registered owner or operator to maintain a record of Airworthiness Directives (AD)

A— listing all ADs pertinent to his or her model.
B— in a logbook for each airframe, engine, propeller, or rotor.
C— listing status of applicable ADs.

Records containing the following information shall be retained and transferred with the aircraft at the time the aircraft is sold.

1. *The total time in service of each airframe, each engine, each propeller, and each rotor.*
2. *The current status of life-limited parts of each airframe, engine, propeller, rotor, and appliance.*
3. *The time since last overhaul of all items installed on the aircraft which are required to be overhauled on a specified time basis.*
4. *The current inspection status of the aircraft, including the time since the last inspection required by the inspection program under which the aircraft and its appliances are maintained.*
5. *The current status of applicable airworthiness directives (AD) including, for each the method of compliance, the AD number, and revision date. If the AD involves recurring action, the time and date when the next action is required.*
6. *Copies of the forms prescribed by §43.9(a) for each major alteration to the airframe and currently installed engines, rotors, propellers, and appliances.*

141. A rebuilt aircraft engine may be granted zero time by

A— the manufacturer.
B— a certified powerplant mechanic.
C— an A and P mechanic who holds an inspection authorization.

The owner or operator may use a new maintenance record, without previous operating history, for an aircraft engine rebuilt by the manufacturer or by an agency approved by the manufacturer.

Answers

137 [B] (031) 14 CFR Part 91
138 [C] (031) 14 CFR Part 91
139 [C] (031) 14 CFR Part 91
140 [C] (031) 14 CFR Part 91
141 [A] (031) 14 CFR Part 91

14 CFR Part 125
Certification and Operations: Airplanes having a seating capacity of 20 or more passengers or a maximum payload capacity of 6,000 pounds or more

This regulation prescribes rules governing the operations of U.S.-registered civil airplanes which have a seating configuration of 20 or more passengers, or a maximum payload capacity of 6,000 pounds or more when common carriage is not involved.

142. Who is responsible for performing maintenance, preventive maintenance, and alteration of an aircraft operating under 14 CFR Part 125?

A— The registered owner of the aircraft.
B— The shop performing the maintenance.
C— The holder of the Part 125 certificate.

14 CFR §125.243 states that the certificate holder is primarily responsible for airworthiness, and the performance of maintenance, preventive maintenance, and alteration in accordance with applicable regulations and the certificate holder's manual.

14 CFR Part 135
Operating Requirements: Commuter and On-Demand Operations

This part of the Federal Regulations prescribes rules governing the commuter or on-demand operations that:

- require an Air Carrier Certificate or Operating Certificate under 14 CFR Part 119,
- are used by a certificate holder conducting operations under this part including the maintenance, preventative maintenance and alteration of an aircraft,
- are involved in the transportation of mail by an aircraft conducted under a postal service contract awarded under 39 U.S.C. 5402c,
- are employed or used by an air carrier or commercial operator under this part to perform training, qualification, or evaluation functions under an Advanced Qualification Program under Subpart Y of 14 CFR Part 121, or
- who conduct nonstop sightseeing flights for compensation or hire that begin and end at the same airport, and are conducted within a 25 statute mile radius of that airport.

143. What is the smallest aircraft operating under 14 CFR Part 135 that is required to have a cockpit voice recorder?

Multiengine, turbine-powered airplane or rotorcraft:

A— having a passenger seating configuration of six or more and for which two pilots are required.
B— having a passenger seating configuration of nine or more.
C— having a passenger seating configuration of 20 or more.

No person may operate a multi-engine, turbine-powered airplane or rotorcraft having a passenger seating configuration of six or more and for which two pilots are required by certification or operating rules unless it is equipped with an approved cockpit voice recorder.

Answers

142 [C] (031) 14 CFR Part 125   143 [A] (031) 14 CFR Part 135

144. Aircraft operating under 14 CFR Part 135 with 9 seats or less are maintained in accordance with Parts 43 and 91. What type of inspection would be required for the aircraft?

A— An inspection in accordance with Section 91.409.
B— An inspection as required by the operations maintenance manual specifications.
C— An inspection in accordance with Part 43, Appendix D.

14 CFR §135.411(a) states that aircraft type certificated for a passenger seating configuration, excluding any pilot seat, of nine seats or less, shall be maintained under Parts 91 and 43 of 14 CFR.

14 CFR §91.409(a)(1) states that except as provided in paragraph (c) of this section, no person may operate an aircraft unless, within the preceding 12 calendar months, it has had an annual inspection in accordance with Part 43 of 14 CFR.

Paragraph (c) and 14 CFR §135.411(a)(1) states that an approved aircraft inspection program "may" be used under §135.419.

145. Aircraft operating under 14 CFR Part 135 that are type certificated for a passenger seating configuration, excluding any pilot seat, of nine seats or less, shall be maintained under what two parts of 14 CFR as well as the applicable sections of Part 135?

A— Part 91 and 121.
B— Part 43 and 91.
C— Part 91 and 125.

Aircraft that are type certificated for a passenger seating configuration, excluding any pilot seat, of nine seats or less, shall be maintained under Parts 91 and 43 of this chapter and §§135.415, 135.417, and 135.421.

146. For aircraft operating under 14 CFR Part 135, the approved maintenance program is determined by the number of seats. For aircraft that can be maintained under 14 CFR Parts 91 and 43 maintenance programs, the number of seats are limited to

A— 9 or less.
B— 20 or less.
C— 50 or less.

Aircraft that are type certificated for a passenger seating configuration, excluding any pilot seat, of nine seats or less, shall be maintained under 14 CFR Parts 91 and 43.

147. An owner brings you his light twin-engine airplane operating under Part 135. Where would you find the appropriate instructions for inspecting the aircraft?

A— Flight Manual
B— Operations Specifications
C— 14 CFR Part 43, Appendix D

When the Administrator finds that the aircraft inspections required or allowed under 14 CFR Part 91 are not adequate to meet this part, or upon application by a certificate holder, the Administrator may amend the certificate holder's operations specifications under §135.17 to require or allow an approved aircraft inspection program for any make and model aircraft of which the certificate holder has exclusive use of at least one aircraft.

148. Who is primarily responsible for the airworthiness of an aircraft that is operated under 14 CFR Part 135?

A— The registered owner of the aircraft.
B— The pilot of the aircraft.
C— The certificate holder of the aircraft.

Each certificate holder is primarily responsible for the airworthiness of its aircraft, including airframes, aircraft engines, propellers, rotors, appliances, and parts and shall have its aircraft maintained under this chapter, and shall have defects repaired between required maintenance under 14 CFR Part 43.

149. Which special inspection provision(s) must be included for single-engine aircraft operated under 14 CFR Part 135 for passenger carrying IFR operations?

A— Engine trend monitoring.
B— Engine oil analysis.
C— Both of the above.

For each single engine aircraft to be used in passenger-carrying IFR operations, each certificate holder must incorporate into its maintenance program the manufacturer's recommended engine trend monitoring program, which includes an oil analysis.

Answers

144 [A] (017) 14 CFR Part 135
145 [B] (017) 14 CFR Part 135
146 [A] (017) 14 CFR Part 135
147 [B] (017) 14 CFR Part 135
148 [C] (017) 14 CFR Part 135
149 [C] (017) 14 CFR Part 135

150. The controlling document for the maintenance and inspection of aircraft operated under 14 CFR Part 135 is

A— the operator's FAA-approved manual.
B— AC 43.13-1B and -2B.
C— the aircraft manufacturer's service manual.

Each certificate holder shall have an inspection program and a program covering other maintenance, preventive maintenance, and alterations, that ensures that maintenance, preventive maintenance, and alterations performed by it, or by any other persons, are performed under the certificate holder's manual.

151. In the maintenance portion of a 14 CFR Part 135 operation the relationship between the persons performing the maintenance and the inspection of the maintenance must be closely monitored. Which statement is true?

A— The person performing the maintenance is responsible for the inspection.
B— The person performing the maintenance is not allowed to inspect it.
C— The inspector is under the authority of the person performing the maintenance.

The maintenance organization of a Part 135 operator must contain instructions to prevent any person who performs any item of work from performing any required inspection of that work.

14 CFR Part 183
Representatives of the Administrator

This part of 14 CFR describes the requirements for designating private persons to act as representatives of the Administrator in examining, inspecting, and testing persons and aircraft for the purpose of issuing airman and aircraft certificates.

152. Certain persons are designated by the Administrator to represent the FAA in developing and approving technical data for alteration and repair of U.S.-certificated aircraft. These persons are known as

A— designated aircraft maintenance inspectors.
B— designated engineering representatives.
C— designated airworthiness representatives.

Designated Engineering Representatives (DERs) are designated by the FAA to approve technical data for alteration and repair of U.S.-certificated aircraft.

152a. A major repair is required in the fuselage of a corporate aircraft. A structural engineer has created a drawing of how to repair the damage. The technician should

A— fix the aircraft per the drawing and make a logbook entry.
B— send the drawing to a structural Designated Engineering Representative (FAA DER) for approval and then complete a 337 for FAA field approval.
C— request an FAA Field approval from the local FSDO before doing any work on the aircraft.

A major repair requires approved data for maintenance activity to correct it. The engineering by itself is not approved data. If it is sent to a DER, that approval is all that is necessary for the repair to be completed. If the technician chooses to, they may request the FSDO field approval, which does not require a DER drawing approval.

152b. Title 14 of the Code of Federal Regulations, Part 183 allows the Federal Aviation Administration to designate certain persons to develop and approve technical data for alteration and repair of U.S. certificated aircraft. These persons are known as the

A— Designated Engineering Representative.
B— Designated Airworthiness Representative.
C— Designated aircraft maintenance inspectors.

DERs are designated by the FAA to approve technical data for alteration and repair of U.S.-certificated aircraft.

Answers

150 [A] (017) 14 CFR Part 135
151 [B] (017) 14 CFR Part 135
152 [B] (031) 14 CFR Part 183
152a [C] (031) AC 43.9-1
152b [A] (031) 14 CFR §183.29

Chapter 4
Airworthiness Directives (ADs)

14 CFR Part 39
Airworthiness Directives

This part of 14 CFR prescribes airworthiness directives (AD) that apply to aircraft, aircraft engines, propellers, or appliances (hereinafter referred to in this part as "products") when an unsafe condition exists in a product, and that condition is likely to exist or develop in another product of the same type design.

Included in this chapter are several sample ADs (starting on Page 4-11) to use when studying the test questions on this topic.

Categories of Airworthiness Directives

Airworthiness Directives are Federal Aviation Regulations that are published in the *Federal Register* as amendments to 14 CFR Part 39. They apply to aircraft, aircraft engines, propellers, or appliances, which are referred to as "products," and are issued when an unsafe condition is found to exist in a product, and when that condition is likely to exist or develop in another product of the same type design. ADs are published in the following four categories.

Notice of Proposed Rulemaking (NPRM)

An NPRM is issued and published in the *Federal Register* when an unsafe condition is discovered in a product. Interested persons are invited to comment on the NPRM by submitting such written data, views, or arguments as they may desire. The comment period is usually 60 days, and proposals contained in the notice may be changed or withdrawn in light of the comments received. When an NPRM is adopted as a final rule, it is published in the *Federal Register*, printed, and distributed by first-class mail to the registered owners of the product affected.

Immediately Adopted Rule

ADs of an urgent nature are adopted without the NPRM process, as immediately adopted rules. These ADs usually become effective less than 30 days after publication in the *Federal Register* and are distributed to the registered owners of the product affected.

Emergency ADs

Emergency ADs are issued when immediate corrective action is required. Emergency ADs are distributed to the registered owners of the product affected by telegram, priority mail, or other electronic methods and are effective upon receipt. Emergency ADs are published in the *Federal Register* as soon as possible after the initial distribution.

ADs Issued to Other than Aircraft

ADs may be issued which apply to engines, propellers, or appliances installed on multiple makes or models of aircraft. When the product can be identified as being installed on a specific make or model aircraft, AD distribution is made to the registered owners of those aircraft. However, there are times when a determination cannot be made, and direct distribution to the registered owner is impossible. For this reason, aircraft owners and operators are urged to subscribe to the Summary of Airworthiness Directives which contain all previously published ADs and a biweekly supplemental service. To access AD notes and other regulatory data, you may also use the internet. The most comprehensive site is:

www.airweb.FAA.gov/rgl

Publication of Airworthiness Directives

Individual ADs are distributed to the owners of the affected products and are also made available to maintenance personnel by subscription from the FAA, as described in AC 00-44 *Status of the Federal Aviation Regulations.*

Printed ADs are published in two volumes, with each volume containing two books:

- **Volume I**, Book 1 contains ADs, applicable to small aircraft and all rotorcraft, that were issued between 1943 and 1979 that are still in effect.
- **Volume I**, Book 2 contains all the ADs, applicable to small aircraft and all rotorcraft, that were issued between 1980 and the present.
- **Volume II** contains two books that have the same breakdown of ADs as in Volume I, except these ADs relate to large aircraft, those with a maximum certificated takeoff weight of more than 12,500 pounds.

The Summary of AD Notes contains two indexes; one is alphabetical by the manufacturer of the product, and the other is a numerical listing of all ADs from the oldest to the most recent. The indexes are updated every six months.

New and revised ADs are compiled by the FAA every two weeks and mailed out to the subscribers of the service in the "Biweekly Listings."

Applicability of ADs

Each AD contains an applicability statement specifying the product (aircraft, aircraft engine, propeller, or appliance) to which it applies. Some aircraft owners and operators mistakenly assume that ADs do not apply to aircraft with other than standard airworthiness certificates, i.e., special airworthiness certificates in the restricted, limited, or experimental category. Unless specifically stated, ADs apply to the make and model set forth in the applicability statement regardless of the classification or category of the airworthiness certificate issued for the aircraft. Type certificate and airworthiness certification information are used to identify the product affected. Limitations may be placed on the applicability by specifying the serial number or number series to which the AD is applicable. When there is no reference to serial numbers, all serial numbers are affected.

Construction of an Airworthiness Directive

Airworthiness Directives are all written to a standard format. All contain the AD number, the amendment number, an applicability statement, a required compliance time or date, the effective date and a compliance statement.

The AD Number

Historically, AD notes were identified by a six-digit number such as 90-08-14. The first two digits (90) identify the year the AD was issued. The second two digits (08) identify the biweekly period in which the AD was issued. This is the fifteenth or sixteenth week of the year. The last two digits (14) is the sequential number of the AD issued during this time period. This was the fourteenth AD issued in the eighth biweekly period. Currently, AD notes are identified by an eight-digit number such as 2004-22-08. This change started with AD notes issued in the year 2000.

If the AD is revised, the letter "R" and the number of revisions will be appended to the AD number. For example, AD 81-23-01 R1 is the first revision to AD 81-23-01. It is not uncommon that a revised AD includes additional maintenance procedures that were not in the original AD. The instructions must be complied with and the revised AD be signed off even though the original AD was a one-time compliance AD.

The letter "T" preceding the AD number means that the AD was telegraphed to the owners/operators of the unsafe aircraft.

The Amendment Number

AD notes are amendments to 14 CFR Part 39 and, therefore carry amendment numbers. This amendment number is changed when the AD is revised. For example, the amendment number of AD 90-08-14 as shown on Page 4-20 is Amendment 39-6563.

Applicability Statement

This statement lists the products to which the AD applies. The applicability is given by product model number and applicable serial numbers.

Compliance Time or Date

There are a number of ways compliance time may be stated:

- Before further flight
- Within the next X hours time in service after the effective date of this AD
- Within the next X landings after the effective date of this AD
- Within the next X cycles of engine operation after the effective date of this AD
- By a certain calendar date

Some ADs are repetitive, and must be complied with at specified intervals after the initial compliance. These repetitive compliance requirements may be dispensed with or terminated when a specified permanent fix is developed and complied with.

Effective Date

Near the end of the AD there is a statement similar to the following: "This amendment (39-6563, AD 90-18-14) becomes effective on May 7, 1990."

Compliance Statement

This statement specifies the action that is required by this AD. A typical compliance statement is as follows: "Compliance: Required as indicated in the body of this AD, unless already accomplished." The body of the AD then describes the action that is required and specifies the method of compliance.

Alternate Method of Compliance (AMOC)

Each AD also includes a short section which indicates that someone in the FAA is authorized to approve "Alternative Methods of Compliance" for this AD. This can be as simple as requesting a variation of the inspection interval, or as complex as fabricating a totally different method for addressing the unsafe condition identified by the AD.

Sample Test Questions

153. When, if ever, are manufacturer's service bulletins considered mandatory during an annual inspection?

A— When incorporated in the maintenance manual as required in 14 CFR Part 21.
B— Service bulletins are optional and never required at annual inspections.
C— When incorporated in 14 CFR Part 39.

Compliance with manufacturer's service bulletins becomes mandatory during an annual inspection when the bulletins are incorporated into an Airworthiness Directive under 14 CFR Part 39.

153a. Of the following statements concerning use of manufacturer's service bulletins and compliance with airworthiness directives during an annual inspection, which is true?

A— Service bulletins always take precedence over airworthiness directives when the aircraft is in the heavy maintenance category.
B— If there is a conflict between a service bulletin and an airworthiness directive (AD), you should always follow the AD.
C— Service bulletins are always mandatory when performing required inspections that take precedence over all airworthiness items.

Compliance with manufacturer's service bulletins becomes mandatory during an annual inspection when the bulletins are incorporated into an Airworthiness Directive under 14 CFR Part 39.

154. Use Airworthiness Directive (AD) 80-10-02 to answer this question.

Known information: Messerschmitt-Bolkow-Blohm Model BO-105 helicopter with tail rotor blade grips P/N 105-31722 installed.

While performing a progressive inspection on this helicopter, you note in the aircraft's records that the last compliance with AD 80-10-02 was at an aircraft time of 5402 hours. The records further indicate that the tail rotor blade grips were replaced at an aircraft time of 4902. What action is required at this inspection with a time of 5502?

A— Compliance is required for paragraph (c)(1)(2).
B— Compliance is required for paragraph (c).
C— Compliance is required for paragraph (b)(d)(e).

It has been 100 hours time in service since the last inspection of the tail rotor blade grips and it has been 600 hours since the blade grips were replaced.

Paragraph (b) requires a dye penetrant inspection of the visible part of the inner surface of the tail rotor grip clevis area at each 100 hours time in service.

Paragraph (d) requires the visual inspection of the inboard end of the tail rotor blade grip in accordance with the "Special Instructions" of the maintenance manual at each 100 hours time in service.

Paragraph (e) requires the dye penetrant inspection of the rotor blade grip in the vicinity of the bore of the laminated pack retaining bolt within the next 600 hours time in service.

155. Use AD 80-15-12 to answer this question.

Known Information: Agusta Helicopter model A109A, serial number 7162, having main rotor mast bearing inner race P/N 109-0404-14 installed.

What action is required at 200 hours time in service after the effective date of AD 80-15-12, if no evidence of oil leaks are found?

A— Install bearing race P/N 109-0404-14-15.
B— Continue compliance with paragraph (d).
C— Replace packing P/N 109-0406-68.

Paragraph (b) states that if no evidence of oil leaks is found, continue in service and comply with paragraph (d) of this AD.

155a. How do we know the revisions to an Airworthiness Directive?

A— By the number at the bottom of the AD.
B— In the first paragraph of the AD.
C— In the AD number.

AD notes are currently identified by an eight-digit number such as 2000-22-08. The first four digits (2000) identify the year in which the AD was issued. The second two digits identify the biweekly period in which the AD was issued. This is the twenty-first or twenty-second week of the year. The last two digits (08) is the sequential number of the AD issued during this time period. This is the eighth AD issued in the twenty-second biweekly period.

Continued

Answers Note: *All Learning Statement Codes (in parentheses) are preceded by "IAR." See explanation on Page 1–10.*

153 [C] (031) 14 CFR Part 39
153a [B] (031) 14 CFR §39.27
154 [C] (020) AD 80-10-02
155 [B] (020) AD 80-15-12
155a [C] (031) AC 39-7C

If the AD is revised, the letter "R" and the number of revisions will be appended to the AD number. For example, AD 81-23-01 R1 is the first revision to AD 81-23-01. The effective date of the revision may be found in the last sentence in the body of the AD.

156. Use AD 81-23-01 R1 to answer this question.

Known Information: Beechcraft model 65-A88 serial number LP-34.

Following the requirements of AD 81-23-01 R1, which of the following hardware is compatible for use with a used preload indicating washer?

A— FN22M-1414.
B— LWB 22-14-31.
C— VEP 220121-14-32.

A used PLI washer can be used with an FN22-1414 nut and a VEP 220121-14-32 bolt.

157. Use AD 82-06-12 to answer this question.

Known Information: Air Tractor model AT-302, serial number 302-6521, equipped with 1-inch-thick (P/N 40007-2) main landing gear struts.

On the effective date of AD 82-06-12, this aircraft had accumulated 703 hours and 897 landings. Which of the following actions would the AD require?

A— Within 500 landings, replace landing gear.
B— Within 100 hours, inspect in accordance with paragraph (c).
C— Within 20 hours, inspect in accordance with paragraph (c).

Model AT-302 airplane: On struts having exceeded, or upon accumulating, 600 hours time-in-service or 3,000 landings, whichever occurs first, within 20 hours time-in-service or 100 landings, whichever comes first, after the effective date of this AD and thereafter at intervals of 100 hours time-in-service or 500 landings, whichever occurs first, inspect the struts and replace as necessary in accordance with paragraph (c).

158. Use AD 82-11-05 to answer this question.

Known Information: Bendix Magneto model D-2000 with blue identification plate, serial number 8121106. Time since new 2,002.5 hours, with 198.2 hours of operation in the preceding year. A review of aircraft records, prior to an annual inspection, revealed no compliance with AD 82-11-05. What action, if any, is required?

A— The blue identification plate found on the magneto indicates that it is a factory overhaul with AD compliance by the manufacturer.
B— Compliance with AD will be by inspection and gear replacement.
C— The magneto must be inspected per Bendix Service Bulletin number 617 within the next 50 hours time in service.

Magnetos with 500 hours or more time in service since new or overhaul must be inspected in accordance with the Detailed Instructions of Bendix Service Bulletin No. 617, dated November 1981, or later FAA-approved revision within the next 50 hours in service. Since this magneto has 2,002.5 hours since new, the distributor gear assembly must be replaced.

159. Use AD 82-11-05 to answer this question.

Known Information: Bendix Magneto model D-2000 with blue identification plate, serial number 8122206. Time since new 2,002.5 hours, with 198.2 hours of operation in the preceding year. A review of aircraft records, prior to an annual inspection, revealed no compliance with AD 82-11-05. What action, if any, is required?

A— Airworthiness Directive does not apply by serial number.
B— Compliance with AD will be by inspection and gear replacement.
C— The magneto must be inspected per Bendix Service Bulletin number 617 within the next 50 hours time in service.

This AD does not apply to this magneto because serial number 8122206 is above the applicable serial number of 8122106.

Answers

156 [C] (020) AD 81-23-01 R1
157 [C] (020) AD 82-06-12
158 [B] (020) AD 82-11-05
159 [A] (020) AD 82-11-05

160. Use AD 90-01-06 to answer this question.

Known Information: Enstrom Helicopter model F-28C-2 equipped with tail rotor gear box P/N 28-13500-1, incorporating a Boston gear XR-137-2YR. Installed gear box was placed in service, in new condition, at total time of 2,104 hours.

During the last compliance with AD 90-01-06 at total time of 3,204 hours, five metal flakes of less than the AD specified size were noted in the gear box oil. The gear box was flushed and reserviced. A test flight was conducted per the requirements of the AD. No further metal was found at that time. What action, if any, may be required during the annual inspection at 3,304 hours concerning this AD?

A— Replace the tail rotor gear box.
B— No further action is required at this time.
C— If metal is found during the inspection, clean gearbox, reservice, and test fly. If metal reappears, replace gearbox.

The gear box was placed in service, in new condition at 2,104 hours. By the next annual inspection at 3,304 hours, the gear box will have 1,200 hours in service. Paragraph (a)(2) states "Remove all -1, -3, or -5 tail rotor gearboxes containing spiral miter gear-set "Boston Gear XR-137-2YR" and "Boston Gear XR-137-2YL," with 1,200 or more hours time in service since the last overhaul and replace with an airworthy gear box."

161. Use AD 90-08-14 to answer this question.

Known Information: Beechcraft model 56TC, serial number TG-28, aircraft total time 1,701.9 hours.

During the current annual inspection, aircraft records indicated AD 90-08-14 was last complied with at 1,500.4 hours. At that time, cracks were found in the wing forward spar carry-through web structure. One crack in the left forward bend radius, measuring .45 inch and one in the right aft bend radius measuring .40 inch, both were stop drilled. At this inspection, both cracks that were previously stop-drilled showed no additional cracking. However, another crack was found in the right aft web face emanating from one huckbolt hole to another. What action should be taken concerning compliance with this AD?

A— Stop-drill the additional crack and reinspect in 100 hours.
B— Install kit P/N 58-4008 within the next 25 hours.
C— Approve for return to service and reinspect within the next 100 hours.

This AD was complied with 201.5 hours previously and is therefore due for a reinspection at this time. The stop-drilled cracks in the bend radius show no additional cracking.

There is a crack in the web face that joins two of the Huckbolt holes. This crack requires that the web face be repaired by the applicable Beech P/N 58-4008 kit within the next 25 hours time in service.

Answers

160 [A] (020) AD 90-01-06

161 [B] (020) AD 90-08-14

162. Use AD 90-08-14 to answer this question.

During the current annual inspection, aircraft records indicated AD 90-08-14 was last complied with at 1,500 hours. At that time, cracks were found in the wing forward spar carry-through web structure. One crack in the left forward bend radius, measuring .45 inch and one in the right aft bend radius measuring .40 inch, both were stop drilled. At this inspection, both cracks that were previously stop-drilled showed no additional cracking. However, another crack was found in the right aft web face that passes through two holes and extends another .78 inches. What action should be taken concerning compliance with this AD?

A— Install Beech kit P/N 58-4008 within the next 25 hours.
B— Approve for return to service, and reinspect in the next 25 hours.
C— Install Beech P/N 58-4008 kit prior to further flight.

The new crack that was found exceeds the dimensions found in paragraph (a)(3)(iii) which states: "If any crack passes through two fastener holes and extends beyond the holes for more than 0.5 inch, prior to further flight repair the web face with the applicable Beech P/N 58-4008 kit as specified in the above SB."

163. Use AD 93-24-03 to answer this question.

Known Information: Beech Bonanza model E33, serial number CD1224, current aircraft total time 3,677 hours.

The last recorded entry for this AD was at aircraft total time of 3,176 hours. At that time, the forward rudder spar was inspected in accordance with Beech Service Bulletin (SB) No. 2333, Revision 1. Following the inspection at that time, Supplemental Type Certificate (STC) SA4899NM was used to install SMP reinforcement bracket. What action is required at the current aircraft total time?

A— Reinspect in accordance with Beech Service Bulletin SB No. 2333, Revision 1.
B— No further action is required by this AD.
C— Replace rudder assembly with Part number 33-630000-167.

The Spacecraft Machine Products (SMP) reinforcement bracket was installed in accordance with STC SA4899NM, and no repetitive action is required by this AD.

163a. Use AD 93-24-03 to answer this question.

Known Information: Beech Bonanza model E33, serial number CD1224, current aircraft total time 3,676 hours.

The last recorded entry for this AD was at aircraft total time of 3,176 hours. At that time, the forward rudder spar was inspected in accordance with Beech Service Bulletin No. 2333, Revision 1. No cracks were found at that time. During the current inspection at 3,676 hours, cracks were found in the area of the upper hinge. What action should be taken concerning compliance with this AD?

A— Reinspect the rudder forward spar for cracks in accordance with the instructions in Beech Service Bulletin No. 2333, Revision 1, at intervals not to exceed 500 hours time in service.
B— Since cracks were found only in the upper hinge area, no further action is required by this AD.
C— Stop-drill the cracks and install an SMP upper-hinge reinforcement bracket in accordance with STC SA4899NM.

For a crack in the upper hinge, the AD states that the corrective action is to stop-drill and install the reinforcement bracket per the referenced STC.

164. Use AD 95-13-08 to answer this question.

Known Information: Pratt & Whitney Canada (PWC) model PT6A-67D turboprop engine. Engine serial number PC-E114029, total time since new 1,572 hours at the effective date of AD 94-10-02, June 15, 1994.

The maintenance record indicates the CT disk and blade assembly was debladed and inspected per the accomplishment instructions of PWC SB No. 14128, Revision 1, at 1,498 hours on December 16, 1992, in accordance with AD 92-27-19. When will the next deblading and inspection become due?

A— 1,548 hours.
B— 1,622 hours.
C— 1,748 hours.

The compressor turbine was debladed and inspected in accordance with PWC SB 14128, Revision 1 prior to receipt of this AD. It was done according to AD 92-27-19 which this AD supersedes.

These blades have a total of 1,572 hours since new, and they must be debladed, inspected, and replaced if necessary within the next 50 hours time in service.

The time in service at the receipt of this AD was 1,572 hours, so the next deblading will be due at 1,622 hours.

Answers

162 [C] (020) AD 90-08-14
163 [B] (020) AD 93-24-03
163a [C] (020) AD 93-24-03
164 [B] (020) AD 95-13-08

165. Use AD 95-13-08 to answer this question.

Known Information: Pratt & Whitney Canada (PWC) model PT6A-67D turboprop engine. Engine serial number PC-E114029, total time since new 1,572 hours at receipt of AD 95-13-08.

The maintenance record indicates the CT disk and blade assembly was debladed and inspected per the accomplishment instructions of PWC SB No. 14128, Revision 1, at 2,087 TIS hours on December 16, 1992, in accordance with AD 92-27-19. The aircraft now has 2,116 TIS. When will the next deblading and inspection become due?

A—2,187 hours—10 hours over due.
B—2,166 hours—due immediately.
C—2,137 hours—21 hours to go.

The compressor turbine was debladed and inspected in accordance with PWC SB 14128, Revision 1 at 2,087 hours time in service (TIS).

AD 95-13-08 states that engines with serial numbers between PC-E114001 and PC-B114044 (this engine's SN is PC-E114029) with blade sets having more than 600 hours TIS since new on the effective date of this AD, must be debladed, inspected and replaced, if necessary within the next 50 hours TIS.

These blades have a total of 1,572 hours since new, so they must be debladed, inspected, and replaced if necessary within the next 50 hours TIS.

The TIS at the receipt of this AD was 2,087 hours, and it now has 2,116 hours TIS. This is 29 hours since the last inspection so the next deblading will be due at 2,137 hours, which is 21 hours to go.

Answers

165 [C] (020) AD 95-13-08

Sample Airworthiness Directives

MESSERSCHMITT-BOLKOW-BLOHM-GMBH AND MESSERSCHMITT-BOLKOW-BLOHM HELICOPTER

80-10-02 MESSERSCHMITT-BOLKOW-BLOHM: Amendment 39-3765. Applies to Model BO-105 series helicopters with tail rotor blade grip P/N 105-31711 or P/N 105-31722 installed, certificated in any category.

To prevent failure of the tail rotor system, accomplish the following:

(a) Within the next 10 hours time in service after the effective date of this AD, unless already accomplished within the last 90 hours time in service, and thereafter at intervals not to exceed 100 hours time in service from the last inspection, inspect the visible part of the inner surface of the tail rotor blade grip clevis area (do not remove blade retaining bolt bushings) for cracks using the dye penetrant method in accordance with Messerschmitt-Bolkow-Blohm BO-105 Alert Service Bulletin No. 18 dated March 15, 1979, or an FAA-approved equivalent.

(b) Within the next 100 hours after installing a replacement tail rotor blade grip in accordance with paragraph (g) of this AD, and thereafter at intervals not to exceed 100 hours time in service from the last inspection, inspect the visible part of the inner surface of the tail rotor blade grip clevis area (do not remove blade retaining bolt bushings) for cracks using the dye penetrant method in accordance with Messerschmitt-Bolkow-Blohm BO-105 Alert Service Bulletin No. 18 dated March 15, 1979, or an FAA-approved equivalent.

(c) Within the next 100 hours time in service after the effective date of this AD—

(1) Visually inspect the inboard end of the tail rotor blade grip for cracks in accordance with paragraph 2.A.1 "Accomplishment Instructions" of Messerschmitt-Bolkow-Blohm Service Bulletin 30-24 dated December 1, 1978, or an FAA-approved equivalent; and

(2) Inspect the tail rotor blade grip in the vicinity of the bore of the laminated pack retaining bolt (on the inner side) for cracks using the dye penetrant method in accordance with paragraph 2.A.2 "Accomplishment Instructions" of Messerschmitt-Bolkow-Blohm BO-105 Service Bulletin 30-24 dated December 1, 1978, or an FAA-approved equivalent.

(d) Within the next 100 hours time in service after accomplishing the inspection required by paragraph (c)(1) of this AD or installing a replacement tail rotor blade grip in accordance with paragraph (g) of this AD, and thereafter at intervals not to exceed 100 hours time in service from the last inspection, visually inspect the inboard end of the tail rotor blade grip for cracks in accordance with "Special Inspections." Chapter 10 of the Messerschmitt-Bolkow-Blohm BO-105 Maintenance and Overhaul Manual or an FAA-approved equivalent.

(e) Within the next 600 hours time in service after accomplishing the inspection required by paragraph (c)(2) of this AD or installing a replacement tail rotor blade grip in accordance with paragraph (g) of this AD, and thereafter at intervals not to exceed 600 hours from the last inspection, inspect the tail rotor blade grip in the vicinity of the bore of the laminated pack retaining bolt (on the inner side) for cracks using the dye penetrant method in accordance with "Special Inspections," Chapter 10, of the Messerschmitt-Bolkow-Blohm BO-105 Maintenance and Overhaul Manual or an FAA-approved equivalent.

(f) If, during any inspection required by this AD, any cracks are found, before further flight, replace the cracked tail rotor blade grip in accordance with paragraph (g) of this AD.

(g) For all replacement tail rotor blade grips installed after the effective date of this AD—

(1) Use a new or used crack-free tail rotor blade grip of the same part number. Before installation of a used tail rotor blade grip, inspect the part using the dye penetrant method to ensure that it is crack-free; and

(2) Comply with the repetitive inspection requirements of paragraphs (b), (d), and (e) of this AD.

NOTE: This AD applies to both tail rotor blade grips installed on the helicopter.

This amendment becomes effective May 1, 1980, as to all persons except those persons to whom it was made immediately effective by the telegram dated March 30, 1979, which contained this amendment.

AGUSTA, COSTRUZIONI
AERONAUTICHE GIOVANNI

80-15-12 COSTRUZIONI AERONAUTICHE GIOVANNI AGUSTA: Amendment 39-3854 Applies to Model A109A series helicopters, certificated in all categories, all serial numbers up to S/N 7165 inclusive, which have main rotor mast bearing inner race P/N 109-0404-14 installed.

Compliance required as indicated.

To prevent failure of the main rotor mast upper thrust bearing, accomplish the following:

(a) Within the next 25 hours time in service after the effective date of this AD, unless already accomplished, visually inspect the area between the swashplate support, P/N 109-0110-05, and the main transmission upper case flange, for evidence of oil leaks in accordance with "ACCOMPLISHMENT INSTRUCTIONS," Part I, paragraph A, of Agusta Service Bulletin No. 109-12, Revision A, dated December 12, 1979 (hereinafter referred to as the Service Bulletin), or an FAA-approved equivalent.

(b) If no evidence of oil leaks is found, continue in service and comply with paragraph (d) of this AD.

(c) If, as a result of the inspection required in paragraph (a) of this AD, or of a repetitive inspection required by paragraph (d) of this AD, evidence of oil leaks is found, raise swashplate support, P/N 109-0110-05, and carefully inspect, using the visual method, the entire exposed surface of the bearing inner race, P/N 109-0404-14, for evidence of damage or cracks.

(i) If no cracks or damage are found, replace the packing P/N 109-0406-68, with new packing in accordance with "ACCOMPLISHMENT INSTRUCTIONS," Part I, paragraph B.2 of the Service Bulletin, or an FAA-approved equivalent.

(ii) If cracks or damage are found, before further flight, except that the helicopter may be flown to a base in accordance with FAR 21.197 and 21.199 where the repairs may be accomplished, replace the bearing inner race with a new part number inner race, P/N 109-0404-14-15, in accordance with "ACCOMPLISHMENT INSTRUCTIONS," Part II, of the Service Bulletin, or an FAA-approved equivalent.

(iii) Upon accomplishment of paragraph (c)(i) or (c)(ii) of this AD, return to service and comply with paragraph (d) of this AD.

(d) After the termination of each flight, conduct the inspection described in paragraph (a) of this AD on all helicopters up to S/N 7165, inclusive.

(e) Within the next 200 hours time in service after the effective date of this AD, unless already accomplished, for all helicopters up to S/N 7165 inclusive, and except for helicopters S/N 7140, 7142, 7148, 7150, 7152, 7158, 7160, 7161, 7162, 7163 and 7164, remove the main rotor mast upper bearing inner race, P/N 109-0404-14, and replace with a new part number bearing inner race, P/N 109-0404-14-15, in accordance with Part II of the Service Bulletin, or an FAA-approved equivalent, and continue to comply with paragraph (d) of this AD.

(f) For all main transmission gearboxes S/N 58 and below, held as spares, replace the main rotor mast upper bearing inner race, P/N 109-0404-14, with a new inner race P/N 109-0404-14-15 in accordance with Part II of the Service Bulletin before release of the gearbox to service.

(g) Upon request of an operator, the Chief, Aircraft Certification Staff, FAA, Europe, Africa, and Middle East Office, c/o American Embassy, Brussels, Belgium, may adjust the compliance time specified in paragraph (d) of this AD provided such requests are made through an FAA maintenance inspector and the request contains substantiating data to justify the request for that operator.

(h) For the purpose of this AD, and FAA-approved equivalent may be approved by the Chief, Aircraft Certification Staff, AEU-100, Europe, Africa, and Middle East Office, Federal Aviation Administration, c/o American Embassy, Brussels, Belgium.

This amendment becomes effective August 7, 1980.

BEECH

81-23-01 R1 BEECH: Amendment 39-4289. Applies to the following model airplanes regardless of the category or categories of airworthiness certification:

MODELS	SERIAL NUMBER (S/N)*
65, A65 & A65-8200	LC-181 through LC-335
70	LB-1 through LB-35
65-A80, 65-A80-8800 & 65-B80	LD-151 through LD-511 and LD-34, LD-46, LD-119
65-A88, 65-88	LP-1 through LP-54
65-90, 65-A90, B90 & C-90	LJ-1 through LJ-929
E90	LW-1 through LW-342
99, 99A, B99	U-1 through U-164
100 & A100	B-1 through B-247
B 100	BE-1 through BE-102, and BE-104
Military:	
L23F* *	**LF-7 through LF-76**
65-A90-1	LM-1 through LM-144
65-A90-2	LS-1, -2, -3
65-A90-3	LT-1, -2
65-A90-4	LU-1 through LU-16
NU-8F	LG-1

*Except that airplanes which have installed BEECHCRAFT Kit No. 90-4077-1 S, BEECHCRAFT Kit No. 99-4023-1 S, or Aviadesign Supplemental Type Certificate SA1178CE or SA1583CE are not affected by this AD.

**Except that Model L23F airplanes which do not have a preload indicating washer assembly (i.e., one with radial holes in a center ring) are not affected by this AD.

COMPLIANCE: Required as indicated, unless already accomplished.

In order to assure integrity of bolts and nuts at the lower forward attachments of outer wing panels to the wing center section, accomplish the following:

A) Prior to next flight, accomplish all of the following:

1. Remove all bolts, washers, and nuts from each lower forward wing attachment and thoroughly clean each removed part. Throughout all action required by this AD:

a. Use procedures in the applicable Beech Maintenance Manual except where other procedures are specified by this AD,

b. Unless different instruction from Beech Aircraft Corporation is obtained and followed, reposition wing, as necessary, to remove or reinstall bolt by hand without using any tool,

c. Keep parts of each preload indicating washer assembly together so that parts of one assembly cannot be intermingled with parts of another assembly,

d. Clean each removed part with naptha or methyl ethyl ketone (MEK) using a bristle brush, and repeat this cleaning as necessary prior to each subsequently specified action until lubricant is applied, and

e. Accomplish all of the specified actions on both (i.e., left and right) sides of the airplane.

2. Visually inspect each bolt and nut for reddish rust. Do not classify copper residue over cadmium plating as rust. For a bolt, rust is acceptable only on the end (including not more than one thread) farthest from the head and within counterboard recess between wrench serrations of the bolt head. For compliance with Paragraph A)6 and C), below, classify a bolt as rusted if rust is found elsewhere. Classify a nut as rusted if rust is found anywhere.

81-23-01 R1 BEECH *(continued)*

3. Visually inspect each bolt and nut for a pit or crack in steel (not cadmium or copper plating) material. Use 10X or stronger magnifying glass. For each bolt, pay particular attention to the fillet and shank, including threads. For each nut, pay particular attention to the chamfer (that faces the bolt head when installed) and perceptible threads adjacent to this chamfer. (Refer to Paragraphs A)6 and C) below.)

4. Bake each bolt and nut continuously for 23 hours at 350 degrees to 400 degrees Fahrenheit and cool in still air.

5. After accomplishment of Paragraph A)4, above, use a magnetic particle method of Advisory Circular AC43.13-1A to inspect each bolt and nut for a crack, paying particular attention to locations specified in Paragraph A)3, above. For each bolt, use a fluorescent particle method with 5250 to 6750 ampereturns in a coil to produce longitudinal magnetization in each bolt. (6,000 ampereturns means 2,000 amperes in a 3-turn coil or 1,000 amperes in a 6-turn coil, etc.) For each nut, use any magnetic particule method with 500 to 700 amperes through a central conductor of at least 0.6-inch diameter through two nuts to produce circular magnetization. Demagnetize each bolt and nut after the above inspection.

6. Replace each rusted, pitted, and/or cracked nut and bolt with a new Part Number (P/N) as follows:

a. If new preload indicating (PLI) washer assembly is to be used in accordance with Paragraph A)9, below, nut P/N is 72789-1414, 72789M-1414, FN22-1414, or FN22M-1414. ("M" in P/N denotes black coating. All eligible nuts have a locking feature which necessitates use of a wrench for full engagement with bolt.)

b. If a used PLI washer assembly is reinstalled in accordance with Paragraph A)9, below, nut P/N is 72789-1414 or FN22-1414.

c. Bolt is P/N LWB 22-14-XX or VEP 220121-14-XX where XX is 31 for airplanes with S/N LD-34, LD-46, LD-119, and LJ-1 through LJ-67, and XX is 32 for all other airplanes affected by this AD.

Replace preload indicating washer with new P/N 61475-14-43.5 assembly (not any other P/N) if this assembly is available. Obtain new parts only from Beech Service Centers or Beech Aircraft Corporation. (Neither baking nor field inspection of new parts is necessary.) Do not replate any part.

7. Clean the bore and recessed washer seat area of the outboard and inboard wing fittings with naptha or methyl ethyl ketone (MEK). Visually inspect these areas for corrosion, burrs, gouges and coining. If any defect is found, contact Beech Aircraft Service Department, 9709 East Central, Wichita, Kansas 67201; telephone (316) 681-7261, 7278, or 7352, for rework disposition. Also, if any defect is found, treat the bore and recessed washer seat areas of the inboard and outboard wing fittings with Alodine 1200, 1200S, or 1201. Allow the alodine coating to dry for 5 minutes. Wash the coating with water and blow dry with air without wiping. Paint treated washer seat areas with zinc chromate primer (obtain locally) and allow primer to dry.

8. Coat the inspected areas of the wing fittings, all of each bolt, all of each nut, and all of each preload indicating washer assembly with either clean MIL-C-16173, Grade 2 corrosion preventative compound or clean General Electric G322L Versilube Silicone Lubricant.

9. Install removed or new parts using standard procedures except as follows:

a. Preload indicating (PLI) washer assembly may be reused with P/N 72789-1414 and/or P/N FN22-1414 nuts, only.

b. Ascertain that a radius of the adjacent washer is next to the fillet under the bolt head and next to the outer edge of the recess in each wing fitting. Position wing as necessary to allow bolt to slide into fitting without use of any tool.

c. Tighten the joint by rotating the nut (do not turn the bolt). Use standard procedure if new PLI washer assembly is installed. If used PLI washer assembly is reinstalled, make necessary correction for any torque wrench adapter and apply 3250 to 3400 inch-pounds torque, but install new PLI washer assembly if center ring of the used assembly turns after 3400 inch-pounds torque is applied. Do not allow wrench to bear against fitting.

BEECH

d. Coat entire portion of bolt that projects beyond nut, using a material that is specified in Paragraph A)8, above.

e. Make aircraft maintenance record entry showing work accomplished, especially procedure used for tightening nut, and whether new or used PLI washer was installed. Indicating washer assembly with either clean MIL-C-16173, Grade 2 corrosion preventative compound or clean General Electric G322L Versilube Silicone Lubricant.

B) Between 90 and 110 hours time-in-service after accomplishment of action specified by Paragraph A) of this AD, check nut tightness, using the same procedure that was used for accomplishment of Paragraph A)9c, above.

C) Within 3 days after replacing a part in accordance with Paragraph A)6, above, or noting a defect when complying with this AD, submit a written report to the Federal Aviation Administration via an FAA M or D Report (FAA Form 8330-2) or a letter to the office specified in Paragraph E), below and send the replaced part(s) to Beech Aircraft Corporation. In the submitted report, please advise date of last previous bolt removal.

D) A special flight permit in accordance with Federal Aviation Regulation 21.197 for flight to the nearest base is permitted in order to accomplish Paragraph A) of this AD. The nearest FAA Flight Standards District Office may be contacted to obtain a telegraphic special flight permit.

E) Any equivalent method of compliance with this AD must be approved by the Chief, Aircraft Certification Program, Federal Aviation Administration, Room 238, Terminal Building 2299, Mid-Continent Airport, Wichita, Kansas 67209; Telephone (316) 269-7000, 7001, or 7002.

This amendment becomes effective on January 4, 1982, to all persons except those to whom it has already been made effective by an airmail letter from the FAA dated October 31, 1981.

AIR TRACTOR, INC.

==================================

82-06-12 AIR TRACTOR: Amendment 39-4350. Applies to Models AT-300 (S/Ns 300-0001 through 300-9999); AT-301 (S/Ns 301-0001 through 301-9999); AT-302 (S/Ns 302-0001 through 302-9999); AT-400 (S/Ns 400-0244 through 400-9999); and AT-400A (S/Ns 400A-0397 through 400A-9999) airplanes certified in any category and equipped with l-inch-thick (P/N 40007-2 or P/N 40058-1) main landing gear struts.

COMPLIANCE: Required as indicated, unless already accomplished.

To prevent possible failure of the P/N 40007-2 or P/N 40058-1 main landing gear struts accomplish the following:

(a) Models AT-300 and AT-301 airplanes:

(1) On struts having exceeded, or upon accumulating, 1,000 hours time-in-service or 5,000 landings, whichever occurs first, within 20 hours time-in-service or 100 landings, whichever occurs first, after the effective date of this AD and thereafter at intervals of 100 hours time-in-service or 500 landings, whichever occurs first, inspect and replace as necessary the landing gear struts in accordance with paragraph (c).

(2) On struts having exceeded, or upon accumulating, 2,000 hours time-in-service or 7,500 landings, whichever occurs first, prior to further flight, replace the struts with new struts of the same part number.

(b) Models AT-302, AT-400 and AT-400A airplanes:

(1) On struts having exceeded, or upon accumulating, 600 hours time-in-service or 3,000 landings, whichever occurs first, within 20 hours time-in-service or 100 landings, whichever occurs first, after the effective date of this AD and thereafter at intervals of 100 hours time-in-service or 500 landings, whichever occurs first, inspect the struts and replace as necessary in accordance with paragraph (c).

(2) On struts having exceeded, or upon accumulating, 1,200 hours time-in-service, or 6,000 landings, whichever occurs first, prior to further flight, replace the struts with new struts of the same part number.

(c) Remove the left and right outboard fuselage clamp blocks. Remove all minor corrosion on both main landing gears by sandblasting. Inspect both main landing gears using dye penetrant or magnetic particle inspection procedures with special attention in the areas of strut contact with the clamp blocks. Replace all parts which are damaged, cracked, or have severe corrosion pitting with new parts of the same part number before further flight. All struts returned to service must be painted.

(d) The aircraft hours and landings may be used as the time-in-service or landings on the struts if time-in-service or landings on the struts cannot be established by the airplane maintenance records.

(e) A special flight permit may be issued in accordance with FAR 21.197 to allow flight of the aircraft to a location where this AD can be accomplished.

(f) An equivalent method of compliance with this AD may be used when approved by the Chief, Aircraft Certification Division, Federal Aviation Administration, 4400 Blue Mound Road, Fort Worth, Texas 76101.

Snow Engineering Company Service Letter No. 45, dated November 1, 1981, covers the subject matter of this AD.

Compliance with this Service Letter within the last 100 hours time-in-service or 500 landings, whichever comes first, satisfies the initial inspection requirement of paragraphs (a) and (b) of this AD.

This amendment becomes effective on March 25, 1982.

BENDIX

= =

82-11-05 **BENDIX**: Amendment 39-4389. Applies to Bendix Engine Products Division D-2000 and D-2200 series magnetos with serial numbers below 35480 (red identification plate) and with serial numbers below 8122106 (blue identification plate), unless identified with an "X" in the upper left corner of the identification plate.

Compliance required as indicated, unless already accomplished.

To reduce the possibility of engine power loss and engine damage resulting from looseness of the distributor gear electrode, accomplish Paragraphs (a) and (b):

(a) Comply with the inspection requirements specified in the "Detailed Instructions" of Bendix Service Bulletin No. 617, dated November 1981, or later FAA-approved revision in accordance with the following schedule:

MAGNETO TIME IN SERVICE SINCE NEW OR OVERHAUL	ACCOMPLISH
Less than 500 hours	Within the next 50 hours in service and every 100 hours in service thereafter up to 550 hours in service.
500 hours or more	Within the next 50 hours in service.

(b) Magnetos with 1900 hours or more in service since new or overhaul: Within the next 100 hours time in service, replace distributor gear assembly with new serviceable gear assembly in accordance with Bendix Service Bulletin No. 617, dated November 1981, or later FAA approved revision.

(c) If the distributor block is contaminated with brass filings or bronze colored dust, inspect the engine as follows:

(1) Observe engine pistons through spark plug hole for evidence of burning.

(2) Check valve dry tappet clearance per engine manufacturer's instructions.

If piston damage, or lower than specified dry tappet clearance, is present, the engine must be inspected in accordance with the engine manufacturer's instructions for continued airworthiness.

Equivalent means of compliance may be approved by the Chief of the New York Aircraft Certification Office, ANE-170, Federal Aviation Administration (FAA), New England Aircraft Certification Division, Federal Building, JFK International Airport, Jamaica, New York 11430. As permitted by FAR 21.197, aircraft may be flown to a base where maintenance required by this AD can be accomplished.

This AD is effective June 9, 1982.

ENSTROM

90-01-06 ENSTROM HELICOPTER CORPORATION: Amendment 39-6457. Docket No. 89-ASW-59.

Applicability: Enstrom Model F-28, F-28A, F-28C, F-28C-2, F-28F, 280, 280C, 280F and 280FX Series Helicopters, equipped with tail rotor gearboxes, P/N 28-13500-1, 28-13525-1, -3, and -5, containing spiral miter gear-set "Boston Gear XR-137-2YR" and "Boston Gear XR-137-2YL."

Compliance: Required as indicated, unless already accomplished.

To prevent the loss of tail rotor thrust and directional control, which could result in loss of the helicopter, accomplish the following:

(a) Within the next five hours' time in service—

(1) Determine from the aircraft log book if tail rotor gearbox, P/N 28-13500-1, 28-13525-1, -3, or -5, is installed in the helicopter;

(2) Remove all -1, -3 or -5 tail rotor gearboxes containing spiral miter gear-set "Boston Gear XR-137-2YR" and "Boston Gear XR-137-2YL," with 1,200 or more hours' time in service since the last overhaul, and replace with an airworthy gearbox; and

(3) For tail rotor gearboxes with less than 1,200-hours' time in service since the last overhaul, remove the magnetic chip detector (plug), drain the oil from the tail rotor gearbox, filter the oil using a white filter paper, and inspect the magnetic plug and the filter paper with a ten-power magnifying glass—

(i) If no metal contaminants are found, return the tail rotor gearbox to service;

(ii) If the inspection required by paragraph (a)(3) above reveals the presence of more than 15 thin metal flakes, splinters, or granular-shaped steel particles greater than 0.005-inches thick or longer than 0.015 inches, remove and replace the tail rotor gearbox with an airworthy gearbox; and

(iii) If metal contaminants are found that are fewer in number and smaller than those described in paragraph (ii) above, conduct further servicing and inspection in accordance with paragraph (a)(4).

(4) Flush the gearbox with clean oil and clean the magnetic plug with a cotton swab and/or an air gun.

NOTE: Do not clean the magnetic plug with a strong magnet. This weakens the magnet on the chip detector.

(i) Refill the tail rotor gearbox with Mil-L-6082B Shell SAE 10W, Mil-L-6082B Texaco SAE 10W, or Mil-L-22851B Phillips SAE 20W-50W lubricant. If any of these lubricants are not available, consult Enstrom Helicopter Corporation, Customer Service Department, for a possible alternative.

(ii) Conduct a serviceability check by flying the helicopter for one hour at various power settings up to full power, and then repeat the inspection required by paragraphs (a)(3) above.

(A) If no metal contaminants are found, return the tail rotor gearbox to service.

(B) If the repeat inspection reveals the presence of any metal contaminants, regardless of size or number, remove and replace the tail rotor gearbox with an airworthy gearbox.

(b) At intervals not to exceed 100 hours' time in service on all gearboxes returned to service after passing the inspections of paragraph (a), remove the magnetic chip detector (plug), drain the oil from the tail rotor gearbox, filter the oil using a white filter paper, and inspect the magnetic plug and the filter paper with a ten-power magnifying glass.

(1) If the inspection reveals the presence of any metal contaminants, regardless of size or number, remove and replace the tail rotor gearbox with an airworthy gearbox.

(2) If no metal contaminants are found return the tail rotor gearbox to service.

(c) Within 1,200 hours' time in service since the last overhaul, remove and replace the tail rotor gearbox with an airworthy gearbox.

(d) An alternate method of compliance with this AD, which provides an equivalent level of safety, may be used when approved by the Manager, Chicago Aircraft Certification Office, FAA, 2300 East Devon Avenue, Room 232, Des Plaines, Illinois 60018.

(e) In accordance with Sections 21.197 and 21.199, flight is permitted to a base where the maintenance required by this AD may be accomplished.

This amendment (39-6457, AD 90-01-06) becomes effective on February 1, 1990.

BEECH

90-08-14 BEECH: Amendment 39-6563. Docket No. 89-CE-26-AD.
Applicability: The following airplanes certificated in any category.

MODELS	SERIAL NUMBERS
95, B95, B95A, D95A, E95	TD-1 through TD-721
95-55, 95-A55, 95-B55 and 95-B55A	TC-1 through TC-2456, except TC-350
95-C55, 95-C55A, D55, D55A, E55 AND E55A	TC-350 and TE-1 through TE-1201
95-B55B (T42A)	TF-1 through TF-70
56TC, A56TC	TG-1 through TG-94
58, 58A	TH-1 through TH-1475

Compliance: Required as indicated in the body of the AD, unless already accomplished.

To prevent cracks in the wing forward spar carry-through web structure from propagating to lengths that could compromise the integrity of the wing attachment to the fuselage, accomplish the following:

(a) Within the next 100 hours time-in-service (TIS), after the effective date of this AD, or upon the accumulation of 1,500 hours total TIS, whichever occurs later, and thereafter at the intervals specified below, inspect the wing forward spar carry-through web structure in accordance with the instructions in Beech Service Bulletin (SB) No. 2269, Revision 1, dated March 1990.

(1) If no cracks are found, repeat the inspection at 500 hour TIS intervals thereafter.

(2) For cracks in the bend radius:

(i) If the crack length is less than 2.25 inches, prior to further flight stop drill the crack in accordance with the instructions in Beech SB No. 2269, Revision 1, and reinspect for crack progression every 200 hours TIS thereafter. Only one stop drilled crack for the left side and one stop drilled crack for the right side of the web structure are permissible.

(ii) If the crack length is greater than 2.25 inches but less than 4.0 inches, prior to further flight stop drill the crack in accordance with the instructions in Beech SB No. 2269, Revision 1, and within the next 100 hours TIS, repair the web structure with the applicable Beech Part Number (P/N) 58-4008 kit as specified in the above SB. After installation of the applicable Beech P/N 58-4008 kit, dye-penetrant inspect this area for cracks within the next 1,500 hours TIS from the time of installation of the applicable kit, and reinspect for cracks at 500 hours TIS intervals thereafter. If cracks are detected in these subsequent inspections, prior to further flight, contact the Wichita Aircraft Certification Office at the address below for disposition.

(iii) If the crack length is greater than 4.0 inches, prior to further flight repair the web structure with the applicable Beech P/N 58-4008 kit as specified in the above SB. After installation of the applicable Beech P/N 58-4008 kit, dye-penetrant inspect this area for cracks within the next 1,500 hours TIS from the time of installation of the applicable kit, and reinspect for cracks at 500 hours TIS intervals thereafter. If cracks are detected in these subsequent inspections, prior to further flight, contact the Wichita Aircraft Certification Office at the address below for disposition.

(3) For cracks in the web face, in the area of the huckbolt fasteners:

(i) If the crack length is less than 1.0 inch, reinspect for crack progression every 100 hours TIS thereafter. Only one crack for the left side and one crack for the right side are permissible, provided neither crack exceeds 1.0 inch in length.

NOTE 1: Do not stop drill these cracks due to the possibility of damaging the structure behind the web face.

(ii) If any crack length is greater than 1.0 inch, or a crack is connecting two fastener holes, within the next 25 hours TIS, repair the web face with the applicable Beech P/N 58-4008 kit as specified in the above SB. After installation of the applicable Beech P/N 58-4008 kit, dye-penetrant inspect this area for cracks within the next 1,500 hours TIS from the time of installation of the applicable kit, and reinspect for cracks at 500 hours TIS intervals thereafter. If cracks are detected in these subsequent inspections, prior to further flight, contact the Wichita Aircraft Certification Office at the address below for disposition.

BEECH

(iii) If any crack passes through two fastener holes and extends beyond the holes for more than 0.5 inch, prior to further flight repair the web face with the applicable Beech P/N 58-4008 kit as specified in the above SB. After installation of the applicable Beech P/N 58-4008 kit, dye-penetrant inspect this area for cracks within the next 1,500 hours TIS from the time of installation of the applicable kit, and reinspect for cracks at 500 hours TIS intervals thereafter. If cracks are detected in these subsequent inspections, prior to further flight, contact the Wichita Aircraft Certification Office at the address below for disposition.

(4) If cracks are found on the same side of the airplane in both the forward and aft web face, or the bend radii, and any of the cracks are more than 1.0 inch long, prior to further flight repair the web structure with the applicable Beech P/N 58-4008 kit as specified in the above SB. After installation of the applicable Beech P/N 58-4008 kit, dye-penetrant inspect this area for cracks within the next 1,500 hours TIS from the time of installation of the applicable kit, and reinspect for cracks at 500 hours TIS intervals thereafter. If cracks are detected in these subsequent inspections, prior to further flight, contact the Wichita Aircraft Certification Office at the address below for disposition.

NOTE 2: If a fuselage skin crack is discovered around the opening for the lower forward carry-through fitting, an external doubler may be required.

(b) Airplanes may be flown in accordance with FAR 21.197 to a location where the AD may be accomplished.

(c) An alternate method of compliance or adjustment of the initial or repetitive compliance times, which provides an equivalent level of safety, may be approved by the Manager, Wichita Aircraft Certification Office, FAA, Room 100, 1801 Airport Road, Wichita, Kansas 67209.

NOTE 3: The request should be forwarded through an FAA Maintenance Inspector, who may add comments and then send it to the Manager, Wichita Aircraft Certification Office.

All persons affected by this directive may obtain copies of the document referred to herein upon request to Beech Aircraft Corporation, Commercial Service, Department 52, P.O. Box 85, Wichita, Kansas 67201-0085; or may examine this document at the FAA, Office of the Assistant Chief Counsel, Room 1558, 601 East 12th Street, Kansas City, Missouri 64106.

This amendment (39-6563, AD 90-08-14) becomes effective on May 7, 1990.

BEECH AIRCRAFT CORPORATION

AIRWORTHINESS DIRECTIVE
SMALL AIRCRAFT

93-24-03 BEECH AIRCRAFT CORPORATION: Amendment 39-8752. Docket No. 93-CE-22-AD. Supersedes AD 92-15-06, Amendment 39-8300 which superseded AD 91-23-07, Amendment 39-8076.

Applicability: The following Beech model and serial numbered airplanes, certificated in any category:

MODELS	SERIAL NUMBERS
35-33, 35-A33, 35-B33, 35-C33, E33, F33, and G33	CD-1 through CD-1304
35-C33A, E33A, and F33A	CE-1 through CE-1425
E33C and F33C	CJ-1 through CJ-179
36 and A36	E-1 through E-2518
A36TC and B36TC	EA-1 through EA-500

Compliance: Required as indicated after the effective date of this AD, unless already accomplished (compliance with superseded AD 92-15-06 or superseded AD 91-23-07).

To prevent separation of the rudder from the airplane caused by cracks in the forward rudder spar, accomplish the following:

(a) Upon the accumulation of 1,000 hours time-in-service (TIS) or within the next 100 hours TIS, whichever occurs later, inspect the rudder forward spar for cracks in accordance with the instructions in Beech Service Bulletin (SB) No 2333, Revision 1, dated November 1991.

(b) If no cracks are found, accomplish one of the following:

(1) Reinspect the rudder forward spar for cracks with the instructions in Beech SB No 2333, Revision 1, dated November 1991, at intervals not to exceed 500 hours TIS until either paragraph (b)(2), (b)(3), or (b)(4) of this AD is accomplished;

(2) Install Kit No 33-6001-1 S in accordance with Beech SB No 2333, Revision 1, dated November 1991;

(3) Install a Spacecraft Machine Products (SMP) rudder spar upper-hinge reinforcement bracket in accordance with Supplemental Type Certificate (STC) SA4899NM; or

(4) Replace the rudder assembly with either part number 33-630000-137, -139, -141, -167, or -169, as applicable, in accordance with the instructions in Beech SB No 2333, Revision 1, dated November 1991.

(c) If cracks are found, prior to further flight, accomplish one of the following:

(1) Replace the rudder assembly with either part number 33-630000-137, -139, -141, -167, or -169, as applicable, in accordance with the instructions in Beech SB No. 2333, Revision 1, dated November 1991;

(2) Install Kit No. 33-6001-1 S in accordance with Beech SB No. 2333, Revision 1, dated November 1991; or

(3) If the cracks are found in the area of the upper hinge, the middle hinge, or both the upper and middle hinge as specified in Beech SB No. 2333, Revision 1, dated November 1991, then stop drill the cracks and install an SMP upper-hinge reinforcement bracket in accordance with STC SA4899NM. For cracks in the middle hinge, install the upper-hinge reinforcement bracket and also install an SMP rudder spar middle-hinge reinforcement bracket in accordance with STC SA5870NM.

93-24-03

(d) If a modification or replacement has been accomplished in accordance with either paragraph (b)(2), (b)(3), (b)(4), (c)(1), (c)(2), or (c)(3) of this AD, then no repetitive inspections are required by this AD.

(e) Special flight permits may be issued in accordance with FAR 21.197 and 21.199 to operate the airplane to a location where the requirements of this AD can be accomplished.

(f) An alternative method of compliance or adjustment of the initial or repetitive compliance times that provides an equivalent level of safety may be approved by the Manager, Wichita Aircraft Certification Office, FAA, 1801 Airport Road, Mid-Continent Airport, Wichita, Kansas 67209. The request shall be forwarded through an appropriate FAA Maintenance Inspector, who may add comments and then send it to the Manager, Wichita Aircraft Certification Office.

Note: Information concerning the existence of approved alternative methods of compliance with this AD, if any, may be obtained from the Wichita Aircraft Certification Office.

(g) The inspections, installations, or replacements required by this AD shall be done in accordance with Beech Service Bulletin No. 2333, Revision 1, dated November 1991. This incorporation by reference was previously approved by the Director of the Federal Register in accordance with 5 U.S.C. 552(a) and 1 CFR Part 51 on August 22, 1992. Copies may be obtained from Beech Aircraft Corporation, P.O. Box 85, Wichita, Kansas 67201-0085. Copies may be inspected at the FAA, Central Region, Office of the Assistant Chief Counsel, Room 1558, 601 E. 12th Street, Kansas City, Missouri, or at the Office of the Federal Register, 800 North Capitol Street, NW., suite 700, Washington, DC.

(h) This amendment (39-8752) supersedes AD 92-15-06, Amendment 39-8300 which superseded AD 91-23-07, Amendment 39-8076.

(i) This amendment (39-8752) becomes effective on January 21, 1994.

**PRATT & WITNEY AIRCRAFT
OF CANADA, INC.**

AIRWORTHINESS DIRECTIVE

95-13-08 Pratt & Whitney Canada: Amendment 39-9288. Docket 95-ANE-33. Supersedes AD 94-10-02, Amendment 39-8909.

Applicability: Pratt & Whitney Canada (PWC) Model PT6A-67D turboprop engines with serial numbers prior to PC-E114100, installed on but not limited to Beech Model 1900D airplanes. NOTE: This airworthiness directive (AD) applies to each engine identified in the preceding applicability provision, regardless of whether it has been modified, altered, or repaired in the area subject to the requirements of this AD. For engines that have been modified, altered, or repaired so that the performance of the requirements of this AD is affected, the owner/operator must use the authority provided in paragraph (o) to request approval from the Federal Aviation Administration (FAA). This approval may address either no action, if the current configuration eliminates the unsafe condition, or different actions necessary to address the unsafe condition described in this AD. Such a request should include an assessment of the effect of the changed configuration on the unsafe condition addressed by this AD. In no case does the presence of any modification, alteration, or repair remove any engine from the applicability of this AD.

Compliance: Required as indicated, unless accomplished previously.

To prevent aircraft handling problems due to imposition of the engine RPM restriction, accomplish the following:

(a) For those operators that have previously complied with AD 94-10-02, this AD requires compliance with only paragraph (n).

(b) Prior to further flight, amend the Beech Model 1900D Aircraft Flight Manual (AFM), Part Number (P/N) 129-590000-3, by inserting the following requirements between pages 2-4 and 2-5:

"ENGINE OPERATING LIMITATIONS

Gas Generator RPM (N1)—Continuous operation of the gas generator between 94.0% and 97.1% is prohibited.

NOTES

1. This limitation does not prohibit the use of N1's between 94.0% and 97.1% when the pilot in command determines that the power setting is required for the safe operation of the airplane. If such occurrences exceed 5 minutes, the engine(s) must be inspected in accordance with Pratt & Whitney Canada Service Bulletin No. 14128, Revision 3, dated April 19, 1993.

2. This limitation does not prohibit the use of static Take-Off Power and Maximum Continuous Power between 94.0% and 97.1% N1 to meet the required Take-Off performance. If such occurrences exceed 5 minutes, the engine(s) must be inspected in accordance with Pratt & Whitney Canada Service Bulletin No. 14128, Revision 3, dated April 19, 1993.

3. Operation at 94.0% and below, and at 97.1% and above are permitted. Continuous operation at 94.1% through 97.0% is prohibited.

4. "Continuous Operation" means time periods exceeding 5 minutes.

5. High Speed Cruise Power Tables found in the Pilot's Operating Manual may produce N1's in the prohibited range. Flights should be planned using Intermediate or Long Range Power settings. 6. The goal of the operator should be to keep the total time of operation in the prohibited range to the absolute minimum, since the effects of operating between N1's of 94.0% and 97.1% are cumulative.

PLACARDS

Located in front of the pilot on the aft edge of the glareshield between the Master Caution annunciator and the fire extinguisher control switch:

CONTINUOUS OPERATION BETWEEN
94.0% AND 97.1% N1 IS PROHIBITED
SEE AFM"

95-13-08

(c) Compliance with the requirements of paragraph (b) of this AD may also be accomplished by inserting a copy of this AD into the Beech Model 1900D AFM.

(d) Prior to further flight, install the placard as specified in paragraph (b) of this AD.

(e) For engines that have not been inspected prior to the effective date of this AD in accordance with PWC SB No. 14128, Revision 1, dated November 13, 1992, or debladed and inspected in accordance with PWC SB No. 14128, Revision 2, dated December 22, 1992, or PWC SB No. 14128, Revision 3, dated April 19, 1993, accomplish the following:

(1) For engines with Serial Numbers PC-E114001 to PC-E114044, within 25 hours time in service (TIS) after the effective date of AD 94-10-02, June 15, 1994, deblade the CT disk, inspect the entire disk surface area and fir tree area of the CT blades for cracking and the trailing edge of the blade airfoil section for irregularities, and replace, if necessary, with serviceable parts, in accordance with the Accomplishment Instructions of PWC SB No. 14128, Revision 3, dated April 19, 1993.

(2) For engines with Serial Numbers PC-E114045 to PC-E114099, within 50 hours TIS after the effective date of AD 94-10-02, June 15, 1994, deblade the CT disk, inspect the entire disk surface area and fir tree area of the CT blades for cracking, and replace, if necessary, with serviceable parts, in accordance with the Accomplishment Instructions of PWC SB No. 14128, Revision 3, dated April 19, 1993.

(f) For engines that have been inspected in accordance with PWC SB No. 14128, Revision 1, dated November 13, 1992, prior to the effective date of this AD, deblade the CT disk, inspect the entire disk surface area and fir tree area of the CT blades for cracking, and replace, if necessary, with serviceable parts, in accordance with the Accomplishment Instructions of PWC SB No. 14128, Revision 3, dated April 19, 1993, as follows:

(1) For blade sets with greater than 600 hours TIS since new on the effective date of AD 94-10-02, June 15, 1994, deblade, inspect, and replace, if necessary, within the next 50 hours TIS after the effective date of AD 94-10-02, June 15, 1994.

(2) For blade sets with greater than or equal to 250 hours TIS, and less than or equal to 600 hours TIS, since new, on the effective date of AD 94-10-02, June 15, 1994, deblade, inspect, and replace, if necessary, within the next 100 hours TIS after the effective date of AD 94-10-02, June 15, 1994.

(3) For blade sets with less than 250 hours TIS since new on the effective date of AD 94-10-02, June 15, 1994, deblade, inspect, and replace, if necessary, within the next 250 hours TIS after the effective date of AD 94-10-02, June 15, 1994.

(g) For uninstalled CT disk and blade assemblies that have not been inspected in accordance with the Accomplishment Instructions of PWC SB No. 14128, Revision 2, dated December 22, 1992, or PWC SB No. 14128, Revision 3, dated April 19, 1993, in the preceding 250 hours TIS from the effective date of AD 94-10-02, June 15, 1994, deblade the CT disk, inspect the entire disk surface area and fir tree area of CT blades for cracking, and replace, if necessary, with serviceable parts, in accordance with the Accomplishment Instructions of PWC SB No. 14128, Revision 3, dated April 19, 1993, prior to installation.

(h) For engines with CT disk and blade assemblies that have been debladed and inspected in accordance with the Accomplishment Instructions of PWC SB No. 14128, Revision 2, dated December 22, 1992, or PWC SB No. 14128, Revision 3, dated April 19, 1993, prior to the effective date of AD 94-10-02, June 15, 1994, within 250 hours TIS since the last deblading and inspection, deblade the CT disk, inspect the entire disk surface area and fir tree area of CT blades for cracking, and replace, if necessary, with serviceable parts, in accordance with the Accomplishment Instructions of PWC SB No. 14128, Revision 3, dated April 19, 1993.

(i) For CT disk and blade assemblies that have been debladed and inspected in accordance with paragraphs (e), (f), (g), and (h) of this AD, deblade the CT disk, reinspect the entire disk surface area and fir tree area of CT blades for cracking, and replace, if necessary, with serviceable parts, in accordance with the Accomplishment Instructions of PWC SB No. 14128, Revision 3, dated April 19, 1993, at intervals not to exceed 250 hours TIS since the last deblading and inspection performed in accordance with the Accomplishment Instructions of PWC SB No. 14128, Revision 3, dated April 19, 1993.

95-13-08

(j) Install a CT stator assembly, a CT shroud housing, and a small exit duct assembly in accordance with PWC SB No. 14132, Revision 1, dated May 12, 1993, at the next shop visit after the effective date of this AD, or within 30 days after the effective date of this AD, whichever occurs first.

(k) Install CT blades and feather seals in accordance with PWC SB No. 14142, Revision 1, dated May 12, 1993, at the next shop visit after the effective date of this AD, or 30 days after the effective date of this AD, whichever occurs first.

(l) For the purpose of this AD, a shop visit is defined as when major engine flanges are separated.

(m) Installation of improved hardware in accordance with paragraphs (j) and (k) of this AD constitutes terminating action for the inspections required by paragraphs (e) through (i) of this AD.

(n) For aircraft equipped with engines that have complied with paragraphs (j) and (k) of this AD, or AD 94-10-02, accomplish the following:

(1) Remove the amendment to the Beech Model 1900D AFM, P/N 129-590000-3, described in paragraphs (b) or (c) of this AD.

(2) Remove the placard described in paragraph (d) of this AD.

(o) An alternative method of compliance or adjustment of the compliance time that provides an acceptable level of safety may be used if approved by the Manager, Engine Certification Office. The request should be forwarded through an appropriate FAA Principal Maintenance Inspector, who may add comments and then send it to the Manager, Engine Certification Office.

NOTE: Information concerning the existence of approved alternative methods of compliance with this AD, if any, may be obtained from the Engine Certification Office.

(p) Special flight permits may be issued in accordance with sections 21.197 and 21.199 of the Federal Aviation Regulations (14 CFR 21.197 and 21.199) to operate the aircraft to a location where the requirements of this AD can be accomplished.

(q) The inspections and modifications shall be done in accordance with the following SBs:

Document No.	Pages	Revision	Date
PWC SB No. 14128 Total pages: 5	1–5	3	April 19, 1993
PWC SB No. 14132 Total pages: 6	1–6	1	May 12, 1993
PWC SB No. 14142 Total pages: 7	1–7	1	May 12, 1993

This incorporation by reference was approved by the Director of the Federal Register in accordance with 5 U.S.C. 552(a) and 1 CFR part 51. Copies may be obtained from Pratt & Whitney Canada, 1000 Marie-Victorin, Longueil, Quebec, Canada J4G 1A1. Copies may be inspected at the FAA, New England Region, Office of the Assistant Chief Counsel, 12 New England Executive Park, Burlington, MA; or at the Office of the Federal Register, 800 North Capitol Street, NW., suite 700, Washington, DC.

(r) This amendment becomes effective on July 12, 1995.

Chapter 5
FAA Order 8130.21

Sample Test Questions

FAA Order 8130.21 (current version) is an FAA internal publication entitled *Procedures for Completion and Use of the Authorized Release Certificate, FAA Form 8130-3, Airworthiness Approval Tag.*

Note: The current guidance material for FAA Order 8130.21 (reprinted in this chapter starting on Page 5–7) was released October 26, 2009 and removed all references concerning the "Class" of a product. (Formerly, export items were classified as Class I, Class II, or Class III.) However, as of the publication of this IA Test Prep, the FAA has not changed the computer test and reference material to reflect this deletion. Therefore, references to both editions of the FAA Order (8130.21F and 8130.21G) are used in this book. Questions dealing with product Class I, II and III are found in 8130.21F; all others are referenced in 8130.21G. (A copy of the previous release of this Order may be downloaded from the IA "Reader Resources" page on the ASA website.)

166. An Authorized Release Certificate (FAA Form 8130-3), may be used for what purpose?

A— Return to service after maintenance.
B— Identification of newly manufactured parts.
C— Both A and B.

An Authorized Release Certificate (FAA Form 8130-3) may be issued for return to service after maintenance, preventive maintenance, repair, or alteration requiring a sign off by persons authorized by 14 CFR §43.7 provided the applicable record keeping requirements of 14 CFR §§43.9, 91.417, 121.369(c), and 135.427(c) are complied with.

The aviation industry has advised the FAA that efforts to ensure part traceability and accountability would be greatly enhanced if the form could also be issued upon completion of the product to identify the product and follow the product in shipment. The FAA has revised the form to permit its use (by an FAA production approval holder or their supplier) in this manner with a signature by either the FAA or its appropriately authorized designee.

167. An Authorized Release Certificate (FAA Form 8130-3) may be used for what purpose?

A— Return to service after maintenance.
B— Export of Class I products.
C— Both A and B.

An Authorized Release Certificate may be used by an FAA-approved repair station to signify approval for return to service of a part after a major repair.

The tag cannot be used for export approval of a Class I product (a complete aircraft, aircraft engine or propeller) but it can be used for export approval of a Class II product (a major component of a Class I product), or a Class III product (a standard part, such as an AN, NAS, or SAE part).

168. An Authorized Release Certificate (FAA Form 8130-3), when used to return to service an aeronautical product,

A— is only used for a major repair.
B— is only used for a major alteration.
C— may be used for either a major repair or alteration.

A maintenance release or Authorized Release Certificate may be used by a certificate holder under 14 CFR Part 121 or 135 who has a continuous airworthiness maintenance program or by a repair station certificated under 14 CFR Part 145 to approve a Class II or Class III product for return to service after a major repair or a major alteration. The use of this form for this purpose is optional.

169. When an 8130-3 form is used to split bulk shipments of new products for domestic use, the original 8130-3 form should be

A— returned to the PAH that supplied it.
B— sent with the shipment of product.
C— copied and sent with a certifying statement.

For those shipments of products required to be split, the original Form 8130-3 will be used to develop the Form 8130-3 that will accompany the new quantity of product to the end-user. Copy the original form and attach it to a written statement, certifying that the copied 8130-3 form is a certified "true copy of the original."

Answers *All answer references in this chapter are to FAA Order 8130.21F and 8130.21G.*

166 [C] (030) FAA Order 8130.21
167 [A] (030) FAA Order 8130.21
168 [C] (030) FAA Order 8130.21
169 [C] (030) FAA Order 8130.21

170. 14 CFR Part 21 defines classes of aeronautical products for export. An Authorized Release Certificate (FAA Form 8130-3) may be used to return to service which of those classes?

A— Class I and II.
B— Class II and III.
C— Both A and B.

The 8130-3 form and its related guidance material continues to evolve. Revisions to the Order are a direct result of the harmonization effort between the FAA and its counterpart airworthiness authorities to clarify, improve and address questions raised regarding various sections of the Order. Export airworthiness approvals of Class II and Class III products are issued in the form of a Form 8130-3.

171. When using an 8130-3 form, which of the following could you approve for export and return to service?

A— Engine.
B— Propeller.
C— Landing gear.

The 8130-3 form and its related guidance material continues to evolve. Revisions to the Order are a direct result of the harmonization effort between the FAA and its counterpart airworthiness authorities to clarify, improve and address questions raised regarding various sections of the Order.

A Class I product is an aircraft, aircraft engine or propeller.

A Class II product is a major component of an aircraft, aircraft engine or propeller, the failure of which could jeopardize the safety of the aircraft, engine or propeller. This includes the landing gear.

A Class III product is any part or component that is not a Class I or Class II product.

171a. The Airworthiness Approval Tag (8130-3) may be used for which of the following?

A— Identification of parts.
B— Domestic Airworthiness Approvals.
C— Both A and B.

The 8130-3 form and its related guidance material continues to evolve. This form is the preferred method for documenting the approval of products, parts and appliances considered approved by the FAA. Chapter 2 of the Order lists "Domestic Airworthiness Approvals"; in Chapter 3, blocks 6 through 13 of the form are explained, which allow the entry of sufficient data to specifically identify the parts being approved.

172. As an IA, you are inspecting an aircraft that has a field coil of the alternator rewound. The 8130-3 Authorized Release Certificate has been properly completed. As an IA, you would

A— give the owner a list of discrepancies.
B— reject the repair.
C— sign block 7 of the 337 Form provided, and return the aircraft to service.

According to FAA Order 8130.21G ¶3-3a, although the use of this form for return to service is optional, the FAA recommends its use. A properly completed form, signed by an authorized individual, is acceptable documentation to be able to release the part to the aircraft, and the aircraft back to service.

173. A starter-generator was rewound as per the manufacturer's manual. An A&P working for a Part 121 certificate holder has signed the paperwork and dated it. He has filled out an 8130-3 form. As an IA, you would

A— reject, as per 14 CFR Part 91.
B— reject, as per FAA Order 8130.21.
C— reject, as per 14 CFR Part 43.

Only those persons authorized as stated in Order 8130.21G ¶3-3a, when authorized by §43.7(c), (d), and (e) may issue a Form 8130-3 for approval for return to service of products and parts that have undergone maintenance, preventive maintenance, rebuilding or alterations. A&P and IAs are excluded from those who are approved to sign for Return to Service.

Answers Note: *All Learning Statement Codes (in parentheses) are preceded by "IAR"—see explanation on Page 1–10.*

170 [B] (030) FAA Order 8130.21
171 [C] (030) FAA Order 8130.21
171a [C] (030) FAA Order 8130.21
172 [C] (030) FAA Order 8130.21
173 [B] (030) FAA Order 8130.21

174. Authorized Release Certificates (FAA Form 8130-3) may be used by which maintenance entity for approving Class I, II and III products for return to service after maintenance or alteration?

A— Inspection Authorizations.
B— 14 CFR Part 145, Certified Repair Stations.
C— Either A or B.

The work must be accomplished by a certificate holder under 14 CFR Part 121 or 135, having a continuous airworthiness maintenance program or by a repair station certificated under Part 145.

175. When authorized, a Part 145 repair station may return products to service by using FAA Form 8130-3 (Authorized Release Certificate). When such a repair station uses this form to return a component to service after a major repair, a maintenance release or FAA Form 337 is

A— not required.
B— a required attachment.
C— not required if the component is installed at the repair station.

The work must be accomplished by a certificate holder under 14 CFR Part 121 or 135, having a continuous airworthiness maintenance program or by a repair station certificated under Part 145. In all cases, the Form 8130-3 must be signed by the appropriately authorized representative of the maintenance facility in accordance with Part 43. Specifically, §43.9 specifies attachments such as work orders, air carrier records, and FAA Form 337 which may be required to accompany Form 8130-3.

175a. A company receives a new part from the U.S. manufacturer of the product. The item does not come with an FAA 8130-3 form for Airworthiness Approval. The inspector should

A— return the product to the manufacturer.
B— create an 8130-3 for the part, and sign it.
C— install the part, an 8130-3 is not required.

Although the FAA encourages the use of 8130-3 forms for domestic use, the form is not required. Also, the FAA does not authorize A&Ps or IA's to sign the form.

175b. A part has been overhauled by a certified repair station. What document should accompany the part when the technician receives it?

A— Logbook entry.
B— A certificate of conformance on company letterhead.
C— An 8130-3 form signed by an authorized representative from the repair station on the lower right side of the form.

Although the FAA encourages the use of 8130-3 forms for domestic use, the form is not required. Also, the FAA does not authorize A&Ps or IA's to sign the form.

Answers

174 [B] (030) FAA Order 8130.21
175 [B] (030) FAA Order 8130.21
175a [C] (030) FAA Order 8130.21
175b [C] (030) FAA Order 8130.21

Sample Forms

U.S. DEPARTMENT OF TRANSPORTATION
FEDERAL AVIATION ADMINISTRATION

National Policy

ORDER 8130.21H CHG 1

01/11/16

SUBJ: Procedures for Completion and Use of the Authorized Release Certificate, FAA Form 8130-3, Airworthiness Approval Tag

1. Purpose. This change revises paragraphs 1-1, 1-2, 1-3, 1-8, 1-9, 1-13, 2-1, 2-3, 2-6, 2-8, 4-1, and 4-5, and adds paragraph 1-13, to reflect amendment 21-98 to Title 14 of the Code of Federal Regulations (14 CFR) 21.137. This change also revises appendix D to reflect updates to Federal Aviation Administration (FAA) records management references.

2. Who This Change Affects. All Washington headquarters branch levels of the Aircraft Certification Service (AIR), Flight Standards Service, and the Regulatory Support Division; the Aviation System Standards office; the branch levels in the AIR directorates and regional Flight Standards Service divisions; all Aircraft Certification Offices; all Manufacturing Inspection District Offices and Manufacturing Inspection Satellite Offices; all Flight Standards District Offices; the Aircraft Certification Branch and Flight Standards Branch at the FAA Academy; all applicable representatives of the FAA; and all international field offices.

3. Explanation of Changes. This change updates chapters 1, 2, and 4 to state that a production approval holder (PAH) with an FAA-approved quality system that includes procedures described in 14 CFR 21.137(o), and persons authorized by such a PAH, may issue FAA Form 8130-3 as an authorized release document.

4. Disposition of Transmittal Paragraph. Retain this transmittal sheet until the directive is canceled by a new directive.

5. Effective Date. This change is effective 01/11/2016.

PAGE CHANGE CONTROL CHART

Remove Pages	Dated	Insert Pages	Dated
ii thru iv	8/01/13	ii thru iv	01/11/16
1-1	8/01/13	1-1	01/11/16
1-4 thru 1-5	8/01/13	1-4 thru 1-6	01/11/16

Distribution: Electronic **Initiated By: AIR-100**

FAA Order 8130.21G

01/11/2016 8130.21H CHG 1

PAGE CHANGE CONTROL CHART (Continued)

Remove Pages	Dated	Insert Pages	Dated
2-1	**8/01/13**	**2-1**	**01/11/16**
2-3 thru 2-4	**8/01/13**	**2-3 thru 2-4**	**01/11/16**
2-6	**8/01/13**	**2-6**	**01/11/16**
2-10	**8/01/13**	**2-10**	**01/11/16**
4-1	**8/01/13**	**4-1**	**01/11/16**
4-3 thru 4-9	**8/01/13**	**4-3 thru 4-9**	**01/11/16**
D-1	**8/01/13**	**D-1**	**01/11/16**

Susan J. M. Cabler
Susan J. M. Cabler
Acting Manager, Design, Manufacturing,
& Airworthiness Division
Aircraft Certification Service

2

U.S. DEPARTMENT OF TRANSPORTATION
FEDERAL AVIATION ADMINISTRATION
National Policy

08/01/13

SUBJ: Procedures for Completion and Use of the Authorized Release Certificate, FAA Form 8130-3, Airworthiness Approval Tag

This order describes the procedures for completion and use of the Federal Aviation Administration (FAA) Authorized Release Certificate, FAA Form 8130-3, Airworthiness Approval Tag. The order describes the use of the form for domestic airworthiness approval, conformity inspections, and prepositioning; airworthiness approval of new products and articles; and splitting bulk shipments of previously shipped products and articles. It also provides guidance for the issuance of the form for approval for return to service of products and articles, the export airworthiness approval of products and articles, and the electronic exchange of the form.

James D. Seipel
Manager
Production and Airworthiness Division, AIR-200

Distribution: Electronic **Initiated by:** AIR-200

Table of Contents

Paragraph *Page*

01/11/2016 8130.21H CHG 1

Table of Contents (Continued)

Table of Contents (Continued)

FAA Order 8130.21H

Chapter 1. General Information

1-1. Purpose of This Order. This order describes the procedures for completion and use of the Federal Aviation Administration (FAA) Authorized Release Certificate, FAA Form 8130-3, Airworthiness Approval Tag. The order describes the use of the form for the following purposes:

a. Domestic airworthiness approval, including conformity inspections, prepositioning of new products or articles pending approval, and splitting bulk shipments of previously produced products and articles;

b. Approval for return to service of products and articles; and

c. Export airworthiness approval of products and articles.

Note 1: For the purposes of this order, the term "product" refers only to aircraft engines and propellers.

Note 2: The use of the word "should" throughout this order refers to a recommended practice. The associated activity is not a requirement; therefore, a record of completion is not required.

Note 3: The samples of FAA Form 8130-3 in appendix A to this order are provided for reference only. To properly complete a form, the originator must follow the procedures within this order.

1-2. Audience. FAA personnel; designees; production approval holders (PAH) with an approved quality system that includes the procedures pursuant to Title 14 of the Code of Federal Regulations (14 CFR) 21.137(o); air agencies certificated under 14 CFR part 145, Repair Stations; U.S. air carrier certificate holders; and distributors.

1-3. Where Can I Find This Order and FAA Form 8130-3?

a. You can find this order at http://www.faa.gov/regulations_policies/orders_notices/.

b. FAA Form 8130-3 may be obtained through normal distribution channels from the FAA Logistics Center, AML-4060, Oklahoma City, Oklahoma. FAA Form 8130-3 can be ordered through the Customer Care Center, AML-31, at 405-954-3793 or toll free at 1-888-322-9824, or may be obtained on the Internet at http://www.faa.gov/aircraft.

1-4. Cancellation. FAA Order 8130.21G, Procedures for Completion and Use of the Authorized Release Certificate, FAA Form 8130-3, Airworthiness Approval Tag, dated October 26, 2009, is canceled.

1-5. Explanation of Policy Changes. This revision—

a. Replaces FAA Form 8130-3 (06-01) with FAA Form 8130-3 (02-14) as part of an effort to harmonize the Authorized Release Certificate among multiple civil aviation authorities (CAA). The changes to the Authorized Release Certificate include—

(1) Changing the term "National Aviation Authority" to "Civil Aviation Authority" in Block 1.

(2) Removing the Eligibility block.

(3) Removing the term "Batch" from Block 10.

(4) Renumbering the various blocks within the Authorized Release Certificate.

(5) Changing the date format to (dd/mm/yyyy).

b. In chapter 1, adds a note that describes the proper use of the word "should" throughout the order, referring to it as a recommended practice.

c. Moves all FAA Form 8130-3 record retention requirements to chapter 1.

d. Moves all procedures concerning lost FAA Form 8130-3 and reissuance because of typographical errors to chapter 1.

e. Removes the term "issuer" and replaces it with "originator."

f. Updates guidance in all chapters concerning copies of FAA Form 8130-3.

g. Updates all the block-by-block instructions and sample figures referring to FAA Form 8130-3 (02-14).

h. In chapter 2, includes guidance concerning commercial parts and FAA Form 8130-3.

i. In chapter 2, updates the documentation procedures for splitting bulk shipments.

j. Harmonizes the implementing instructions for Block 7 in all chapters.

k. Adds specific guidance to all chapters concerning batch or lot numbers in Block 12 in the block-by-block instructions.

l. Moves all sample figures of FAA Form 8130-3 from the chapters to a separate appendix.

m. In chapter 3, adds a new section concerning articles marked with dual or multiple part numbers.

n. In chapter 3, adds a new section concerning return-to-service information relevant to the European Union.

o. In chapter 3, the terms "Inspected" and/or "Tested" have been updated in the Status/Work block to coincide with the Maintenance Annex Guidance (MAG) Between the Federal Aviation Administration for the United States of America and the European Aviation Safety Agency for the European Union.

p. In chapter 4, adds the words "and Articles" in the section title and rewords this section to correspond with 14 CFR 21.331(c) and (d).

q. In chapter 4, the terms "NEW" and "USED" have been updated in the Status/Work block in accordance with § 21.331, and two notes have been added to clarify their use.

r. In chapter 4, adds a new section in the Remarks block (Block 12) when exporting a product or article to a country or jurisdiction that does not have a bilateral agreement with the United States.

s. In appendix A, adds a sample figure that shows how to complete FAA Form 8130-3 when shipping a rebuilt engine to a country within the European Union.

t. Removes the sample figures containing extensible markup language (XML) fragments from chapter 5.

u. Removes the required use of the watermark "Printed from Electronic File" from the electronic FAA Form 8130-3.

1-6. Effective Date. This order is effective on February 1, 2014.

1-7. Purposes for Which FAA Form 8130-3 Cannot Be Used.

a. FAA Form 8130-3 is not a delivery or shipping document.

b. FAA Form 8130-3 may not be issued by organizations or individuals other than those approved/authorized by the FAA.

c. Aircraft are not to be released using FAA Form 8130-3.

d. FAA Form 8130-3 does not constitute approval to install a product or article on a particular aircraft, aircraft engine, or propeller; however, it does assist the end user in determining the airworthiness approval status of a product or article.

e. A mixture of production- and maintenance-released products and articles is not permitted on the same FAA Form 8130-3.

f. A mixture of products and articles released against approved and non-approved design data is not permitted on the same FAA Form 8130-3.

g. FAA Form 8130-3 cannot be used to export a prototype product or article.

1-8. Authorization to Issue FAA Form 8130-3. The following persons may issue FAA Form 8130-3 in accordance with the appropriate chapter of this order:

a. FAA ASIs;

b. Persons with the appropriate function codes in accordance with FAA Order 8000.95, Designee Management Policy, when authorized by Certificate Letter of Authority (CLOA); or in accordance with FAA Order 8100.8, Designee Management Handbook when authorized by their Certificate of Authority (COA);

c. Persons authorized in accordance with FAA Order 8100.15, Organization Designation Authorization Procedures;

d. Persons authorized in accordance with paragraph 3-1 of this order to issue FAA Form 8130-3 to approve a product or article for return to service; and

e. Persons authorized in accordance with a PAH's approved quality system that includes the procedures described in § 21.137(o).

1-9. Record Retention Requirements for FAA Form 8130-3. Refer to FAA Order 0000.1, FAA Standard Subject Classification System; FAA Order 1350.14, Records Management; or your office Records Management Officer (RMO)/Directives Management Officer (DMO) for guidance regarding retention or disposition of records.

a. When issued for domestic airworthiness and export airworthiness approvals for a new product or article (to include conformity inspections, prepositioning, and splitting of bulk shipments), the originator must retain FAA Form 8130-3 for no less than 5 years for products and articles and 10 years for critical parts.

b. When issued for return to service, the following statements describe how long the originator must retain FAA Form 8130-3, unless regulatory requirements stipulate otherwise:

(1) If FAA Form 8130-3 is issued as an approval for return to service by an appropriately certificated organization (that is, a PAH or 14 CFR part 121, 135, or 145 organization), the originator should retain a copy of FAA Form 8130-3 for a period of 2 years after the work is approved for return to service, unless the work is repeated or superseded. An air carrier's manual requirements may require a longer retention period.

(2) If a certificated repair station uses FAA Form 8130-3 as the approval for return to service for a major repair in accordance with 14 CFR part 43, Maintenance, Preventive Maintenance, Rebuilding, and Alteration, the repair station should retain a copy of the document for 2 years.

1-10. Lost FAA Form 8130-3 Issued for Domestic Airworthiness Approvals, Return To Service, and Export Airworthiness Approvals. If a copy of an FAA Form 8130-3 is requested, correlation must be established between FAA Form 8130-3 and the applicable product(s) or article(s). The originator must retain a copy of each FAA Form 8130-3 issued to allow verification of the original data. There is no restriction on the number of copies of FAA Form 8130-3 that may be sent to the customer or retained by the originator.

1-11. Reissuance Because of Typographical Errors. The originator may reissue FAA Form 8130-3 to correct typographical errors.

a. The recipient of the incorrect FAA Form 8130-3 must provide a written request and a copy of the incorrect FAA Form 8130-3 to the originator.

b. The request for a corrected FAA Form 8130-3 may be honored without reverification of the product or article condition. The reissued FAA Form 8130-3 is not a statement of current condition and must refer to the FAA Form 8130-3 being corrected. Include this reference in Block 12 using the following statement: "This FAA Form 8130-3 corrects the error(s) in Block(s) [enter block number(s) corrected] of the FAA Form 8130-3 [enter form tracking number] dated [enter issuance date] and does not cover conformity/condition/release to service." The reissued form must be marked as such. Both forms must be retained according to the retention period.

1-12. Information Systems and Automation.

a. FAA Form 8130-3 may be computer-generated for local reproduction, but must duplicate the format of the original Government-printed form. The overall form as designed must not be changed, nor may any words be added or deleted (with the exception of filling in the blanks). White is the preferred color for the paper; however, if another color is used, the information contained on the form must be legible. Text on FAA Form 8130-3 that is required by this order may be preprinted. The size of blocks, in relationship to each other, may vary slightly, but all blocks must remain in their original location. FAA Form 8130-3 also may be reduced in overall size to reduce paper consumption, but not to the extent it is no longer easily readable and readily recognizable. The details to be entered on the form may be either machine/computer-printed or handwritten using block letters and must be easy to read, with limited use of abbreviations. All entries on the form must be made in permanent ink and be in English. If a deviation to FAA Form 8130-3 becomes necessary, the FAA employee involved should ensure the deviations are substantiated, documented, and concurred with by the appropriate supervisor. The deviations must be submitted to AIR-200 for review and approval.

b. Approval holders should develop procedures for managing information systems consistent with Advisory Circular (AC) 21-43, Issuance of Production Approvals Under Subparts G, K, & O. These procedures should include a secured electronic auditing system that reflects all system changes and a secured monitoring system that records all transactions by items such as part number, serial number (when applicable), and quantity shipped.

c. Automation and use of an electronic signature on FAA Form 8130-3 is allowed by all persons who issue the form; however, using automation and electronic signature does not relieve the designee or person authorized to issue FAA Form 8130-3 from verifying the product or article conforms to FAA-approved design data and is in a condition for safe operation.

1-13. PAHs Issuing Authorized Release Documents Using FAA Form 8130-3.
FAA Form 8130-3 is considered an airworthiness approval, as defined by § 21.1, only when issued by the FAA or an FAA designee. When FAA Form 8130-3 is issued by a PAH with an approved quality system that includes the procedures described in § 21.137(o), the form is considered an authorized release document. A PAH may authorize its personnel to issue authorized release documents, using FAA Form 8130-3, for a new or used aircraft engine, propeller, or articles manufactured by the PAH itself. An authorized release document issued by a PAH is a certifying statement that a given aircraft engine, propeller, or article conforms to its approved design and is in a condition for safe operation at the time of examination and release of the document, unless otherwise specified (refer to § 21.331). As described in § 21.137(o), a PAH that intends to issue these documents must detail the appropriate procedures in its quality manual. AC 21-43 states that a PAH should use FAA Form 8130-3 when issuing an authorized release document.

> **Note:** Because this order provides instructions to FAA employees and designees, it frequently describes FAA Form 8130-3 as an airworthiness approval. As described above, however, FAA Form 8130-3 may, under various circumstances, be referred to as an airworthiness approval, Authorized Release Certificate (ARC), airworthiness approval tag, or authorized release document. This form constitutes an FAA approval only when issued by the FAA or its designee.

Chapter 2. Domestic Airworthiness Approvals

2-1. General Information on Domestic Airworthiness Approvals.

a. FAA Form 8130-3 is the preferred method for documenting the approval of products and articles approved by the FAA. The FAA recommends each PAH include FAA Form 8130-3 with all eligible product and article shipments, while also minimizing the quantity of forms used for bulk shipments (for example, 500 turbine blades shipped with 1 form vs. 500 forms). This will help aviation authorities and industry to ensure complete traceability and ease the movement of products and articles through the aviation system. Issuing FAA Form 8130-3 with all eligible product and article shipments enables end users to determine the airworthiness approval status of the products and articles. Only an FAA ASI, authorized designee, or person authorized by a PAH's FAA-approved quality system that includes the procedures described in § 21.137(o) is authorized to issue FAA Form 8130-3 for the domestic purposes described in this chapter. Except as provided in paragraphs 2-2 and 2-6 of this order, products and articles not produced under an FAA production approval are not eligible to receive an FAA Form 8130-3. FAA Form 8130-3 does not constitute approval to install a product or article on a particular aircraft, aircraft engine, or propeller.

b. FAA Form 8130-3 must be completed as described in paragraph 2-8 of this order.

c. FAA Form 8130-3 must be correlated with the shipment. Additional copies of FAA Form 8130-3 may be provided upon request. The forms may not be delivered before the product or article is shipped.

d. If FAA Form 8130-3 is issued as an airworthiness approval of a new product or article (this is to include conformity inspections, prepositioning, and splitting of bulk shipments), the originator must retain the FAA Form 8130-3 for no less than 5 years for products and articles and 10 years for critical parts.

e. The copies of FAA Form 8100-1, Conformity Inspection Report, and FAA Form 8130-3 may be retained in paper format or in a secure database, provided the database contains all of the information required on FAA Form 8130-3. An acceptable means of compliance is provided in AC 21-43 or AC 120-78, Acceptance and Use of Electronic Signatures, Electronic Recordkeeping Systems, and Electronic Manuals (when applicable). Duplicates of FAA Form 8130-3, including signatures retained in a database, do not need to be graphic images of the documents.

f. Unique identification is required to enable or provide product or article traceability. The preferred method is a unique form tracking number in Block 3. However, if traceability is provided through other information on the form combined with a number in Block 3, this is also acceptable.

g. The signature of the person authorized to issue FAA Form 8130-3 may be applied electronically to Block 13b from domestic or international locations. With the exception of paragraphs 1-10 and 1-11, at the time the signature is authorized to be placed on FAA Form 8130-3, the person whose signature appears on the form must have direct access

FAA Order 8130.21H

to the product or article to verify it conforms to FAA-approved design data and is in a condition for safe operation.

> **Note:** The time and location of the authorization of the form issuance may be different from the time and location of the printing of the form.

h. In the case where a product or article is presented for inspection for the issuance of FAA Form 8130-3, and the product or article is sealed in a package that does not afford a visible inspection, the authorized person must obtain objective evidence to determine the appropriate inspections were conducted and approved before the issuance of FAA Form 8130-3.

i. Products or articles received without an FAA Form 8130-3 must not be mixed with those received with FAA Form 8130-3. This is to preclude shipment of products and articles that were not received with an FAA Form 8130-3. When more than one product or article is listed on a supplemental FAA Form 8130-3, the product or article does not need to be from the same quantity or shipment, as long as it was received with an FAA Form 8130-3 and traceability has been maintained.

j. The User/Installer Responsibilities statements may be placed on either side of the form. If the statements are placed on the back side of the form, a note in Block 12 must reference that fact. When copies of the forms are generated, these statements must be provided with the copies.

2-2. Conformity Inspections. When requested on FAA Form 8120-10, Request for Conformity, FAA Form 8130-3 is used to ship a prototype product or article. Any nonconformities/deviations relative to the product or article conformity inspection must have prior aircraft certification office (ACO)/designated engineering representative (DER), organization designation authorization (ODA), or FAA project manager acknowledgement of disposition. Before signing FAA Form 8130-3, any nonconformities/deviations must be appropriately acknowledged and dispositioned and be annotated in Block 12. Only an FAA ASI or authorized designee is permitted to sign the form. (Refer to appendix A, figure A-2, to this order.) When the request for conformity includes a quantity of articles in excess of those articles subject to the required certification, the tag for those excess articles should indicate they are prepositioned.

> **Note:** Prototype product(s)/article(s) pending certification under an FAA project number are not eligible for installation on in-service, type-certificated aircraft. Upon approval of the applicable design data and completion of an inspection to validate conformity to that approved design data and condition for safe operation, that product or article may be considered new.

2-3. Domestic Airworthiness Approval of New Products (Aircraft Engines and Propellers).

a. FAA Form 8130-3 can be issued for domestic shipments to identify the airworthiness approval status of new products produced under 14 CFR part 21, Certification Procedures for Products and Parts. The use of FAA Form 8130-3 for this purpose is optional, but the FAA recommends its use. When used for an airworthiness approval for new products (engines or propellers), the following statement must be entered: "Airworthiness approval – Engine [or Propeller]." (Refer to appendix A, figure A-3, to this order.)

b. Persons eligible to issues FAA Form 8130-3 for domestic shipments of new products include: authorized FAA ASIs, persons authorized by a CLOA in accordance with FAA Order 8000.95, persons authorized by a COA in accordance with FAA Order 8100.8, persons authorized in accordance with FAA Order 8100.15, and persons authorized by the FAA-approved quality system of a PAH, if that quality system includes the procedures described in § 21.137(o). Any person issuing FAA Form 8130-3 must determine the product meets the FAA-approved design data and is in a condition for safe operation before issuing the form. FAA Form 8100-1 should be used to document conformity inspections.

c. An FAA Form 8130-3 for domestic shipments of products to identify airworthiness approval does not constitute an export approval. Exporters must meet the applicable requirements of part 21, subpart L, Export Airworthiness Approvals (refer to chapter 4 of this order).

2-4. Domestic Airworthiness Approval of New Articles.

a. FAA Form 8130-3 can be issued for domestic shipments to identify the airworthiness approval status of new articles produced under part 21. The use of FAA Form 8130-3 for this purpose is optional, but the FAA recommends its use. (Refer to appendix A, figure A-4, to this order.)

b. A person must determine the article meets the FAA-approved design data and is in a condition for safe operation before issuing FAA Form 8130-3. The FAA managing office must determine if an FAA Form 8100-1 has to be completed for each FAA Form 8130-3 issued based on the PAH's quality system's health and/or the designee's previous history, experience, or performance, or if the information can be stored and retrieved in another format (for example, electronic database).

c. When produced under an FAA production approval, a standard or commercial part is eligible for issuance of an FAA Form 8130-3 airworthiness approval. Use of FAA Form 8130-3 for this purpose is recommended, but not mandatory. The inclusion of FAA Form 8130-3 helps document the airworthiness and traceability of these parts. Refer to FAA Order 8110.118, Commercial Parts, for further information concerning commercial parts.

d. Issuance of FAA Form 8130-3 as an airworthiness approval does not constitute an export approval, because compliance with a specific country's or jurisdiction's special import requirements may not have been verified.

e. FAA Form 8130-3 may be issued to document airworthiness approvals at PAH facilities, including PAH suppliers and associate facilities identified in the PAH's approved procedures. The form also may be issued by a designated person at PAH suppliers with direct shipment authorization or associate facilities outside the United States, if the FAA finds there is no undue burden associated with the form's issuance.

f. FAA Form 8130-3 will not be issued by non-PAH suppliers for products or articles shipped to their PAH's facilities for use on production products or for proof of the PAH's source inspection requirements at suppliers. If, however, the supplier has its own production approval for the products and articles, and the products and articles are part of another PAH's higher level design, then FAA Form 8130-3 may be issued.

2-5. Domestic Airworthiness Approval of New Products and Articles at Distributors, Part 121 and Part 135 Certificate Holders, and Part 145 Certificated Repair Stations.

a. New products and articles at distributors, repair stations certificated under part 145 (Repair Stations), the holder of a U.S. air carrier certificate operating under part 121 (Operating Requirements: Domestic, Flag, and Supplemental Operations), or part 135 (Operating Requirements: Commuter and On Demand Operations and Rules Governing Persons On Board Such Aircraft), with an approved continued airworthiness maintenance program, may be eligible to have an FAA Form 8130-3 issued as a domestic airworthiness approval. All other approvals must be issued in accordance with the appropriate chapter of this order.

b. When completing FAA Form 8130-3, the name and address of the organization where FAA Form 8130-3 was issued must be documented in Block 4, along with the PAH's name in Block 12. (Refer to appendix A, figure A-5, and paragraph 2-8 of this order.)

2-6. Prepositioned Products and Articles.

a. General. FAA Form 8130-3 may be used to identify airworthiness approval status of prepositioned products or articles before type certificate (TC)/supplemental type certificate (STC) approval. Use of the form for this purpose is allowed.

b. Applicability. Eligible products and articles are production products and articles conformed as part of an FAA certification project, and produced under a production certificate (PC) holder's FAA-approved quality system in accordance with part 21.

c. System Requirements. The PC holder must have a procedure that tracks the configuration of the product or article from the manufacturer through shipment until the TC/STC is issued. The procedures must be adequate to ensure the requirements of § 21.146(b) and (c) are met.

d. Completion of FAA Form 8130-3 for a Prepositioned Product or Article. The following persons may issue an FAA Form 8130-3 for a prepositioned product or article:

(1) An FAA ASI,

(2) A DMIR,

(3) An ODA, or

(4) An authorized DAR employed by the PC holder.

e. Certification Issuance. After the TC/STC is issued, but before installation, the conforming product or article should be verified as incorporated into the design.

2-7. Splitting Bulk Shipments of Previously Shipped New Products and Articles.

a. General.

(1) After an FAA Form 8130-3 is issued for a bulk shipment, a new FAA Form 8130-3 may be used to split bulk shipments of previously shipped new products or articles.

(2) Products or articles received without an FAA Form 8130-3 must not be mixed with those received with FAA Form 8130-3. This is to preclude shipment of products and articles that were not received with FAA Form 8130-3. When more than one product or article is listed on a supplemental FAA Form 8130-3, each product or article does not need to be from the same shipment, as long as it was received with FAA Form 8130-3 and traceability has been maintained.

b. Eligibility and System Requirements.

(1) Splitting bulk shipments is permitted when the specific products or articles were produced under an FAA production approval to include PAH associate facilities and PAH-approved suppliers having direct shipment authorization.

(2) The facilities authorized to split bulk shipments are PAHs, PAH associate facilities, distributors, PAH-approved suppliers having direct shipment authorization, and those certificate holders described in paragraph 3-1a of this order. This may include PAH associate facilities and PAH-approved suppliers that have direct shipment authorization located outside the United States. (This is not considered an export; the act of exporting is when the product or article is found to be airworthy, meets the special conditions of the importing country or jurisdiction, and is transferred from one CAA's regulatory authority to another CAA's regulatory authority.)

(3) An authorized facility as described in paragraph 2-7b(2) above must have a written procedure in place for controlling products or articles when splitting bulk shipments.

(4) An authorized facility may split a bulk shipment of previously shipped new products or articles as many times as the original quantity listed in Block 9 allows.

c. Procedures and Documentation for Splitting Bulk Shipments. For those shipments of products or articles required to be split, the following procedure will be used if an approved electronic system to issue supplemental FAA Forms 8130-3 for this purpose is not in place.

(1) Make a copy of the FAA Form 8130-3 received with the original shipment of products or articles that identifies the bulk shipment. (Refer to appendix A, figure A-7a, to this order.)

> **Note:** If the user responsibilities statements are placed on the back side of the form, the copies of the forms must include these statements.

(2) Include the following written statement (an example) or similar statement: "(Company name) certifies this/the attached document is a copy of FAA Form 8130-3. The prior FAA Form 8130-3 received by our facility is maintained on file pursuant to our document retention standards. That prior FAA form tracking number is [ACE-549]. The new tracking number for this portion of the split bulk shipment is [S1-054321]. The number of products or articles being shipped under this approval is [500]. Signed [quality control/assurance manager or authorized representative] Dated [month day year]" (Refer to appendix A, figure A-7b, to this order for an example.) A quality control/assurance manager or authorized representative from that facility must sign and date the written statement. You can include this statement by attaching a separate sheet of paper to the copy of the FAA Form 8130-3.

(3) Maintain the prior FAA Form 8130-3 and a copy of the written statement on file.

2-8. Block-By-Block Instructions for Completing FAA Form 8130-3 for Domestic Airworthiness Approvals.

a. Block 1. Approving Civil Aviation Authority/Country. FAA/United States. (Preprinted.)

b. Block 2. Authorized Release Certificate, FAA Form 8130-3, Airworthiness Approval Tag. (Preprinted.)

c. Block 3. FAA Form Tracking Number.

(1) Enter the unique number established by the numbering system. (Refer to paragraph 2-1f of this order.)

(2) The organization that splits bulk shipments of previously shipped products or articles received from a PAH must establish a new tracking number and enter that number on the certifying statement or on the supplemental FAA Form 8130-3. (Refer to paragraph 2-7c of this order.)

d. Block 4. Organization Name and Address.

(1) Enter the full name and physical address (no post office box numbers) of the organization or facility for which the form is being issued, and the PAH certificate or project number (for example, Certificate No. LI1R 123K or Certificate No. PQ1234NM), as appropriate. A logo or other identification of the organization is permitted if it can be contained within the block.

> **Note:** If an FAA Form 8130-3 is issued at a PAH's extension facility with its own project number (assigned by the geographic managing office), the facility's project number, along with its full name and address, must be used.

(2) When a supplier has direct shipment authorization from a PAH, or conformity inspections are performed on behalf of a PAH/applicant at the supplier's facility, the following information must be entered:

(a) PAH/applicant name and address, and

(b) Supplier name and address.

(3) If a supplier to a PAH produces and ships a product or article, the supplier must either have direct shipment authorization from a PC/PAH holder or hold a production approval for each article shipped. If the supplier holds its own production approval, and the products and articles were manufactured and are being shipped under that approval, the information required in paragraph 2-8d(1) must be listed.

(4) When completing FAA Form 8130-3 at a distributor's facility, enter the name and the address of that facility.

e. Block 5. Work Order/Contract/Invoice Number. To facilitate customer traceability of the product or article, enter the work order number, contract number, invoice number, or similar reference number, and state the number of pages attached to the form, including dates, if applicable. If the shipment list contains the information required in Blocks 7 through 11, these respective blocks may be left blank if a list is attached to the form. In this case, the following statement must be entered in Block 12: "This is the certification statement for the products and articles listed on the attached document dated _________, containing pages ______ through ______." In addition, the shipping list must cross-reference the form tracking number located in Block 3.

f. Block 6. Item. When FAA Form 8130-3 is issued, a single item number or multiple item numbers (for example, same item with different serial numbers) may be used for the same part number. Multiple items must be numbered in sequence, although not necessarily beginning with the number one (for example, 0040, 0050, 0062, 0063). If a separate listing is used, enter "List Attached" (refer to paragraph 2-8e of this order for further instructions).

g. Block 7. Description. Enter the name or description of the product or article. Preference should be given to the term used in the instructions for continued airworthiness or maintenance data (for example, illustrated parts catalog, aircraft maintenance manual, or service bulletin).

h. Block 8. Part Number. Enter each part number of the product or article. In the case of an aircraft engine or propeller, the model designation may be used.

i. Block 9. Quantity. Enter the quantity of each product or article shipped.

j. Block 10. Serial Number. If the product or article is required by 14 CFR part 45, Identification and Registration Marking, to be identified with a serial number, enter it here. Additionally, any other serial number not required by regulation also may be entered. If no serial number is entered in this block, enter "N/A." If a specific batch or lot number is used, refer to the instructions for Block 12.

k. Block 11. Status/Work. The following table describes what to enter in a specific situation. Only one term may be entered in Block 11, which should reflect the majority of the work performed. The use of upper or lower case in this block does not matter.

Table 1—FAA Form 8130-3 for Domestic Airworthiness Approvals Block 11 Terms

Enter—	*For—*
"NEW"	The production of a new product or article in conformity with the approved design data.
"PROTOTYPE"	The production of a new product or article in conformity with the non-approved design data.

l. Block 12. Remarks. The use of upper or lower case in this block does not matter.

(1) State any information in this block, either directly or by reference to supporting documentation, necessary for the user or installer to determine the airworthiness of the product or article. If necessary, a separate sheet may be used and referenced from the main FAA Form 8130-3. Each statement must clearly identify the product or article in Block 6 to which it relates.

(2) Below are examples of conditions that could necessitate a statement in this block. These statements may or may not be appropriate depending on the form's purpose.

(a) When the request for conformity includes a quantity of product(s)/article(s) in excess of those product(s)/article(s) subject to the required certification, the tag for those excess product(s)/article(s) should indicate they are prepositioned. Include the following statement in this block: "Prototype product(s)/article(s) pending certification under FAA project number [enter number] are not eligible for installation on in-service, type-certificated aircraft. Upon approval of the applicable design data and completion of an inspection to validate conformity to that approved design data and condition for safe operation, that product or article may be considered new." Block 13a will be marked as "Non-approved design data specified in Block 12." (Refer to appendix A, figure A-6, to this order.)

(b) The purpose of the form (for example, airworthiness approval, conformity, or prepositioning).

(c) Attachment when used. Attachments should include the form tracking number of the corresponding FAA Form 8130-3.

(d) Compliance with airworthiness directives (AD) or service bulletins.

(e) For technical standard order (TSO) articles, enter the applicable TSO number.

(f) Information on life-limited parts (for example, total time, total cycles, or time since new).

(g) If a specific batch or lot number is used to control or trace the product or article, enter the batch or lot number in this block.

(h) Shelf-life data.

(i) Drawing number and revision level.

(j) Information needed to support shipment with shortages or reassembly after delivery.

(k) Any data not appropriate in other blocks.

(l) When used for conformity, the words "Conformity Inspection" must be entered. In addition, an explanation of the product or article use (for example, pending approved data, TC pending, for test only) must be provided. Information concerning a conformity inspection such as design data, revision level, date, project number, and special instructions as shown on FAA Form 8120-10 must be entered in this block.

(m) When issued at a supplier facility with direct shipment authorization from the PAH, the words "Airworthiness approval – Direct shipment authorization" must be entered in Block 12, and the information from paragraph 2-8d(2) of this order must be entered in Block 4. (Refer to appendix A, figure A-8, to this order.)

(n) When FAA Form 8130-3 is issued at a distributor, part 121 or part 135 certificate holders, or a part 145 certificated repair stations, enter the following statement: "The product(s)/article(s) shipped under this approval were produced by [insert PAH's name]." (Refer to appendix A, figure A-5, to this order.)

(o) When used for an airworthiness approval for new products (engines or propellers), the following statement must be entered: "Airworthiness approval – Engine [or Propeller]." (Refer to paragraph 2-3 of this order.)

(p) When used for prepositioning, the following statement must be made: "Prepositioned product(s)/article(s) were conformed to design data under FAA project number [enter number], for the issuance of a TC/STC modification of [enter make and model number]. Product(s)/article(s) conforming to design at issuance of the TC/STC is/are certified as airworthy and is/are in a condition for safe operation without further showing." (Refer to appendix A, figure A-6, to this order.)

(q) When used for airworthiness approval for a new subcomponent of a Parts Manufacturer Approval (PMA)/TSO authorization article higher assembly, complete FAA Form 8130-3 with the subcomponent information, and enter a statement in Block 12 indicating the article is a subcomponent of a PMA or TSO authorization (for example, "This [insert article description] is a subcomponent of a(n) [FAA PMA article/TSO authorization].") (Refer to appendix A, figure A-9, to this order.)

m. Block 13a. Airworthiness Approval.

(1) Place a check in the "Approved design data and are in a condition for safe operation" box if the products and articles were manufactured using FAA-approved design data and found to be in a condition for safe operation. Checking this box and signing Block 13b means the products and articles listed on the form meet the FAA-approved design data and are in a condition for safe operation.

(2) Place a check in the "Non-approved design data specified in Block 12" box when FAA Form 8130-3 is used for—

(a) Conformity of a prototype product or article certification program, or

(b) Prepositioning products or articles before the issuance of a TC/STC.

n. Block 13b. Authorized Signature. This block must be signed by the authorized person issuing the form. The block may be signed by an FAA ASI, authorized designee, or person authorized under a PAH's approved quality system that includes the procedures described in § 21.137(o). A person signing Block 13b may use an alternative to a handwritten signature (for example, a computer-generated signature; refer to appendix C to this order for definition) only when authorized by the FAA. The approval signature must be applied manually at the time and place of issuance, except as provided in paragraph 2-1g of this order.

o. Block 13c. Approval/Authorization No. Enter the approval/authorization number of the authorized representative/organization identified in Block 13b. If signed by an FAA inspector, the authorization number is the applicable office identifier. If signed by a PAH or a person authorized by a PAH's quality system, enter the applicable production approval number (for example, PT1234CE or PQ1234NM).

p. Block 13d. Name (Typed or Printed). Enter the typed or printed name of the authorized representative or organization whose signature appears in Block 13b.

q. Block 13e. Date (dd/mmm/yyyy). Enter the date on which Block 13a is completed, or in the case of electronically generated forms, the date the conformity determination is made and the form is authorized to be issued. The date must be in the following format: two-digit day, first three letters of the month, and four-digit year, for example, 03 Feb 2008. This does not need to be the same as the printing or shipping date, which may occur later. The use or omission of slashes, hyphens, or spaces in the date does not matter.

r. Blocks 14a through 14e. Shade, darken, or otherwise mark to preclude inadvertent or unauthorized use.

2-9. Reissuance by a PAH for Returned Products and Articles.

a. The new products and articles returned to a PAH may be eligible for a new FAA Form 8130-3 if—

(1) The new products and articles were produced under the PAH's production approval.

(2) The PAH maintains a procedure to accept products and articles back into its quality system.

(3) Tests and inspections are performed in accordance with procedures contained in the PAH's quality system to determine the returned product or article still meets the original type design under which it was produced and is still in a condition for safe operation.

b. If the conditions in paragraph 2-9a(1) through (3) are met, a new FAA Form 8130-3 in accordance with chapter 2 of this order may be issued.

c. If FAA Form 8130-3 is returned with the products and articles, the originator should retain that form on file with (or have reference to) the new FAA Form 8130-3.

Chapter 3. Approval for Return To Service of Products and Articles

3-1. General Information on Approval for Return To Service.

a. Air agencies certificated under part 145, or the holder of a U.S. air carrier certificate operating under part 121 or part 135, with an approved continued airworthiness maintenance program may issue an FAA Form 8130-3 for approval for return to service for a product or article maintained or altered under part 43.

> **Note:** The restriction in this order relating to the issuance of the form does not apply when the form is used as a maintenance record and approval for return to service. Copies of the form when used as a maintenance record or an approval for return to service may be provided to the owner/operator or others who require copies of maintenance records as prescribed by the applicable CFRs.

b. A PAH may issue an FAA Form 8130-3 for approval for return to service after rebuilding, altering, or inspecting its product in accordance with §§ 43.3(j) and 43.7(d). The use of FAA Form 8130-3 for this purpose is optional, but the FAA recommends its use. This will help aviation authorities and industry to ensure complete traceability and ease the movement of products and articles through the aviation system. (Refer to paragraph 3-2a(2) and appendix A, figure A-10, of this order.)

c. FAA Form 8130-3 does not constitute approval to install a product or article on a particular aircraft, aircraft engine, or propeller.

d. Blocks 14a through 14e on FAA Form 8130-3 are used to indicate approval for return to service (along with the information contained in Blocks 1 through 12).

e. FAA Form 8130-3 must be completed as outlined in the Block-by-Block instructions in paragraph 3-6 of this order.

f. FAA Form 8130-3 must be correlated with the shipment. Additional copies of FAA Form 8130-3 may be provided upon request. The forms may not be delivered before the product or article is shipped.

g. FAA Form 8130-3 may be retained in paper format or in a secure database, provided the database contains all the information required on FAA Form 8130-3, complies with AC 120-78 (when applicable), and is available for FAA review upon request. When FAA Form 8130-3 is issued for approval for return to service in accordance with this chapter, a copy of FAA Form 8130-3 that accompanied each shipment, product, or article, must comply with the recordkeeping requirements of parts 43, 91, 121, 135, and 145. These forms must be retained by the facility where FAA Form 8130-3 is issued. Duplicates of FAA Form 8130-3, including signatures retained in a database, do not need to be graphic images of the documents. However, when a supplemental FAA Form 8130-3 is issued as described by this order, traceability back through a system that ensures the products and articles were received with FAA Form 8130-3 must be possible.

h. Many part numbers are applied in a nonpermanent manner (for example, ink stamp or paper label). In other cases, maintenance is required in areas where articles are permanently identified. During the maintenance process, these part numbers may be removed or otherwise obscured. If during maintenance the part number is removed or obscured, the persons performing the maintenance must document the part number and, if applicable, serial number, total time and cycles, heat code (if applicable), and any and all part markings on maintenance documents before performing the work. The article information must be reapplied after maintenance per acceptable practices. FAA Form 8130-3, when completed in accordance with this order, may be considered the article's identification.

i. Unique identification is required to enable or provide product or article traceability. The preferred method is a unique form tracking number in Block 3. However, if traceability is provided through other information on the form combined with a number in Block 3, this is also acceptable.

j. The signature of the person authorized to issue FAA Form 8130-3 may be applied electronically to Block 14b from domestic and international locations. At the time the signature is authorized to be placed on FAA Form 8130-3, the person whose signature appears on the form must have direct access to the products, articles, forms, and other data to monitor the process, perform spot-checks, and ensure the work specified on the form was accomplished in accordance with part 43 and, in respect to that work, the items are approved for return to service.

k. The User/Installer Responsibilities statements may be placed on either side of the form. If the statements are placed on the back side of the form, a note in Block 12 must reference that fact. When copies of the forms are generated, these statements must be provided with the copies.

l. When an article marked with dual or multiple part numbers is received, the article will be processed in the following manner:

(1) Articles must be maintained using the instructions for the part number referenced in the customer's work request.

(2) Part markings applied under part 45 may only be modified as defined in FAA-approved or -accepted data (for example, maintenance manuals, ADs, manufacturer drawings, technical data, and service bulletins).

(3) All related records pertaining to the maintenance performed must reflect the same part number for the maintenance of a particular article as referenced in the customer's request. The following describes some, but not all, scenarios that could be encountered when articles are received. Typically, the part number on the customer's work request must match the part number on the return to service. In all cases, the maintainer provides records according to § 43.9. An FAA Form 8130-3 may be issued per this order. (Refer to appendix A, figure A-11, to this order.)

(a) If the article is marked with the PC part number, the return to service will reference that number.

(b) If the article is marked with a PMA and/or technical standard order authorization (TSOA) number, the return to service created will reference that number.

(c) If the article is marked with more than one PAH number (that is, PC, PMA, or TSOA), request guidance from the customer to determine which part number to use to support the customer's work request.

(d) If the customer is an air carrier (part 121 or part 135), commercial operator (part 125), or foreign air carrier with N-registered aircraft (14 CFR 129.14), follow the air carrier/operator's maintenance programs.

3-2. Approval for Return To Service After Maintenance, Preventive Maintenance, Rebuilding, and Alteration—Products and Articles.

a. Only those persons described in paragraph 3-1a and b, when authorized by § 43.7(c), (d), and (e), may issue an FAA Form 8130-3 for approval for return to service of products and articles that have undergone maintenance, preventive maintenance, rebuilding, or alteration, provided the applicable recordkeeping requirements of §§ 43.9, 91.417, 91.421, 121.380, 135.439, or 145.219 are met. The use of FAA Form 8130-3 for this purpose is optional, but the FAA recommends its use. This will help aviation authorities and industry to ensure complete traceability and ease the movement of products and articles through the aviation system. (Refer to appendix A, figure A-11, to this order.)

(1) All work must be performed under the control of part 121 or part 135 certificate holders having a continued airworthiness maintenance program or an air agency certificated under part 145. This applies to all FAA-certificated repair stations, both domestic and foreign.

(2) A PAH may use FAA Form 8130-3 for approval for return to service of products and articles as set forth in §§ 43.3(j) and 43.7(d). The completion of Blocks 14a through 14e will be used when the PAH rebuilds or alters any product manufactured by it under a TC or PC, TSO authorization, PMA, or product and process specification issued by the FAA. The PAH completes Block 14a by checking the appropriate box "Other regulation specified in Block 12." Refer to paragraph 3-6l.

(a) Documentation as outlined in § 43.9 ensures a PAH has in place a method for tracking the rebuild and/or alteration work performed and who performed it. This documentation method should become part of the FAA-approved quality system.

(b) As a minimum, the PAH quality system should address the PAH's procedures for rebuild and alteration that—

1 Dictate the data used for rebuilding and alteration. Section 43.7(d) requires that except for minor alterations, products or articles must be worked under technical data approved by the FAA. It is acceptable to rebuild using the same FAA-approved design data used for manufacturing. The PAH may alternatively develop data specifically for rebuilding, as long as that data is FAA-approved.

2 Identify by name and job title all persons authorized to return rebuilt or altered products and articles to service, to include signing of return-to-service documents.

3 Identify the records required for approval for return to service and how to complete them in compliance with § 43.9(a). Concerning the name and signature of the person approving the product(s)/article(s) for approval for return to service, the certificate type and number of the approving person must be documented as well. In the case of PAHs rebuilding their own products and articles, the certificate number is the assigned FAA project number (under part 21, subpart F, Production Under Type Certificate Only; K, Parts Manufacturer Approvals; or O, Technical Standard Order Approvals) or the PC number (under part 21, subpart G, Production Certificates).

(c) Section 43.7(d) authorizes the PAH to return to service any item worked on under § 43.3(j). Any employee of the PAH may therefore issue approval for return-to-service documents, but the PAH should deem them qualified and authorized in writing—the approval for return-to-service documents are signed as part of their approval. Issuing approval for return-to-service documents for rebuild and alteration activities is not a designee function. While the person issuing approval for return-to-service documents may also be an FAA designee, that person must not perform approval for return to service in a designee capacity or record a designee number on any approval for return-to-service document.

(3) When FAA Form 8130-3 is used as an approval for return to service to meet the terms and conditions of a bilateral agreement's maintenance implementation procedures (MIP), the air agency or air carrier must check the two boxes in Block 14a stating "14 CFR 43.9 Return to Service" and "Other regulations specified in Block 12" and provide the appropriate information in Blocks 11 and 12. This is considered to be a dual release FAA Form 8130-3. (Refer to appendix A, figure A-13, to this order.)

(4) If another authority's approved maintenance data are used to maintain products and articles and those data are not addressed in the provisions of a MIP, FAA Form 8130-3 should not be used.

b. In all cases, an appropriately authorized representative of the air agency, air carrier, or PAH in accordance with § 43.7(c), (d), or (e) must make the approval for return to service of products and articles.

3-3. Approval for Return To Service—Products and Articles.

a. Products and articles may be inspected and approved for return to service by persons described in paragraph 3-1a and b of this order. Issuance of FAA Form 8130-3 for this purpose is optional, but the FAA recommends its use. This will help aviation authorities and industry to ensure complete traceability and ease the movement of products and articles through the aviation system. When used for this purpose, an air agency or air carrier must accomplish the inspection. FAA Form 8130-3 can be used for this purpose, provided the applicable recordkeeping requirements of §§ 43.9, 91.417, 121.380, and 135.439 are met and the quality system includes the following:

(1) Traceability to an FAA-approved source of manufacture of new products and articles.

(2) Monitoring of the current status of the product and article in relation to shelf life and AD compliance. Each functional test/inspection must be performed in accordance with the standards set forth by § 43.13.

(3) Provisions for the retention of all records that may be necessary as part of the airworthiness documentation required by either part 21, 43, 91, 121, 135, or 145 for approval for return to service (for example, AD compliance).

(4) Provisions for documentation (FAA Form 8130-3, Block 12 or an attachment) clearly states the process used to determine airworthiness, including each reference to invoices, manufacturer maintenance manuals, or other instructions for continued airworthiness and FAA-approved/acceptable technical data. Attachments should include the form tracking number of the corresponding FAA Form 8130-3.

b. In all cases, FAA Form 8130-3 must be signed by the appropriately authorized representative of an FAA-approved air agency, air carrier, or PAH.

3-4. Issuance of FAA Form 8130-3 for Used Products and Articles Removed from a U.S.-Certificated Aircraft for Installation on Another U.S.-Certificated Aircraft.

a. FAA Form 8130-3 may be issued for approval for return to service of those products and articles removed from a U.S.-certificated aircraft (under an operating certificate in accordance with part 121 or part 135) for use on another aircraft operated under the same air carrier certificate. The removal and installation of products and articles must be accomplished in accordance with the air carrier's approved maintenance program or other acceptable methods, techniques, and practices; or FAA-approved/accepted data that is acceptable to the air carrier's approved maintenance program. The use of FAA Form 8130-3 for this purpose is optional.

b. Those products and articles removed from a U.S.-certificated aircraft other than those referenced in paragraph 3-4a must have an airworthiness determination made in accordance with § 43.13(a) and (b) by an FAA-approved air agency or U.S. air carrier. This also includes compliance with applicable ADs, modification status, and total time/cycles for those products and articles as required by §§ 91.417, 121.380, and 135.439. The use of FAA Form 8130-3 for this purpose is optional, but the FAA recommends its use.

3-5. Approval for Return To Service Information Relevant to the European Union.

a. Approval for Return To Service After Maintenance, Preventive Maintenance, Rebuilding, and Alteration—Products and Articles.

(1) Approval. Aviation Authorities (AA) in the European Union may recognize an approval for return-to-service FAA Form 8130-3 only from part 145 repair stations or air carriers that also obtained a European Aviation Safety Agency (EASA) part 145 approval appropriately rated for the product or article at the time the product or article was approved for return to service.

(2) Completing FAA Form 8130-3. A product or article approved for return to service with a dual release on FAA Form 8130-3 is eligible for installation on a United States or European Union registered aircraft. For a dual release, check both boxes in Block 14a and include the following statement in Block 12: "Certifies that the work specified in Block 11/12 was carried out in accordance with EASA Part 145 and in respect to that work the [product/article] is considered ready for release to service under EASA Part 145 approval no. [insert number: EASA 145-XXX]." The FAA approval/certification number must be entered in Block 14c. A dual release certificate can only be issued by facilities that are both FAA and EASA approved located in the United States or Europe listed in Appendix 2 of the safety agreement.

(3) Guidance for the use of FAA Form 8130-3 and EASA Form 1 single release for special cases is referenced in the MAG under the aviation safety agreement between the United States and the European Union. Refer to MAG section B, appendix 1 for the United States, and MAG section C, appendix 1 for the European Union.

b. Rebuilt Engines. The following information pertains to and must be followed when issuing FAA Form 8130-3 to export a rebuilt engine to the European Union:

(1) Background. On May 1, 2011, the aviation safety agreement between the United States and the European Union went into effect. This agreement streamlined the regulatory cooperation between the United States and the European Union on certification and continued operational safety. It allows for the reciprocal acceptance of FAA and EASA certification and oversight of civil aviation products and repair stations. With the signing of this agreement and the supporting Technical Implementation Procedures (TIP), EASA recognizes the term "Rebuilt Engines" as a manufacturing certification practice, not a maintenance release by the FAA. The TIP places the same import requirements on rebuilt engines that are on new aircraft engines.

(2) Completing FAA Form 8130-3. The appropriate term to be entered in Block 11 will be "See Block 12." A comment will be added to Block 12 stating, "Rebuilt to original PAH's specifications." "Approved design data and are in a condition for safe operation" will be marked in Block 13a once the aircraft engine is rebuilt to the manufacturer's approved design specifications. Block 13b will be completed by an authorized person at the PAH, and Block 13c will be the PAH's certificate number. (Refer to appendix A, figure A-12, to this order.)

3-6. Block-By-Block Instructions for Completing FAA Form 8130-3 for Approval for Return To Service.

a. Block 1. Approving Civil Aviation Authority/Country. FAA/United States. (Preprinted.)

b. Block 2. Authorized Release Certificate, FAA Form 8130-3, Airworthiness Approval Tag. (Preprinted.)

c. Block 3. FAA Form Tracking Number. Enter the unique number established by the numbering system. (Refer to paragraph 3-1i of this order.)

d. Block 4. Organization Name and Address. Enter the full name and physical address (no post office box numbers) of the organization or facility for which the form is being issued, and the facility's certificate number (for example, certificate No. LI1R 123K or X9MA123H), as appropriate. A logo or other identification of the organization is permitted if it can be contained within the block.

e. Block 5. Work Order/Contract/Invoice Number.

(1) Fill in the work order number, contract number, and/or invoice number related to the shipment list, or maintenance release authorization number, and state the number of pages attached to the form, including dates, if applicable. If the shipment list contains the information required in Blocks 6 through 11, the respective blocks may be left blank if a list is attached to the form. In this case, the following statement must be entered in Block 12: "This is the certification statement for the articles listed on the attached document dated _________, containing pages ______ through ______." In addition, the shipping list must cross-reference the form tracking number located in Block 3. (Refer to appendix A, figure A-4, to this order.)

(2) If a work order/contract/invoice number is not available, enter "N/A."

f. Block 6. Item. When FAA Form 8130-3 is issued, a single item number or multiple item numbers (for example, same item with different serial numbers) may be used for the same part number. Multiple items must be numbered in sequence, although not necessarily beginning with the number one (for example, 0040, 0050, 0062, 0063). If a separate listing is used, enter "List Attached" (refer to paragraph 3-6e of this order for further instructions).

g. Block 7. Description. Enter the name or description of the product or article. Preference should be given to the term used in the instructions for continued airworthiness or maintenance data (for example, illustrated parts catalog, aircraft maintenance manual, or service bulletin).

h. Block 8. Part Number. Enter each part number of the product or article. In case of an aircraft engine or propeller, the model designation may be used. If the article being worked is a subassembly that does not have a part number of its own, enter the next higher assembly number followed by the word "subassembly."

i. Block 9. Quantity. Enter the quantity of each product or article shipped.

j. Block 10. Serial Number. If the product or article is required by part 45 to be identified with a serial number, enter it here. Additionally, any other serial number not required by regulation also may be entered. If no serial number is entered in this block, enter "N/A." If a specific batch or lot number is used, refer to the instructions for Block 12.

k. Block 11. Status/Work. The following table describes what to enter in a specific situation. The term entered in Block 11 should reflect the majority of the work performed by the organization. The use of upper or lower case in this block does not matter.

Table 2—FAA Form 8130-3 for Approval for Return To Service Block 11 Terms

Enter—	*For—*
"OVERHAULED"	A process that ensures the product or article is in complete conformity with the applicable service tolerances specified in the type certificate holder's or equipment manufacturer's instructions for continued airworthiness, or in the data approved or accepted by the authority. The product or article will be at least disassembled, cleaned, inspected, repaired as necessary, reassembled, and tested in accordance with the approved or accepted data.
"See Block 12"	Products or articles rebuilt or altered by authorized PAHs in accordance with § 43.3(j). Refer to paragraph 3-6l(3).
"REPAIRED"	Repair of defect(s) using an applicable standard.
"INSPECTED" and/or "TESTED"	Examination or measurement in accordance with an applicable standard (for example, visual inspection, functional testing, or bench testing).
"MODIFIED"	Alteration of a product or article to conform to an applicable standard.

Note: The applicable standard must be described in Block 12.

l. Block 12. Remarks. The use of upper or lower case in this block does not matter.

(1) Describe the work identified in Block 11 and associated results necessary for the user or installer to determine the airworthiness of the product or article in relation to the work being certified. This can be done either directly or by reference to supporting documentation. If necessary, a separate sheet may be used and referenced from the main FAA Form 8130-3. Each statement must clearly identify which product or article in Block 6 it relates to.

(2) Below are examples of conditions that could necessitate a statement in this block. These statements may or may not be appropriate depending on the form's purpose.

(a) Data required by § 43.9, including the reference and revision status. If other documents such as work orders, shop travelers, or FAA Form 337, Major Repair and Alteration (Airframe, Powerplant, Propeller, or Appliance), are used by the certificate holder to comply with §§ 43.9 and 43.11, they must be specifically referenced in this block.

(b) Compliance with ADs or service bulletins.

(c) Repairs carried out.

(d) Modifications carried out.

(e) Replacement articles installed.

(f) Life-limited parts status (for example, total time, total cycles, time since new).

(g) If a specific batch or lot number is used to control or trace the product or article, enter the batch or lot number in this block.

(h) Deviations from the customer work order.

(i) Release statements to satisfy a CAA maintenance requirement.

(j) Information needed to support shipment with shortages or re-assembly after delivery.

> **Note:** Examples in paragraph 3-6l(2)(i) show the possibility of dual release against both part 43 and another CAA's maintenance requirement or the single release by a part 145-approved maintenance facility against a CAA maintenance requirement. However, care should be taken to check the relevant box(es) in Block 14a to validate the release. A dual release requires the approved data to be approved/accepted by both the FAA and appropriate CAA.

(3) When an authorized person completes Blocks 14a through 14e for the purpose of rebuilding or altering a product they hold the approval for in accordance with § 43.3(j), the term "See Block 12" will be entered in Block 11, and one of the following statements will be entered in Block 12: "Rebuilt to original PAH's specifications" or "Altered to original PAH's specifications."

m. Blocks 13a through 13e. Shade, darken, or otherwise mark to preclude inadvertent or unauthorized use.

n. Block 14a. Approval for Return To Service. Mark the appropriate box(es) indicating which regulations apply to the completed work. If the box "Other regulations specified in Block 12" is marked, then the regulations of the other CAA(s) must be identified in Block 12. At least the left box must be marked, or both boxes may be marked, as appropriate.

(1) The regulations of the other CAA must be specifically identified in Block 12. The completed work can be accomplished in accordance with the regulations of the FAA, or the regulations of the FAA and another CAA. The data used to complete the work must be clearly stated in Block 12 or attached to the form and the attachment identified in Block 12. If the work has been done in accordance with both the regulations of the FAA and another CAA, both boxes

must be checked. (Refer to paragraph 3-2a(3) of this order for dual release instructions.) Attachments should include the form tracking number of the corresponding FAA Form 8130-3.

(2) The phrase "Rebuilt (altered or inspected) to original PAH's specifications" will be entered in Block 12 when a PAH rebuilds, alters, or inspects their product in accordance with § 43.3(j) or § 43.7(d).

o. Block 14b. Authorized Signature. This space will be completed with the signature of the authorized person. Only persons specifically authorized are permitted to sign this block. The approval signature must be applied at the time and place of issuance and manually applied, except as provided in paragraph 3-1j of this order.

p. Block 14c. Approval/Certificate No. Enter the PAH, air agency, or air carrier certificate number (for example, OTWR165K or PQ1234NM).

q. Block 14d. Name (Typed or Printed). Enter the typed or printed name of the authorized representative whose signature appears in Block 14b.

r. Block 14e. Date (dd/mmm/yyyy). The date to be entered in Block 14e for approval for return to service will be the date on which the original work was completed (refer to § 43.9). The date must be in the following format: two-digit day, first three letters of the month, and four-digit year, for example, 03 Feb 2008. This does not need to be the same as the printing or shipping date, which may occur later. The use or omission of slashes, hyphens, or spaces in the date does not matter.

Chapter 4. Export Airworthiness Approvals of Aircraft Engines, Propellers, or Articles

4-1. General Information on Export Airworthiness Approvals.

a. An export occurs when a product or article is found to be airworthy, meets any available special conditions of the importing country or jurisdiction, and is transferred from one CAA's regulatory authority to another CAA's regulatory authority. Part 21, subpart L contains the regulatory requirements regarding the application for and issuance of an export airworthiness approval for aircraft engines, propellers and articles manufactured under U.S. regulations. An FAA ASI, authorized designee, or person authorized by a PAH's approved quality system pursuant to § 21.137(o) may determine (1) the product or article conforms to its FAA-approved design data, (2) whether the importing country or jurisdiction requires any special conditions, and (3) for a product, whether the product is in a condition for safe operation. If a PAH knows that a product or article will be installed on a non-U.S.-registered aircraft, or on an aircraft registered in a country where the CAA requires an export airworthiness approval, the FAA ASI or authorized designee should issue an export approval regardless of the aircraft's location. FAA Form 8130-3 does not constitute approval to install a product or article on a particular aircraft, aircraft engine, or propeller. For policies and procedures concerning the export of products and articles, as well as compliance with special requirements of importing countries, refer to FAA Order 8130.2, Airworthiness Certification of Products and Articles.

b. The country or jurisdiction of import may have a requirement that the FAA certify the exported product(s)/article(s) conform to that country's or jurisdiction's CAA-approved design approval. This is similar to the requirement placed on a CAA to certify product(s)/article(s) exported to the United States meet the FAA-approved type design in accordance with part 21, subpart N, Acceptance of Aircraft Engines, Propellers, and Articles for Import. The check in Block 13a ("Approved design data and are in a condition for safe operation") indicates the product(s)/article(s) meet the CAA- and FAA-approved design and are in a condition for safe operation.

(1) It is the exporter's responsibility to meet the special import requirements of the country or jurisdiction to which the product(s)/article(s) are being shipped. When any requirements, including the special requirements determined necessary by the importing country/jurisdiction for its certification basis (for example, changes to meet environmental conditions), cannot or will not be satisfied, the exporter, through coordination with the FAA, must obtain a written statement from the CAA of the importing country/jurisdiction indicating acceptance of the specific nonconformity (or nonconformities, as applicable). All nonconformities to the CAA-approved design must be noted in Block 12, Remarks, of FAA Form 8130-3.

(2) The requirements for a specific country or jurisdiction may be found in one or both of the following: (1) a bilateral agreement or (2) a specific document submitted to the FAA for publication that contains import requirements. The FAA Web site http://www.faa.gov/aircraft/air_cert/international contains a listing of the bilateral agreements as well as a listing of requirements submitted to the FAA by importing countries and jurisdictions. Refer to AC 21-2, Complying with the Requirements of Importing Countries or

Jurisdictions When Exporting U.S. Products, Articles, or Parts, now maintained online at http://www.faa.gov/aircraft/air_cert/international/export_aw_proc/sp_req_import/.

Note: Refer to annex 1 of the Agreement Between the United States of America and the European Union on Cooperation in the Regulation of Civil Aviation Safety and its Annexes for a listing of European countries included in this agreement.

(3) If a statement is requested by the country or jurisdiction of import to document that country's or jurisdiction's design approval data, and no such corresponding design approval data is available, a statement to that effect must be written in Block 12.

(4) If a written statement of acceptance has been received from the importing CAA regarding noncompliance to its approved design, a copy of this written statement of acceptance must be included with FAA Form 8130-3 to meet the requirements of § 21.331.

(5) The following instructions are to be followed before issuing an export airworthiness approval:

(a) Review. FAA Order 8130.2, Airworthiness Certification of Aircraft and Related Products, provides instructions for completing FAA Form 8130-1, Application for an Export Certificate of Airworthiness. Although an applicant may complete FAA Form 8130-1 for aircraft engines, propellers, and articles, these applications may also be made orally. An oral application or request should be made to the FAA or designated FAA representative authorized to issue those approvals. Designees will maintain records of the inspection and issuance or denial of FAA Form 8130-3. These records must be made available for review and evaluation as requested by FAA personnel. FAA Form 8130-1 may be documented electronically instead of completing, printing, signing, and retaining it in the paper format.

(b) Inspection. When the application is determined acceptable, the product or article must be inspected to the extent necessary to ensure that it conforms to the FAA-approved design data, is in a condition for safe operation, is properly identified, and meets any design or special requirements of the importing country or jurisdiction. The FAA managing office must determine whether an FAA Form 8100-1 has to be completed for each FAA Form 8130-3 issued for export based on the PAH's quality system's health and/or the designee's previous history, experience, or performance, or if the information can be stored and retrieved in another format (for example, electronic database). If required by the FAA managing office responsible for the designee/designee organization, each designee authorized to issue approvals for export will document the inspection results on FAA Form 8100-1 for periodic review and evaluation by the FAA.

1 When documenting the "nomenclature of item inspected" in Block 9 of the FAA Form 8100-1, also include the form tracking number (Block 3) and item number (Block 6) from the FAA Form 8130-3 completed for the product export airworthiness approval.

2 When applicable, FAA Form 8100-1 must include the results of the inspection, date of issuance, country of destination, description of product, and manufacturer's invoice or shipping document number.

(c) In the case where a product or article is presented for inspection for the issuance of FAA Form 8130-3, and the product or article is sealed in a package that does not afford a visible inspection, the authorized person must obtain the objective evidence to determine the appropriate inspections were conducted and approved before the issuance of FAA Form 8130-3.

(6) A PAH with an approved quality system that includes the procedures described in § 21.137(o) may use FAA 8130-3 to issue an authorized release document for a new or used aircraft engine, propeller, or article manufactured by the PAH itself. An authorized release document is a certifying statement by a PAH that, at the time of the release of the document, a given aircraft engine, propeller, or article conforms to its approved design and is in a condition for safe operation, unless otherwise specified (refer to § 21.331). When a PAH uses an authorized release document for the purpose of export, § 21.137(o) requires the PAH to comply with the procedures applicable to the export of new and used aircraft engines, propellers, and articles specified in § 21.331 and the responsibilities of exporters specified in § 21.335.

c. Splitting of previously exported bulk shipments by a PAH or a PAH's associate facility is not within the control or jurisdiction of the FAA. Therefore, once products or articles are exported, those items would be under the control or jurisdiction of the receiving authority.

d. FAA Form 8130-3 may be issued for products or articles outside the United States if the FAA finds no undue burden in administering the applicable requirements in accordance with § 21.325(c).

e. FAA Form 8130-3 must be completed as described in paragraph 4-5 of this order.

f. FAA Form 8130-3 must be correlated with the shipment. Additional copies of FAA Form 8130-3 may be provided upon request. The forms may not be delivered before the product or article is shipped.

g. FAA Form 8100-1 and FAA Form 8130-3 may be retained in paper format or in a secure database, provided the database contains all of the information required on FAA Form 8130-3. An acceptable means of compliance is provided in AC 21-43 or AC 120-78 (when applicable), and is available for FAA review upon request. Duplicates of FAA Form 8130-3, including signatures retained in a database, do not need to be graphic images of the documents.

h. Unique identification is required to enable or provide product or article traceability. The preferred method is a unique form tracking number in Block 3. However, if traceability is provided through other information on the form combined with a number in Block 3, this is also acceptable.

i. The signature of the person authorized to issue FAA Form 8130-3 may be applied electronically to Block 13b. With exception of paragraphs 1-10 and 1-11, at the time the signature is authorized to be placed on FAA Form 8130-3, the person whose signature appears on the form must have direct access to the product or article to verify it conforms to FAA-approved design data and is in a condition for safe operation, or any special conditions required by the importing country or jurisdiction are met.

j. An FAA Form 8130-3 issued subsequent to the original finding of airworthiness is considered a recurrent airworthiness approval, for example, a PMA or TSO authorization article that left the PAH's quality/inspection system and is being presented for export.

4-2. New Products and Articles. Export airworthiness approvals for aircraft engines, propellers, and articles are issued in accordance with § 21.331.

4-3. Used Products and Articles. Export airworthiness approvals for used aircraft engines, propellers, or articles are issued in accordance with § 21.331(c). If a used aircraft engine or propeller does not meet a requirement of § 21.331(c), § 21.331(d) allows for a deviation from the requirement if the CAA of the importing country or jurisdiction accepts the deviation in a form and manner acceptable to the FAA. The differences between the used aircraft engine or propeller and its approved design must be listed in Block 12 as exceptions.

Note: A deviation cannot be obtained for a used article.

4-4. PMA Articles. The following applies when exporting PMA articles using FAA Form 8130-3:

a. Various bilateral agreements with countries have specific additional requirements for the acceptance of U.S. PMA articles into those countries. The applicable bilateral agreement should be reviewed for the specific provisions associated with PMA articles.

b. When a particular bilateral agreement requires such a specific provision for PMA articles, statements must be entered in Block 12, if applicable.

c. The determination of a PMA article's criticality, as required to be entered in Block 12 when exported, can only be determined by the actual design approval holder (that is, the FAA-PMA holder).

d. The text of all bilateral agreements can be found at http://www.faa.gov/aircraft/air_cert/international/bilateral_agreements.

4-5. Block-By-Block Instructions for Completing FAA Form 8130-3 for Export Airworthiness Approvals.

a. Block 1. Approving Civil Aviation Authority/Country. FAA/United States. (Preprinted.)

b. Block 2. Authorized Release Certificate, FAA Form 8130-3, Airworthiness Approval Tag. (Preprinted.)

c. Block 3. FAA Form Tracking Number. Enter the unique number established by the numbering system. (Refer to paragraph 4-1h of this order.)

d. Block 4. Organization Name and Address.

(1) Enter the full name and physical address (no post office box numbers) of the organization or facility for which the form is being issued, and the facility's certificate number (for example, Certificate No. LI1R123K or Certificate No. PQ1234NM), as appropriate. A logo or other identification of the organization is permitted if it can be contained within the block.

> **Note:** If an FAA Form 8130-3 is issued at a PAH's extension facility with its own project number (assigned by the geographic managing office), the facility's project number, along with its full name and address, must be used.

(2) When a supplier has direct shipment authorization from a PAH, the following information must be entered:

(a) PAH name and address.

(b) Supplier name and address.

(c) PAH certificate or project number (for example, certificate No. PC 700 or PQ0123CE). If the supplier is unsure what number to use, consult the PAH for assistance.

(3) If a supplier produces a product or article as a replacement product or article, the supplier must either have direct shipment authorization or hold a production approval (PMA/TSO authorization) for each replacement product or article shipped. If the supplier holds its own production approval, and the products or articles were manufactured and are being shipped under that approval, the information required in paragraph 4-5d(1) must be listed.

e. Block 5. Work Order/Contract/Invoice Number.

(1) Fill in the work order number, contract number, and/or invoice number related to the shipment list, or maintenance release authorization number, and state the number of pages attached to the form, including dates, if applicable. If the shipment list contains the information required in Blocks 7 through 11, these respective blocks may be left blank if a list is attached to the form. In this case, the following statement must be entered in Block 12: "This is the certification statement for the products or articles listed on the attached document dated _________, containing pages ______ through ______." (Refer to appendix A, figure A-4, to this order.)

(2) In addition, the shipment list must cross-reference the form tracking number located in Block 3. The shipment list may contain more than one item, but it is the responsibility of the shipper to determine whether the CAA of the importing country or jurisdiction will accept bulk shipments under a single FAA Form 8130-3. If the CAA does not permit bulk shipments under a single form, Blocks 6 through 11 of each form must be filled in for each product or article shipped.

(3) If work order/contract/invoice number is not available, enter "N/A."

f. Block 6. Item. When FAA Form 8130-3 is issued, a single item number or multiple item numbers (for example, same item with different serial numbers) may be used for the same part number. Multiple items must be numbered in sequence, although not necessarily beginning with the number one (for example, 0040, 0050, 0062, 0063). If a separate listing is used, enter "List Attached" (refer to paragraph 4-5e of this order for further instructions).

g. Block 7. Description. Enter the name or description of the product or article. Preference should be given to the term used in the instructions for continued airworthiness or maintenance data (for example, illustrated parts catalog, aircraft maintenance manual, or service bulletin).

h. Block 8. Part Number. Enter each part number of the product or article. In the case of an aircraft engine or propeller, the model designation may be used.

i. Block 9. Quantity. Enter the quantity of each product or article shipped.

j. Block 10. Serial Number. If the product or article is required by part 45 to be identified with a serial number, enter it here. Additionally, any other serial number not required by regulation also may be entered. If no serial number is entered in this block, enter "N/A." If a specific batch or lot number is used to control or trace the product or article, enter the batch or lot number in Block 12

k. Block 11. Status/Work. The following table describes what to enter in a specific situation. Only one term may be entered in Block 11, which should reflect the majority of the work performed. The use of upper or lower case in this block does not matter.

Table 3—FAA Form 8130-3 for Export Airworthiness Approvals Block 11 Terms

Enter—	*For—*
"NEW"	The production of a new product or article in conformity with the approved design data.
"USED"	The aircraft engine, propeller, or article having any time in service.

Note 1: If the aircraft engine, propeller, or article is considered "USED," do not issue FAA Form 8130-3 for the purpose of export to the European Union. The European Union only accepts FAA Form 8130-3 issued for the purpose of return to service by a repair station/maintenance organization when used products or articles are being shipped to a European Union member country. Refer to paragraph 3-5 of this order for related policy and procedures.

Note 2: Before exporting a product or article, all special import requirements must be met, or that product or article may be rejected by the importing country or jurisdiction.

l. Block 12. Remarks. The use of upper or lower case in this block does not matter.

(1) State any information in this block, either directly or by reference to supporting documentation, necessary for the user or installer to determine the airworthiness of the product or article. Bilateral agreements may require certain statements to be added to the export form; those statements should be entered in this block. If necessary, a separate sheet may be used and referenced from the main FAA Form 8130-3. Each statement must be clearly identified as to the product or article in Block 6 to which it relates. If there is no statement, state "none."

(2) Below are examples of conditions that could necessitate a statement in this block. These statements may or may not be appropriate depending on the form's purpose.

(a) If the term "USED" is in Block 11, make reference to other substantiating airworthiness data in this block. Examples include a "Return To Service" FAA Form 8130-3, logbooks, airworthiness directive compliance, or status of life-limited parts.

(b) When a new FAA Form 8130-3 is issued to correct errors, the following statement must be entered: "This FAA Form 8130-3 corrects the error(s) in block(s) [enter block numbers corrected] of the FAA Form 8130-3 [enter form tracking number] dated [enter issuance date] and does not cover conformity/condition/release to service."

(c) Attachments when used. Attachments should include the form tracking number of the corresponding FAA Form 8130-3.

(d) When FAA Form 8130-3 is issued at a distributor, part 121 or part 135 certificate holders, or a part 145 certificated repair stations, enter the following statement: "The product(s)/article(s) shipped under this approval were produced by [insert PAH's name]."

(e) For TSO articles, enter the applicable TSO number.

(f) Information on life-limited parts (for example, total time, total cycles, time since new).

(g) Shelf-life data.

(h) Shortages or outstanding work, for example, missing parts on an assembly or reassembly after shipment.

(i) If a specific batch or lot number is used to control or trace the product or article, enter the batch or lot number in this block.

(j) Any data not appropriate in other blocks.

(3) When used by authorized suppliers with properly documented direct shipment authorization from the PAH, the words "Direct shipment authorization" must be entered in Block 12, and the information from paragraph 4-5d(2) of this order must be entered in Block 4. (Refer to appendix A, figure A-16, to this order.)

(4) When used for export approval for used products and articles returned to service based on the requirements of part 43, the words "Used product(s)/article(s), shipped per

CAA statement of acceptance for used product(s)/article(s)" must be entered. Refer to paragraph 4-3, which stipulates the importing authority must submit a written statement accepting used products and articles. Refer to applicable bilateral agreements.

(5) If a written statement of acceptance has been received from the importing CAA regarding a noncompliance to its approved design, the noncompliance to the CAA-approved design must be entered in Block 12. This does not apply to used articles according to § 21.331(d).

(6) When used for an export for a new subcomponent of a PMA/TSO authorization article higher assembly, complete FAA Form 8130-3 with the subcomponent information, and enter a statement in Block 12 indicating the part or article is a subcomponent of a PMA or TSO authorization (for example, "This part is a subcomponent of a PMA/TSO authorization"). (Refer to appendix A, figure A-15, to this order.)

(7) If a statement is requested by the country to which the product or article is being exported that documents that country's design approval data, and no such corresponding design approval data is available, a statement to that effect must be written in Block 12.

(8) If the PAH holds the type design data for replacement articles produced under an STC, "Produced by the STC design approval holder" must be entered in Block 12.

(9) If the originator has found that the product or article meets the special import requirements of the importing country or jurisdiction, Block 12 should indicate: "Export airworthiness approval – This article meets the special requirements of (enter country)." If the importing country or jurisdiction does not have special import requirements applicable to the product or article, then the statement does not need to be included. The finding of compliance with a particular country's or jurisdiction's special import requirements should not be interpreted to preclude export or re-export to another country, but the exporter/re-exporter should confirm compliance with the special import requirements of the destination country or jurisdiction. (Refer to appendix A, figure A-14, to this order.)

(10) If the country or jurisdiction to which a product or article is exported does not have a bilateral agreement with the United States, or has not stated any special import requirements, FAA Form 8130-3 may still be issued as an export approval. In this case, Block 12 should indicate "Export airworthiness approval. No special import requirements for [enter name of country or jurisdiction] stated at time of issuance."

(11) For exported aircraft engines or propellers, include total time and, if applicable, time since overhauled.

m. Block 13a. Airworthiness Approval. Place a check in the "Approved design data and are in a condition for safe operation" box if the products and articles were manufactured using FAA-approved design data and found to be in a condition for safe operation. Checking this box and signing Block 13b means the products and articles listed on the form meet the FAA-approved design data, are in a condition for safe operation, and, in the case of export, meet the importing country's or jurisdiction's design approval and meet the special requirements of that importing country or jurisdiction.

n. Block 13b. Authorized Signature. This block must be signed by the authorized person issuing the form. The block must be signed by an FAA ASI, authorized designee, or person authorized under a PAH's approved quality system that includes the procedures described in § 21.137(o). A person signing Block 13b may use an alternative to a handwritten signature (for example, a computer-generated signature) only when authorized by the FAA. The approval signature must be applied manually at the time and place of issuance, except as provided in paragraph 4-1i of this order.

o. Block 13c. Approval/Authorization No. Enter the approval/authorization number of the authorized representative/organization identified in Block 13b. If signed by an FAA inspector, the authorization number is the applicable office identifier. If signed by a PAH, or a person authorized by a PAH's quality system, enter the applicable production approval number (for example, PT1234CE or PQ1234NM).

p. Block 13d. Name (Typed or Printed). Enter the typed or printed name of the authorized representative/organization whose signature appears in Block 13b.

q. Block 13e. Date (dd/mmm/yyyy). The date must be in the following format: two-digit day, first three letters of the month, and four-digit year, for example, 03 Feb 2008. This does not need to be the same as the printing or shipping date, which may occur later. The use or omission of slashes, hyphens, or spaces in the date does not matter.

r. Blocks 14a through 14e. Shade, darken, or otherwise mark to preclude inadvertent or unauthorized use.

4-6. Reissuance by a PAH for Returned Products and Articles.

a. The new products and articles returned to a PAH may be eligible for a new FAA Form 8130-3 if—

(1) The new products and articles were produced under the PAH's production approval.

(2) The PAH maintains a procedure to accept products and articles back into their quality system.

(3) Tests and inspections are performed in accordance with procedures contained in the PAH's quality system to determine the returned product or article still meets the original type design it was produced under and still is in a condition for safe operation.

b. If the conditions in paragraphs 4-6a(1) through (3) are met, a new FAA Form 8130-3 in accordance with chapter 4 of this order may be issued.

c. If FAA Form 8130-3 is returned with the products and articles, the originator should retain that form on file with (or have reference to) the new FAA Form 8130-3.

Chapter 5. Electronic Use of the Authorized Release Certificate, FAA Form 8130-3, Airworthiness Approval Tag

5-1. Purpose of This Chapter. This chapter provides guidance on the acceptance and use of the electronic exchange of FAA Form 8130-3, for those entities that elect to comply with the required standards and guidance that governs the use of such electronic documentation for aircraft products and articles.

5-2. Background on Electronic FAA Form 8130-3.

a. The Government Paperwork Elimination Act (GPEA), Public Law 105-277, Title XVII, and the Electronic Signatures in Global and National Commerce Act (E-Sign), Public Law 106-229, encourage use of electronic signatures.

b. Before the enactment of E-Sign on June 30, 2000, the regulations on signatures acknowledging satisfaction of manufacturing and maintenance requirements did not reflect current advances in information storage and retrieval technology. These earlier rules were developed when use of electronic media for the storage and retrieval of data was neither available to, nor contemplated by, the aviation industry or the FAA.

c. As the complexity of aircraft design, manufacturing, and maintenance processes increased, the number of records and documents generated and required to be retained by aircraft manufacturers, owners, operators, and repair facilities expanded dramatically. Electronic information storage and retrieval systems have enhanced significantly the aviation industry's ability not only to meet FAA record-retention requirements, but also to manufacture, operate, and maintain today's highly complex aircraft and aircraft systems in a demanding operational environment.

d. The Office of Management and Budget (OMB), Executive Office of the President, has issued OMB Circular A-130, Management of Federal Information Resources. OMB Circular A-130 directs the FAA and other Government agencies to recognize the limitations on electronic recordkeeping systems due to restrictions on the use of electronic signatures. The FAA recognizes this limitation and will now permit the use of electronic signatures on the electronic FAA Form 8130-3. Manufacturers, owners, operators, and maintenance personnel may now use complete electronic recordkeeping systems because the requirement to authenticate documents with non-electronic signatures has been eliminated. Such systems may be used to generate FAA Form 8130-3 that can be properly authenticated with an electronic signature.

e. As a result of the above (GPEA, enactment of E-Sign, and OMB Circular A-130), the FAA and industry formed the Electronic Documentation Project Team (EDPT) to develop an industry specification to enable the electronic exchange of FAA Form 8130-3 for aircraft products and articles. The requirements contained in this chapter for the use of the electronic version of FAA Form 8130-3 and the specifications contained in ATA Spec 2000 are the direct result of the efforts put forth by that team. Not only were the requirements of FAA Form 8130-3 developed, but corresponding forms used by other authorities (for example, EASA or Transport Canada Civil Aviation (TCCA)) were considered.

f. The use and acceptance of electronic FAA Form 8130-3 (and other corresponding EASA and TCCA forms) offers several distinct advantages over the current paper format:

(1) Through the use of standard data semantics and structures contained in ATA Spec 2000, a higher degree of data reliability and consistency will be achieved.

(2) Through adoption of common, widely available digital security technologies, it is considerably more difficult to forge or alter data without being detected, and the data can more easily be traced directly to the source.

(3) Identifying a document signer (signatory) will be easier through the elimination of traceability difficulties associated with illegible handwritten entries and the deterioration of paper documents.

(4) The frequency of lost, damaged, and unreadable documents can be significantly reduced.

(5) The automated processes for generating, transmitting, and processing data will significantly reduce costly human errors.

(6) The cost and difficulty to store, retrieve, and analyze information can be substantially reduced.

5-3. General Procedures for the Use of Electronic FAA Form 8130-3.

a. The use of the electronic transfer FAA Form 8130-3 procedures is strictly voluntary when issuing FAA Form 8130-3 for its intended purpose as specified in chapters 2, 3, and 4 of this order. If authorized persons elect to implement the following procedures, it must be understood that both the originator and recipient of the electronic form must comply with the procedures in this chapter. If for whatever reason the data recipient is unable to accept the electronic form, the issuance of the form must be in paper format in accordance with the appropriate chapter of this order.

b. Those authorized persons who elect to issue the electronic FAA Form 8130-3 for products and articles must comply with the guidelines in this chapter, specific block-by-block instructions contained in chapters 2, 3, and 4 as appropriate, and the standardized set of data formats, data requirements, business guidelines, and reference documents in the ATA Spec 2000. ATA Spec 2000 is available by contacting Airlines for America at 202-626-4000, or via email at a4a@airlines.org.

5-4. Use of the Electronic FAA Form 8130-3.

a. Each time an electronic FAA Form 8130-3 is issued for a product or article (for example, new, export, conformity, or approval for return to service), a new electronic FAA Form 8130-3 must be generated by the originator for each item. To maintain traceability for all other electronic FAA Form 8130-3s issued for a particular product or article, all available electronic FAA Form 8130-3s from prior transfers/returns to service should be attached as reference for historical purposes.

(1) The digitally signed electronic FAA Form 8130-3 generated for a given transaction is considered to be an original document.

(2) Unlike the paper format of FAA Form 8130-3 that is manually signed by the originator, the unaltered electronic FAA Form 8130-3 may be transmitted multiple times. These transmittals would only be used if the end user has lost the first data transaction or a typographical error was found (for whatever reason). When the data is sent again because the original data was lost or damaged after being transmitted to the end user, a new form tracking number would NOT be established. If data is resent because of a typographical error in the first transmission, a new form tracking number must be established for the data transmittal.

b. A separate electronic FAA Form 8130-3 must be issued for each product or article part number. A quantity greater than one may be listed on the electronic FAA Form 8130-3 for the part number if it is not serialized in accordance with the applicable part 45. This order is not applicable if the product or article does not have a part number.

c. A separate electronic FAA Form 8130-3 must be issued for each product or article identified with a serial number that is required to be applied to products and articles in accordance with part 45.

d. The originator and the receiver of the electronic FAA Form 8130-3 must archive the digitally signed file, including any attached previous references, for the required time as mentioned in this order or as stipulated in the record retention requirements for that organization.

e. One of the advantages of using an electronic FAA Form 8130-3 is the ability to more easily integrate data across various other systems, such as inventory, maintenance, or spare parts management. To facilitate this, it may be necessary to include additional information in the electronic format that is not necessary in the paper format, for example, adding a Manufacturer's CAGE/NCAGE Code in Block 12 to unambiguously identify the part. (Although CAGE/NCAGE Codes are not required by the FAA, these codes have been accepted in the commercial aviation industry as a standard means of identifying entities.)

f. If the electronic system is acceptable between two trading partners, the FAA cannot require that a dual system be implemented for the same product or article. That is, for a given product or article, only one Authorized Release Certificate is allowed, in either a paper or an electronic format. This is to protect the integrity of the data, because it would not be correct to have two documents (paper and electronic) for the same item. If the electronic system is inoperative/ineffective for any reason, the paper format must be used until such time the electronic system can prove to be effective.

g. The electronic implementation of FAA Form 8130-3 is the legal and official document that uses the data in the XML file, not the PDF/paper copy generated from the XML file. If a PDF/paper copy is generated from the XML file, it is permissible to have additional information when viewing that form as a paper copy or on a computer screen version. This information should appear in Block 12 (unless it is necessary to present it in another block to clarify the contents of that block) to provide information to the end user on the airworthiness of the product or article. Examples of additional information that may be on the electronic version and not required by the paper format as defined in chapters 2, 3, and 4 may be viewed in appendix A, figures A-17 through A-19, to this order. Please note that these figures are provided as examples only.

5-5. Specific Requirements Other Than Those Listed in ATA Spec 2000. When constructing an electronic FAA Form 8130-3 transfer system to meet the requirements in this order, the following must be considered and addressed in the organization's manual or in the directions for the operating system. This information must be made available to each individual responsible for using the operating system.

a. Security.

(1) The electronic system should protect confidential information.

(2) The system should provide a means to identify if the data has changed so that appropriate action may be taken.

(3) A corresponding policy and management structure should support the computer hardware and software that delivers the information to establish each FAA Form 8130-3 for issuance, and the computer hardware and software to issue (receive) FAA Form 8130-3 for each product or article shipped.

b. Operating Procedures. Before introducing an electronic transfer system for FAA Form 8130-3 for products or articles shipped, procedures must be established, incorporated, and maintained for the operating system to include the following:

(1) Procedures describing how companies will effectively implement these procedures with their trading partners.

> **Note:** If authorized persons elect to implement the following procedures, it must be understood that both the originator and recipient of the electronic FAA Form 8130-3 must comply with the procedures in this chapter. If for whatever reason the data recipient is unable to accept the electronic form, the issuance of the form must be in paper format in accordance with the appropriate chapter of this order.

(2) Procedures for making the data available to the FAA upon request. Each person who elects to use the electronic transfer system will make available (at the FAA's request) an authorized employee or representative to access and explain, if necessary, the data for each electronically transferred FAA Form 8130-3. The individual must be familiar with the computer system and assist in accessing the necessary computerized information. This system must be capable of producing paper copies of FAA Form 8130-3 and of the viewed information at the request of the FAA.

(3) Procedures describing how the electronic FAA Form 8130-3 will be stored and retrieved. The archive system must ensure the integrity of the stored data, regardless of the storage medium, and that no unauthorized changes can be made to the data.

(4) Procedures for obtaining, maintaining, and controlling digital security certificates for the individual(s)/organization authorized to sign an electronic FAA Form 8130-3.

(5) Procedures for reviewing the computerized personal identification codes system to ensure the system will not permit password duplication.

(6) Procedures for periodically auditing the computer system to ensure the integrity of the system. A record of the audit should be completed and retained on file as part of the person's record retention requirements. This audit may be a computer program that automatically audits itself. In addition to the computer generated audit, a manual audit should be conducted annually to verify the integrity of the system.

(7) Audit procedures to ensure the integrity of each computerized workstation. If the workstations are server-based and contain no inherent attributes that enable or disable access, there is no need for each workstation to be audited.

(8) A description of the training procedure and requirements necessary to authorize access to the computer hardware and software system. (Recognizing the details will vary with the different individuals who need access, the training description may simply be part of the position description. Its location should be referenced in the manual or work instructions.)

(9) Procedures describing the method of identifying the product or article to the receivable electronic FAA Form 8130-3 and how that product or article is identified while in storage. Procedures must include how the necessary information from the electronic FAA Form 8130-3 is provided to the user/installer in order to complete the appropriate maintenance record after installation as required by part 43.

5-6. User/Installer Responsibilities. Because FAA Form 8130-3 is being issued electronically when complying with this chapter of the order, the User/Installer Responsibilities referenced at the bottom of each hard or screen copy of the form are not visible in the XML format. Therefore, the following statements are provided as a reminder of the user/installer responsibilities per the applicable regulations:

a. It is important to understand the existence of this document alone does not automatically constitute authority to install the aircraft engine, propeller, or article.

b. Where the user/installer performs work in accordance with the national regulations of an airworthiness authority different than the airworthiness authority of the country specified in Block 1, it is essential the user/installer ensures their airworthiness authority accepts aircraft engines, propellers, or articles from the airworthiness authority of the country or jurisdiction specified in Block 1.

c. Statements in Blocks 14a and 14e do not constitute installation certification. In all cases, aircraft maintenance records must contain an installation certification issued in accordance with the national regulations by the user/installer before the aircraft may be flown.

5-7. Sample Uses of Electronic FAA Form 8130-3. In appendix A are examples of various uses of the electronic FAA Form 8130-3. Each example depicts a sample PDF/paper copy of the data.

> **Note:** Any PDF/paper copies produced from a valid electronic form must meet all requirements described in this order, but may vary somewhat in layout and format.

5-8. Intent to Use Electronic FAA Form 8130-3. Those authorized persons who elect to issue the electronic FAA Form 8130-3 for transmittal of products or articles are required to notify their geographic FAA office of their intent before implementation. Figure 5-1 is provided as an example of a letter of intent.

Figure 5-1. Sample Letter of Intent for an Electronic Certification System

Note: Use company's letterhead.

To: [Responsible FAA office (manufacturing or flight standards having geographic jurisdiction responsibility over the requester's facility)]

From: [Company name]

Date: [Date]

Subject: Use of Electronic Transfer System for FAA Form 8130-3

This letter is to inform you that [company name] intends to use an electronic transfer system for FAA Form 8130-3 for products or articles shipped to [name of receiver of certificate] in accordance with our documented instructions. This system has been established using the guidelines outlined in FAA Order 8130.21, chapter 5.

This organization intends to implement the electronic transfer system on [date].

Our company facilities, equipment, and personnel are available for your review and/or inspection at [address]. Please contact [name] at [telephone number] if you have any questions regarding the implementation of the FAA Form 8130-3 electronic transfer system.

Sincerely,

[Requester signature]
[Requester name]

08/01/2013

8130.21H
Appendix A

Appendix A. Samples of FAA Form 8130-3 (For Reference Only)

Figure A-1. FAA Form 8130-3 (Blank)

1. Approving Civil Aviation Authority/Country: FAA/United States	2. AUTHORIZED RELEASE CERTIFICATE FAA Form 8130-3, AIRWORTHINESS APPROVAL TAG				3. Form Tracking Number:
4. Organization Name and Address:					5. Work Order/Contract/Invoice Number:
6. Item:	7. Description:	8. Part Number:	9. Quantity:	10. Serial Number:	11. Status/Work:
12. Remarks:					
13a. Certifies the items identified above were manufactured in conformity to: ☐ Approved design data and are in a condition for safe operation. ☐ Non-approved design data specified in Block 12.			14a. ☐ 14 CFR 43.9 Return to Service ☐ Other regulation specified in Block 12 Certifies that unless otherwise specified in Block 12, the work identified in Block 11 and described in Block 12 was accomplished in accordance with Title 14, Code of Federal Regulations, part 43 and in respect to that work, the items are approved for return to service.		
13b. Authorized Signature:		13c. Approval/Authorization No.:	14b. Authorized Signature:		14c. Approval/Certificate No.:
13d. Name (Typed or Printed):		13e. Date (dd/mmm/yyyy):	14d. Name (Typed or Printed):		14e. Date (dd/mmm/yyyy):
User/Installer Responsibilities					

It is important to understand that the existence of this document alone does not automatically constitute authority to install the aircraft engine/propeller/article.

Where the user/installer performs work in accordance with the national regulations of an airworthiness authority different than the airworthiness authority of the country specified in Block 1, it is essential that the user/installer ensures that his/her airworthiness authority accepts aircraft engine(s)/propeller(s)/article(s) from the airworthiness authority of the country specified in Block 1.

Statements in Blocks 13a and 14a do not constitute installation certification. In all cases, aircraft maintenance records must contain an installation certification issued in accordance with the national regulations by the user/installer before the aircraft may be flown.

FAA Form 8130-3 (02-14) NSN: 0052-00-012-9005

A-1

FAA Order 8130.21H

08/01/2013

8130.21H
Appendix A

Figure A-2. Sample FAA Form 8130-3 for a Conformity Inspection

1. Approving Civil Aviation Authority/Country: FAA/United States	2. AUTHORIZED RELEASE CERTIFICATE FAA Form 8130-3, AIRWORTHINESS APPROVAL TAG				3. Form Tracking Number: AP54321
4. Organization Name and Address: Anyone's Aviation, 1104 Wing Avenue, Anyplace, TX 72212 (AP54321)					5. Work Order/Contract/Invoice Number: WO 99987
6. Item:	7. Description:	8. Part Number:	9. Quantity:	10. Serial Number:	11. Status/Work:
1	Flap Track	B9876-1	8	N/A	PROTOTYPE

12. Remarks:

Conformity Inspections

Detail article conformity for FAA Project AP54321, dated 10 Feb 2008, Drawing No. 12345-001, Revision G1, dated 1 Oct 2007 requested.

1. **Request for Conformity FAA 8120.10, #06-09222, dated 19 Feb 2008 reviewed.**
2. **Copy of FAA 8130-9, Statement of Conformity, dated 3 May 2007 provided, reviewed, and attached.**
3. **Part No. B9876-1 Flap Track (8 ea.), inspected to engineering to include Drawing No. 12345-001, Revision G1, dated 1 Oct 2007.**

DEVIATION: 8 ea. Flap Tracks, Part No. B9876-1 holes should be ".250 +or- .005" Holes are oversized by ".020." DER Disposition: Oversized holes does not affect static testing and parts can be used as is per DER-888002-SW, A. Engineer, dated 11 Feb 2008.

13a. Certifies the items identified above were manufactured in conformity to: ☐ Approved design data and are in a condition for safe operation. ☒ Non-approved design data specified in Block 12.		14a. ☐ 14 CFR 43.9 Return to Service ☐ Other regulation specified in Block 12. Certifies that unless otherwise specified in Block 12, the work identified in Block 11 and described in Block 12 was accomplished in accordance with Title 14, Code of Federal Regulations, part 43 and in respect to that work, the items are approved for return to service.	
13b. Authorized Signature: *A. Inspector*	13c. Approval/Authorization No.: DARF-1234567-SW	14b. Authorized Signature:	14c. Approval/Certificate No.:
13d. Name (Typed or Printed): A. Inspector	13e. Date (dd/mmm/yyyy): 03 Mar 2008	14d. Name (Typed or Printed):	14e. Date (dd/mmm/yyyy):

User/Installer Responsibilities

It is important to understand that the existence of this document alone does not automatically constitute authority to install the aircraft engine/propeller/article.

Where the user/installer performs work in accordance with the national regulations of an airworthiness authority different than the airworthiness authority of the country specified in Block 1, it is essential that the user/installer ensures that his/her airworthiness authority accepts aircraft engine(s)/propeller(s)/article(s) from the airworthiness authority of the country specified in Block 1.

Statements in Blocks 13a and 14a do not constitute installation certification. In all cases, aircraft maintenance records must contain an installation certification issued in accordance with the national regulations by the user/installer before the aircraft may be flown.

FAA Form 8130-3 (02-14) NSN: 0052-00-012-9005

A-2

08/01/2013

8130.21H
Appendix A

Figure A-3. Sample FAA Form 8130-3 for Domestic Airworthiness Approval for an Engine

1. Approving Civil Aviation Authority/Country:	2.				3. Form Tracking Number:
FAA/United States	**AUTHORIZED RELEASE CERTIFICATE** FAA Form 8130-3, AIRWORTHINESS APPROVAL TAG				5648944
4. Organization Name and Address: Big Engine Manufacturing Co., 5 Aviation Way, Small Town, KS 67021 (B7EM123T)					5. Work Order/Contract/Invoice Number: BR549
6. Item:	7. Description:	8. Part Number:	9. Quantity:	10. Serial Number:	11. Status/Work:
1	Engine	TSIO-550B1D1	1	P222264	NEW

12. Remarks:

Airworthiness approval – Engine

13a. Certifies the items identified above were manufactured in conformity to: ☒ Approved design data and are in a condition for safe operation. ☐ Non-approved design data specified in Block 12.		14a. ☐ 14 CFR 43.9 Return to Service ☐ Other regulation specified in Block 12 Certifies that unless otherwise specified in Block 12, the work identified in Block 11 and described in Block 12 was accomplished in accordance with Title 14, Code of Federal Regulations, part 43 and in respect to that work, the items are approved for return to service.	
13b. Authorized Signature: *A. Inspector*	13c. Approval/Authorization No.: DARF-54123-SW	14b. Authorized Signature:	14c. Approval/Certificate No.:
13d. Name (Typed or Printed): A. Inspector	13e. Date (dd/mmm/yyyy): 28 Aug 2007	14d. Name (Typed or Printed):	14e. Date (dd/mmm/yyyy):

User/Installer Responsibilities

It is important to understand that the existence of this document alone does not automatically constitute authority to install the aircraft engine/propeller/article.

Where the user/installer performs work in accordance with the national regulations of an airworthiness authority different than the airworthiness authority of the country specified in Block 1, it is essential that the user/installer ensures that his/her airworthiness authority accepts aircraft engine(s)/propeller(s)/article(s) from the airworthiness authority of the country specified in Block 1.

Statements in Blocks 13a and 14a do not constitute installation certification. In all cases, aircraft maintenance records must contain an installation certification issued in accordance with the national regulations by the user/installer before the aircraft may be flown.

FAA Form 8130-3 (02–14) NSN: 0052-00-012-9005

A-3

FAA Order 8130.21H

08/01/2013

8130.21H
Appendix A

Figure A-4. Sample FAA Form 8130-3 for Domestic Airworthiness Approval for a New Product or Article (Packing List)

1. Approving Civil Aviation Authority/Country: FAA/United States	2. **AUTHORIZED RELEASE CERTIFICATE** FAA Form 8130-3, AIRWORTHINESS APPROVAL TAG				3. Form Tracking Number: 991004327
4. Organization Name and Address: Parts Manufacturing Corporation, 6210 Wing Avenue, Anyplace, TX (PQ02469SW)					5. Work Order/Contract/Invoice Number: V234ZY 6 pages attached dated 12 Oct 2005
6. Item:	7. Description:	8. Part Number:	9. Quantity:	10. Serial Number:	11. Status/Work:
List Attached					

12. Remarks:

Airworthiness approval

This is the certification statement for the articles listed on the attached document dated 12 Oct 2005, containing pages 1 through 6 with the Form Tracking Number 991004327 on each of the pages.

13a. Certifies the items identified above were manufactured in conformity to: ☒ Approved design data and are in a condition for safe operation. ☐ Non-approved design data specified in Block 12.		14a. ☐ 14 CFR 43.9 Return to Service ☐ Other regulation specified in Block 12. Certifies that unless otherwise specified in Block 12, the work identified in Block 11 and described in Block 12 was accomplished in accordance with Title 14, Code of Federal Regulations, part 43 and in respect to that work, the items are approved for return to service.	
13b. Authorized Signature: *A. Inspector*	13c. Approval/Authorization No.: DARF-761104-NM	14b. Authorized Signature:	14c. Approval/Certificate No.:
13d. Name (Typed or Printed): A. Inspector	13e. Date (dd/mmm/yyyy): 12 Oct 2007	14d. Name (Typed or Printed):	14e. Date (dd/mmm/yyyy):

User/Installer Responsibilities

It is important to understand that the existence of this document alone does not automatically constitute authority to install the aircraft engine/propeller/article.

Where the user/installer performs work in accordance with the national regulations of an airworthiness authority different than the airworthiness authority of the country specified in Block 1, it is essential that the user/installer ensures that his/her airworthiness authority accepts aircraft engine(s)/propeller(s)/article(s) from the airworthiness authority of the country specified in Block 1.

Statements in Blocks 13a and 14a do not constitute installation certification. In all cases, aircraft maintenance records must contain an installation certification issued in accordance with the national regulations by the user/installer before the aircraft may be flown.

FAA Form 8130-3 (02-14) NSN: 0052-00-012-9005

A-4

08/01/2013 — 8130.21H Appendix A

Figure A-5. Sample FAA Form 8130-3 for Airworthiness Approval When Issued at a Distributor Facility

1. Approving Civil Aviation Authority/Country: FAA/United States	2. AUTHORIZED RELEASE CERTIFICATE FAA Form 8130-3, AIRWORTHINESS APPROVAL TAG				3. Form Tracking Number: ACE235
4. Organization Name and Address: Ace Aircraft Parts Distribution Co., 100 Lake Drive, San Antonio, TX 78007					5. Work Order/Contract/Invoice Number: PO #451960
6. Item:	7. Description:	8. Part Number:	9. Quantity:	10. Serial Number:	11. Status/Work:
1	Exhaust Valve	GE637781	5 ea.	N/A	NEW
12. Remarks: Airworthiness approval The product(s)/article(s) shipped under this approval were produced by Sample Engines, Incorporated.					
13a. Certifies the items identified above were manufactured in conformity to: ☒ Approved design data and are in a condition for safe operation. ☐ Non-approved design data specified in Block 12.			14a. ☐ 14 CFR 43.9 Return to Service ☐ Other regulation specified in Block 12 Certifies that unless otherwise specified in Block 12, the work identified in Block 11 and described in Block 12 was accomplished in accordance with Title 14, Code of Federal Regulations, part 43 and in respect to that work, the items are approved for return to service.		
13b. Authorized Signature: *A. Inspector*		13c. Approval/Authorization No.: DARF-000243-SW	14b. Authorized Signature:		14c. Approval/Certificate No.:
13d. Name (Typed or Printed): A. Inspector		13e. Date (dd/mmm/yyyy): 30 May 2007	14d. Name (Typed or Printed):		14e. Date (dd/mmm/yyyy):

User/Installer Responsibilities

It is important to understand that the existence of this document alone does not automatically constitute authority to install the aircraft engine/propeller/article.

Where the user/installer performs work in accordance with the national regulations of an airworthiness authority different than the airworthiness authority of the country specified in Block 1, it is essential that the user/installer ensures that his/her airworthiness authority accepts aircraft engine(s)/propeller(s)/article(s) from the airworthiness authority of the country specified in Block 1.

Statements in Blocks 13a and 14a do not constitute installation certification. In all cases, aircraft maintenance records must contain an installation certification issued in accordance with the national regulations by the user/installer before the aircraft may be flown.

FAA Form 8130-3 (02-14) — NSN: 0052-00-012-9005

A-5

08/01/2013

8130.21H
Appendix A

Figure A-6. Sample FAA Form 8130-3 for Identification of Prepositioned Products or Articles

1. Approving Civil Aviation Authority/Country: FAA/United States	2. AUTHORIZED RELEASE CERTIFICATE FAA Form 8130-3, AIRWORTHINESS APPROVAL TAG				3. Form Tracking Number: BR549
4. Organization Name and Address: Executive Airplanes, 337 Modification Way, Anyplace, TX 75000 (PC123)					5. Work Order/Contract/Invoice Number: WO 87800
6. Item:	7. Description:	8. Part Number:	9. Quantity:	10. Serial Number:	11. Status/Work:
001	Coffee Maker	EA 6451-2	1	02346	PROTOTYPE
002	Galley Cabinet Door	EA 5471-2	1	77759	
003	PCU Panel	EA 7500-1	1	99999	
004	Coat Closet Door	EA 98700	1	66654	

12. Remarks:

Prepositioned articles

Prototype product(s)/article(s) pending certification under FAA project number [enter number] not eligible for installation on in-service, type-certificated aircraft. Upon approval of the applicable design data and completion of an inspection to validate conformity to that approved design data and condition for safe operation, that product or article may be considered new.

13a. Certifies the items identified above were manufactured in conformity to: ☐ Approved design data and are in a condition for safe operation. ☒ Non-approved design data specified in Block 12.		14a. ☐ 14 CFR 43.9 Return to Service ☐ Other regulation specified in Block 12. Certifies that unless otherwise specified in Block 12, the work identified in Block 11 and described in Block 12 was accomplished in accordance with Title 14, Code of Federal Regulations, part 43 and in respect to that work, the items are approved for return to service.	
13b. Authorized Signature: *A. Inspector*	13c. Approval/Authorization No.: DMIR-003486-SW	14b. Authorized Signature:	14c. Approval/Certificate No.:
13d. Name (Typed or Printed): A. Inspector	13e. Date (dd/mmm/yyyy): 10 Oct 2007	14d. Name (Typed or Printed):	14e. Date (dd/mmm/yyyy):

User/Installer Responsibilities

It is important to understand that the existence of this document alone does not automatically constitute authority to install the aircraft engine/propeller/article.

Where the user/installer performs work in accordance with the national regulations of an airworthiness authority different than the airworthiness authority of the country specified in Block 1, it is essential that the user/installer ensures that his/her airworthiness authority accepts aircraft engine(s)/propeller(s)/article(s) from the airworthiness authority of the country specified in Block 1.

Statements in Blocks 13a and 14a do not constitute installation certification. In all cases, aircraft maintenance records must contain an installation certification issued in accordance with the national regulations by the user/installer before the aircraft may be flown.

FAA Form 8130-3 (02-14) NSN: 0052-00-012-9005

A-6

Figure A-7a. Sample FAA Form 8130-3 for Splitting Bulk Shipments

<table>
<tr><td colspan="2">1. Approving Civil Aviation Authority/Country:
FAA/United States</td><td colspan="3">2.
AUTHORIZED RELEASE CERTIFICATE
FAA Form 8130-3, AIRWORTHINESS APPROVAL TAG</td><td>3. Form Tracking Number:
ACE-549</td></tr>
<tr><td colspan="5">4. Organization Name and Address:
OEM Airplane Company, 110 Stunt Flyer Road, Memphis, TN 76005 (PC023)</td><td>5. Work Order/Contract/Invoice Number:
WO 5678</td></tr>
<tr><td>6. Item:</td><td>7. Description:</td><td>8. Part Number:</td><td>9. Quantity:</td><td>10. Serial Number:</td><td>11. Status/Work:</td></tr>
<tr><td>1</td><td>Flap Track Roller</td><td>65B9999-1</td><td>1000</td><td>N/A</td><td>NEW</td></tr>
<tr><td colspan="6">12. Remarks:

Airworthiness approval</td></tr>
<tr><td colspan="3">13a. Certifies the items identified above were manufactured in conformity to:
☒ Approved design data and are in a condition for safe operation.
☐ Non-approved design data specified in Block 12.</td><td colspan="3">14a. ☐ 14 CFR 43.9 Return to Service ☐ Other regulation specified in Block 12
Certifies that unless otherwise specified in Block 12, the work identified in Block 11 and described in Block 12 was accomplished in accordance with Title 14, Code of Federal Regulations, part 43 and in respect to that work, the items are approved for return to service.</td></tr>
<tr><td colspan="2">13b. Authorized Signature:
A. Inspector</td><td>13c. Approval/Authorization No.:
DMIR-650987-NM</td><td colspan="2">14b. Authorized Signature:</td><td>14c. Approval/Certificate No.:</td></tr>
<tr><td colspan="2">13d. Name (Typed or Printed):
A. Inspector</td><td>13e. Date (dd/mmm/yyyy):
25 May 2007</td><td colspan="2">14d. Name (Typed or Printed):</td><td>14e. Date (dd/mmm/yyyy):</td></tr>
<tr><td colspan="6">User/Installer Responsibilities</td></tr>
<tr><td colspan="6">It is important to understand that the existence of this document alone does not automatically constitute authority to install the aircraft engine/propeller/article.

Where the user/installer performs work in accordance with the national regulations of an airworthiness authority different than the airworthiness authority of the country specified in Block 1, it is essential that the user/installer ensures that his/her airworthiness authority accepts aircraft engine(s)/propeller(s)/article(s) from the airworthiness authority of the country specified in Block 1.

Statements in Blocks 13a and 14a do not constitute installation certification. In all cases, aircraft maintenance records must contain an installation certification issued in accordance with the national regulations by the user/installer before the aircraft may be flown.</td></tr>
</table>

FAA Form 8130-3 (02-14) NSN: 0052-00-012-9005

Figure A-7b. Sample FAA Form 8130-3 for Splitting Bulk Shipments (Separate Sheet of Paper)

AIRPLANE COMPANY COPY STATEMENT

(Company name) certifies the attached document is a copy of FAA Form 8130-3. The prior FAA Form 8130-3 received by our facility is maintained on file pursuant to our document retention standards. That prior FAA form tracking number is **ACE-549**. The new tracking number for this portion of the split bulk shipment is **S1-054321**. The number of products or articles being shipped under this approval is **500**.

A. Quality Manager Oct 24 2007

A. Quality Manager Date

FAA Order 8130.21H

08/01/2013

8130.21H
Appendix A

Figure A-8. Sample FAA Form 8130-3 for a Direct Shipment Authorization

1. Approving Civil Aviation Authority/Country: FAA/United States	2. AUTHORIZED RELEASE CERTIFICATE FAA Form 8130-3, AIRWORTHINESS APPROVAL TAG				3. Form Tracking Number: 991004327
4. Organization Name and Address: Original Parts Manufacturing Corporation, 6210 Wing Avenue, Anyplace, AL (PQ02269CE) Everybody's Aircraft Supply Co., 810 Red Baron Way, Hooterville, OK 74032					5. Work Order/Contract/Invoice Number: WO 2020
6. Item:	7. Description:	8. Part Number:	9. Quantity:	10. Serial Number:	11. Status/Work:
1	Wing Tip	AE637781-1	5 ea.	N/A	NEW

12. Remarks:

Airworthiness approval – Direct shipment authorization

13a. Certifies the items identified above were manufactured in conformity to: ☒ Approved design data and are in a condition for safe operation. ☐ Non-approved design data specified in Block 12.		14a. ☐ 14 CFR 43.9 Return to Service ☐ Other regulation specified in Block 12 Certifies that unless otherwise specified in Block 12, the work identified in Block 11 and described in Block 12 was accomplished in accordance with Title 14, Code of Federal Regulations, part 43 and in respect to that work, the items are approved for return to service.	
13b. Authorized Signature: *A. Inspector*	13c. Approval/Authorization No.: DMIR-00243-CE	14b. Authorized Signature:	14c. Approval/Certificate No.:
13d. Name (Typed or Printed): A. Inspector	13e. Date (dd/mmm/yyyy): 13 Apr 2008	14d. Name (Typed or Printed):	14e. Date (dd/mmm/yyyy):

User/Installer Responsibilities

It is important to understand that the existence of this document alone does not automatically constitute authority to install the aircraft engine/propeller/article.

Where the user/installer performs work in accordance with the national regulations of an airworthiness authority different than the airworthiness authority of the country specified in Block 1, it is essential that the user/installer ensures that his/her airworthiness authority accepts aircraft engine(s)/propeller(s)/article(s) from the airworthiness authority of the country specified in Block 1.

Statements in Blocks 13a and 14a do not constitute installation certification. In all cases, aircraft maintenance records must contain an installation certification issued in accordance with the national regulations by the user/installer before the aircraft may be flown.

FAA Form 8130-3 (02-14) NSN: 0052-00-012-9005

A-8

Figure A-9. Sample FAA Form 8130-3 for Airworthiness Approval for a New Subcomponent for a PMA Article

1. Approving Civil Aviation Authority/Country: FAA/United States	2. AUTHORIZED RELEASE CERTIFICATE FAA Form 8130-3, AIRWORTHINESS APPROVAL TAG				3. Form Tracking Number: Smith 007-1
4. Organization Name and Address: Sample Engines Inc., 49 Timber Lane, San Antonio, TX 75005 (PQ0000SW)					5. Work Order/Contract/Invoice Number: WO 671960
6. Item:	7. Description:	8. Part Number:	9. Quantity:	10. Serial Number:	11. Status/Work:
1	Exhaust Valve	GE1637781	5 ea.	N/A	NEW
12. Remarks: Airworthiness approval This exhaust valve is a subcomponent of an FAA PMA article.					
13a. Certifies the items identified above were manufactured in conformity to: ☒ Approved design data and are in a condition for safe operation. ☐ Non-approved design data specified in Block 12.			14a. ☐ 14 CFR 43.9 Return to Service ☐ Other regulation specified in Block 12 Certifies that unless otherwise specified in Block 12, the work identified in Block 11 and described in Block 12 was accomplished in accordance with Title 14, Code of Federal Regulations, part 43 and in respect to that work, the items are approved for return to service.		
13b. Authorized Signature: *A. Inspector*		13c. Approval/Authorization No.: DMIR-00007-SW	14b. Authorized Signature:		14c. Approval/Certificate No.:
13d. Name (Typed or Printed): A. Inspector		13e. Date (dd/mmm/yyyy): 14 Oct 2007	14d. Name (Typed or Printed):		14e. Date (dd/mmm/yyyy):

User/Installer Responsibilities

It is important to understand that the existence of this document alone does not automatically constitute authority to install the aircraft engine/propeller/article.

Where the user/installer performs work in accordance with the national regulations of an airworthiness authority different than the airworthiness authority of the country specified in Block 1, it is essential that the user/installer ensures that his/her airworthiness authority accepts aircraft engine(s)/propeller(s)/article(s) from the airworthiness authority of the country specified in Block 1.

Statements in Blocks 13a and 14a do not constitute installation certification. In all cases, aircraft maintenance records must contain an installation certification issued in accordance with the national regulations by the user/installer before the aircraft may be flown.

FAA Form 8130-3 (02-14) NSN: 0052-00-012-9005

08/01/2013

8130.21H
Appendix A

Figure A-10. Sample FAA Form 8130-3 for a Rebuilt Product or Article

1. Approving Civil Aviation Authority/Country: FAA/United States	2. **AUTHORIZED RELEASE CERTIFICATE** FAA Form 8130-3, AIRWORTHINESS APPROVAL TAG				3. Form Tracking Number: ACME-12345
4. Organization Name and Address: Acme Airplane Company, 110 Aviation Place, Somewhere, OK (PC62)					5. Work Order/Contract/Invoice Number: WO 98765
6. Item:	7. Description:	8. Part Number:	9. Quantity:	10. Serial Number:	11. Status/Work:
1	Fuel Control	PW54667	1	N/A	See Block 12
12. Remarks: Rebuilt (altered) to original PAH's specifications.					
13a. Certifies the items identified above were manufactured in conformity to: ☐ Approved design data and are in a condition for safe operation. ☐ Non-approved design data specified in Block 12.			14a. ☐ 14 CFR 43.9 Return to Service ☒ Other regulation specified in Block 12. Certifies that unless otherwise specified in Block 12, the work identified in Block 11 and described in Block 12 was accomplished in accordance with Title 14, Code of Federal Regulations, part 43 and in respect to that work, the items are approved for return to service.		
13b. Authorized Signature:	13c. Approval/Authorization No.:		14b. Authorized Signature: A. Inspector		14c. Approval/Certificate No.: PC #62
13d. Name (Typed or Printed):	13e. Date (dd/mmm/yyyy):		14d. Name (Typed or Printed): A. Inspector		14e. Date (dd/mmm/yyyy): 30 Apr 2008

User/Installer Responsibilities

It is important to understand that the existence of this document alone does not automatically constitute authority to install the aircraft engine/propeller/article.

Where the user/installer performs work in accordance with the national regulations of an airworthiness authority different than the airworthiness authority of the country specified in Block 1, it is essential that the user/installer ensures that his/her airworthiness authority accepts aircraft engine(s)/propeller(s)/article(s) from the airworthiness authority of the country specified in Block 1.

Statements in Blocks 13a and 14a do not constitute installation certification. In all cases, aircraft maintenance records must contain an installation certification issued in accordance with the national regulations by the user/installer before the aircraft may be flown.

FAA Form 8130-3 (02-14) NSN: 0052-00-012-9005

A-10

Figure A-11. Sample FAA Form 8130-3 for Approval for Return To Service

1. Approving Civil Aviation Authority/Country: FAA/United States	2. **AUTHORIZED RELEASE CERTIFICATE** FAA Form 8130-3, AIRWORTHINESS APPROVAL TAG				3. Form Tracking Number: 2004-664
4. Organization Name and Address: Anyone's Repair Station, 1104 Wing Avenue, Anyplace, TX 22212 (PW8RW813J)					5. Work Order/Contract/Invoice Number: W 8851
6. Item:	7. Description:	8. Part Number:	9. Quantity:	10. Serial Number:	11. Status/Work:
010	Actuator	69A321	1	3384-L	REPAIRED

12. Remarks:

"The work specified has been accomplished in accordance with [*insert type of manual or specification, number, and revision date*]."

13a. Certifies the items identified above were manufactured in conformity to: ☐ Approved design data and are in a condition for safe operation. ☐ Non-approved design data specified in Block 12.		14a. ☒ 14 CFR 43.9 Return to Service ☐ Other regulation specified in Block 12. Certifies that unless otherwise specified in Block 12, the work identified in Block 11 and described in Block 12 was accomplished in accordance with Title 14, Code of Federal Regulations, part 43 and in respect to that work, the items are approved for return to service.	
13b. Authorized Signature:	13c. Approval/Authorization No.:	14b. Authorized Signature: A. Inspector	14c. Approval/Certificate No.: PW8RW813J
13d. Name (Typed or Printed):	13e. Date (dd/mmm/yyyy):	14d. Name (Typed or Printed): A. Inspector	14e. Date (dd/mmm/yyyy): 12 Oct 2007

User/Installer Responsibilities

It is important to understand that the existence of this document alone does not automatically constitute authority to install the aircraft engine/propeller/article.

Where the user/installer performs work in accordance with the national regulations of an airworthiness authority different than the airworthiness authority of the country specified in Block 1, it is essential that the user/installer ensures that his/her airworthiness authority accepts aircraft engine(s)/propeller(s)/article(s) from the airworthiness authority of the country specified in Block 1.

Statements in Blocks 13a and 14a do not constitute installation certification. In all cases, aircraft maintenance records must contain an installation certification issued in accordance with the national regulations by the user/installer before the aircraft may be flown.

FAA Form 8130-3 (02-14) NSN: 0052-00-012-9005

FAA Order 8130.21H

08/01/2013 8130.21H
Appendix A

Figure A-12. Sample FAA Form 8130-3 for a Rebuilt Engine to a Country Within the European Union

1. Approving Civil Aviation Authority/Country: FAA/United States	2. **AUTHORIZED RELEASE CERTIFICATE** FAA Form 8130-3, AIRWORTHINESS APPROVAL TAG			3. Form Tracking Number: ACME-12345
4. Organization Name and Address: ACME Airplane Company, 110 Aviation Place, Somewhere, OK (PC62)				5. Work Order/Contract/Invoice Number: WO 2468

6. Item:	7. Description:	8. Part Number:	9. Quantity:	10. Serial Number:	11. Status/Work:
1	Engine	PW-00001	1	AB100001	See Block 12

12. Remarks:

Export Airworthiness Approval – This product meets the special requirements of [enter country]. Rebuilt to original PAH's specifications.

13a. Certifies the items identified above were manufactured in conformity to: ☒ Approved design data and are in a condition for safe operation. ☐ Non-approved design data specified in Block 12.		14a. ☐ 14 CFR 43.9 Return to Service ☐ Other regulation specified in Block 12. Certifies that unless otherwise specified in Block 12, the work identified in Block 11 and described in Block 12 was accomplished in accordance with Title 14, Code of Federal Regulations, part 43 and in respect to that work, the items are approved for return to service.	
13b. Authorized Signature: *A. Inspector*	13c. Approval/Authorization No.: PC #62	14b. Authorized Signature:	14c. Approval/Certificate No.:
13d. Name (Typed or Printed): A. Inspector	13e. Date (dd/mmm/yyyy): 30 Aug 2011	14d. Name (Typed or Printed):	14e. Date (dd/mmm/yyyy):

User/Installer Responsibilities

It is important to understand that the existence of this document alone does not automatically constitute authority to install the aircraft engine/propeller/article.

Where the user/installer performs work in accordance with the national regulations of an airworthiness authority different than the airworthiness authority of the country specified in Block 1, it is essential that the user/installer ensures that his/her airworthiness authority accepts aircraft engine(s)/propeller(s)/article(s) from the airworthiness authority of the country specified in Block 1.

Statements in Blocks 13a and 14a do not constitute installation certification. In all cases, aircraft maintenance records must contain an installation certification issued in accordance with the national regulations by the user/installer before the aircraft may be flown.

FAA Form 8130-3 (02–14) NSN: 0052-00-012-9005

A-12

Figure A-13. Sample FAA Form 8130-3 for Dual Release Approval for Return To Service

1. Approving Civil Aviation Authority/Country: FAA/United States	2. AUTHORIZED RELEASE CERTIFICATE FAA Form 8130-3, AIRWORTHINESS APPROVAL TAG				3. Form Tracking Number: 2004-1009
4. Organization Name and Address: Anyone's Repair Station, 1104 Wing Avenue, Anyplace, TX 22212 (OC2R025L)					5. Work Order/Contract/Invoice Number: W 13884
6. Item:	7. Description:	8. Part Number:	9. Quantity:	10. Serial Number:	11. Status/Work:
001	Antenna	12342	1	AN-223-H	OVERHAULED

12. Remarks:

Overhauled in accordance with CMM 12342, section 2A3B, revision 23, S/B and FAA AD XYZ-2001 complied with. Full details of work carried out per work order no. W 13884.

Certifies that the work specified in Block 11/12 was carried out in accordance with EASA Part 145 and in respect to that work the [product/article] is considered ready for release to service under EASA Part 145 approval no. [insert number: EASA 145-XXX].

13a. Certifies the items identified above were manufactured in conformity to: ☐ Approved design data and are in a condition for safe operation. ☐ Non-approved design data specified in Block 12.		14a. ☒ 14 CFR 43.9 Return to Service ☒ Other regulation specified in Block 12 Certifies that unless otherwise specified in Block 12, the work identified in Block 11 and described in Block 12 was accomplished in accordance with Title 14, Code of Federal Regulations, part 43 and in respect to that work, the items are approved for return to service.	
13b. Authorized Signature:	13c. Approval/Authorization No.:	14b. Authorized Signature: A. Inspector	14c. Approval/Certificate No.: OC2R025L
13d. Name (Typed or Printed):	13e. Date (dd/mmm/yyyy):	14d. Name (Typed or Printed): A. Inspector	14e. Date (dd/mmm/yyyy): 13 Oct 2005

User/Installer Responsibilities

It is important to understand that the existence of this document alone does not automatically constitute authority to install the aircraft engine/propeller/article.

Where the user/installer performs work in accordance with the national regulations of an airworthiness authority different than the airworthiness authority of the country specified in Block 1, it is essential that the user/installer ensures that his/her airworthiness authority accepts aircraft engine(s)/propeller(s)/article(s) from the airworthiness authority of the country specified in Block 1.

Statements in Blocks 13a and 14a do not constitute installation certification. In all cases, aircraft maintenance records must contain an installation certification issued in accordance with the national regulations by the user/installer before the aircraft may be flown.

FAA Form 8130-3 (02-14) NSN: 0052-00-012-9005

08/01/2013 8130.21H
Appendix A

Figure A-14. Sample FAA Form 8130-3 for Export Airworthiness Approval

1. Approving Civil Aviation Authority/Country: FAA/United States	2. **AUTHORIZED RELEASE CERTIFICATE** FAA Form 8130-3, AIRWORTHINESS APPROVAL TAG				3. Form Tracking Number: BE5432987
4. Organization Name and Address: Dave's Aircraft Parts Manufacturing, 2010 Falcon Way, Somewhere, OK (PQ5410SW)					5. Work Order/Contract/Invoice Number: WO 2185
6. Item:	7. Description:	8. Part Number:	9. Quantity:	10. Serial Number:	11. Status/Work:
1	Flap	C 54321	1	9876543	NEW

12. Remarks:

Export Airworthiness Approval – This article meets the special requirements of (enter country).

13a. Certifies the items identified above were manufactured in conformity to: ☒ Approved design data and are in a condition for safe operation. ☐ Non-approved design data specified in Block 12.		14a. ☐ 14 CFR 43.9 Return to Service ☐ Other regulation specified in Block 12. Certifies that unless otherwise specified in Block 12, the work identified in Block 11 and described in Block 12 was accomplished in accordance with Title 14, Code of Federal Regulations, part 43 and in respect to that work, the items are approved for return to service.	
13b. Authorized Signature: *A. Inspector*	13c. Approval/Authorization No.: DMIR-000011-SW	14b. Authorized Signature:	14c. Approval/Certificate No.:
13d. Name (Typed or Printed): A. Inspector	13e. Date (dd/mmm/yyyy): 23 Oct 2007	14d. Name (Typed or Printed):	14e. Date (dd/mmm/yyyy):

User/Installer Responsibilities

It is important to understand that the existence of this document alone does not automatically constitute authority to install the aircraft engine/propeller/article.

Where the user/installer performs work in accordance with the national regulations of an airworthiness authority different than the airworthiness authority of the country specified in Block 1, it is essential that the user/installer ensures that his/her airworthiness authority accepts aircraft engine(s)/propeller(s)/article(s) from the airworthiness authority of the country specified in Block 1.

Statements in Blocks 13a and 14a do not constitute installation certification. In all cases, aircraft maintenance records must contain an installation certification issued in accordance with the national regulations by the user/installer before the aircraft may be flown.

FAA Form 8130-3 (02-14) NSN: 0052-00-012-9005

A-14

08/01/2013

8130.21H
Appendix A

Figure A-15. Sample FAA Form 8130-3 for Export Airworthiness Approval for a New Subcomponent for a TSO Authorization Article

1. Approving Civil Aviation Authority/Country: FAA/United States	2. **AUTHORIZED RELEASE CERTIFICATE** FAA Form 8130-3, AIRWORTHINESS APPROVAL TAG				3. Form Tracking Number: ACE 2345
4. Organization Name and Address: Ace Instrument Company, 1224 Wiley Post Drive, Oklahoma City, OK (PT0906SW)					5. Work Order/Contract/Invoice Number: WO 2020
6. Item:	7. Description:	8. Part Number:	9. Quantity:	10. Serial Number:	11. Status/Work:
1	Gimbal Ring	RI 4586	1	N/A	NEW
12. Remarks: Export airworthiness approval – This article meets the special requirements of (enter country). This [Gimbal Ring] is a subcomponent of a TSO authorization.					
13a. Certifies the items identified above were manufactured in conformity to: ☒ Approved design data and are in a condition for safe operation. ☐ Non-approved design data specified in Block 12.			14a. ☐ 14 CFR 43.9 Return to Service ☐ Other regulation specified in Block 12. Certifies that unless otherwise specified in Block 12, the work identified in Block 11 and described in Block 12 was accomplished in accordance with Title 14, Code of Federal Regulations, part 43 and in respect to that work, the items are approved for return to service.		
13b. Authorized Signature: *A. Inspector*		13c. Approval/Authorization No.: DMIR-003333-SW	14b. Authorized Signature:		14c. Approval/Certificate No.:
13d. Name (Typed or Printed): A. Inspector		13e. Date (dd/mmm/yyyy): 25 Oct 2007	14d. Name (Typed or Printed):		14e. Date (dd/mmm/yyyy):

User/Installer Responsibilities

It is important to understand that the existence of this document alone does not automatically constitute authority to install the aircraft engine/propeller/article.

Where the user/installer performs work in accordance with the national regulations of an airworthiness authority different than the airworthiness authority of the country specified in Block 1, it is essential that the user/installer ensures that his/her airworthiness authority accepts aircraft engine(s)/propeller(s)/article(s) from the airworthiness authority of the country specified in Block 1.

Statements in Blocks 13a and 14a do not constitute installation certification. In all cases, aircraft maintenance records must contain an installation certification issued in accordance with the national regulations by the user/installer before the aircraft may be flown.

FAA Form 8130-3 (02-14) NSN: 0052-00-012-9005

A-15

FAA Order 8130.21H

08/01/2013

8130.21H
Appendix A

Figure A-16. Sample FAA Form 8130-3 for a Direct Shipment Authorization for Export

1. Approving Civil Aviation Authority/Country: FAA/United States	2. **AUTHORIZED RELEASE CERTIFICATE** FAA Form 8130-3, AIRWORTHINESS APPROVAL TAG				3. Form Tracking Number: 991004327
4. Organization Name and Address: Original Parts Manufacturing Corporation, 6210 Wing Avenue, Anyplace, AL (PQ02269CE) Everybody's Aircraft Supply Co., 810 Red Baron Way, Hooterville, OK 74032					5. Work Order/Contract/Invoice Number: WO 2020
6. Item:	7. Description:	8. Part Number:	9. Quantity:	10. Serial Number:	11. Status/Work:
1	Wing Tip	AE637781-1	5 ea.	N/A	NEW

12. Remarks:

Export airworthiness approval – This article meets the special requirements of (enter country).

Direct shipment authorization

13a. Certifies the items identified above were manufactured in conformity to: ☒ Approved design data and are in a condition for safe operation. ☐ Non-approved design data specified in Block 12.		14a. ☐ 14 CFR 43.9 Return to Service ☐ Other regulation specified in Block 12 Certifies that unless otherwise specified in Block 12, the work identified in Block 11 and described in Block 12 was accomplished in accordance with Title 14, Code of Federal Regulations, part 43 and in respect to that work, the items are approved for return to service.	
13b. Authorized Signature: *A. Inspector*	13c. Approval/Authorization No.: DMIR-00243-CE	14b. Authorized Signature:	14c. Approval/Certificate No.:
13d. Name (Typed or Printed): A. Inspector	13e. Date (dd/mmm/yyyy): 13 Apr 2008	14d. Name (Typed or Printed):	14e. Date (dd/mmm/yyyy):

User/Installer Responsibilities

It is important to understand that the existence of this document alone does not automatically constitute authority to install the aircraft engine/propeller/article.

Where the user/installer performs work in accordance with the national regulations of an airworthiness authority different than the airworthiness authority of the country specified in Block 1, it is essential that the user/installer ensures that his/her airworthiness authority accepts aircraft engine(s)/propeller(s)/article(s) from the airworthiness authority of the country specified in Block 1.

Statements in Blocks 13a and 14a do not constitute installation certification. In all cases, aircraft maintenance records must contain an installation certification issued in accordance with the national regulations by the user/installer before the aircraft may be flown.

FAA Form 8130-3 (02-14) NSN: 0052-00-012-9005

A-16

08/01/2013

8130.21H
Appendix A

Figure A-17. Sample FAA Form 8130-3 for an Electronic Export Airworthiness Approval

1. Approving Civil Aviation Authority/Country: FAA/United States	2. AUTHORIZED RELEASE CERTIFICATE FAA Form 8130-3, AIRWORTHINESS APPROVAL TAG				3. Form Tracking Number: 04040608
4. Organization Name and Address: Aircraft Manufacturing Co., 106 Shady Pines Drive, Any Town, CA 34567, United States; **Supplier Code:** 63321; **Production Certificate Number:** PC 777					5. Work Order/Contract/Invoice Number: **Customer Order No.:** TS4567 **Customer ID Code:** XYZ
6. Item:	7. Description:	8. Part Number:	9. Quantity:	10. Serial Number:	11. Status/Work:
1	Bearing	16-44784-1	100 ea.	N/A	NEW
12. Remarks: **Manufacturer Code:** 73489 EXPORT: United Kingdom					
13a. Certifies the items identified above were manufactured in conformity to: ☒ Approved design data and are in a condition for safe operation. ☐ Non-approved design data specified in Block 12.			14a. ☐ 14 CFR 43.9 Return to Service ☐ Other regulation specified in Block 12. Certifies that unless otherwise specified in Block 12, the work identified in Block 11 and described in Block 12 was accomplished in accordance with Title 14, Code of Federal Regulations, part 43 and in respect to that work, the items are approved for return to service.		
13b. Authorized Signature: Digital signature on file.		13c. Approval/Authorization No.: DARF-54123-SW	14b. Authorized Signature:		14c. Approval/Certificate No.:
13d. Name (Typed or Printed): A. Inspector		13e. Date (dd/mmm/yyyy): 19 Dec 2007	14d. Name (Typed or Printed):		14e. Date (dd/mmm/yyyy):

User/Installer Responsibilities

It is important to understand that the existence of this document alone does not automatically constitute authority to install the aircraft engine/propeller/article.

Where the user/installer performs work in accordance with the national regulations of an airworthiness authority different than the airworthiness authority of the country specified in Block 1, it is essential that the user/installer ensures that his/her airworthiness authority accepts aircraft engine(s)/propeller(s)/article(s) from the airworthiness authority of the country specified in Block 1.

Statements in Blocks 13a and 14a do not constitute installation certification. In all cases, aircraft maintenance records must contain an installation certification issued in accordance with the national regulations by the user/installer before the aircraft may be flown.

FAA Form 8130-3 (02-14) NSN: 0052-00-012-9005

A-17

FAA Order 8130.21H

08/01/2013

8130.21H
Appendix A

Figure A-18. Sample FAA Form 8130-3 for an Electronic Conformity Airworthiness Approval

1. Approving Civil Aviation Authority/Country: FAA/United States	2. AUTHORIZED RELEASE CERTIFICATE FAA Form 8130-3, AIRWORTHINESS APPROVAL TAG				3. Form Tracking Number: 08-3456-LT-NMO-01
4. Organization Name and Address: Aircraft Manufacturing Co., 106 Shady Pines Drive, Any Town, CA 34567, United States **Supplier Code:** 73489 **Production Certificate Number:** PC 777			Exinol, Inc. Bldg. 5B, 1 Exinol Way Small Town, OK 74747, United States **Supplier Code:** 16754		5. Work Order/Contract/Invoice Number: **Work Order No.:** 150374 **Customer ID Code:** 73489
6. Item:	7. Description:	8. Part Number:	9. Quantity:	10. Serial Number:	11. Status/Work:
1	Right ASG section 2 & amp; 3 feeder assy, container, CW640	XX1675-536	1 ea.	N/A	PROTOTYPE

12. Remarks:

Manufacturer Code: 16754
CONFORMITY
DRO Log No.: 2006-0436D
Drawing and Revision Level: XX1675-536, Rev. A, 02 Oct 2006, MFR: 37952
Conformity Project Number: CT8196ET-S
Airworthiness Deviation Text: NONE

Prototype product(s)/article(s) pending certification under FAA project number [enter number] not eligible for installation on in-service, type-certificated aircraft. Upon approval of the design data the product(s)/article(s) listed above are considered NEW and conform to approved design data and are in a condition for safe operation without further showing.

13a. Certifies the items identified above were manufactured in conformity to: ☐ Approved design data and are in a condition for safe operation. ☒ Non-approved design data specified in Block 12.		14a. ☐ 14 CFR 43.9 Return to Service ☐ Other regulation specified in Block 12. Certifies that unless otherwise specified in Block 12, the work identified in Block 11 and described in Block 12 was accomplished in accordance with Title 14, Code of Federal Regulations, part 43 and in respect to that work, the items are approved for return to service.	
13b. Authorized Signature: Digital signature on file.	13c. Approval/Authorization No.: DMIR-54123-SW	14b. Authorized Signature:	14c. Approval/Certificate No.:
13d. Name (Typed or Printed): Alfred R. Gibson	13e. Date (dd/mmm/yyyy): 19 Dec 2007	14d. Name (Typed or Printed):	14e. Date (dd/mmm/yyyy):

User/Installer Responsibilities

It is important to understand that the existence of this document alone does not automatically constitute authority to install the aircraft engine/propeller/article.

Where the user/installer performs work in accordance with the national regulations of an airworthiness authority different than the airworthiness authority of the country specified in Block 1, it is essential that the user/installer ensures that his/her airworthiness authority accepts aircraft engine(s)/propeller(s)/article(s) from the airworthiness authority of the country specified in Block 1.

Statements in Blocks 13a and 14a do not constitute installation certification. In all cases, aircraft maintenance records must contain an installation certification issued in accordance with the national regulations by the user/installer before the aircraft may be flown.

FAA Form 8130-3 (02-14) NSN: 0052-00-012-9005

A-18

Figure A-19. Sample FAA Form 8130-3 for an Electronic Approval for Return To Service

1. Approving Civil Aviation Authority/Country:	2.				3. Form Tracking Number:
FAA/United States	**AUTHORIZED RELEASE CERTIFICATE** FAA Form 8130-3, AIRWORTHINESS APPROVAL TAG				99999
4. Organization Name and Address: American Repair Services, 3434 Harbor Drive, Suite 12B, Dallas, TX 76645, United States **Supplier Code:** 12345 **Certificate Number:** RX333K					5. Work Order/Contract/Invoice Number: **Work Order No.:** RO16754 **Customer ID Code:** 53111
6. Item:	7. Description:	8. Part Number:	9. Quantity:	10. Serial Number:	11. Status/Work:
1	Air Motor	C48401-302	1 ea.	64654564	OVERHAULED

12. Remarks:

Manufacturer Code: 84848

Unit was disassembled, cleaned, and inspected. Articles were reworked/replaced as required to return unit to a serviceable condition.

Repair Manual 55-11-22, Rev. 1, 22 Mar 2001

Notice: An airworthiness directive may apply to the unit described herein. The installed is responsible for ensuring complete compliance with any applicable airworthiness directives.

13a. Certifies the items identified above were manufactured in conformity to: ☐ Approved design data and are in a condition for safe operation. ☐ Non-approved design data specified in Block 12.		14a. ☒ 14 CFR 43.9 Return to Service ☐ Other regulation specified in Block 12 Certifies that unless otherwise specified in Block 12, the work identified in Block 11 and described in Block 12 was accomplished in accordance with Title 14, Code of Federal Regulations, part 43 and in respect to that work, the items are approved for return to service.	
13b. Authorized Signature:	13c. Approval/Authorization No.:	14b. Authorized Signature: Digital signature on file	14c. Approval/Certificate No.: S3RX333K
13d. Name (Typed or Printed):	13e. Date (dd/mmm/yyyy):	14d. Name (Typed or Printed): Mary Jones	14e. Date (dd/mmm/yyyy): 10 Jan 2008

User/Installer Responsibilities

It is important to understand that the existence of this document alone does not automatically constitute authority to install the aircraft engine/propeller/article.

Where the user/installer performs work in accordance with the national regulations of an airworthiness authority different than the airworthiness authority of the country specified in Block 1, it is essential that the user/installer ensures that his/her airworthiness authority accepts aircraft engine(s)/propeller(s)/article(s) from the airworthiness authority of the country specified in Block 1.

Statements in Blocks 13a and 14a do not constitute installation certification. In all cases, aircraft maintenance records must contain an installation certification issued in accordance with the national regulations by the user/installer before the aircraft may be flown.

FAA Form 8130-3 (02-14) NSN: 0052-00-012-9005

Appendix B. Acronyms

The following acronyms are used in this order:

14 CFR	Title 14 of the Code of Federal Regulations
AA	Aviation Authority
AC	advisory circular
ACO	aircraft certification office
AD	airworthiness directive
AN	Air Force-Navy Aeronautical Standard
ARC	Authorized Release Certificate
ASI	aviation safety inspector
ATA	Air Transport Association of America, Inc.
CAA	civil aviation authority
COA	Certificate of Authority
DAR	designated airworthiness representative
DER	designated engineering representative
DMIR	designated manufacturing inspection representative
EASA	European Aviation Safety Agency
FAA	Federal Aviation Administration
MAG	Maintenance Annex Guidance
MIP	maintenance implementation procedures
MS	Military Standard
NAS	National Aerospace Standards
ODA	organization designation authorization
PAH	production approval holder
PC	production certificate
PMA	Parts Manufacturer Approval
SAE	Society of Automotive Engineers
STC	supplemental type certificate
TC	type certificate
TCCA	Transport Canada Civil Aviation
TIP	Technical Implementation Procedures
TSO	technical standard order
TSOA	technical standard order authorization
XML	extensible markup language

Appendix C. Definitions

1. Applicable Standard. A manufacturing/design/maintenance/quality standard, method, technique, or practice approved by or acceptable to a civil aviation authority (CAA).

2. Approved Design Data. Applicable design data that has been granted an approval (for example, type certificate, supplemental type certificate, technical standard order authorization, parts manufacturer approval, or equivalent) by the relevant CAA.

> **Note:** For the purposes of this definition, the European Aviation Safety Agency is considered to be a CAA.

3. Article. A material, part, component, process, or appliance.

4. Authentication. The means by which a system validates the identity of an authorized user. This may include a password, personal identification number, cryptographic key, badge, stamp, or combination thereof, or any other method of identifying an authorized user.

5. Commercial Part. An article that is listed on a Federal Aviation Administration (FAA)-approved commercial parts list included in a design approval holder's instructions for continued airworthiness.

6. Computer Hardware. A computer and the associated physical equipment directly involved in the performance of communications or data processing functions.

7. Computer Software. Written, printed, or other technologically accepted media such as programs, routines, and symbolic languages used in the operation of computers.

8. Deliverable Software. Computer software with a part number that meets FAA standards for software design and use.

9. Digital Certificate. A digitally signed statement that binds the identifying information of a user, computer, or service to a public/private key pair.

10. Digital Signature. A secure digital means of conveying the same meaning as an individual's handwritten signature in an electronic document, which when printed may or may not contain an exact copy of the originating handwritten signature.

11. Direct Shipment Authorization. The written authorization granted by a production approval holder (PAH) with responsibility for the airworthiness of a product or article, to a supplier, to ship articles produced in accordance with the PAH's quality/inspection system directly to end users without the articles being processed through the PAH's own facility.

12. Distributor. Any person engaged in the sale or transfer of products and articles for installation in type-certificated aircraft, aircraft engines, or propellers, and that conducts no manufacturing activities.

13. Electronic Recordkeeping System or Manual. A system of record processing in which records or manuals are entered, stored, and retrieved electronically by a computer system.

14. Electronic Signature. An exact copy of a handwritten signature that is securely produced by electronic means.

15. End User. For the purpose of this order, means the person taking possession of the product or article.

16. Export. When a product or article is found to be airworthy, meets the special conditions of the importing country or jurisdiction, and is transferred from one CAA's regulatory authority to another CAA's regulatory authority.

17. Installation Eligibility. Acceptability of an article for installation on type-certificated product(s) based on airworthiness data and the configuration of the product.

18. Originator. The issuer (FAA-authorized individuals or organization designation authorization holder) who began the approval process signing FAA Form 8130-3, Airworthiness Approval Tag.

19. Password. An identification code or device required to access stored material, intended to prevent information from being viewed, edited, or printed by unauthorized persons.

20. Product. Per Title 14 of the Code of Federal Regulations 21.1, "product" is defined as an aircraft, aircraft engine, or propeller. However, for the purposes of this order, "product" refers only to aircraft engines and propellers.

21. Public Key Infrastructure. The method for verifying the validity and status of the digital certificate of a message sender.

22. Quality System. A documented organizational structure containing responsibilities, procedures, processes, and resources that implement a management function to determine and enforce quality principles.

23. Receiver of the Electronic FAA Form. The entity who receives the form electronically from the originator. If the entity that receives the form electronically isn't the end user, the receiver of the electronic form must either transfer the form electronically (in accordance with chapter 5 of this order) or issue a paper FAA Form 8130-3 in accordance with the appropriate chapter contained in this order.

24. Record. Information inscribed on a tangible medium or stored in an electronic or other medium that is retrievable in perceivable form.

25. Recurrent Airworthiness Approval. Issuance of FAA Form 8130-3 for products or articles based on a prior finding by a PAH that the product or article was airworthy, and a current finding that the product or article remains airworthy.

26. Signature. Any form of identification used to acknowledge completion of an act and authenticate a record entry. A signature must be traceable to the individual making the entry, and it must be handwritten or be part of an electronic signature system or other form acceptable to the FAA.

27. Standard Part. A part manufactured in complete compliance with an established government or industry-accepted specification that contains design, manufacturing, and uniform identification requirements. The specification must include all information necessary to produce and conform the part, and must be published so that any person/organization may manufacture the part.

> **Note:** Examples of specifications include, but are not limited to, National Aerospace Standards (NAS), Air Force-Navy Aeronautical Standard (AN), Society of Automotive Engineers (SAE), SAE Aerospace Standard (AS), and Military Standard (MS).

28. Trading Partners. A person transmitting FAA Form 8130-3 and a person capable of receiving FAA Form 8130-3 in the form of paper or electronic data.

01/11/2016

8130.21H CHG 1
Appendix D

Appendix D. Administrative Information

1. Distribution. This order is distributed to the Washington Headquarters division levels of the Aircraft Certification Service and Flight Standards Service; to the branch levels of the Aircraft Certification Service; to the branch levels in the regional Flight Standards Divisions and Aircraft Certification Directorates; to all Flight Standards District Offices and International Field Offices; to all Aircraft Certification Offices; to all Certificate Management Offices and all Manufacturing Inspection District and Satellite Offices; and to the Aircraft Certification and Airworthiness Branches at the FAA Academy.

2. Deviations. Adherence to the procedures in this order is necessary for uniform administration of this directive material. Any deviations from this guidance material must be coordinated and approved by the Design, Manufacturing, and Airworthiness Division, AIR-100. If a deviation becomes necessary, the FAA employee involved should ensure the deviations are substantiated, documented, and concurred with by the appropriate supervisor. The deviation must be submitted to AIR-100 for review and approval. The limits of Federal protection for FAA employees are defined in § 2679 of Title 28 of the United States Code.

3. Suggestions for Improvements. Please forward all comments on deficiencies, clarifications, or improvements regarding this order to:

Aircraft Certification Service
Planning and Program Management Division, AIR-510
ATTN: Directives Management Officer
800 Independence Avenue SW.
Washington, DC 20591

FAA Form 1320-19, Directive Feedback Information, is located in appendix E to this order for your convenience. FAA employees may also use the automated directive feedback system (DFS) to submit comments at http://avsdfs.avs.faa.gov/. If you require an immediate interpretation, please contact your local FAA managing office. If further interpretation is required, that FAA managing office will contact AIR-100 at (202) 267-1575. However, you also should complete FAA Form 1320-19 as a followup to the conversation.

4. Records Management. Refer to FAA Order 0000.1, FAA Standard Subject Classification System; FAA Order 1350.14, Records Management; or your office Records Management Officer (RMO)/Directives Management Officer (DMO) for guidance regarding retention or disposition of records.

D-1

Appendix E. FAA Form 1320-19, Directive Feedback Information

U.S. Department
of Transportation

**Federal Aviation
Administration**

Directive Feedback Information

Please submit any written comments or recommendations for improving this directive, or suggest new items or subjects to be added to it. Also, if you find an error, please tell us about it.

Subject: Order 8130.21H

To: Directive Management Officer, 9-AWA-AVS-AIR-DMO@faa.gov

(Please check all appropriate line items)

An error (procedural or typographical) has been noted in paragraph __________ on page ________.

☐ Recommend paragraph _____________ on page _____________ be changed as follows:
(attach separate sheet if necessary)

☐ In a future change to this directive, please include coverage on the following subject *(briefly describe what you want added)*:☐ Other comments:

☐ I would like to discuss the above. Please contact me.

Submitted by: ___________________________________ Date: ___________________

Telephone Number: ___________________ Routing Symbol: _________________

FAA Form 1320-19 (10-98)

1. Approving Civil Aviation Authority/Country: FAA/United States	2. **AUTHORIZED RELEASE CERTIFICATE** FAA Form 8130–3, AIRWORTHINESS APPROVAL TAG				3. Form Tracking Number:
4. Organization Name and Address:					5. Work Order/Contract/Invoice Number:
6. Item:	7. Description:	8. Part Number:	9. Quantity:	10. Serial Number:	11. Status/Work:
12. Remarks:					

13a. Certifies the items identified above were manufactured in conformity to: ☐ Approved design data and are in a condition for safe operation. ☐ Non-approved design data specified in Block 12.		14a. ☐ 14 CFR 43.9 Return to Service ☐ Other regulation specified in Block 12 Certifies that unless otherwise specified in Block 12, the work identified in Block 11 and described in Block 12 was accomplished in accordance with Title 14, Code of Federal Regulations, part 43 and in respect to that work, the items are approved for return to service.	
13b. Authorized Signature:	13c. Approval/Authorization No.:	14b. Authorized Signature:	14c. Approval/Certificate No.:
13d. Name (Typed or Printed):	13e. Date (dd/mmm/yyyy):	14d. Name (Typed or Printed):	14e. Date (dd/mmm/yyyy):

User/Installer Responsibilities

It is important to understand that the existence of this document alone does not automatically constitute authority to install the aircraft engine/propeller/article.

Where the user/installer performs work in accordance with the national regulations of an airworthiness authority different than the airworthiness authority of the country specified in Block 1, it is essential that the user/installer ensures that his/her airworthiness authority accepts aircraft engine(s)/propeller(s)/article(s) from the airworthiness authority of the country specified in Block 1.

Statements in Blocks 13a and 14a do not constitute installation certification. In all cases, aircraft maintenance records must contain an installation certification issued in accordance with the national regulations by the user/installer before the aircraft may be flown.

FAA Form 8130–3 (02–14) NSN: 0052-00-012-9005

FAA Form 8130-3

Chapter 6
Advisory Circulars

The Advisory Circular (AC) System

The FAA issues advisory circulars to inform the aviation public of nonregulatory material of interest. An AC is issued to provide guidance and information in its designated subject area or to show a method acceptable to the Administrator for complying with a related Federal Aviation Regulation. Advisory circulars are issued in a number system that **corresponds to the subject areas** of 14 CFR. Therefore, the AC number usually reflects the section of the regulatory material it is seeking to explain. For example: AC 39-7 deals with 14 CFR Part 39, AC 43-9 refers to §43.9, and AC 91.67 to Part 91. The ACs covered in this chapter are those most applicable and useful to the IA, and which contain information that may appear in the FAA IA Knowledge Test questions.

Some Advisory Circulars have been replaced with:

Computerized Testing Supplements FAA-CT-8080-
Practical Test Standards FAA-S-8081-
Knowledge Test Guides FAA-G-8082-
FAA Aeronautical Handbooks FAA-H-8083-

AC 39-7D
Airworthiness Directives

This AC provides guidance and information to owners and operators of aircraft concerning their responsibility for complying with airworthiness directives (ADs) and recording AD compliance in the appropriate maintenance records.

176. An aircraft originally certificated in one of the standard categories has been altered and is now certificated in the restricted category. Airworthiness Directives (ADs) subsequently issued for the aircraft, if applicable,

A— must be complied with.
B— may be complied with as a matter of good practice.
C— must be complied with only if additionally specifying restricted category.

Unless specifically stated, ADs apply to the make and model set forth in the applicability statement, regardless of the classification or category of the airworthiness certificate issued for the aircraft.

177. Airworthiness Directives (ADs) may be issued to aircraft with which of the following categories?

A— Experimental and Primary.
B— Restricted and Standard.
C— Both A and B.

Unless specifically stated, ADs apply to the make and model set forth in the applicability statement, regardless of the classification or category of the airworthiness certificate issued for the aircraft.

178. The applicability statement of an airworthiness directive (AD) states "applies to Martin model 50 aircraft." This statement would cause the AD to apply to which of the following classification of aircraft?

A— Any Martin model 50 aircraft.
B— Only Martin model 50 aircraft with standard airworthiness certificates.
C— Any Martin model 50 aircraft, except those with experimental certificates.

Unless specifically stated, ADs apply to the make and model set forth in the applicability statement, regardless of the classification or category of the airworthiness certificate issued for the aircraft.

Answers

176 [A] (031) AC 39-7
177 [C] (031) AC 39-7
178 [A] (031) AC 39-7

179. Airworthiness Directives have amendment numbers. That number is considered an amendment to what?

A— The Type Certificate Data Sheet.
B— The Airworthiness Certificate.
C— The Federal Aviation Regulation Part 39.

Airworthiness Directives are Federal Aviation Regulations, and each numbered amendment is an amendment to 14 CFR Part 39.

180. 14 CFR Section 91.417(a)(2)(v), requires the revision dates of applicable airworthiness directives (AD) be made part of the aircraft records. How is the revision date determined? The revision date is

A— listed in the opening sentence of all ADs.
B— the effective date of the latest amendment.
C— noted in the AD number, such as, AD 90-05-01R2.

The revision date required by §91.417(a)(2)(v) is the effective date of the latest amendment to the AD and is found in the last sentence of the body of each AD.

181. Where do you find the revision date of an AD?

A— At the beginning of the AD.
B— In the last sentence of the body of the AD.
C— It will be located beside the Amendment Number of the AD.

The revision date required by §91.417(a)(2)(v) is the effective date of the latest amendment to the AD and is found in the last sentence of the body of each AD.

AC 43-4A
Corrosion Control for Aircraft

This AC is a summary of currently available data regarding identification and treatment of corrosive attack on aircraft structure and engine materials. It is available as a free "PDF" download from the FAA's Regulatory and Guidance Library on their website at this address: **http://www.airweb.faa.gov/rgl**

181a. What severity is corrosion that goes to a depth of 0.001 inch?

A— Severe.
B— Light.
C— Moderate.

Light corrosion is characterized by discoloration or pitting to a depth of approximately 0.001 inch maximum. Moderate corrosion has pitting depths as deep as 0.010 inch. Severe corrosion has pitting deeper than 0.010 inch.

Answers

179 [C] (031) AC 39-7
180 [B] (031) AC 39-7
181 [B] (031) AC 39-7
181a [B] (026) AC 43-4

AC 43-9C
Maintenance Records

This AC describes methods, procedures and practices that have been determined to be acceptable means of showing compliance with the general aviation maintenance record making and record keeping requirements of 14 CFR Parts 43 and 91. This material is not mandatory, nor is it regulatory and acknowledges that the FAA will consider other methods that may be presented. It is issued for guidance purposes and outlines several methods of compliance with the regulations.

Note: The information in AC 43-9 does not apply to air carrier maintenance records made and retained in accordance with 14 CFR Part 121.

182. Who is responsible for keeping and maintaining the maintenance records for an aircraft operated under 14 CFR Part 91?

A— The pilot who flies the aircraft.
B— The maintenance personnel who does the work.
C— The owner or operator of the aircraft.

14 CFR §91.417 states that an aircraft owner/operator shall keep and maintain aircraft maintenance records. 14 CFR §§43.9 and 43.11 state that maintenance personnel, however, are required to make the record entries.

183. How long should an FAA Form 337 for a major alteration to an airframe be retained?

A— For one year or until the work is repeated or superceded.
B— It becomes a part of the permanent maintenance record and it is transferred with the aircraft when it is sold.
C— For two years or until the work is repeated or superceded.

In the past the owner or operator has been permitted to maintain a list of current major alterations to the airframe, engine(s), propeller(s), rotor(s), or appliances. This procedure did not produce a record of value to the owner/operator or to maintenance persons in determining the continued airworthiness of the alteration since such a record was not in sufficient detail. This section of the rule has been changed. It now prescribes that copies of FAA Form 337, issued for the alteration, be made a part of the permanent maintenance record which is transferred with the aircraft when it is sold.

184. When a repair station makes a major repair, the repair may be approved for return to service on an FAA Form 337, customer's work order, or tag attached to the product. 14 CFR Part 43, Appendix B, requires a maintenance release statement be printed on which of these documents?

A— FAA Form 337.
B— Tag attached to the product.
C— Both A and B.

The maintenance release must contain the information specified in paragraph (b)(1), (2) and (3) of Appendix B of Part 43, be made a part of the aircraft maintenance record, and retained by the owner/operator as specified in §91.417. The maintenance release is usually a special document (normally a tag) and is attached to the product when it is approved for return to service. The maintenance release may, however, be on a copy of the work order written for the product. When this is done (for major repairs only) the entry on the work order must meet paragraph (b)(1), (2) and (3) of the appendix.

Answers

182 [C] (030) AC 43-9 183 [B] (030) AC 43-9 184 [B] (030) AC 43-9

AC 43.9-1F
Instructions for Completion of FAA Form 337

This AC provides instructions for completing FAA Form 337, *Major Repair and Alteration (Airframe, Powerplant, Propeller, or Appliance)*. (*See* Pages 6–89 and 6–90 of this *IA Test Prep* for an example of FAA Form 337.)

185. When completing FAA Form 337, item 1, information for make, model, and serial number should be taken from the

A— aircraft logbook.
B— aircraft manufacturer's identification plate.
C— aircraft Form 8050-3, Certificate of Aircraft Registration.

The information to complete the "Make," "Model," and "Serial Number" blocks will be found on the aircraft manufacturer's identification plate. The "Nationality and Registration Mark" is the same as shown on AC Form 8050-3 Certificate of Aircraft Registration.

186. Data used for approving major repairs must be FAA approved prior to its use for that purpose. Which of the following is approved data for major repairs?

A— Airworthiness Directive.
B— Previously approved FAA Form 337 for a like model aircraft.
C— Supplemental type certificate data.

Data used as a basis for approving major repairs or alterations for return to service must be FAA approved prior to its use for that purpose. This includes: 14 CFR (e.g. Airworthiness Directives), ACs (e.g. AC 43.13-1B under certain circumstances), TSOs, parts manufacturing approval (PMA), FAA-approved manufacturer's instructions, kits and service handbooks, Type Certificate Data Sheets, and Aircraft Specifications. Although STCs are approved data, they are always an alteration to the TCDS.

187. Major alteration data may come from various sources however, in each case the data must

A— be manufacturer-approved.
B— result from engineering authorization.
C— be traceable to an approved source.

Data used as a basis for approving major repairs or alterations for return to service must be FAA approved prior to its use for that purpose.

188. When considering major repairs, data presented in an advisory circular (AC) is

A— approved by the Administrator if so stated in the text of the AC.
B— approved by the Administrator.
C— acceptable to the Administrator.

Certain Advisory Circulars, such as AC 43.13-1B and 43.13-2A, are approved by the Administrator (under certain circumstances) and are so noted in the text of the AC.

189. As the holder of an inspection authorization, you have just completed an inspection of a major repair by an A&P mechanic on a small aircraft. The repair was found to be airworthy and eligible for return to service. The maintenance record contains a properly completed FAA Form 337. The signature of the A&P mechanic on the 337 constitutes

A— approval for return to service.
B— conformity to approved data.
C— airworthiness of the repair.

A signature, in block 6 of the FAA Form 337, certifies that the work was done in conformity with approved data.

190. A signature by an FAA airworthiness inspector in block 3 of FAA Form 337 indicates

A— approval of data.
B— approval for return to service.
C— conformity of the work performed.

The signature of an FAA inspector in block 3 signifies that the data to be used in performing this major alteration or major repair complies with accepted industry practices and all applicable parts of 14 CFR. This signature in Block 3 makes the 337 a "Field Approved" Form 337, and is necessary only when the data in Block 8 is ***not*** *previously approved data.*

Answers

185 [B] (030) AC 43.9-1
186 [A] (030) AC 43.9-1
187 [C] (030) AC 43.9-1
188 [A] (030) AC 43.9-1
189 [B] (021) AC 43.9-1
190 [A] (030) AC 43.9-1

191. The completion and signing of block 6 (conformity statement) of FAA Form 337 indicates

A— FAA approval of the work only.
B— agency approval for return to service.
C— the work was accomplished to FAA-approved data.

The completion and signing of block 6 certifies that the work was done in conformity with FAA-approved data.

192. The approval of non-previously approved data described in item 8 (Description of Work Accomplished) of FAA Form 337 is indicated by

A— marking of the approved box and signature in item 7 (Approval for Return to Service) of the FAA Form 337.
B— an approval statement and signature in item 3 (For FAA Use Only) of the FAA Form 337.
C— signature of Authorized Individual in item 6 (Conformity Statement) of the FAA Form 337.

If the data has not been previously approved, the applicant will make both copies of FAA Form 337 available to the local FAA district office. When the inspector determines that the major repair or major alteration data complies with applicable regulations and is in conformity with accepted industry practices, the inspector will record data approval by entering the appropriate statement in block 3 on both the original and duplicate copy and return both forms to the applicant. This is a "Field Approval" 337.

193. A statement and signature of an FAA Representative in item 3 (For FAA Use Only) of FAA Form 337 indicates

A— approval of data to accomplish the work.
B— approval for return to service.
C— conformity of work performed.

The signature of the FAA inspector in block 3 indicates that the data has been examined and found to comply with applicable regulations and is in conformity with accepted industry practices.

194. When submitting an FAA Form 337 to the FAA, using non-previously-approved data, which item on the form should be left blank?

A— Item 8...... "Description of Work Accomplished"
B— Item 7...... "Approval For Return to Service"
C— Item 6...... "Conformity Statement"

If the data has not been previously approved by the FAA, the Form 337 is completed with all the information regarding the repair or alteration, including Items 6 and 8, but Item 7 is left blank. Note: *Enter all data required in Item 6, including your name,* ***but do not sign Item 6 at this time****. The FAA inspector will examine the form and the data and if it is approved, he or she will note the approval in Item 3 "For FAA Use Only" and return the form to the submitter. After the work is completed, Item 6 is signed. Then the aircraft is inspected by an IA, Item 7 is signed, and the aircraft is returned to service.*

Answers

191 [C] (030) AC 43.9-1
192 [B] (030) AC 43.9-1
193 [A] (030) AC 43.9-1
194 [B] (030) AC 43.9-1

AC 43.13-1B
Acceptable Methods, Techniques and Practices— Aircraft Inspection and Repair

This AC contains methods, techniques, and practices acceptable to the Administrator for inspection and repair to civil aircraft, only when there is no manufacturer repair or maintenance instructions. This data generally pertains to minor repairs. It may be used as a basis for FAA data approval for major repairs. This data may be used as approved data when: (1) The user has determined that it is appropriate to the product being repaired; (2) directly applicable to the repair being made; and (3) not contrary to manufacturer's data.

195. What is the maximum permissible free play of an elevator "full-span" trim tab?

Given: Trim tab chord = 4.5 inches

A—.140
B—.090
C—.045

If the tab span equals or exceeds 35 percent of the span of the supporting control surface, the total free play at the tab trailing edge should not exceed 1 percent of the distance from the tab hinge line to the trailing edge of the tab perpendicular to the tab hinge line.

The tab span is 100% of the supporting control surface and its distance from the hinge line to the trailing edge is 4.5 inches. It is allowed to have a free play of 1% of 4.5 or 0.045 inch.

196. What is the maximum permissible free play of the elevator trim tab?

Given: Elevator span (length) = 5 ft.
Trim tab span (length) = 1 ft.
Trim tab chord = 3 inches

A—.040
B—.060
C—.020

If the tab span does not exceed 35 percent of the span of the supporting control surface, the total free play at the tab trailing edge should not exceed 2 percent of tab chord.

The tab span is 20% of the supporting control surface and its chord is 3 inches. It is allowed to have a free play of 2% of 3 or 0.060 inch.

197. Which of the following would be an excessive amount of free play of an elevator "full span" trim tab?

Given: Trim tab chord = 4.5 inches

A—.045
B—.060
C—.020

The tab span is 100% of the supporting control surface and its distance from the hinge line to the trailing edge is 4.5 inches. It is allowed to have a free play of 1% of 4.5 or 0.045 inch. Since .060 is larger than this maximum allowable, it is excessive.

198. If a piece of metal is bent at 135 degree open-angle bend, the metal would have a bend angle of

A—90°.
B—45°.
C—35°.

To form a 135° open-angle bend, the metal must be bent through a bend angle of 45°.

199. If a piece of metal is bent to 45 degrees closed-angle bend, the metal would have a bend angle of

A—90°.
B—135°.
C—155°.

To form a 45° closed-angle bend, the metal must be bent through a bend angle of 135°. This is 45° beyond 90°.

Answers

195 [C] (013) AC 43.13-1
196 [B] (013) AC 43.13-1
197 [B] (013) AC 43.13-1
198 [B] (006) AC 43.13-1
199 [B] (006) AC 43.13-1

200. How much metal is used in a 90° bend in 0.064-inch 2024-T aluminum alloy sheet using a 3/16-inch bend radius?

A— 0.250 inch.
B— 0.340 inch.
C— 0.312 inch.

The bend allowance for a 90° bend may be found by using this formula:

Bend allowance =

$[(0.01743 \cdot R) + (0.0078 \cdot T)] \cdot \text{degree of bend}$

$= (0.01743 \cdot 0.1875) + (0.0078 \cdot 0.064) \cdot 90$

$= (0.0033 + 0.000499) \cdot 90$

$= 0.0038 \cdot 90$

$= 0.342$ *inch*

201. What would be the flat layout dimension (before-bending dimension) of a U-shaped channel, 1" x 5" x 1", 2024-T6, .064 thick, using minimum bend radius?

A— 7.0
B— 6.28
C— 6.68

Using Table 4-6 on Page 4-14 of AC 43.13-1B, find the minimum bend radius for .064-inch 2024T6 to be 3t or 0.192 inch.

The bend allowance for a 90° bend may be found by using this formula:

$$\text{Bend Allowance} = \frac{2\pi\,(R + 1/2T)}{4} = \frac{6.28\,(.192 + .032)}{4} = 0.352 \text{ inch}$$

Setback for a 90° bend is the metal thickness + the bend radius = .064 + .192 = 0.256

The flat for each of the 1" sides is 1.00 – .256 = .744 × 2 = 1.488

The flat for the 5" base is 5.00 – 2(.256) = 5.00 – .512 = 4.488 inch

The two bend allowances is 2 × .352 = .704

The total width of the material for the channel is 1.488 + 4.488 + .704 = 6.68 inches.

202. Determine the location of the second sight line of a U-shaped channel, 1" × 5" × 1", 2024-T6, .064 thickness using minimum bend radius.

A— 6"
B— 5.45"
C— 5.77"

A sight line is a line drawn on a sheet metal layout that is inside the bend allowance and is one bend radius from the bend tangent line under the radius bar. The sight line is lined up directly below the nose of the radius bar on a cornice brake. When the metal is clamped in this position, the bend tangent line is in the correct position for the start of the bend.

The minimum bend radius found in Table 4-6 on Page 4-14 of AC 43.13-1B for .064-inch 2024T6 is 3t or 0.192 inch.

Measure from one end and make these lines on the metal:

First bend tangent line
0.744 inches

First sight line
(0.744 + 0.192): 0.936 inches

Second bend tangent line
(0.744 + 0.352): 1.096 inches

Third bend tangent line
(0.744 + 0.352 + 4.448: 5.584 inches

Second sight line
(0.744 + 0.352 + 4.488 + 0.192): 5.776 inches

Fourth bend tangent line
(0.744 + 0.352 + 4.488 + 0.352): 5.936 inches

Total width
(0.744 + 0.352 +4.488 +0.352 + 0.744):
6.680 inches

Note: *The use of a sight line is discussed in detail in Chapter 2 of the ASA* Airframe Volume 1: Structures, *in the Aviation Maintenance Technician Series.*

Answers

200 [B] (006) AC 43.13-1
201 [C] (006) AC 43.13-1
202 [C] (006) AC 43.13-1

203. How far from the mold line of a piece of 0.051-inch aluminum alloy should the jaws of the brake be set back if the bend is 120° and the bend radius is 3/16-inch (0.1875)?

A— 0.238 inch.
B— 1.732 inch.
C— 0.413 inch.

1. *Setback = K(BR + MT)*
2. *Find the value of K for 120° bend in Table 4-7 on Page 4-15 of AC 43.13-1B. K = 1.732*
3. *Add the bend radius (0.1875) plus the metal thickness (0.051). BR + MT = 0.2385*
4. *Multiply BR + MT by K. 1.732 · 0.2385 = 0.413*

The jaw of the brake must be set back 0.413 inch from the mold line.

203a. As an IA holder you are asked to inspect a major repair made with approved data. The repair required the mechanic to fabricate a sheet metal bracket made of 2024-T3 aluminum alloy, .032 inch in thickness. The bracket was bent using the lowest-value bend radii to a closed angle of 37°. If the bracket was constructed correctly and the metal in the bend was not over stressed, how much setback should the mechanic have used in the brake?

A— .032 inch.
B— .096 inch.
C— .287 inch.

Table 4-6 in AC 43.13-1B indicates the bend radii for a 90° bend is 2t – 4t. The question states that the bend was made using the lowest value radii. Since the thickness is .032, a 2t dimension would be the minimal bend: 2 × .032 = .064; this would be the setback for a 90° bend. For a 143° bend (180 – 37 = 143), refer to Table 4-7 to get the K factor of 2.988. Multiplying this K-factor by the .064 result above gives a setback of 0.287.

204. What is the diameter of an AN470AD4-4 rivet?

A— 4/16 (1/4)
B— 4/32 (1/8)
C— 4/8 (1/2)

This is the meaning of the rivet identification numbers:

AN470AD4-4	*Complete part number*
AN	*Air Force–Navy standard number*
MS	*Military standard number*
470 or 20470	*Universal head rivet*
4	*Diameter in 1/32-inch increments 4/32 (1/8) inch*
4	*Length in 1/16-inch increments 4/16 (1/4) inch*

205. A sheet metal repair is to be made using two pieces of .040 inch aluminum riveted together. All rivet holes are drilled for 3/32 inch rivets. The length of the rivets to be used will be

A— 1/8 inch.
B— 1/4 inch.
C— 5/16 inch.

To form a proper shop head, the rivet shank should extend for 1-1/2 diameters from the sheet. For a 3/32 (0.0938) rivet, this would be 0.140. Two thicknesses of the metal would be 0.080 inch. The length of the rivet would have to be 0.080 + 0.140 = 0.220. The nearest standard rivet to this is 1/4 inch or 0.250.

206. A sheet metal repair is to be made using two pieces of .0625 inch aluminum riveted together. All rivet holes are drilled for 1/8 inch rivets. The length of the rivets to be used will be

A— 5/32 inch.
B— 3/16 inch.
C— 5/16 inch.

To form a proper shop head, the rivet shank should extend for 1-1/2 diameters from the sheet. For a 1/8 (0.125) rivet, this would be 0.1875. Two thicknesses of the metal would be 0.125 inch. The length of the rivet would have to be 0.1875 + 0.125 = 0.3125 or 5/16 inch.

Answers

203 [C] (006) AC 43.13-1
205 [B] (006) AC 43.13-1
203a [C] (014) AC 43.13-1
206 [C] (006) AC 43.13-1
204 [B] (032) AC 43.13-1

207. What rivet should be selected to join two sheets of .032 and .020 aluminum?

A— AN470AD 4-3
B— AN470AD 4-4
C— AN470AD 5-3

The diameter of the rivet should be preferably about three times the thickness of the thicker sheet. 3 × .032 = .096 inch. This is slightly larger than 3/32 inch. A 1/8-inch rivet (-4) rivet would be chosen. It should protrude 1-1/2 diameters from the metal so its length would be 0.052 + 0.1875 = 0.239 inch. The nearest rivet to this would be 1/4 inch. The proper rivet is an AN470AD 4-4.

208. What rivets should be selected to join two sheets of .032 aluminum?

A— MS20425D 4-3
B— MS20470AD 4-4
C— MS20455DD 5-3

The diameter of the rivet should be preferably about three times the thickness of the thicker sheet. 3 × .032 = .096 inch. This is slightly larger than 3/32 inch. A 1/8-inch rivet (-4) rivet would be chosen. It should protrude 1-1/2 diameters from the metal so its length would be 0.064 + 0.1875 = 0.2515 inch. The nearest rivet to this would be 1/4 inch. The proper rivet is an MS20470AD 4-4.

208a. How is a rivet made from 2024 material visually identified?

A— "DD" material code in the part number.
B— Raised dot on the rivet head.
C— Two raised dashes on the rivet head.

Refer to AC 43.13-1B Figure 4-4 (this figure will be available during the test as Exhibit I in Section III of the Computer Testing Supplement). This figure shows the Head Marking of this rivet is two raised dashes. Although "DD" is the material code for this rivet, it is not visually identifiable as is the marking on the rivet head.

208b. How is the length of a universal rivet head measured?

A— From the top of the head to the end of the rivet stem.
B— From the underside of the head to the end of the rivet stem.
C— In 32nds of an inch.

RIVET IDENTIFICATION

The material can be identified by the head marking

Rivet	Material Code	Head Marking	Material
	A	PLAIN (Dyed)	1100
	AD	DIMPLED	2117
	D	RAISED DOT	2017T
	DD	TWO RAISED DASHES	2024
	B	RAISED CROSS (Dyed)	5056
	E	RAISED CIRCLE	7050
	M	TWO DOTS	Monel

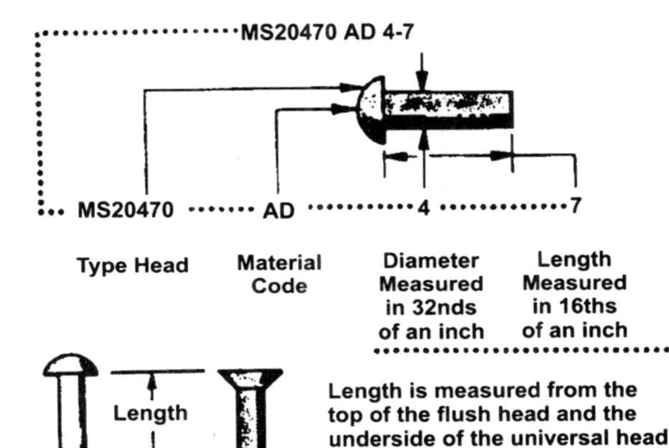

Figure 4-4. Rivet identification and part number breakdown.

Refer to AC 43.13-1B Figure 4-4 (this figure will be available during the test as Exhibit I in Section III of the Computer Testing Supplement). The length of a rivet is measured from the top of a flush head rivet and from the underside of the head of a universal rivet.

208c. A rivet identified as an MS20470AD5-8 would be

A— 5/16ths in diameter.
B— 8/32nds in length.
C— 5/32nds in diameter.

Refer to AC 43.13-1B Figure 4-4 (this figure will be available during the test as Exhibit I in Section III of the Computer Testing Supplement). The first digit after the material code (AD in this example) indicates the rivet diameter is measured in 32nds of an inch.

Answers

207 [B] (032) AC 43.13-1
208 [B] (032) AC 43.13-1
208a [C] (032) AC 43.13-1
208b [B] (032) AC 43.13-1
208c [C] (032) AC 43.13-1

208d. What is the shear strength of a 2117T rivet?

A— 10,000 psi.
B— 30,000 psi.
C— 100,000 psi.

Refer to AC 43.13-1B Table 4-8 (this figure will be available during the test as Exhibit II in Section III of the Computer Testing Supplement). In this table, the various rivet materials are arranged in vertical columns, and the shear strength and bearing strength are in horizontal rows at the bottom. The table entry at the intersection of shear strength and a 2117T rivet is 30,000 psi.

208e. Which of the following rivets must be heat treated before using it?

A— 2117T.
B— 2017T.
C— 2017T-HD.

Refer to AC 43.13-1B Table 4-8 (this figure will be available during the test as Exhibit II in Section III of the Computer Testing Supplement). In this table, the various rivet materials are arranged in vertical columns, and the "Heat Treat Before Using" requirement is shown in a horizontal row at the bottom. The table entry at the intersection of "Heat Treat Before Using" and a 2017T rivet is "Yes."

208f. Which of the following rivets is a flathead style?

A— AN 427.
B— AN 441.
C— AN 456.

Refer to AC 43.13-1B Table 4-8 (this figure will be available during the test as Exhibit II in Section III of the Computer Testing Supplement). In this table, the various rivet materials are arranged in vertical columns, and rivet styles are shown in horizontal rows. Reading down the first column, there are two flathead rivets listed, AN 441 and AN 442.

208g. What is intergranular corrosion?

A— Corrosion integrated inside the metal.
B— The formation of corrosion along the grain boundaries within a metal alloy.
C— The formation of corrosion against the grain boundaries within a metal alloy.

The formation of corrosion along the grain boundaries within a metal alloy is called "intergranular corrosion."

208h. What is the maximum allowable voltage drop in the main power wires for a 28-volt aircraft system, when the generator is carrying rated current?

A— 0.5 volt.
B— 1 amp.
C— 1 volt.

The voltage drop in the main power wires from the generator source or the battery to the bus should not exceed 2 percent of the regulated voltage when the generator is carrying rated current, or when the battery is being discharged at the 5-minute rate. See *AC 43.13-1B Table 11-6 for acceptable voltage drop.*

Nominal system voltage	Allowable voltage drop continuous operation	Intermittent operation
14	0.5	1
28	1	2
115	4	8
200	7	14

Table 11-6. Tabulation chart (allowable voltage drop between bus and utilization equipment ground).

209. What is the dimension for the height of the shop head of an AN470AD4-4 rivet?

A— .0625 (thick).
B— .1875 (thick).
C— .0525 (thick).

The shop head of a universal head rivet (AN470 or MS20470) is 1/2 diameter thick and 1-1/2 diameter wide. The rivet shank is 1/8 (.125) so the shop head should be 0.0625 thick.

Answers

208d [B] (032) AC 43.13-1
208e [B] (032) AC 43.13-1
208f [B] (032) AC 43.13-1
208g [B] (032) AC 43.13-1
208h [C] (032) AC 43.13-1
209 [A] (032) AC 43.13-1

Material	1100	2117T	2017T	2017T-HD	2024T	5056T	7075-T73
Head Marking	Plain	Dimpled	Raised Dot	Raised Dot	Raised Double Dash	Raised Cross	Three Raised Dashes
AN Material Code	A	AD	D	D	DD	B	
AN425 78□ Counter-Sunk Head	X	X	X	X	X		X
AN426 100□ Counter-Sunk Head MS20426	X	X	X	X	X	X	X
AN427 100□ Counter-Sunk Head MS20427							
AN430 Round Head MS20470	X	X	X	X	X	X	X
AN435 Round Head MS20613 MS20615							
AN 441 Flat Head							
AN 442 Flat Head MS20470	X	X	X	X	X	X	X
AN 455 Brazier Head MS20470	X	X	X	X	X	X	X
AN 456 Brazier Head MS20470	X	X	X	X	X	X	X
AN 470 Universal Head MS20470	X	X	X	X	X	X	X
Heat Treat Before Using	No	No	Yes	No	Yes	No	No
Shear Strength psi	10000	30000	34000	38000	41000	27000	
Bearing Strength psi	25000	100000	113000	126000	136000	90000	

Table 4-8a. Aircraft rivet identification. *(Continued on the next page…)*

Material	Carbon Steel	Corrosion-Resistant Steel	Copper	Monel	Monel Nickel-Copper Alloy	Brass	Titanium
Head Marking	Recessed Triangle	Recessed Dash	Plain	Plain	Recessed Double Dots	Plain	Recessed Large and Small Dot
AN Material Code		F	C	M	C		
AN425 78□ Counter-Sunk Head							
AN426 100□ Counter-Sunk Head MS20426							MS 20426
AN427 100□ Counter-Sunk Head MS20427	X	X	X	X			
AN430 Round Head MS20470							
AN435 Round Head MS20613 MS20615	X MS20613	X MS20613	X		X MS20615	X MS20615	
AN 441 Flat Head	X		X	X			X
AN 442 Flat Head MS20470							
AN 455 Brazier Head MS20470							
AN 456 Brazier Head MS20470							
AN 470 Universal Head MS20470							
Heat Treat Before Using	No	No	No	No	No	No	No
Shear Strength psi	35000	65000	23000	49000	49000		95000
Bearing Strength psi	90000	90000					

(...continued from previous page)

Table 4-8b. Aircraft rivet identification.

210. During a sheet metal repair, a row of 2017 rivets installed by the aircraft manufacturer was replaced by an equal number of 2117 rivets of the same diameter by the person making the repairs. Would you, as an IA, approve the repair if it is otherwise eligible for return to service?

A— Yes, the strength of 2117 rivets is greater than that of 2017 rivets.
B— No, the strength of 2117 rivets is less than that of 2017 rivets.
C— Yes, the strength of 2117 rivets is equal to that of 2017 rivets (2117 being the recommended replacement for 2017).

It is acceptable to replace 2017-T3 rivets of 3/16-inch diameter or less, and 2024-T4 rivets of 5/32-inch diameter or less with 2117-T3 rivets for general repairs, provided the replacement rivets are 1/32 inch greater in diameter than the rivets they replace, and the edge distances and spacing are not less than the allowable minimums.

211. Brazing of steel joints in aircraft primary structural members may be acceptable under certain conditions. What is the most important consideration in brazing a primary structural member?

A— The brazing alloy used is the same as the original.
B— Brazing was the original method of construction.
C— The steel is not heated until it loses its heat treating.

Brazing may be used for repairs to primary aircraft structure only if brazing was originally approved for the particular application.

212. Repairs to tubular engine mounts may be repaired by using a larger diameter replacement tube telescoped over the stub of the original member and welded in place. How is the replacement tube preferably cut?

A— Diagonally cut at 45°.
B— Fishmouth cut at 30°.
C— Perpendicular cut at 90°.

Engine-mount members should preferably be repaired by using a larger diameter replacement tube telescoped over the stub of the original member and using a 30° fishmouth cut and rosette welds.

213. A lower horizontal stabilizer streamlined brace is to be repaired by welding. The brace size is 1-1/4 inch. The repair should be accomplished using which of the following materials?

A— A round insert tube of the same material, one gauge thicker than the original streamlined tube and a minimum of 5.01 inches.
B— An outside sleeve of at least the same gauge with a minimum length of 9.128 inches.
C— An inside sleeve of the same streamlined tubing as original with a maximum insert length of 6.43 inches.

According to Figure 4-43 of AC 43.13-1B, a streamline tube splice using split sleeve (applicable to wing and tail surface brace struts and other members), and a repair to a 1-1/4 inch streamline tube should be repaired with an outside sleeve of the same streamlined tubing as the original and at least the same gauge. The minimum length of the splice would be 9.128 inches.

214. A major repair to a structural component is presented to you for inspection and approval. You find that a number of rivets were installed in place of several spot welds that had been opened during the repair. With regard to the replacement of the spot welds with rivets, the repair would

A— require 50 percent more rivets than the original number of welds.
B— require specific approval by an authorized representative of the FAA.
C— not be acceptable; spot welds may not be rejoined using rivets.

With regard to the repair of a stainless steel structure, substitution of bolted or riveted connections for spot-welded joints are to be specifically approved by a DER or the FAA.

Answers

210	[B]	(006)	AC 43.13-1	211	[B]	(032)	AC 43.13-1	212	[B]	(023)	AC 43.13-1
213	[B]	(018)	AC 43.13-1	214	[B]	(018)	AC 43.13-1				

215. What is considered an acceptable radius for a 90-degree bend using 0.064-inch thick aluminum alloy 2024-T3?

A— 3 to 5 times sheet thickness (3 x 0.064 = 0.192).
B— 6 to 8 times sheet thickness (6 x 0.064 = 0.384).
C— 9 to 10 times sheet thickness (9 x 0.064 = 0.576).

Using Table 4-6 on Page 4-14 of AC 43.13-1B, follow the column for 0.064-inch thick sheet down to the row for 2024-T3 alloy and temper. The intersection is at 3t-5t. This means that the minimum bend radius for a 90-degree bend in 0.064-inch thick 2024-T3 aluminum alloy is 0.192 inch.

215a. Where would the location of the sight line be when bending sheet metal?

A— Neutral line plus the bend radius.
B— Bend tangent line plus the bend radius.
C— Bend allowance plus the thickness.

The sight line is a line drawn on a sheet metal layout that is one bend radius from the bend-tangent line. The sight line is lined up directly below the nose of the radius bar in a cornice brake. When the metal is clamped in this position, the bend tangent line is in the correct position for the start of the bend.

215b. The setback used for a 90 degree bend using 2024-T6, 0.032 inch thickness would be

A— insufficient for a 45 degree closed-angle bend.
B— excessive for a 45 degree closed-angle bend.
C— insufficient for a 135 degree closed-angle bend.

The setback for a 90° bend is the bend radius plus the metal thickness. When the metal is bent through more than 90°, a K-factor must be applied. For a closed-angle of 45° the metal must be bent through an angle of 135° (180 – 45 = 135). The K-factor for 135° is 2.414.

The setback for a 45° closed angle is bend radius + metal thickness × 2.414. This is greater than that required for a 90° bend.

215c. Can a 0.032-inch bend radius be used for a 90-degree bend using 5052-H32 aluminum alloy, 0.128 inch thick?

A— May be used.
B— The metal must be annealed prior to bending.
C— Radius cannot be used.

Table 4-6 on Page 4-14 of AC 43.13-1B lists the recommended bend radii for 5052-H32 of 0.128-inch thickness as being between 1/2T and 1-1/2T. This would be between 0.064 and 0.192 inch. A bend radius of 0.032 inch would be too small for this metal.

215d. The sight line drawn on material to be bent in a cornice brake is located at what position from the bend tangent line?

A— The setback measurement.
B— The bend radius length.
C— The bend allowance length.

The definition of "sight line (sheet metal layout)" is a line, drawn on a sheet-metal layout, one bend radius from the bend tangent line. The sight line is placed so it is directly below the nose of the radius bar in a leaf brake. When the metal is clamped in this position, the bend tangent line is at the correct position for the beginning of the bend.

215e. To determine the length of a flat in a sheet metal project, which measurement is subtracted from the mold line length?

A— The bend radius.
B— The setback dimension.
C— The bend allowance.

Setback is a measurement used in sheet metal layout. It is the distance the jaws of a brake must be setback from the mold line to form a bend. For a 90° bend, the point is back from the mold line to a distance equal to the bend radius plus the metal thickness.

215f. What is the setback for a straight-line bend in 2024-T3 sheet metal of.032 inch thickness formed to a bend angle of 30°?

A— .0343.
B— .1280.
C— .2679.

Table 4-6 in AC 43.13-1B indicates the bend radii for a 90° bend is2t – 4t. Since the thickness is .032, a 3t dimension would be the nominal bend: 3 × .032 = .096; .096 + .032 = .128. This would be the setback for a 90° bend. For a 30° bend, refer to Table 4-7 to get the K factor of 0.2679. Multiplying this K-factor by the .128 result above gives a setback of 0.0343.

Answers

215 [A] (027) AC 43.13-1
215c [C] (027) AC 43.13-1
215a [B] (027) AMT-STRUCT
215d [B] (027) AC 43.13-1
215b [A] (027) AMT-STRUCT
215e [B] (027) AC 43.13-1
215f [A] (006) AC 43.13-1

216. Using minimum bend radius, which of the following would require the most material for a 90-degree bend?

Given: Bend Allowance $= \dfrac{2\pi\,(R + 1/2T)}{4}$

A— 2024-T3, .032 thick
B— 5052-H32, .016 thick
C— 7075-T6, .032 thick

Using Table 4-6 on Page 4-14 of AC 43.13-1B find the minimum bend radius for each of the metals given.

.032-inch 2024-T3 = 2t = .064 — BA = 0.125
.016-inch 5052-H32 = 0 — BA = 0.0125
.032-inch 7075-T6 = 3t = .096 — BA = 0.176

217. What is the minimum bend radius for a 90-degree bend for 0.064-inch thick 7075-T6 aluminum alloy?

A— 0.250 inch.
B— 0.187 inch.
C— 0.312 inch.

Using Table 4-6 on Page 4-14 of AC 43.13-1B, follow the column for 0.064-inch thick sheet down to the row for 7075-T6 alloy and temper. The intersection is at 4t-6t. Note 1 says that an alclad sheet "may be bent over slightly smaller radii than the corresponding tempers of uncoated alloy." This means that the minimum bend radius for a 90-degree bend in 0.064-inch thick 7075-T6 aluminum alloy is slightly less than 0.256 inch.

218. What strength rating of a rivet do you use when joining two pieces of sheet metal?

A— rivets equal to the strength of the metal.
B— rivets slightly greater than the strength of the metal.
C— rivets slightly less than the strength of the metal.

The rivet chosen should have a shear strength slightly less than the bearing strength of the sheet metal being joined. If the riveted joint should fail, the rivets should shear rather than the sheet tearing.

In Table 4-9 on Page 4-37 of AC 43.13-1B, the rivet sizes below the lines in each column will fail in shear before the sheet fails in bearing.

219. A sheet metal repair is presented to you for inspection. The repair involves a splice of bare 2024-T3 sheet aluminum .032-inches thick using a single lap sheet joint and joined with 3/32-inch 2117-AD protruding-head rivets. How many rivets per inch of splice width are required?

A— 2.9.
B— 8.3.
C— 11.1.

Using Table 4-9 on Page 4-37 of AC 43.13-1B, follow the column under 3/32-inch diameter rivets to the row for 0.032-inch thickness metal. The number 11.1 is the number of rivets per inch needed. This is below the line in this column, indicating that the joint will fail in shear, as it should. The rivets will shear before the sheet tears.

Note c. specifies that for single lap sheet joints, 75% of the number shown may be used. (11.1 × .75 = 8.32).

220. A sheet metal repair is presented to you for inspection. The repair involves a splice on an intermediate frame using a single-lap joint, 2017 ALCLAD sheet aluminum, 0.040-inch thick with 1/8-inch 2117-AD protruding-head rivets. What is the minimum number of rivets per inch required for the splice?

A— 3.7.
B— 4.0.
C— 6.2.

Using Table 4-10 on Page 4-38 of AC 43.13-1B, follow the column under 1/8-inch diameter rivets to the row for 0.040-inch thickness metal. The number 6.2 is the number of rivets per inch needed. This is below the line in this column, indicating that the joint will fail in shear, as it should, rather than in bearing. The repair is for an intermediate frame, and according to note b. 60% of the number may be used. (6.2 × .60 = 3.72).

Answers

216 [C] (015) AC 43.13-1
217 [A] (015) AC 43.13-1
218 [C] (006) AC 43.13-1
219 [B] (006) AC 43.13-1
220 [A] (027) AC 43.13-1

221. A sheet metal repair is presented to you for inspection. The repair was to a single-lap sheet joint of 5052-H-14 sheet aluminum .091-inches thick. The repair involved the replacement of 2117-AD protruding-head rivets with AN-3 bolts. How many bolts should be installed per inch of width?

A— 2.4.
B— 3.2.
C— 4.0.

Using Table 4-11 on Page 4-39 of AC 43.13-1B, follow the row for 0.091-inch thick metal to the column for AN-3 bolts. This shows that there are 3.2 bolts per inch of repair needed.

Note c. states that for a single lap sheet joint, 75% of the number shown may be used. (3.2 × .75 = 2.4).

222. A sheet metal repair is presented to you for inspection. The repair involves the installation of four bolts in place of the required number of 2117-AD protruding-head rivets for a single lap joint splice in 5052-H14 sheet aluminum .064-inches thick. If the repair was otherwise eligible for return to service, how would you proceed?

A— Disapprove the repair, bolts may not be used to replace rivets in this thickness of aluminum sheet.
B— Approve the repair, the substitution of the rivets with the bolts exceeds the strength of the rivets.
C— Disapprove the repair, the number of bolts per inch of width is too few.

This repair would have to be disapproved, because Table 4-11 on Page 4-39 of AC 43.13-1B does not allow the use of bolts in 5052 aluminum alloy of less than 0.081-inch thickness. If bolts are used in thinner material, the joint will fail in bearing.

222a. How would you identify compression failure in a wooden aircraft spar?

A— Cracks running parallel to the grain of the wood.
B— Separation of the wood fibers near the end.
C— Streaks perpendicular to the wood fiber.

Compression failures are characterized by a buckling of the fibers that appear as streaks on the surface of the wood substantially at right angles to the grain.

222b. When performing an annual inspection on a wooden spar, you find an area with dark spots in the wood. How can you determine if this is acceptable damage?

A— Coin tap test.
B— Bromide solution test.
C— Probe the area with a sharp metal tool.

Discoloration is not necessarily an indication of decay or dry rot. Probe the area in question, if accessible, with a sharp metal tool. The wood structure should be solid and firm. If the suspect area feels soft and mushy it can be assumed that it has rotted.

223. Delamination and debonding in composite structures can be detected by which of the following methods?

A— Coin tap test
B— Visual inspection
C— Both A and B

Tap testing is widely used for a quick evaluation of any accessible aircraft surface to detect the presence of delamination or debonding.

223a. What is likely to result if a delamination of a composite structure is not repaired?

A— Broken fibers.
B— Misalignment of fibers.
C— Wicking.

When layers of laminated structure delaminate there is no matrix material to bond the fibers together and the fibers may break.

223b. What NDT method would ***not*** be suitable for detecting cracks in a composite structure?

A— Dye penetrant.
B— Ultrasonic.
C— Tap test.

Penetrant inspection is used on nonporous metal and nonmetal; components to find material discontinuities that are open to the surface. Composite materials are made in layers and only cracks in the surface ply could be detected by dye penetrant. Ultrasonics and tap tests are used on composite structures.

Answers

221 [A] (027) AC 43.13-1
222b [C] (018) AC 43.13-1
222 [A] (024) AC 43.13-1
223 [A] (024) AC 43.13-1
222a [C] (018) AC 43.13-1
223a [A] (032) ASA-AMT-STRUC
223b [A] (024) AC 43.13-1

223c. The tap test (coin tap test) is one of the simplest methods of finding faults in honeycomb materials. What types of faults can be detected?

A— Delaminations and surface scratches.
B— Impact damage and structural flaws to laminated structure.
C— Delamination and debonding.

Tap testing is widely used for a quick evaluation of any accessible aircraft surface to detect the presence of delamination or debonding.

223d. What is an approved method of inspection on composite materials?

A— X-ray.
B— Dye penetrant.
C— Thermography.

Radiographic inspection using X-rays or gamma rays can be used to inspect the inside of a composite material for delaminations and for moisture and corrosion in honeycomb material.

223e. In composite fiber construction, thread orientation is important in producing maximum strength. Warp is what direction on the fabric roll?

A— Warp is at 90°.
B— Warp is at 0°.
C— Warp is at 45°.

The warp threads run the length of the fabric as it comes off of the fabric roll. The warp direction is designated as 0°.

223f. What kind of defect is detected when using longitudinal magnetization?

A— Longitudinal.
B— Transverse.
C— Lateral.

Longitudinal magnetization occurs when the part is subjected to circular current. This circular current causes the magnetic lines of force to flow through the length of the part in a longitudinal path. A "transverse" crack, oriented 90 degrees to the magnetic flow, will appear because the crack disrupts the magnetic flow through the part and attracts the magnetic inspection particles.

223g. What kind of defect is detected when using circular magnetization?

A— Longitudinal.
B— Transverse.
C— Lateral.

Circular magnetization occurs when the part is subjected to longitudinal current. This longitudinal current causes the magnetic lines of force to flow around the part in a circular path. A "longitudinal" crack, oriented 90 degrees to the magnetic flow, will appear because the crack disrupts the magnetic flow through the part and attracts the magnetic inspection particles.

223h. Which type of crack can be detected by magnetic particle inspection using either circular or longitudinal magnetization?

A— 45 degrees.
B— Longitudinal.
C— Transverse.

Longitudinal magnetization produces a magnetic field that extends lengthwise in the material. It is used to detect faults that extend across the part, perpendicular to the lines of magnetic flux. Circular magnetization produces a magnetic field that extends across the material. It can detect faults that are oriented along the length of the part. Either type of magnetization can detect a fault that runs at 45 degrees to the length of the part.

223i. The term "resin rich" refers to a composite material that is

A— filled with resin but lacks sufficient reinforcing fiber.
B— filled with reinforcing fiber but lacks sufficient resin to thoroughly wet the fiber.
C— cured without using external heating.

A localized area in a composite layup that is filled with resin but lacks sufficient reinforcing fiber is called a "resin rich" area. A resin rich laminate is usually more brittle than a layup with the proper amount of resin. It will also weigh more.

Answers

223c [C] (024) AC 43.13-1	223d [A] (024) AMT-G	223e [B] (024) AMT-STRUCT
223f [B] (024) AC 43.13-1	223g [A] (024) AC 43.13-1	223h [A] (024) AMT-G
223i [A] (032) ASA-DAT		

223j. For detection of which of the following could a coin or ring tap test be used in a composite structure?

A— Voids and occlusions.
B— Delaminations and occlusions.
C— Delamination, debonding, and voids.

Tap testing is widely used for a quick evaluation of any accessible aircraft surface to detect the presence of delamination or debonding. A void is an empty area in the composite laminate, and this term is interchangeable with delamination.

223k. Which NDT method should not be used to detect cracks in advanced composite structures?

A— Dye penetrant.
B— Ultrasonic.
C— Both A and B.

Penetrant inspection is used on nonporous metal and nonmetal components in order to find material discontinuities that are open to the surface. Composite materials are constructed in layers, so only cracks in the surface ply could be detected by dye penetrant. Ultrasonics and tap tests are used on composite structures.

223l. When working with composites, the warp direction is designated at zero degrees. In what direction would the tracer run?

A— 0 degrees.
B— 45 degrees.
C— 90 degrees.

Warp tracers are threads of a different color from the warp threads that are woven into a material in order to identify the direction of the warp threads.

224. Corrosion under paint that looks "worm-like" is

A— intergranular.
B— filiform.
C— fretting.

Filiform corrosion is a special form of oxygen concentration cell occurring on metal surfaces that have an organic coating system. It is recognized by its characteristic worm-like trace of corrosion products beneath the paint film.

224a. A concentration of what occurs under a faying surface where the solution is stagnant?

A— Metal ion.
B— Oxygen.
C— Active-Passive.

The solution may consist of water and ions of the metal which is in contact with the water. A high concentration of metal ions will normally exist under a faying surface where the solution is stagnant.

224b. Galvanic corrosion is also known as

A— dissimilar metals corrosion.
B— uniform surface corrosion.
C— intergranular corrosion.

Galvanic corrosion, also known as dissimilar metals corrosion, occurs when two dissimilar metals make contact in the presence of an electrolyte.

224c. Corrosion can be described as the

A— polishing of one surface by sliding contact with another harder surface.
B— breakdown of metal surfaces due to excessive friction between two parts.
C— loss of metal from the surface by chemical or electrochemical action.

Corrosion is an electrolytic action which takes place within a metal or on its surface. The metal reacts with an electrolyte, and part of the metal is changed into a porous salt. This salt absorbs and holds water in contact with the metal and causes more corrosion to form.

224d. Corrosion is

A— a natural occurrence that attacks metal by physical action.
B— a natural occurrence that attacks metal by chemical action.
C— an unnatural occurrence that attacks metal by chemical action.

Corrosion is a natural occurrence that attacks metal by chemical or electrochemical action and converts it back to a metallic compound.

Answers

223j [C] (024) AC 43.13-1
224 [B] (024) AC 43.13-1
223k [A] (024) AC 43.13-1
224a [A] (024) AC 43.13-1
224c [C] (026) FAA-H-8083-31
223l [A] (024) ASA-DAT
224b [A] (026) AC 43.13-1
224d [B] (026) AC 43.13-1

224e. Which of the following conditions is necessary for electrochemical corrosion to exist?

A— Presence of continuous conductive path.
B— Electrical contact between similar metals.
C— An oxygen depleted environment.

Four conditions must exist before electrochemical corrosion can occur. They are: (1) A metal subject to corrosion (anode); (2) A dissimilar conductive material (cathode), which has less tendency to corrode; (3) Presence of a continuous, conductive liquid path (electrolyte); and (4) Electrical contact between the anode and the cathode (usually in the form of metal-to-metal contact such as rivets, bolts, and corrosion).

224f. Which factors affect the rate of corrosion?

A— Type of metal, heat treatment, and grain direction.
B— Fiber orientation and lamination.
C— Improper adhesive pot life and prolonged assembly time.

Some factors that influence metal corrosion and the rate of corrosion are: (1) type of metal; (2) heat treatment and grain direction; (3) presence of a dissimilar, less corrodible metal; (4) anodic and cathodic surface areas (in galvanic corrosion); (5) temperature; (6) presence of electrolytes (hard water, salt water, battery fluids, etc.); (7) availability of oxygen; (8) presence of biological organisms; (9) mechanical stress on the corroding metal; (10) time of exposure to a corrosive environment; and (11) lead/graphite pencil marks on aircraft surface metals.

225. Cracks and other flaws within composite structure materials can be detected by which of the following methods?

A— Coin tap test.
B— Ultrasonic inspection.
C— Both A and B.

Ultrasonic inspection is finding increasing application in aircraft bonded construction and repair.

226. Applying zinc chromate primer to an aluminum surface has what effect?

A— It prevents further corrosion by insulating the surface against electrolytic action.
B— It forms an organic layer on the surface.
C— It causes the aluminum alloy core to corrode rapidly.

Zinc chromate primer, enamels, chlorinated rubber compounds, etc., are organic coatings commonly used to protect metals.

226a. Applying zinc chromate primer to an aluminum surface has what effect?

A— It inhibits the formation of corrosion by ionizing the surface.
B— It provides a film that protects the metal from water.
C— It helps clean the surface from fingerprints and traces of oil.

Zinc chromate primer is an inhibitive primer that releases chromate ions and holds them on the surface of the metal. This ionized surface prevents the electrolytic action necessary for corrosion to form.

226b. Water on an Alclad aluminum surface creates what type of surface?

A— Neutral.
B— Cathodic.
C— Anodic.

Water in an open area of an aluminum surface readily absorbs oxygen from the air. In this process, two molecules of water (H_2O) combine with one molecule of oxygen (O_2) and take four electrons from the surface of the aluminum metal, leaving the surface anodic (an area that has lost some of its electrons).

227. Which of the following abrasives could not be used to remove corrosion from aluminum alloys?

A— Aluminum oxide
B— Silicon Carbide
C— Garnet

According to Table 6-1 on Page 6-22 of AC 43.13-1B, aluminum oxide and garnet can be used to remove corrosion from aluminum alloys. Silicon carbide is not recommended.

Answers

224e	[A]	(026)	AC 43.13-1	224f	[A]	(026)	AC 43.13-1	225	[B]	(024)	AC 43.13-1
226	[B]	(026)	AC 43.13-1	226a	[A]	(026)	AMT-STRUCT	226b	[C]	(026)	AMT-G
227	[B]	(026)	AC 43.13-1								

227a. Certain types of abrasive paper or cloth may be used to remove corrosion from ferrous alloys having a strength of 220,000 PSI. Which of the following cannot be used on this alloy?

A— Aluminum oxide.
B— Garnet.
C— Silicon carbide.

Table 6-1 on Page 6-22 of AC 43.13-1B lists the abrasives that may be used to remove corrosion from ferrous alloys. Aluminum oxide and silicon carbide are allowed but garnet is not.

227b. Ferrous alloys having a strength of 220,000 PSI should have corrosion removed with which of the following materials?

A— Garnet.
B— Acid-based stripper.
C— Silicon carbide.

Table 6-1 on Page 6-22 of AC 43.13-1B lists the abrasives that may be used to remove corrosion from ferrous alloys. Silicon carbide is allowed.

228. The formation of an oxide film on the surface of aluminum has what effect?

A— The film insulates the aluminum from any electrolyte (gas or liquid), and will not itself further react with the oxygen.
B— A porous or interrupted film is produced and the metal will continue to react with the oxygen in the air until the metal is completely eaten away.
C— It has no effect on the aluminum surface.

A thin, continuous, air-tight film of aluminum oxide can be chemically or electrolytically formed on the surface of the aluminum alloy. This film prevents air or moisture from reaching the metal and prevents the formation of corrosion. This film will not react with the oxygen.

228a. What effect would oxidation have on an aluminum surface?

A— Promotes corrosion.
B— Slows corrosion.
C— Stops corrosion.

The formation of a tightly adhering oxide film offers increased resistance under most corrosive conditions. It slows the formation of corrosion.

228b. Fretting is a major cause of aircraft repairs. What does fretting cause?

A— Prevents aluminum oxides from forming.
B— Allows moisture to be trapped between the metals.
C— Allows oxidation to form between the metals.

Fretting corrosion occurs at the interface of two highly-loaded surfaces that are not supposed to move against one another. When vibration causes the surfaces to rub together, the protective oxide film is rubbed off and the oxides act as abrasives increasing the amount of damage and preventing new oxides from forming on the surface.

228c. How does corrosion cause aluminum alloy to lose its structural strength?

A— Aluminum loses electrons and becomes an anode-type material.
B— Aluminum loses electrons and becomes a cathode-type material.
C— Aluminum gains electrons and becomes an anode-type material.

Aluminum, when covered with an electrolyte, becomes anodic. It has lost electrons and attracts negative ions from the electrolyte to form the salts of corrosion and weaken the metal.

229. Which of the following materials, once oxidized, corrodes further?

A— Iron
B— Aluminum
C— Both A and B

One of the most familiar kinds of corrosion is red iron rust. Red iron rust results from atmospheric oxidation of steel surfaces. Some metal oxides protect the underlying base metal but red rust is not a protective coating. Its presence actually promotes additional attack by attracting moisture from the air, acting as a catalyst to promote additional corrosion.

Answers

227a [B] (026) AC 43.13-1
228a [B] (026) AC 43.13-1
227b [C] (026) AC 43.13-1
228b [C] (026) AC 43.13-1
228 [A] (026) AC 43.13-1
228c [A] (026) AMT-G
229 [A] (026) AC 43.13-1

230. The formation of red rust on a steel surface has what effect?

A— It forms a film that acts as a protective coating on the steel surface.
B— Its presence actually promotes additional attack by attracting moisture from the air, acting as a catalyst to promote additional corrosion.
C— It has no effect on the steel surface.

One of the most familiar kinds of corrosion is red iron rust. Red iron rust results from atmospheric oxidation of steel surfaces. Some metal oxides protect the underlying base metal but red rust is not a protective coating. Its presence actually promotes additional attack by attracting moisture from the air, acting as a catalyst to promote additional corrosion.

231. The formation of red rust on which of the following surfaces would actually promote additional attack by attracting moisture from the air and acting as a catalyst to promote additional corrosion?

A— Aluminum
B— Steel
C— Magnesium

One of the most familiar kinds of corrosion is red iron rust. Red iron rust results from atmospheric oxidation of steel surfaces. Some metal oxides protect the underlying base metal but red rust is not a protective coating. Its presence actually promotes additional attack by attracting moisture from the air, acting as a catalyst to promote additional corrosion.

232. As the holder of an IA, you are inspecting a major repair to an aircraft where the work was accomplished using previously approved data based on the manufacturer's structure repair manual. However, the person making the repair deviated from the data. This was done by substituting a row of 24 AN470-4 rivets that attaches the rudder skin to the rudder ribs with 24 mechanically-locking blind rivets. In view of this change to the repair, what action would you take?

A— Approve the repair, the use of the blind rivets would be a minor deviation by itself.
B— Reject the repair, the use of the blind rivets would be a major repair by itself.
C— Require additional approval of the rivet substitution using AC 43.13-1B for the repair.

AC 43.13-1B, Chapter 7, ¶7-2j discusses the limited use of blind rivets, then references earlier ¶4-57(f) which says, "Blind rivets shall not be used...on aircraft control surfaces...CAUTION: For sheet metal repairs to the airframe, the use of blind rivets must be specifically authorized by the airframe manufacturer or approved by a representative of the FAA."

The substitution of blind rivets therefore would be considered a major deviation and the repair should be rejected.

233. Close-tolerance (AN-173 through AN-186) bolts are used in high-performance aircraft in applications where the bolted joint is subject to severe load reversals and vibration. Which of the following bolts may be substituted for close-tolerance bolts?

A— No substitutions may be made.
B— AN-3 through AN-20 Hex-head bolts.
C— AN-21 through AN-36 Clevis bolts.

Do not substitute for close-tolerance fasteners without specific instructions from the aircraft manufacturer or the FAA.

233a. A major repair has a bolt pattern drilled with holes that are .3125 inch in diameter. What size AN bolts would be required?

A— AN3.
B— AN4.
C— AN5.

In an example of an Air Force-Navy standard bolt designation, the numeral that follows the "AN" indicates the diameter in sixteenths of an inch. Since .3125 = 5/16, AN5 would be the correct size needed.

234. You have been asked to inspect a major repair which involves the fabrication and replacement of a 3/32-inch 7x7 control cable. The cable is fabricated using 3/32-inch AN667 swaged terminal ends. What should you use to check the diameter of the terminal?

A— Go/No-Go gauge.
B— Micrometer.
C— Either A or B.

After the swaging operation is complete, use a Go/No-Go gauge or a micrometer to check the terminal barrel diameter. The shank diameter of a properly swaged terminal is given in Table 7.5 on Page 7-31 of AC 43.13-1B.

Answers

230 [B] (026) AC 43.13-1
231 [B] (026) AC 43.13-1
232 [B] (019) AC 43.13-1
233 [A] (013) AC 43.13-1
233a [C] (005) FAA-H-8083-30
234 [C] (024) AC 43.13-1

235. When inspecting turnbuckles, what is the maximum number of threads that may be exposed on either side of the turnbuckle-barrel?

A— 2.
B— 3.
C— 6.

Adjust the turnbuckle to the correct cable tension so that no more than three threads are exposed on either side of the turnbuckle barrel.

236. You have been asked to inspect a major repair which involves the fabrication and replacement of a 3/32-inch 7x7 control cable. The cable is fabricated using 3/32-inch AN667 swaged terminal ends. What is the diameter of a properly swaged terminal of this size?

A— .190.
B— .250.
C— .500.

Using Table 7-5 on Page 7-31 of AC 43.13-1B, follow the row for 3/32-inch 7x7 cable to the column for shank diameter after swaging. The diameter of the shank should be reduced to 0.190 inch when the terminal is properly swaged.

237. What type of wrap would you consider acceptable when using .040-inch stainless steel safety wire to safety turnbuckles on 5/32 cable?

A— Single wrap.
B— Double wrap.
C— .040-inch stainless steel safety wire is not approved to safety turnbuckles on 5/32-inch cable.

The double-wrap method is acceptable for safetying a turnbuckle in a 5/32-inch cable using 0.040-inch stainless steel wire.

238. A turnbuckle on a 5/32 cable, safetied with a single wrap of stainless steel wire, must have wire of at least what diameter?

A— .030.
B— .040.
C— .057.

When using a single-wrap method for safetying a turnbuckle in a 5/32-inch cable using stainless steel wire, the wire would have to be at least 0.057-inch diameter.

239. A hole is drilled to .3750. What size bolt is required to fit this hole?

A— AN4.
B— AN5.
C— AN6.

The diameter of an aircraft bolt is specified in increments of 1/16 inch by its AN number. An AN6 bolt has a diameter of 6/16 or 3/8 inch (0.3750).

240. A hole is drilled to .18750. What size bolt is required to fit this hole?

A— AN5.
B— AN4.
C— AN3.

The diameter of an aircraft bolt is specified in increments of 1/16 inch by its AN number. An AN3 bolt has a diameter of 3/16 inch (0.18750).

241. Turnbuckles must have a minimum of how many turns of safety wire wrapping?

A— 3.
B— 4.
C— 5.

When safetying a turnbuckle, wrap the ends of the safety wire at least four turns around the shank and cut it off.

242. What is the maximum differential pressure allowed when performing a compression check on a reciprocating engine?

A— 30%.
B— 25%.
C— 20%.

The compression testing of aircraft engine cylinders is a test to determine the internal condition of the combustion chamber cylinder assembly by ascertaining if any appreciable internal leakage is occurring. If a cylinder has less than a 60/80 reading (a 25% leakage) on the differential test gauges on a hot engine, the cylinder must be removed and inspected.

Answers

235	[B]	(024)	AC 43.13-1
236	[A]	(024)	AC 43.13-1
237	[B]	(015)	AC 43.13-1
238	[C]	(015)	AC 43.13-1
239	[C]	(014)	AC 43.13-1
240	[C]	(014)	AC 43.13-1
241	[B]	(015)	AC 43.13-1
242	[B]	(024)	AC 43.13-1

243. A proposed airframe alteration will require a section of MIL-H-8788-10 hydraulic hose to flex through 60° of travel. The system will operate at 210° centigrade and 1,200 psi. What is the minimum bend radius for this installation?

A— 3-1/4 inches.
B— 5-1/2 inches.
C— 7-1/2 inches.

According to the no-flexing table in Figure 9-10 on Page 9-24 of AC 43.13-1B, the minimum bend radius for MIL-H-8788-10 hose is 6.50 inches.

The flexing-range table in this figure shows that for 60° of travel for this hose, a bend factor of 1.16 should be used. Multiply the bend radius by the bend factor. 6.5 × 1.16 = 7.54 or 7-1/2 inches.

244. You are inspecting an installation involving the use of MIL-H-8794-5 hose. The system operates at a maximum of 1,000 psi and 210° temperature. The hose must travel through a flexing range of 60 degrees. The minimum bend radii to which the hose may be subjected is

A— 1.8-inch minimum bend radii.
B— 2.03-inch minimum bend radii.
C— 3.37-inch minimum bend radii.

According to the bend radius chart in Figure 9-10 on Page 9-24 of AC 43.13-1B, the minimum bend radius for MIL-H-8794-5 hose operating at 1,000 psi is 1.75 inches.

The flexing range table in this figure shows that for 60° of travel for this hose, a bend factor of 1.16 should be used. Multiply the bend radius by the bend factor. 1.75 × 1.16 = 2.03 inches.

244a. How would the minimum bend radius for a section of MIL-H-8794-5 hydraulic hose used in a flexing installation of 90°, and operating at 2,500 psi, compare with the same hose used in a non-flexing installation at maximum operating pressure? The flexing installation would

A— have a greater minimum bend radius.
B— have a smaller minimum bend radius.
C— be equal to the non-flexing installation.

Refer to AC 43.13-1B, Figure 9-10. The -5 hose operating at 2,500 psi with no flexing requires a minimum bend radius of 3 inches. A factor of 1.25 (N) is used to multiply the minimum non-flexing bend radius when 90° of flexing occurs; this results in a minimum bend radius of 3.75. Comparing this number with the bend radius for non-flexing hose at maximum operating pressure of 3,000 psi, which is 3.25, the flexing installation will have a greater minimum bend radius.

245. Where is the datum located on a single-engine rotorcraft certificated under 14 CFR Part 27?

A— Tip of rotor.
B— Plane of rotor.
C— Anywhere the manufacturer determines it to be.

The datum is an imaginary vertical plane from which all horizontal measurements are taken for balance purposes with the aircraft in level flight attitude. The datum may be located anywhere the manufacturer chooses and it is normally specified in the Aircraft Specifications or Type Certificate Data Sheets.

245a. The location of the datum on an aircraft is determined by the

A— FAA.
B— aircraft manufacturer.
C— aircraft owner.

The datum is an imaginary vertical plane, chosen by the aircraft manufacturer, from which all horizontal distances are measured for balance purposes with the aircraft in level flight attitude.

246. For airplanes that do not have an approved aircraft flight manual, equipment required by the certification regulation to be on the airplane during calculation of the certificated empty weight, may be found in which document?

A— Manufacturer's equipment list.
B— Type Certificate Data Sheets.
C— Master minimum equipment list.

For aircraft that do not have an approved aircraft flight manual, the required equipment items are listed in the Aircraft Specifications, or for aircraft that have Type Certificate Data Sheets, in an equipment list furnished by the aircraft manufacturer.

Answers

243	[C]	(005)	AC 43.13-1	244	[B]	(005)	AC 43.13-1	244a	[A]	(013)	AC 43.13-1
245	[C]	(029)	AC 43.13-1	245a	[B]	(029)	AC 43.13-1	246	[A]	(018)	AC 43.13-1

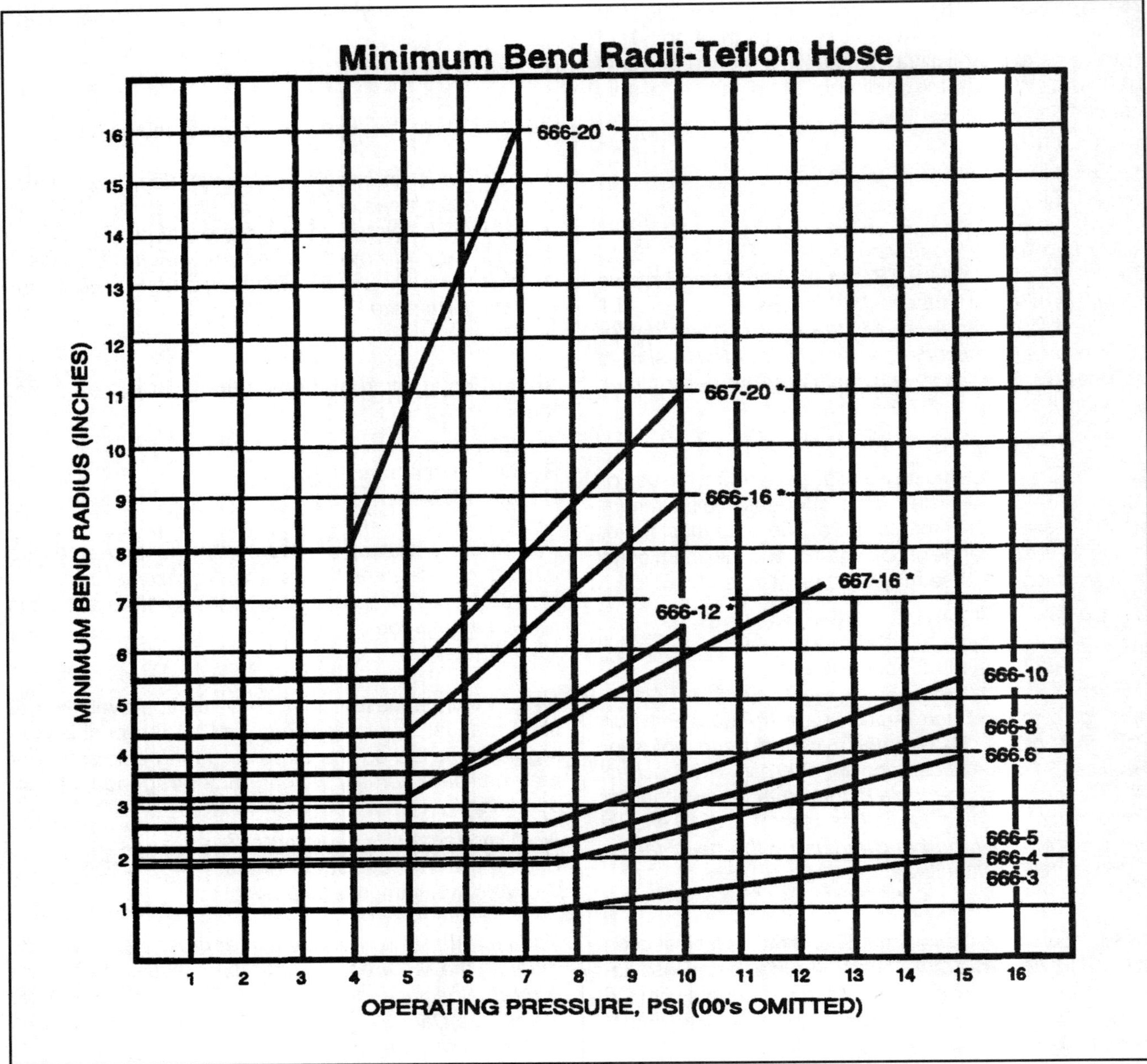

Question 247

247. *(Refer to figure on opposite page.)* What is the minimum bend radius that can be used on a 666-5 Teflon hose that will operate with 1,100 psi pressure?

A— 1 inch.
B— 1-1/2 inches.
C— 1-3/4 inches.

1. *Use the chart for the Minimum Bend Radius for Teflon Hose.*
2. *Follow the vertical line for an operating pressure of 1,100 psi upward until it intersects the diagonal line for 666-5 hose.*
3. *Project a horizontal line to the left to the Minimum Bend Radius (Inches) index.*
4. *The minimum bend radius for 666-5 hose under a pressure of 1,100 psi is 1-1/2 inches.*

248. During an annual inspection, the pilot's seatbelt is found to be in a questionable condition. The belt is identified on the label as TSO-C22 but does not show the TSO rating. The owner wishes to have the belt tested rather than replaced. At what tensile load should the belt be tested?

A— 1,250 pounds.
B— 1,500 pounds.
C— 1,750 pounds.

Airworthy one-person type certificated belts should be able to withstand a tensile load of 525 pounds, and TSO belts should withstand the rated tensile load indicated on the belt label. Most one-person TSO belts are rated for 1,500 pounds.

249. You are inspecting an installation involving the use of Mil-H-8794-5 hydraulic hose. The system operates at a maximum of 2000 PSI and 210° temperature. The hose must travel through a flexing range of 90°. The minimum bend radii to which the hose may be subjected is

A— 2.75-inch minimum bend radii.
B— 3.24-inch minimum bend radii.
C— 6.50-inch minimum bend radii.

Using Figure 9-10 on Page 9-24 of AC 43.13-1B, follow these steps:

1. *Find the bend factor N by following a vertical line for the 90° flexing range upward to the diagonal line. From this point project a line to the left and find that N = 1.25.*
2. *In the bend radius chart follow the 2,000 psi vertical line until it intersects the curve for -5 hose. From this point project a line to the left and find that the no-flexing bend radius is 2.6 inches.*
3. *Find the flexing minimum bend radius by multiplying the no-flexing bend radius by the bend factor:*

Minimum bend radius = N · no-flex bend radius
= 1.25 · 2.6
= 3.25 inches

250. When a non-flex hose angle-of-bend increases, what happens to the minimum bend radius?

A— increases
B— decreases
C— remains constant

According to AC 43.13-1B Figure 9-10, the minimum allowable bend radius for non-flex hose is determined by the diameter of the hose, not by the hose angle.

251. What is the relationship of the bend radius of a MIL-H-8794-5 hose operating at 3,000 psi with a 90-degree flexing range, compared to the same hose at maximum operating psi with no flexing? The flexing bend radius

A— increases.
B— decreases.
C— remains constant.

According to AC 43.13-1B Figure 9-10, a MIL-H-8794-5 hose operating at 3,000 psi with no flexing has a minimum allowable bend radius of 3-3/8 inches. If this hose is operated at the same pressure but has a flexing range of 90 degrees, the minimum bend radius must be multiplied by a bend factor N of 1.25. The minimum bend radius under these conditions is increased to 4.2 inches (3.375 · 1.25 = 4.2).

Answers

247	[B]	(018)	AC 43.13-1	248	[B]	(024)	AC 43.13-1	249	[B]	(013)	AC 43.13-1
250	[C]	(028)	AC 43.13-1	251	[A]	(028)	AC 43.13-1				

252. Which -5 hose has a larger minimum bend radius at 3,000 psi with a 90-degree flexing range?

A— Non-flex hose
B— Flexible hose
C— Minimum bend radius would be the same for non-flex and flexible hoses

According to AC 43.13-1B Figure 9-10, a MIL-H-8794-5 hose that is operated in a flexing installation (flexible hose) is required to have a larger minimum bend radius than the same hose in a non-flexing installation. The amount larger the bend radius must be is determined by the bend factor "N" chart.

253. When computing weight and balance, the empty weight includes the weight of the airframe, engine(s) and all items of operating equipment permanently installed. Empty weight also includes

A— the unusable fuel, hydraulic fluid, and undrainable oil or, in some aircraft all of the oil.
B— all usable fuel, maximum oil, hydraulic fluid, but does not include the weight of pilot, passengers, or baggage.
C— all usable fuel and oil, but does not include any radio equipment or instruments that were installed by someone other than the manufacturer.

14 CFR Parts 23 and 25 both specify that the empty weight includes optional and special equipment, fixed ballast, full reservoirs of hydraulic fluid, and engine lubricating oil but it includes only unusable, or residual, fuel. Other parts of the Federal Regulations include only the undrainable oil.

253a. Usable fuel is included in what measurement?

A— Useful load.
B— Empty weight.
C— Both A and B.

The useful load is the empty weight subtracted from the maximum weight of the aircraft. The useful load consists of the pilot, crew (if applicable), maximum oil, fuel, passengers and baggage unless otherwise noted.

254. What determines whether the value of the moment is preceded by a plus (+) or a minus (–) sign in aircraft weight and balance?

A— The moment of the weight in reference to the datum.
B— The result of a weight being added or removed and the location relative to the datum.
C— The location of the datum in reference to the aircraft CG.

A moment is a force that causes rotation about a point, and in order to specify the direction of the rotation, signs (+) and (–) are assigned to the moment. In aircraft weight and balance, a positive moment is one that causes the aircraft nose to go up, and a negative moment is one that causes the nose to go down.

Since a moment is the product of weight and the distance from the datum, and both of these are signed values, we have four choices for the sign of the moment:

1. *A positive weight (weight added) and a positive arm (arm behind the datum) gives a positive moment.*
2. *A positive weight and a negative arm (arm ahead of the datum) gives a negative moment.*
3. *A negative weight (weight removed) and a positive arm gives a negative moment.*
4. *A negative weight and a negative arm gives a positive moment.*

255. The empty weight CG of an airplane lies within the empty weight CG limits,

A— it is necessary to calculate the CG extremes.
B— it is not necessary to calculate the CG extremes.
C— minimum fuel should be used in both forward and rearward CG checks.

If the empty-weight CG of an airplane lies within the empty-weight CG limits, it is not necessary to calculate the CG extremes. The airplane cannot be legally loaded in such a way that either its forward or aft CG limits can be exceeded.

Answers

252	[B]	(028)	AC 43.13-1
253	[A]	(029)	AC 43.13-1
253a	[A]	(029)	AC 43.13-1
254	[B]	(029)	AC 43.13-1
255	[B]	(029)	AC 43.13-1

255a. What would the licensed empty weight CG be if an aircraft was weighed with the following?

Nose gear.................................650 lbs at –31 inches
Main gear...............................1,820 lbs at +84 inches
Fixed ballast ....10 lbs at +114 inches
Fuel (avgas).................................30 gal at +55 inches
Oil ....3 gal at –39 inches

A— 52.22
B— 53.14
C— 51.09

First find the CG as the aircraft was weighed:

Item	**Weight**	**Arm**	**Moment**
Nose gear	*650*	*–31*	*–20,150*
Main gear	*1,820*	*+84*	*152,880*
Totals	*2,470*		*132,730*

The CG as weighed is 132,730 ÷ 2,470 = +53.74

The weight of the ballast, fuel and oil are already included in these weights listed for nose and main. Now the weight of the fuel must be subtracted out, but the weight of the ballast (which is listed as "Fixed") and oil stay in:

	2,470		*132,730*
Fuel	*–180*	*+55*	*9,900*
Totals	*2,290*	*+53.6*	*122,830*

The EWCG is 122,830 ÷ 2,290 = 53.6 inches aft of datum.

256. When computing the maximum forward loaded CG of an aircraft, minimum weights, arms, and moments should be used for items of useful load that are located aft of the

A— rearward CG limit.
B— forward CG limit.
C— datum.

When computing the maximum forward-loaded CG of an aircraft, you should use maximum weight of all items of useful load located ahead of the forward CG limit and the minimum weight for all items of useful load located behind the forward CG limit.

257. When making a rearward weight and balance check to determine that the CG will not exceed the rearward limit during extreme conditions, the items of useful load which should be computed at their minimum weights are those located forward of the

A— forward CG limit.
B— datum.
C— rearward CG limit.

When making a rearward weight and balance check to determine that the loaded CG cannot fall behind the rearward CG limit, you should use the maximum of all items of the useful load whose CG is behind the rear limit and the minimum weights for all items that are forward of the rearward CG limit.

258. An aircraft with an empty weight of 2,100 pounds and an empty weight CG of +32.5 was altered as follows:

1. two 18-pound passenger seats located at +73 were removed.
2. structural modifications were made at +77, increasing weight by 17 pounds.
3. a seat and a safety belt weighing 25 pounds were installed at +74.5.
4. radio equipment weighing 35 pounds was installed at +95.

What is the new empty weight CG?

A— +34.01
B— +33.68
C— +34.65

Item	**Weight**	**Arm**	**Moment**
Aircraft	*2,100*	*32.5*	*68,250*
Seats (remove)	*36 (-)*	*73*	*2,628 (-)*
Modification	*17*	*77*	*1,309*
Seat	*25*	*74.5*	*1,862.5*
Radio	*35*	*95*	*3,325*
Totals	*2,141*	*33.68*	*72,118.5*

The new empty weight is 2,141 pounds and the CG is located at fuselage station 33.68.

Answers

255a [B] (008) FAA-H-8083-1	256 [B] (029) AC 43.13-1	257 [C] (029) AC 43.13-1
258 [B] (008) AC 43.13-1		

259. An aircraft has an empty weight of 4,954 pounds at a CG of +30.5 inches. The CG range is +32.0 inches to +42.1 inches. Find the minimum weight of the ballast, mounted at +162 inches, necessary to bring the CG within the CG range.

A— 61.98 pounds.
B— 30.58 pounds.
C— 57.16 pounds.

The CG of the aircraft is out of allowable range by 1.5 inches. Its CG is at fuselage station 30.5 and the forward CG limit is at station 32.0.

To find the amount of ballast needed to be attached at fuselage station 162, multiply the empty weight of the aircraft by the distance the CG is to be moved and divide this by the distance between the ballast location and the desired CG location.

$$\text{Ballast weight} = \frac{\text{Empty weight} \cdot \text{Distance out}}{\text{Distance ballast to new CG}}$$

$$= \frac{4{,}954 \cdot 1.5}{162 - 32}$$

$$= 57.16 \text{ pounds}$$

Attaching a 57.16 pound ballast at fuselage station 162 will move the empty-weight CG to fuselage station 32.0.

260. Determine the cable size of a 40-foot length of single cable in free air, with a continuous rating, running from a bus to the equipment in a 28 volt system with a 15-ampere load and a 1-volt drop.

A— No. 10.
B— No. 12.
C— No. 18.

Refer to the continuous flow conductor chart, Figure 11-2 on Page 11-30 of AC 43.13-1B:

1. *Follow the 15-ampere diagonal line down until it intersects the horizontal line for 40 feet in the 28-volt column.*
2. *This intersection occurs between the vertical lines for 12-gage and 10-gage wire.*
3. *You would need to use a 10-gage wire (always use the larger of the two wires) to carry a 15-ampere load for 40 feet without exceeding a 1-volt drop.*

260a. What is the maximum amount of time a circuit can be in operation and still be an intermittent duty circuit?

A— Three minutes.
B— Two minutes.
C— One minute.

Referring to Figures 11-2 and 11-3 in AC 43.13-1B, there is a significant difference in the current-carrying capacity of the same diameter wire, depending on whether or not the current flow is intermittent or continuous. However, there is no definition given for understanding what length of time can be used to determine this. This information is located in paragraph 11-68.d., where it states "...or intermittent operation (maximum two minutes)."

261. What size wire would be used for a starter motor if you have 28 volts, 20 feet of wire, and 200 amps?

A— No. 2
B— No. 1
C— No. 4

Refer to the intermittent flow conductor chart, Figure 11-3 on Page 11-31 of AC 43.13-1B:

1. *Follow the 200-ampere diagonal line down to its end, the vertical line for a 1-gauge wire. This is the smallest wire that can carry this amount of current without overheating.*
2. *Using Table 11-9 on Page 11-23, find the resistance of 20 feet of this wire to be 0.003 ohm.*
3. *The voltage drop caused by 200 amps flowing through a resistance of 0.003 ohm is 0.6 volt which is less than the voltage drop allowed in Table 11-6 on Page 11-21 for 28-volt intermittent service.*

Answers

259 [C] (008) AC 43.13-1
260 [A] (003) AC 43.13-1
260a [B] (003) AC 43.13-1
261 [B] (003) AC 43.13-1

262. Determine the minimum wire size of a single cable carrying a continuous current of 20-amperes 10 feet from the bus to the equipment in a 28-volt system with an allowable 1-volt drop.

A— No. 12.
B— No. 14.
C— No. 16.

Refer to the continuous flow conductor chart, Figure 11-2 on Page 11-30 of AC 43.13-1B:

1. *Follow the 20-ampere diagonal line down until it intersects the horizontal line for 10 feet in the 28-volt column.*
2. *This intersection occurs at the vertical lines for 16-gage wire.*
3. *A 16-gage wire will carry a 20-ampere load for 10 feet without exceeding a 1-volt drop.*

263. Determine the maximum length of a No. 12 single cable that can be used between a 28-volt bus and a component utilizing 20 amperes continuous load in free air with a maximum acceptable 1-volt drop.

A— 23.5 feet.
B— 26.5 feet.
C— 12.5 feet.

Refer to the continuous flow conductor chart, Figure 11-2 on Page 11-30 of AC 43.13-1B:

1. *Follow the 20-ampere diagonal line down until it intersects the vertical line for No. 12 wire size.*
2. *This intersection occurs on the horizontal line for 23.5 feet in the 1-volt drop column.*
3. *You can use up to 23.5 feet of No. 12 cable to carry a 20-ampere load without exceeding a 1 volt drop.*

264. What would be the maximum amperage when using an AN14-gage wire with a temperature rating of 150°C in a 28-volt system with a continuous load?

A— 14 amps.
B— 10 amps.
C— 7.5 amps.

Using Table 11-9 on Page 11-23 of AC 43.13-1B, follow the 150°C column down until it meets the row for 14-gage wire. This intersection occurs at 14 amperes. An AN14-gage wire with a temperature rating of 150°C can carry a continuous current of 14 amps.

265. Determine the maximum length of a No. 16 cable to be installed from a bus to the equipment in a 28-volt system with a 25-ampere intermittent load and a 1-volt drop.

A— 8 feet.
B— 10 feet.
C— 12 feet.

Refer to the intermittent flow conductor chart, Figure 11-3 on Page 11-31 of AC 43.13-1B:

1. *Draw a line midway between the 20- and the 30-ampere diagonal line.*
2. *Follow this line down until it intersects the vertical line for No. 16 wire size.*
3. *This intersection occurs on the horizontal line for 16 feet in the 2-volt drop column. Half the cable length will create half the voltage drop, therefore 8 feet will produce a 1-volt drop.*
4. *You cannot exceed 8 feet of No. 16 cable carrying a 25-ampere load without exceeding a 1 volt drop.*

266. A mechanic has installed electric wiring in accordance with an STC, with the exception of adding 2 feet of conduit that passes through the firewall. The STC calls for 16-gage wire with a 150°C temperature rating, 28 volt, 14 amps, single cable in free air. The total length of the wire is 14 feet. Is this installation acceptable with the addition of the conduit?

A— No, there is too much amperage for this length of wire.
B— Yes, the installation is acceptable.
C— No, the voltage drop does not allow this installation.

According to Table 11-9 on Page 11-23 of AC 43.13-1B, a 16-gage wire with a temperature rating of 150°C can carry only 11 amps in a conduit without overheating. The installation would have to be disapproved.

Answers

262 [C] (003) AC 43.13-1
263 [A] (003) AC 43.13-1
264 [A] (003) AC 43.13-1
265 [A] (003) AC 43.13-1
266 [A] (003) AC 43.13-1

267. An electric motor draws 14.5 amps, using a 9-wire bundle. What gage wire having a 200°C temperature rating should be used, and what size circuit breaker would be used?

A— 16-gage wire, 15 amp circuit breaker.
B— 14-gage wire, 20 amp circuit breaker.
C— 12-gage wire, 30 amp circuit breaker.

Using Table 11-9 on Page 11-23 of AC 43.13-1B, follow the current values in the 200°C column to the first value larger than 14.5 amps. This is 18 amps. Follow this row to the left and see that a 14-gage wire will carry 18 amps continuously in a bundle.

In Table 11-3 on Page 11-15 you will find that a 14-gage wire should be protected with a 20-amp circuit breaker.

268. You have a 28-volt system, AN20 wire in a bundle and 14 feet of wire. What are the maximum amps?

A— 7 amps.
B— 10 amps.
C— 5 amps.

Using Figure 11-2 on Page 11-30 of AC 43.13-1B, follow the horizontal line for 14 feet in the 28-volt column to the right until it intersects the vertical line for AN20 wire. The diagonal line for 7 amps also meets at this point. An AN20 wire can carry 7 amps for 14 feet in a 28-volt system without exceeding the allowable voltage drop or current limitation.

269. What is the location of the CG of a nose wheel airplane with these specifications?

Datum is aft of the main wheels
Main wheels are at -75.0
Nose wheel is at -153.0
Weight of nose wheel is 340 pounds
Total net weight of airplane is 2,006 pounds

A— -67.8
B— +82.8
C— -88.2

D = Datum to main wheels = -75
F = Weight of the nose wheel = 340 pounds
L = Distance between main and nose wheels = 153.0 – 75.0 = 78 inches
W = Net weight of airplane = 2,006 pounds

Use the formula:

$$CG = -(D + \frac{F \cdot L}{W})$$

$$= -(75 + \frac{340 \cdot 78}{2{,}006})$$

$$= -88.2 \text{ inches}$$

The CG is 88.2 inches forward of the datum.

269a. When selecting an electrical conductor that will operate at an elevated temperature (above 20°C), the T2 factor requires what value for its determination?

A— The constant 234.5°C.
B— L2 (maximum allowable run length).
C— I_{MAX} (maximum allowable current).

When the estimated or measured conductor temperature (T2) exceeds 20 °C, such as in areas having elevated ambient temperatures or in fully loaded power-feed wires, the maximum allowable run length (L2), must be shortened from L1 (the 20 °C value). There are two formulas, one for copper conductor, and one for aluminum. Although the copper formula uses 234.5, the aluminum formula uses 238.1 degrees, therefore Answer [A] is incorrect. Answer [C] is incorrect because the maximum allowable current (I_{MAX}) is only used when the wires are carrying a high percentage of their current capability at elevated temperature.

269b. Determine the estimated conductor temperature (T2) for Mil-W-22759/10 under the following conditions:

1. Ambient temperature is 43° Celsius (110° Fahrenheit).
2. Conductor rating temperature is 150° Celsius.
3. Circuit current is 5 amperes.
4. Max current is 11 amperes.

A— 72 degrees Celsius.
B— 115 degrees Celsius.
C— 201 degrees Celsius.

Refer to AC 43.13-1B Chapter 11, Section 5 "Electrical Wire Rating," paragraph 11-66 d.(6). The formula is:

$$T_2 = T_1 + (T_R - T_1)(\sqrt{I_2 \div I_{MAX}})$$

Filling in the given data, $T_1 = 43$ (in 2 places), $T_R = 150$; $I_2 = 5$, and $I_{MAX} = 11$. Therefore, $T_2 = 43 + 107 \times .67 = 114.69$, which is closest to answer B.

Answers

267 [B] (003) AC 43.13-1
268 [A] (003) AC 43.13-1
269 [C] (008) AC 43.13-1
269a [B] (001) AC 43.13-1
269b [B] (003) AC 43.13-1

270. What is the location of the CG of a tail wheel airplane with these specifications?

Datum is leading edge of the wing
Right main wheel weighs 564 pounds at +3.0
Left main wheel weighs 565 pounds at +3.0
Tail wheel is at +225.0
Weight of tail wheel is 40 pounds
(67 pounds – 27 pounds of tare)

A— 12.6
B— 10.6
C— 15.4

Weighing point	***Location***	***Scale reading***	***Tare***	***Net weight***	***Moment***
Right wheel	*+3*	*564*	*0*	*564*	*1,692*
Left wheel	*+3*	*565*	*0*	*565*	*1,695*
Tail wheel	*+225*	*67*	*27*	*40*	*9,000*
				1,169	*12,387*

Find the CG by dividing the total moment by the total weight:

$$CG = \frac{Total\ Moment}{Total\ Weight} = \frac{12{,}387}{1{,}169} = 10.6 \text{ inches aft of the datum}$$

The CG is 10.6 inches aft of the datum.

271. What is the location of the CG of a nose wheel airplane with these specifications?

Datum is leading edge of the wing
Right main wheel weighs 604 pounds at +34.0
(609# – 5# tare)
Left main wheel weighs 615 pounds at +34.0
(620# – 5# tare)
Nose wheel is at -33.8
Weight of nose wheel is 454 pounds
(464# – 10# tare)

A— 15.4
B— 14.8
C— 16.7

Weighing point	***Location***	***Scale reading***	***Tare***	***Net weight***	***Moment***
Right wheel	*+34*	*609*	*5*	*604*	*20,536*
Left wheel	*+34*	*620*	*5*	*615*	*20,910*
Nose wheel	*–33.8*	*464*	*10*	*454*	*–15,345*
				1,673	*26,101*

Find the CG by dividing the total moment by the total weight:

$$CG = \frac{Total\ Moment}{Total\ Weight} = \frac{26{,}101}{1{,}673} = 15.6 \text{ inches aft of the datum}$$

The CG is 15.6 inches aft of the datum.

Answers

270 [B] (008) AC 43.13-1
271 [A] (008) AC 43.13-1

272. The CG of an aircraft may be determined by

A— dividing the total arms by the total moments.
B— dividing the total moment by the total weight.
C— multiplying the total weight by the total moment.

The CG is found by dividing the total of all moments by the total weight.

273. What is the most forward adverse loaded CG for a tail wheel airplane with these specifications?

Airplane empty weight .... 1,169 pounds
Empty weight CG.. +10.6
Forward CG limit.. +8.5
Pilot.. 170 pounds at +16.0
Fuel tanks.. +22
METO horsepower 160
Minimum fuel .... 80 pounds at +22

A— +11.9
B— +8.5
C— +13.2

No passengers or items behind the forward CG limit will be carried.

The fuel is behind the forward CG limit so minimum fuel will be carried. Minimum pounds of fuel is METO horsepower divided by 2.

Item	*Weight*	*Arm*	*Moment*
Airplane (empty)	*1,169*	*+10.6*	*12,391*
Pilot	*170*	*+16.0*	*2,720*
Fuel	*80*	*+22.0*	*1,760*
Totals	*1,419*		*16,871*

Find the most forward loaded CG by dividing the total moment by the total weight:

$$CG = \frac{\text{Total Moment}}{\text{Total Weight}}$$

$$= \frac{16{,}871}{1{,}419}$$

$$= 11.9 \text{ inches aft of the datum}$$

The most forward CG limit is +8.5 and the adverse loaded CG is +11.9 so the forward CG limit cannot be exceeded.

274. What is the most rearward adverse loaded CG for a tail wheel airplane with these specifications?

Airplane empty weight .... 1,169 pounds
Empty weight CG.. +10.6
Rearward CG limit .... +21.9
Pilot.. 170 pounds at +16.0
Fuel tanks.. +22
Fuel quantity 40 gallons........................... 240 pounds
Passengers
2 in rear seats .... 340 pounds at +48
Baggage (max. placarded) .... 100 pounds at +75.5

A— +23.8
B— +20.7
C— +21.9

All passengers and items behind the rearward CG limit will be carried.

The fuel is behind the rearward CG limit so full fuel will be carried.

Item	*Weight*	*Arm*	*Moment*
Airplane (empty)	*1,169*	*+10.6*	*12,391*
Pilot	*170*	*+16.0*	*2,720*
Fuel	*240*	*+22.0*	*5,280*
Passengers (2)	*340*	*+48.0*	*16,320*
Baggage	*100*	*+75.5*	*7,550*
Totals	*2,019*		*44,261*

Find the most rearward loaded CG by dividing the total moment by the total weight:

$$CG = \frac{\text{Total Moment}}{\text{Total Weight}}$$

$$= \frac{44{,}261}{2{,}019}$$

$$= 21.9 \text{ inches aft of the datum}$$

The most rearward CG limit is +21.9 and the adverse loaded CG is +21.9 so the rearward CG limit cannot be exceeded.

Answers

272 [B] (029) AC 43.13-1
273 [A] (008) AC 43.13-1
274 [C] (008) AC 43.13-1

275. Given:
Empty weight .... 15,000 lbs
2 front passengers .... + 85.5
2 rear passengers .... +118.1
2 (two) 50-gal fuel tanks .... +95.0
Baggage 100 lbs .... +110.0

Empty weight CG ....98 inches
Maximum weight .... 16,200 lbs
Forward CG limit .... +90.0
Aft CG limit .... +110.0

As loaded, this aircraft would be

A— overweight, but within the CG limits.
B— underweight and outside the aft CG limit.
C— overweight and outside the forward CG limit.

Item	*Weight*	*Arm*	*Moment*
Aircraft	*15,000*	*98.0*	*1,470,000*
Front passengers	*340*	*85.5*	*29,070*
Rear passengers	*340*	*118.1*	*40,154*
Fuel	*600*	*95.0*	*57,000*
Baggage	*100*	*110.0*	*11,000*
Totals	*16,380*	*98.1*	*1,607,224*

Divide the total moment of 1,607,224 pound-inches by the total weight of 16,380 pounds to find the loaded CG of 98.1 inches aft of the datum. This is more than the allowable weight of 15,000 pounds, but is within the CG limits of +90 to +110 inches.

276. Given:
Empty Weight .... 1,095 lbs
Empty Weight CG .... +33.8
Pilot .... + 39.0
Passenger .... +39.0
Fuel 135 lbs .... +42.0
Oil 11 lbs .... –12.0
Baggage 120 lbs .... +64.0

Forward CG limit 34.1
Aft CG limit 38.0
Maximum weight .... 1,750 lbs
100 hp

(The aircraft was certificated under CAR 3 regulations.)

When making a forward and rearward extreme condition check on this aircraft, which of the following would be correct?

A— Rearward check, 1,721 lbs; 37.9
B— Rearward check, 1,701 lbs; 37.3
C— Forward check, 1,350 lbs; 34.4

Rearward extreme CG check:

Include every item that will move the CG behind the aft limit of +38.0. This includes passenger, full fuel, and baggage. The pilot and oil are essential and must be included.

Item	*Weight*	*Arm*	*Moment*
Aircraft	*1,095*	*33.8*	*37,011*
Pilot	*170*	*39.0*	*6,630*
Passenger	*170*	*39.0*	*6,630*
Fuel	*135*	*42.0*	*5,670*
Baggage	*120*	*64.0*	*7,680*
Oil	*11*	*–12.0*	*–132*
Totals	*1,701*	*37.3*	*63,489*

The rearward extreme condition check is 1,701 pounds at 37.3 inches aft of the datum. This is within the allowable weight and CG range.

Forward extreme CG check:

Include every item that will move the CG ahead of the forward limit of +34.1. The pilot and oil are essential and must be included. Include only the minimum fuel which is the horsepower divided by 2, or 50 pounds. No baggage is included.

This includes passenger, full fuel, and baggage.

Item	*Weight*	*Arm*	*Moment*
Aircraft	*1,095*	*33.8*	*37,011*
Pilot	*170*	*39.0*	*6,630*
Fuel	*50*	*42.0*	*2,100*
Oil	*11*	*–12.0*	*–132*
Totals	*1,326*	*34.4*	*45,609*

The forward extreme condition check is 1,326 pounds at 34.4 inches aft of the datum. This is also within the allowable weight and CG range.

Answers

275 [A] (008) AC 43.13-1
276 [B] (008) AC 43.13-1

276a. While performing a forward extreme conditions CG check on a twin-engine aircraft with 300-horsepower Continental engines, what would be the quantity of fuel required if the fuel tanks are aft of the rear CG limits?

A— 25 gallons.
B— 50 gallons.
C— 100 gallons.

When performing a forward extreme CG check, minimum weights must be used for all items located behind the forward CG limit (anything that would move the CG aft of the forward limit). This airplane has a total of 600 horsepower and minimum fuel for weight and balance purposes is 1/12 gallon for each METO horsepower. 50 gallons of fuel is required for this check (600 ÷ 12 = 50).

277. When inspecting a composite structure, what direction is the ply orientation in the upper layer of laminate?

A— 0°.
B— 40°.
C— 90°.

When the selvage-edge direction is not known, the direction of the surface ply may be considered to be 0° for the placement of the warp clock.

278. A letter "E" drill bit is the same as a 1/4-inch fraction. What is the decimal equivalent?

A— .3750
B— .2500
C— .1250

An E-letter drill and a 1/4-inch fraction drill are both 0.2500 inch in diameter.

278a. What is the diameter of a hole drilled with a 0.3125-inch drill?

A— 19/64
B— 11/32
C— 5/16

The common fraction that is equivalent to 0.3125 is 5/16 inch.

278b. What is 28 degrees Celsius in Fahrenheit?

A— 86.2 degrees Fahrenheit.
B— 81.4 degrees Fahrenheit.
C— 82.4 degrees Fahrenheit.

To change °C to °F, multiply °C by 1.8 and then add 32:

$$(28°C \times 1.8) + 32 = 82.4°F$$

278c. If the aircraft CG is located at 27.3% of MAC, the LEMAC is at station 1022, and the TEMAC is at station 1198, at what station is the CG?

A— 48.
B— 1022.
C— 1070.

Set up a formula in which 27.3 is to 100 as X is to 176 (distance between LEMAC and TEMAC or 1198 – 1022). The aircraft CG is given as a percentage of MAC, so its ratio to 100% will be the same as the ratio of the unknown number to the total distance between LEMAC and TEMAC. This number must be added to the LEMAC location to obtain the actual CG location with reference to the aircraft datum.

$$\frac{27.3}{100} = \frac{X}{176}$$

$$X = 48$$

$$48 + 1022 = 1070$$

278d. A decimal fraction of .1875 would relate to which of the following drill sizes?

A— 3/32.
B— 3/16.
C— 3/8.

Using basic mathematics, divide each fraction numerator by the fraction denominator to determine the decimal equivalent.

278e. A D-size drill is.004 inch smaller than a 1/4-inch fractional size. What is the decimal equivalent size of a D-drill?

A— .2460
B— .2480
C— .2500

1/4 inch is .2500 decimal. Subtracting .004 from .2500 provides .2460 as the answer.

Answers

276a [B] (008) AC 43.13-1
278a [C] (032) ASA-MHB
277 [A] (022) AMT-STRUC
278b [C] (032) ASA-MHB
278d [B] (005) ASA-MHB
278 [B] (032) ASA-MHB
278c [C] (032) ASA-MHB
278e [A] (005) ASA-MHB

278f. A technician has put a swaged end on a 1/8th cable, and has brought it to you for inspection. The shank diameter should be

A—.250
B—.141
C—.219

Swage-type terminals, manufactured in accordance with AN, are suitable for use in civil aircraft up to, and including, maximum cable loads. When swaging tools are used, it is important that all the manufacturers' instructions, including "go and no-go" dimensions, be followed in detail to avoid defective and inferior swaging. Observance of all instructions should result in a terminal developing the full-rated strength of the cable. Critical dimensions, both before and after swaging, are shown in AC 43.13-1B, Table 7-5.

278g. If the swage dimension is correct, then a proof load test should be conducted. For a 3/16th cable, the applied load should be:

A—2,000 lbs
B—2,200 lbs
C—2,520 lbs

Test the cable by proof-loading it to 60 percent of its rated breaking strength.

278h. How many degrees Fahrenheit is equal to -5° Celsius?

A—0 degrees.
B—23 degrees.
C—33 degrees.

To change °C to °F, multiple °C by 1.8, and then add 32:

(-5 × 1.8) + 32 = 23°F.

278i. What is 76° Fahrenheit equal to on the Celsius scale?

A—25 degrees.
B—37 degrees.
C—44 degrees.

To change °F to °C, subtract 32 degrees and then divide by 1.8:

(76 – 32) ÷ 1.8 = 25°C.

278j. A major repair requires the installation of AN3 bolts. What fractional size drill would be required for this installation?

A—3/16.
B—3/32.
C—3/64.

AN bolt dimensions are referenced in their fractional relationship to sixteenths of an inch. An AN3 bolt is 3/16-inch in diameter (decimal range .186 to .189 inches). An AN4 bolt is 4/16- inch in diameter (decimal .246 to .249 inches). A decimal drill of 0.1875 has a fractional value of 3/16 inch.

Answers

278f [C] (005) AC 43.13
278g [B] (005) AC 43.13-1
278h [B] (007) ASA-MHB
278i [A] (007) ASA-MHB
278j [A] (005) AC 43.13-1, ASA-AMT-G

AC 43.13-2A
Acceptable Methods, Techniques, and Practices— Aircraft Alteration

This AC contains methods, techniques, and practices acceptable to the Administrator for use in altering civil aircraft. ***Note:*** Be aware that although the FAA released the updated AC 43.13-2B in March 2008, the applicable text and reference exhibits in the most recent Computer Test Supplement (FAA-CT-8080-8D) have not yet been updated and therefore still reflect the information contained in AC 43.13-2A.

279. What publication is an acceptable document for obtaining load factors when computing static load tests for equipment installations in U.S.-certificated aircraft?

A— AC 43.13-2A.
B— Aircraft Flight Manual.
C— Type Certificate Data Sheet.

Chapter 1 of AC 43.13-2A Acceptable Methods, Techniques, and Practices — Aircraft Alterations *gives the required load factors for equipment installation in an aircraft.*

280. Static test load factors are the ultimate load factors multiplied by prescribed casting, fitting, bearing, and/or other special factors. Where no special factors apply, the static load factors are equal to the

A— limit load factors.
B— ultimate load factors.
C— critical load factors.

Chapter 1.2.c. of AC 43.13-2A Acceptable Methods, Techniques, and Practices — Aircraft Alterations *states that "Where no special factors apply, the static load factors are equal to the ultimate load factors."*

281. When installing additional equipment in an aircraft, if not otherwise specified, the ultimate load factor used in the static load test is

A— four times the weight of the equipment.
B— variable, depending on the direction of the applied force.
C— the limit load factor multiplied by 1.5.

Chapter 1.2.b. of AC 43.13-2A Acceptable Methods, Techniques, and Practices — Aircraft Alterations *states that "Ultimate load factors are the limit load factors multiplied by a prescribed factor of safety." 14 CFR §23.303 states that "Unless otherwise provided, a factor of safety of 1.5 must be used."*

282. A piece of equipment weighing 6.5 pounds is to be installed aft of the crew compartment on a rotorcraft certificated under 14 CFR Part 27. What is the forward static load test to ensure structural integrity for this piece of equipment?

Forward static load is

A— 26 pounds.
B— 39 pounds.
C— 58.5 pounds.

When a piece of equipment is installed in a rotorcraft aft of the crew compartment, its installation must withstand a static load of 4.0 g. If the equipment weighs 6.5 pounds, the installation must withstand a load of 26 pounds.

283. What document could be referenced to determine if an aircraft is eligible for ski installation?

A— Aircraft airworthiness certificate.
B— Type certificate data sheets.
C— Advisory Circular 43.13-2A.

Only aircraft approved for operation on skis are eligible for ski installations in accordance with this chapter. Eligibility can be determined by referring to the Aircraft Specifications, Type Certificate Data Sheets, Aircraft Listings, Summary of Supplemental Type Certificates, or by contacting the manufacturer.

284. At what static test loads should a banner tow hitch be tested to ensure structural integrity of the aircraft structure?

A— 1.5 times the operating weight of the banner.
B— 2.0 times the operating weight of the banner.
C— 2.5 times the operation weight of the banner.

Install the hitch to support a limit load equal to at least two times the operating weight of the banner.

Answers

279	[A]	(011)	AC 43.13-2	280	[B]	(011)	AC 43.13-2	281	[C]	(011)	AC 43.13-2
282	[A]	(011)	AC 43.13-2	283	[B]	(011)	AC 43.13-2	284	[B]	(011)	AC 43.13-2

285. What are the minimum requirements to operate a gyroscopic instrument at the proper speed?

A— Minimum required vacuum and rated CFM.
B— A flow requirement of .50 cubic feet per minute and a pressure drop of 2.00 in. Hg.
C— A flow requirement of 2.30 cubic feet per minute and a pressure drop of 4.00 in. Hg.

Gyroscopic instruments require optimum value of airflow to produce their rated rotor speed. It is important that the air pump and the plumbing provide the required vacuum and the rated air flow in cubic feet per minute at the instrument.

286. What are the minimum requirements to operate a gyroscopic instrument at the proper speed?

A— Enough vacuum required to operate the maximum number of instruments and positive pressure.
B— Maximum negative pressure to run total instruments, plus pressure drop in plumbing.
C— Enough vacuum required to operate the minimum number of instruments and positive pressure.

Gyroscopic instruments require optimum value of airflow to produce their rated rotor speed. It is important that the air pump and the plumbing provide the required vacuum and the rated air flow in cubic feet per minute at the instrument.

287. The vacuum pump of a vacuum instrumentation system must be able to deliver what?

A— Enough vacuum required to operate the maximum number of instruments and sufficient negative pressure.
B— Maximum negative pressure to run total instruments plus pressure drop in plumbing.
C— System minimum pressure differential at required airflow rate per minute.

Gyroscopic instruments require optimum value of airflow to produce their rated rotor speed. It is important that the air pump and the plumbing provide the required vacuum and the rated air flow in cubic feet per minute at the instrument.

288. You are required to install an instrument vacuum system in an airplane. There will be a bank and pitch indicator (gyro horizon), a directional gyro, and a turn and slip indicator. The installation will require 20 feet of 1/2-inch OD aluminum tubing having a wall thickness of 0.042 inch, and four 90° elbows.

The vacuum pump required for this installation must be capable of producing a negative pressure of

A— 4.00 inches of mercury.
B— 5.53 inches of mercury.
C— 11.53 inches of mercury.

1. Determine the amount of negative pressure (suction) and the number of cubic feet of air required for each instrument. This is found in the manufacturer's specifications for the instruments:

bank and pitch indicator... 2.30 CFM at 4.00 in. hg.
directional gyro .... 1.30 CFM at 4.00 in. hg.
turn and slip indicator ...... 0.50 CFM at 2.00 in. hg.

2. Find the total air flow required:

bank and pitch indicator.......................... 2.30 CFM
directional gyro .... 1.30 CFM
turn and slip indicator .... 0.50 CFM
Total air flow required.............................. 4.10 CFM

3. Find the equivalent length of tubing that accounts for the pressure drop in the four right-angle elbows. Table 11.2 on Page 86 of AC 43.13-2A shows that each 90° elbow has a pressure drop equivalent to 0.62 feet of straight tubing. The equivalent length of tubing accounting for four elbows is:

20 feet + 4 × 0.62 = 20 + 2.48 = 22.48 feet

4. Find the pressure drop through the tubing installation. Using the chart in Figure 11.1 in AC 43.13-2A. Follow these steps:

i. This chart gives the pressure drop per 10-foot length of tubing so the equivalent length of the tubing in feet is divided by 10.

22.48 ÷ 10 = 2.248

ii. Locate the airflow of 4.10 cubic feet per minute across the bottom of the chart and follow a line vertically upward until it intersects the curve for 1/2 × 0.042 tubing.

iii. From this intersection, project a horizontal line to the left to find the pressure drop per 10-foot length of tubing. This shows that there is a pressure drop of 0.68 inches of mercury for each 10-foot length of tubing. The pressure drop for the entire installation is:

0.68 · 2.248 = 1.53 inches of mercury

Continued

Answers

285 [A] (011) AC 43.13-2
286 [A] (011) AC 43.13-2
287 [C] (011) AC 43.13-2
288 [B] (011) AC 43.13-2

5. *Find the minimum suction, or pressure differential, the pump must be capable of producing:*

   Both the bank and pitch indicator and the directional gyro require 4.0 in. hg, and at the maximum flow, there is a pressure drop of 1.53 in. hg. in the tubing. The pump, therefore must be capable of producing a differential pressure of 5.53 in. hg.

289. A vacuum-driven gyro system was installed in accordance with an STC, with the exception that a 90-degree elbow was added. The system utilized a 1/2-inch, .042 line and 4 CFM. What would be the pressure drop across the elbow?

A— .062 in. Hg.
B— .067 in. Hg.
C— .042 in. Hg.

Find the pressure drop through the tubing installation. Using the chart in Figure 11.1 on Page 86 in AC 43.13-2A follow these steps:

1. *Follow the diagonal line for 1/2 × .042 until it intersects the vertical line for 4 cubic feet per minute (CFM). This intersection is at 0.67 inches of mercury drop for 10 feet.*
2. *The pressure drop for 1 foot is 0.067 inch of mercury.*

Find the equivalent length of tubing that accounts for the pressure drop in a 90° elbow. Table 11.2 on Page 86 of AC 43.13-2A shows that each 90° elbow has a pressure drop equivalent to 0.62 feet of straight tubing.

Multiply the pressure drop per foot (0.067) by 0.62 to find the pressure drop across the 90° elbow:

0.067 × 0.62 = 0.042 inch of mercury

289a. In reviewing the installation data for a proposed STC, you note that there will be additional vacuum supply demand. The customer's aircraft will require the following additional vacuum system plumbing to make the installation: 2 feet of 1/2 inch × .042 inch tubing, and two 90° elbows. This plumbing will supply an additional instrument that requires 1.20 CFM airflow. The existing vacuum system requires 4.80 CFM of airflow. The optimum instrument pressure differential will remain at 4.0 in. Hg. How much additional pressure drop will result from the added plumbing? (Answer in inches of mercury (in. Hg.))

A— .4455 in Hg.
B— .6000 in Hg.
C— .6455 in Hg.

Existing parameters are 4.8 CFM airflow, and optimum pressure differential is 4.0 in.Hg. The newly-added instrument requires an additional 1.2 CFM, making a total new system demand of 6.0 CFM. From Table 11-3, determine that the pressure drop associated with a ½ by 0.042 90° elbow is the equivalent of 0.62 feet of straight tubing. The 2 elbows to be added make the total 1.24 feet. Next, refer to Figure 11-2 and find the slant line labeled ½ × .042. Since the new requirements are for 6 CFM, find the vertical line associated with 6 CFM, and follow it until it intersects the ½ by .042. Next, read horizontally over to "Pressure drop per ten-foot length inches of mercury"; the number is 1.3 inches Hg per ten feet. The total tubing line length added is 3.24 (2 plus 1.24); 3.24 divided by 10 is .324. That number × 1.3 = 0.4212, closest to [A].

290. When installing an instrument filter, it should be installed

A— in the manifold intake port.
B— at the instrument intake port.
C— either A or B if the instruments are interconnected.

Filters are used to prevent dust, lint, and other foreign matter from entering the instrument and vacuum system. Filters may be located in the instrument intake port or at the manifold intake port when the instruments are interconnected.

Answers

289 [C] (011) AC 43.13-2
289a [A] (009) AC 43.13-2B, Ch 11
290 [C] (011) AC 43.13-2

FAA-H-8083-1
Aircraft Weight and Balance Handbook

This *Aircraft Weight and Balance Handbook* supersedes AC 91-23A *Pilot's Weight and Balance Handbook*. Its object is twofold:

- To provide the aviation maintenance technician with the method of determining the empty weight and empty-weight center of gravity of an aircraft
- To furnish the flight crew with information on loading and operating the aircraft to ensure its weight is within the allowable limit and the center of gravity is within the allowable range.

291. In a normally certificated aircraft, the "CG of the airfoil" (the aerodynamic center of the wing) is normally

A— before the aircraft CG.
B— coincidental of the aircraft CG.
C— aft of the aircraft CG.

Longitudinal stability is maintained by ensuring that the CG is slightly ahead of the center of lift.

291a. What is the advantage of locating the datum line forward of the nose of the aircraft?

A— To have a reference point to make all measurements.
B— To allow for all positive computations.
C— To allow both positive and negative computations.

The reference datum is an imaginary vertical plane from which all horizontal distances are measured for balance purposes. When the datum is ahead of the aircraft, all of the arms are positive and computational errors are minimized.

291b. What is meant by zero fuel weight?

A— Standard weight plus crew.
B— Operational aircraft empty weight.
C— Operational weight including payload minus fuel load.

Zero fuel weight is the weight of an aircraft with all of the useful load except the fuel on board.

291c. After a major repair that increased the structural weight of the airframe, a new empty weight and center of gravity were calculated. To ensure the airplane has acceptable flight characteristics, a maximum forward and rearward center of gravity check was made. If the repair has been properly completed, when the aircraft is loaded for the extreme rearward check, the center of lift will be

A— located behind the center of gravity.
B— coincidental with the center of gravity.
C— located forward of the center of gravity.

An airplane is designed to have stability that allows it to be trimmed so it will maintain straight and level flight with hands off of the controls. Longitudinal stability is maintained by ensuring the CG is slightly ahead of the center of lift. This produces a fixed nose-down force independent of the airspeed.

292. Normally, an aircraft will have acceptable flight characteristics if the CG is located at what percentage of the average chord?

A— 40 percent.
B— 25 percent.
C— 60 percent.

Longitudinal stability is maintained by ensuring that the CG is slightly ahead of the center of lift. This places the CG at approximately 25 percent of the mean aerodynamic chord (MAC) for best longitudinal stability.

Answers

291 [C] (029) FAA-H-8083-1
291a [B] (029) FAA-H-8083-1
291b [C] (029) ASA-DAT
291c [A] (029) FAA-H-8083-1
292 [B] (029) FAA-H-8083-1

292a. For most airplanes, if the center of gravity is located at 25% of the mean aerodynamic chord, the center of lift will

A— be forward of the center of gravity.
B— coincide with the center of gravity.
C— be behind the center of gravity.

It is important for longitudinal stability that the CG be located ahead of the center of lift of a wing. Since the center of lift is expressed as percent MAC, the location of the CG is expressed in the same terms. An aircraft has acceptable flight characteristics if the CG is located somewhere near the 25 percent average chord point. This means the CG is located one-fourth of the distance back from the LEMAC to the TEMAC. Such a location places the CG forward of the aerodynamic center for most airfoils.

293. If the aircraft weight and balance data states that the MAC is from stations 860.2 to 1040.9, and a weight and balance computation determines that the CG is located at station 910.2, the location expressed in percentage of MAC is

A— 27.7% MAC.
B— 14.9% MAC.
C— 24.7% MAC.

The length of the mean aerodynamic chord (MAC) is 1,040.9 – 860.2 = 180.7 inches.

Leading edge of the mean aerodynamic chord (LEMAC) to CG = 910.2 – 860.2 = 50 inches.

Use the formula:

$$CG = \frac{\text{Distance aft of LEMAC} \cdot 100}{MAC}$$

$$= \frac{50 \cdot 100}{180.7}$$

$$= 27.7\%\ MAC$$

293a. Find the CG in percent of MAC.

Given:

CG ....840.5 inches aft of datum
LEMAC ....786 inches aft of datum
TEMAC ....1,029 inches aft of datum
Empty weight.. 7,201 pounds

A— 21.79 % MAC
B— 22.43 % MAC
C— 28.94 % MAC

First, find the length of the mean aerodynamic chord (MAC) by subtracting the station of the leading edge of the mean aerodynamic chord (LEMAC) from the station of the trailing edge of the mean aerodynamic chord (TEMAC).

MAC = TEMAC – LEMAC = 1029 – 786 = 243 inches

Then, find the distance of the CG from the LEMAC:

CG from LEMAC = 840.5 – 786 = 54.5 inches

Then, use the formula:

$$CG\ in\ \%MAC = \frac{\text{CG in inches from LEMAC} \cdot 100}{MAC}$$

$$= \frac{54.5 \cdot 100}{243}$$

$$= 22.43\%\ MAC$$

294. Given:

Nose wheel weighs 20,500 lbs ....+120.0
Right wheel weighs 70,000 lbs........................+600.0
Left wheel weighs 70,500 lbs ....+600.0
LEMAC ....+500.0
TEMAC ....+690.0

Find the CG expressed in %MAC for this aircraft.

A— 20.46%
B— 21.5%
C— 18.0%

The CG is found by dividing the total moment by the total weight:

Item	*Weight*	*Arm*	*Moment*
Nose wheel	*20,500*	*120.0*	*2,460,000*
Right wheel	*70,000*	*600.0*	*42,000,000*
Left wheel	*70,500*	*600.0*	*42,300,000*
	161,000	*538.9*	*86,760,000*

The length of the mean aerodynamic chord (MAC) is 690.0 – 500.0 = 190 inches. Leading edge of the mean aerodynamic chord (LEMAC) to CG = 538.9 – 500.0 = 38.9 inches.

Use the formula:

$$CG = \frac{\text{Distance aft of LEMAC} \cdot 100}{MAC}$$

$$= \frac{38.9 \cdot 100}{190}$$

$$= 20.46\%\ MAC$$

Answers

292a [C] (002) FAA-H-8083-1
293 [A] (008) FAA-H-8083-1
293a [B] (008) FAA-H-8083-1
294 [A] (008) FAA-H-8083-1

295. A mechanic has just installed a new reinforcement bracket on an airplane. The bracket weighs 18 lbs and was installed at station 352. With the given information, which of the following statements is true?

Given:

Empty weight of A/C 4,836
LEMAC location .... +115.28
200 lbs of baggage .... +19.0
Empty weight CG .... +124.13
2 pilot seats .... +95.0
Minimum fuel (350 lbs) .... +126.8
Forward CG limit... +120
Aft CG limit .... +132.5

A— The new CG remains within CG limits.
B— The new CG is located at 4.34 inches forward of LEMAC.
C— Ballast will have to be added.

The CG is found by dividing the total moment by the total weight:

Item	*Weight*	*Arm*	*Moment*
Airplane E.W.	*4,836*	*124.13*	*600,292.7*
Bracket	*18*	*352*	*6,336.0*
Baggage	*200*	*19.0*	*3,800.0*
Pilots (170 × 2)	*340*	*95.0*	*32,300.0*
Fuel	*350*	*126.8*	*44,380.0*
	5,744	*119.62*	*687,108.7*

The CG is 119.62. Ballast will have to be added to return the aircraft to an acceptable CG.

295a. A major alteration has been completed to the empennage of a light-twin engine airplane. Additional reinforcement was added at +355 inches, weighing 20 pounds. Working as the holder of an inspection authorization, you examine the weight and balance condition using the following information. If the items of useful load listed below are placed on the aircraft, the aircraft would:

A— require ballast.
B— exceed the rearward CG limit.
C— be within the loaded CG limits.

Two pilots weighing a total of 380 pounds at +95 inches.
Baggage weighing 200 pounds at +110 inches.
Minimum fuel weighing 350 pounds at +125.8 inches.
Known information (before repair):
Empty weight 4,855 pounds.
Empty weight center of gravity +124.13 inches.
CG range (+126.0) to (+135.0) at 7,000 pounds.
(+122.0) to (+135.0) at 6,200 pounds.
(+120.0) to (+135.0) at 5,200 pounds.

296. Which statement is correct regarding the amount of fuel to be in the aircraft when making a check for the forward adverse loaded center of gravity?

A— Full fuel if the tank is forward of the forward CG limit.
B— Full fuel if the tank is aft of the forward CG limit.
C— Full fuel if the tank is aft of the rearward CG limit.

When making a forward adverse-loaded CG check, include full fuel if the tank is located forward of the forward CG limit. If the tank is aft of the forward CG limit, use the minimum fuel of 1/12 gallon for each maximum except takeoff (METO) horsepower.

297. An airplane is loaded with a ramp weight of 3,650 pounds with a CG of 94.0. Approximately how much baggage would have to be moved from the rear baggage compartment at station 180 to the forward baggage compartment at station 40 to move the CG to 92.0?

A— 78.14 pounds.
B— 62.14 pounds.
C— 52.14 pounds.

The weight is to be shifted from station 180 to station 40, a distance of 140 inches. The CG is to be shifted from 94.0 to 92.0, a distance of 2 inches.

Use the formula:

$$\text{Weight to be shifted} = \frac{\Delta CG \cdot \text{Total weight}}{\text{Distance weight is shifted}} = \frac{2 \cdot 3{,}650}{140} = 52.14 \text{ pounds}$$

298. What is the name of the maximum permissible weight of an aircraft with no disposable fuel?

A— Maximum ramp weight.
B— Maximum landing weight.
C— Maximum zero fuel weight.

The maximum zero fuel weight means the maximum permissible weight of an aircraft with no disposable fuel.

Answers

295 [C] (008) FAA-H-8083-1
295a [C] (008) FAA-H-8083-1
296 [A] (029) FAA-H-8083-1
297 [C] (008) FAA-H-8083-1
298 [C] (029) FAA-H-8083-1

FAA-G-8082-11C
Inspection Authorization Knowledge Test Guide

This study guide provides guidance for persons who conduct annual and progressive inspections and approve major repairs and/or major alterations of aircraft. The guide is primarily intended for mechanics who hold or who are preparing to take the test for inspection authorization (IA). The guide stresses the important role that certificated mechanics who hold an IA have in air safety.

299. A person holding an inspection authorization is responsible for ensuring certain levels of safety in aeronautical products. When inspecting a major alteration for conformity and approval for return to service, the holder of an inspection authorization must determine that the data used is

A— acceptable to the Administrator.
B— acceptable to the manufacture.
C— compatible with other installations.

Inspect the configuration of the repair or alteration for conformity to the approved data and the performance standards of 14 CFR Part 43. At the same time, the aircraft should still comply with applicable airworthiness requirements and the repair or alteration should be compatible with all other installations.

300. Which of the following data could be used as approved data without any further approval?

A— Data developed by an owner and approved by a mechanic holding an inspection authorization.
B— Data developed for and incorporated in a supplemental type certificate.
C— Data developed by a commercial operator under a continuous maintenance program.

Approved data to be used for major repairs and major alterations may be: Supplemental Type Certificates (STCs).

301. When considering a major alteration to be made in accordance with a Supplemental Type Certificate (STC), what must the approving mechanic with inspection authorization ensure?

A— The STC was legally obtained.
B— The STC is compatible with other systems.
C— The STC is in compliance with the certification rule.

During inspection of major repairs and alterations, the holder of an IA must also determine that they are compatible with previous repairs and modifications that have been made to the aircraft.

302. A small aircraft with a standard airworthiness certificate has been modified per supplemental type certificate (STC) that is designed for a restricted, special purpose operation. The work was performed to 14 CFR Part 43 standards and was compatible with all other installations. As the IA, who inspects the alteration you would

A— reject the FAA Form 337 because the aircraft no longer meets the requirements for its standard airworthiness certificate.
B— approve the aircraft for return to service.
C— hold the FAA Form 337 until the aircraft has been recertified with the appropriate airworthiness certificate.

Because the work was done according to approved data (the STC), was performed to 14 CFR Part 43 standards and was compatible with all other installations, the IA is cleared to approve the aircraft for return to service.

303. For an appliance major repair, which of the following data could be used without further approval?

A— Advisory Circular.
B— Manufacturer's manual.
C— Previously approved FAA Form 337.

Approved data to be used for major repairs and major alterations may be: Appliance manufacturer's manuals (excluding installation instructions).

304. When considering a major alteration to be made in accordance with a supplemental type certificate (STC), the approving mechanic with inspection authorization must ensure

A— the STC is compatible with other previous alterations.
B— the STC is applicable with make and model.
C— Both A and B.

Answers

299	[C]	(011)	FAA-G-8082-11
300	[B]	(011)	FAA-G-8082-11
301	[B]	(011)	FAA-G-8082-11
302	[B]	(010)	FAA-G-8082-11
303	[B]	(011)	FAA-G-8082-11
304	[C]	(010)	FAA-G-8082-11

During inspection of a major alteration made in accordance with an STC, the holder of an IA should not only verify that the STC is authorized for the make and model aircraft currently being modified, but must also determine that it is compatible with previous alterations that have been made to the aircraft.

305. For an appliance, the manufacturer's manual could be used as approved data for which of the following?

A— Major repairs.
B— Major alterations.
C— Both A and B.

Approved data to be used for major repairs and major alterations may include Appliance Manufacturer's Manuals (excluding Installation Instructions).

306. For an appliance major alteration, which of the following could be used as approved data without further approval?

A— Advisory Circulars.
B— Manufacturer's manual.
C— Neither A or B.

Approved data to be used for major repairs and major alterations may include Appliance Manufacturer's Manuals (excluding Installation Instructions).

307. The owner of an aircraft wants to know the feasibility of performing a major alteration to her aircraft. The alteration would be based on a field-approved FAA Form 337, from another modifier of a like aircraft. Which of the following statements would be correct in this situation?

A— If the proposed alteration does not conflict with any previous modification, the data could be used without further approval.
B— The data could be used as the basis in seeking a field approval.
C— The data could be used without further approval if the FAA Form 337 stated "approval issued for duplication of identical aircraft."

An FAA field approval (FAA Form 337) with an FAA signature in Block 3 issued for duplication of identical aircraft may be used as approved data only when the identical alteration is performed on an aircraft of identical make, model, and series ***by the original modifier****. It may, however, be used by others as the basis for seeking a field approval.*

308. As an IA, you are performing an annual inspection on an aircraft and notice that the aircraft does not have an N number. The owner says that he will not be flying the aircraft until the N number is on the aircraft. As an IA, what is your responsibility in this situation?

A— Give the owner a list of discrepancies and find the aircraft unairworthy.
B— Approve the aircraft for return to service.
C— Sign the aircraft off as per 14 CFR Part 43, App. A (c)(9).

Required aircraft identification markings are discussed in 14 CFR Part 45. It is the owner's or operator's responsibility to have the nationality and registration markings properly displayed on the aircraft (14 CFR §91.9(c)). The holder of an IA can and should offer advisory service to owners and operators in regard to any deficiencies in markings; however, such deficiencies are not cause to report an aircraft in an "unairworthy" condition.

Answers

305 [C] (011) FAA-G-8082-11
306 [B] (011) FAA-G-8082-11
307 [B] (010) FAA-G-8082-11
308 [B] (017) FAA-G-8082-11

AC 91-67
Minimum Equipment Requirements for General Aviation Operations Under 14 CFR Part 91

This AC describes acceptable methods for the operation of aircraft under Title 14 of the Federal Regulations (14 CFR) Part 91 with certain inoperative instruments and equipment which are not essential for safe flight. This AC was cancelled on Nov 3, 2017, as it is no longer aligned with current ICAO standards. However, applicants may still see questions pertaining to this AC and should monitor **www.faa.gov** to stay informed of a new edition.

309. A pilot delivers his 14 CFR Part 91 airplane to you, the IA, for an annual inspection. He informs you that the autopilot has become inoperative on the flight in. He requests that you postpone repair under the MEL. On consulting the MEL, the remarks section states "*(M) As required by 14 CFR." What document should next be referenced in determining if this item of equipment could be deferred?

A— 14 CFR Part 91.
B— Kinds of Operations List (KOL).
C— Procedures Document.

Procedures Document, as referred to in this AC, pertains to a separate document containing the O and M procedures developed by the operator and any other operating information applicable to operation with an MEL, such as the "as required by 14 CFR" items that list 14 CFR by part and section or stipulate the operating conditions.

310. Operational and Maintenance procedures (O and M), and/or references for disabling or rendering inoperative items of equipment are found in what document of an 14 CFR Part 91 operator's MEL?

A— Procedures Document.
B— Letter of Authorization (LOA).
C— Master Minimum Equipment List.

Procedures Document, as referred to in this AC, pertains to a separate document containing the O and M procedures developed by the operator and any other operating information applicable to operation with an MEL, such as the "as required by 14 CFR" items that list 14 CFR by part and section or stipulate the operating conditions.

311. The owner of a light-twin, operated under 14 CFR Part 135, consults you, as the holder of an inspection authorization, about the feasibility of installing an STC on his aircraft. What effect could a proposed modification have on the aircraft's approved MEL?

A— The installation would not affect the MEL, however, since the STC is not on the MMEL, the STC's function would not be deferrable.
B— The installation may invalidate the MMEL (Master MEL).
C— The items on the MMEL are not amendable by the IA, however, the STC's function, if allowable, could be deferred under 14 CFR Section 91.213(d).

An MEL is a precise listing of instruments, equipment, and procedures that allows an aircraft to be operated under specific conditions with inoperative equipment. The MMEL, as part of the MEL, by nature does not cover equipment installed or modified under other STCs. Any STC or other major modification may make the MMEL invalid for a particular modified aircraft.

312. Items of inoperative equipment that are deferred under an 14 CFR Part 91 MEL may remain deferred

A— until the annual inspection becomes due, or the progressive inspection has completed a complete inspection period.
B— so long as it is inspected and determined not to pose a hazard to further operation.
C— until the time period specified by the A, B, C, codes of the MEL require repair or replacement.

If an owner does not want specific inoperative equipment repaired, then the maintenance person must check each item to see if it conforms to the requirements of 14 CFR §91.213(d). The operator and maintenance personnel should also assess how permanent removal of the item could affect safe operation of the aircraft. Each item of inoperative equipment that is to remain inoperative must be appropriately placarded.

Answers

309 [C] (025) AC 91-67
310 [A] (025) AC 91-67
311 [B] (025) AC 91-67
312 [B] (025) AC 91-67

313. How may discrepancies to a Part 135 aircraft with an MEL be disposed of during periods of Part 91 operation?

A— The provisions of the 14 CFR Part 91 MMEL may be used until the aircraft is returned to 14 CFR Part 135 operation.
B— Any discrepancies must be disposed of by the provisions of the aircraft's MEL.
C— The provisions of 14 CFR Section 91.213(d) could be used during periods of 14 CFR Part 91 operations.

A person authorized to use an approved Minimum Equipment List issued for a specific aircraft under Part 121, 125, or 135 of this chapter shall use that Minimum Equipment List in connection with operations conducted with that aircraft under this part without additional approval requirements.

314. A small single-engine turboprop airplane that is operated under 14 CFR Part 91 is presented to you for an annual inspection. During the inspection, you observed two items of non-essential equipment have been previously deferred. Further inspection reveals the items are deactivated and placarded properly. The aircraft does not have a minimum equipment list. How should you proceed?

A— Complete the annual inspection, but inform the owner that the items should be repaired as soon as possible.
B— Inspect the items and continue to defer if safe to do so.
C— List the items as unairworthy.

A turboprop airplane is required to have an approved MEL. If there is no MEL on the aircraft, there is no authorization to operate with deferred equipment. Therefore the items must be listed as unairworthy.

315. During the annual inspection of an 14 CFR Part 91, small, single-engine, turbo-propeller airplane, an item of nonessential inoperable equipment was noted. If the airplane does not have an approved minimum equipment list, how must this discrepancy be handled?

A— Deactivate and placard the item, make a maintenance record entry.
B— Make a maintenance record entry listing the item as unairworthy.
C— The discrepancy may be disposed of by either A or B.

A turboprop airplane is required to have an approved MEL. If there is no MEL on the aircraft, there is no authorization to operate with deferred equipment. Therefore the items must be listed as unairworthy.

316. As the holder of an inspection authorization, you have performed an inspection on a light, reciprocating twin-engine airplane, used in 14 CFR Part 135 air carrier operation. During the inspection, an inoperative instrument was discovered. The instrument is not listed on the approved MEL; however, this instrument is required by the aircraft TCDS. The pilot points out that the flight to a location where the aircraft would be repaired is under 14 CFR Part 91. She suggests you defer the maintenance under 14 CFR Section 91.213. How would you proceed?

A— Determine that the instrument is nonessential, if so, deactivate and placard, then return to service.
B— Make a maintenance record entry listing the instrument as an unairworthy item.
C— If the instrument is not on the Kinds of Operations List (KOL) as necessary for the flight, the airplane could be returned to service.

If an aircraft is operating as a Part 135 air carrier, it must comply with the MEL guidance and cannot "flip" back to being a Part 91 operation. Therefore, the fact that there is an MEL drives the conclusion. All components on an aircraft can be put into 1 of 3 categories (based on their airworthiness requirements):

1. *Items clearly required for the aircraft to be airworthy (engines, wings, tires, struts, etc.).*
2. *Items clearly NOT required for the aircraft to be airworthy (curtains, flight phones, entertainment systems, etc.).*
3. *Items which do not clearly fall into either of the two previous categories.*

MEL items are those that fall into category 3. The MEL does not include every piece of equipment or system on the aircraft. When no specific mention of a unit or system is made in the MEL and it does not fall into the non-essentials category, then it is necessary that the equipment be in place and operative.

Answers

313 [B] (025) AC 91-67 and Part 91
314 [C] (025) AC 91-67 and Part 91
315 [B] (025) AC 91-67 and Part 91
316 [B] (025) AC 91-67 and Part 91

317. The letter designators in column 1 (ITEM) of an MEL have what significance to an 14 CFR Part 91 operator during aircraft maintenance?

A— Establishes the repair time at which that item, if deferred, must be repaired.
B— Indicates who may defer the item, the pilot or mechanic.
C— This designator has no significance to the 14 CFR Part 91 operator.

The Parts A, B, C, and D codes, listed in column 1 of the MMEL, apply only to operations conducted under 14 CFR Parts 121, 125, 129, and 135, and have no significance to a Part 91 operator.

318. What significance does the asterisk * symbol in column 4 of an MEL have for maintenance personnel?

A— The maintenance person should see the reference section for the corresponding Federal Aviation Regulation.
B— It indicates the item of equipment was added by the operator and not originally on the Master MEL.
C— The item of equipment, if inoperative, must be placarded.

The asterisk "" symbol in column 4 indicates that the listed item, if inoperative, must be placarded to inform and remind the crewmembers and maintenance personnel of the equipment condition.*

319. The owner of an airplane asks you to perform an annual inspection on the airplane he has just purchased. The previous owner transferred all the aircraft's records including his MEL and procedures document with the aircraft. The new owner informs you that the propeller deicing system is not working and asks you to defer repair of it under the MEL. The MEL lists the equipment and states in the remarks section, "May be inoperable provided flight is not operated in known or forecast icing conditions." How should you proceed?

A— Inform the owner that the equipment cannot be deferred under this MEL.
B— Determine that the equipment is nonessential, if so, deactivate and placard before approving for return to service.
C— If the equipment is not required by the aircraft's certification rule, then defer it per the MEL, ensuring any (M) maintenance procedures are accomplished.

The MEL and LOA are not transferable. The MEL and LOA must be surrendered to the FSDO exercising jurisdiction. The new owner must decide if he/she wants to operate with an MEL or under the provisions of 14 CFR §91.213(d). If the owner elects to operate with an MEL, he/she must apply for one at the appropriate FSDO as described in this AC.

Answers

317 [C] (025) AC 91-67
318 [C] (025) AC 91-67
319 [A] (025) AC 91-67

Reprints of Advisory Circulars

U.S. Department of Transportation
Federal Aviation Administration

Advisory Circular

Subject: Airworthiness Directives | **Date:** 3/2/12 | **AC No:** 39-7D
Initiated by: AFS-300 | **Change:**

1. PURPOSE. This advisory circular (AC) provides guidance and information to owners and operators of aircraft concerning their responsibility for complying with Airworthiness Directives (AD) and recording AD compliance in the appropriate maintenance records.

2. CANCELLATION. This AC cancels AC 39-7C, ADs, dated November 16, 1995.

3. RELATED REGULATIONS. Title 14 of the Code of Federal Regulations (14 CFR):

- Part 39, Airworthiness Directives,
- Part 43, Maintenance, Preventive Maintenance, Rebuilding, and Alteration,
    - Section 43.9, Content, Form, and Disposition of Maintenance, Preventive Maintenance, Rebuilding, and Alteration Records (except inspections performed in accordance with part 91, part 125, § 135.411(a)(1), and § 135.419 of this chapter), and
    - Section 43.11, Content, Form, and Disposition of Records for Inspections Conducted Under Parts 91 and 125 and §§ 135.411(a)(1) and 135.419 of This Chapter;
- Part 91, General Operating and Flight Rules,
    - Section 91.403, General,
    - Section 91.417, Maintenance Records, and
    - Section 91.419, Transfer of Maintenance Records.

4. RELATED READING MATERIAL (current editions).

- FAA-IR-M-8040.1, Airworthiness Directives Manual, and
- FAA Order 8110.103, Alternative Methods of Compliance (AMOC).

5. BACKGROUND. The authority for the role of the Federal Aviation Administration (FAA) regarding the promotion of safe flight for civil aircraft may be found in Title 49 of the United State Code (49 U.S.C.) § 44701 et. seq. (formerly, Title VI of the Federal Aviation Act of 1958 and related statutes). Pursuant to 49 U.S.C. § 44709(a), the Administrator of the FAA may reinspect and reexamine, at any time, a civil aircraft, aircraft engine, propeller, or appliance. One way the FAA has implemented its authority is through part 39. Pursuant to its authority, the FAA issues ADs when an unsafe condition exists in a product (aircraft, aircraft engine, propeller, or appliance) and is likely to exist or develop in other products of the same type design. ADs are issued by the FAA to notify aircraft owners and operators of an unsafe condition and to require action(s) to resolve the unsafe condition. ADs prescribe the conditions and limitations, including inspection, repair, or alteration under which the product may continue to be operated. ADs are authorized under part 39 and issued in accordance with the public rulemaking procedures of the Administrative Procedure Act (APA), Title 5 of the United States Code (5 U.S.C.) § 553, and FAA procedures in 14 CFR part 11.

6. PRINCIPAL CHANGES. References to specific 14 CFR parts have been updated and text reworded for clarification throughout this AC.

7. AD CATEGORIES. ADs are published in the Federal Register (FR) as amendments to part 39 when an unsafe condition is found to exist in a product. Depending on the urgency, ADs are issued as follows:

a. Notice of Proposed Rulemaking (NPRM) followed by a Final Rule. This is the most common type of AD. An NPRM is issued whenever safety considerations do not require the immediate imposition of action under an AD. Anyone is invited to comment on the NPRM by submitting written comments. After the comment period closes, the final rule is prepared, taking into account the comments received. Actions proposed in the notice may be changed or withdrawn in light of the comments. When the final rule, resulting from the NPRM, is adopted, it is published in the FR, and is also available electronically to subscribers at the FAA's Regulatory and Guidance Library (RGL) Web site, http://rgl.faa.gov.

b. Final Rule; Request for Comment (FRC) (commonly referred to as an "Immediately Adopted Rule"). This is used in certain cases, when an unsafe condition warrants the immediate adoption of a rule without prior notice. In general, we issue an FRC only when it is impractical to complete the prior notice requirement procedure because the compliance time for the required action is shorter than the time necessary for the public to comment and for the FAA to publish the final rule. Comment dispositions are only published when a comment warrants a change to the FRC, or when a significant issue is raised that might have wide or continuing interest among members of the affected public.

c. Emergency ADs. These ADs are of an urgent nature (e.g., immediate safety of flight) and cannot wait for publication in the FR. When an Emergency AD is issued, the AD applies only to the people who receive "actual notice" by First-Class mail and/or fax to the registered owners of those aircraft. Therefore, a followup AD is published in the FR normally as an FRC. The FR version contains the "good cause" findings required by §§ 553(b)(3)(B) and 553(d) of the APA and makes the AD effective to all persons. Other than very minor corrections (such as obvious typographical errors) and the addition of certifications and analyses required by statute, standard formatting required for FR publication, and material required for incorporation by reference, the version published in the FR is identical to the Emergency AD.

8. ELECTRONIC DISTRIBUTION OF ADS.

a. AD Copies. Official copies of any AD are available in the FR at the Web site, http://www.gpoaccess.gov/fr/index.html. The FAA only distributes Emergency ADs by

First-Class Mail and/or fax. All other final rule ADs are available by signing up for electronic delivery via the FAA RGL Web site, http://rgl.faa.gov. On this Web site, you will find a link entitled "Subscribe for email delivery of ADs and SAIBs." Subscribers must enter their e-mail address and pick the aircraft/engine/propeller makes and models they want to receive information for. Subscribers will automatically receive all applicable ADs and Special Airworthiness Information Bulletins (SAIB) by e-mail.

b. Appliance AD. A final rule AD related to an appliance is distributed using the electronic delivery described above for the aircraft model(s) selected.

c. NPRMs. The FAA does not electronically distribute NPRMs. To find NPRMs, you must visit the FAA's RGL at the Web site, http://rgl.faa.gov, or the FR Web site listed above.

9. APPLICABILITY OF ADs. Each AD contains an applicability statement specifying the product (aircraft, aircraft engine, propeller, or appliance) to which it applies. Unless stated otherwise (see subparagraph 9b of this AC), ADs only apply to type-certificated (TC) aircraft, including ADs issued for an engine, propeller, and appliance.

a. TC'd Aircraft, Engines, and Propellers. For ADs issued against aircraft, engines, and propellers certified under 14 CFR part 21, the Type Certificate Data Sheet (TCDS) is used to identify the affected product. Limitations may be placed on applicability by specifying the serial number or number series to which the AD is applicable. When there is no reference to serial numbers, all serial numbers are affected. The following are examples of AD applicability statements for TC'd products:

(1) "This AD applies to Hawker Beechcraft Corporation (Type Certificate previously held by Raytheon Aircraft Company) Model 1900, 1900C, and 1900D airplanes, certificated in any category." This statement makes the AD applicable to all airplanes of the model listed, regardless of the type of airworthiness certificate issued to the TC'd aircraft.

(2) "This AD applies to Hawker Beechcraft Corporation (Type Certificate previously held by Raytheon Aircraft Company) Model 1900D airplane, Serial Numbers UE-1 through UE-439, certificated in any category." This statement specifies certain aircraft by serial number within a specific model, regardless of the type of airworthiness certificate issued to the TC'd aircraft.

(3) "This AD applies to Aerotek II, Inc. Models B-1 and B-1A airplanes, certificated in any category except restricted." This statement makes the AD applicable to all TC'd airplanes except those issued a special airworthiness certificate in the restricted category.

(4) "This AD applies to Learjet Inc. (Type Certificate Previously Held by Gates Learjet Corporation) Model 23, 24, 24A, 24B, 24B-A, 24D, 24D-A, 24E, 24F, 25, 25A, 25C, 25D, and 25F airplanes, certificated in any category, modified by Supplemental Type Certificate SA1731SW, SA1669SW, or SA1670SW." This statement makes the AD applicable to all TC'd airplanes listed when altered by the Supplemental Type Certificate (STC) listed, regardless of the type of airworthiness certificate issued to the TC'd aircraft.

(5) "This AD applies to Lycoming Engines Models AEIO-360-A1A and IO-360-A1A." This statement makes the AD applicable to the engine models listed that are installed on TC'd aircraft.

b. Non-TC'd Aircraft and Products Installed Thereon. Non-TC'd aircraft (e.g., amateur-built aircraft, experimental exhibition) are aircraft for which the FAA has not issued a TC under part 21. The AD applicability statement will identify if the AD applies to non-TC'd aircraft or engines, propellers, and appliances installed thereon. The following are examples of applicability statements for ADs related to non-TC'd aircraft:

(1) "This AD applies to Honeywell International Inc. Auxiliary Power Unit (APU) models GTCP36-150(R) and GTCP36-150(RR). These APUs are installed on, but not limited to, Fokker Services B.V. Model F.28 Mark 0100 and F.28 Mark 0070 airplanes, and Mustang Aeronautics, Inc. Model Mustang II experimental airplanes. This AD applies to any aircraft with the listed APU models installed." This statement makes the AD applicable to the listed auxiliary power unit (APU) models installed on TC'd aircraft, as well as non-TC'd aircraft.

(2) "This AD applies to Lycoming Engines Models AEIO-360-A1A and IO-360-A1A. This AD applies to any aircraft with the listed engine models installed." This statement makes the AD applicable to the listed engine models installed on TC'd and non-TC'd aircraft.

c. Changed Products. An AD applies to each product identified in the applicability statement, regardless of whether it has been modified, altered, or repaired in the area subject to the requirements of the AD. For products that have been modified, altered, or repaired so that performance of the requirements of the AD is affected, the owner/operator must use the alternative methods of compliance (AMOC) provision of the AD (see paragraph 12) to request approval from the FAA. This approval may address either no action if the current configuration eliminates the unsafe condition, or different actions necessary to address the unsafe condition described in the AD. In no case does the presence of any alteration, modification, or repair remove any product from the applicability of this AD.

10. AD COMPLIANCE. ADs are regulations issued under part 39. Therefore, no person may operate a product to which an AD applies, except in accordance with the requirements of that AD. Owners and operators should understand that to "operate" not only means piloting the aircraft, but also causing or authorizing the product to be used for the purpose of air navigation, with or without the right of legal control as owner, lessee, or otherwise. Compliance with Emergency ADs can be a problem for operators of leased aircraft because they may not be aware of the AD and safety may be jeopardized.

11. COMPLIANCE TIME OR DATE.

a. Specified Compliance Time. The belief that AD compliance is only required at the time of a required inspection (e.g., at a 100-hour or annual inspection) is not correct. The required compliance time is specified in each AD, and no person may operate the affected product after expiration of that stated compliance time without an AMOC approval for a change in compliance time.

b. Requirements. Compliance requirements specified in ADs are established for safety reasons and may be stated in various ways. Some ADs are of such a serious, time-critical nature that they require compliance before further flight (e.g., to prevent uncommanded engine shutdown with the inability to restart the engine; the compliance statement may be written as "Prior to further flight, inspect...."). Other ADs express compliance time in terms of a specific number of hours in operation (e.g., "Within the next 50 hours time-in-service after the effective date of this AD, visually inspect...."). Compliance times may also be expressed in operational terms (e.g., "Within the next 10 landings after the effective date of this AD...."). For turbine engines, compliance times are often expressed in terms of cycles. A cycle normally consists of an engine start, takeoff, operation, landing, and engine shutdown.

c. Expression of Time. When a direct relationship between airworthiness and calendar-time is identified, compliance time may be expressed as a calendar date. For example, if the compliance time is specified as "Within 12 months after the effective date of this AD..." with an effective date of July 15, 1995, the deadline for compliance is July 15, 1996.

d. Special Flight Permits. In some instances, you may need to fly an aircraft to a repair facility to do the work required by an AD. Unless the AD states otherwise, you may apply to the FAA for a special flight permit following the procedures in part 21, § 21.199.

12. AMOCs.

a. AMOC Definition. An AD contains the required method for resolving an unsafe condition. An AMOC is a different way, other than the one specified in an AD, to address the unsafe condition on an aircraft, aircraft engine, propeller, or appliance. The term AMOC is used to define a FAA-approved AMOC to the specific requirements of an AD, including a change in the required time to accomplish the AD. An AMOC must ensure the unsafe condition is resolved by providing an acceptable level of safety.

b. Submitting an AMOC Request. In accordance with part 39, § 39.19, AMOC requesters should send their AMOC proposal to their principal inspector (PI). The PI may add comments and must forward a copy of the AMOC proposal to the manager of the FAA office identified in the AD. The requester may, at the same time they send it to their PI, send a copy of the proposal to the manager of the office identified in the AD. If the requester doesn't have a PI (such as a design approval holder (DAH)), we advise them to send the proposal directly to the manager of the FAA office identified in the AD.

c. FAA Approval. Any AMOC must provide an acceptable level of safety and be substantiated and approved by the Aircraft Certification Office (ACO) before it may be used.

13. RESPONSIBILITY FOR AD COMPLIANCE AND RECORDATION. The owner or operator of an aircraft is primarily responsible for maintaining that aircraft in an Airworthy condition, including compliance with ADs.

a. Means of Accomplishment. This responsibility may be met by ensuring that properly certificated and appropriately rated maintenance person(s) accomplish the requirements of the AD and properly record this action in the appropriate maintenance records. This action must be accomplished within the compliance time specified in the AD or the aircraft may not be operated.

b. Other Inspections. Maintenance persons may also have direct responsibility for AD compliance, aside from the times when AD compliance is the specific work contracted by the owner or operator. When a 100-hour, annual, progressive, or any other inspection required under 14 CFR part 91, 121, 125, or 135 is accomplished, § 43.15(a) requires the person performing the inspection to determine that all applicable airworthiness requirements are met, including compliance with ADs.

c. Progressive Inspections. Maintenance persons should note that even though an inspection of the complete aircraft is not made, if the inspection conducted is a progressive inspection, determination of AD compliance is required for those portions of the aircraft inspected.

d. Continuous Inspection Programs. For aircraft being inspected in accordance with a continuous inspection program (§ 91.409), the person performing the inspection must ensure that an AD is complied with only when the portion of the inspection program being handled by that person involves an area covered by a particular AD. The program may require a determination of AD compliance for the entire aircraft by a general statement, compliance with ADs applicable only to portions of the aircraft being inspected, or it may not require compliance at all. This does not mean AD compliance is not required at the compliance time or date specified in the AD. It only means that the owner or operator has elected to handle AD compliance apart from the inspection program. The owner or operator remains fully responsible for AD compliance.

e. Required Entries into Records. The person accomplishing the AD is required by § 43.9 to record AD compliance. The entry must include those items specified in § 43.9(a)(1) through (a)(4). The owner or operator is required by § 91.405 to ensure that maintenance personnel make appropriate entries and, by § 91.417, to maintain those records. Owners and operators should note that there is a difference between the records required to be kept by the owner under § 91.417 and those that § 43.9 requires maintenance personnel to make. In either case, the owner or operator is responsible for maintaining proper records.

14. RECURRING/PERIODIC ADs. Some ADs require repetitive or periodic inspection. In order to provide for flexibility in administering such ADs, an AD may provide for adjustment of the inspection interval to coincide with inspections required by part 91, or other regulations. The conditions and approval requirements under which adjustments may be allowed are stated in the AD. Any other modification or adjustment of the compliance time of the AD must be requested through the AMOC process as described in paragraph 13.

15. DETERMINING REVISION DATES.

a. Revision Date. The revision date required by § 91.417(a)(2)(v) is the effective date of the latest amendment to the AD and may be found in paragraph (a) of the body of each AD. For example, "This airworthiness directive (AD) is effective April 12, 2005."

b. Emergency ADs. Similarly, the revision date for an Emergency AD is the date it was issued. For example, "Emergency airworthiness directive (AD) 2006-17-51, issued

August 15, 2006, is effective immediately upon receipt." Each emergency AD is followed by an FRC version published in the FR that will reflect the amendment number of the regulation, including effective date. For example, "This AD is effective September 18, 2006, to all persons except those persons to whom it was made immediately effective by Emergency AD 2006-17-51, issued on August 15, 2006, which contained the requirements of this amendment."

16. SUMMARY.

a. Owner/Operator Responsibility. The registered owner or operator of an aircraft is responsible for compliance with ADs for the airframe, engine, propeller, and appliance as stated in the applicability statement of the AD for all aircraft it owns or operates.

b. Maintenance Personnel Responsibility. Maintenance personnel are responsible for determining that all applicable airworthiness requirements are met when they accomplish an inspection in accordance with part 43.

ORIGNAL SIGNED by
/s/ Raymond Towles for

John M. Allen
Director, Flight Standards Service

FAA-G-8082-11C

Inspection Authorization Knowledge Test Guide

U.S. Department of Transportation
Federal Aviation Administration

Inspection Authorization Knowledge Test Guide

July 2010

U.S. Department of Transportation

Federal Aviation Administration

Flight Standards Service

FAA-G-8082-11C

CONTENTS

INTRODUCTION

FAA-G-8082-11C, Inspection Authorization Knowledge Test Guide, provides information for preparing to take the following knowledge test. This document supersedes FAA-G-8082-11B, dated 2008.

TEST NAME	TEST CODE
Inspection Authorization	IAR

The Federal Aviation Administration (FAA) initiated the issuance of the inspection authorization (IA) more than 35 years ago. This system of allowing qualified mechanics the privilege of performing certain inspections has served well in the maintenance of the U.S. civil fleet. The attainment of an IA and performance of the duties thereof greatly enhance the privileges and responsibilities of the aircraft mechanic. The IA permits the airframe and powerplant (A&P) mechanic to perform a greater variety of maintenance and alterations than any other single maintenance entity.

The determination of airworthiness during an inspection is a serious responsibility. For many general aviation aircraft, the annual inspection could be the only in-depth inspection it receives throughout the year. In view of the wide ranging authority conveyed with the authorization, the test examines a broader field of knowledge than required for the A&P certificate and reflects the emphasis placed on perpetuating air safety.

This guide is not offered as an easy way to obtain the necessary information for passing the inspection authorization knowledge test. Rather, the intent of this guide is to define and narrow the field of study to the required knowledge areas included in the test. Federal Aviation Administration (FAA) airman knowledge tests are effective instruments for aviation safety and regulation measurement. However, these tests can only sample the vast amount of knowledge every aviation maintenance technician needs.

Comments may be e-mailed to AFS630Comments@faa.gov.

KNOWLEDGE TEST ELIGIBILITY REQUIREMENTS

Eligibility is established at the local FAA Flight Standards District Office (FSDO) or International Field Office (IFO) prior to taking the inspection authorization knowledge test.

You are eligible for the inspection authorization knowledge test if you meet the requirements of Title 14 of the Code of Federal Regulations (14 CFR) part 65, section 65.91(c):

"To be eligible for an inspection authorization, an applicant must—

(1) Hold a currently effective mechanic certificate with both an airframe rating and a powerplant rating, each of which is currently effective and has been in effect for a total of at least 3 years;

(2) Have been actively engaged, for at least the 2-year period before the date of application, in maintaining aircraft certificated and maintained in accordance with this chapter;

(3) Have a fixed base of operations at which he or she may be located in person or by telephone during a normal working week, but it need not be the place where he or she will exercise his inspection authority;

(4) Have available the equipment, facilities, and inspection data necessary to properly inspect airframes, powerplants, propellers, or any related part or appliance; and

(5) Pass a written test on the ability to inspect according to safety standards for returning aircraft to service after major repairs and major alterations and annual and progressive inspection performed under part 43 of this chapter."

KNOWLEDGE AREAS ON THE TEST

The inspection authorization knowledge test is comprehensive, as it must test your knowledge in many subject areas. When applying for an IA you should review 14 CFR part 65, section 65.91(c)(5), for the knowledge areas on the test.

DESCRIPTION OF THE TEST

All test questions are the objective, multiple-choice-type. Each question can be answered by the selection of a single response. Each test question is independent of other questions; therefore, a correct response to one does not depend upon or influence the correct response to another. **The minimum passing score is 70 percent.**

The Inspection Authorization test contains 50 questions, and you are allowed 3 hours to complete the test. (Note: Occasionally, 51 to 53 questions appear on the test. The additional questions are validation questions, new questions being tested, that will not count against you if missed. Extra time is allotted for completion of any additional questions.)

INTERVIEW AND TEST REGISTRATION PROCESS

The first step in taking the inspection authorization knowledge test is to contact your local FSDO or IFO to make an appointment to interview with an aviation safety inspector (ASI) (airworthiness) to determine eligibility before registering for the knowledge test. At the interview, the inspector will ask you to complete two copies of FAA Form 8610-1, Mechanic's Application for Inspection Authorization and provide proof of identity. An acceptable identification document includes a recent photograph, signature, and actual residential address, if different from the mailing address. This information may be presented in more than one form of identification.

Acceptable forms of identification include, but are not limited to, driver license, government identification card, passport, alien residency (green) card, and military

2

identification card. Some applicants may not possess the identification documentation described. In any case, you should always check with your local FSDO or IFO if you are unsure of the kind of identification to bring to the interview.

During the interview, you will be asked to demonstrate to the inspector's satisfaction that you meet the requirements for the authorization as specified in 14 CFR part 65, section 65.91(c)(1) through (4).

The inspector will interview to the extent necessary to determine that you clearly understand the inspection authorization privileges, limitations, and responsibilities. Once your qualifications have been demonstrated, the inspector will sign both of the Form 8610-1 you completed. You must present one copy of the form at the test site in order to take the test; the other copy will remain on file at the FSDO or IFO.

The next step is the actual registration process, which is accomplished in one of two ways. You may contact the computer testing designees (CTDs) through their national 1-800 numbers (refer to telephone numbers following this paragraph), or call directly to a local site. A complete listing of test centers may be found on the Internet at the FAA Web site at www.faa.gov under the heading "Training and Testing." You will then need to schedule a test date and make financial arrangements for test payment. You may cancel your appointment according to the cancellation policy of the CTD. If you do not follow the CTD's cancellation policies, you could be subject to a cancellation fee.

Computer Assisted Testing Service (CATS)
1801 Murchison Drive, Suite 288
Burlingame, CA 94010
Applicant inquiry and test registration: 1-800-947-4228
From outside the U.S. (650) 259-8550

LaserGrade Computer Testing
16821 SE McGillivray Blvd., Suite 201
Vancouver, WA 98683
Applicant inquiry and test registration: 1-800-211-2753 or 1-800-211-2754
From outside the U.S. (360) 896-9111

PROCESS FOR TAKING A KNOWLEGE TEST

At the test site, you will again be asked to provide proper identification and the completed FAA Form 8610-1. Testing center personnel will not begin the test until the required items are verified.

Before you take the actual test, you will have the option to review a tutorial that demonstrates test navigation and formatting. It also provides some sample questions. The actual test is time limited; however, you should have sufficient time to complete and review your test.

3

Communication between individuals through the use of words is a complicated process. In addition to being an exercise in the application and use of aeronautical knowledge, a knowledge test is also an exercise in communication since it involves the use of the written language. Since the tests involve written rather than spoken words, communication between the test writer and the person being tested may become a difficult matter if both parties do not exercise care. Consequently, considerable effort is expended to write each question in a clear, precise manner. Carefully read the instructions given with each test, as well as the statements in each test item.

The inspection authorization knowledge test has been mistakenly considered by some to be an open book test because the use of reference material is permitted during the test. To view the test in this manner is a misconception. There has always been a core knowledge requirement for which no reference material was provided. Therefore, it should be noted that, during the tests, there are subject areas for which reference material is not included in the test supplement. These areas will draw on skills acquired as an airframe and powerplant mechanic and which are necessary to properly inspect work performed by others.

The IA test supplement provides appropriate segments of 14 CFR regulations, charts, graphs, and technical data necessary to solve problems contained in the test. Prior to taking the test, you should take a few minutes to look through the supplement to determine what is included.

You should carefully read the information and instructions provided with the test, as well as the statements in each test item.

When taking a test, you should keep the following points in mind:

- Answer each question in accordance with the latest regulations and guidance publications.
- Read each question carefully before looking at the answer options. You should clearly understand the problem before attempting to solve it.
- After formulating an answer, determine which answer option most nearly corresponds with your answer. The answer you choose should completely resolve the problem.
- From the answer options given, it may appear that there is more than one possible answer; however, there is only one answer that is correct and complete. The other answers are either incomplete, erroneous, or derived from popular misconceptions.

4

- If a certain question is difficult for you, it is best to mark it for review and proceed to the next question. After you answer the less difficult questions, return to those you marked for review and answer them. The review marking procedure will be explained to you prior to starting the test. Although the computer should alert you to unanswered questions, make sure every question has an answer recorded. This procedure will enable you to use the available time to the maximum advantage.
- When solving a calculation problem, select the answer that most nearly matches your solution. The problem has been checked by various individuals and with different types of calculators; therefore, if you have solved it correctly, your answer will be closer to the correct answer than any of the other choices.

USE OF TEST AIDS AND MATERIALS

You may use aids, reference materials, and test materials within the guidelines listed below, if actual test questions or answers are not revealed. All models of aviation-oriented calculators may be used, including small electronic calculators that perform only arithmetic functions (add, subtract, multiply, and divide) may be used. Simple programmable memories, which allow addition to, subtraction from, or retrieval of one number from the memory, are permissible. Also, simple functions such as square root and percent keys are permissible.

The following guidelines apply:

1. You may use any reference materials provided with the test. In addition, you may use scales, straightedges, protractors, plotters, and electronic or mechanical calculators that are directly related to the test.
2. Manufacturer's permanently inscribed instructions on the front and back of such aids (e.g., formulas, conversions, and weight and balance formulas) are permissible.
3. Testing centers may provide a calculator to you and/or deny use of your personal calculator based on the following limitations:
    a. Prior to and upon completion of the test while in the presence of the proctor, you must actuate the ON/OFF switch and perform any other function that ensures erasure of any data stored in memory circuits.
    b. The use of electronic calculators incorporating permanent or continuous type memory circuits without erasure capability is prohibited. The proctor may refuse the use of your calculator when unable to determine the calculator's erasure capability.
    c. Printouts of data must be surrendered at the completion of the test if the calculator incorporates this design feature.

   d. The use of magnetic cards, magnetic tapes, modules, computer chips, or any other device upon which prewritten programs or information related to the test can be stored and retrieved is prohibited.
   e. You are not permitted to use any booklet or manual containing instructions related to use of test aids.
4. Dictionaries are not allowed in the testing area.
5. The proctor makes the final determination relating to test materials and personal possessions you may take into the testing area.

DYSLEXIC TESTING PROCEDURES

If you are a dyslexic applicant, you may request approval from the local Flight Standards District Office (FSDO) or International Field Office (IFO) to take an airman knowledge test using one of the three options listed in preferential order:

Option 1. Use current testing facilities and procedures whenever possible.

Option 2. You may use a Franklin Speaking Wordmaster® to facilitate the testing process. The Wordmaster® is a self-contained electronic thesaurus that audibly pronounces typed in words and presents them on a display screen. It has a built-in headphone jack for private listening. The headphone feature must be used during testing to avoid disturbing others.

Option 3. If you do not choose to use the first or second option, you may request a proctor to assist in reading specific words or terms from the test questions and supplement material. In the interest of preventing compromise of the testing process, the proctor must be someone who is non-aviation oriented. The proctor must provide reading assistance only, with no explanation of words or terms. When this option is requested, the FSDO or IFO inspector must contact the Airman Testing Standards Branch (AFS-630) for assistance in selecting the test site and proctor.

Prior to approval of any option, the FSDO or IFO inspector must advise you of the regulatory certification requirement of being able to read, write, speak, and understand the English language.

CHEATING OR OTHER UNAUTHORIZED CONDUCT

Computer testing centers must follow strict security procedures to avoid test compromise. These procedures are established by the FAA and are covered in FAA Order 8080.6 (as amended), Conduct of Airman Knowledge Tests. The FAA has directed testing centers to terminate a test at any time a test proctor suspects a cheating incident has occurred. An FAA investigation will then be conducted. If the investigation determines that cheating or unauthorized conduct has occurred, then any airman certificate or rating that you hold may be revoked, and you will be prohibited for 1 year from applying for or taking any test for a certificate or rating under 14 CFR part 65.

AIRMAN KNOWLEDGE TEST REPORTS

Upon completion of the knowledge test, you will receive your Airman Knowledge Test Report (with the testing center's embossed seal), which reflects your score. The test site retains the FAA Form 8610-1, Mechanic's Application for Inspection Authorization.

The Airman Knowledge Test Report lists the learning statement codes for questions answered incorrectly. The total number of learning statement codes shown on the Airman Knowledge Test Report is not necessarily an indication of the total number of questions answered incorrectly.
The Learning Statement Reference Guide for Airman Knowledge Testing, found at www.faa.gov, contains the listings of reference materials, learning statement codes, and learning statements. The learning statement codes, as used in airman testing, refer to a measurable statement of knowledge that a student should be able to demonstrate following a defined element of training. You should match the codes on your Airman Knowledge Test Report to the codes in the Learning Statement Reference Guide to review your areas of deficiency.

Listings of reference materials have been prepared by the FAA to establish references for all knowledge standards. The listings contain reference materials to be used when preparing for all airman knowledge tests.

After passing the test, present your Airman Test Report to an ASI at the FSDO or IFO where you interviewed. It is best to return to the original interviewer if possible; however, any available airworthiness ASI can complete the authorization process. At that time, the ASI will again review your application and discuss any questions you have. When the ASI is satisfied that all requirements are met, the certificate will be issued.

Should you require a duplicate Airman Knowledge Test Report due to loss or destruction of the original, send a signed request accompanied by a check or money order for $1 payable to the FAA. Your request should be sent to:

Federal Aviation Administration
Airman Certification Branch, AFS-760
P.O. Box 25082
Oklahoma City, OK 73125

Airman Knowledge Test Reports are valid for the 24-calendar-month period preceding the month you complete the test. If the Airman Knowledge Test Report expires before you finalize the inspection authorization, you must retake the knowledge test.

RETESTING PROCEDURES

If you receive a score lower than 70 percent, you may not apply for retesting until 90 days after the date that you failed the test. Any attempt to retest prior to the 90-day waiting period is contrary to 14 CFR part 65, and could result in revocation of any airman certificates that you hold. Whether retesting after a failed examination or simply retesting for a better score, you will need to present your test report in order to retest.

TRAINING AND TESTING PUBLICATIONS AND GENERAL INFORMATION

Most of the current Flight Standards Service airman training and testing publications can be obtained in electronic format from the FAA Web site, www.faa.gov. The training and testing publications and general information can be found on the opening page of that Web site under the Training and Testing tab. If a publication is not available in electronic format, there are instructions for obtaining paper copies. Information found on the Web site includes the following:

- Advisory Circulars
- Knowledge testing sites
- Knowledge Test statistics
- Knowledge testing supplements
- Other testing information
- Practical Test Standards
- Training handbooks
- Learning Statement Reference Guide
- Code of Federal Regulations
- Airworthiness Directives
- Type Certificate Data Sheets

Airman Knowledge Test Questions

Sample questions are contained in the Sample Test Questions and Answers section of this test guide and represent the types of questions included in the actual test banks. Practicing these questions will help you become familiar with similar questions on the airman knowledge tests. The knowledge test is not designed to intimidate any prospective airman; it is designed to measure understanding of the rules and regulations required to receive an FAA certificate. The list of reference materials contained in this test guide is provided to ensure that instructors and students are able to determine the importance of the subject matter to be taught and learned.

8

Knowledge Testing Supplements

The knowledge testing supplements contain the graphics, legends, and maps that are needed to successfully respond to certain knowledge test items. These supplements will be provided by CTD test center personnel during the airman knowledge test.

Airman Knowledge Test Statistics

Test statistics for all airman knowledge tests are contained in a series of tables organized by year and subject area. Individual tables are provided for the following subject areas: test volume, pass rates, average test scores, countries, regions, and district offices.

Learning Statement Reference Guide

Learning statement codes replace the old subject matter codes and are noted on the test report. They refer to measurable statements of knowledge that a student should be able to demonstrate following a defined element of training. The learning statement corresponding to the learning statement code on the test report can be located in the Learning Statement Reference Guide on the Web site.

Suggestions for Studying for the IA Test

1. Be familiar with the parts of 14 CFR, as listed in the Publications and Technical Data section.

2. Study 14 CFR parts 91 and 135 aircraft maintenance and inspection requirements.

3. Be familiar with aircraft type certificate data sheets and specifications. This should include the differences and history of these documents. Applicant should know how revisions are noted.

4. Study 14 CFR part 43, appendixes A, B, and D, for detailed information regarding major repairs, major alterations, and annual inspections.

5. Learn to use the graphs and tables in AC 43.13-1B, (or most current revision) Acceptable Methods, Techniques and Practices—Aircraft Inspection and Repair; and AC 43.13-2A, (or most current revision) Acceptable Methods, Techniques, and Practices—Aircraft Alterations.

6. Be familiar with airworthiness directives for small aircraft and rotorcraft. This should include knowledge of 14 CFR part 39.

7. Be familiar with the completion of FAA Form 337 (Major Repair and Alteration—Airframe, Powerplant, Propeller, or Appliance). Guidance in this area is provided in AC 43.9-1F, Instructions for Completion of FAA Form 337, Major Repair and Alteration (Airframe, Powerplant, Propeller, or Appliance).

8. Know the requirements for maintenance and inspection record entries for 14 CFR parts 43 and 91. Guidance in this area is provided in AC 43.9C, Maintenance Records, and AC 39-7C, Airworthiness Directives.

9. Be familiar with minimum equipment lists for general aviation aircraft. Guidance in this area is provided in AC 91-67, Minimum Equipment Requirements for General Aviation Operations under 14 CFR part 91.

10. Be familiar with all aspects of weight and balance computations. Applicant must be able to:

    a. calculate basic empty weight and center of gravity in both inches and percent of mean aerodynamic chord (MAC).

    b. conduct adverse loading checks for extreme forward and rearward CG positions.

    Applicants should practice making changes to an aircraft weight and balance report by simulating installation or removal of equipment and then computing the forward, aft, and empty-weight center of gravity (CG). Guidance in this area is provided in AC 65-9, Airframe and Powerplant Mechanics General Handbook and FAA-H-8083-1, Aircraft Weight and Balance Handbook. Also, many commercial publications are available on this subject.

NOTE: You should use the most current versions of the referenced documents.

10

SAMPLE TEST QUESTIONS AND ANSWERS

1. What ignition system is approved for a Lycoming engine model 0-540-A4A5?

A) Bendix magneto model D6LN-3031.

B) Slick magneto models 662 and 663.

C) Bendix magneto models S6LN-20 and S6LN-21.

Answer: C. Learning Statement: IAR013, Determine design specific.

Note: Old Subject Matter Knowledge Code: Y303. Type Certificate Data Sheet No. E-295, Note 8.

2. A lower horizontal stabilizer streamlined brace is to be repaired by welding. The brace size is 1¼ inch.

The repair should be accomplished using which of the following materials?

A) A round insert tube of the same material, one gauge thicker than the original streamlined tube and a minimum length of 5.01 inch.

B) An outside sleeve of at least the same gauge with a minimum length of 9.128 inches.

C) An inside sleeve of the same streamlined tubing as original with a maximum insert length of 6.43 inches.

Answer: B. Learning Statement: IAR019, Determine repair requirements.

Note: Old Subject Matter Knowledge Code: K49. Advisory Circular (AC) 43.13-1B, chapter 2, paragraph 81; and figure 2.13.

3. Use Airworthiness Directive (AD) AD 80-10-02 to answer this question.

Known Information: Messerschmitt-Bolkow-Blohm Model BO-105 helicopter with tail rotor blade grips P/N 105-31722 installed.

While performing a progressive inspection on this helicopter, you note in the aircraft's records that the last compliance with AD 80-10-02 was at an aircraft time of 5402 hours. The records further indicate that the tail rotor blade grips were replaced at an aircraft time of 4902. What action is required at this inspection with a time of 5502?

A) Compliance is required for paragraph (c)(1)(2).

B) Compliance is required for paragraph (e).

C) Compliance is required for paragraphs (b)(d) and (e).

Answer: C. Learning Statement: IAR020, Interpret data.

Note: Old Subject Matter Knowledge Code: A14. AD 80-10-02.

11

4. Where can the major items to be inspected be found that must be included in a checklist used while performing an annual inspection on a fixed-wing aircraft?

A) FAA Form 8130-10.

B) 14 CFR part 43, appendix D.

C) AC 43.13-1B.

Answer: B. Learning Statement: IAR016, Determine regulatory requirement.

Note: Old Subject Matter Knowledge Code: K49. 14 CFR part 43, section 43.15(c) states: "Sec. 43.15 Additional performance rules for inspections....

(c) Annual and 100-hour inspections.

(1) Each person performing an annual or 100-hour inspection shall use a checklist while performing the inspection. The checklist may be of the person's own design, one provided by the manufacturer of the equipment being inspected, or one obtained from another source. This checklist must include the scope and detail of the items contained in appendix D to this part and paragraph (b) of this section...."

5. The sight line drawn on material to be bent in a cornice brake is located at what position from the bend tangent line?

A) The setback measurement.

B) The bend radius length.

C) The bend allowance length.

Answer: B. Learning Statement: IAR027, Recall principles of sheet metal forming.

Note: Old Subject Matter Knowledge Code K05. Order 8130.21C. The work must be accomplished by a certificate holder under 14 CFR part 121 or 135, having a continuous airworthiness maintenance program or by a repair station certificated under part 145.

6. When installing additional equipment in an aircraft, if not otherwise specified, the ultimate load factor used in the static load test is

A) four times the weight of the equipment.

B) variable, depending on the direction of applied force.

C) the limit load factor multiplied by 1.5.

Answer: C. Learning Statement: IAR022, Recall alteration/design fundamentals.

Note: Old Subject Matter Knowledge Code: K50. AC 43.13-2A, chapter 1, paragraph 3. Ultimate load factors are limit load factors multiplied by a 1.5 safety factor.

12

7. Which of the following locations in 14 CFR provides for the fabrication of aircraft replacement and modification parts?

A) 14 CFR part 21, section 21.303.

B) 14 CFR part 23, appendix B.

C) 14 CFR part 45, section 45.21.

Answer: A. Learning Statement: IAR030, Recall regulatory requirements.

Note: Old Subject Matter Knowledge Code A112. 14 CFR part 21, subpart K, section 21.303, defines who may produce modification and replacement parts for sale and those persons to which the part does not apply.

8. A proposed airframe alteration will require a section of Mil-H-8788-10 hydraulic hose to flex through 60° of travel. The system will operate at 210 °Centigrade and 1,200 psi. What is the minimum bend radius for this installation?

A) 3¼ inches.

B) 5½ inches.

C) 7¾ inches.

Answer: C. Learning Statement: IAR013, Determine design specific.

Note: Old Subject Matter Knowledge Code: K49. Acceptable Methods, Techniques, and Practices—Aircraft Inspection and Repair, chapter 10, paragraph d; and figure 10.5.

9. Where would you find the marking and placards required for Cessna Model 208, serial number 20800044?

A) Type Certificate Data Sheet No. A37CE.

B) Airplane Flight Manual, Cessna P/N D1286-13PH.

C) Model 208 Series Maintenance Manual.

Answer: B. Learning Statement: IAR011, Determine correct data.

Note: Old Subject Matter Knowledge Code: A157. 14 CFR Part 23, Subpart G, Operating Limitations and Information.

13

10. Which of the following aircraft, operating under 14 CFR part 91, could the holder of an inspection authorization approve for return-to-service after a major alteration has been made in accordance with technical data approved by the administrator?

A) A commuter category, multiengine, turbopropeller airplane.

B) A transport category, multiengine, turbojet airplane.

C) Either A or B.

Answer: C. Learning Statement: IAR031, Recall regulatory specific.

Note: Old Subject Matter Knowledge Code: A45. 14 CFR part 65, section 65.95(a). "Sec. 65.95 Inspection authorization: privileges and limitations.

(a) The holder of an inspection authorization may—

(1) Inspect and approve for return to service any aircraft or related part or appliance (except any aircraft maintained in accordance with a continuous airworthiness program) after a major repair or major alteration to it in accordance with Part 43 of this chapter, if the work was done in accordance with technical data approved by the Administrator; and

(2) Perform an annual or perform or supervise a progressive inspection according to §§ 43.13 and 43.15 of this chapter."

14

PUBLICATIONS AND TECHNICAL DATA

The following publications and technical data provide information for aircraft inspection.

FAA-G-8082-19, Inspection Authorization (IA) Guide

This information manual provides guidance for persons who conduct annual and progressive inspections and approve major repairs and/or major alterations of aircraft. The information is primarily intended for mechanics that hold and IA. This manual stresses the important role that certificated mechanics that hold an IA have in air safety.

Title 14 of the Code of Federal Regulations (14 CFR)

The Code of Federal Regulations is a codification of the general and permanent rules published in the Federal Register by the Executive departments and agencies of the Federal Government. The Code is divided into 50 titles, which represent broad areas subject to Federal regulation. Each title is divided into chapters, which usually bear the name of the issuing agency. Title 14—Aeronautics and Space is composed of four chapters. Chapter I of this title is the Federal Aviation Administration, Department of Transportation (DOT). This chapter contains parts 1–199.

The following CFR parts are of particular interest to the holder of an inspection authorization.

PART	TITLE
1	Definitions and Abbreviations
11	General Rulemaking Procedures
21	Certification Procedures for Products and Parts
23	Airworthiness Standards: Normal, Utility, Acrobatic, and Commuter Category Airplanes
25	Airworthiness Standards: Transport Category Airplanes
27	Airworthiness Standards: Normal Category Rotorcraft
29	Airworthiness Standards: Transport Category Rotorcraft
31	Airworthiness Standards: Manned Free Balloons
33	Airworthiness Standards: Aircraft Engines
35	Airworthiness Standards: Propellers
39	Airworthiness Directives
43	Maintenance, Preventive Maintenance, Rebuilding, and Alteration
45	Identification and Registration Marking
47	Aircraft Registration

65	Certification: Airmen Other Than Flight Crewmembers
91	General Operating and Flight Rules
119	Certification: Air Carriers and Commercial Operators
125	Certification and Operations: Airplanes Having a Seating Capacity of 20 or More Passengers or a Maximum Payload Capacity of 6,000 Pounds or More
135	Operating Requirements: Commuter and On-Demand Operations and Rules Governing Persons on Board Such Aircraft
183	Representatives of the Administrator

The official FAA copy of the Code of Federal Regulations may be obtained from the Regulatory and Guidance Library at the Federal Aviation Administration Web site at www.faa.gov or the U.S. Government Printing Office (GPO) Web site at www.access.gpo.gov.

TYPE CERTIFICATE DATA SHEETS AND SPECIFICATIONS

Type Certificate Data Sheets and Specifications (TCDS) set forth essential factors and other conditions, which are necessary for U.S. airworthiness certification. Aircraft, engines, and propellers which conform to a U.S. type certificate (TC) are eligible for U.S. airworthiness certification when found to be in a condition for safe operation and ownership requisites are fulfilled. There are two kinds of certification documents contained in the TCDS file:

1. Type Certificate Data Sheets
2. Specifications

Type Certificate Data Sheets were originated and first published in January 1958. 14 CFR part 21, section 21.41, indicates they are part of the type certificate. As such, a type certificate data sheet is evidence the product has been type certificated. Generally, type certificate data sheets are compiled from details supplied by the type certificate holder; however, FAA may request and incorporate additional details when conditions warrant.

Specifications were originated during implementation of the Air Commerce Act of 1926. Specifications are FAA recordkeeping document issued for both type certificated and non-type-certificated products which have been found eligible for U.S. airworthiness certification. Although they are no longer issued, specifications remain in effect and will be further amended. Specifications covering type certificated products may be converted to type certificate data sheets at the option of the type certificate holder. However, to do so requires the type certificate holder to provide an equipment list. A specification is not part of a type certificate.

The official FAA copy is available on the Internet at the FAA Web site titled Regulatory and Guidance Library at rgl.faa.gov. This is a free service.

16

SUMMARY OF AIRWORTHINESS DIRECTIVES FOR SMALL AIRCRAFT AND ROTORCRAFT

An airworthiness directive (AD) contains information regarding an unsafe condition that exists in an aircraft, aircraft engine, propeller, or appliance when that condition is likely to exist or develop in other products of the same type design. No person may operate a product to which an AD applies, except in accordance with the requirements of the AD. All ADs are summarized and issued by the FAA. New and revised ADs are published bi-weekly and mailed to registered owners of effected equipment and subscription holders. Airworthiness directives are issued in two weight categories:

1. Small aircraft with a maximum certificated takeoff weight aircraft of 12,500 pounds or less, and all rotorcraft regardless of weight
2. Large aircraft over 12,500 pounds maximum certificated takeoff weight

Each of these categories is presented in three books. Included in these books are the airframe ADs and the ADs applicable to the engines, propellers, and appliances of the category.

The official FAA copy is available on the Internet at the FAA Web site titled Regulatory and Guidance Library (RGL) at rgl.faa.gov.

The ADs are totally searchable and easily located. The individual airworthiness directives and the AD bi-weeklies on the RGL Web site are considered official FAA copy and may be used in lieu of purchasing paper copies. This is a free service. Questions concerning the RGL may be directed to the Delegation & Airworthiness Programs Branch (AIR-140) at (405) 954-4103.

ADVISORY CIRCULARS

The Federal Aviation Administration issues advisory circulars to inform the aviation public in a systematic way of nonregulatory material. Unless incorporated into a regulation by reference, the contents of an advisory circular are not binding on the public. Advisory circulars are issued in a numbered-subject system corresponding to the numerical part of the subject regulation (AC 39-7 would therefore deal with a subject related to 14 CFR part 39 or Airworthiness Directives).

An advisory circular is issued to provide guidance and information in a designated subject area or to show a method acceptable to the Administrator for complying with a related 14 CFR part. Electronic versions (as revised) are available on the Internet at the FAA Web site.

- AC 39-7, Airworthiness Directives
- AC 43-4, Corrosion Control for Aircraft
- AC 43-11, Reciprocating Engine Overhaul Terminology and Standards

17

- AC 43.13-1, Acceptable Methods, Techniques and Practices—Aircraft Inspection and Repair
- AC 43.13-2, Acceptable Methods, Techniques, and Practices—Aircraft Alterations
- AC 43-9, Maintenance Records
- AC 43.9-1, Instructions for Completion of FAA Form 337 Major Repair and Alteration (Airframe, Powerplant, Propeller, or Appliance)
- AC 43-210, Standardized Procedures for Requesting Field Approval of Data, Major Alterations, and Repairs
- AC 91-67, Minimum Equipment Requirements for General Aviation Operations Under FAR Part 91

OTHER PUBLICATIONS

Additional information of particular interest to the holder of an inspection authorization can be found in FAA-H-8083-1, Aircraft Weight and Balance Handbook.

ADDITIONAL SOURCES OF INSPECTION DATA

Several commercial publisher offer subscription services that include the Airworthiness Directives, Advisory Circulars, and Type Certificate Data Sheets along with other inspection data. They may be found in aviation trade paper and magazines.

FAA-G-8082-19

INSPECTION AUTHORIZATION INFORMATION GUIDE

U.S. Department of Transportation
Federal Aviation Administration

Inspection Authorization Information Guide

2010

U.S. DEPARTMENT OF TRANSPORTATION
FEDERAL AVIATION ADMINISTRATION
Flight Standards Service

PREFACE

This publication provides guidance for persons who conduct annual and progressive inspections and approve major repairs and/or major alterations of aircraft holding or eligible to hold a U.S. type certificate. The Federal Aviation Administration (FAA) intends this information for mechanics who hold an Inspection Authorization (IA). This manual stresses the important role that certificated mechanics who hold an inspection authorization have in air safety.

Since inspection requirements and regulations change, the FAA recommends that you contact your local Flight Standards District Office (FSDO) for help with questions regarding inspection, alteration, and repair of aircraft.

Title 14 of the Code of Federal Regulations (14 CFR) part 65, Certification: Airmen Other Than Flight Crewmembers, describes the privileges of mechanics holding an inspection authorization. 14 CFR part 43, Maintenance, Preventive Maintenance, Rebuilding, and Alteration, describes maintenance rules and standards of performance.

This publication is available free of charge for download, in PDF format on the FAA website at www.faa.gov or may also be purchased from http://bookstore.gpo.gov.

Send comments in email form to: afs630comments@faa.gov.

TABLE OF CONTENTS

Chapter 1
RENEWAL AND CHANGE OF FIXED BASE

NOTIFICATION OF CHANGE OF FIXED BASE

When the holder of an inspection authorization (IA) moves the base of operation to a different Flight Standards District Office (FSDO) or International Field Office (IFO) area, the receiving office must be notified in writing before exercising the privileges of the authorization.

EXPIRATION OF AN INSPECTION AUTHORIZATION

The IA expires every odd-numbered year (2015, 2017, etc.) on March 31. An IA holder must continue to meet the yearly requirements of Title 14 Code of Federal Regulations (14 CFR) part 65, section (§) 65.93 in order to retain the authorization. Yearly means April 1 through March 31.

RENEWAL OF INSPECTION AUTHORIZATION

Renewal Application Paperwork

Application for renewal requires the following:

1. Evidence the applicant still meets the requirements of 14 CFR § 65.91(c)(1) through (4).
2. Completed Federal Aviation Administration (FAA) Form 8610-1, Mechanic's Application for Inspection Authorization, in duplicate. (Refer to appendix 1, figure 1.)
3. Evidence the applicant meets the requirements of 14 CFR § 65.93(a) for both the first and second year in the form of an activity sheet or log, training certificates, and/or oral test results, as applicable.

1

4. Verification of the applicant's activities by the jurisdictional FSDO or IFO for renewal at an office other than the jurisdictional office.
5. Evidence that an applicant employed by a repair station personally inspected, performed, and returned aircraft to service (in addition to the signed application).

Eligibility

Meeting the requirements of 14 CFR § 65.93(a) does not mean that the applicant must meet all 5 of the listed requirements. To be eligible for renewal of an IA for a 2-year period, "the applicant must show completion of one of the activities" by March 31 of the first year and completion of one during the second year. For instance, the applicant may show evidence of having "performed at least one annual inspection for each 90 days" during the first year and meet the same requirement for the second year for a total of eight annuals prior to the renewal date to qualify for renewal. The same logic applies to major repairs and major alterations or training. The number of annual inspections, major repairs, and major alterations performed cannot be mixed to meet a *single year's requirement* simply because 14 CFR § 65.93(a) does not provide for such combinations. However, the applicant can meet one requirement of 14 CFR § 65.93(a) for the first year and a different requirement for the second year to qualify for renewal.

NOTE: An inspection program required under 14 CFR part 91, § 91.409(e) is not acceptable as IA activity. Partial inspections such as phases or events on more than one aircraft are not acceptable as activity. A progressive inspection is a complete inspection on one identified aircraft that completes a cycle each 12 calendar months.

2

Approved Training Renewal Option

Successful completion of an 8-hour refresher course, acceptable to the Administrator, in one of the 12-month periods preceding the renewal application includes the following requirements:

1. The refresher course must contain subjects directly related to aircraft maintenance, inspection, repairs, and alterations. In addition, some nontechnical subjects, such as human factors or professionalism as they relate to aviation maintenance personnel, may be acceptable. Training must not be used to promote a new or existing product.
2. The instructional requirements of § 65.93(a)(4) may be met by accumulating at least 8 hours of maintenance training each year.
3. Each person who intends to use 8 hours of instruction each year to meet § 65.93(a)(4) must, at the time of renewal, provide proof of attendance for instruction received. Acceptable proof of attendance consists of a certificate of training or similar document showing the name of the course, name of attendee, course identification number, expiration date, description of the course content, time in hours, the date, location, and course instructor's name and signature.

   All FSDOs and IFOs must accept, without further showing maintenance, technical training conducted by a manufacturer or its authorized representative on its type-certificated (TC), Supplemental Type Certificate (STC), Technical Standard Order (TSO), or Parts Manufacturer Approval (PMA) product, component, or accessory that is considered acceptable to the Administrator and in compliance with this policy.

Oral Test Requirement

If an IA holder does not meet the renewal requirements at the end of the first year, the holder must take and pass an oral test

3

administered at their local Flight Standards District Office or International Field Office prior to exercising the privileges of their certificate in the second year.

The oral test given by an aviation safety inspector (ASI) is to ensure that the applicant's knowledge of regulations and standards are current (requires a passing grade of 70 percent). A failure of the oral test will result in nonrenewal of the IA. The ASI administering the oral test will provide the IA with evidence of passing or failing the test in the form of written documentation. The applicant should retain the oral test results.

Renewal Notes

1st YEAR NOTE: The holder of an IA issued less than 90 days before March 31 of the first year (even year) need not comply with § 65.93(a)(1) through (5) for the first year of the 2-year authorization period.

2nd YEAR NOTE: The holder of an IA issued less than 90 days before the expiration date March 31 (odd year) does not need to comply with § 65.93(a)(1) through (5) for that quarter, but the IA holder still needs to apply for a renewal.

A completed example of the IA authorization card, FAA Form 8310-5, is shown in appendix 1, figure 2.

4

Chapter 2
BASIC FUNCTIONS OF THE INSPECTION AUTHORIZATION

GENERAL

The basic functions of the holder of an inspection authorization (IA) are set forth in Title 14 of the Code of Federal Regulations (14 CFR) part 65, § 65.95. With the exception of aircraft maintained in accordance with a Continuous Airworthiness Maintenance Program under part 121, an IA may inspect and approve for return to service any aircraft or related part or appliance after a major repair or major alteration. Also, the holder of an IA may perform an annual inspection and he or she may supervise or perform a progressive inspection.

APPROVING MAJOR REPAIRS AND MAJOR ALTERATIONS

What To Look for During an Inspection

A primary responsibility of the holder of an IA is to determine airworthiness by inspecting repairs or alterations for conformity to approved data, and assuring that the aircraft is in a condition for safe operation. During inspection of major repairs or major alterations, the holder of an IA must also determine that they are compatible with previous repairs and alterations that have been made to the aircraft.

The holder of an IA must personally perform the inspection. The regulations do not provide for delegation of this responsibility. Approving major repairs and major alterations is a serious responsibility. The approval action should consist of a detailed investigation to establish at least that:

1. All replacement parts installed conform to approved design and/or have traceability to the original equipment manufacturer (OEM).

2. As installed, the installation conforms to approved data that is applicable to the installation.
3. Workmanship meets the requirements of 14 CFR part 43, § 43.13 (the aircraft or product is equal to its original or properly altered condition).
4. The data used is appropriate to the aircraft certification rule (CAR 3, 14 CFR part 23, etc.).
5. Work is complete and compatible with other structures or systems.

The holder of an IA *cannot* approve the *data* for major repairs or major alterations. He or she may, however, inspect to see that alterations conform to data previously approved by the Administrator (14 CFR part 65, § 65.95). This means the holder of an IA ensures that approved data is available and is used as the basis for the approval. This availability determination should be made prior to beginning the repair or alteration. If data is unavailable, or if the holder of an IA is unsure of the acceptability of the available data, the local Aviation Safety Inspector (ASI) should be consulted. The ASI may, as the circumstances warrant, be able to:

1. Establish an acceptable basis for approval,
2. Approve the data, or
3. Recommend application for a supplemental type certificate.

Quite often major repairs are performed that are eventually covered by fabric, metal skin, or another structure. When this situation exists, the holder of an IA should have a clear understanding with the mechanic performing the repair that a precover inspection is necessary. The inspection should assure that the repair was made in accordance with acceptable methods, techniques, and practices prescribed by 14 CFR part 43 and that the structure to be covered is free from defects, corrosion, or wood rot, and is protected from the elements. In

6

addition, the holder of an IA should inspect other affected areas for hidden damage if the aircraft has been involved in an accident or incident. An entry is required to be made in the maintenance record and FAA Form 337, Major Repair and Alteration, must be completed. (Refer to appendix 1, figure 3, showing typical entries on the front and back of FAA Form 337.)

Minor deviation from approved data is permissible if the change is one that could be approved as a minor alteration when considered alone. Be sure to list the deviations on FAA Form 337 and make an entry in the maintenance record when completing the aircraft records. When in doubt, contact the local ASI who may decide the change is not minor and would need specific approval or an amendment of the original approval.

Approved Data

Substantiating and descriptive technical data sources used to make a major repair or alteration that is approved by the Administrator include those on the following list:

1. Type Certificate Data Sheets (TCDS).
2. Supplemental Type Certificate (STC) data, provided it specifically applies to the item being repaired/altered. Such data may be used in whole or part as included within the design data associated with the STC.
3. Appliance manufacturer's manuals or instructions, unless specifically not approved by the Administrator, are approved for major repairs.
4. Airworthiness Directives (ADs).
5. FAA Form 337, which has been used to approve multiple identical aircraft (only by the original modifier).

   NOTE: Aviation safety inspectors (ASI) no longer approve data for use on multiple aircraft.

7

6. U.S. Civil Aviation Authority (CAA) Form 337, dated before October 1, 1955.
7. FAA-approved portions of SRMs.
8. Designated Engineering Representative (DER)-approved data, only when approval is authorized under his or her specific delegation.
9. Organization Designation Authorization (ODA)-approved data, when the major alteration is performed specific to the authorization granted.
10. Data in the form of an Appliance Type Approval issued by the Minister of Transport Canada for those parts or appliances for which there is no current Technical Standard Order (TSO) available. The installation manual provided with the appliance includes the Transport Canada certificate as well as the date of issuance and an environmental qualification statement.
11. Foreign bulletins, for use on U.S.-certificated foreign aircraft, when approved by the foreign authority.
12. Data describing an article or appliance used in an alteration which is FAA-approved under a TSO. As such, the conditions and tests required for TSO approval of an article are minimum performance standards. The article may be installed only if further evaluation by the operator (applicant) documents an acceptable installation which may be approved by the Administrator.
13. Data describing a part or appliance used in an alteration which is FAA-approved under a Parts Manufacturer Approval (PMA). An STC may be required to obtain a PMA as a means of assessing Airworthiness and/or performance of the part.

    NOTE: Installation eligibility for subsequent installation or reinstallation of such part or appliance in a type-certificated (TC) aircraft, other than the aircraft for which airworthiness was originally demonstrated, is acceptable, provided the part or appliance meets its performance requirements and is

8

operationally compatible for installation. The operator/applicant must provide evidence of previously approved installation by TC, STC, or field approval on FAA Form 337 that will serve as a basis for follow-on field approval.

14. Any FAA-approved Service Bulletins (SB) and letters or similar documents, including DER approvals.
15. Foreign bulletins as apply to a U.S.-certificated product made by a foreign manufacturer located within a country with whom a Bilateral Agreement (BA) is in place and by letter of specific authorization issued by the foreign civil air authority. The Bilateral Web site is located at: http://www.faa.gov/aircraft/air_cert/international/bilateral_agreements/baa_basa_listing.
16. Other data approved by the Administrator.
17. Advisory Circular (AC) 43.13-1, current edition, for FAA-approved major repairs on nonpressurized areas of aircraft only when the user determines that it is:
    a. Appropriate to the product being repaired;
    b. Directly applicable to the repair being made; and
    c. Not contrary to the airframe, engine, propeller, product, or appliance manufacturer's data.
18. AC 43.13-2, current edition, for FAA-approved major alterations on nonpressurized areas of aircraft 12,500 pounds gross weight or less only when the user determines that it is:
    a. Appropriate to the product being repaired;
    b. Directly applicable to the alteration being made; and
    c. Not contrary to the airframe, engine, propeller, product, or appliance manufacturer's data.
19. Service and repair data provided by small airplane manufacturers (although, in most cases, not specifically approved) has provided for continued airworthiness of their product. Service experience in using this data when performing major repairs to nonpressurized small airplanes that were

9

FAA-G-8082-19

originally type certificated before January 1, 1980, has proven to be very reliable if followed with no deviations. When the data is used in this manner, the manufacturer's data (with page, paragraph, etc.) must be referenced on FAA Form 337 in block 8.

Components of the Inspection

Inspecting repairs or alterations consists of these basic operations:

1. Determine that the repair or alteration data has FAA approval.
2. Inspect the configuration of the repair or alteration for conformity to the approved data and the performance standards of 14 CFR part 43. At the same time, the aircraft should still comply with applicable airworthiness requirements, and the repair or alteration should be compatible with all other installations.
3. Ensure that all operating limitations affected by an alteration are appropriately revised. Sometimes, limitations are in the form of flight manual supplements, instrument range markings, placards, or combinations of these. See the local ASI for limitations on changes which can be made.
4. Determine that aircraft record entries have been made and the weight and balance data and equipment list have been revised, when appropriate. There should be a statement on the FAA Form 337 to the effect that the weight and balance data and equipment list have been revised. When an alteration results in a change in the center-of-gravity (CG) position, the affected CG limit should be investigated under adverse loading conditions unless the new CG falls within an approved empty CG range. For instance, if the CG has shifted aft, the loading conditions should be computed to see that the aircraft does not exceed the aft CG limit. It is the pilot's responsibility to have the aircraft correctly loaded.

10

However, when approving an alteration, it is the IA's responsibility to see that weight and balance data have been revised. The aircraft record entries may refer to the FAA Form 337 for details, such as: "Installed STOL kit in accordance with STC SA 940 CE drawing number 5084 dated April 24, 1996. See FAA Form 337, this date, for details."

5. Indicate approval in block 7 of FAA Form 337, and return both copies to the person who performed the work, for disposition in accordance with 14 CFR part 43, appendix B.

ANNUAL AND PROGRESSIVE INSPECTIONS

The procedures and scope for annual inspections are set forth in 14 CFR part 43, appendix D, and should be followed in detail. The scope and detail for a progressive inspection is established by the owner or operator in accordance with 14 CFR part 91, § 91.409(d). There are additional requirements for annual and progressive inspections listed in 14 CFR part 43, § 43.15. The scope and detail of 100-hour and annual inspections are the same. Record entries are very important as they are the only evidence an aircraft owner has to show compliance with the inspection requirements of 14 CFR part 91, § 91.409. (Refer to appendix 1, figure 4, of this manual.)

The following reminders should help determine aircraft compliance with all airworthiness requirements. (Refer to 14 CFR part 43, § 43.15.)

Configuration

The aircraft should conform to the aircraft specification or type certificate data sheet, any changes by supplemental type certificates, and/or its properly altered condition. When the aircraft does not conform, use the procedures for "unairworthy" items listed in 14 CFR part 43, § 43.11(a)(5).

11

1. Alterations to the product may have changed some of the operating limitations.
2. Unrecorded alterations or repairs may have been made in the past and warrant one of the following:
    a. Contacting the owner for pertinent information
    b. Conducting an inspection and personally approving for return to service by completing FAA Form 337, if approved data is available
    c. Contacting the local ASI for assistance
3. The aircraft specification or type certificate data sheet indicates when a flight manual is required. It also identifies limitations which must be displayed in the form of markings and placards.
4. Unlike the specifications, type certificate data sheets do not contain a list of equipment approved for a particular aircraft. The list of required and optional equipment can be found in the equipment list furnished by the manufacturer of the aircraft. Sometimes a later issue of the list is needed to cover recently approved items. Serial number eligibility should always be considered.

Condition

The holder of an IA may use the checklist in 14 CFR part 43, appendix D, the manufacturer's inspection sheets, or a checklist designed by the holder of an IA, that includes the scope and detail of the items listed in appendix D, to check the condition of the entire aircraft. This includes checks of the various systems listed in 14 CFR part 43, § 43.15.

Routine servicing is *not* a part of the annual inspection. The inspection itself is essentially a visual evaluation of the condition of the aircraft and its components and certain operational checks. The manufacturer may recommend certain services to be performed at various operating intervals. These services can

12

often—in fact, should—be done conveniently during an annual inspection, but are not considered to be a part of the inspection itself.

It is very important that the holder of an IA be familiar with the manufacturer's service manuals, bulletins, and letters for the product being inspected. Use these publications to avoid overlooking problem areas.

AC 43-16A, Aviation Maintenance Alerts, is also an important source of service experience. The articles for the alerts are taken from selected service difficulties reported to the FAA on Form 8010-4, Malfunction or Defect Reports. Monthly copies of the alerts are provided on the Internet at www.faa.gov under the Aircraft tab, and then under "Aircraft Safety." Comments may be sent by letter, with name and address typed or legibly printed to:

Federal Aviation Administration
Aviation Data Systems Branch, AFS-620
P.O. Box 25082
Oklahoma City, OK 73125

When the holder of an IA approves an aircraft for return to service, he or she will be held responsible for the condition of the aircraft *as of the time of approval*.

Minimum Equipment List

The minimum equipment list (MEL) is intended to permit operations with certain inoperative items of equipment for the minimum period of time necessary until repairs can be accomplished. It is important that repairs are accomplished at the earliest opportunity in order to return the aircraft to its design level of safety and reliability. Be mindful of the following points with respect to MELs:

1. When inspecting aircraft operating with an MEL, the holder of an IA should review the document where inoperative

items are recorded (aircraft maintenance record, logbook, discrepancy record, etc.) to determine the state of airworthiness with regard to those recorded discrepancies. Inspections of aircraft with approved MELs will be in accordance with 14 CFR under which the MEL was issued.

2. Those MELs specifying repair intervals through the use of A, B. C, D codes require repairs of deferred items at or prior to the repair times established by the letter designated category. In such instances, some items previously deferred may not be eligible for continued deference at the inspection or may require additional maintenance. Where repair intervals are not specified by codes in the MEL, all MEL-authorized inoperative instruments and/or equipment should be repaired or inspected and deferred before approval for return to service.
3. Aircraft established on a progressive inspection program require that all MEL-authorized inoperative items be repaired or inspected and deferred at each inspection whether or not the item is encompassed in that particular segment.
4. When inspecting aircraft operating without an MEL, 14 CFR part 91, § 91.213(d), allows certain aircraft not having an approved MEL to be flown with inoperative instruments and/or equipment. These aircraft may be presented for annual or progressive inspection with such items previously deferred or may have inoperative instruments and equipment deferred during an inspection. In either case, the holder of an IA is required by 14 CFR part 43, § 43.13(b) to determine that:
    a. The deferrals are eligible within the guidelines of that rule.
    b. All conditions for deferral are met, including proper recordation in accordance with 14 CFR part 43, sections (§§) 43.9 and 43.11.
    c. Deferral of any item or combination of items will not affect the intended function of any other operable instruments and/or equipment, or in any manner constitute

14

a hazard to the aircraft. When these requirements are met, such an aircraft is considered to be in a properly altered condition with regard to those deferred items.

Airworthiness Directives

The holder of an IA is required by 14 CFR part 43, § 43.13, to determine that all applicable ADs for aircraft, powerplants, propellers, instruments, and appliances have been accomplished. You must consider the following:

1. If the maintenance records indicate compliance with an AD, the holder of an IA should make a reasonable attempt to verify the compliance. It is not uncommon for a component to have compliance with an AD accomplished and properly recorded then later be replaced by another component on which the AD has not been accomplished. The holder of an IA is not expected to disassemble major components (cylinders, crankcases, etc.) if adequate records of compliance exist.
2. When the maintenance records *do not* contain indications of AD compliance, the holder of an IA should:
   a. Make the AD an item on a discrepancy list provided to the owner, in accordance with 14 CFR part 43, § 43.11(b);
   b. With the owner's concurrence, do whatever disassembly is required to determine the status of compliance; or
   c. Obtain concurrence of the owner to comply with the AD.
3. Often, an AD calls for an inspection, with a modification or inspection required at a later date. It is very important to identify, in the maintenance record entry, the portion of the AD complied with and the exact method of compliance.
4. 14 CFR part 91, § 91.417(a)(2)(v) requires each registered owner or operator to keep a record of the current status of applicable ADs. This status includes for each the method of compliance, AD number, and revision date. If the AD involves

15

recurring action, the time and date should be recorded when the next action is required. As a vital part of the services performed, the holder of an IA may wish to provide the owner with information he or she is expected to keep. (Refer to appendix 1, figure 5.)

5. The owner should also be informed of any subsequent requirements of an AD or whether a reinspection is required at operating intervals other than at annual inspections. Often, the subsequent requirements are at 100-hour intervals and will need to be done whether or not the aircraft is required to have 100-hour inspections. Where a progressive inspection is involved, the approved program should state how and when the AD review will be accomplished. However, as a mechanic or IA, you should be aware of an AD that is pending or due, and is not in the area you are inspecting. It is good customer relations to inform the owner or pilot of the situation.

Malfunction or Defect Reports

All malfunctions or defects that come to the attention of the holder of an IA should be reported on FAA Form 8010-4. (Refer to appendix 1, figure 6.) Copies of the self-addressed form are available at all Flight Standards District Offices (FSDOs). It is easy to complete and requires no postage. The form can also be completed on line at www.faa.gov under the Aircraft tab, and then under Report Safety Issues. Prompt reporting will contribute much toward improving air safety by helping correct unsafe conditions.

Paperwork Review

The owner or operator is responsible for maintaining the equipment list, CG and weight distribution computations, and loading schedules, if necessary. The following items must be considered:

16

1. The holder of an IA as required by 14 CFR part 43, § 43.13, determines that the required placards and documents set forth in the aircraft specification or type certificate data sheet are available and current. The aircraft should be reported as being in an unairworthy condition if these placards and documents are not available. A missing, incorrect, or improperly located placard is considered an unairworthy item, and the owner or operator should be informed that, under the requirements of 14 CFR part 91, § 91.9, the aircraft may not be operated until a correct and properly placed placard is available.
2. The holder of an IA should refer to the registration and airworthiness certificates for the owner's name and address; the aircraft make, model, registration, and serial numbers needed for recording purposes. Be sure not to use manufacturer trade names as they do not always coincide with the actual model designation (Cessna Skylane model designation is 182, Piper Seneca III is PA 34 220T, etc.). If registration and airworthiness certificates are not available, the aircraft does not need to be reported in unairworthy condition; however, the owner or operator should be informed that the documents required by 14 CFR part 91, § 91.203(a)(2), should be in the aircraft and the airworthiness certificate displayed *when the aircraft is operated.*
3. On aircraft for which no approved flight manual is required, the operating limitations prescribed during original certification, and as required by 14 CFR part 91, § 91.9, must be carried in or be affixed to the aircraft. Range markings on the instruments, placards, and listings must be worded and located as specified in the type certificate data sheet. (Refer to appendix 1, figure 7.)

Aircraft Markings

Required aircraft identification markings are discussed in 14 CFR part 45. It is the owner's or operator's responsibility to have the nationality and registration markings properly displayed on the aircraft (14 CFR part 91, § 91.9(c)). The holder of an IA can and should offer advisory service to owners and operators in regard to any deficiencies in markings; however, such deficiencies are not cause to report an aircraft in "unairworthy" condition.

Aircraft With Discrepancies or Unairworthy Conditions

If the aircraft is not approved for return to service after a required inspection, use the procedures specified in 14 CFR part 43, § 43.11. This will permit an owner to assume responsibility for having the discrepancies corrected prior to operating the aircraft. Discrepancies or unairworthy conditions can be resolved in the following ways:

1. The discrepancies can be cleared by a person who is authorized by 14 CFR part 43 to do the work. Preventive maintenance items could be cleared by a pilot who owns or operates the aircraft, provided the aircraft is not used under 14 CFR part 121, 129, or 135; except that approval may be granted to allow a pilot operating a rotorcraft in a remote area under 14 CFR part 135 to perform preventive maintenance.
2. The owner may want the aircraft flown to another location to have repairs completed, in which case the owner should be advised that the issuance of FAA Form 8130-7, Special Flight Permit, is required. This form is commonly called a ferry permit and is detailed in 14 CFR part 21, § 21.197. The certificate may be obtained in person or by fax at the local FSDO or from a Designated Airworthiness Representative.
3. If the aircraft is found to be in an unairworthy condition, an entry will be made in the maintenance records that the inspection was completed and a list of unairworthy items was provided to the owner. When all unairworthy items are corrected by a person authorized to perform maintenance and that person makes an entry in the maintenance record

18

for the correction of those items, the aircraft is approved for return to service. (Refer to appendix 1, figures 8 and 9.)

Incomplete Inspection

If an annual inspection is not completed, the holder of an IA should:

1. Indicate any discrepancies found in the aircraft records.
2. *Not* indicate that an annual inspection was completed.
3. Indicate in the aircraft records the extent to which the inspection was completed and all work accomplished.

19

Chapter 3
MAINTENANCE RECORDS

REQUIRED RECORDATION

The holder of an IA and other maintenance personnel or agencies are required to record maintenance, inspections, or alterations performed or approved in accordance with the requirements of 14 CFR part 43, §§ 43.9 and 43.11. The owner or operator is required by 14 CFR part 91, § 91.417 to keep maintenance records. The holder of an IA is also required to indicate the total aircraft time in service when a required inspection is done.

Responsibility for maintenance work performed rests with the person whose signature and certificate number is entered on the appropriate maintenance record and/or forms. The responsibility for annual and progressive inspections and approval for return to service after major repairs or major alterations is assumed by the holder of an IA whose signature and certificate number appears on the appropriate maintenance records.

FAA FORM 337

FAA Form 337, Major Repair and Alteration (Airframe, Powerplant, Propeller, or Appliance), serves two purpose. It provides:

1. Owners and operators with a record of major repairs and major alterations, indicating details and approval.
2. The FAA with a copy for the aircraft records.

An example of a typical completed FAA Form 337 is provided in appendix 1, figure 3.

20

Completion

The person who performed or supervised the major repair or major alteration prepares the original FAA Form 337 (two originals). Instructions for the completion of FAA Form 337 appear in AC 43.9-1 (as revised), Instructions for Completion of FAA Form 337, Major Repair and Alteration (Airframe, Powerplant, Propeller, or Appliance). The holder of an IA then further processes the forms when they are presented for approval. If the IA holder finds a major alteration or a major repair to be in conformity with FAA-approved data, the IA holder must review the FAA Form 337 for completeness and accuracy, and complete item 7. The IA holder should ensure that the duplicate is an exact and legible reproduction of the original. Signatures should be original in ink, and not carbon copies.

Disposition

In accordance with 14 CFR part 43, the person performing a major repair or major alteration must make the proper entry in the maintenance records and distribute the originals of the completed and signed FAA Form 337 to the:

1. Aircraft owner and,
2. FAA in one of the following two forms:
   a. One completed hardcopy form within 48 hours to:

      Federal Aviation Administration
      Aircraft Registration Branch, AFS-751
      PO Box 25724
      Oklahoma City, OK 73125

   b. An electronic form (with electronic signature) automatically through the Web site at eformservice.faa.gov/eForm337.aspx.

If the FAA Form 337 is completed for extended-range fuel tanks installed within the passenger compartment or a baggage compartment, the person who performs the work and the person

21

authorized to approve the work by 14 CFR part 43, § 43.7, must execute an FAA Form 337 in at least triplicate, as required by 14 CFR part 43, appendix B. One copy of the FAA Form 337 must be placed on board the aircraft, as specified in 14 CFR part 91, § 91.417(d). The remaining copies are distributed as previously mentioned.

If FAA Form 337 has been completed for engines, propellers, spare parts or components, both copies of the form, with the approval portion completed, should be attached to the part or component until it is installed on an aircraft. The mechanic who makes the installation will, in accordance with 14 CFR part 43, § 43.9(a)(4), complete both copies of FAA Form 337 by filling in blocks 1 and 2, and sign for the installation in the aircraft records, making reference to the FAA Form 337 in the record entry. The copies are distributed as previously mentioned.

WEIGHT AND BALANCE

Weight and balance data is not required on the FAA Form 337. However, it is imperative that weight and balance checks be made very carefully. Since aircraft manufacturers use varying methods of weight and balance control, it not feasible to provide a universally adaptable method. The example provided in appendix 1, figure 10, of this guide is general in nature and can be modified to suit the aircraft involved. When revising weight and balance data, the following general guidelines should be used:

1. The weight and balance data should be kept together in the aircraft records.
2. When making revisions, use a permanent, easily identified method with full-size sheets of paper large enough to contain complete computations and to minimize the possibility of becoming detached or lost.
3. Each page should be identified with the aircraft by make, model, serial number, and registration number.

22

4. The pages should be signed and dated by the person making the revision.
5. The nature of the weight change should be described.
6. The old weight and balance data should be marked "superseded" and dated.
7. A new page should show the date of the old figures it supersedes.
8. Appropriate fore and/or aft extreme loading conditions should be investigated and the computations shown.
9. Example loading computations may be helpful.
10. On large aircraft, be careful to distinguish between empty weight and operating weights that may include commissary supplies, spare parts, lavatory water, etc.
11. On small aircraft, it is often convenient to post a placard in the aircraft indicating the empty weight, useful load, empty CG, and example loadings or general instructions to cover the most likely loading conditions. (Refer to 14 CFR § 91.9(b)(2).) AC 120-27 (as revised), Aircraft Weight and Balance Control, and FAA-H-8083-1 (as revised), Aircraft Weight and Balance Handbook, contain useful information applicable to the functions performed by the holder of an IA on general aviation aircraft.

23

Chapter 4
SUGGESTIONS FOR DEVELOPING GOOD OWNER/IA RELATIONS

Be sure to come to a mutual agreement with the aircraft owner concerning exactly what work is to be performed. Misunderstandings usually result from a lack of clear communication. Get it straight. Attention to the following details will usually avoid the ill will a later disagreement could generate.

1. Itemize the work to be done so the owner will have a clear understanding of the work order.
2. Establish a firm understanding about the cost, or range of cost, anticipated for the job.
3. If an annual inspection is involved, indicate that certain maintenance is required to perform the inspection, such as:
   a. Removing cowling and fairing, and opening inspection plates.
   b. Cleaning the aircraft and engine.
   c. Disassembling wheels and other components to determine their condition.
4. Advise the owner that an annual inspection involves determination of compliance with aircraft specifications and ADs.
5. Agree whether routine servicing is to be included as part of the inspection or if it is to be performed separately. Such servicing is not a part of the inspection, but may be conveniently done while conducting the inspection. Items might include:
   a. Cleaning spark plugs,
   b. Servicing landing gear shock struts,
   c. Changing oil,

24

d. Making minor adjustments,

e. Servicing brakes,

f. Dressing nicked propeller blades,

g. Lubricating where necessary, or

h. Stop-drilling small cracks and minor patching of cowling and baffles.

6. The owner should be made aware that the annual or progressive inspection does not include correction of discrepancies or unairworthy items and that such maintenance will be additional to the inspection. Maintenance and repairs may be accomplished simultaneously with the inspection by a person authorized to perform maintenance if the owner and the IA holder agree. This method would result in an aircraft that is approved for return to service upon completion of the inspection. A written list of discrepancies and unairworthy items not repaired concurrently with the inspection must be made and given to the owner. Record uncorrected discrepancies and unairworthy items in the maintenance records. The owner must make arrangement for correction or deferral of items on the list of discrepancies and unairworthy items with a person authorized to perform maintenance prior to returning the aircraft to service. The holder of the IA ensures that any item permitted to be inoperative by a MEL or under 14 CFR part 91, § 91.213(d)(2) is properly placarded and any maintenance for deferral has been performed. Any deferred items are to be included on the list of discrepancies and unairworthy items. The owner should be informed that the aircraft should not be operated until the discrepancies and unairworthy items are corrected or are appropriately deferred.

7. Establish a reasonable time frame to accomplish the inspection.

8. Request the owner to supply the complete aircraft records (airframe, engines, and propellers) for study, review, and entries.

Point out that this is necessary to conduct an annual inspection properly.

9. Complete the inspection as soon as practicable. An aircraft can sit idle in a shop waiting for parts, even though the inspection has actually been completed. In this case, it is advisable to officially report the aircraft unairworthy. (Refer to 14 CFR part 43, § 43.11(a)(5).) When the parts arrive, the repairs can be completed and the aircraft approved for return to service in the usual manner by the person who makes the repairs. The time lapse may represent several weeks or months, and conditions can deteriorate on the aircraft. Also, there is the chance that an AD involving some part of the aircraft may have been issued in the interim. In these cases, it might be unwise to complete the repairs originally intended and sign off the aircraft as airworthy without doing another complete inspection.
10. Complete the aircraft record entries as required by 14 CFR part 43, §§ 43.9 and 43.11 and provide sufficient information for the owner to comply with 14 CFR part 91, § 91.417(a)(2)(i). Make adequate descriptions of repairs or alterations if accomplished along with the inspection.
11. Record compliance with all ADs actually accomplished. Provide sufficient information for the owner to comply with 14 CFR part 91, § 91.417(a)(2)(v). A general statement such as "All ADs complied with" is *not* an adequate entry and should be avoided. Many owners keep a separate record of AD compliance in the back of the logbook or in a section specifically provided for this record. This is a good place to identify the ADs of a recurring nature and show when the next compliance is required. (Refer to appendix 1, figures 12 and 13, for typical entries.)
12. When approving repairs and alterations, the IA holder should be available as work progresses on major jobs; affected areas and structures can be seen and repairs can be inspected and improved more readily than after completion of the entire job.

26

13. Remind the owners or operators that they are responsible for operational requirements, such as:

    a. Very high frequency (VHF) omnidirectional range (VOR) equipment checked in accordance with 14 CFR part 91, § 91.171.

    b. Altimeter and altitude reporting equipment test and inspections in accordance with 14 CFR part 91, § 91.411.

    c. Air traffic control (ATC) transponder inspection in accordance with 14 CFR part 91, § 91.413. These tests and inspections are not part of the annual inspection.

27

Appendix 1
SAMPLE FORMS AND RECORDS

No certificate may be issued unless a completed application form has been received (14 CFR 65)

U. S. DEPARTMENT OF TRANSPORTATION
FEDERAL AVIATION ADMINISTRATION
MECHANIC'S APPLICATION FOR INSPECTION AUTHORIZATION-PRIVACY ACT

Form Approved: OMB No. 2120-0022 02/28/2011

1. NAME *(Last, First, Middle)*		2. MECHANIC CERTIFICATE NO.
Doe, John J.		A&P 12345678
3. MAILING ADDRESS *(Number, Street, City, State/County, Zip Code) (Place at which you desire to receive Airworthiness Directives, etc.)*	4a. FIXED BASE OF OPERATIONS	4b. TELEPHONE NO.
	PLACE AT WHICH YOU MAY BE LOCATED IN PERSON DURING NORMAL WORKING WEEK	PLACE AT WHICH YOU MAY BE LOCATED BY TELEPHONE DURING NORMAL WORKING WEEK
1450 E Cheltenham Ave Cleveland County Oklahoma City, OK 73098	Meridian Aviation, Downtown Airpark 5060 S Western Ave Oklahoma City OK 73452	(405) 555-1875

	YES	NO
5. HAVE YOU HELD A MECHANIC CERTIFICATE WITH BOTH AIRFRAME AND POWERPLANT RATINGS FOR THE 3 YEARS PRECEDING THE DATE OF THIS APPLICATION ?	☑	☐
6. HAVE YOU BEEN ACTIVELY ENGAGED, FOR AT LEAST THE 2-YEAR PERIOD BEFORE THE DATE OF APPLICATION IN MAINTAINING AIRCRAFT CERTIFICATED AND MAINTAINED IN ACCORDANCE WITH THE CFRs ?	☑	☐
7. HAS YOUR MECHANIC CERTIFICATE AND/OR RATINGS BEEN REVOKED OR SUSPENDED DURING THE 3-YEAR PERIOD PRECEDING THIS APPLICATION ?	☐	☑
8. HAS AN INSPECTION AUTHORIZATION BEEN DENIED YOU WITHIN 90 DAYS PREVIOUS TO THIS APPLICATION ? IF ANSWER IS "YES", EXPLAIN IN REMARKS.	☐	☑
9. HAVE YOU MET THE MINIMUM REQUIREMENTS FOR RENEWAL OF INSPECTION AUTHORIZATION ? *(For Renewal Only)*	☐	☐

10. BASIS FOR RENEWAL *(Number Performed Per Renewal Period)*

ALTERATIONS		REPAIRS		ANNUAL INSP.		PROGRESSIVE INSP.		RECENT ISSUANCE – IN EFFECT LESS THAN 90 DAYS BEFORE EXPIRATION DATE. ☐
First Year Period	*Second Renewal Period*	*First Year Period*	*Second Renewal Period*	*First Year Period*	*Second Renewal Period*	*First Year Period*	*Second Renewal Period*	

FAA ACCEPTED COURSE/SEMINAR NO., LOCATION, AND DATE *(First Year Period)*	FAA ACCEPTED COURSE/SEMINAR NO., LOCATION, AND DATE *(Second Renewal Period)*

11. AIRCRAFT MAINTENANCE ACTIVITY DURING LAST 2 YEARS

DATES	NAME AND ADDRESS OF REPAIR STATION, FACILITY, MANUFACTURER, OPERATOR, ETC.	DESCRIPTION OF ACTIVITY
FROM June 12, 20XX TO PRESENT	Meridian Aviation, Downtown Airpark 5077 S Western Ave Oklahoma City, OK 73458	Inspection, repair, and overhaul of single-engine and multiengine aircraft
FROM TO		
FROM TO		

12. REMARKS

Endorsement expires in 30 days.

13. CERTIFICATION: *I certify that the statements made above and in all attachments hereto are correct and true.*

DATE	SIGNATURE OF APPLICANT
March 22, 20XX	*John J. Doe*

14. RECORD OF ACTION *(For FAA use only)*

	DATE	INSPECTOR'S SIGNATURE	OFFICE IDENTIFICATION
■ ENDORSEMENT	March 22, 20XX	*John Milford* John Milford	ASW 25
☐ ISSUANCE ☐ RENEWAL ☐ VOLUNTARY SURRENDER			

FAA Form 8610-1 (02-2010) SUPERCEDES PREVIOUS EDITION

Figure 1. FAA Form 8610-1, Mechanic's Application for Inspection Authorization.

A-1

FAA-G-8082-19

UNITED STATES OF AMERICA
DEPARTMENT OF TRANSPORTATION
FEDERAL AVIATION ADMINISTRATION

INSPECTION AUTHORIZATION

This certifies that Robert D. Burge

holder of Mechanic Certificate No. 222442345 has been authorized to exercise the privileges of Federal Aviation Regulation 65.95.

This authority expires March 31, 20XX unless sooner revoked by the Administrator of the Federal Aviation Administration or extended by endorsement on the reverse of this card.

DATE ISSUED	SIGNATURE, FLT. STDS. INSPECTOR
3/16/20XX	*Mike Johnson* Mike Johnson

SIGNATURE OF AUTHORIZED MECHANIC: *Robert D. Burge*

FAA FORM 8310-5 (8-80) SUPERSEDES PREVIOUS EDITION

SAMPLE

front

Authority to exercise the privileges of FAR 65.95 has been endorsed or renewed to expire on the date shown below.

EXPIRATION DATE	ENDORSED BY INSPECTOR	FAA OFFICE
3/31/20XX	*Mike Johnson*	SW-FSDO-2

back

Figure 2. FAA Form 8310-5, Inspection Authorization.

A-2

US Department of Transportation Federal Aviation Administration

MAJOR REPAIR AND ALTERATION
(Airframe, Powerplant, Propeller, or Appliance)

Form Approved OMB No. 2120-0020 2/28/2011 | Electronic Tracking Number | For FAA Use Only

INSTRUCTIONS: Print or type all entries. See Title 14 CFR §43.9, Part 43 Appendix B, and AC 43.9-1 (or subsequent revision thereof) for instructions and disposition of this form. This report is required by law (49 U.S.C. §44701). Failure to report can result in a civil penalty for each such violation. (49 U.S.C. §46301(a))

1. Aircraft	Nationality and Registration Mark: N12345J	Serial No.: 721-43566	
	Make: FleetWing	Model: FW200	Series: 80
2. Owner	Name (*As shown on registration certificate*): Mike J. Urbach	Address (*As shown on registration certificate*): Address 2414 N. Lincoln; City Milltown; State OK; Zip 73122; Country	

3. For FAA Use Only

The technical data identified herein has been found to comply with applicable airworthiness requirements and is hereby approved for use only on the above aircraft, subject to conformity inspection by a person authorized in 14 CFR part 43, section 43.7.

Maria Johnson Maria Johnson, ASI

4. Type		5. Unit Identification			
Repair	Alteration	Unit	Make	Model	Serial No.
☑	☐	AIRFRAME	———	(*As described in Item 1 above*)	———
☐	☐	POWERPLANT			
☐	☐	PROPELLER			
☐	☐	APPLIANCE	Type / Manufacturer		

6. Conformity Statement

A. Agency's Name and Address	B. Kind of Agency	
Name: Eugene Henson	✓ U. S. Certificated Mechanic	☐ Manufacturer
Address: 212 SW 66th Street	☐ Foreign Certificated Mechanic	C. Certificate No. A&P 1709665
City: Milltown State: OK	☐ Certificated Repair Station	
Zip: 73122 Country:	☐ Certificated Maintenance Organization	

D. I certify that the repair and/or alteration made to the unit(s) identified in item 5 above and described on the reverse or attachments hereto have been made in accordance with the requirements of Part 43 of the U.S. Federal Aviation Regulations and that the information furnished herein is true and correct to the best of my knowledge.

Extended range fuel per 14 CFR Part 43 App. B ☐	Signature/Date of Authorized Individual: *Eugene Henson* *March 2, 20xx* Eugene Henson

7. Approval for Return to Service

Pursuant to the authority given persons specified below, the unit identified in item 5 was inspected in the manner prescribed by the Administrator of the Federal Aviation Administration and is ☑ Approved ☐ Rejected

BY				
	☐ FAA Flt. Standards Inspector	☐ Manufacturer	☐ Maintenance Organization	☐ Persons Approved by Canadian Department of Transport
	☐ FAA Designee	☐ Repair Station	✓ Inspection Authorization	Other (*Specify*)

Certificate or Designation No.	Signature/Date of Authorized Individual
A&P 9486717 IA	*Martin M. Sawyer* *April 2, 20xx* Martin M. Sawyer

FAA Form 337 (10-06)

Figure 3. FAA Form 337, Major Repair and Alteration (Airframe, Powerplant, Propeller, or Appliance) (front view). Note the FAA inspector's data approval for a major repair (block 3).

A-3

FAA-G-8082-19

NOTICE

Weight and balance or operating limitation changes shall be entered in the appropriate aircraft record. An alteration must be compatible with all previous alterations to assure continued conformity with the applicable airworthiness requirements.

8. Description of Work Accomplished
(If more space is required, attach additional sheets. Identify with aircraft nationality and registration mark and date work completed.)

N12345J	03/02/20XX
Nationality and Registration Mark	**Date**

Aircraft Total Time 6,210 hours

1. Removed horizontal stabilizer from aircraft and opened top and bottom skin at rear spar. Removed cracked rear spar and replaced with new spar (part number FW10204-56) in accordance with FleetWing structural repair manual No. 410, chapter 2, and figure 9-12. Original rivet pattern and type (MS20470AD3-4) were maintained.

   DATE: 02/25/20XX, inspected repair work to interior of horizontal stabilizer prior to closure of top skin. Found repair to be in accordance with data indicated and ready for final closure. An inspection of the complete interior of the stabilizer for hidden damage and condition at this time revealed no damage and good structural condition.

Martin M. Sawyer

Martin M. Sawyer, A&P 9486717 IA

2. Primed interior of stabilizer and closed upper skin. Installed on aircraft, rigged elevator and operationally checked in accordance with manufacturer's maintenance manual (FW4490).

3. No change in weight and balance.

END

☐ **Additional Sheets Are Attached**

FAA Form 337 (10-06)

Figure 3 (cont'd). FAA Form 337, Major Repair and Alteration (Airframe, Powerplant, Propeller, or Appliance) (back view).

A-4

March 22, 20XX

Total Aircraft Time 1,502.0 Hours

Tach Time 972.4 Hours

I certify that this aircraft has been inspected in accordance with an annual inspection as per Air Tractor AT502 owner's manual and was determined to be in an airworthy condition.

Joseph P. Kline
A&P 123467899 IA

NOTE: This is an example of a record entry for an **annual inspection** determining the aircraft to be in airworthy condition. The date, aircraft total time, and tachometer (tach) or recorder reading are included. The tach or recorder readings should not be confused with the total time and should only be shown in **addition** to the total time entry. The mechanic's certificate number is suffixed by the letters "IA" indicating that the mechanic is the holder of an inspection authorization. Maintenance done in conjunction with the inspection should be entered as a separate entry.

Figure 4. Example of an airworthy annual inspection maintenance record entry.

A-5

Page 1 of 1 Date 12/02/2000
Registration Number N937JM Aircraft Make/Model FleetWing FW-25-200 Serial Number 2842015
Aircraft Certification Date 07/12/2000
Engine Model Lycoming 0-320-D3G Serial Number L-7656-38A

AD Number	Revision Date	Applicable S.B. Number and Subject	Date and Hours at Compliance	Method of Compliance	One Time	Recurring	Next Compliance at Hours/Date	Authorized Signature, Type, and Number
2008-26-13		Inspect Oil Cooler Hose	5/27/1996 3,102 hours	Replaced hose assembly with TSO 53a, type D hose. 100-hour recurrent inspection or longer required.			NA	Bill Jenkins A&P 23453322 IA
2005-20-R1	Oct. 10, 2005	Inspect fuel cells IAW SB 1134	12/14/2005 2,823 hours	Inspected IAW FleetWing service bulletin 1134 sections A and B.	X		No further action required	Joe Kline A&P 123467899 IA
2001-02-03		Fuel quantity indicators.	02/15/2001 502 hours	Replaced right and left fuel quantity indicators per AD paragraph B2.	X		No further action required	Jimmy Miller A&P 23244411
2000-26-01		Inspect flap jackscrew IAW SB 1002	02/15/2001 502 hours	Inspected IAW FleetWing SB 1002.		X	Inspection required each 3,000 hours	Jimmy Miller A&P 23244411

Figure 5. Airworthiness Directive Compliance Record (suggested format).

A-6

OMB No. 2120-0003
08/31/2008

DEPARTMENT OF TRANSPORTATION FEDERAL AVIATION ADMINISTRATION MALFUNCTION OR DEFECT REPORT	OPER. Control No.	
	ATA Code	8120
	1. A/C Reg. No.	N- 6696J

Enter pertinent data	MANUFACTURER	MODEL/SERIES	SERIAL NUMBER
2. AIRCRAFT	Cessna	421B	421B79485
3. POWERPLANT	Continental	GTSIO520L	C216977
4. PROPELLER	McCauley	3AF34C92	42279
5. SPECIFIC PART (of component) CAUSING TROUBLE			
Part Name	MFG. Model or Part No.	Serial No.	Part/Defect Location.
Wastegate shaft	Garrett PN 4166952	NA	Left engine wastegate
6. APPLIANCE/COMPONENT (Assembly that includes part)			
Comp/Appl Name	Manufacturer	Model or Part No.	Serial Number
Wastegate	Garrett	480164-10	1121

Part TT	Part TSO	Part Condition	7. Date Sub.
1,222 hrs	NA	warped	1/22/20XX

8. Comments (Describe the malfunction or defect and the circumstances under which it occurred. State probable cause and recommendations to prevent recurrence.)

Pilot reported loss in aircraft's critical altitude. Inspection revealed the left engine's wastegate shaft warped and binding. The shaft's freedom of travel was also found to be partially restricted due to carbon buildup in the bearings. This is possibly a contributing factor in the warping. Recommend lubricating wastegate valve with approved lubricant such as Mouse Milk or WD-40 when shaft is cool.

Optional Information:

Check a box below, if this report is related to an aircraft

☐ Accident; Date ______ ☐ Incident; Date ______

DISTRICT OFFICE | OPERATOR DESIGNATOR | OTHER | COMMUTER | FAA | MFG. | AIR TAXI | MECH. X | OPER. | REP. STA.

SUBMITTED BY: [signature]

TELEPHONE NUMBER: (429) 555 — 6219

FAA FORM 8010-4 (10-92) SUPERSEDES PREVIOUS EDITIONS

NOTE: This is a typical FAA Form 8010-4 (revised 10-92). The holder of an IA is urged to use this form for all malfunctions or defects that cannot be attributed to poor maintenance procedures. Provide the information requested on the form. Note that item 8 requests information concerning how the defect can be corrected. The form may be obtained at the local Flight Standards District Office or online at the FAA's web site (FAA.gov). At the web site the form is found under the Aircraft tab and under the Advisories & Guidance—Service Difficulty Reports (SDR) heading. The form may be e-mailed, faxed, or mailed to the addresses or telephone fax number noted under Service Difficulty Reporting System (SDRS) Submissions on the web site.

Figure 6. FAA Form 8010-4, Malfunction or Defect Report.

A-7

FAA-G-8082-19

Operating Limitations:	**Zeph-Air 63-1A N40023**
RPM	Do not exceed 2,300
Oil temperature	212 °F max.
Airspeed limits—do not exceed:	
Level flight or climb	95 knots
Glide or dive	130 knots
Gross weight	1,200 lb
Empty CG	14.4 inches aft of datum
Useful load	453 lb
Kinds of operation	VFR—day

Figure 7. Operating limitations placard.

A-8

March 22, 20XX

Total Aircraft Time 3,202.5 hours

Hobbs Meter Reading 975.5 hours

I certify that this aircraft has been inspected in accordance with an annual inspection, and a list of discrepancies and unairworthy items dated March 22, 20XX, have been provided for the aircraft owner.

Joseph P. Kline
A&P 1123456789 IA

Figure 8. Example of an unairworthy annual inspection maintenance record entry.

A-9

Academy Aviation
Hangar 4
North Philadelphia Airport
Philadelphia, PA 19114

Mr. Morris McCall
1450 W. Cheltenham Ave.
Philadelphia, PA 19125

Mr. McCall:

This is to certify that on March 22, 20XX, I completed an annual inspection on your aircraft, Condor 191B, S/N 3945, N1234, and found the following unairworthy items:

Compression in No. 3 cylinder read 30 over 80, which is below the manufacturer's recommended limits.

The muffler has a broken baffle plate which is blocking the engine exhaust outlet.

There is a 6-inch crack on bottom of the left wing just aft of the main landing gear attach point.

Jospeh P. Kline
A&P 123456789 IA

Figure 9. Discrepancy list to be provided to an aircraft owner when reporting an aircraft with unairworthy items after completing an annual inspection.

A-10

Weight and Balance Revision — **Date: 05/06/20XX**

N44933 Cessna 182L

Serial Number 18234329

Supersedes Computations found on FAA Form 337, dated 10/22/20XX.

	Weight	Arm	Moment
Removed the following equipment:			
1. Turn coordinator P/N C661003-0211	2.50 lb	15.0	37.50
2. Directional gyro P/N 0760099	+3.12	13.5	+42.12
Total	5.62		79.62
	1,709.60	35.26	60,282.20
	–5.62		–79.62
Aircraft after removal	1,703.98	35.20	60,202.58

	Weight	Arm	Moment
Installed the following equipment:			
1. Vector2 Autopilot system including turn coordinator and directional gyro.	13.0 lb	32.7	425.13
	1,703.98		60,202.20
	+13.00		+425.13
***REVISED LICENSED EMPTY WEIGHT**	**1,716.98**		**60,627.71**

***NEW USEFUL LOAD: 1,083.02**

Forward Limit Check (Limit +38.4)				Rearward Limit Check (Limit +47.4)			
	Wt	Arm	Moment		Wt	Arm	Moment
A/C Empty	1,716.98	35.31	60,627.21	**A/C Empty**	1,716.98	35.31	60,627.21
Fwd Seats	170.00	36.00	6,120.00	**Fwd Seats**	170.00	36.00	6,120.00
Aft Seats				**Aft Seats**	340.00	71.00	24,140.00
Fuel (min.)	115.00	48.00	5,520.00	**Fuel (max.)**	360.00	48.00	17,280.00
Oil	22.00	–15.00	–330.00	**Oil**	22.00	–15.00	–330.00
Baggage				**Baggage**	120.00	97.00	11,640.00
	2,023.98	35.50	71,937.71		2,728.98	43.78	119,477.71

Joseph P Kline

Joseph P Kline
A&P 123456789 IA

NOTE: Computations are shown. Form is signed, dated, and identifies the computations or figures it supersedes. It is recommended that the manufacturer's weight and balance data forms be used for specific aircraft.

Figure 10. Weight and balance revision for a typical light, single-engine aircraft.

A-11

FAA-G-8082-19

July 12, 20XX

Aircraft Total Time: 1,566 hours

Complied with Airworthiness Directive (AD) 20XX-12-10R1, effective date June 30, 20XX. Modified the airplane by compliance with paragraph 2(b) of AD. Installed FleetWing Service Kit SK 1910 as required by AD. No recurring action required.

Bill Quinlan
A&P 143298671

Figure 11. One-time Airworthiness Directive compliance entry.

May 23, 20XX

Engine Total Time: 720 hours

Complied with Airworthiness Directive (AD) 20XX-10-12, Alcon Turbo Chargers by inspection as required by paragraphs (b) through (g) of AD. Turbine housing found satisfactory, next inspection due at 920 hours.

Joe Knight
A&P 279387792

Figure 12. Recurrent Airworthiness Directive compliance entry.

A-12

Appendix 2
PUBLICATIONS AND TECHNICAL DATA

The following publications and technical data provide information for aircraft inspection.

1. TITLE 14 OF THE CODE OF FEDERAL REGULATIONS

The Code of Federal Regulations is a codification of the general and permanent rules published in the Federal Register by the Executive departments and agencies of the Federal Government. The Code is divided into 50 titles, which represent broad areas subject to Federal regulation. Each title is divided into chapters, which usually bear the name of the issuing agency. Title 14—Aeronautics and Space (14 CFR) is composed of four chapters. Chapter 1 of this title is the Federal Aviation Administration (FAA), Department of Transportation (DOT). This chapter contains parts 1–199. The following 14 CFR parts are of particular interest to the holder of an Inspection Authorization.

14 CFR PART NUMBER	TITLE
1	Definitions and Abbreviations
11	General Rulemaking Procedures
21	Certification Procedures for Products and Parts
23	Airworthiness Standards: Normal, Utility, Acrobatic, and Commuter Category Airplanes
25	Airworthiness Standards: Transport Category Airplanes
27	Airworthiness Standards: Normal Category Rotorcraft
29	Airworthiness Standards: Transport Category Rotorcraft
31	Airworthiness Standards: Manned Free Balloons
33	Airworthiness Standards: Aircraft Engines
35	Airworthiness Standards: Propellers
39	Airworthiness Directives

43	Maintenance, Preventive Maintenance, Rebuilding, and Alteration
45	Identification and Registration Marking
47	Aircraft Registration
65	Certification: Airmen Other Than Flight Crewmembers
91	General Operating and Flight Rules
119	Certification: Air Carriers and Commercial Operators
125	Certification and Operations: Airplanes Having a Seating Capacity of 20 or More Passengers or a Maximum Payload Capacity of 6,000 Pounds or More
135	Operating Requirements: Commuter and On-Demand Operations and Rules Governing Persons on Board Such Aircraft
183	Representatives of the Administrator

The Code of Federal Regulations may be obtained from the Regulatory and Guidance Library at the FAA web site, www.faa.gov, for the official FAA copy or the U.S. Government Printing Office (GPO) web site www.gpo.gov.

2. TYPE CERTIFICATE DATA SHEETS AND SPECIFICATIONS

Type Certificate Data Sheets and Specifications (TCDS) set forth essential factors and other conditions, which are necessary for U.S. airworthiness certification. Aircraft, engines, and propellers which conform to a U.S. type certificate (TC) are eligible for U.S. airworthiness certification when found to be in a condition for safe operation and ownership requisites are fulfilled.

There are two kinds of certification documents contained in the TCDS file: (1) Type Certificate Data Sheets and (2) Specifications.

Type Certificate Data Sheets were originated and first published in January 1958. 14 CFR part 21, section 21.41, indicates they are part of the type certificate. As such, a type certificate data sheet is evidence the product has been type certificated. Generally, type certificate data sheets are compiled from details supplied by the type certificate holder; however, the FAA may

A-14

request and incorporate additional details when conditions warrant.

Specifications were originated during implementation of the Air Commerce Act of 1926. Specifications are FAA recordkeeping document issued for both type certificated and non-type certificated products which have been found eligible for U.S. airworthiness certification. Although they are no longer issued, specifications remain in effect and will be further amended. Specifications covering type-certificated products may be converted to type certificate data sheets at the option of the type certificate holder. However, to do so requires the type certificate holder to provide an equipment list. A specification is not part of a type certificate.

The official FAA copy is available on the Internet at the FAA web site under Regulatory and Guidance Library at http://rgl.faa.gov. This is a free service.

3. SUMMARY OF AIRWORTHINESS DIRECTIVES FOR SMALL AIRCRAFT AND ROTORCRAFT

An airworthiness directive (AD) contains information regarding an unsafe condition that exists in an aircraft, aircraft engine, propeller, or appliance when that condition is likely to exist or develop in other products of the same type design. No person may operate a product to which an AD applies, except in accordance with the requirements of the AD.

All ADs are summarized and issued by the FAA. New and revised ADs are published biweekly and mailed to registered owners of affected equipment and to subscription holders. ADs are issued in two weight categories:

1. Small aircraft with a maximum certificated takeoff weight aircraft of 12,500 pounds or less, and all rotorcraft regardless of weight.

2. Large aircraft over 12,500 pounds maximum certificated takeoff weight.

The official FAA copy is available on the Internet at the FAA web site under Regulatory and Guidance Library (RGL) at http://rgl.faa.gov.

The ADs are totally searchable and easily located. The individual airworthiness directives and the AD biweeklies on the website are considered approved FAA copy and may be used in lieu of purchasing paper copies. This is a free service. Questions concerning the RGL may be directed to AIR-140, Delegation and Airworthiness Programs Branch, at (405) 954-4103.

4. ADVISORY CIRCULARS

The Federal Aviation Administration issues Advisory Circulars (ACs) to inform the aviation public in a systematic way of nonregulatory material. Unless incorporated into a regulation by reference, the contents of an Advisory Circular are not binding on the public. Advisory Circulars are issued in a numbered-subject system corresponding to the numerical part of the subject regulation (FAA AC 39-7 would deal with a subject related to 14 CFR part 39, Airworthiness Directives).

An Advisory Circular is issued to provide guidance and information in a designated subject area or to show a method acceptable to the Administrator for complying with a related Federal Aviation Regulation. Electronic versions are available on the Internet at the FAA web site.

The following Advisory Circulars are of interest to mechanics:

AC 39-7, Airworthiness Directives

AC 43-4, Corrosion Control for Aircraft

AC 43-11, Reciprocating Engine Overhaul Terminology and Standards

A-16

AC 43.13-1, Acceptable Methods, Techniques and Practices—Aircraft Inspection and Repair

AC 43.13-2, Acceptable Methods, Techniques, and Practices—Aircraft Alterations

AC 43-9, Maintenance Records

AC 43.9-1, Instructions for Completion of FAA Form 337, Major Repair and Alteration (Airframe, Powerplant, Propeller, or Appliance)

AC 43-18, Fabrication of Aircraft Parts by Maintenance Personnel

AC 43-210, Standardized Procedures for Requesting Field Approval of Data, Major Alterations, and Repairs

AC 91-67, Minimum Equipment Requirements for General Aviation Operations Under FAR Part 91

Additional information of particular interest to the holder of an inspection authorization is found in FAA-H-8083-1, Aircraft Weight and Balance Handbook.

5. ADDITIONAL SOURCES OF INSPECTION DATA

Several commercial publishers offer subscription services that include the Airworthiness Directives, Advisory Circulars, and Type Certificate Data Sheets, along with other inspection data. They may be found in aviation trade paper and magazines.

A-17

FAA-G-8082-19

U.S. Department
of Transportation

Federal Aviation Administration

Advisory Circular

Subject: **MINIMUM EQUIPMENT REQUIREMENTS FOR GENERAL AVIATION OPERATIONS UNDER FAR PART 91**

Date: **6/28/91**
Initiated by: **AFS-820**

AC No: **91-67**
Change:

1. PURPOSE. This advisory circular (AC) describes acceptable methods for the operation of aircraft under Federal Aviation Regulations (FAR) Part 91 with certain inoperative instruments and equipment which are not essential for safe flight.

a. These acceptable methods of operation are:

(1) Operation of aircraft with a Minimum Equipment List (MEL), as authorized by FAR §91.213(a).

(2) Operation of aircraft without an MEL under FAR §91.213(d).

b. This AC also explains the reprocess for obtaining Federal Aviation Administration (FAA) approval of an MEL.

2. RELATED FAR SECTIONS. The following FAR provide additional information on operations with or without a FAR Part 91 MEL:

a. FAR § 43.9: Content, form, and disposition of maintenance, preventive maintenance, rebuilding, alterations, and alteration records (except inspections performed in accordance with FAR Parts 91, 123, and 125 and FAR §§135.411(a)(1) and 135.419).

b. FAR §43.11: Content, form, and disposition of the records for inspections conducted under FAR Parts 91 and 125 and FAR §§135.411(a)(1) and 135.419.

c. FAR §91.205: Powered civil aircraft with standard category U.S. airworthiness certificates: Instrument and equipment requirements.

d. FAR §91.405: Maintenance required.

3. FORMS AND REPORTS. The FAA Flight Standards District Office (FSDO) contacted by an MEL applicant provides the applicant a Master Minimum Equipment List (MMEL) for the applicant's particular aircraft.

4. RELATED READING MATERIAL. Users of this AC will find detailed background and in-depth information in the Federal Register Vol. 53, No. 239, December 13, 1988. The public may obtain copies of this issue of the Federal Register from the FAA, Office of Public Affairs, Public Inquiry Center, APA-230, 800 Independence Ave. SW, Washington, DC 20591.

5. BACKGROUND. Except as provided in FAR §91.213, all instruments and equipment installed on an aircraft must be operative in order for the operator to operate it. However, the FAA recognized that safe flight can be conducted under the MEL concept and under specific conditions with inoperative instruments and equipment.

a. Regulatory History. Until the most recent change to FAR §91.213, the MEL concept applied only to air carrier and commercial operations and general aviation operators of multiengine aircraft for which FAA had developed an MMEL. Operators of aircraft for which FAA had not developed an MMEL had to comply with FAR §91.405. This section required that all aircraft discrepancies occurring between required inspections had to be repaired in accordance with FAR Part 43 before the aircraft could be operated. This meant that all the aircraft's instruments and equipment, regardless of whether they were essential or not to the flight operation conducted, had to be operative. This requirement often placed a burden on operators.

b. Amendments to FAR Part 91. Over the past decade, the FAA initiated several rule making projects to alleviate the regulatory burden of FAR §91.405. Before the issuance of a final rule change, FAA encouraged public and industry participation, accepted and reviewed public comments, and conducted public hearings which were attended by other Government agencies and the industry.

(1) The FAA briefly suspended FAR §91.213 and allowed issuance of MEL's by exemption. During this period, the FAA gained valuable information on the usefulness and safety aspects of using MEL's in general aviation.

(2) Further, general aviation operators have a history of safe operations using FAR §91.205 as the sole reference for determining the instrument and equipment requirements for a particular flight.

(3) However, operators indicated the need for relief from FAR §91.405, and the FAA agreed that the FAR should reflect current operational practices. Consequently, the FAA amended FAR Parts 43 and 91 in December 1988.

c. New Regulatory Requirements. The amendment to FAR Parts 43 and 91 provides a regulatory basis for the operation of aircraft with inoperative instruments and equipment. Operators conduct these operations within a framework of a controlled program of maintenance inspections, repairs, and parts replacement. However, operators must exercise good judgment and have, at each required inspection, any inoperative instrument or equipment repaired or inspected or the maintenance deferred, as appropriate.

6. DEFINITIONS.

a. Aircraft Evaluation Group (AEG). The AEG is the FAA office responsible for the development and publication of an approved MMEL for those aircraft within its area of responsibility.

b. Aircraft Flight Manual (AFM). The AFM is the source document for operational limitations and performance for an aircraft. The term AFM can apply to either an airplane flight manual or a rotorcraft flight manual. FAA requires an AFM for type certification. The responsible FAA Aircraft Certification Office (ACO) approves an AFM.

c. Aircraft Maintenance Manual (AMM). The AMM is the source document for maintenance procedures for an aircraft. The term AMM can apply to either an airplane maintenance manual or a rotorcraft maintenance manual. FAA requires the AMM for type certification.

d. Airworthiness Directive (AD). An AD is a mandatory airworthiness requirement for a particular make and model aircraft or installed equipment. An AD is supplementary to the aircraft original airworthiness approval.

e. Air Transportation Association (ATA) Numbering System. The standard ATA numbering system refers to systems on different aircraft in a standardized manner. MMEL's use the ATA num-bering system.

f. Calendar Days include all days, with no exclusion for weekends and holidays.

g. Deactivation means to make a piece of equipment or an instrument unusable to the pilot/crew by preventing its operations.

h. Deferred Maintenance is the postponement of the repair or replacement of an item of equipment or an instrument.

i. Equipment List is an inventory of equipment installed by the manufacturer or operator on a particular aircraft.

j. Flight Operations Evaluation Board (FOEB). The FOEB is composed of FAA personnel who are operations, avionics, airworthiness, and aircraft certification specialists. The FOEB develops an MMEL for a particular aircraft type under the direction of the AEG.

k. Inoperative means that a system and/or component has malfunctioned to the extent that it does not accomplish its intended purpose and/or is not consistently functioning normally within its approved operating limits or tolerances.

l. Kinds of Operations List (KOL). The KOL specifies the kinds of operations (e.g., visual flight rules (VFR), instrument flight rules (IFR), day, or night) in which the aircraft can be operated. The KOL also indicates the installed equipment that may affect any operating limitation. Although the certification rules require this information, there is no standard format; consequently, the manufacturer may furnish it in various ways.

m. Letter of Authorization (LOA). The FSDO issues an LOA to the operator when the FSDO authorizes the operator to operate under the provisions of an MEL. Together, the LOA, the procedures document (paragraph v. following, and the MMEL constitute a Supplemental Type Certificate (STC). The operator must carry the STC in the aircraft during its operation.

n. Maintenance is the inspection, overhaul, repair, preservation, or replacement of parts. This definition excludes preventive maintenance (see paragraph u. following). After a mechanic performs maintenance, other than preventive maintenance, a properly certificated maintenance person must approve the aircraft for return to service.

o. MMEL. An MMEL contains a list of items of equipment and instruments that may be inoperative on a specific type of aircraft (e.g., BE-200, Beechcraft model 200). It is also the basis for the development of an individual operator's MEL.

p. MEL. The MEL is the specific inoperative equipment document for a particular make and model aircraft by serial and registration numbers; e.g., BE-200, N12345. A FAR Part 91 MEL consists of the MMEL for a particular type aircraft, the MMEL's preamble, the procedures document, and a LOA. The FAA considers the MEL as an STC. As such, the MEL permits operation of the aircraft under specified conditions with certain equipment inoperative.

q. Next Required Inspection is the one required under either an FAA-approved inspection program, a 100-hour inspection, or an annual inspection, as appropriate.

r. Operations (O) and Maintenance (M) procedures in the MMEL refer to the specific maintenance procedures the operator uses to disable or render items of equipment inoperative and to specific operating conditions and limitations, as appropriate.

(1) An O symbol in column 4 of the MMEL indicates that a specific operations procedure must be accomplished before or during operation with the listed item of equipment inoperative. Normally, the flightcrew accomplishes these procedures; however, other personnel, such as maintenance personnel, may be qualified and authorized to perform the procedure.

(2) An M symbol in column 4 of the MMEL indicates that a specific maintenance procedure must be accomplished before beginning operation with the listed item of equipment inoperative. Normally, maintenance personnel accomplish these procedures; however, other personnel, such as the flightcrew, may be qualified and authorized to perform certain functions. Qualified maintenance personnel must perform procedures requiring specialized knowledge, skills, or the use of tools or test equipment.

s. Operator refers to an individual or company (corporation, entity, etc.). As used in this AC, operator applies to those who are applicants for, or holders of, authority to conduct operations under the provisions of a FAR Part 91 MEL.

t. Placard is a decal or label with letters at least 1/8-inch high. The operator or mechanic must place the placard on or near inoperative equipment or instruments so that it is visible to the pilot or flightcrew and alerts them to the inoperative equipment.

u. Preventive Maintenance. The term preventive maintenance refers to simple or minor preservation operations and/or the replacement of small standard parts not involving complex assembly. FAR Part 43, Appendix A(c), contains a list of preventive maintenance items. Qualified mechanics or certificated pilots may accomplish preventive maintenance and approve the aircraft for return to service.

v. Procedures Document as referred to in this AC pertains to a separate document containing the O and M procedures developed by the operator and any other operating information applicable to operation with an MEL, such as the "as required by the FAR" items that list the FAR by part and section or stipulate the operating conditions.

w. Proposed Master Minimum Equipment List (PMMEL). The PMMEL is the working document used as the basis for development of the MMEL. Normally, the manufacturer proposes it during the certification process. However, an operator of a unique type aircraft, for which an MMEL does not exist, may submit a PMMEL for FAA approval.

x. ***Return to Service.*** Return to service has two applications. An appropriately certificated person approves an aircraft for return to service after an inspection or after maintenance. A certificated pilot, in fact, returns the aircraft to service after the pilot conducts an appropriate preflight and accepts the aircraft for an intended flight.

y. ***Small Aircraft*** means aircraft with a maximum certificated takeoff weight of 12,500 pounds or less.

z. ***STC.*** An STC is a major change in type design not great enough to require a new application for a type certificate under FAR §21.19. An example would be installation of a powerplant different from what was included the original type certificate.

aa. ***Type Certificate Data Sheets (TCDS)*** and Specifications are documents issued by the FAA which describes the aircraft's airworthiness requirements relating to a specific type, make, and model of aircraft. These documents are available at a FSDO.

David R. Harrington
Acting Director, Flight Standards Service

7. COMMENTS INVITED. Comments regarding this publication should be directed to:

Federal Aviation Administration
Field Programs Division, AFS-500
Advisory Circular Staff
P.O. Box 20034, Gateway Building
Dulles International Airport
Washington, DC 20041-2034

Every comment will not necessarily generate a direct acknowledgment to the commenter. Comments received will be considered in the development of upcoming revisions to AC's or other related technical material.

(intentionally left blank)

vi

6/28/91 AC 91-67

CONTENTS

vii

6/28/91 AC 91-67

CHAPTER 1. GENERAL

1. APPLICABILITY. This AC provides guidance for the operation of the following aircraft under FAR Part 91:

a. Aircraft for which no MMEL has been developed by the FOEB:

(1) Rotorcraft.

(2) Nonturbine-powered airplanes.

(3) Gliders.

(4) Lighter-than-air aircraft.

b. Aircraft for which an MMEL has been developed but for which the FSDO has not authorized operation with an MEL.

(1) Small rotorcraft.

(2) Nonturbine-powered small single and multiengine airplanes.

c. All other aircraft which have an MEL or for which an operator seeks MEL authorization under FAR § 91.213.

d. An operator may operate an aircraft for which FAA has issued an original Experimental airworthiness certificate in accordance with FAR § 91.213 only when authorized in that certificate's operating limitations.

e. ***This AC does not apply to*** operators holding certificates issued under FAR Parts 121, 125, 129, and 135.

f. Holders of letters of full deviation authority from FAR Part 125 and operating under FAR Part 91, Subpart F, may apply for authorization to operate with a FAR Part 91 MEL.

2. MEL VS. FAR § 91.213(d). Although FAA amended FAR Part 91 to provide relief to operators under the MEL concept, some operators may find it less burdensome or less complicated to operate under the provisions of FAR § 91.213(d). The applicant should discuss the requirements of each method with FSDO inspectors to decide which method of compliance better suits the particular operation. Appendix 3 contains a list of commonly asked questions which may assist in the decision.

a. An MEL is a precise listing of instruments, equipment, and procedures that allows an aircraft to be operated under specific conditions with inoperative equipment. The MMEL, as part of the MEL, by nature does not cover equipment installed or modified under other STC's. Any STC or other major modification may make the MMEL invalid for a particular modified aircraft.

b. The FAR require that all equipment installed on an aircraft in compliance with the airworthiness standards and operating rules be operative. The FAA-approved MMEL includes those items of equipment and other items which the FAA finds may be inoperative and yet maintain an acceptable level of safety. Obviously, the MMEL does not contain required items such as wings, flaps, rudders, etc. When a FAR Part 91 operator uses an MMEL as an MEL, all instruments and equipment not covered in the MMEL must be operative at all times regardless of the operation conducted, unless:

(1) They are newly installed and are not instruments or equipment specifically required by the airworthiness rule under which the aircraft is type certificated, required by AD, or required for specific operations under FAR § 91.213(b)(1)-(3), such as Traffic Alert and Collision Avoidance System (TCAS), an extra piece of navigational equipment, a windshear detection device, a ground proximity warning system, a radar altimeter, passenger convenience items, etc.;

Chap 1
Par 1 1

(2) The operator has developed procedures for disabling or rendering them inoperative; and

(3) The operator has contacted the FSDO having oversight within 10 calendar days following an installation and requested that the equipment be added to the MMEL.

(i) The operator must furnish the following information:

(A) A copy of the STC or FAA Form 337, Major Repair and Alteration, that approved each equipment installation and the associated limitations listed in the AFM supplement or on the 337. The FOEB needs this information to account for installation differences as well as for maintaining MMEL relief that is consistent with the limitations.

(B) A system description that details sufficiently the interface of the equipment with the crew; i.e., location, controls, operations, how it is used, etc.

(C) A statement that describes the transfer of function when the equipment is inoperative; i.e., not required for the flight, as per crew procedures, because of alternate systems, etc.

(ii) If the FAA determines that the equipment has been previously considered by the FOEB for inclusion in the MMEL and denied, or if the FOEB convenes and denies inclusion, the FAA will not grant relief. The equipment must be operational before aircraft can take off.

(iii) If the FOEB determines that the equipment should be added to the MMEL, the operator will receive the updated MMEL and must prepare O and M procedures for that piece of equipment.

c. If FAA has not authorized operating with an MEL for an operator's specific aircraft, the operator may apply for an MEL (Chapter 3, paragraph 20). However, the operator can always elect to operate without an MEL under the provisions of FAR § 91.213(d).

(1) FAR § 91.213(d) requires only those instruments listed in FAR § 91.213(d)(2) to be operative.

(2) The operator can operate the aircraft with those instruments and equipment not listed in FAR § 91.213(d)(2) inoperative.

3. RELATIONSHIP BETWEEN THE PMMEL, THE MMEL, AND THE MEL. When an aircraft is first manufactured, the FOEB determines the minimum operative instruments and equipment required for safe flight in that aircraft type in each authorized operating environment. During the type certification process, the manufacturer submits a PMMEL to the FOEB. Based on its determinations, the FOEB reviews the PMMEL and develops an MMEL from it. Once the FOEB approves the MMEL, a copy is available to each FSDO via an automated system that allows the FSDO to download the MMEL onto a diskette or hard copy. The FSDO provides MMEL's to applicant's to use along with the procedures document, preamble, and LOA, as an MEL.

a. As technology changes and new equipment becomes available, the FOEB will reconvene to develop new MMEL's or to revise and update existing ones.

b. When an FOEB makes a change to an MMEL, all operators using that MMEL as their MEL will receive a postcard advising them of the revised MMEL. The FSDO provides operators copies of the revised MMEL. The operator then makes the necessary changes to the procedures document through the normal revision process (Chapter 3, paragraph 22).

4. SINGLE- AND MULTIENGINE MEL's. The FAA has developed MMEL's for most of the FAA type certificated aircraft in general service today. All multiengine airplanes have an MMEL that is specific to the type design; e.g., Beech Baron, BE-58. The FAA has developed a generic, single-engine MMEL to provide to operators of single-engine aircraft.

5. AIRCRAFT FOR WHICH NO MMEL HAS BEEN DEVELOPED.

a. If an FOEB has not developed an MMEL for a certain type of rotorcraft, nonturbine-powered airplane, glider, or lighter-than-air aircraft, that aircraft may be operated with inoperative equipment under the provisions of FAR § 91.213(d).

b. In those cases where an operator has an older or rare design aircraft that has no MMEL, the operator may submit a PMMEL to the appropriate FOEB for evaluation. Once the AEG approves the MMEL, the operator could use it as the MEL along with the other required documents.

6. MEL RESTRICTIONS. Operators of small rotorcraft, nonturbine-powered small single- and multiengine airplanes, and other aircraft for which a MMEL has been developed, may elect to operate with a MEL or under the provisions of FAR § 91.213(d). However, the latter option does not apply if the aircraft has an MEL approved under FAR Parts 121, 125, 129, or 135. For example, an owner has leased an aircraft to an air carrier operator, and the air carrier operator has applied for and received an approved MEL for FAR Part 135 operations. Compliance with such an MEL is mandatory, even during FAR Part 91 operations. If the operator wants to operate under FAR § 91.213(d), the operator would have to surrender the MEL authorization.

7. REMOVAL OR DEACTIVATION. When an operator elects to operate without an MEL, any inoperative instrument or equipment must either be removed (FAR § 91.213(d)(3)(i)) or deactivated (FAR § 91.213(d)(3)(ii)), then placarded.

a. Removal of any item of equipment that affects the airworthiness of an aircraft requires following an approved procedure. A properly certificated maintenance person must record the removal in accordance with FAR § 43.9. A person authorized by FAR § 43.7 must make the appropriate adjustments to the aircraft's weight and balance information and the equipment list, fill out and submit FAA Form 337, and approve the aircraft for return to service.

b. The operator must evaluate any proposed deactivation to assure there is no adverse effect that could render another system less than fully capable of its intended function.

(1) A certificated pilot can accomplish deactivation involving routine pilot tasks, such as turning off a system. However, for a pilot to deactivate an item or system, that task must come under the definition of preventive maintenance in FAR Part 43, Subpart A.

(2) If the deactivation procedures do not fall under preventive maintenance, a properly certificated maintenance person must accomplish the deactivation. The maintenance person must record the deactivation in accordance with FAR § 43.9 (figure 1 – Sample Maintenance Record Entries.).

c. Placarding can be as simple as writing the word "inoperative" on a piece of masking tape and attaching it to the inoperative equipment or to its cockpit control. Placarding is essential since it reminds the pilot that the equipment is inoperative. It also ensures that future flightcrews and maintenance personnel are aware of the discrepancy.

8. INOPERATIVE EQUIPMENT AND REQUIRED INSPECTIONS. An operator may defer maintenance on inoperative equipment that has been deactivated or removed and placarded inoperative.

a. When the aircraft is due for inspection in accordance with the FAR, the operator should have all inoperative items repaired or replaced.

b. If an owner does not want specific inoperative equipment repaired, then the maintenance person must check each item to see if it conforms to the requirements of FAR § 91.213(d). The operator and maintenance personnel should also assess how permanent removal of the item could affect safe operation of the aircraft.

(1) The repair interval categories (A, B, C, D, etc.) in the MMEL do not apply to FAR Part 91 MEL's.

(2) The maintenance person must furnish the owner/operator with a signed and dated list of all discrepancies not repaired.

(3) The maintenance person must ensure that each item of inoperative equipment that is to remain inoperative is placarded appropriately.

Placard (Minimum 1/8-inch high letters)

Landing Light Inoperative:

PREVENTIVE MAINTENANCE ENTRY:

(DATE) Total time ________ hours, Landing light bulb removed in accordance with (manufacturer) maintenance manual, Chapter ________ , page ________ . Landing light switch placarded inoperative.

______________________________ ____________________

Pilot's Signature Certificate Number

Placard (Minimum 1/8-inch high letters)

Aircraft Heater Inoperative:

MAINTENANCE ENTRY (FAR §43.9):

(DATE) Total time ________ hours. Aircraft heater and control switch deactivated by capping heater fuel lines in accordance with (manufacturer) maintenance manual, Chapter ________ , page ________ . Heater control switch placarded inoperative.

______________________________ ____________________

Mechanic's Signature Certificate Number

Figure 1. Sample Maintenance Record Entries

9.-12. RESERVED

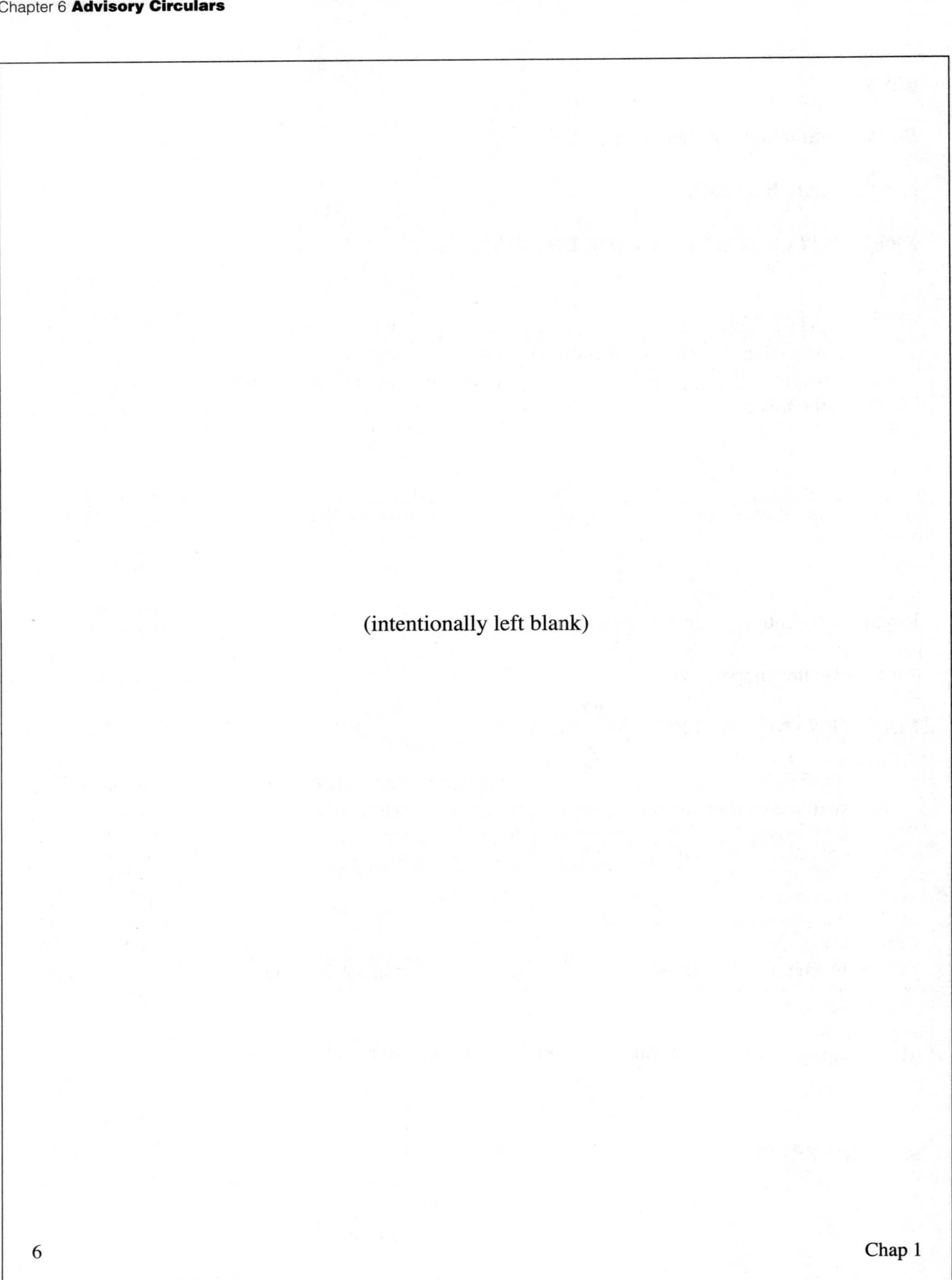

(intentionally left blank)

6 Chap 1

CHAPTER 2. CONDUCTING OPERATIONS WITHOUT AN MEL

13. APPLYING FAR § 91.213(d). This chapter provides guidance for operators who elect to conduct flight operations under the provisions of FAR § 91.213(d). Operating under FAR § 91.213(d) requires no application to or approval from FAA. An operator, after operating under FAR § 91.213(d), may elect at any time to apply for authorization to operate under an MEL (Chapter 3).

14. THE DECISION SEQUENCE. Figure 2 is a flow chart depicting the typical sequence of events a pilot or operator, operating under FAR § 91.213(d), should follow when the pilot or operator discovers inoperative equipment. For example, during a preflight inspection for a VFR-day, cross-country flight, the pilot discovers at the number 2 automatic direction finder (ADF) head is inoperative.

a. The pilot checks the aircraft's equipment list or KOL to see if the number 2 ADF is a required item (FAR § 91.213(d)(2)(ii)). If the number 2 ADF is required in the equipment list or KOL, the aircraft is not airworthy. The operator must have the number 2 ADF replaced or repaired before operating the aircraft. In this example, the number 2 ADF is not a required item on the equipment list.

b. Next, the pilot checks the airworthiness regulation under which the aircraft was certificated to determine if the number 2 ADF is part of the VFR-day type certificate (FAR § 91.213(d)(2)(i)). (These requirements are summarized in a TCDS, copies of which are available at FSDO's or from qualified maintenance personnel.) If the number 2 ADF is required as part of the VFR-day type certification, the aircraft is not airworthy. The operator must have the number 2 ADF replaced or repaired before operating the aircraft In this example, the number 2 ADF is not required by type certificate.

c. Next, the pilot checks to see if an AD requires the number 2 ADF. The pilot can accomplish this by checking the aircraft's maintenance log to see if the number 2 ADF was installed as a result of an AD. However, it may be necessary for the pilot to consult a qualified maintenance person to determine AD compliance. If an AD requires the number 2 ADF to be operative, the aircraft is not airworthy. The operator must have the number 2 ADF replaced or repaired before operating the aircraft. In this example, there is no AD require the number 2 ADF to be operative.

d. Next, the pilot checks to see the number 2 ADF is required by FAR §§ 91.215, 91.205, or 91.207. The pilot can accomplish this by checking those sections of the FAR or by consulting with a maintenance technician or FSDO personnel. If any of those sections of the FAR require a number 2 ADF, then the aircraft would not be airworthy with the number 2 ADF inoperative. The operator must have the number 2 ADF replaced or repaired before operating the aircraft. In this example, those sections of the FAR do not require the number 2 ADF to be operative.

e. At this point the inoperative number 2 ADF must either be removed from the aircraft (FAR § 91.213(d)(3)(i)) or deactivated (FAR § 91.213(d)(3)(ii)). The person removing or deactivating the number 2 ADF must placard it inoperative in the appropriate location. (A pilot should consult maintenance personnel before deactivating or having maintenance personnel remove any item of equipment.)

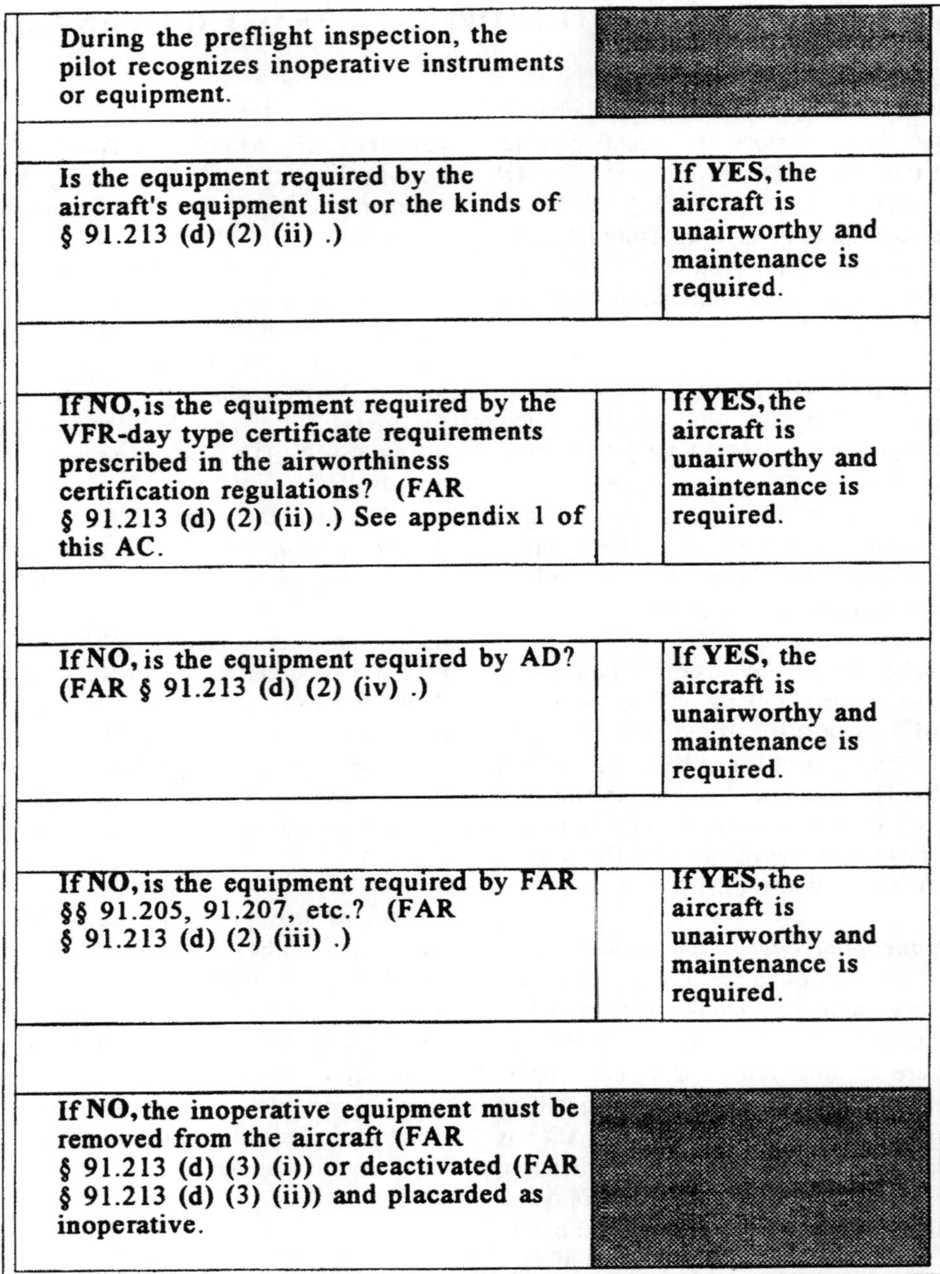

During the preflight inspection, the pilot recognizes inoperative instruments or equipment.

Is the equipment required by the aircraft's equipment list or the kinds of § 91.213 (d) (2) (ii) .)	If **YES**, the aircraft is unairworthy and maintenance is required.
If **NO**, is the equipment required by the VFR-day type certificate requirements prescribed in the airworthiness certification regulations? (FAR § 91.213 (d) (2) (ii) .) See appendix 1 of this AC.	If **YES**, the aircraft is unairworthy and maintenance is required.
If **NO**, is the equipment required by AD? (FAR § 91.213 (d) (2) (iv) .)	If **YES**, the aircraft is unairworthy and maintenance is required.
If **NO**, is the equipment required by FAR §§ 91.205, 91.207, etc.? (FAR § 91.213 (d) (2) (iii) .)	If **YES**, the aircraft is unairworthy and maintenance is required.

If **NO**, the inoperative equipment must be removed from the aircraft (FAR § 91.213 (d) (3) (i)) or deactivated (FAR § 91.213 (d) (3) (ii)) and placarded as inoperative.

At this point the pilot shall make a final determiniation to confirm that the inoperative instrument/equipment does not constitute a hazard under the anticipate operational conditions before release for departure.

Figure 2. Pilot Decision Sequence When Operating Without An MEL.

f. Finally, the pilot should decide whether the inoperative number 2 ADF creates a hazard for the anticipated conditions of the flight, e.g., VFR-day.

15.-18. RESERVED

CHAPTER 3. OPERATING AIRCRAFT

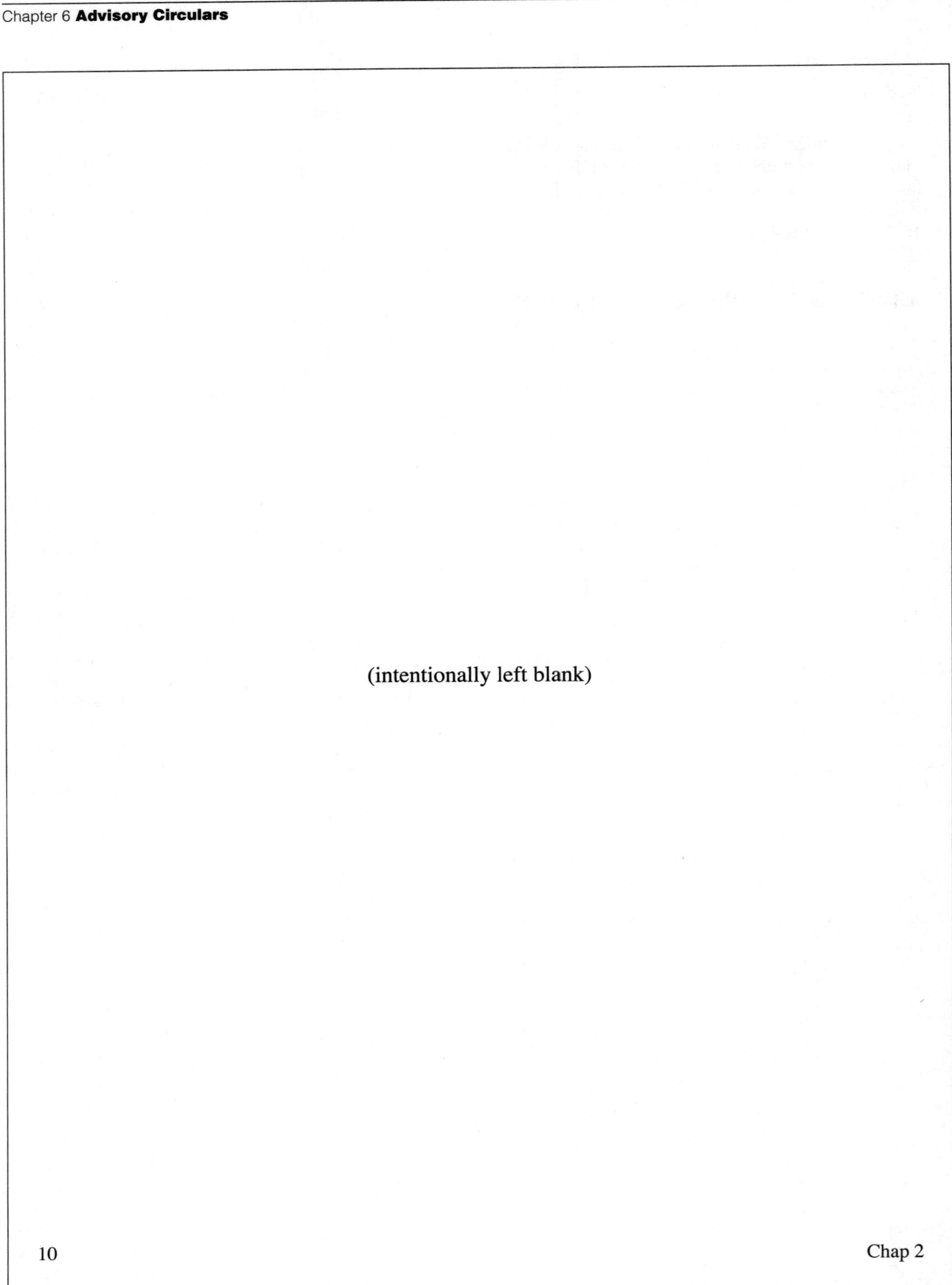
(intentionally left blank)

10 Chap 2

WITH AN MEL

19. APPLICABILITY. This chapter provides guidance for operators who want to conduct flight operations under the provisions of an MEL.

20. APPLYING FOR MEL APPROVAL. FAA has only one procedure for the issuance of FAR Part 91 MEL's, and it is the procedure the FSDO will follow for FAR Part 91 MEL authorizations. The operator who wishes to conduct operations with an MEL must contact the FSDO which has jurisdiction over the geographic area where the aircraft is based and make an appointment.

a. FAR Part 91 operators who received MEL authorization under the approval system in place before July 5, 1990, have letters of authorization that will expire. Those operators may continue to operate as usual; however, at least 30 days before the letter is due to expire, the operator should contact the issuing FSDO so that the FSDO can issue a new LOA.

b. For FAR Part 91 operators seeking MEL authorization under the current approval system, the FSDO will assign a Flight Standards inspector to advise the applicant about FAR requirements pertinent to using an MEL. During the initial appointment, the applicant will likely be dealing with a team of inspectors from the operations, airworthiness, and avionics units.

c. The inspector will provide the applicant with a copy of the appropriate MMEL, a copy of this AC, and a copy of the preamble to the MMEL. If the operator has installed items of equipment that are not on the MMEL, the operator must request that the MMEL be amended to include those items of equipment. This request is made to the FSDO.

d. The operator and the team of inspectors discuss the requirements for the procedures document. When FSDO personnel believes that the operator understands the requirements for operating with an MEL, the FSDO issues the operator the LOA (appendix 4).

(1) The LOA contains the legal name of the operator and the address of the operator's principal base of operations.

(2) Both the FAA inspector and the operator (or the operator's bona fide representative) sign the LOA.

e. If, after meeting with FSDO personnel and discussing MEL operational considerations, an inspector believes that the applicant does not have a good understanding of the requirements, the FSDO will not issue the LOA. If the LOA is still desired, the applicant should obtain the necessary understanding of the requirements from appropriate sources. After obtaining and understanding the requirements, the applicant can again request the LOA from the FSDO. The applicant could also elect to operate under FAR § 91.213(d).

f. Once the FSDO issues the LOA, the applicant is then responsible for developing a document that contains O and M procedures for disabling or rendering inoperative items of equipment in accordance with FAR Parts 43, 91, or 145 (if a repair station accomplishes the activity), as appropriate. No further FAA approval is necessary, and the operator can begin flight operations. The MMEL, preamble, LOA, and the procedures document are now considered an MEL.

(1) The operator should develop the O and M procedures using guidance contained in the manufacturer's aircraft flight and/or maintenance manuals, the manufacturer's recommendations, engineering specifications, and other appropriate sources. The operator may consult FSDO airworthiness inspectors for advice or clarification, but the operator is responsible for preparing the document.

(2) The operator must consider the following when preparing the procedures documents.

(i) The operator's procedures document may be more restrictive than the MMEL either by the applicant's choice or because of AD's or operating rules. The operator's procedures document may not be less restrictive than the MMEL.

(ii) The title page of the procedures document must contain the following statement:

> This MEL is applicable to FAR Part 91 operations only and may not be used for operations conducted under FAR Parts 121, 125, 129, or 135.

(iii) The operator must use the ATA numbering system for equipment and instruments, as is used in all MMEL's (appendix 1). The operator must use the ATA numbering system in sequence when describing O and M procedures, including the numbers for equipment installed in all aircraft. When equipment is not installed in a specific aircraft, the applicant need not develop O and M procedures for those items of equipment.

(iv) Operators must ensure that the procedures document lists the items of equipment that are actually installed on the specific aircraft. This provides guidance to a pilot as to which items of equipment may be inoperative for a particular operation.

(v) Equipment specifically required by the airworthiness rule under which the aircraft is type certificated, equipment required by AD, and equipment required for specific operations under FAR § 91.213(b)(1), (2) and (3) must be operative. It is important to note that all items related to the airworthiness of the aircraft that are not included on the MMEL must be operative.

(vi) The Parts A, B, C, and D codes, listed in column 1 of the MMEL, apply only to operations conducted under FAR Parts 121, 125, 129, and 135.

(vii) Where the MMEL states "as required by FAR," the procedures document should list the particular FAR by part and section, or describe the actual FAR requirement applicable to the operator's particular operation. For example, where the FAR requires a clock for IFR flight, the operator's procedures document should say, "May be inoperative for VFR."

(viii) The procedures document must specify suitable limitations in the form of placards, maintenance procedures, crew operating procedures, and offer restrictions to ensure an acceptable level of safety.

(ix) The procedures document must specify those conditions under which an item may be inoperative. The remarks must also identify required maintenance or operational tasks. The symbol "O" or "M", placed in column 4 of the MMEL (appendix 1), indicates that an O and M procedure is applicable to that item. Indicating O and M procedures in the procedures document provides flightcrews and ground support personnel with a single procedural reference document.

(x) If the O and M procedures are already stated in the AFM, the maintenance manual, or other available FAA-approved source, the operator needs to show only the reference, e.g., O:AFM, pp. 3-8 through 3-10, para. 347. If the operator uses this reference format in the procedures document, the referenced source must be readily available to the ground support personnel, and a copy of the references source must be carried in the aircraft and be readily available to the flightcrew.

(xi) If the O and M procedures are not in the AFM, the maintenance manual, or other available FAA-approved source, or if the operator wishes to use a different procedure, then the operator must list the procedure in the procedures document.

(xii) The procedures document may not conflict with the AFM limitations, emergency procedures, AD's, or the AMM.

(3) An operator may begin operations before completion of the procedures document. If the operator has not yet developed a procedure for an item, that item must be operative. When an instrument or item of equipment becomes inoperative, the operator must follow the procedure indicated in the procedures document or the operator could be in noncompliance with the FAR.

21. MEL AUTHORIZATION. The MEL applies only to a particular aircraft make, model, serial number, and registration number. Also, it applies only to the operator who received the authorization.

a. When more than one operator has operational control of a specific aircraft, all operators must meet with inspectors from the issuing FSDO to discuss MEL operational considerations, as described in paragraph 20. The FSDO may find it appropriate to list all operators on the LOA. Each operator must sign the "Statement of Operator" on the LOA.

b. The FSDO may issue operators who use several aircraft of the same type a single LOA that lists each aircraft by serial and registration numbers. The FSDO will issue separate letters for different types of aircraft.

(1) When operators add or delete aircraft of the same type from their fleet, they must notify the FSDO having oversight within 10 calendar days following the change. The FSDO will reissue the LOA containing the new information. Again, both the operator and the inspector must sign the new LOA.

(2) The operator must surrender the previous letter upon reissuance of a new one. The FSDO should place the old letter in the operator's file.

c. At any time after operating with a FAR Part 91 MEL, an operator may elect to operate under FAR § 91.213(d). The operator must surrender the LOA to the issuing FSDO and must conform to all provision of FAR § 91.213(d) during operations.

22. REVISIONS. The operator may have to revise the procedures document under several conditions. The AEG may authorize an FOEB to revise the MMEL, the operator may add equipment, or the FOEB may develop a type specific MMEL for a single-engine aircraft.

a. When the FOEB revises an MMEL, the FAA automated, national MMEL data base notifies operators who have MEL authorizations by mail. The operator is then responsible for obtaining a copy of the revised MMEL from the FSDO that issued the authorization. Within 30 calendar days of notification, the operator must replace the superseded revision of the MMEL with the current revision and add or delete procedures to the procedures document, as applicable.

b. Within 10 calendar days of installing new equipment not on the MMEL, the operator may request that the MMEL be amended.

(1) If the items of newly installed equipment are not instruments or equipment specifically required by the airworthiness rule under which the aircraft is type certificated, an AD, or for specific operations under FAR § 91.213(b)(1), (2) or (3); and exceed what is listed on the MMEL; and the FSDO has determined that the equipment has not previously been denied for inclusion in an MMEL, the operator may petition the FOEB for inclusion of the newly installed equipment in the MMEL. All petitions, with appropriate supporting information, will be forwarded by the FSDO to the appropriate FOEB. Then the operator may add the equipment temporarily to the MMEL and develop appropriate O and M procedures for the equipment. The operator may then operate with the equipment inoperative pending a decision by the FOEB on the operator's request for an MMEL revision to include the equipment.

(2) If the FOEB has previously denied the inclusion of the equipment, or if the equipment is safety related, or if the equipment was previously installed or is "factory original," the operator may still petition the FOEB through the FSDO for inclusion of the equipment in the MMEL. However, the operator may not gain relief for the equipment by adding the equipment to the MMEL temporally and adding procedures to the procedures document pending the FOEB's decisions. The equipment must be operative before operating the aircraft.

c. Although FAA has developed a generic MMEL for operators of single-engine aircraft, an FOEB may decide that a complex, turbine-powered single-engine aircraft requires a type-specific MMEL. For example, an FOEB has developed a type-specific MMEL for the Cessna 208, Caravan.

(1) When an FOEB develops a specific MMEL for a single-engine aircraft, the FAA will notify all holders of MEL's for that aircraft under the generic MMEL that the specific MEL is available.

(2) Within 30 calendar days of notification, the operator must obtain the MMEL from the FSDO and begin the process for a new LOA. Only by issuing a new LOA will the FSDO be assured that the operator has and is using the type-specific MMEL.

(3) Once the FSDO issues the new LOA, the operator must develop, within an additional 30 calendar days, a new procedures document that conforms to the requirements of the type-specific MMEL. The operator will find that most of the procedures that were acceptable under the generic MMEL will transfer to the new procedures document. If equipment becomes inoperative while the operator is developing the new procedures document, the operator may still use the previous procedures, as appropriate.

23. CONDUCTING OPERATIONS WITH AN MEL. In addition to carrying the documents that comprise the MEL onboard the aircraft, the operator must have onboard any technical manuals needed to accomplish O and M procedures. Figure 3 illustrates the sequence of events involved in applying the MEL to inoperative equipment.

a. Inoperative Items Before Flight. During a preflight inspection for a VFR-day flight, the pilot discovers a navigation light is inoperative.

(1) The pilot checks the aircraft's MEL to determine under what, if any, flight conditions the aircraft could be operated without operator navigation lights. The MEL indicates that the aircraft may be operated during daylight hours without operable navigation lights.

(2) The pilot checks the procedures document and deactivates the navigation lights by pulling the correct circuit breaker and having it collared by an appropriately certificated person.

(3) The pilot places a placard which indicates that the lights are inoperative near the navigation light control.

(4) The pilot examines the conditions of the proposed flight and determines that the flight can be conducted safely without navigation lights.

b. In-Flight Failures. An MEL applies only to the takeoff of an aircraft with inoperative instruments or equipment. The pilot's operating handbook or the AFM indicate procedures to follow for instrument or equipment failure in flight. The pilot in command (PIC) should handle the in-flight failure in accordance with those procedures. As soon as possible after landing safely, the PIC must enter a notation of the Inoperative equipment in the aircraft's maintenance records, logbooks, or discrepancy record. Before the next takeoff, the pilot must apply the MEL to inoperative equipment as per the procedures in paragraph a. above. An MEL allows the PIC to defer maintenance on many items under the following conditions:

(1) The aircraft is in a condition for safe flight, and

(2) For the inoperative item, the pilot has followed the specific conditions, limitations, and procedures in the procedures document.

c. Deactivation or Removal and Placarding. See: Chapter 1, paragraph 7.

d. Correcting MEL Inoperative Items. The MEL permits operations with inoperative items of equipment for the minimum period of time necessary until the equipment is repaired. It is important that operators have repairs done at the earliest opportunity in order to return the aircraft to its design level of safety and reliability. In all cases, inoperative equipment must be repaired or the maintenance deferred at the aircraft's next required inspection (FAR § 91.405(c)).

(1) Operators shall establish procedures to correct those inoperative items authorized within specified time requirements.

(2) Owners of aircraft operated under FAR Part 91 may opt to use one of several types of airworthiness inspection systems, depending upon the operator's use of the aircraft. Therefore, the tie between required inspections or inspection segments will vary.

(3) Items of inoperative equipment, authorized by the MEL to be inoperative, must be inspected or repaired by qualified maintenance personnel, or maintenance deferred, at the next 100 hour, annual, progressive, or unscheduled inspection. However, if FAR § 91.213 requires that an item be repaired, the item cannot be deferred.

e. Recordkeeping Requirements. A record of inoperative equipment must remain in the aircraft so pilots will be aware of all discrepancies.

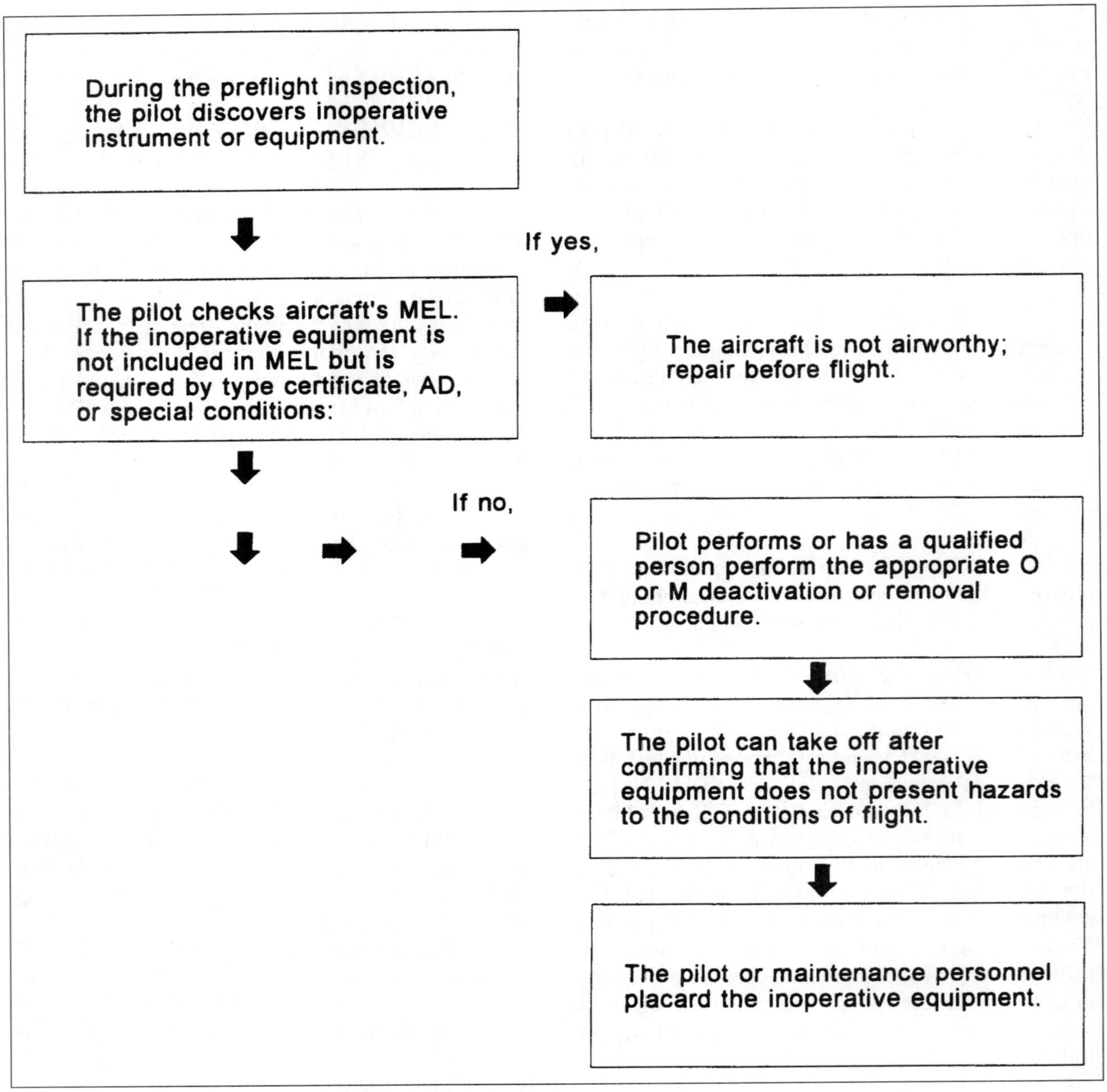

During the preflight inspection, the pilot discovers inoperative instrument or equipment.

↓

The pilot checks aircraft's MEL. If the inoperative equipment is not included in MEL but is required by type certificate, AD, or special conditions:

If yes, → The aircraft is not airworthy; repair before flight.

↓

If no, → Pilot performs or has a qualified person perform the appropriate O or M deactivation or removal procedure.

↓

The pilot can take off after confirming that the inoperative equipment does not present hazards to the conditions of flight.

↓

The pilot or maintenance personnel placard the inoperative equipment.

Figure 3. Pilot Decision Sequence When Operating With An MEL

(1) Since some operators do not carry aircraft logbooks in the aircraft, a discrepancy record or log (figure 4) is a good alternative. When an operator uses this type of discrepancy log in lieu of the aircraft's maintenance records, the operator must retain the log as a part of the aircraft's records as per FAR § 91.417(b).

(2) If the operator elects to use the aircraft maintenance record to log inoperative items, that portion of the record must be carried onboard the aircraft during all operations.

(3) Corrective actions and maintenance procedures shall be accomplished and recorded in accordance with FAR §§ 43.9, 91.405, and 91.417.

(4) Failure to record an inoperative item may result in an operation of the aircraft contrary to the FAR because subsequent pilots would not be able to determine the airworthiness of the aircraft.

f. Aircraft Used in Multiple Operations. FAR § 91.213(c) allows a person who has an approved MEL under FAR Parts 125, 129, or 135 to use that MEL for FAR Part 91 operations. The FAR Parts 121, 125, 129, or 135 MEL must specify requirements for authorized FAR Part 91 operators to comply with the more restrictive provisions established in the approved MEL. It is important that operators be capable of conducting operations in accordance with the MEL. This includes, but is not limited to, accomplishing required maintenance in accordance with the certificate holder's requirements.

(1) The use of a leased aircraft creates a situation where several persons may be operating the same aircraft under different regulations. For example, a Cessna 340 could be operated by an approved school under FAR Part 91, by an air carrier under FAR Parts 135, and by a rental pilot under FAR Part 91. FAA will not approve multiple MEL's, which would create pilot confusion, with discrepancy lists and sets of procedures for the same aircraft. in the example, the aircraft would operate under the FAR Part 135 MEL, including the A, B, C, and D codes, with approval from the FSDO for other users to conduct operations under other regulations.

(2) FAA will grant operators approval for multiple users of an MEL under FAR Parts 121, 125, 129, or 135 MEL, subject to the following conditions:

(i) The operator is responsible for training all persons in the MEL's use, including the logging and clearing of discrepancies and the use of the A, B, C, and D codes.

(ii) Operators shall maintain a complete, current list of all persons trained and authorized to use the MEL.

(iii) The operator is responsible for determining the aircraft's maintenance status on its return from a FAR Part 91 operation. The operator must accomplish this before the aircraft is put back into FAR Parts 121, 125, 129, or 135 service.

(iv) FAA Principal Operations Inspectors shall verify that operators have established procedures that ensure an acceptable level of safety before authorizing persons to use the MEL under FAR Part 91.

AC 91-67 6/28/91

COMPANY or OPERATOR'S NAME: ______________________________

N ______________________________

LOG SHEET NUMBER: ______________

DATE: ______________________________

LOCATION: ______________________________

DISCREPANCY: ______________________________

____

CORRECTIVE ACTION: ______________________________

____

____

____

______________	______________	______________
Signature	Certificate Number	Date

Figure 4. Sample Equipment Discrepancy Record

18 Chap 3

APPENDIX 1. SAMPLE MMEL DOCUMENTS

FEDERAL AVIATION ADMINISTRATION MASTER MINIMUM EQUIPMENT LIST

Page:
Revision:
Date:

(AIRCRAFT TYPE)

Preamble – PART 91 ONLY

This preamble is applicable to, and will be included in, Master Minimum Equipment Lists (MMEL) issued under the provisions of Section 91.213(a)(2). It is not applicable to MMELs issued under the provisions of Parts 121,125,129, and 135 of the FAR.

Except as provided in Section 91.213(d), or under the provisions of an approved MMEL, all equipment installed on an aircraft in compliance with the airworthiness standards or operating rules must be operative. Experience has shown that with the various levels of redundancy designed into modern aircraft, operation of every system or component installed may not be necessary when the remaining equipment can provide an acceptable level of safety.

An MMEL is developed by the FAA, with participation by the aviation industry, to improve aircraft utilization and thereby provide more convenient and economic air transportation for the public. The FAA-approved MMEL includes only those items of equipment which the Administrator finds may be inoperative and yet maintain an acceptable level of safety by appropriate conditions and limitations. The MMEL and FAA-issued letter of authorization are used as an MEL by an operator and permit operation of the aircraft with inoperative equipment.

The MMEL includes all items of installed equipment that are permitted to be inoperative. Equipment required by the FAR, and optional equipment in excess of FAR requirements, are included with appropriate conditions and limitations. For each listed item, the installed equipment configuration considered to be normal for the aircraft is specified. Items of equipment installed on aircraft (except for passenger convenience items such as galley equipment and passenger entertainment devices), such as "TCAS," wind shear detection devices, and ground proximity warning systems (GPWS) that are in excess of what is required, and are not listed on the MMEL must be operational for dispatch unless MMEL relief is sought through the FSDO having jurisdiction for the operator. If MMEL relief is sought, the operator

APPENDIX 1. SAMPLE MMEL DOCUMENTS (Continued)

must notify the FSDO who will make a request of the FOEB to convene and consider adding the equipment to the MMEL. The operator may then dispatch with the equipment disabled, or rendered inoperative, in accordance with all FAR. It is incumbent on the operator to endeavor to determine if O and/or M procedures for that equipment must be developed. If so, any procedures developed must comply with all FAR. Procedures developed to use the MMEL must not conflict with either the Aircraft Flight Manual limitations, Emergency Procedures, or with Airworthiness Directives (AD), all of which take precedence over the MMEL and those procedures. Suitable conditions and limitations in the form of placards, maintenance procedures, crew operating procedures, and other restrictions, as necessary, are required to be accomplished by the operator to ensure that an acceptable level of safety is maintained. Those procedures should be developed from guidance provided in the manufacturer's aircraft flight and/or maintenance manuals, manufacturer's recommendations, engineering specifications, and other appropriate sources. Procedures must not be contrary to any FAR. Wherever the statement "was required by FAR" appears in the MMEL, the operator must either list the specific FAR by Part and Section and carry the FAR on board the aircraft, or specify the requirements and/or limitations to conduct the flight in accordance with the appropriate FAR.

The MMEL is intended to permit operations with inoperative items of equipment for the minimum period of time necessary until repairs can be accomplished. It is important that repairs be accomplished at the earliest opportunity in order to return the aircraft to its design level of safety and reliability. Inoperative equipment in all cases must be repaired, or Inspected and deferred, by qualified maintenance personnel at the next required inspection Section 91.405(c). The repair intervals indicated by the Letters A, B, and C inserted adjacent to column 2 are NOT applicable to this MMEL.

The MMEL provides for release of the aircraft for flight with inoperative equipment. When an item of equipment is discovered to be inoperative, it is reported by making an entry in the aircraft maintenance records. The item is then either repaired, or deferred per the MMEL or other approved means acceptable to the Administrator, prior to further operation. In addition to the specific MMEL conditions and limitations, determination by the operator that the aircraft is

APPENDIX 1. SAMPLE MMEL DOCUMENTS (Continued)

in condition for safe operations under anticipated flight conditions must be made for all items of inoperative equipment. When these requirements are met, the aircraft may be considered airworthy and returned to service. Operators are responsible for exercising the necessary operational control to ensure that an acceptable level of safety is maintained. When operating with multiple inoperative items, the interrelationship between those items, and the effect on aircraft operation and crew workload, must be considered. Operators are expected to establish a controlled and sound repair program, including the parts, personnel, facilities, procedures, and schedules to ensure timely repair.

WHEN USING THE MMEL, COMPLIANCE WITH THE STATED INTENT OF THE PREAMBLE, DEFINITIONS, CONDITIONS, AND LIMITATIONS SPECIFIED IN THE MMEL IS REQUIRED.

APPENDIX 1. SAMPLE MMEL DOCUMENTS (Continued)

FEDERAL AVIATION ADMINISTRATION MASTER MINIMUM EQUIPMENT LIST

TYPE AIRCRAFT

Definitions

1. System Definitions. System numbers are based on the ATA Specification Number 100 and items are numbered sequentially.

a. "Item" (column 1) lists the equipment, system, component, or function in a column.

b. "Number Installed" (column 2) is the number of items normally installed in the aircraft. This number represents the aircraft configuration considered in developing this MMEL. Should the number be a variable (e.g., passenger cabin items), a number is not required.

c. "Number Required for Dispatch" (column 3) is the minimum number (quantity) of items required for operation provided the conditions specified in column 4 are met.

d. "Remarks or Exceptions" (column 4) in this column include a statement either prohibiting or permitting operation with a specific number of items inoperative, provisos (conditions and limita-tions) for such operation, and appropriate notes.

e. A vertical bar (change bar) in the margin indicates a change, additions or deletion in the adjacent text for the current revision of that page only. The change bar is dropped at the next revision of the page.

2. "Airplane Flight Manual or Rotorcraft Flight Manual" (AFM/RFM) is the document required for type certification and approved by the responsible FAA ACO. The FAA approved AFM/RFM for the specific aircraft is listed on the applicable TCDS.

3. "As required by FAR" means that the listed item is subject to certain provisions (restrictive or permissive) expressed in the FAR operating rules. The number of items required by the FAR must be operative. Items installed that are in excess of the FAR requirements may be permitted to be inoperative if not otherwise required by the MMEL.

4. The asterisk "*" symbol in column 4 indicates the listed item if inoperative, must be placarded to inform and remind the crewmembers and mainten-ance personnel of the equipment condition.

Note: To the extent practical placards should be located adjacent to the control or indicator for the item affected; however, unless otherwise specified, placard wording and location is determined by the operator.

5. The dash "-" symbol in column 2 and/or column 3 indicates a variable number (quantity) of the item(s) installed.

APPENDIX 1. SAMPLE MMEL DOCUMENTS (Continued)

6. "Deleted" in the remarks column after a sequence item indicates that the item was previously listed but is now required to be operative if installed in the aircraft

7. "ER" refers to extended range operations of a two-engine airplane which has a type design approval for ER operations and complies with the provisions of Advisory Circular 120-42A.

8. "Federal Aviation Regulations" (FAR) are the applicable portions of the Federal Aviation Act and Federal Aviation Regulations.

9. "Flight Day" refers to a 24 hour period (from midnight to midnight) either Universal Coordinated Time (UCT) or local time, as established by the operator, during which at least one flight is initiated for the affected aircraft

10. "Icing Conditions" indicate an atmospheric environment causing ice to form on the aircraft or in the engine(s).

11. Alphabetical symbol in column 4 indicates a proviso (condition or limitation) for operation with the listed item inoperative.

12. "Inoperative" indicates a system and/or component malfunction that does not accomplish its intended purpose and/or is not consistently functioning normally within its approved operating limit(s) or tolerance(s).

13. "Notes:" in column 4 provide additional information for crewmember or maintenance consideration. Notes are used to identify applicable material intended to assist with compliance, not to relieve the operator of the responsibility for compliance. Notes are not a part of the provisos.

14. Inoperative components of an inoperative system are components directly associated with and having no other function than to support that system. (Warning/caution systems associated with the inoperative system must be operative unless relief is specifically authorized per the MMEL.)

15. "(M)" symbol indicates a requirement for a specific maintenance procedure which must be accomplished prior to operation with the listed item inoperative. Normally these procedures are accomplished by maintenance personnel; however, other personnel may be qualified and authorized to perform certain functions. Procedures requiring specialized knowledge or skill, or requiring the use of tools or test equipment should be accomplished by maintenance personnel. The satisfactory accomplishment of all maintenance procedures regardless of who performs them, is the responsibility of the operator. Appropriate procedures are required to be published as part of the operator's manual.

16. "(O)" symbol indicates a requirement for a specific operations procedure which must be accomplished in planning for and/or operating with the listed item inoperative. Normally these procedures are accomplished by the flight crew; however, other personnel may be qualified and authorized to perform certain functions. The operator

APPENDIX 1. SAMPLE MMEL DOCUMENTS (Continued)

is responsible for the satisfactory accomplishment of all procedures, regardless of who performs them. Appropriate procedures are required to be published as a part of the operator's manual.

NOTE: The (M) and (O) symbols require the operator to develop procedures for the removal, disabling, or rendering inoperative of items of equipment, in accordance with FAR Part 91, Part 145, or Part 43, as appropriate.

17. "Deactivated" and "Secured" suggest that the specified component is in an acceptable condition for safe flight. The operator will provide an acceptable method of securing or deactivating.

18. "Visual Flight Rules" (VFR) are as defined in FAR Part 91. This precludes a pilot from filing an Instrument Flight Rules (IFR) flight plan.

19. "Visual Meteorological Conditions" (VMC) indicate an atmospheric environment that allows a flight to proceed under the visual flight rules.

20. Three days are 3 consecutive calendar days (72 hours), excluding the day the malfunction was recorded in the aircraft maintenance record/logbook. For example if it were recorded at 10 a.m. on January 26th, the 3-day interval would begin at midnight the 26th and end at midnight the 29th.

21. Category C. Items in this category shall be repaired within 10 consecutive calendar days (240 hours), excluding the day the malfunction was recorded in the aircraft maintenance record/logbook. For example, if it were recorded at 10 a.m. on January 26th, the 10-day interval would begin at midnight the 26th and end at midnight February 5th. The letter designators are inserted adjacent to column 2.

22. Engine Indicating Crew Alerting System (EICAS), Electronic Centralized Aircraft Monitoring System (ECAM) or similar systems that provide electronic messages refer to a system capable of providing different priority levels of systems information messages (e.g., Warning, Caution, Advisory Status and Maintenance). Any airplane discrepancy message that affects dispatchability will normally be at status message level (e.g., Advisory Status) or higher.

23. The three asterisk "em" symbol in column indicates an item which is not required by regulation but which may have been installed on some models of aircraft covered by this MMEL. It should be noted that neither this definition nor the use of this symbol provide authority to install or remove an item from an aircraft.

APPENDIX 2. SAMPLE LETTER OF AUTHORIZATION

Flight Standards District Office
Portland-Hillsboro Airport
3355 NE. Cornell Road
Hillsboro, OR 97124

July 25, 1991

Mr. John Dough, President
John Dough Enterprises
Hangar 9, Suite 203
Portland-Hillsboro Airport
Hillsboro, OR 97124

Dear Mr. Dough:

This letter is issued under the provisions of FAR § 91.213(a)(2) of the Federal Aviation Regulations (FAR) and authorizes John Dough Enterprises only to operate Cessna Citation 500, N81149, Serial No. 12345, under the Master Minimum Equipment List (MMEL), using it as a Minimum Equipment List (MEL).

This letter of authorization and the MMEL constitute a Supplemental Type Certificate for the aircraft and must be carried on board the aircraft as prescribed by FAR § 91.213(a)(2).

Operations must be conducted in accordance with the MMEL. Operations and Maintenance (O and M) procedures for the accomplishment of rendering items of equipment inoperative must be developed by the operator. Whose procedures should be developed from guidance provided in the manufacturer's aircraft flight and/or maintenance manuals manufacturer's recommendations, engineering specifications, and other appropriate sources. Such operations or maintenance procedures must be accomplished in accordance with the provisions and requirements of FAR Part 91, Part 145, or Part 43.

A means of recording discrepancies and corrective actions must be in the aircraft at all times and available to the pilot in command. Failure to perform O and M procedures in accordance with Part 91, Part 145, or Part 43 as appropriate, or to comply with the provisions of the MMEL, preamble, O and M procedures and other related documents, is contrary to FAR and invalidates this letter. All MMEL items that contain the statement "as required by FAR" must either state the FAR by part and section (e.g., 91.205) with the appropriate FAR carried aboard the aircraft, or the operational requirements/limitations required for dispatch must be clearly stated. When the MMEL is revised by the Flight Operations Evaluation Board (FOEB), John Dough Enterprises will be notified by post card of the revision. John Dough Enterprises must then obtain a copy of the revision from this Flight Standards District Office (FSDO), or the FSDO having jurisdiction, and incorporate any changes as soon as practicable including O and M procedures as required.

John Dough Enterprises must develop O and M procedures that correspond with those listed in the MMEL. John Dough Enterprises must also list the "as required by FAR" by specific FAR part and section, or state the operational requirements/limitations for aircraft dispatch. These items must be contained in a procedures document that is

APPENDIX 2. SAMPLE LETTER OF AUTHORIZATION (Continued)

separate from the MMEL and must accompany the MMEL, preamble, and letter of authorization (LOA). They must all be onboard the aircraft anytime it is operated.

Equipment installed on this aircraft (other than passenger convenience items such as galley equipment and passenger entertainment devices) that are in excess of what is required, and are not listed on the MMEL, must be operational for dispatch unless a request is made to this FSDO (or subsequent FSDO that has jurisdiction) to seek relief from the FOEB, through a revision to the MMEL, at the earliest opportunity for the FOEB to convene. If MMEL relief is sought, this FSDO (or subsequent FSDO) must be notified within 10 calendar days (including weekends and holidays) following installation. The operator may then conduct operations with the equipment inoperative for dispatch provided it is disabled, or rendered inoperative, in accordance with all the FAR. It is the responsibility of John Dough Enterprises to determine if O and/or M procedures must be developed for disabling, or rendering inoperative, the equipment. If so, any procedures that are developed must comply with all FAR. If MMEL relief is not sought, the FSDO need not be notified following installation of the equipment.

Should John Dough Enterprises relocate its principal base of operations (address), they must notify in writing both this FSDO and the new FSDO that will have jurisdiction, within 10 calendar days following relocation.

This letter is issued without an expiration date and will remain valid until voluntarily surrendered by John Dough Enterprises, John Dough Enterprises ceases to be the operator of N81149, or it is suspended or revoked for cause by the Federal Aviation Administration (FAA). In any case, should it become invalid, it must be returned to this office or the FSDO having jurisdiction within 10 calendar days from the date it becomes invalids

Sincerely,

Principal Operations Inspector

STATEMENT OF OPERATOR

As evidenced by my signature below, I certify that John Dough Enterprises will operate Cessna 500, N81149, in compliance with the authorizations, provisions, and limitations incumbent with the utilization of this LOA issued in accordance with FAR § 91.213(a)(2). A copy of this letter will be made a part of the MEL file maintained by this FSDO of John Dough Enterprises.

____

Signature Title Date

APPENDIX 3. SAMPLE TITLE PAGE

JOHN DOUGH ENTERPRISES

MINIMUM EQUIPMENT LIST
PROCEDURES DOCUMENT

Cessna 500 N81149

This minimum equipment list is applicable to FAR Part 91 operations only and may not be used for operations conducted under FAR Parts 121, 125, or 135

APPENDIX 4. COMMONLY ASKED QUESTIONS ABOUT MEL's

1. Can the operator make changes to an MEL document without changes having been made to the MMEL? How do they get approved?

To make changes to the MEL, the operator must write to the FSDO exercising jurisdiction over its operation that it wishes to have the MEL revised. This would apply to newly installed equipment that is not required by type certification rules, operating rules, and/or is in excess of what is required and is not listed on the MMEL. The FSDO will contact the FOEB and request the equipment be considered at the next meeting of the FOEB. During the interim, the aircraft may be operated with the items of equipment inoperative provided the operator has developed O and M procedures (as applicable) that comply with all the FAR.

2. What happens if my aircraft is destroyed in an accident? Do I need to return the MEL and Letter of Authorization (LOA) to the issuing FSDO?

If the MEL and LOA survive in a readable form, they must be surrendered to the issuing FSDO, or the FSDO having jurisdiction for the operator, with an official notification of the aircraft's destruction in an accident. A National Transportation Safety Board's indication of the aircraft's destruction is sufficient evidence if the aircraft was destroyed outside of the appropriate FSDO's jurisdiction.

3. What if an FAA inspector asks to see my MEL, procedures document, and LOA?

Because the FAR requires that the MEL, procedures documents and LOA be carried onboard the aircraft, the operator must show an FAA inspector, or other authorized representative of the Administrator, the documents when requested.

4. What happens when the original MEL is no longer appropriate?

This would depend on the conditions that caused the MEL to become inappropriate, since MEL's must be revised when MMEL's are revised.

5. Does the FAA perform any type of surveillance after approval of an MEL? If so, how often?

FAA inspectors do not specifically survey or inspect operators using an MEL. However, as part of a ramp inspection, inspectors will check to determine if an aircraft is operating with an MEL or under the provisions of FAR § 91.213(d).

6. What happens to the MEL if the aircraft is sold?

The MEL and LOA are not transferable. The MEL and LOA must be surrendered to the FSDO exercising jurisdiction. The new owner must decide if he/she wants to operate with an MEL or under the provisions of FAR § 91.213(d). If the owner elects to operate with an MEL, he/she must apply for one at the appropriate FSDO as described in this AC.

APPENDIX 4. COMMONLY ASKED QUESTIONS ABOUT MEL's (Continued)

7. Can an operator request withdrawal of an approved MEL and elect to operate under FAR 91.213(d)?

Both provisions of FAR § 91.213 offer relief to operators. Operators will find more relief operating with an MEL. However, an operator can surrender an MEL and LOA by submitting them to the issuing FSDO with a letter indicating that the operator no longer wishes to operate with an MEL. As of the date the MEL and LOA were surrendered, the aircraft must be operated under FAR § 91.213(d) provided it can meet the requirements of FAR § 91.213(d). If the FSDO determines it cannot, it must continue to operate under the MEL.

8. Can the applicant operate the aircraft under FAR § 91.213(d) while waiting for an approval to a proposed MEL?

Under the regulation, the operator would have no choice except to operate under FAR § 91.213(d), to whatever extent he/she can, until authorization to operate under an MEL is received and the LOA issued.

9. If an MEL LOA is issued in one FSDO's jurisdiction, do I have to have it reissued if I'm operating in another area of jurisdiction?

No. The FAA considers an MEL LOA issued by one FSDO as sufficient for use in any other FSDO's jurisdiction.

10. How do I transfer my MEL and LOA if I move out of the jurisdiction of the issuing FSDO?

For a FAR Part 91 operation, the operator must notify both the FSDO exercising oversight, and the FSDO that will exercise oversight, of the new location of the aircraft within 10 calendar days following the relocation. The previous FSDO will then forward the operator's MEL file to the acquiring FSDO through the FAA's regional office having jurisdiction for the new location. The acquiring FSDO will enter the new location information into the national MMEL data base for revision and update.

11. There are a number of items on the Beech 58P Baron MMEL that need clarifying. For example, the MMEL states that you can take off with one fuel quantity indicator inoperative provided that an approved reliable means is established to determine there is enough fuel required by regulation. What is an example of "an approved means"?

The pilot can visually check the fuel and, if it were full, know how much fuel was onboard for the flight. A dipstick calibrated for that aircraft or any other means that provides a positive measurement would be acceptable.

12. Does an operator have to use the sample discrepancy record provided in this AC?

The sample is the preferred method, but the operator may devise one of their own choosing; however, it must contain at least the information indicated in the sample in this AC.

APPENDIX 4. COMMONLY ASKED QUESTIONS ABOUT MEL's (Continued)

13. If I hold a Part 125 Deviation, may I receive authorization to conduct operations under a Part 91 MEL?

If an operator holds a Deviation to Part 125 and **does not hold a Part 125 Operating Certificate**, he/she may be issued a LOA to conduct operations under the provisions of a FAR Part 91 MEL.

14. If I currently have an approved MEL issued under Part 91, how does this new procedure affect me and my operation under that MEL?

You may continue to operate under your present LOA until it expires at which time you would be reissued the new letter. You could also choose to surrender your current letter, prior to its expiration date, in exchange for a new letter along with the MEL. Either way, you may continue to use your MEL document containing O and M procedures as the procedures document referenced in this AC.

15. Since I no longer have to submit my O and M procedures to the FSDO for approval prior to receiving the LOA, can I accomplish the activity by mail?

No. It is important that Aviation Safety Inspectors (ASI) from the issuing FSDO meet with you (or a bonafide representative of you or your organization/company, etc. having signature authority) to discuss MEL operating procedures prior to issuance of the LOA. This is necessary to ascertain your ability to operate in accordance with the provisions of an MEL. All MEL, therefore, can only be issued in person.

16. Who is a bonafide representative?

It can be anyone with signature authority; i.e., the chief pilot, director of operations, director of maintenance, or other company officer. In the case where none of the above are applicable, a letter on company letterhead introducing the individual as a bona fide representative and signed by a company officer may suffice.

17. Must I make my request for a meeting with the FSDO inspectors in writing?

There is no specified requirement that the request be made in writing, however, it is the prerogative of the FSDO to make such a request at their option.

18. Must ASI from all three disciplines; operations, maintenance, and avionics, be available for the meeting to discuss MEL operating procedures?

It is preferred that all three disciplines be represented; however, it is not necessary if due to work constraints they will not be available within a reasonable period of time. It is important that an operations inspector be involved in the discussion since the MEL is an operating document.

Chapter 7
Type Certificate Data Sheets, Aircraft Specifications and Listings

TCDS Background Information

Type Certificate Data Sheets (TCDS) and Specifications set forth essential factors and other conditions which are necessary for U.S. airworthiness certification. Aircraft, engines, and propellers which conform to a U.S. Type Certificate (TC) are eligible for U.S. airworthiness certification when found to be in a condition for safe operation, and ownership requisites are fulfilled.

Type Certificate Data Sheets which were originated and first published in January of 1958 are, according to 14 CFR §21.41, a part of the type certificate, and as such are evidence that the product has been type-certificated. Generally, TCDS's are compiled from details supplied by the type certificate holder; however, the FAA may request and incorporate additional details when conditions warrant.

Specifications are *not* a part of a type certificate; they were originated during implementation of the Air Commerce Act of 1926. They are FAA recordkeeping documents issued for both type-certificated and non-type-certificated products which have been found eligible for U.S. airworthiness certification. Although they are no longer issued, specifications remain in effect and will be further amended. Specifications covering type-certificated products may be converted to Type Certificate Data Sheets at the option of the type certificate holder. However, to do so requires the type certificate holder to provide an equipment list.

Specifications are subdivided into five major groups:

1. **Type Certificated Aircraft, Engines, and Propellers**

   This covers standard, restricted, and limited types issued for domestic, foreign, and military surplus products.

2. **Group II: Aircraft, Engine, and Propeller Approvals**

   This covers domestic, foreign, and military surplus products constructed or modified between October 1, 1927 and August 22, 1938, all of which have met minimum airworthiness requirements without formal type certification. Such products are eligible for standard airworthiness certification as though they were type-certificated products.

3. **Group III: Aircraft, Engine, and Propeller Approvals**

   This covers domestic products manufactured prior to October 1, 1927, and foreign products manufactured prior to June 20, 1931, and certain military surplus engines and propellers, all of which have met minimum airworthiness requirements of the Air Commerce Act of 1926 and implementing Air Commerce Regulations without formal type certification. Such products are eligible for standard airworthiness certification as though they were type-certificated products.

4. **Group IV: Engine Ratings**

   This covers unapproved engines rated for maximum power and speed only, their use being limited to specific aircraft with maximum gross weights less than 1,000 pounds. Such engines are not eligible for independent airworthiness certification. These ratings are no longer issued.

5. **Group V: Engine Approvals**

   This covers military surplus engines meeting Civil Air Regulations (CAR) 13 design requirements without formal type certification. Such engines are eligible for airworthiness certification as though they were type-certificated engines.

Note: Most products found in Groups II, III, and IV were approved prior to 1938. Although such products may still be eligible for U.S. airworthiness certification, they may require issuance of specific operating limitations. Specifications covering Groups II, III, IV, and V products may be recognized in two ways:

1. An approval number which begins with 2- (sometimes A-2- or G-2-), 3-, 4-, or 5E-.
2. The words Group 2, Group 3, Group 4, or Group 5E in lieu of the specification number.

Specifications have also been used to record the approval of major alterations performed on any of the products for which they were issued. Such approvals are presently recorded on a "Supplemental Type Certificate" (STC). STCs are not published in data sheet format. However, they are listed in the "Summary of Supplemental Type Certificates" when the holder indicates that parts (kits), data, and design rights are available to the public (*see* the latest revision of Advisory Circular 00-21.5 for ordering instructions). The best reference for researching STC data is found on the FAA website:

www.airweb.FAA.gov/RGL

TCDS Availability

Nowadays the method of looking up TCDS is all online, yet historically, "Type Certificate Data Sheets, Specifications, and Listings" were available in six volumes in both paper and mircofiche format from the Superintendent of Documents; some independent companies also make the data available in electronic format:

Volume I, *Single-engine airplanes.* Contained documents for all single-engine fixed-wing airplanes regardless of takeoff weight.

Volume II, *Small multiengine airplanes.* Contained documents for multi-engine airplanes of 12,500 pounds or less maximum takeoff weight.

Volume III, *Large multiengine airplanes.* Contained documents for multi-engine airplanes of more than 12,500 pounds maximum takeoff weight.

Volume IV, *Rotorcraft, gliders, and balloons.* Contained documents on all rotorcraft, gliders, and piloted balloons.

Volume V, *Aircraft engines and propellers.* Contained documents for engines and propellers of all types and models.

Volume VI, *Aircraft Listing and Aircraft Engine and Propeller Listing.* Contained information pertaining to older aircraft models of which there are 50 or less on the FAA Aircraft Registry. It also contained engine and propeller information for which approvals have expired or for which the manufacturer no longer holds a production certificate.

Volumes I through V were available on a one-year subscription basis, with monthly supplementary service included with the subscription. Volume VI was seldom revised, and was available on a single-sale basis.

The historically used method of looking up TCDS is still referred to on the FAA test and therefore covered here. Because Type Certificate Data Sheets are periodically revised by the FAA, you must have the latest revision to make a valid airworthiness decision. The best way to ensure that you are reviewing the most current data is to access information from the Regulatory and Guidance library found on the FAA website (**http://rgl.faa.gov**).

Coded Entries

Many Aircraft and Engine Specifications and some Type Certificate Data Sheets carry coded information to describe the general characteristics of the product. These may be found in the model caption line or a separate line entry entitled "Type" or "Designation."

Aircraft Codes

An example of an aircraft code (Designation) is:

4PCLM

The first number followed by the letter P is the number of seats, for both passengers and crew.

The next letter or compound letters designate the cockpit or cabin design:

O .... open cockpit
C .... closed cabin
O–C .... convertible

The next letter designates the basic kind of aircraft:

L .... landplane
S .... seaplane
L–S .... convertible
Am .... amphibian
Fb .... flying boat
Ag .... autogiro
H .... helicopter

The last letter designates the wing design:

M .... monoplane
B .... biplane

Engine Codes

An example of an engine code (Type) is:

6HOA

The first number is the number of cylinders.

The first letter or letters designate the cylinder arrangement:

L .... in-line
V .... vee
R .... radial
HO .... horizontally opposed
I .... inverted

The second letter designation is for the type of coolant:

A .... air
W .... liquid

Additional letters represent modifications of the original designation.

Sample Test Questions

Note: Any TCDS sources referenced in the answer lines below that are not contained within this IA Test Prep can be found in the Computer Testing Supplement (FAA-CT-8080-8D), in the TCDS section of that document ("Section IV"), which is available as a separate downloadable file on the Reader Resources page for this book at **www.asa2fly.com/reader/ia**.

320. Type certificate data sheets were originated and first published in what year?

A— July 1950.
B— January 1958.
C— December 1962.

Type Certificate Data Sheets were originated and first published in January 1958.

321. If a complete listing of required equipment is not found in the type certificate data sheet, what document could it be found in?

A— Aircraft Part Catalog.
B— Aircraft Maintenance Manual.
C— Approved Flight Manual.

Aircraft Specifications contain a complete listing of the required equipment. Type Certificate Data Sheets do not include this information, but they specify as required equipment an FAA-approved Flight Manual. This manual includes a list of the required equipment.

321a. While performing an annual inspection on a general aviation aircraft, you cannot find the operating limitations in the Type Certificate Data Sheet. In what document will it be found?

A— Aircraft Maintenance Manual.
B— Aircraft Parts Catalog.
C— Airplane Flight Manual.

The Aircraft Maintenance Manual, as the title indicates, explains how to maintain the aircraft. The Aircraft Parts Catalog (or Illustrated Parts Catalog) shows a pictorial breakdown of almost all components with their part numbers, allowing the user to order parts. Of these three, only the "Airplane Flight Manual" (AFM) will include operating limitations for the aircraft.

322. Type Certificate Data Sheets (TCDS) that were converted from Aircraft Specifications will list additional information not found in other TCDS's. What is that information?

A— Optional engines.
B— Notes section.
C— Required equipment.

Specifications covering type certificated products may be converted to Type Certificate Data Sheets at the option of the type certificate holder. However, to do so requires the type certificate holder to provide an equipment list.

323. Concerning the Type Certificate Data Sheet, why would it be important for the IA to know the Certification Basis for an aircraft? Because it determines if

A— a flight manual is required for the aircraft.
B— there are instructions for Continued Airworthiness.
C— Both A and B.

The certification basis determines whether a flight manual or Instructions for Continued Airworthiness are required. Aircraft certificated under 14 CFR Part 23 require both. See 14 CFR §23.1581 and Appendix G. Aircraft certified under CAR Part 3 are not required to have an AFM or an ICA.

324. All aircraft manufactured after March 1, 1979, are required to have an approved flight manual (AFM). Concerning aircraft manufactured prior to that date, which of the following is true?

A— Aircraft not having an AFM are operated per instructions of the aircraft specifications sheet or the type certificate data sheet.
B— Only multiengine aircraft had Approved Flight Manuals prior to the 1979 date.
C— Both A and B are true.

Aircraft Specifications and TCDS's have the operating limitations that must be adhered to if the aircraft does not have an Approved Flight Manual.

Answers Note: *All Learning Statement Codes (in parentheses) are preceded by "IAR." See explanation on Page 1–10.*

320 [B] (031) TCDS
321 [C] (011) TCDS
321a [C] (011) TCDS
322 [C] (031) TCDS
323 [C] (017) TCDS
324 [A] (017) TCDS

324a. Helicopters manufactured after September 16, 1992 are required to have

A— an anti-torque warning light.
B— safety belts and shoulder harnesses.
C— an emergency locator transmitter.

For each rotorcraft manufactured after September 16, 1992, each applicant must show that each occupant's seat is equipped with a safety belt and shoulder harness.

325. In Type Certificate Data Sheets, an aircraft designation of 4PCLM means what?

A— 4 cylinder, propeller, cabin, light, monoplane.
B— 4 passenger and crew, landplane, monocoque.
C— 4 passenger and crew, closed, landplane, monoplane.

The coded entry 4PCLM found in a TCDS identifies the aircraft as being a four-place, closed, landplane, monoplane.

326. If you as the holder of an inspection authorization were approached by an aircraft owner and asked to replace the engine on the aircraft with one of a different make and horsepower rating, which of the following would you consult first in determining acceptability?

A— Manufacturer's maintenance manual.
B— Type certificate data sheet.
C— Summary of supplemental type certificates.

First, check the Type Certificate Data Sheets for the aircraft in which the engine is to be installed. If this engine is listed as approved for this aircraft in the TCDS, its installation would be a minor alteration.

327. The latest revision to individual pages in type certificate data sheets are indicated

A— by a dash number following the type certificate data sheet number in the heading block of page 1.
B— by the word "revision" at the top of the page followed by the number of the revision.
C— by the number of the revision listed below the corresponding page number in a box at the bottom of page 1.

The revision number for each page may be determined by looking at the bottom of Page 1 of the TCDS. Immediately below each page number is the revision number for that page.

327a. What does the date in the box on the opening page of a Type Certificate Data Sheet indicate?

A— Number 1, the date of the latest revision.
B— Number 2, the date of page 1 revision.
C— Both #1 and #2.

The matrix shown at the bottom of the first page of a TCDS shows two things: (1) the total number of pages in the document, and (2) the revision status of each page. Since a change to any page of the TCDS will require a change to the status box in the upper right corner of the TCDS first page, that revision status should also be reflected in the matrix table box reflecting status of page 1. However, the editor has seen numerous TCDS documents where the revision status shown in this matrix box is outdated and does not correlate to the actual revision status in the upper right corner box.

328. What is the certification basis used for the issuance of a type certificate for a Piper model PA18A-150 restricted aircraft?

A— CAR 3.
B— CAR 8.10(b).
C— CAR 4a.

The Piper PA18A-150 (restricted) was certificated under CAR 8.10(b) which covers the restricted category aircraft.

329. Which one of the following aircraft has continued airworthiness instructions?

A— 421C.
B— AA1A.
C— PA-32-300.

14 CFR §23.1529 requires that all aircraft certificated under Part 23 have instructions for continued airworthiness.

The Cessna 421C and Piper PA-32-300 are both certificated under CAR 3. The AA1A is certificated under 14 CFR Part 23 and is the only one of the three that requires the instructions for continued airworthiness.

Answers

324a	[B]	(031)	14 CFR Part 27
325	[C]	(011)	TCDS
326	[B]	(011)	TCDS
327	[C]	(011)	TCDS
327a	[C]	(011)	TCDS
328	[B]	(011)	TCDS
329	[B]	(031)	TCDS

330. What is the certification basis and category(s) of Cessna model A152, serial number A1520824?

A— CAR Part 3, standard.
B— 14 CFR Part 25, utility.
C— CAR 3, acrobatic.

The Cessna A152 is the acrobatic version of the Cessna 152, and is licensed under CAR 3, acrobatic category.

331. What is the certification basis for a Cessna 188 operated in the restricted category?

A— CAR Part 3.
B— 14 CFR Part 21.
C— 14 CFR Part 23.

The Cessna 188 is certificated in the restricted category under 14 CFR Part 21 "Certification Procedures for Products, Articles and Parts."

332. What is the type certificate number and certification basis for a Cessna model C421B?

A— A7CE, CAR Part 3.
B— A7CA, CAR Part 4.
C— 2A13, CAR Part 3.

Cessna model 421B was certificated under CAR Part 3, and its specifications are covered in TCDS A7CE.

333. What is the certification basis for a Piper PA-28-235, serial number 28-11062?

A— CAR Part 3.
B— 14 CFR Part 21.
C— 14 CFR Part 23.

Piper PA-28-235 is certificated under Civil Air Regulations (CAR) Part 3.

334. Where would you find the marking and placards required for Cessna Model 208, serial number 20800044?

A— Type Certificate Data Sheet No. A37CE.
B— Airplane Flight Manual, Cessna P/N D1286-13PH.
C— Model 208 Series Maintenance Manual.

Cessna Model 208 is certificated under 14 CFR Part 23, and is required to have an Airplane Flight Manual. The required markings and placards are included in the AFM.

334a. What is the year of manufacture for Cessna serial number 21062194?

A— 1966.
B— 1977.
C— 1980.

Since neither the aircraft model nor the series is provided, this question requires some additional aviation knowledge. The first few digits in the aircraft serial number usually refer to the aircraft model. So this will lead to referencing the TCDS for Cessna 210 series aircraft, 3A21. The last entry in each section in a TCDS provides a list (or a range) of the serial numbers applicable to that specific model. S/N 21062194 can be found in Section XIV Model 210M/T210M, and was produced in 1977.

Serial Nos. Eligible Models 210M/T210M: 21061574 through 21062273 (1977 Model) 21061042, 21062274 through 21062954 (1978 Model).

335. 14 CFR §43.16, "Airworthiness Limitations" applies to inspections in which of the following aircraft?

A— Cessna 207.
B— Cessna T210.
C— Cessna 421B.

Airplanes certificated under 14 CFR Part 23 are required to have "Instructions for Continued Airworthiness" which includes an Airworthiness Limitations Section.

Of these airplanes only the Cessna 207 is certificated under 14 CFR Part 23. The Cessna T210 and 421B are both certificated under Civil Air Regulations Part 3.

335a. Which of the following aircraft must be maintained in accordance with the "Airworthiness Limitations" section of the manufacturer's maintenance manual or Instructions for Continued Airworthiness?

A— Cessna 177B.
B— Cessna 421C.
C— Piper PA-32-260.

14 CFR 23 Appendix G23.4 states that maintenance according to the Airworthiness Limitations section of the Instructions for Continued Airworthiness applies only to aircraft certificated under 14 CFR Part 23. The Cessna 177B is certificated under this part but both the Cessna 421C and the Piper PA-32-260 are certificated under CAR 3.

Answers

330 [C] (011) TCDS
333 [A] (011) TCDS
331 [B] (011) TCDS
334 [B] (017) TCDS
335 [A] (017) TCDS
332 [A] (011) TCDS
334a [B] (013) TCDS
335a [A] (017) TCDS

336. You are inspecting a PA-28-235, SN 28-10300. The aircraft has 90 pounds of oil on board. How much of this oil is considered "unusable"?

A— 5.7 lbs.
B— 0 lbs.
C— 2.4 lbs.

PA-28-235, SN 28-10300 specifications are listed under TCDS 2A13. On Page 9 we find that the aircraft has an oil capacity of 12 quarts, and in Note 1 on Page 38 we see that this model of PA-28 has unusable oil of 2.4 pounds.

337. The approved engines for the Piper PA-28-235, SN 28-10005, are the Lycoming O-540-B2B5, O-540-B1B5, or O-540-B4B5. Which of the following propellers are approved for use with the O-540-B4B5 engine only?

A— McCauley 1IP235PFA80.
B— Hartzell HC-C2YK-1/8468A-4.
C— Sensenich M80BMM.

On Page 8 of TCDS 2A13 we see that this particular PA-28 is approved for a Sensenich M80BMM or 80BM8, if it is equipped with the Lycoming O-540-B4B5 engine.

337a. What is the basis of certification for Hartzell propeller number HC-12X?

A— CAR 3.
B— CAR 14.
C— FAR 35.

The Hartzell HC-12X propeller is certificated under Civil Air Regulations Part 14 (CAR 14).

337b. According to TCDS 2A13, Note 18 refers to the use of two bushings for the McCauley propeller on a PA-28S-160 — SN 28-1136. Which of the following would be a true statement?

A— They are for use with a fixed pitch propeller.
B— They are for use only with a constant speed propeller.
C— They must be used with a Lycoming O-540 engine.

Note 18 requires two propeller flange bushings to be replaced with Lycoming #72068S bushings in order to use the McCauley propeller. This is a fixed-pitch propeller. This can be determined because no governor assembly is listed with the propeller information.

338. What is the center-of-gravity range for Piper aircraft, model PA-28-235, serial number 28-7200078 at 1,950 pounds?

A— (+82.5) to (93.0).
B— (+81.0) to (94.5).
C— (+85.1) to (93.5).

The CG range for this airplane at 2,100 pounds or less is (+85.1) to (93.5).

338a. The PA-28-235 is certified under CAR Part 3. However, a limited number of systems are also required under FAR 23. What requirement must some of these aircraft meet in Part 23?

A— Cockpit controls.
B— Fuel flow requirements.
C— Landing gear.

Effective May 3, 1962 the PA-28-235 had to meet requirements of 14 CFR §23.955 (fuel flow requirements) and §23.959 (unusable fuel supply).

338b. The Piper PA-28-235 is certified under CAR Part 3. However, a limited number of systems are also required under which of the following?

A— 14 CFR Part 25.
B— 14 CFR Part 23.
C— 14 CFR Part 21.

14 CFR Part 23, according to Page 34 of TCDS 2A13, "Data Pertinent to all models, Certification Basis" for the PA-28-235.

339. How many engine/propeller combinations are approved for a Piper PA-28-235, serial number 28-7310005?

A— Four.
B— Three.
C— One.

The only engine approved for this particular PA-28-235 is a Lycoming O-540-B4B5 with carburetor setting IO-5404, and a Hartzell HC-C2YK-1()F/F8468A-4 propeller.

Answers

336	[C]	(017)	TCDS
337	[C]	(013)	TCDS
337a	[B]	(017)	TCDS
337b	[A]	(013)	TCDS
338	[C]	(013)	TCDS
338a	[B]	(013)	TCDS
338b	[B]	(013)	TCDS
339	[C]	(013)	TCDS

340. What year was the Cessna A150L, serial number A1500293 manufactured?

A—1971.
B—1970.
C—1972.

The design of the Cessna A150L was approved on June 8 of 1970, but serial number A1500293 is a 1972 model.

341. What is the maximum permissible static RPM for the L model of a 1974 Cessna 150, for utility and acrobatic categories?

A—Not over 2600, not under 2500.
B—Not over 2560, not under 2460.
C—Not over 2475, not under 2375.

The 1974 Cessna 150L has a Continental O-200-A engine. In the utility category, there are three propeller options for a 1974 L model:

1A102/HCM 6948
1A102/PCM 6948
1A102/OCM 6948

In the acrobatic category, only the 1A102/OCM 6948 propeller is authorized for use on a 1974 model. Therefore, the maximum permissible throttle setting is not over 2560 and not under 2460.

342. Which propeller is applicable for a Cessna L model 150, certified for the utility and acrobatic category manufactured in 1974?

A—1A101/HCM.
B—1A101/PCM.
C—1A102/OCM.

The 1974 Cessna 150L, certificated in both the Utility and Acrobatic categories, is approved for a McCauley 1A102/OCM 6948 propeller.

343. You are performing an annual inspection on a Cessna 150L, Serial Number 15074861. The propeller, recently installed on the airplane, is a McCauley 1A101/PCM. Which of the following statements would be correct?

A—This propeller is applicable only to the A150L Model.
B—This propeller is applicable to all "L" models of 150s.
C—Neither A or B.

The McCauley1A101/PCM propeller is not approved for the A150L and it is approved only for the 1974 model 150L certificated in the utility category.

344. A Lycoming model O-540-J2B5D engine should have which magneto(s) installed?

A—Bendix D6LN-3230.
B—Slick 6024 and 6025.
C—Bendix S6LN-20 and S6LN-21.

Note 8 on Page 7 of TCDS E-295 shows that for this -J2B5D engine, a Bendix dual magneto D6LN-3230 is required.

345. What is the certification basis for the engine of a Piper PA-28-235, SN 28-7210005?

A—CAR Part 3.
B—14 CFR Part 23.
C—CAR Part 13.

The Piper PA-28-235, SN 28-7210005 has a Lycoming O-540-B4B5 engine that is certificated under Civil Air Regulations Part 13.

346. What is one characteristic of the engine for the Piper PA-28-235? It has a crankshaft that

A—requires Lycoming Flange bushing 720685.
B—has one fifth-order damper and one sixth-order damper.
C—has larger oil passage at rear main.

The PA-28-235 is approved for a Lycoming O-540-B2B5, O-540-B1B5, or O-540-B4B5. These engines refer to TCDS E-295, Notes 5 and 6 regarding the crankshaft dampers.

Note 6 states: These engines incorporate crankshafts with two sixth-order dampers unless a "5" is part of the model designation, i.e., -A1A5. Engines so designated have one fifth-order damper and one sixth-order damper instead of the two sixth-order dampers.

Answers

340 [C] (013) TCDS
343 [C] (013) TCDS
341 [B] (013) TCDS
344 [A] (011) TCDS
342 [C] (013) TCDS
345 [C] (011) TCDS
346 [B] (011) TCDS

347. What is the retirement life of the heated windshield of Cessna model 421C, serial number 421C1110?

A—8,000 hours.
B—4,000 hours.
C—13,200 hours.

Note 3 on Page 23 (Service Information) of TCDS A7CE shows that heated windshield P/N 9910460-1 installed in a Cessna 421C S/N 421C1110 has a mandatory structural retirement life of 13,200 hours.

348. Which of the following Cessna 182 models requires an approved flight manual?

A—182A, Serial Number 33870.
B—182L, Serial Number 18259080.
C—182R, Serial Number 18268542.

In the Equipment section of "Data Pertinent to Model Items I through XII" (Page 20 of TCDS 3A13), "The basic required equipment as prescribed in the applicable airworthiness requirements (see Certification Basis) must be installed in the aircraft for certification. This equipment must include a current Airplane Flight Manual effective S/N 18266591 through 18268586 and R18200584 through R18202041."

349. What are the required travel limits of the elevator of Cessna model 182L, serial number 18258907?

A—Up 20° ±2°, Down 15° ±2°.
B—Up 26° ±1°, Down 17° ±1°.
C—Up 24° ±1°, Down 24° ±1°.

Section VI, Page 9 of TCDS 3A13 contains the specifications for Cessna 182L. Control Surface Movements shows that the elevator travel relative to stabilizer is "Up 26° ±1°, Down 17° ±1°."

350. What are the travel limits of the ailerons of Cessna model 182L?

A—Up 25° ±2°, Down 15° ±1°.
B—Up 20° ±2°, Down 15° ±2°.
C—Up 26° ±1°, Down 17° ±1°.

Section VI, Page 9 of TCDS 3A13 contains the specifications for Cessna 182L. Control Surface Movements shows that the aileron travel is "Up 20° ±2°, Down 15° ±2°."

351. What are the travel limits of the rudder of Cessna model 182L?

A—Right 25° ±2°, Left 15° ±1°.
B—Right 20° ±2°, Left 15° ±2°.
C—Right 24° ±1°, Left 24° ±1°.

Section VI, Page 9 of TCDS 3A13 contains the specifications for Cessna 182L. Control Surface Movements shows that the rudder travel is "Right 24° ±1°, Left 24° ±1°."

352. What are the minimum number of placards required for a Cessna model 182L?

A—3
B—5
C—4

The required placards are listed in the section of TCDS 3A13 entitled "Applicable to Models 182E, 182F, 182G, 182H, 182J, 182K, 182L, 182M."

1. *In full view of the pilot:*

   "This airplane must be operated as a normal category airplane in compliance with operating limitations stated in the form of placards, markings and manuals. No acrobatic maneuvers including spins approved.

   Flight Maneuvering Load Factors

   Flaps Up +3 −1.52
   Flaps Down +3.5
   Maximum design weight 2,800 lbs

   Reference weight and balance data for loading instructions."

2. *On the fuel selector valve plate:*

   "Both off. Left tank level flight only 31 gal. Both on for landing and takeoff all flight attitudes 60 gal. Right tank level flight only 31 gal."

3. *On the control lock:*

   "Control lock—Remove before starting engine."

4. *On the baggage door:*

   "120 lb. maximum baggage and/or auxiliary seat passengers. For additional loading instructions, see weight and balance data."

Answers

347	[C]	(011)	TCDS	348	[C]	(012)	TCDS	349	[B]	(012)	TCDS
350	[B]	(013)	TCDS	351	[C]	(020)	TCDS	352	[C]	(020)	TCDS

353. Which of the following specifications would you refer to during the annual inspection for the engine data of a Cessna model 182L?

A— Type Certificate Data Sheet No. 3A13.
B— Type Certificate Data Sheet No. P3EA.
C— Type Certificate Data Sheet No. E-273.

TCDS 3A13 specifies that Cessna Model 182L has a Continental O-470-R engine. The engine data for this engine is found in TCDS E-273.

354. You are performing an annual inspection on Piper model PA22-150, serial number 22-3401, certificated in normal category only. Which of the following engine/propeller combinations is correct according to the aircraft specifications?

A— Lycoming O-320-A1A engine and Sensenich M74DM propeller.
B— Lycoming O-320-A1B engine and Hartzell hub HC82XG-6 with 7636D-4 blades.
C— Lycoming O-320-B2A engine and Sensenich M76AM-2 propeller.

The normal engine for the PA22-150 is a Lycoming O-320-A2A or O-320-A2B.

Item 106 on Page 11 of TCDS 1A6 lists Lycoming O-320-A1A and O-320-A1B as approved alternate engines.

In the section under "Propellers and Propeller Accessories," item 6 lists the approved constant-speed controllable propeller for a Lycoming O-320-A1A or O-320-A1B as a Hartzell hub HC82XG-6, with 7636D-4 blades.

355. What is the maximum weight and center-of-gravity range (at maximum weight) of Piper model PA-34-200T, serial number 34-7970136?

A— (+101.2) to (+105.2) at 4690 lbs.
B— (+90.6) to (+94.6) at 4570 lbs.
C— (+89.1) to (95.4) at 4500 lbs.

The gross takeoff weight for this aircraft is 4,570 pounds. For this weight, the CG range, with the gear extended, is (+90.6) to (+94.6).

356. Where would the holder of an inspection authorization find a listing of required placards for a Gulfstream American model AA-1A?

A— Aircraft Flight Manual.
B— Type Certificate Data Sheet.
C— Aircraft Parts Catalog.

Note 2 in the "Data Pertinent to All Models" on Page 6 of TCDS A11EA contains a listing of all of the placards that are required for this airplane.

357. Which of the following engines would be acceptable for Piper model PA22-150, serial number 22-3389, certificated in normal and utility category?

A— Lycoming O-320-A2A.
B— Lycoming O-320-A1A.
C— Lycoming O-320-A1B.

The Lycoming O-320-A2A or O-320-A2B engines are approved for the PA22-150 certificated in the normal and utility categories.

Note 3 on Page 15 of TCDS 1A6 requires that when operating in the normal and utility categories, item 407 (control modification kit) be installed. When this kit is installed, propeller item 6 (Hartzell constant-speed) is not eligible.

Propeller item 5 is required. This is a fixed-pitch Sensenich M74DM propeller, which is eligible for a Lycoming O-320-A2A or O-320-A2B.

The only engine-propeller combination listed that is acceptable for the Piper PA22-150 S/N 22-3389 operating in the normal and utility category is the Lycoming O-320-A2A with a Sensenich M74DM propeller.

358. To be eligible to operate in either the normal or utility category, Piper model PA22-150, serial number 22-4450, must have which of the following installed?

A— A control modification kit.
B— Hartzell propeller with hub HC82XG-6, with 7636D-4 blades.
C— Placard, limiting rear seat capacity to a maximum of 170 pounds.

Note 3 on Page 15 of TCDS 1A6 specifies that Serial Nos. 22-3218, 22-3387 and up are eligible to be operated as a Normal or Utility Category Airplane in compliance with the approved Airplane Flight Manual provided item 407 (control modification kit) is installed.

Answers

353	[C]	(020)	TCDS	354	[B]	(020)	TCDS	355	[B]	(020)	TCDS
356	[B]	(011)	TCDS	357	[A]	(013)	TCDS	358	[A]	(013)	TCDS

359. What is the diameter of a McCauley propeller with a hub model 3AF32C509 and blades L82NF[X]-2?

A— 78 inches.
B— 80 inches.
C— 82 inches.

According to Note 2 on Page 3 of TCDS P57GL, blade L82NF[X]-2 has a basic diameter of 82 inches and it has been reduced by 2 inches (see Page 2). Therefore, the diameter is now 80 inches.

360. What is the minimum diameter of a Hartzell propeller with a HC-C2YF-1 hub and 7068 blades?

A— 60 inches.
B— 70 inches.
C— 68 inches.

The diameter limits for 7068 blades is from 70 inches to 60 inches. The minimum diameter is 60 inches.

361. How many occupants is the Twin Commander Model 680 certificated to carry?

A— 5
B— 6
C— 7

The Twin Commander is licensed as a 7PCLM Normal Category airplane. The meanings of these numbers are:

7P = 7 place (occupants)
C = Closed
L = Landplane
M = Monoplane

Also, on Page 2, the number of seats is listed as 7, and further data is given concerning the location reference to the datum: 2 at +95, 2 at +128, and 3 at +168.

362. How many cylinders does a Lycoming IGSO-480 have?

A— 8.
B— 6.
C— 4.

The IGSO-480 is a 6HOA engine:

I .... Fuel injected
G .... Geared
S .... Supercharged
O .... Opposed cylinders
480 .... 480 cubic inches cylinder displacement
6 .... 6 cylinders
HO .... Horizontally opposed cylinders
A .... Air cooled

Also, Page 3, note 7 clearly describes the basic model (GSO-480-A1A6) as a six-cylinder.

363. What is the maximum oil inlet temperature allowed for a Lycoming IGSO-480-A1G6?

A— 225°F.
B— 245°F.
C— 215°F.

Note 1 of TCDS E-284 lists the maximum oil inlet temperatures of the GSO-480-A1A6, -A2A6, and -A1C6 as 225°F. For all other -480 engines, the maximum temperature is 245°F.

364. Which engine is certified with the model 188B, serial number 18801823?

A— Continental O-470-R
B— Continental O-470-S
C— Continental IO-520-D

Section V applies to the 188B model. Although the S/N range shown on Page 8 includes the reference aircraft serial number, it is not definitive for the engine. That clarification is found on Page 6, where the engine models (O-470-R and O-470-S) are listed with appropriate aircraft S/N.

365. Which of the following statements is true regarding the maximum weight of the various aircraft identified in TCDS A9CE?

A— All aircraft have a limit of 3300 pounds.
B— All models have the same weight
C— The maximum weight varies depending on the category use of the aircraft.

Note 3 of the "Data Pertinent to All Models" section states, "When operating in the restricted category, operators may approve higher maximum weights as permitted by FAA Advisory Circular 20-33B..."

Answers

359 [B] (013) TCDS
360 [A] (013) TCDS
361 [C] (020) TCDS
362 [B] (020) TCDS
363 [B] (020) TCDS
364 [A] (020) TCDS
365 [C] (020) TCDS

366. Where is the datum located for a Cessna 172A model, with a serial number of 47743?

A— Front face of firewall.
B— Lower front face of firewall.
C— Upper doorsill.

Under the "Data Pertinent to All Models" section, the first information listed is the "Datum." There are two locations for the datum on 172 model aircraft, depending on aircraft serial number. For S/Ns 28000 through 47746, the datum is located at the "front face of firewall."

367. Airplane Flight Manuals are required equipment in which Cessna 172 models?

A— 172 N and later.
B— Only those with Lycoming engines.
C— Those with S/N prior to 17271035.

Under the "Data Pertinent to All Models 172 through 172Q" section, AFM is required equipment for S/N 17271035 and on. Compare that S/N to specific models and you find it listed in Section VIII for the Cessna 172N model.

368. What is the largest diameter propeller authorized for a Cessna 207?

A— 82 inches.
B— 80 inches.
C— 78 inches.

The TCDS specifies that a diameter not over 82 inches is to be used on the standard Cessna model 207.

369. Which statement is true regarding the electrical system certified for the 207A model aircraft?

A— All models are 28 volts.
B— Only models built in 1977 have a 14-volt system.
C— 28-volt systems start with S/N 20700414.

The TCDS spells out the two different electrical systems used on the 207 series. A cross-check of the S/N for the 14-volt system and the S/N used on the different 207A models indicates that only the 1977 model had the 14-volt system.

370. What year did Cessna reduce the maximum flap deflection on the 210 series aircraft?

A— 1964.
B— 1965.
C— 1966.

The following Cessna 210 models all had 40 degrees of flaps authorized: the initial model 210 approved in 1959, the model 210A approved in 1960, models 210B and C introduced in 1961 and 1962 respectively, the 210D approved in 1963, the 210E approved in 1964, and the 210F and T210F released in 1965. Starting with the introduction of the 210G and T210G (with a TCDS approval August 23, 1966), all subsequent 210 models have a maximum flap-down movement of 30 degrees.

371. What was the last 210 model to require the stall warning indicator specified on Cessna drawing 0511062-4?

A— 210F.
B— T210G.
C— 210G.

The TCDS section titled "Data Pertinent to All Models" in the "Equipment" section specifies a stall warning indicator per Cessna drawing 0511062-4 as required equipment for aircraft S/N 21057001 through 21058818. The last S/N listed is the last one applicable for the 210F model.

372. Which items on the P210 model are subject to a mandatory retirement life?

A— Alternator.
B— Main landing gear.
C— Windshield.

In Note 5 under the "Data Pertinent to All Models" section of TCDS 3A21, the windshield, the rear cabin top windows, side windows, and ice detector light lens are all listed as having a retirement life of 13,000 hours.

Answers

366	[A]	(020)	TCDS	367	[A]	(020)	TCDS	368	[A]	(020)	TCDS
369	[B]	(020)	TCDS	370	[C]	(020)	TCDS	371	[A]	(020)	TCDS
								372	[C]	(020)	TCDS

373. Which of the following Piper PA-32 aircraft are **not** authorized to have Club Seat Installation?

A— PA-32-300.
B— PA-32R-300.
C— PA-32S-300.

In TCDS A3SO, Note 11 calls out almost all models of the PA-32, however, a review reveals that the 32S model is ***not*** *included in the list of models for which this seating arrangement is allowed. Also reference Section III, Model PA32S-300 of the TCDS. "Optional Club Seats" is not an option for this model.*

373a. How many club seats may be installed in Piper PA32-301T, serial number 8221552?

A— 1.
B— 2.
C— 4.

Referencing the TCDS for the Piper PA32 (A3SO), Section XI, model PA32-301T, indicates that 2 club seats are optional at location +119.1 (page 19 of 31).

374. What is the report number for the Pilot Operating Handbook (POH) for the PA32-301?

A— VB-1080.
B— VB-1070.
C— VB-1060.

The TCDS lists various AFMs and POHs as required equipment. The table on Pages 28–29 gives the specific aircraft model, the document title (AFM or POH), the report number, the date it was approved, and the aircraft serial number affected.

375. What is the part number for the approved governor assembly for the PA-32R-301, S/N 3213029?

A— F-4-11.
B— F-4-11B.
C— V-5-4.

Sections VII and VIII of the TCDS both apply to the PA-32R-301 aircraft. However, the referenced S/N is listed in Section VIII, where the approved governor is listed as the Hartzell V-5-4.

375a. What is the maximum ramp weight for a Cessna 208, S/N 20800065?

A— 7335 lbs.
B— 7800 lbs.
C— 8035 lbs.

The maximum ramp weight for aircraft with serial numbers 20800061 and up is 8,035 lbs.

375b. What is the unusable fuel and CG location for a Cessna 208B, S/N 208B0076, that has been modified with SK208-52 "Wing Tank External Sump Installation"?

A— 20.1 lbs @ +185.7.
B— 24.1 lbs @ +186.4.
C— 24.1 lbs @ +206.4

Any Cessna model 208B (S/N 208B0001 through 208B0089) modified with SK208-52 "Wing Tank External Sump Installation" has unusable fuel of 24.1 lbs at +206.4.

375c. What statement is required to accompany a PT6A-65B, when it is exported to the U.S. for installation on a U.S.-registered aircraft?

A— The engine conforms to its United States type design (Type certificate No. E4EA) and is in condition for safe operation.
B— This engine has been subjected by the manufacturer to a final operational check and is in a proper state of airworthiness.
C— Both A and B.

Both statements A and B express the import requirements, which state that "to be considered eligible for installation on U.S.-registered aircraft, each engine to be exported (from Canada) to the United States shall be accompanied by a Certificate of Airworthiness for export or certifying statement endorsed by the exporting cognizant civil airworthiness authority."

Answers

373	[C]	(020)	TCDS
373a	[B]	(020)	TCDS
374	[C]	(020)	TCDS
375	[C]	(020)	TCDS
375a	[C]	(020)	TCDS
375b	[C]	(020)	TCDS
375c	[C]	(020)	TCDS

375d. What is the usable oil tank capacity for a PT6A-52?

A— Not given.
B— 2.5 gallons.
C— 1.5 gallons.

The PT6A-52 engine model is shown in the 6th column in Table II. The entry in the "Usable Oil Tank Capacity" is shown as "—". Page 1 of this TCDS has a legend stating that the symbol, "—" indicates "same as preceding model." The preceding two columns also have "—"; the last numerical entry is 1.5 gal listed in the column for the PT6-60 and -60A models.

375e. If a PT6A-20 engine is used in conjunction with a reversing propeller application, what governor type must be used?

A— Woodward Propeller Governor Type X210XXX.
B— Honeywell Propeller Governor Type X210XXX.
C— This engine cannot be utilized with a reversing propeller.

For reversing applications, the PT6A-6A and PT6A-20 engines must be equipped with Woodward Propeller Governor Type X210XXX.

375f. Of what material is the propeller blade on a Hartzell HC-B3M constructed?

A— Alloy steel.
B— Aluminum.
C— Aramid composite.

The table on Page 1 of the TCDS has a column titled "Blade Construction"; under that heading "Aramid Composite" is the only entry.

375g. How is left-hand rotation designated for a Hartzell HC-B3M propeller blade?

A— "L" in the last position of the hub model designation.
B— "L" in the last position of the blade model designation.
C— Neither A nor B.

In the blade model designation LM100838, the first letter in the series "L" denotes left-hand rotation.

375h. How many blades does the McCauley 3GFR34C(7--) propeller have?

A— 2.
B— 3.
C— 4.

The "3" in the hub model designation code indicates the number of blades on this propeller.

375i. What is the minimum diameter for a McCauley blade 106G used with hub model 3GFR34C703?

A— 100 inches.
B— 106 inches.
C— Neither A nor B.

The table in the TCDS indicates the 106 propeller has a minimum allowable diameter of 100 inches.

375j. What is the magneto timing for a Continental C90-12F?

A— 26 top, 28 bottom.
B— 24 top, 24 bottom.
C— Not given.

The table in the TCDS showing the magneto timing (in degrees before Top Dead Center) indicates 26 top and 28 bottom for the C90-8F model. In the C90-12F column, the table shows "—" which the legend says "indicates same as preceding model." Therefore, the -12F has the same timing as the -8F model.

375k. What is the weight of a Continental C90-8FJ?

A— 184 lbs.
B— 188 lbs.
C— 190 lbs.

The standard C90-8F engine weighs 184 lbs. However, according to Note 4, the letter "J" denotes incorporation of Model B-46 Ex-Cell-O fuel injector, P/N 530499, or American Bosch Model PSC-4A-95A2, P/N 534505, at a weight increase of 4 pounds over the corresponding carburetor-equipped engine.

Answers

375d [C] (020) TCDS
375g [C] (020) TCDS
375e [A] (020) TCDS
375h [B] (020) TCDS
375j [A] (020) TCDS
375f [C] (020) TCDS
375i [A] (020) TCDS
375k [B] (020) TCDS

375l. What is the unusable fuel weight and CG location for a Cessna 402B, with TSIO-520-E engines operating at 2700 RPM?

A— 6 pounds at +152.0.
B— 12 pounds at +152.0.
C— 18 pounds at +164.0.

In TCDS A7CE, the fuel capacity for this model of Cessna is stated as 102 gallons (2 wing tip tanks, 51 gallons each, 50 gallons usable) at +152.0. This is confirmed by referring to Note 1, paragraph (a) on Page 23.

375m. According to TCDS P-920, the prefix "L" for blade model LC7666D means:

A— Large pitch change knob.
B— L.H. tractor.
C— L.H. pusher.

Refer to Note 2 of the TCDS, which states "L prefix denotes L.H. pusher."

375n. How much unusable fuel and what is the location of the unusable fuel for serial number 21062775?

A— 60 lbs at station +46.
B— 6 lbs at station +23.
C— 9 lbs at station +46.

Comparison of the given S/N to the S/N range of the various Cessna 210 models in TCDS 3A21 indicates that this S/N is for a Model 210M (on Page 17 of the TCDS). For fuel capacity, the TCDS refers to Note 1 (Page 25), where it states that the 210 Model M has unusable fuel of 6 lbs at (+23).

375o. For weight and balance purposes, what is the location of any unusable fuel on Cessna model 210 serial number 21058713?

A— +23.
B— +38.
C— +46.

Reference TCDS 3A21, Section VI, model 210E (s/n range 2105811 through 21058715) and on page 7 of 45, it indicates the following:

Fuel Capacity 65 gal. (63.4 gal. usable); two 32.5 gal. tanks in wings at +48.

See Note 1 for data on unusable fuel.

NOTE 1. Current weight and balance report including list of equipment included in certificated empty weight, and loading instructions when necessary must be provided for each aircraft at the time of original certification. The certificated empty weight and corresponding center of gravity location must include unusable fuel of 60 lbs at (+46) on Models 210 and 210A, 9 lbs at (+46) on the 210B, 210C, 210D, 210E, 210-5(205) 210-5A(205A); 12 lbs at (+46) on the 210F, T210F; and 6 lbs at (+23) on the 210G, T210G, 210H, T210H, 210J, T210J, 210K, T210K, 210L, T210L, 210M, T210M, 210N, T210N, P210N through S/N's 21064535 and P21000760; and 18 lbs at (+38) on S/N's 21064536 and up, and P21000761 and up; and undrainable oil of 0 lbs at (-19) through S/N 21061039 and full oil of 18.8 lbs at (-12.5) S/N 21061040 and up, and S/N P21000001 and up.

375p. What is the total weight of unusable fuel on a Twin Aero Commander, model 695, serial number 95039?

A— 15.5 lbs.
B— 33.6 lbs.
C— 53.6 lbs.

Note 1 in each Aircraft Type Certificate Data Sheet contains information concerning the certificated empty weight and corresponding center of gravity location. This information must include unusable (undrainable) fuel.

Answers

375l [B] (020) TCDS
375m [C] (020) TCDS
375n [B] (020) TCDS
375o [C] (029) TCDS
375p [B] (020) TCDS

Type Certificate Data Sheets

2A13
Revision 44
PIPER

PA-28-140	PA-28-151
PA-28-150	PA-28-161
PA-28-160	PA-28-181
PA-28-180	PA-28R-201
PA-28-235	PA-28R-201T
PA-28S-160	PA-28-236
PA-28S-180	PA-28RT-201
PA-28R-180	PA-28RT-201T
PA-28R-200	PA-28-201T

October 15, 1997

TYPE CERTIFICATE DATA SHEET NO. 2A13

This data sheet, which is a part of Type Certificate 2A13, prescribes conditions and limitations under which the product for which the type certificate was issued meets the airworthiness requirements of the Federal Aviation Regulations.

Type Certificate Holder — The New Piper Aircraft, Inc.
2926 Piper Drive
Vero Beach, Florida 32960

I - Model PA-28-160 (Cherokee), 4 PCLM (Normal Category), Approved October 31, 1960, for S/N 28-03; 28-1 through 28-4377; and 28-1760A.

Engine — Lycoming O-320-B2B or O-320-D2A with carburetor setting 10-3678-32

Fuel — 91/96 minimum grade aviation gasoline

Engine Limits — For all operations, 2700 r.p.m. (160 hp)

Propeller and Propeller Limits — Sensenich M74DM or 74DM6 on S/N 28-1 through 28-1760, and 28-1760A.
Sensenich M74DMS or 74D6S5 on S/N 28-1761 through 28-4377.
Static r.p.m. at maximum permission throttle setting not over 2425 r.p.m., not under 2325 r.p.m.
No additional tolerance permitted.
Diameter: Not over 74", not under 72.5".

Propeller Spinner — Piper P/N 14422-00 on S/N 28-1 through 28-1760, and 28-1760A.
Piper P/N 63760-04 or P/N 65805-00 on S/N 28-1761 through 28-4377.
See NOTE 11.

Airspeed Limits

Never exceed	171 mph	(148 knots)	CAS
Maximum structural cruising	140 mph	(121 knots)	CAS
Maneuvering	129 mph	(112 knots)	CAS
Flaps Extended	115 mph	(100 knots)	CAS

Page No.	1	2	3	4	5	6	7	8	9	10	11	12	13	14	15	16	17	18
Revision No.	44	43	43	43	44	43	44	43	44	43	44	44	44	43	43	44	44	44
Page No.	19	20	21	22	23	24	25	26	27	28	29	30	31	32	33	34	35	36
Revision No.	43	43	44	43	44	43	44	43	44	44	44	43	44	43	44	44	43	43
Page No.	37	38	39	40	41	42												
Revision No.	43	43	43	44	44	43												

Center of Gravity Range	(+84.0) to (+95.9) at 1650 lb. or less (+85.9) to (+95.9) at 1975 lb. (+89.2) to (+95.9) at 2200 lb. Straight line variation between points given.
Empty Weight C. G. Range	None
Maximum Weight	2200 lb.
No. of Seats	4 (2 at +85.5, 2 at +118.1)
Maximum Baggage	125 lb. at (+142.8) on S/N 28-1 through 28-1760, and 28-1760A. See NOTE 8. 200 lb. at (+142.8) on S/N 28-1761 through 28-4377.
Fuel Capacity	50 gallons at (+95) (2 wing tanks) See NOTE 1 for data on system fuel.
Oil Capacity	8 quarts at (+32.5), 6 quarts usable See NOTE 1 for data on system oil.

Control Surface Movements

Wing flaps	(± 2°)	Up	0°	Down	40°
Ailerons	(± 2°)	Up	30°	Down	15°
Rudder	(± 2°)	Left	27°	Right	27°
Stabilator	(± 2°)	Up	18°	Down	2°
Stabilator Tab	(± 1°)	Up	3°	Down	12°

Nose Wheel Travel

(±1°) Left 30° Right 30°
(Effective on S/N 28-1 through 28-3377, and 28-1760A)
(±1°) Left 22° Right 22°
(Effective S/N 28-3378 through 28-4377)

Manufacturer's Serial Nos. 28-03; 28-1 through 28-4377; and 28-1760A.

II - Model PA-28-150 (Cherokee), 4 PCLM (Normal Category), Approved June 2, 1961, for S/N 28-03; 28-1 through 28-4377; and 28-1760A.

Engine	Lycoming O-320-A2B or O-320-E2A with carburetor setting 10-3678-32
Fuel	80/87 minimum grade aviation gasoline
Engine Limits	For all operations, 2700 r.p.m. (150 hp)
Propeller and Propeller Limits	Sensenich M74DM or 74DM6 on S/N 28-1 through 28-1760, and 28-1760A. Sensenich M74DMS or 74DM6S5 on S/N 28-1761 through 28-4377. Static r.p.m. at maximum permissible throttle setting not over 2375 r.p.m., not under 2275 r.p.m. No additional tolerance permitted. Diameter: Not over 74", not under 72.5".
Propeller Spinner	Piper P/N 14422-00 on S/N 28-1 through 28-1760, and 28-1760A. Piper P/N 63760-04 or 65805-00 on S/N 28-1761 through 28-4377. See NOTE 11.

Airspeed Limits

Never exceed	171 mph	(148 knots)	CAS
Maximum structural cruising	140 mph	(121 knots)	CAS
Maneuvering	129 mph	(112 knots)	CAS
Flaps Extended	115 mph	(100 knots)	CAS

Center of Gravity Range	(+84.0) to (+95.9) at 1650 lb. or less (+85.9) to (+95.9) at 1975 lb. (+88.4) to (+95.9) at 2150 lb. Straight line variation between points given.
Empty Wt. C. G. Range	None
Maximum Weight	2150 lb.
No. of Seats	4 (2 at +85.5, 2 at +118.1)
Maximum Baggage	125 lb. at (+142.8) on S/N 28-1 through 28-1760, and 28-1760A. See NOTE 8. 200 lb. at (+142.8) on S/N 28-1761 through 28-4377.
Fuel Capacity	50 gallons at (+95) (2 wing tanks) See NOTE 1 for data on system fuel.
Oil Capacity	8 quarts at (+32.5) (6 quarts usable) See NOTE 1 for data on system oil.

Control Surface Movements

Wing flaps	(± 2°)	Up	0°	Down	40°
Ailerons	(± 2°)	Up	30°	Down	15°
Rudder	(± 2°)	Left	27°	Right	27°
Stabilator	(± 1°)	Up	18°	Down	2°
Stabilator Tab	(± 1°)	Up	3°	Down	12°

Nose Wheel Travel
(± 2°) Left 30° Right 30°
(Effective on S/N 28-03; 28-1 through 28-3377; and 28-1760A)
(± 2°) Left 22° Right 22°
(Effective on S/N 28-3378 through 28-4377)

Manufacturer's Serial Nos. 28-03; 28-1 through 28-4377; and 28-1760A.

III - Model PA-28-180 (Cherokee), 4 PCLM (Normal Category), Approved August 3, 1962; 2 PCLM (Utility Category), Approved December 6, 1966, for S/N 28-03; 28-671 through 28-5859; and 28-7105001 through 28-7205318.

Engine	Lycoming O-360-A3A or 0-360-A4A with carburetor setting 10-3878 or 10-4164-1
Fuel	91/96 minimum grade aviation gasoline
Engine Limits	S/N 28-671 through 28-1760, and 28-1760A (except S/N 28-1571 and S/N 28-1573) (See NOTE 4): Maximum permissible takeoff, 2475 r.p.m. For all other operations, 2700 r.p.m. (180 hp) S/N 28-1571; 28-1573; 28-1761 through 28-5859; and 28-7105001 through 28-7205318: For all operations, 2700 r.p.m. (180 hp)
Propeller and Propeller Limits	Sensenich M76EMM or 76EM8 on S/N 28-671 through 28-1760, and 28-1760A (except S/N 28-1571 and S/N 28-1573). Sensenich M76EMMS or 76EM8S5 on S/N 28-1571, 28-1573; 28-1761 through 28-5859; and 28-7105001 through 28-7205318. Static r.p.m. at maximum permissible throttle setting not over 2450 r.p.m., not under 2275 r.p.m. No additional tolerance permitted. Diameter: Not over or under 76". See NOTE 10.

Propeller Spinner

Piper P/N 14422-00 on S/N 28-671 through 28-1760, and 28-1760A.
Piper P/N 63760-04 or 65805-00 on S/N 28-1761 through 28-5859; and 28-7105001 through 28-7205318.
See NOTE 11.

Airspeed Limits

Never exceed	171 mph	(148 knots)	CAS
Maximum structural cruising	140 mph	(121 knots)	CAS
Maneuvering	129 mph	(112 knots)	CAS
Flaps Extended	115 mph	(100 knots)	CAS

Center of Gravity Range

Utility Category (See NOTE 9)

(+84.0)	to	(+86.5)	at	1650 lb. or less
(+85.8)	to	(+86.5)	at	1950 lb.

Normal Category (See NOTE 15)
(S/N 28-671 through 28-5859)

(+84.0)	to	(+95.9)	at	1650 lb. or less
(+85.9)	to	(+95.9)	at	1975 lb.
(+89.2)	to	(+95 9)	at	2200 lb.
(+92.1)	to	(+94.5)	at	2400 lb.

Normal Category
(S/N 28-7105001 through 28-7205318)

(+84.0)	to	(+95.9)	at	1650 lb. or less
(+87.0)	to	(+95.9)	at	2150 lb.
(+87.8)	to	(+95.9)	at	2200 lb.
(+91.0)	to	(+94.5)	at	2400 lb.

Straight Line Variation Between Points Given

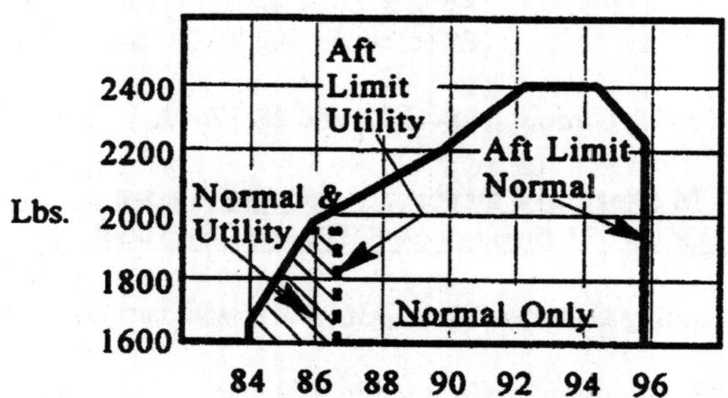

(S/N 28-671 thru 28-5859)

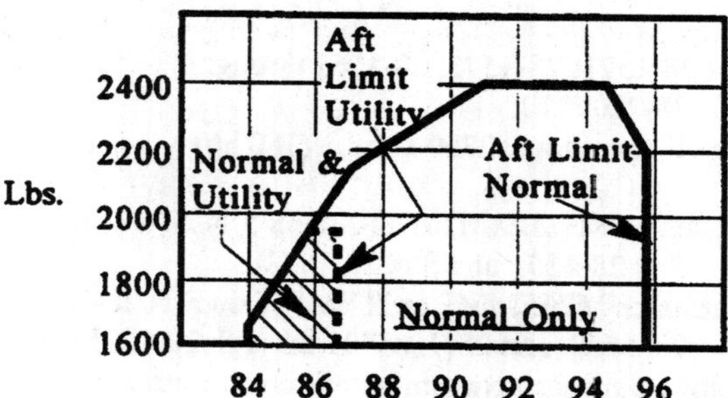

(S/N 28-7105001 thru 28-7205318)

Empty Weight C. G. Range

None

Maximum Weight	Normal Category: 2400 lb. Utility Category: 1950 lb.
No. of Seats	Normal Category: 4 (2 at +85.5, 2 at +118.1) Utility Category: 2 (2 at +85.5)
Maximum Baggage	Eligible Normal Category Only: 125 lb. at (+142.8) on S/N 28-671 through 28-1760, and 28-1760A. See NOTE 8. 200 lb. at (+142.8) on S/N 28-1761 through 28-5859, and 28-7105001 through 28-7205318.
Fuel Capacity	50 gallons at (+95) (2 wing tanks) See NOTE 1 for data on system fuel.
Oil Capacity	8 quarts at (+32.5) (6 quarts usable) See NOTE 1 for data on system oil.

Control Surface Movements

Wing flaps	(± 2°)	Up	0°	Down	40°
Ailerons	(± 2°)	Up	30°	Down	15°
Rudder	(± 2°)	Left	27°	Right	27°
Stabilator	(± 1°)	Up	18°	Down	2°
Stabilator Tab	(± 1°)	Up	3°	Down	12°

Nose Wheel Travel

(± 2°) Left 30° Right 30°
(Effective on S/N 28-671 through 28-3377)
(± 2°) Left 22° Right 22°
(Effective on S/N 28-3378 through 28-5859, and 28-7105001 through 28-7205318)

Manufacturer's Serial Nos.

28-03; 28-671 through 28-5859; and 28-7105001 through 28-7205318.
The manufacturer is authorized to issue airworthiness certificates for airplane serial numbers:

28-4704	28-4745	28-4754	28-4763	28-4776
28-4791	28-4795	28-4826	28-4834	28-4859
28-4875	28-4879	28-4891	28-4907	28-4919
28-4922	28-4935	28-4945	28-4946	28-4947
28-4955	28-4959	28-4961	27-4964	28-4967
28-4968	28-4971	28-4975	28-4977	28-4985
28-4995	28-4999	28-5004	28-5007	28-5015
28-5017	28-5018	28-5019	28-5020	28-5023
28-5026	28-5027	28-5028	28-5031	28-5039
28-5041	28-5046	28-5051	28-5053	28-5057
28-5060	28-5061	28-5062	28-5063	28-5064

28-5066 through 28-5859, and 28-7105001 through 28-7205318 under the delegation option provisions of FAR 21. See NOTE 17 and 20.

IV - Model PA-28S-160 (Cherokee), 4 PCSM (Normal Category), Approved February 25, 1963, for S/N 28-1 through 28-1760; and S/N 28-1760A.

Engine	Lycoming O-320-D2A with carburetor setting 10-3678-32 (See NOTE 18)
Fuel	100/130 minimum grade aviation gasoline
Engine Limits	For all operations, 2700 r.p.m. (160 hp)
Propeller and Propeller Limits	McCauley 1A175-GM Static r.p.m. at maximum permissible throttle setting not over 2360 r.p.m., not under 2260 r.p.m. No additional tolerance permitted. Diameter: Not over 79", not under 78".

Propeller Spinner	Piper P/N 14422-00 spinner required.

Airspeed Limits

Never exceed	153 mph	(133 knots)	CAS
Maximum structural cruising	140 mph	(121 knots)	CAS
Maneuvering	129 mph	(112 knots)	CAS
Flaps Extended	115 mph	(100 knots)	CAS

Center of Gravity

(+85.1)	to	(+93.5)	at	1850 lb. or less
(+87.0)	to	(+93.5)	at	2100 lb.
(+87.9)	to	(+93.5)	at	2140 lb.

Straight line variation between points given.

Empty Weight C. G. Range	None
Maximum Weight	2140 lb.
No. of Seats	4 (2 at +85.5, 2 at +118.1)
Maximum Baggage	125 lb. at (+142.8)
Fuel Capacity	50 gallons at (+95) (2 wing tanks) See NOTE 1 for data on system fuel.
Oil Capacity	8 quarts at (+32.5) (6 quarts usable) See NOTE 1 for data on system oil.

Control Surface Movements

Wing flaps	(±2°)	Up	0°	Down	40°
Ailerons	(±2°)	Up	30°	Down	15°
Rudder	(±2°)	Left	27°	Right	27°
Stabilator	(±1°)	Up	18°	Down	2°
Stabilator Tab	(±1°)	Up	3°	Down	12°

Manufacturer's Serial Nos.	28-03; 28-1 through 28-1760; and 28-1760A.

V - Model PA-28S-180 (Cherokee), 4 PCSM (Normal Category), Approved May 10, 1963, for S/N 28-671 through 28-5859, and 28-7105001 through 28-7105234.

Engine	Lycoming O-360-A3A or 0-360-A4A with carburetor setting 10-4164-1 See NOTE 19.
Fuel	100/130 minimum grade aviation gasoline
Engine Limits	S/N 28-671 through 28-1760, and 28-1760A (except S/N 28-1571 and S/N 28-1573): Maximum permissible takeoff, 2350 r.p.m. For all other operations, 2700 r.p.m. (180 hp) See NOTE 4. S/N 28-1571; 28-1573; 28-1761 through 28-5859; and 28-7105001 through 28-7105234: For all operations, 2700 r.p.m. (180 hp)
Propeller and Propeller Limits	McCauley 1A200-FA8248 on S/N 28-671 to 28-1760, and 28-1760A. McCauley 1A200-DFA8248 on S/N 28-1761 through 28-5859, and 28-7105001 through 28-7105234. Static r.p.m. at maximum permissible throttle setting not over 2190 r.p.m., not under 2140 r.p.m. No additional tolerance permitted. Diameter: Not over 82", not under 81".

Propeller Spinner	Spinner required. Piper P/N 14422-00 on S/N 28-671 through 28-1760, and 28-1760A. Piper P/N 63760-04 or 65805-00 on S/N 28-1761 through 28-5859, and 28-7105001 through 28-7105234.

Airspeed Limits

Never exceed	153 mph	(133 knots)	CAS
Maximum structural cruising	140 mph	(121 knots)	CAS
Maneuvering	129 mph	(112 knots)	CAS
Flaps Extended	115 mph	(100 knots)	CAS

Center of Gravity

(+85.1)	to	(+92.5)	at	1850 lb. or less
(+87.0)	to	(+92.5)	at	2100 lb.
(+89.8)	to	(+92.5)	at	2222 lb.

Straight line variation between points given.

Empty Weight C. G. Range None

Maximum Weight 2222 lb.

No. of Seats 4 (2 at +85.5, 2 at +118.1)

Maximum Baggage 125 lb. at (+142.8)

Fuel Capacity 50 gallons at (+95) (2 wing tanks)
See NOTE 1 for data on system fuel.

Oil Capacity 8 quarts at (+32.5) (6 quarts usable)
See NOTE 1 for data on system oil.

Control Surface Movements

Wing flaps	(±2°)	Up	0°	Down	40°
Ailerons	(±2°)	Up	30°	Down	15°
Rudder	(±2°)	Left	27°	Right	27°
Stabilator	(±1°)	Up	18°	Down	2°
Stabilator Tab	(±1°)	Up	3°	Down	12°

Manufacturer's Serial Nos. 28-671 through 28-5859, and 28-7105001 through 28-7105234. See NOTE 3. The manufacturer is authorized to issue airworthiness certificates for airplane serial numbers:

28-4704	28-4745	28-4754	28-4763	28-4776
28-4791	28-4795	28-4826	28-4834	28-4859
28-4875	28-4879	28-4891	28-4907	28-4919
28-4922	28-4935	28-4945	28-4946	28-4947
28-4955	28-4959	28-4961	27-4964	28-4967
28-4968	28-4971	28-4975	28-4977	28-4985
28-4995	28-4999	28-5004	28-5007	28-5015
28-5017	28-5018	28-5019	28-5020	28-5023
28-5026	28-5027	28-5028	28-5031	28-5039
28-5041	28-5046	28-5051	28-5053	28-5057
28-5060	28-5061	28-5062	28-5063	28-5064

28-5066 through 28-5859, and 28-7105001 through 28-7105234 under the delegation option provisions of FAR 21. See NOTE 17 and 20.

VI - Model PA-28-235 (Cherokee Pathfinder), 4 PCLM (Normal Category), Approved July 15, 1963, for S/N 28-10001 through 28-11378, and 28-7110001 through 28-7210023.

Engine	Lycoming O-540-B2B5, O-540-B1B5, or O-540-B4B5 with carburetor setting 10-4404, 10-5042, or 10-5054. (Baffle P/N 68759 required with 10-5054 setting.)
Fuel	80/87 minimum grade aviation gasoline
Engine Limits	For all operations, 2575 r.p.m. (235 hp)
Propeller and Propeller Limits	McCauley 1P235PFA80 Static r.p.m. at maximum permissible throttle setting not over 2300 r.p.m., not under 2125 r.p.m. No additional tolerance permitted. Diameter: Not over 80", not under 78.5". or Hartzell HC-C2YK-1/8468A-4 or HC-C2YK-1( )F/F8468A-4 Pitch: High 27° ± 2°, Low 13.5° ± .2° at 30" station. Diameter: Not over 80", not under 80". Governor assembly: Hartzell F-4-3 ( ) or F-4-13 See NOTE 21. or Approved for Use with O-540-B4B5 Engine Only: Sensenich M80BMM or 80BM8 Pitch from 69" to 71". Static r.p.m. at maximum permissible throttle setting not over 2300 r.p.m., not under 2150 r.p.m. No additional tolerances permitted. Diameter: Not over 80", not under 78.5".
Propeller Spinner	Piper P/N 65209-00 or P/N 63760-03 with fixed pitch propeller. Spinner required. Piper P/N 65435-0 or P/N 68713 or P/N 66785 spinner tip and P/N 66786 spinner shell or P/N 67790-0 spinner, P/N 67791-0 bulkhead, P/N 67793-0 bulkhead and P/N 99499-0 plate. Two each P/N 67794-0 cuff, or Kit 760 452V with constant speed propeller. See NOTE 14.

Airspeed Limits

Never exceed	197 mph	(171 knots)	CAS
Maximum structural cruising	156 mph	(136 knots)	CAS
Maneuvering	138 mph	(120 knots)	CAS
Flaps Extended	115 mph	(100 knots)	CAS

Center of Gravity Range

S/N 28-10001 through 28-11378 (See NOTE 16):

(+81.5)	to	(+93.5)	at	2100 lb. or less
(+91.5)	to	(+93.5)	at	2900 lb.

S/N 28-7110001 through 28-7210023:

(+85.1)	to	(+93.5)	at	2100 lb. or less
(+86.0)	to	(+93.5)	at	2600 lb.
(+91.5)	to	(+93.5)	at	2900 lb.

Straight line variation between points given.

Empty Weight C. G. Range	None
Maximum Weight	2900 lb.
No. of Seats	4 (2 at +85.5, 2 at +118.1)
Maximum Baggage	200 lb. at (+142.8)

Fuel Capacity	84 gallons at (+95) (50 gallons in 2 wing tanks, 34 gallons in 2 tip tanks). See NOTE 1 for data on system fuel.
Oil Capacity	12 quarts at (+34.1)(9 ¼ quarts usable) See NOTE 1 for data on system oil.

Control Surface Movements

Wing flaps	(±2°)	Up	0°	Down	40°
Ailerons	(±2°)	Up	30°	Down	15°
Rudder	(+2°)	Left	27°	Right	27°
Stabilator	(±1°)	Up	18°	Down	2°
Stabilator Tab	(±1°)	Up	3°	Down	12°

Nose Wheel Travel

(±2°) Left 30° Right 30°
(Effective on S/N 28-10001 through 28-11039)
(±2°) Left 22° Right 22°
(Effective on S/N 28-11040 through 28-11378, and 28-7110001 through 28-7210023)

Manufacturer's Serial Nos. 28-10001 through 28-11378, and 28-7110001 through 28-7210023. The manufacturer is authorized to issue airworthiness certificates for airplane serial numbers 28-11063, 28-11064, 28-11070, 28-11072 through 28-11378, and 28-7110001 through 28-7210023 under the delegation option provisions of FAR 21. See NOTE 17 and 20.

VII - Model PA-28-140 (Cherokee Cruiser), 2 PCLM (Utility or Normal Category); 1950 lb. Maximum Weight, Approved February 14, 1964; 2150 lb. Maximum Weight, Approved June 17, 1965; for S/N 28-20001 through 28-26946, and 28-7125001 through 28-7725290.

Engine Lycoming O-320-E2A with carburetor setting 10-3678-32 or O-320-E3D with carburetor setting 10-5009

Fuel 80/87 minimum grade aviation gasoline

Engine Limits For all operations 2700 r.p.m (150 hp)

Propeller and Propeller Limits

For 1950 lb. maximum weight - Normal Category; S/N 28-20001 through 28-20939; or Utility Category, S/N 28-20001 through 28-26946, and 28-7125001 through 28-7725290:
Sensenich M74DM or 74DM6
Static r.p.m. at maximum permissible throttle setting not over 2425 r.p.m., not under 2150 r.p.m.
No additional tolerance permitted.
Diameter: Not over 74", not under 72.5".

For 2150 lb. maximum weight - Normal Category; S/N 28-20940 through 28-26946, and 28-7125001 through 28-7725290:
Sensenich M74DM or 74DM6
Static r.p.m. at maximum permissible throttle setting not over 2425 r.p.m., not under 2275 r.p.m.
No additional tolerance permitted.
Diameter: Not over 74", not under 72.5".

Propeller Spinner Piper P/N 14422-00.
See NOTE 11.

Airspeed Limits

Never exceed	171 mph	(148 knots)	CAS
Maximum structural cruising	140 mph	(121 knots)	CAS
Maneuvering	129 mph	(112 knots)	CAS
Flaps Extended	115 mph	(100 knots)	CAS

Center of Gravity Range	Utility Category					
	(+84.0)	to	(+86.5)	at	1650 lb. or less	
	(+85.8)	to	(+86.5)	at	1950 lb.	
	Normal Category					
	(+84.0)	to	(+95.9)	at	1650 lb. or less	
	(+85.9)	to	(+95.9)	at	1975 lb.	
	(+88.4)	to	(+95.9)	at	2150 lb.	

Straight line variation between points given.

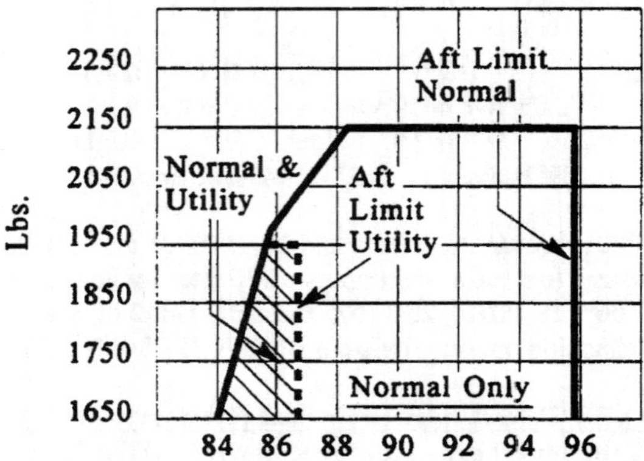

Empty Weight C. G. Range — None

Maximum Weight —
Normal Category: 1950 lb. on S/N 28-20001 through 28-20939 (See NOTE 6).
2150 lb. on S/N 28-20940 through 28-26946, and 28-7125001 through 28-7725290.
Utility Category: 1950 lb. on S/N 28-20001 through 28-26946, and 28-7125001 through 28-7725290.

No. of Seats — 2 at (+85.5)

Maximum Baggage —
Eligible Normal Category Only:
100 lb. at (+117) on S/N 28-20001 through 28-20939 (See NOTE 12).
200 lb. at (+117) on S/N 28-20940 through 28-26946, and 28-7125001 through 28-7725290.
300 lb. at (+117 and +133) on S/N 28-20940 through 28-26946, and 28-7125001 through 28-7725290 (See NOTE 13).

Fuel Capacity —
50 gallon at (+95) (2 wing tanks)
See NOTE 1 for data on system fuel.

Oil Capacity —
8 quarts at (+32.5) (6 quarts usable)
See NOTE 1 for data on system oil.

Control Surface Movements

Wing flaps	(±2°)	Up	0°	Down	40°
Ailerons	(±2°)	Up	30°	Down	15°
Rudder	(±2°)	Left	27°	Right	27°
Stabilator	(±1°)	Up	18°	Down	2°
Stabilator Tab	(±1°)	Up	3°	Down	12°

Nose Wheel Travel —
(±2°) Left 30° Right 30°
(Effective on S/N 28-20001 through 28-21845; 28-21931 through 28-21934; and 28-7425001 through 28-7725290)
(±2°) Left 22° Right 22°
(Effective on S/N 28-21846 through 28-21930; 28-21935 through 28-26946; and 28-7125001 through 28-7325674)

Manufacturer's Serial Nos.	28-20001 through 28-26946; and 28-7125001 through 28-7725290. The manufacturer is authorized to issue airworthiness certificates for airplane serial numbers 28-24677, 28-24682, 28-24697, 28-24698, 28-24700, 28-24703, 28-24704, 28-24705, 28-24706, 28-24709, 28-24710, 28-24712, 28-24713, 28-24714, 28-24715 through 28-26946, and 28-7125001 through 28-7725290 under the delegation option provisions of FAR 21. See NOTE 17 and 20.

VIII - Model PA-28-140 (Cherokee Cruiser), 4 PCLM (Normal Category), 2 PCLM (Utility Category), Approved June 17, 1965, for S/N 28-20001 through 28-26946, and 28-7125001 through 28-7725290.

Engine	Lycoming O-320-E2A with carburetor setting 10-3678-32 or 10-5009 or O-320-E3D with carburetor setting 10-5009
Fuel	80/87 minimum grade aviation gasoline
Engine Limits	For all operations 2700 r.p.m. (150 hp)
Propeller and Propeller Limits	Sensenich M74DM or 74DM6 Static r.p.m. at maximum permissible throttle setting not over 2425 r.p.m., not under 2275 r.p.m. No additional tolerance permitted. Diameter: Not over 74", not under 72.5".
Propeller Spinner	Piper P/N 14422-00. See NOTE 11.

Airspeed Limits

Never exceed	171 mph	(148 knots)	CAS
Maximum structural cruising	140 mph	(121 knots)	CAS
Maneuvering	129 mph	(112 knots)	CAS
Flaps Extended	115 mph	(100 knots)	CAS

Center of Gravity Range

Utility Category

(+84.0)	to	(+86.5)	at	1650 lb. or less
(+85.8)	to	(+86.5)	at	1950 lb.

Normal Category

(+84.0)	to	(+95.9)	at	1650 lb. or less
(+85.9)	to	(+95.9)	at	1975 lb.
(+88.4)	to	(+95 9)	at	2150 lb.

Straight line variation between points given.

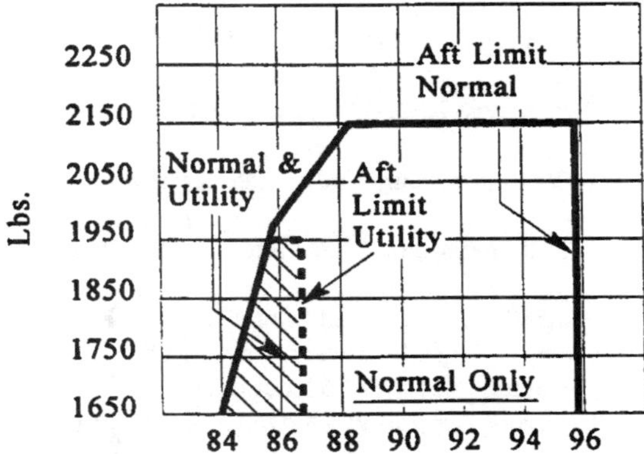

Empty Weight C. G. Range	None
Maximum Weight	Normal Category: 2150 lb. Utility Category: 1950 lb.
No. of Seats	Normal Category: 4 (2 at +85.5, 2 at +117) Utility Category: 2 (2 at +85.5)
Maximum Baggage	Eligible Normal Category only: 100 lb. at (+117) on S/N 28-20001 through 28-20939 (See NOTE 12). 200 lb. at (+117) on S/N 28-20940 through 28-26946; and 28-7125001 through 28-7725290. 300 lb. at (+117 and +133) on S/N 28-20940 through 28-26946; and 28-7125001 through 28-7725290 (See NOTE 13).
Fuel Capacity	50 gallons at (+95) (2 wing tanks) See NOTE 1 for data on system fuel.
Oil Capacity	8 quarts at (+32.5) (6 quarts usable) See NOTE 1 for data on system oil.

Control Surface Movements

Wing flaps	(±2°)	Up	0°	Down	40°
Ailerons	(±2°)	Up	30°	Down	15°
Rudder	(±2°)	Left	27°	Right	27°
Stabilator	(±1°)	Up	18°	Down	2°
Stabilator Tab	(±1°)	Up	3°	Down	12°

Nose Wheel Travel

(±2°) Left 30° Right 30°
(Effective on S/N 28-20940 through 28-21845; 28-21931 through 28-21934; and 28-7425001 through 28-7725290)

(± 2°) Left 22° Right 22°
(Effective on S/N 28-21846 through 28-21930; 28-21935 through 28-26946; and 28-7125001 through 28-7325674)

Manufacturer's Serial Nos.

28-20001 through 28-26946, and 28-7125001 through 28-7725290. The manufacturer is authorized to issue airworthiness certificates for airplane serial numbers 28-24677, 28-24682, 28-24697, 28-24698, 28-24700, 28-24703, 28-24704, 28-24705, 28-24706, 28-24709, 28-24710, 28-24712, 28-24713, 28-24714, 28-24715 through 28-26946, and 28-7125001 through 28-7725290 under the delegation option provisions of FAR 21.
See NOTE 17 and 20.

IX - Model PA-28R-180 (Arrow), 4 PCLM (Normal Category), Approved June 8, 1967, for S/N 28R-30002 through 28R-31270, and 28R-7130001 through 28R-7130013.

Engine	Lycoming IO-360-B1E
Injector	Bendix type RSA-5ADI Parts List No. 2524297
Fuel	100/130 minimum grade aviation gasoline
Engine limits	For all operations, 2700 r.p.m. (180 hp)
Propeller and Propeller Limits	Hartzell constant speed Model HC-C2YK-( )/7666A-0 or HC-C2YK-1( )F/F7666A Pitch: High 29.0° ± 1°, Low 13.0° ± .2° at 30" Station. Diameter: Not over 76", not under 74.5". Governor Assembly: Hartzell F-2-2 ( ) or F-2-7 ( ) Avoid continuous operation between 2000 - 2200 r.p.m.

Propeller Spinner	Piper P/N 68713 or P/N 66785 spinner tip and P/N 66786 spinner shell, or P/N 67790-0 spinner, P/N 67791-0 bulkhead, P/N 67793-0 bulkhead, and P/N 99499-0 plate. Two each P/N 67794-0 cuff or Kit 760 410V. See NOTE 11.

Airspeed Limits

Never exceed	214 mph	(186 knots)	CAS
Maximum structural cruising	170 mph	(148 knots)	CAS
Maneuvering	134 mph	(116 knots)	CAS
Flaps extended	125 mph	(109 knots)	CAS
Maximum gear extension	150 mph	(130 knots)	CAS
Maximum gear retraction	125 mph	(109 knots)	CAS

Center of Gravity Range

(+81.0) to (+95.9) at 1925 lb. or less
(+91.0) to (+95.9) at 2500 lb.
Straight line variation between points given.
Moment due to retracting of landing gear (+819 in-lb.)

Empty Weight C. G. Range None

Maximum Weight 2500 lb.

No. of Seats 4 (2 at +85.5, 2 at +118.1)

Maximum Baggage 200 lb. at (+142.8)

Fuel Capacity 50 gallons at (+95) (2 wing tanks)
See NOTE 1 for data on system fuel.

Oil Capacity 8 quarts at (+29.5) (6 quarts usable)
See NOTE 1 for data on system oil.

Control Surface Movements

Wing flaps	(±2°)	Up 0°	Down 40°
Ailerons	(±2°)	Up 30°	Down 15°
Rudder	(±2°)	Left 27°	Right 27°
Stabilator	(±1°)	Up 18°	Down 2°
Stabilator Tab	(±1°)	Up 3°	Down 12°

Nose Wheel Travel (±2°) Left 30° Right 30°

Manufacturer's Serial Nos.

28R-30002 through 28R-31270, and 28R-7130001 through 28R-7130013. The manufacturer is authorized to issue airworthiness certificates for airplane serial numbers:

28R-30538	28R-30546	28R-30559	28R-30586	28R-30587
28R-30602	28R-30603	28R-30605	28R-30624	28R-30627
28R-30638	28R-30639	28R-30642	28R-30684	28R-30697
28R-30708	28R-30726	28R-30739	28R-30740	28R-30747
28R-30750	28R-30752	28R-30759	28R-30760	28R-30766
28R-30776	28R-30779	28R-30785	28R-30787	28R-30795
28R-30801	28R-30809	28R-30815	28R-30819	28R-30821
28R-30824	28R-30827	28R-30832	28R-30835	28R-30838
28R-30842	28R-30845	28R-30849	28R-30853	28R-30857
28R-30860	28R-30865	28R-30866	28R-30867	28R-30868
28R-30869	28R-30872	28R-30874	28R-30875	28R-30877

through 28R-31270, and 28R-7130001 through 28R-7130013 under the delegation option provisions of FAR 21. See NOTE 17 and 20.

X - Model PA-28R-200 (Arrow), 4 PCLM (Normal Category), Approved January 16, 1969, S/N 28R-35001 through 28R-35820 and 28R-7135001 through 28R-7135229.

Engine	Lycoming IO-360-C1C
Injector	Bendix Type RSA-5AD1, Parts List Number 2524450
Fuel	100/130 minimum grade aviation gasoline
Engine Limits	For all operations, 2700 r.p.m. (200 hp)
Propeller and Propeller Limits	Hartzell constant speed Model HC-C2YK-1 ( )/7666A-2 or HC-C2YK-1 ( )F/F7666A Pitch: High 29.0° ±2°, Low 14.0° ±2° at 30 " Station Diameter: Not over 74", not under 72.5" Governor Assembly: Hartzell F-2-7 ( ) Avoid continuous operation between 2000 - 2350 r.p.m.
Propeller Spinner	Piper P/N 66785 spinner tip and P/N 66786 spinner shell or P/N 67790-0 spinner, P/N 67791-0 bulkhead, P/N 67793-0 bulkhead, and P/N 99499-0 plate. Two each P/N 67794-0 cuff or Kit 760 410V. See NOTE 11.

Airspeed Limits

Never exceed	214 mph	(186 knots)	CAS
Maximum structural cruising	170 mph	(148 knots)	CAS
Maneuvering	134 mph	(116 knots)	CAS
Flaps Extended	125 mph	(109 knots)	CAS
Maximum gear extension	150 mph	(130 knots)	CAS
Maximum gear retraction	125 mph	(109 knots)	CAS

Center of Gravity Range	(+81.0) to (+95.9) at 1925 lb. or less (+90.0) to (+95.9) at 2600 lb. Straight line variation between points given. Moment due to retracting of landing gear (+819 in-lb.)
Empty Weight C. G. Range	None
Maximum Weight	2600 lb.
No. of Seats	4 (2 at +85.5, 2 at +118.1)
Maximum Cargo	200 lb. (at +142.8)
Fuel Capacity	50 gallons at (+95)(2 wing tanks) See NOTE 1 for data on system fuel.
Oil Capacity	8 quarts at (+29.5) (6 quarts usable) See NOTE 1 for data on system oil.

Control Surface Movements

Wing flaps	(±2°)	Up	0°	Down	40°
Ailerons	(±2°)	Up	30°	Down	15°
Rudder	(±2°)	Left	27°	Right	27°
Stabilator	(±1°)	Up	18°	Down	2°
Stabilator Tab	(±1°)	Up	3°	Down	12°

Nose Wheel Travel (±2°) Left 30° Right 30°

Manufacturer's Serial Numbers: 28R-35001 through 28R-35820, and 28R-7135001 through 28R-7135229. The manufacturer is authorized to issue airworthiness certificates for airplanes serial numbers 28R-35001 through 28R-35820, and 28R-7135001 through 28R-7135229 under the delegation option provisions of FAR 21.

XI - Model PA-28R-200 (Arrow II), 4 PCLM (Normal Category), Approved December 2, 1971, for S/N 28R-7235001 through 28R-7635545.

This series differs from the basic PA-28R-200 (Item X) by the addition of a five-inch fuselage extension, larger horizontal tail, wing span increase, gross weight increase, and other minor changes.

Engine	Lycoming IO-360-C1C (See NOTE 22) Lycoming IO-360-C1C6 (See NOTE 23)
Injector	Bendix Type RSA-5AD1, Part List Number 2524450
Fuel	100/130 minimum grade aviation gasoline
Engine Limits	For all operations, 2700 r.p.m. (200 hp)
Propeller and Propeller Limits	Hartzell Constant Speed Model HC-C2YK-1 ( ) or HC-C2YK-1( ) F Blade Model 7666A-2 or F7666A-2 (See NOTE 22) Pitch: High 29.0° ± 2°, Low 14.0° ± .2° at 30" Station. Diameter: Not over 74", not under 72.5". Governor Assembly: Hartzell F-2-7( ) Avoid continuous operation between 2000 - 2350 r.p.m. or McCauley Constant Speed Model B2D34C213, Blade Model 90DHA-16 (See NOTE 23) Pitch: High 27.5° ± .5°; Low 12.5° ± .2° at 30" Station. Diameter: Not over 74", not under 73". Governor Assembly: Hartzell F-2-7 ( ) Avoid continuous operation between 1500 and 1950 r.p.m. below 15" manifold pressure.
Propeller Spinner	For the Hartzell Propeller: Piper P/N 66785-00 spinner tip, P/N 66786 spinner shell and P/N 68734-0 bulkhead or P/N 99374-0 spinner installation (same as Kit No. 760 410V). See NOTE 11. For the McCauley Propeller: Piper P/N 66785 spinner tip and P/N 66786 spinner shell or P/N 67790-0 spinner, P/N 67791-0 bulkhead, P/N 67793-0 bulkhead, and P/N 99499-0 plate. Two each P/N 67794-0 cuff, or Kit 760 410V. Spinner and attachment plate installation P/N 35828-2. See NOTES 11 and 23.

Airspeed Limits

Never exceed	214 mph	(186 knots)	CAS
Maximum structural cruising	170 mph	(148 knots)	CAS
Maneuvering	131 mph	(114 knots)	CAS
Flaps Extended	125 mph	(109 knots)	CAS
Maximum gear extension	150 mph	(130 knots)	CAS
Maximum gear retraction	125 mph	(109 knots)	CAS

Center of Gravity Range

(+80.0)	to	(+93.0)	at	1800 lb. or less
(+82.0)	to	(+93.0)	at	2300 lb.
(+87.3)	to	(+93.0)	at	2650 lb.

Empty Weight C. G. Range	None
Maximum Weight	2650 lb.
No. of Seats	4 (2 at +80.5, 2 at +118.1)
Maximum Cargo	200 lb. (at +142.8)
Fuel Capacity	50 gallons at (+95) (2 wing tanks) See NOTE 1 for data on system fuel.
Oil Capacity	8 quarts at (+24.5) (6 quarts usable) See NOTE 1 for data on system oil.

Control Surface Movements	Wing flaps	(±2°)	Up	0°	Down	40°
	Ailerons	(±2°)	Up	30°	Down	15°
	Rudder	(±2°)	Left	27°	Right	27°
	Stabilator	(±1°)	Up	16°	Down	2°
	Stabilator Tab	(±1°)	Up	3°	Down	12°
Nose Wheel Travel		(±2°)	Left	30°	Right	30°

Manufacturer's Serial Numbers — 28R-7235001 through 28R-7635545. The manufacturer is authorized to issue airworthiness certificates for airplane serial numbers 28R-7235001 through 28R-7635545 under the delegation option provisions of FAR 21. See NOTE 20.

XII - Model PA-28-180 (Archer), 4 PCLM (Normal Category), 2 PCLM (Utility Category), Approved May 22, 1972, for S/N 28- E13, and 28-7305001 through 28-7505260.

This series differs from the basic PA-28-180 (Item III) by the addition of a five inch fuselage extension, wing span increase, larger horizontal tail, gross weight increase and other minor changes.

Engine — Lycoming O-360-A4A or O-360-A4M with carburetor settings 10-3878 or 10-5193

Fuel — 100/130 minimum grade aviation gasoline

Engine Limits — For all operations, 2700 r.p.m. (180 hp)

Propeller and Propeller Limits — Sensenich or 76EM8S5 or M76EMMS
Static r.p.m. at maximum permissible throttle setting not over 2425 r.p.m., not under 2325 r.p.m.
No additional tolerance permitted.
Diameter: Not over or under 76".

Propeller Spinner — Piper P/N 65805-00.
See NOTE 11.

Airspeed Limits			
Never exceed	171 mph	(148 knots)	CAS
Maximum structural cruising	140 mph	(121 knots)	CAS
Maneuvering	127 mph	(110 knots)	CAS
Flaps Extended	115 mph	(100 knots)	CAS

Center of Gravity Range

Normal Category
(+82.0) to (+93.0) at 2050 lb. or less
(+87.4) to (+93.0) at 2450 lb.

Utility Category
(+82.0) to (+86.5) at 1950 lb. or less
Straight line variation between points given.

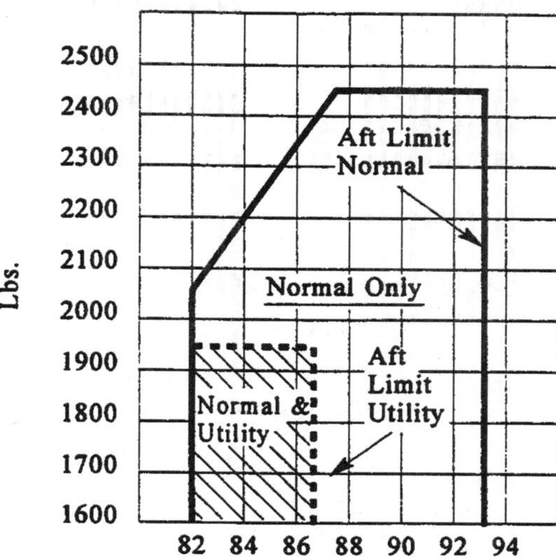

Empty Weight C. G. Range	None
Maximum Weight	Normal Category: 2450 lb. Utility Category: 1950 lb.
No. of Seats	Normal Category: 4 (2 at +80.5 2 at +118.1) Utility Category: 2 (2 at +80.5)
Maximum Baggage	200 lb. at (+142.8)
Fuel Capacity	50 gallons at (+95) (2 wing tanks) See NOTE 1 for data on system fuel.
Oil Capacity	8 quarts at (+27.5) (6 quarts usable) See NOTE 1 for data on system oil.

Control Surface Movements

Wing flaps	(±2°)	Up	0°	Down	40°
Ailerons	(±2°)	Up	30°	Down	15°
Rudder	(±2°)	Left	27°	Right	27°
Stabilator	(±1°)	Up	14°	Down	2°
Stabilator Tab	(±1°)	Up	3°	Down	12°

Nose Wheel Travel

(±2°) Left 22° Right 22°
(S/N 28-E13, 28-7305001 through 28-7305601)
(±2°) Left 30° Right 30°
(S/N 28-7405001 through 28-7505260)

Manufacturer's Serial Numbers: 28-E13, and 28-7305001 through 28-7505260. The manufacturer is authorized to issue airworthiness certificates for airplanes serial numbers 28-7305001 through 28-7505260 under the delegation option provisions of FAR 21. See NOTE 20.

XIII - Model PA-28-235 (Cherokee Pathfinder), 4 PCLM (Normal Category), Approved June 9, 1972, for S/N 28E-11, and 28-7310001 through 28-7710089.

This series differs from the basic PA-28-235 (Item VI) by the addition of a five inch fuselage extension, larger horizontal tail, gross weight increase, and other minor changes.

Engine	Lycoming O-540-B4B5 with carburetor setting 10-5404
Fuel	80/87 minimum grade aviation gasoline

Engine Limits	For all operations, 2575 r.p.m. (235 hp)
Propeller and Propeller Limits	Hartzell HC-C2YK-1( )F/F 8468A-4 Pitch: High 27° ± 2°, Low 13.5 ° ± .2° at 30" station. Diameter: Not over 80", not under 80". Governor Assembly: Hartzell F-4-3( ) or F-4-13 ( ). See NOTE 21.
Propeller Spinner	P/N 99374 Spinner Installation. Spinner required.
Airspeed Limits	Never exceed 197 mph (171 knots) CAS Maximum structural cruising 156 mph (135 knots) CAS Maneuvering 138 mph (119 knots) CAS Flaps Extended 115 mph (99 knots) CAS
Center of Gravity Range	(+79.0) to (+91.5) at 1900 lb. or less (+82.0) to (+91.5) at 2500 lb. (+88.0) to (+91.5) at 3000 lb. Straight line variation between points given.
Empty Weight C. G. Range	None
Maximum Weight	3000 lb.
No. of Seats	4 (2 at +80.5, 2 at +118.1)
Maximum Baggage	200 lb. at (+142.8)
Fuel Capacity	84 gallons (50 gallons in 2 wing tanks at (+95) and 34 gallons in 2 tip tanks at (+95)) See NOTE 1 for data on system fuel.
Oil Capacity	12 quarts at(+29.1) (9¼ quarts usable) See NOTE 1 for data on system oil.
Control Surface Movements	Wing flaps (±2°) Up 0° Down 40° Ailerons (±2°) Up 30° Down 15° Rudder (±2°) Left 27° Right 27° Stabilator (±2°) Up 16° Down 2° Stabilator Tab (±1°) Up 3° Down 12°
Nose Wheel Travel	(±2°) Left 22° Right 22° (S/N 28-E11, 28-7310001 through 28-7310176) (±2°) Left 30° Right 30° (S/N 28-7410001 through 28-7710089)
Manufacturer's Serial Numbers	28-E11, and 28-7310001 through 28-7710089. The manufacturer is authorized to issue airworthiness certificates for airplane serial numbers 28-E11, and 28-7310001 through 28-7710089 under the delegation option provisions of FAR 21. See NOTE 20.

XIV - Model PA-28-151 (Cherokee Warrior), 4 PCLM (Normal Category), 2 PCLM (Utility Category), Approved August 9, 1973, for S/N 28-7415001 through 28-7715314.

Engine	Lycoming O-320-E30 with carburetor setting 10-5009, or 10-5009N, or 10-5135
Fuel	80/87 minimum grade aviation gasoline
Engine Limits	For all operations, 2700 r.p.m. (150 hp)

Propeller and Propeller Limits	Sensenich M74DM6 Static r.p.m. at maximum permissible throttle setting not over 2375 r.p.m., not under 2275 r.p.m. No additional tolerance permitted. Diameter: Not over 74", not under 72". or McCauley 1C160 EGM 7653 Static r.p.m. at maximum permissible throttle setting not over 2400 r.p.m., not under 2300 r.p.m. No additional tolerance permitted. Diameter: Not over 76", not under 74.5".
Propeller Spinner	Piper P/N 35323. See NOTE 11.
Airspeed Limits	Never exceed 176 mph (153 knots) CAS Maximum structural cruising 140 mph (122 knots) CAS Maneuvering 124 mph (108 knots) CAS Flaps Extended 125 mph (109 knots) CAS (S/N 28-7415001 through 28-7515449) Flaps Extended 115 mph (100 knots) CAS (S/N 28-7615001 through 28-7715314)
Center of Gravity Range	Normal Category (+83.0) to (+93.0) at 1950 lb. or less (+87.0) to (+93.0) at 2325 lb. Utility Category (+83.0) to (+86.5) at 1950 lb. or less Straight line variation between points given.
Empty Weight C. G. Range	None
Maximum Weight	Normal Category: 2325 lb. Utility Category: 1950 lb.
No. of Seats	Normal Category: 4 (2 at +80.5, 2 at +118.1) Utility Category: 2 (2 at +80.5)
Maximum Baggage	Eligible Normal Category only: 200 lb. at (+142.8)
Fuel Capacity	50 gallons at (+95) (2 wing tanks) See NOTE 1 for data on system fuel.
Oil Capacity	8 quarts at (+27.5) (6 quarts usable) See NOTE 1 for data on system oil.

Control Surface Movements

Wing Flaps	(±2°)	Up	0°	Down	40°
Ailerons	(±2°)	Up	23°	Down	17°
	(S/N 28-7415001 through 28-7515449)				
Ailerons	(±2°)	Up	25°	Down	12.5°
	(S/N 28-7615001 through 28-7715314)				
Rudder	(±2°)	Left	27°	Right	27°
Stabilator	(±1°)	Up	14°	Down	2°
Stabilator Tab	(±1°)	Up	3°	Down	12°

Nose Wheel Travel (±1°) Left 30° Right 30°

Manufacturer's Serial Numbers	28-7415001 through 28-7715314. The manufacturer is authorized to issue airworthiness certificates for airplanes serial numbers 28-7415001 through 28-7715314 under the delegation option provisions of FAR 21.

XV - A.- Model PA-28-181 (Archer II), 4 PCLM (Normal Category), 2 PCLM (Utility Category), Approved July 8, 1975, for S/N 28-7690001 through 28-8690056; 28-8690061; 28-8690062; and 2890001 through 2890205.

Engine — Lycoming O-360-A4M with carburetor settings 10-3878 or 10-5193 or Lycoming O-360-A4A with carburetor setting 10-5193.

Fuel — 100/130 minimum grade aviation gasoline

Engine Limits —
Applicable to S/N 28-7690001 through 28-7990589:
For all operations, 2700 r.p.m. (180 hp)
Applicable to S/N 28-8090001 through 28-8690056; 28-8690061; 28-8690062; and 2890001 through 2890205:
For takeoff 5 minutes at 2700 r.p.m. (180 hp)
For maximum continuous operation, 2650 r.p.m. (178 hp)

Propeller and Propeller Limits —
Sensenich 76EM8S5
For S/N 28-7690001 through 28-7790607:
Static r.p.m. at maximum permissible throttle setting, not over 2425 r.p.m., not under 2325 r.p.m. at sea level, ISA conditions. (Reference aircraft Maintenance Manual for test procedure to determine approved static r.p.m. under nonstandard conditions.)
No additional tolerance permitted.
Diameter: Not over or under 76".
For S/N 28-7890001 through 28-8690056; 28-8690061; 28-8690062; and 2890001 through 2890205:
Static r.p.m. at maximum permissible throttle setting, not over 2340 r.p.m., not under 2240 r.p.m. at sea level, ISA conditions. (Reference aircraft Maintenance Manual for test procedure to determine approved static r.p.m. under nonstandard conditions.)
No additional tolerance permitted.
Diameter: Not over or under 76".

Propeller Spinner — Piper P/N 65805-00.
See NOTE 11.

Airspeed Limits

Never exceed	171 mph	(148 knots)	CAS
Maximum structural cruising	140 mph	(121 knots)	CAS
Maneuvering	124 mph	(108 knots)	CAS
Flaps Extended	115 mph	(100 knots)	CAS

Center of Gravity Range

Normal Category

(+82.0)	to	(+93.0)	at	2050 lb. or less
(+88.6)	to	(+93.0)	at	2550 lb.

Utility Category

(+82.0)	to	(+93.0)	at	2050 lb. or less
(+83.0)	to	(+93.0)	at	2130 lb.

Straight line variation between points given.

Empty Weight C. G. Range — None

Maximum Weight	Normal Category: Ramp - 2558 lb. * Takeoff - 2550 lb. Utility Category: Ramp - 2138 lb. * Takeoff - 2130 lb. * - Ramp weights for S/N 28-8090001 through 28-8690056; 28-8690061; 28-8690062; and 2890001 through 2890205 only.
No. of Seats	Normal Category: 4 (2 at +80.5, 2 at +118.1) Utility Category: 2 (2 at +80.5)
Maximum Baggage	200 lb. at (+142.8)
Fuel Capacity	50 gallons at (+95)(2 wing tanks) See NOTE 1 for data on system fuel.
Oil Capacity	8 quarts at (+27.5) (6 quarts usable) See NOTE 1 for data on system oil.

Control Surface Movements

Wing flaps	(±2°)	Up	0°	Down	40°
Ailerons	(±2°)	Up	25°	Down	12.5°
Rudder	(±2°)	Left	27°	Right	27°
Stabilator	(±1°)	Up	14°	Down	2°
Stabilator Tab	(±1°)	Up	3°	Down	12°

Nose Wheel Travel (±2°) Left 30° Right 30°

Manufacturer's Serial Numbers 28-7690001 through 28-8690056; 28-8690061; 28-8690062; and 2890001 through 2890205. The manufacturer is authorized to issue airworthiness certificates for airplane serial numbers 28-7690001 through 28-8690056; 28-8690061; 28-8690062; and 2890001 through 2890205 under the delegation option provisions of FAR 21. See NOTE 20.

XV - B.- Model PA-28-181 (Archer III), 4 PCLM (Normal Category), 2 PCLM (Utility Category), Approved August 30, 1994, for S/N 2890206 through 2890231, and 2843001 and up.

Engine	Lycoming O-360-A4M with carburetor settings 10-5193
Fuel	100/130 minimum grade aviation gasoline
Engine Limits	For all operations, 2700 r.p.m. (180 hp)
Propeller and Propeller Limits	Sensenich 76EM8S14-0-62 Static r.p.m. at maximum permissible throttle setting, not over 2340 r.p.m., not under 2240 r.p.m. at sea level, ISA conditions. (Reference aircraft Maintenance Manual for test procedure to determine approved static r.p.m. under nonstandard conditions.) No additional tolerance permitted. Diameter: Not over or under 76".
Propeller Spinner	Piper P/N 83349-02

Airspeed Limits

Never exceed	171 mph	(148 knots)	CAS
Maximum structural cruising	140 mph	(121 knots)	CAS
Maneuvering	124 mph	(108 knots)	CAS
Flaps Extended	115 mph	(100 knots)	CAS

Center of Gravity Range	Normal Category (+82.0) to (+93.0) at 2050 lb. or less (+88.6) to (+93.0) at 2550 lb. Utility Category (+82.0) to (+93.0) at 2050 lb. or less (+83.0) to (+93.0) at 2130 lb. Straight line variation between points given.
Empty Weight C. G. Range	None
Maximum Weight	Normal Category: Ramp - 2558 lb. Takeoff - 2550 lb. Utility Category: Ramp - 2138 lb. Takeoff - 2130 lb.
No. of Seats	Normal Category: 4 (2 at +80.5, 2 at +118.1) Utility Category: 2 (2 at +80.5)
Maximum Baggage	200 lb. at (+142.8)
Fuel Capacity	50 gallons at (+95)(2 wing tanks) See NOTE 1 for data on system fuel.
Oil Capacity	8 quarts at (+27.5) (6 quarts usable) See NOTE 1 for data on system oil.

Control Surface Movements						
Wing flaps	(±2°)	Up	0°	Down	40°
Ailerons	(±2°)	Up	25°	Down	12.5°
Rudder	(±2°)	Left	27°	Right	27°
Stabilator	(±1°)	Up	14°	Down	2°
Stabilator Tab	(±1°)	Up	3°	Down	12°

Nose Wheel Travel	(±2°)	Left	30°	Right	30°

Manufacturer's Serial Numbers	2890206 through 2890231, and 2843001 and up. The manufacturer is authorized to issue airworthiness certificates for airplane serial numbers 2890206 through 2890231, and 2843001 and up under the delegation option provisions of FAR 21.

XVI - A. - Model PA-28-161 (Warrior II), 4 PCLM (Normal Category), 2 PCLM (Utility Category), Approved November 2, 1976, for S/N 28-7716001 through 28-8216300, and 2841001 through 2841365 (Cadet only)

Engine	Lycoming O-320-D3G with carburetor setting 10-5135, 10-5009 or 10-5217, or Lycoming O-320-D2A with carburetor setting 10-5135 or 10-5217.
Fuel	100 octane minimum grade aviation gasoline
Engine Limits	For all operations, 2700 r.p.m. (160 hp)
Propeller and Propeller Limits	Sensenich 74DM6-0-60 Static r.p.m. at maximum permissible throttle setting not over 2430 r.p.m., not under 2330 r.p.m., at sea level, ISA conditions. (Reference aircraft Maintenance Manual for test procedure to determine approved static r.p.m. under nonstandard conditions.) No additional tolerance permitted. Diameter: Not over 74", not under 72".

Propeller and Propeller Limits	or Sensenich 74DM6-0-58 Static r.p.m. at maximum permissible throttle setting not over 2465 r.p.m., not under 2365 r.p.m., at sea level, ISA conditions. (Reference aircraft Maintenance Manual for test procedure to determine approved static r.p.m. under nonstandard conditions.) No additional tolerance permitted. Diameter: Not over 74", not under 72".
Propeller Spinner	Piper P/N 35323 or P/N 36850. See NOTE 11.
Airspeed Limits	Never exceed 160 KIAS Maximum structural cruising 126 KIAS Maneuvering at 2325 lb. gross weight 111 KIAS Maneuvering at 1531 lb. gross weight 88 KIAS Flaps Extended 103 KIAS
Center of Gravity Range	Normal Category (+83.0) to (+93.0) at 1950 lb. or less (+87.0) to (+93.0) at 2325 lb. See NOTE 27. Utility Category (+83.0) to (+93.0) at 1950 lb. or less (+83.8) to (+93.0) at 2020 lb. Straight line variation between points given.
Empty Weight C.G. Range	None
Maximum Weight	Normal Category: 2325 lb. Utility Category: 2020 lb. Ramp: 2332 lb. (Cadet only) See NOTE 27.
No. of Seats	Normal Category: 4 (2 at +80.5, 2 at +118.1) Utility Category: 2 (+2 at +80.5)
Maximum Baggage	Eligible Normal Category only: 200 lb. at(+142.8) 50 lb. (Cadet only)
Fuel Capacity	50 gallons at (+95) (2 wing tanks) See NOTE 1 for data on system fuel.
Oil Capacity	8 quarts at (+27.5) (6 quarts usable) See NOTE 1 for data system oil.

Control Surface Movements

Wing flaps	(±2°)	Up	0°	Down	40°
Ailerons	(±2°)	Up	25°	Down	12.5°
Rudder	(±2°)	Left	27°	Right	27°
Stabilator	(±1°)	Up	14°	Down	2°
Stabilator Tab	(±1°)	Up	3°	Down	12°

Nose Wheel Travel (±1°) Left 30° Right 30°

Manufacturer's Serial Numbers 28-7716001 through 28-8216300, and 2841001 through 2841365 (Cadet only). The manufacturer is authorized to issue airworthiness certificates for airplane serial numbers 28-7716001 through 28-8216300, and 2841001 through 2841365 under the delegation option provisions of FAR 21. See NOTE 20.

XVI - B. Model PA-28-161 (Warrior II), 4 PCLM (Normal Category), 2 PCLM (Utility Category), Approved July 1, 1982, for S/N 28-8316001 through 28-8616057, and 2816001 through 2816109.

Engine	Lycoming O-320-D3G with carburetor setting 10-5135, 10-5009 or 10-5217, or Lycoming O-320-D2A with carburetor setting 10-5135 or 10-5217.
Fuel	100 octane minimum grade aviation gasoline
Engine Limits	For all operations, 2700 r.p.m. (160 hp)
Propeller and Propeller Limits	Sensenich 74DM6-0-60 Static r.p.m. at maximum permissible throttle setting not over 2430 r.p.m., not under 2330 r.p.m. at sea level, ISA conditions. (Reference aircraft Maintenance Manual for test procedure to determine approved static r.p.m. under nonstandard conditions.) No additional tolerance permitted. Diameter: Not over 74", not under 72". or Sensenich 74DM6-0-58 Static r.p.m. at maximum permissible throttle setting not over 2465 r.p.m., not under 2365 r.p.m., at sea level, ISA conditions. (Reference aircraft Maintenance Manual for test procedure to determine approved static r.p.m. under nonstandard conditions.) No additional tolerance permitted. Diameter: Not over 74", not under 72".
Propeller Spinner	Piper P/N 36850. See NOTE 11.
Airspeed Limits	Never exceed 160 KIAS Maximum structural cruising 126 KIAS Maneuvering at 2440 lb. gross weight 111 KIAS See NOTE 26. Maneuvering at 1531 lb. gross weight 88 KIAS Flaps Extended 103 KIAS
Center of Gravity Range	Normal Category (+83.0) to (+93.0) at 1950 lb. or less (+88.3) to (+93.0) at 2440 lb. See NOTE 26. Utility Category (+83.0) to (+93.0) at 1950 lb. or less (+83.8) to (+93.0) at 2020 lb. Straight line variation between points given
Empty Weight C.G. Range	None
Maximum Weight	Normal Category: Ramp - 2447 lb. Takeoff - 2440 lb. See NOTE 26. Utility Category: Ramp - 2027 lb. Takeoff - 2020 lb.
No. of Seats	Normal Category: 4 (2 at +80.5, 2 at +118.1) Utility Category: 2 (2 at +80.5)
Maximum Baggage	Eligible Normal Category only: 200 lb. at (+142.8)
Fuel Capacity	50 gallons at (+95) (2 wing tanks) See NOTE 1 for data on system fuel.

Oil Capacity	8 quarts at (+27.5) (6 quarts usable) See NOTE 1 for data on system oil.

Control Surface Movements						
Wing flaps	(±2°)	Up	0°	Down	40°
Ailerons	(±2°)	Up	25°	Down	12.5°
Rudder	(±2°)	Left	27°	Right	27°
Stabilator	(±1°)	Up	14°	Down	2°
Stabilator Tab	(±1°)	Up	3°	Down	12°
Nose Wheel Travel	(±1°)	Left	30°	Right	30°

Manufacturer's Serial Nos. 28-8316001 through 28-8616057, and 2816001 through 2816109. The manufacturer is authorized to issue airworthiness certificates for airplane serial numbers 28-8316001 through 28-8616057, and 2816001 through 2816109 under the delegation option provisions of FAR 21. See NOTE 20.

XVI - C. Model PA-28-161 (Warrior III), 4 PCLM (Normal Category), 2 PCLM (Utility Category), Approved July 1, 1994, for S/N 2816110 through 2816119, and 2842001 and up.

Engine: Lycoming O-320-D3G with carburetor setting 10-5135, 10-5009 or 10-5217

Fuel: 100 octane minimum grade aviation gasoline

Engine Limits: For all operations, 2700 r.p.m. (160 hp)

Propeller and Propeller Limits: Sensenich 74DM6-0-60
Static r.p.m. at maximum permissible throttle setting not over 2430 r.p.m., not under 2330 r.p.m., at sea level, ISA conditions. (Reference aircraft Maintenance Manual for test procedure to determine approved static r.p.m. under nonstandard conditions.)
No additional tolerance permitted.
Diameter: Not over 74", not under 72".

Propeller Spinner: Piper P/N 36850.
See NOTE 11.

Airspeed Limits:

Never exceed	160 KIAS	
Maximum structural cruising	126 KIAS	
Maneuvering at 2440 lb. gross weight	111 KIAS	See NOTE 26.
Maneuvering at 1531 lb. gross weight	88 KIAS	
Flaps Extended	103 KIAS	

Center of Gravity Range:

Normal Category
(+83.0) to (+93.0) at 1950 lb. or less
(+88.3) to (+93.0) at 2440 lb.
See NOTE 26.

Utility Category
(+83.0) to (+93.0) at 1950 lb. or less
(+83.8) to (+93.0) at 2020 lb.
Straight line variation between points given

Empty Weight C.G. Range: None

Maximum Weight:

Normal Category: Ramp - 2447 lb.
Takeoff - 2440 lb. See NOTE 26.

Utility Category: Ramp - 2027 lb.
Takeoff - 2020 lb.

No. of Seats	Normal Category: 4 (2 at +80.5, 2 at +118.1) Utility Category: 2 (2 at +80.5)
Maximum Baggage	Eligible Normal Category only: 200 lb. at (+142.8)
Fuel Capacity	50 gallons at (+95) (2 wing tanks) See NOTE 1 for data on system fuel.
Oil Capacity	8 quarts at (+27.5) (6 quarts usable) See NOTE 1 for data on system oil.

Control Surface Movements

Wing flaps	(±2°)	Up	0°	Down	40°
Ailerons	(±2°)	Up	25°	Down	12.5°
Rudder	(±2°)	Left	27°	Right	27°
Stabilator	(±1°)	Up	14°	Down	2°
Stabilator Tab	(±1°)	Up	3°	Down	12°

Nose Wheel Travel (±1°) Left 30° Right 30°

Manufacturer's Serial Nos. 2816110 through 2816119, and 2842001 and up. The manufacturer is authorized to issue airworthiness certificates for airplane serial numbers 2816110 through 2816119, and 2842001 and up under the delegation option provisions of FAR 21.

XVII - Model PA-28R-201 (Arrow III), 4 PCLM (Normal Category), Approved November 2, 1976, for S/N 28R-7737002 through 28R-7837317; 2837001 through 2837061; and 2844001 and up.

Engine	Lycoming IO-360-C1C6
Injector	Bendix Type RSA-5AD1, Part List Number 2524450
Fuel	100/130 minimum grade aviation gasoline
Engine Limits	For all operations, 2700 r.p.m. (200 hp)
Propeller and Propeller Limits	McCauley Constant Speed Hub Model B2D34C213, Blade Model 90 DHA-16 Pitch: High 27.5° ± .5°, Low 12.5° ± .2° at 30" station. Diameter: Not over 74", not under 73". Governor Assembly: Hartzell Model F-2-7 ( ) Avoid continuous operation between 1500 and 1950 r.p.m. below 15" manifold pressure. or Hartzell Constant Speed Hub Model HC-C2YK-1( )F, Blade Model F7666A-2R Pitch: High 29.0° ± 2°, Low 14.0° ± .2° at 30" station. Diameter: Not over 74", not under 72". Governor Assembly: F-2-7 ( )
Propeller Spinner	For McCauley propeller: Piper P/N 35838-2 For Hartzell propeller: Piper P/N 99374 See NOTE 11.

Airspeed Limits

Never exceed	183 KIAS
Maximum structural cruising	146 KIAS
Maneuvering	118 KIAS
Flaps Extended	103 KIAS
Maximum Gear Extension	129 KIAS
Maximum Gear Retraction	107 KIAS

Center of Gravity Range	(+82.0) to (+91.5) at 2375 lb. or less (+88.9) to (+91.5) at 2750 lb. Straight line variation between points given. Moment due to retraction of gear (+819 in-lb.)
Empty Weight C.G. Range	None
Maximum Weight	2750 lb.
No. of Seats	4 (2 at +80.5, 2 at +118.1)
Maximum Baggage	200 lb. at (+142.8)
Fuel Capacity	77 gallons at (+95) (2 wing tanks) See NOTE 1 for data on system fuel.
Oil Capacity	8 quarts at (+24.5) (6 quarts usable) See NOTE 1 for data on system oil.

Control Surface Movements						
Wing flaps	(±2°)	Up	0°	Down	40°
Ailerons	(±2°)	Up	25°	Down	12.5°
Rudder	(±1°)	Left	28°	Right	28°
Stabilator	(±1°)	Up	16°	Down	2°
Stabilator Tab	(±1°)	Up	3°	Down	12°
Nose Wheel Travel	(±2°)	Left	30°	Right	30°

Manufacturer's Serial Numbers	28R-7737002 through 28-7837317; 2837001 through 2837061; and 2844001 and up. The manufacturer is authorized to issue airworthiness certificates for airplanes serial numbers 28R-7737002 through 28-7837317; 2837001 through 2837061; and 2844001 and up under the delegation option provisions of FAR 21. See NOTE 20.

XVIII - Model PA-28R-201T (Turbo Arrow III), 4 PCLM (Normal Category), Approved November 2, 1976, for S/N 28R-7703001 through 28R-7803374, and 2803001 through 2803012.

Engine	Continental TSIO-360-F or TSIO-360-FB
Fuel	100/130 minimum grade aviation gasoline
Engine Limits	For all operations, 2575 r.p.m. at 41" Hg. manifold pressure (200 hp)
Propeller and Propeller Limits	1 Hartzell Hub Model BHC-C2YF-1BF, Blade Model F8459A-8R Pitch Setting at 30" Station: High: 29° ± 1.0°, Low: 14.4° ± 0.2°. Diameter: Not over 76", not under 75". Governor: Hartzell E-5 or Woodward G210681 Avoid continuous operation between 2000 and 2200 r.p.m. with engine manifold pressure above 32" Hg. Avoid continuous ground operation in cross and tail winds of over 10 knots between 1700 and 2100 r.p.m.
Propeller Spinner	Hartzell P/N C3568 Spinner Assembly. See NOTE 11.

Airspeed Limits	Never exceed	183 KIAS				
	Maximum structural cruising	146 KIAS				
	Maneuvering	119 KIAS				
	Flaps Extended	103 KIAS				
	Maximum Gear Retraction	107 KIAS				
	Maximum Gear Extension	129 KIAS				
	Maximum Gear Extended	129 KIAS				

Center of Gravity Range

(+86.0) to (+90.0) at 2900 lb.
(+78.0) to (+90.0) at 2240 lb. or less
Straight line variation between points given.
Moment due to retraction of landing gear (+819 in-lb.)

Empty Weight C. G. Range: None

Maximum Weight: Ramp: 2912 lb.
Takeoff: 2900 lb.

No. of Seats: 4 (2 at +80.5, 2 at +118.1)

Maximum Baggage: 200 lb. at (+142.8)

Fuel Capacity: 77 gallons at (+95)(2 wing tanks)
See NOTE 1 for data on system fuel.

Oil Capacity: 8 quarts at (+13.5) (5 quarts usable)
See NOTE 1 for data on system oil.

Maximum Operating Altitude: 20,000 feet

Control Surface Movements						
Wing flaps	(±2°)	Up	0°	Down	40°
Ailerons	(±2°)	Up	25°	Down	12.5°
Rudder	(±1°)	Left	28°	Right	28°
Stabilator	(±1°)	Up	16°	Down	2°
Stabilator Tab	(±1°)	Up	3°	Down	12°

Nose Wheel Travel: (±2°) Left 30° Right 30°

Manufacturer's Serial Numbers: 28R-7703001 through 28R-7803374, and 2803001 through 2803012. The manufacturer is authorized to issue airworthiness certificates for airplanes serial numbers 28R-7703001 through 28R-7803374, and 2803001 through 2803012 under the delegation option provisions of FAR 21. See NOTE 20.

XIX - Model PA-28-236 (Dakota), 4 PCLM (Normal Category), Approved June 1, 1978, for S/N 28-7911001 through 28-8611008; 2811001 through 2811050; and 2845001 and up.

Engine: Lycoming O-540-J3A5D with carburetor setting 10-5054

Fuel: 100/130 minimum grade aviation gasoline

Engine Limits: For all operations, 2400 r.p.m. (235 hp)

Propeller and Propeller Limits: Hartzell HC-F2YR-1( )F/F 8468A-4R
Pitch: High 32° ±2°, Low 16.25° ± ¼°.
Diameter: Not over 80", not under 78".
Governor Assembly: Hartzell F-4-21( )

Propeller Spinner	Hartzell P/N C3568 Spinner Assembly. See NOTE 11.
Airspeed Limits	Never exceed 197 mph (171 knots) CAS Maximum structural cruising 156 mph (135 knots) CAS Maneuvering at 3000 lb. 140 mph (122 knots) CAS Maneuvering at 1761 lb. 108 mph (94 knots) CAS Flaps Extended 115 mph (100 knots) CAS
Center of Gravity Range	(+79.8) to (+92.0) at 1900 lb. or less (+82.5) to (+92.0) at 2500 lb. (+88.5) to (+92.0) at 3000 lb. Straight line variation between points given.
Empty Weight C. G. Range	None
Maximum Weight	3000 lb.
Number of Seats	4 (2 at +80.5, 2 at +118.1)
Maximum Baggage	200 lb. at (+142.8)
Fuel Capacity	77 gallons at (+95)(2 wing tanks) See NOTE 1 for data on system fuel.
Oil Capacity	12 quarts at (+29.1) (9 1/2 quarts usable) See NOTE 1 for data on system oil.
Control Surface Movements	Wing flaps (±2°) Up 0° Down 40° Ailerons (±2°) Up 25° Down 12.5° Rudder (±1°) Left 28° Right 28° Stabilator (±1°) Up 16° Down 2° Stabilator Tab (±1°) Up 3° Down 12°
Nose Wheel Travel	(±1°) Left 30° Right 30°
Manufacturer's Serial Numbers	28-7911001 through 28-8611008; 2811001 through 2811050; and 2845001 and up. The manufacturer is authorized to issue airworthiness certificates for airplane serial numbers 28-7911001 through 28-8611008; 2811001 through 2811050; and 2845001 and up under the delegation option provisions of FAR 21. See NOTE 20.

XX - A. Model PA-28RT-201 (Arrow IV), 4 PCLM (Normal Category), Approved November 13, 1978, for S/N 28R-7918001 through 28R-7918267.

Engine	Lycoming IO-360-C1C6
Injector	Bendix Type RSA-5AD1, Part List Number 2524450
Fuel	100/130 minimum grade aviation gasoline
Engine Limits	For all operations, 2700 r.p.m. (200 hp)
Propeller and Propeller Limits	McCauley Constant Speed Hub Model B2D34C213, Blade Model 90 DHA-16 Pitch: High 27.5° ±.5°, Low 12.5° ±.2° at 30" station. Diameter: Not over 74", not under 73". Governor Assembly: Hartzell Model F-2-7 ( ) Avoid continuous operation between 1500 and 1950 r.p.m. below 15" manifold pressure.

Propeller and Propeller Limits	or Hartzell Constant Speed Hub Model HC-C2YK-1( )F, Blade Model F7666A-2R Pitch: High 29.0° ± 2°, Low 14.0° ± .2° at 30" station. Diameter: Not over 74", not under 72". Governor Assembly: Hartzell Model F-2-7( )
Propeller Spinner	For the McCauley propeller: Piper P/N 35828-2 For the Hartzell propeller: Piper P/N 99374 See NOTE 11.
Airspeed Limits	Never exceed 190 KIAS Maximum structural cruising 149 KIAS Flaps extended 108 KIAS Maximum gear extension 130 KIAS Maximum gear retraction 109 KIAS Maximum gear extended 130 KIAS Maneuvering at 2750 lb. 121 KIAS Maneuvering at 1863 lb. 96 KIAS
Center of Gravity Range	(+85.5) to (+93.0) at 2400 lb. or less (+90.0) to (+93.0) at 2750 lb. Straight line variation between points given. Moment due to retraction of gear (+819 in-lb.)
Empty Weight C. G. Range	None
Maximum Weight	2750 lb.
No. of Seats	4 (2 at +80.5, 2 at +118.1)
Maximum Baggage	200 lb. at (+142.8)
Fuel Capacity	77 gallons at (+95) (2 wing tanks) See NOTE 1 for data on system fuel.
Oil Capacity	8 quarts at (+24.5) 6 quarts usable See NOTE 1 for data on system oil.

Control Surface Movements

Wing flaps	(±2°)	Up	0°	Down	40°
Ailerons	(±2°)	Up	25°	Down	12.5°
Rudder	(±1°)	Left	33°	Right	33°
Stabilator	(±1°)	Up	14°	Down	10°
Stabilator Tab		Up	2.5° (±1°)	Down	10° (±.5°)

Nose Wheel Travel (±2°) Left 30° Right 30°

Manufacturer's Serial Numbers 28R-7918001 through 28R-7918267. The manufacturer is authorized to issue airworthiness certificates for airplane serial numbers 28R-7918001 through 28R-7918267 under the delegation option provisions of FAR 21.

XX - B. Model PA-28RT-201 (Arrow IV), 4 PCLM (Normal Category), Approved November 13, 1978, for S/N 28R-8018001 through 28R-8218026.

Engine	Lycoming IO-360-C1C6
Injector	Bendix Type RSA-5AD1, Part List Number 2524450
Fuel	100/130 minimum grade aviation gasoline

Engine Limits	For 5-minute takeoff, 2700 r.p.m. (200 hp) For maximum continuous operation, 2650 r.p.m. (196 hp)
Propeller and Propeller Limits	McCauley Constant Speed Hub Model 2D34C215, Blade Model 90 DJA-14E Pitch: High 27.5° ± .5°; Low 12.5° ± .2° at 30" station. Diameter: Not over 76", not under 75". Governor Assembly: Hartzell Model F-2-7 ( ) Avoid continuous operation between 1400 and 1750 r.p.m. below 15" manifold pressure.
Propeller Spinner	Piper P/N 35828-2. See NOTE 11.

Airspeed Limits

Never exceed	190 KIAS
Maximum structural cruising	149 KIAS
Flaps Extended	108 KIAS
Maximum gear extension	130 KIAS
Maximum gear retraction	109 KIAS
Maximum gear extended	130 KIAS
Maneuvering at 2750 lb. gross weight	121 KIAS
Maneuvering at 1863 lb. gross weight	96 KIAS

Center of Gravity Range	(+85.5) to (+93.0) at 2400 lb. or less (+90.0) to (+93.0) at 2750 lb. Straight line variation between points given. Moment due to retraction of gear (+819 in-lb.)
Empty Weight C. G. Range	None
Maximum Weight	2750 lb.
Number of Seats	4 (2 at +80.5, 2 at +118.1)
Maximum Baggage	200 lb. at (+142.8)
Fuel Capacity	77 gallons at (+95) (2 wing tanks) See NOTE 1 for data on system fuel.
Oil Capacity	8 quarts at (+24.5) (6 quarts usable) See NOTE 1 for data on system oil.

Control Surface Movements

Wing flaps	(±2°)	Up	0°	Down	40°
Ailerons	(±2°)	Up	25°	Down	12.5°
Rudder	(±1°)	Left	33°	Right	33°
Stabilator	(±1°)	Up	14°	Down	10°
Stabilator Tab		Up	2.5° (±1°)	Down	10° (±.5°)

Nose Wheel Travel	(±2°)	Left	30°	Right	30°

Manufacturer's Serial Numbers	28R-8018001 through 28R-8218026. The manufacturer is authorized to issue airworthiness certificates for airplane serial numbers 28R-8018001 through 28R-8218026 under the delegation option provisions of FAR 21. See NOTE 20.

XXI - Model PA-28RT-201T (Turbo Arrow IV), 4 PCLM (Normal Category), Approved November 13, 1978, for S/N 28R-7931001 through 28R-8631005, and 2831001 through 2831038.

Engine	Continental TSIO-360-FB
Fuel	100/130 minimum grade aviation gasoline

Engine Limits	For all operations, 2575 r.p.m., 41" Hg. manifold pressure (200 hp)
Propeller and Propeller Limits	1 Hartzell Hub Model BHC-C2YF-1( )F, Blade Model F8459A-8R Pitch: High 29° ± 1.0°, Low 14.4° ± .2° at 30" station. Diameter: Not over 76", not under 75". Governor: Hartzell E-5 or Woodward G210681 Avoid continuous operation between 2000 and 2200 r.p.m. with engine manifold pressure above 32" Hg. Avoid continuous ground operation in cross and tail winds of over 10 knots between 1700 and 2100 r.p.m. or 1 Hartzell Hub Model PHC-C3YF-1( )F, Blade Model F7663-2R Pitch: High 33° ± 1°, Low 13.2° ± .2°. Diameter: Not over 76", not under 72". Governor: Hartzell E-5, Woodward G210681 or G210776
Propeller Spinner	For the Hartzell Hub Model BHC-C2YF-1( )F: Hartzell P/N C3568 Spinner Assembly For the Hartzell Hub Model PHC-C3YF-1( )F: Piper PS50077-80 Spinner Assembly (Hartzell C3570) See NOTE 11.

Airspeed Limits

Never exceed	193 KIAS
Maximum structural cruising	152 KIAS
Maneuvering at 2900 lb.	124 KIAS
Maneuvering at 1893 lb.	97 KIAS
Flaps Extended	108 KIAS
Maximum Gear Retraction	111 KIAS
Maximum Gear Extension	133 KIAS
Maximum Gear Extended	133 KIAS

Center of Gravity Range	(+89.0) to (+93.0) at 2900 lb. (+85.0) to (+93.0) at 2240 lb. or less Straight line variation between points given. Moment due to retraction of landing gear (+819 in-lb.)
Empty Weight C. G. Range	None
Maximum Weight	Ramp: 2912 lb. Takeoff: 2900 lb.
No. of Seats	4 (2 at +80.5, 2 at +118.1)
Maximum Baggage	200 lb. at (+142.8)
Fuel Capacity	77 gallons at (+95)(2 wing tanks) See NOTE 1 for data on system fuel.
Oil Capacity	8 quarts at (+13.5) (5 quarts usable) See NOTE 1 for data on system oil.
Maximum Operation Altitude	20,000 feet

Control Surface Movements

Wing flaps	(±2°)	Up	0°	Down	40°
Ailerons	(±2°)	Up	25°	Down	12.5°
Rudder	(±1°)	Left	33°	Right	33°
Stabilator	(±1°)	Up	14°	Down	10°
Stabilator Tab		Up	2.5° (±1°)	Down	10° (±.5°)

Nose Wheel Travel (±2°) Left 30° Right 30°

Manufacturer's Serial Numbers	28R-7931001 through 28R-8631005, and 2831001 through 2831013. The manufacturer is authorized to issue airworthiness certificates for airplane serial numbers 28R-7931001 through 28R-8631005, and 2831001 through 2831038 under the delegation option provisions of FAR 21. See NOTE 20.

XXII - Model PA-28-201T (Turbo Dakota), 4 PCLM (Normal Category), Approved December 14, 1978, for S/N 28-7921001 through 28-7921095.

Engine	Continental TSIO-360-FB
Fuel	100/130 minimum grade aviation gasoline
Engine Limits	For all operations, 2575 r.p.m., 41" Hg. manifold pressure (200 hp)
Propeller and Propeller Limits	1 Hartzell Hub Model BHC-C2YF-1( )F, Blade Model F8459A-8R Pitch: High 29° ± 1.0°, Low 14.4° ± .2° at 30" station. Diameter: Not over 76", not under 75". Governor: Hartzell E-5 or Woodward G210681 Avoid continuous operation between 2000 and 2200 r.p.m. with engine manifold pressure above 32" Hg. Avoid continuous ground operation in cross and tail winds of over 10 knots between 1700 and 2100 r.p.m.
Propeller Spinner	Hartzell P/N C3568 Spinner Assembly. See NOTE 11.
Airspeed Limits	Never exceed 169 KIAS Maximum structural cruising 140 KIAS Maneuvering at 2900 lb. 122 KIAS Maneuvering at 1841 lb. 96 KIAS Flaps Extended 102 KIAS
Center of Gravity Range	(+86.0) to (+90.0) at 2900 lb. (+78.0) to (+90.0) at 2240 lb. or less Straight line variation between points given.
Empty Weight C. G. Range	None
Maximum Weight	2900 lb.
No. of Seats	4 (2 at +80.5, 2 at +118.1)
Maximum Baggage	200 lb. at (+142.8)
Fuel Capacity	77 gallons at (+95)(2 wing tanks) See NOTE 1 for data on system fuel.
Oil Capacity	8 quarts at (+13.5) (5 quarts usable) See NOTE 1 for data on system oil.
Maximum Operation Altitude	20,000 feet

Control Surface Movements						
Wing flaps	(±2°)	Up	0°	Down	40°
Ailerons	(±2°)	Up	25°	Down	12.5°
Rudder	(±2°)	Left	27°	Right	27°
Stabilator	(±1°)	Up	16°	Down	2°
Stabilator Tab	(±1°)	Up	3°	Down	12°
Nose Wheel Travel	(±1°)	Left	30°	Right	30°

Manufacturer's Serial Numbers — 28-7921001 through 28-7921095. The manufacturer is authorized to issue airworthiness certificates for airplane serial numbers 28-7921001 through 28-7921095 under the delegation option provisions of FAR 21. See NOTE 20.

DATA PERTINENT TO ALL MODELS

Datum — 78.4" forward of wing leading edge (straight wing only).
78.4" forward of inboard intersection of straight and tapered sections (semi-tapered wings).

Leveling Means — Two screws left side fuselage below window.

Certification Basis — Type Certificate No. 2A13 issued October 31, 1960.
Date of Application for Type Certificate, February 14, 1958.

Delegation Option Authorization granted per FAR 21, Subpart J, July 17, 1968.

PA-28-140 and PA-28-151: CAR 3 effective May 15, 1956, including Amendments 3-1, 3-2, and 3-4; paragraphs 3.304 and 3.705 of Amendment 3-7 effective May 3, 1962; FAR 23.955 and 23.959 as amended by Amendment 23-7 effective September 14, 1969; and FAR 23.1327 and 23.1547 as amended by Amendment 23-20 effective September 1, 1977.

PA-28-150, PA-28-160, PA-28-180, PA-28-235, PA-28S-160, PA-28S-180, PA-28R-180, and PA-28R-200: CAR 3 effective May 15, 1956, including Amendments 3-1 and 3-2; paragraphs 3.304 and 3.705 of Amendment 3-7 effective May 3, 1962; FAR 23.955 and 23.959 as amended by Amendment 23-7 effective September 14, 1969; and FAR 23.1327 and 23.1547 as amended by Amendment 23-20 effective September 1, 1977.

PA-28-161: CAR 3 effective May 15, 1956, including Amendments 3-1 and 3-2; paragraph 3.387(d) of Amendment 3-4; paragraphs 3.304 and 3.705 of Amendment 3-7 effective May 3, 1962; FAR 23.955 and 23.959 as amended by Amendment 23-7 effective September 14, 1969; FAR 23.1557(c)(1) as amended by Amendment 23-18 effective May 2, 1977; FAR 23.1327 and 23.1547 as amended by Amendment 23-20 effective September 1, 1977; and FAR 36 effective December 1, 1969, through Amendment 36-4.

PA-28-181: CAR 3 effective May 15, 1956, including Amendments 3-2 and 3-4; paragraphs 3.304 and 3.705 of Amendment 3-7 effective May 3, 1962; FAR 23.207, 23.221, 23.955 and 23.959 as amended by Amendment 23-7 effective September 14, 1969; FAR 23.1557(c)(1) as amended by Amendment 23-18 effective May 2, 1977; and FAR 23.1327 and 23.1547 as amended by Amendment 23-20 effective September 1, 1977. FAR 36, Appendix G, Amendment 36-16 for the PA-28-181 (Archer III), S/N 2890206 through 2890231, and 2843001 and up.

PA-28R-201: CAR 3 effective May 15, 1956, including Amendments 3-1 and 3-2; paragraphs 3.304 and 3.705 of Amendment 3-7 effective May 3, 1962; FAR 23.965 of FAR 23 effective February 1, 1965; FAR 23.221, 23.955, 23.959, and 23.1091 as amended by Amendment 23-7 effective September 14, 1969; FAR 23.967(e)(2) as amended by Amendment 23-14 effective December 20, 1973; FAR 23.1093 as amended by Amendment 23-15 effective October 31, 1974; FAR 23.1327 and 23.1547 as amended by Amendment 23-20 effective September 1, 1977; and FAR 36 effective December 1, 1969, through Amendment 36-4 (no acoustical change).

PA-28R-201T: CAR 3 effective May 15, 1956, through Amendment 3-2 including paragraphs 3.304 and 3.705 of Amendment 3-7 effective May 3, 1962; FAR 23.965 of FAR 23 effective February 1, 1965; FAR 23.221, 23.901, 23.909, 23.955, 23.959, 23.1041, 23.1043, 23.1047, 23.1143, and 23.1527 as amended by Amendment 23-7 effective September 14, 1969; FAR 23.1441 as amended by Amendment 23-9 effective June 17, 1970; FAR 23.967(e)(2) as amended by Amendment 23-14 effective December 20, 1973; FAR 23.1305 as amended by Amendment 23-15 effective October 31, 1974; FAR 23.1327 and 23.1547 as amended by Amendment 23-20 effective September 1, 1977; and FAR 36 effective December 1, 1969, through Amendment 36-4.

PA-28-236: CAR 3 effective May 15, 1956, through Amendment 3-2; paragraphs 3.304 and 3.705 of Amendment 3-7 effective May 3, 1962; FAR 23.221, 23.955, 23.959 and 23.1091 as amended by Amendment 23-7 effective September 14, 1969; FAR 23.1093 as amended by Amendment 23-17 effective February 1, 1977; FAR 23.1557(c)(1) as amended by Amendment 23-18 effective May 2, 1977; FAR 23.1327 and 23.1547 as amended by Amendment 23-20 effective September 1, 1977; FAR 23.1581(b)(2) as amended by Amendment 23-21 effective March 1, 1978; and applicable portions of FAR 36, as amended by Amendment 36-9 effective April 3, 1978.

PA-28RT-201: CAR 3, effective May 15, 1956, through Amendment 3-2; paragraphs 3.304 and 3.705 of Amendment 3-7 effective May 3, 1962; FAR 23.965 of FAR 23 effective February 1, 1965; FAR 23.207, 23.221, 23.955, 23.959, and 23.1091 as amended by Amendment 23-7 effective September 14, 1969; FAR 23.201, 23.203, 23.427(c), and 23.967(e)(2) as amended by Amendment 23-14 effective December 20, 1973; FAR 23.1093 as amended by Amendment 23-15 effective October 31, 1974; FAR 23.1557(c)(1) as amended by Amendment 23-18 effective May 2, 1977; FAR 23.1327 and 23.1547 as amended by Amendment 23-20 effective September 1, 1977; FAR 23.1581(b)(2) as amended by Amendment 23-21 effective March 1, 1978; and applicable portions of FAR 36 as amended by Amendment 36-10 effective July 31, 1978.

PA-28RT-201T: CAR 3 effective May 15, 1956, through Amendment 3-2; paragraphs 3.304 and 3.705 of Amendment 3-7 effective May 3, 1962; FAR 23.207, 23.221, 23.901, 23.909, 23.955, 23.959, 23.1041, 23.1043, 23.1047, 23.1091, 23.1143, and 23.1527 as amended by Amendment 23-7 effective September 14, 1969; FAR 23.201, 23.203, 23.427(c), and 23.967(e)(2) as amended by Amendment 23-14 effective December 20, 1973; FAR 23.1093 and 23.1305 as amended by Amendment 23-15 effective October 31, 1974; FAR 23.1557(c)(1) as amended by Amendment 23-18 effective May 2, 1977; FAR 23.1327 and 23.1547 as amended by Amendment 23-20 effective September 1, 1977; FAR 23.1581(b)(2) as amended by Amendment 23-21 effective March 1, 1978; and applicable portions of FAR 36 as amended by Amendment 36-10 effective July 31, 1978.
Compliance with FAR 23.1441 as amended by Amendment 23-9 effective June 17, 1970, will be established with optional oxygen equipment.

PA-28-201T: CAR 3 effective May 15, 1956, through Amendment 3-2; paragraphs 3.304 and 3.705 of Amendment 3-7 effective May 3, 1962; FAR 23.965 of FAR 23 effective February 1, 1965; FAR 23.207, 23.221, 23.901, 23.909, 23.955, 23.959, 23.1041, 23.1043, 23.1047, 23.1091, and 23.1527 as amended by Amendment 23-7 effective September 14, 1969; FAR 23.201 and 23.203 as amended by Amendment 23-14 effective December 20, 1973; FAR 23.1305 as amended by Amendment 23-15 effective October 31, 1974; FAR 23.1093 and 23.1143 as amended by Amendment 23-17 effective February 1, 1977; FAR 23.1557(c)(1) as amended by Amendment 23-18 effective May 2, 1977; FAR 23.1327 and 23.1547 as amended by Amendment 23-20 effective September 1, 1977; FAR 23.1581(b)(2) as amended by Amendment 23-21 effective March 1, 1978; and applicable portions of FAR 36 as amended by Amendment 36-10 effective July 31, 1978.
Compliance with FAR 23.1441 as amended by Amendment 23-9 effective June 17, 1970, will be established with optional oxygen equipment.

Equivalent Safety Finding: CAR 3.757 for Models PA-28-161, PA-28R-201, PA-28R-201T, PA-28-236, PA-28RT-201, PA-28RT-201T, and PA-28-201T only.

Production Basis

Production Certificate No. 206 issued and the manufacturer authorized to issue airworthiness certificates under the delegation option provisions of FAR 21.

Equipment

The basic required equipment as prescribed in the applicable airworthiness regulation (see Certification Basis) must be installed in the aircraft for certification.
In addition, the following documents are required:

MODEL	AFM/POH	REPORT NO.	APPROVED	SERIAL EFFECTIVITY
PA-28-140	AFM	VB-160	2/14/64	28-20001 through 28-26946, and 28-7125001 through 28-7125641
	AFM	VB-339	7/21/71	28-7225001 through 28-7325674
	AFM	VB-557	5/14/73	28-7425001 through 28-7625275
	POH	VB-770	6/16/76	28-7725001 through 28-7725290
PA-28-150	AFM	VB-166	6/2/61	28-1 through 28-4377
PA-28-151	AFM	VB-573	7/25/73	28-7415001 through 28-7615435
	POH	VB-780	6/18/80	28-7715001 through 28-7715314
PA-28-160	AFM	VB-168	10/25/60	28-1 through 28-4377, and 28-1760A
PA-28S-160	AFM	VB-177	2/25/63	28-1 through 28-1760, and 28-1760A
PA-28-161	POH	VB-880	12/16/76	28-7716001 through 28-8216300
	POH	VB-1180	7/1/82	28-8316001 through 28-8616057, and 2816001 through 2816119
	POH	VB-1610	7/12/95	2842001 and up
	POH Supp.	VB-1546	6/30/92	28-8316001 through 28-8616057, and 2816001 through 2816119 (See NOTE 28)
	POH	VB-1360	9/9/88	2841001 through 2841365
	POH Supp.	VB-1545	5/29/92	2841001 through 2841365 (See NOTE 28)
	POH	VB-1565	7/1/94	2816110 through 2816119
PA-28-180	AFM	VB-163	8/3/62	28-671 through 28-5600
	AFM	VB-210	4/22/69	28-5601 through 28-5859, and 28-7105001 through 28-7205091
	AFM	VB-355	9/1/71	28-7205092 through 28-7205318
	AFM	VB-437	5/22/72	28-7305001 through 28-7305601 and 28-E13
	AFM	VB-558	5/14/73	28-7405001 through 28-7505260
PA-28S-180	AFM	VB-179	5/10/63	28-671 through 28-5859, and 28-7105001 through 28-7105234
PA-28-181	POH	VB-760	8/15/75	28-7690001 through 28-7690467
	POH	VB-790	6/18/76	28-7790001 through 28-7990589
	POH	VB-1120	7/2/79	28-8090001 through 28-8690056, 28-8690061, 28-8690062, and 2890001 through 2890231
	POH	VB-1611	7/12/95	2843001 and up
	POH	VB-1563	8/19/94	2890206 through 2890231
PA-28R-180	AFM	VB-173	6/8/67	28R-30001 through 28R-31270, and 28R-7130001 through 28R-7130013
PA-28R-200	AFM	VB-175	1/9/69	28R-35001 through 28R-35820, and 28R-7135001 through 28R-7135229
	AFM	VB-343	10/14/71	28R-7235001 through 28R-7335446
	AFM	VB-560	5/14/73	28R-7435001 through 28R-7635545
PA-28R-201	POH	VB-870	12/21/76	28R-7737001 through 28R-7837317
	POH	VB-1365	9/15/88	2837001 through 2837061
	POH	VB-1612	7/12/95	2844001 and up

(Continued)

MODEL	*AFM/POH*	*REPORT NO.*	*APPROVED*	*SERIAL EFFECTIVITY*
PA-28R-201T	POH	VB-800	12/20/76	28R-7703001 through 28R-7803374
	POH	VB-1370	11/9/89	2803001 through 2803012

PA-28-235	AFM	VB-170	7/15/63	28-10001 through 28-11378, and 28-7110001 through 28-7210023
	AFM Supp.	VB-357	8/25/71	28-10001 through 28-11378, and 28-7110001 through 28-7110023
	AFM	VB-442	6/9/72	28-7310001 through 28-7310176 and 28-E11
	AFM	VB-559	5/14/73	28-7410001 through 28-7610202
	POH	VB-810	1/21/77	28-7710001 through 28-7710089
PA-28-236	AFM	FT-124, App E.	6/1/78	28-7911001 through 28-8611008, and 2811001 through 2811050
	POH	VB-910	6/1/78	28-7911001 through 28-8611008, and 2811001 through 2811050
	POH	VB-1613	7/12/95	2845001 and up
PA-28RT-201	AFM	FT-121, App C.	11/7/78	28R-7918001 through 28R-8218026
	POH	VB-930	11/30/78	28R-7918001 through 28R-7918267
	POH	VB-1130	9/14/79	28R-8018001 through 28R-8218026
PA-28RT-201T	AFM	FT-130, App E.	11/7/78	28R-7931001 through 28R-8631005, and 2831001 through 2831013
	POH	VB-940	11/30/78	28R-7931001 through 28R-8631005, and 2831001 through 2831013
PA-28-201T	AFM	FT-126, App E.	12/14/78	28-7921001 through 28-7921095
		VB-920	1/25/79	28-7921001 through 28-7921095

NOTE 1: Current weight and balance report, including list of equipment included in certification empty weight and loading instructions, when necessary, must be provided for each aircraft at the same time of original certification.

The certificated empty weight and corresponding center of gravity location must include undrainable system oil (not included in the oil capacity) and unusable fuel as noted below.

Unusable Fuel and Oil Quantity	***Applicable Models and Serial Numbers***
Fuel 12.0 lb. at (+103.0)	PA-28R-180, PA-28R-200: all Serial Nos. PA-28-180: S/N 28-E13, and 28-7305001 through 28-7505260
Fuel 12.0 lb. at (+103.0)	PA-28-235: S/N 28-E11, and S/N 28-7310001 through 28-7710089
Fuel 12.0 lb. at (+103.0)	PA-28-151: S/N 28-7415001 through 28-7715314
Fuel 2.2 lb. at (+103.0)	PA-28-140, PA-28-150, PA-28-160: all Serial Nos.
Fuel 2.2 lb. at (+103.0)	PA-28-180: S/N 28-03, S/N 28-671 through 28-5859, and 28-7105001 through 28-7205318
Oil 1.8 lb. at (+27.5)	PA-28-140, PA-28-150, PA-28-160, PA-28-180: S/N 28-03, 28-1 through 28-1760, and 28-1760A
Oil 1.8 lb. at (+27.5)	PA-28-151: S/N 28-7415001 through 28-7715314

Unusable Fuel and Oil Quantity	***Applicable Models and Serial Numbers***
Oil 1.8 lb. at (+40.5)	PA-28-150, PA-28-160: S/N 28-1761 through 28-4377 PA-28-180: S/N 28-1761 through 28-5859, and 28-7105001 through 28-7205318
Oil 1.8 lb. at (+35.5)	PA-28-180: S/N 28-E13, 28-7305001 through 28-7505260
Oil 1.8 lb. at (+36.5)	PA-28R-180: all Serial Nos.

Oil 3.9 lb. at (+35.6)	PA-28R-200: S/N 28R-35001 through 28R-35820, and 28R-7135001 through 28R-7135229
Fuel 2.3 lb. at (+103.0) Oil 2.4 lb. at (+41.0)	PA-28-235: S/N 28-10001 through 28-11378, and 28-7110001 through 28-7210023
Oil 2.4 lb. at (+36.0)	PA-28-235: S/N 28-E11, and 28-7310001 through 28-7710089
Oil 3.9 lb. at (+30.6)	PA-28R-200: S/N 28R-7235001 through 28R-7635545
Oil 1.8 lb. at (+35.5) Fuel 12.0 lb. at (+103.0)	PA-28-181: S/N 28-7690001 through 28-8690056, 28-8690061, 28-8690062, and 2890001 through 2890231
Fuel 30.0 lb. at (+103.0) Oil 3.9 lb. at (+30.6)	PA-28R-201: S/N 28R-7737001 through 28R-7837317, 2837001 through 2837061, and 2844001 and up
Fuel 30.0 lb. at (+103.0)	PA-28R-201T: S/N 28R-7703001 through 28R-7803369, 2831001 through 2831013
Oil 6.0 lb. at (+19.1)	PA-28-161 Cadet: S/N 2841001 through 2841365
Fuel 12.0 lb. at (+103.0) Oil 1.8 lb. at (+27.5)	PA-28-161: S/N 28-7716001 through 28-8616057, and 2816001 through 2816119
Fuel 30.0 lb. at (+103.0) Oil 5.2 lb. at (+36.0)	PA-28-236: S/N 28-7911001 through 28-8611008, 2811001 through 2811050, and 2845001 and up
Fuel 30.0 lb. at (+103.0) Oil 3.9 lb. at (+30.6)	PA-28RT-201: S/N 28R-7918001 through 28R-8218026
Fuel 30.0 lb. at (+103.0) Oil 6.0 lb. at (+19.1)	PA-28RT-201T: S/N 28R-7931001 through 28R-8631005, 2831001 through 2831013
Fuel 30.0 lb. at (+103.0) Oil 6.0 lb. at (+19.1)	PA-28-201T: S/N 28-7921001 through 28-7921095

NOTE 2 The following placards must be displayed in clear view of the pilot:

In Normal Category Aircraft
"THIS AIRPLANE MUST BE OPERATED AS A NORMAL CATEGORY AIRPLANE IN COMPLIANCE WITH OPERATING LIMITATIONS STATED IN THE FORM OF PLACARDS, MARKINGS, AND MANUAL."

In aircraft certificated in both Normal and Utility Categories
"THIS AIRPLANE MAY BE OPERATED AS A NORMAL OR UTILITY CATEGORY AIRPLANE IN COMPLIANCE WITH OPERATING LIMITATIONS STATED IN THE FORM OF PLACARDS, MARKINGS, AND MANUAL."

Reference AFM for additional required placards.

NOTE 3 The Models PA-28-160 and PA-28-180, S/N 28-508 to 28-1760, and 28-1760A may be converted to the seaplane configuration, PA-28S-160 and PA-28S-180, in accordance with Piper Drawing No. 62008.

The Model PA-28-180, S/N 28-1761 through 28-5859, and 28-7105001 through 28-7205318, may be converted to the seaplane configuration, PA-28S-180, in accordance with Piper Drawing No. 65680.

NOTE 4 Takeoff r.p.m. for Models PA-28-180 and PA-28S-180, S/N 28-671 through 28-1760, and 28-1760A, restricted due to fuel flow capability of the emergency pump.

NOTE 5 The Models PA-28-150, PA-28-160, PA-28-180; S/N 28-03, 28-1 through 28-5859, and 28-7105001 through 28-7205318 and PA-28-235; S/N 28-10001 through 28-11378, and 28-7110001 through 28-7210023, may be operated with the door removed in accordance with the FAA approved Airplane Flight Manual Supplement dated September 3, 1963.

The Model PA-28-140 may be operated with the door removed in accordance with the FAA approved Airplane Flight Manual Supplement dated August 12, 1965.

NOTE 6 The Model PA-28-140, 2 PCLM (Normal Category Only), S/N 28-20001 through 28-20939 may be converted:

(a) To a maximum weight of 2150 lb. by the installation of Piper Kit 756 962 and Sensenich propeller M74DM58.
(b) To the four place, 4 PCLM (See Item VIII), configuration in accordance with Piper Drawing 65599.

NOTE 7 The Model PA-28-140, 2 PCLM, S/N 28-20940 through 28-26946, and 28-7125001 through 28-7725290, may be converted to the four place, 4 PCLM (See Item VIII), configuration by the installation of Piper Kit 756 941 and appropriate seats.

NOTE 8 The maximum cargo allowable of 125 lb. for S/N 28-1 through 28-1760, and 28-1760A may be increased to 200 lb. in accordance with Piper Service Spares Letter No. 242.

NOTE 9 The Model PA-28-180 (Normal Category), S/N 28-671 through 28-3832, may be operated in Utility Category in accordance with Service Spares Letter No. 258.

NOTE 10 All PA-28 models with Lycoming O-360-A3A engine and Sensenich propeller Model M76EMM-0, M76EMMS-0, 76EM8S5-0, or 76EM8-0 must avoid continuous operation between 2150 and 2350 r.p.m. Placards must be installed in accordance with Piper Service Letter No. 526, and Airplane Flight Manual Supplement No. 1, dated April 22, 1969.

NOTE 11 The Models PA-28-140, PA-28-150, PA-28-151, PA-28-160, PA-28-180; S/N 28-03, 28-1 through 28-5859, and 28-7105001 through 28-7205318; PA-28R-180 and PA-28R-200 may be operated with the spinner dome removed, or with the spinner dome and rear bulkhead removed. The PA-28-151, S/N 28-7415001 through 28-7715314, may be operated with the spinner dome removed, or with the spinner dome and front and rear bulkheads removed. The PA-28-180, S/N 28-7305001 through 28-7505260, and the PA-28-181; S/N 28-7690001 through 28-8690062, and 2890001 through 2890205, may be operated with the spinner dome removed. The PA-28R-201; S/N 28R-7737002 through 28R-7837317, 2837001 through 2837061, and 2844001 and up, may be operated with the spinner dome removed. The PA-28R-201T; S/N 28R-7703001 through 28R-7803374, and 2803001 through 2803012, may be operated with the spinner dome removed. The PA-28-161, S/N 28-7716001 through 28-8216300 may be operated with the spinner dome and front and rear bulkheads removed. The PA-28-161; S/N 28-8316001 through 28-8616057, 2816001 through 2816119, and PA-28-161 (Cadet), S/N 2841001 through 2841365, may be operated with the spinner dome removed, or with the spinner dome and front and rear bulkheads removed. The PA-28-236; S/N 28-7911001 through 28-8611008, 2811001 through 2811050, and 2845001 and up, may be operated with the spinner dome removed. The PA-28RT-201, S/N 28R-7918001 through 28R-8218026, may be operated with the spinner dome removed. The PA-28RT-201T; S/N 28R-7931001 through 28R-8631005, and 2831001 through 2831013, may be operated with the spinner dome removed. The PA-28-201T, S/N 28-7921001 through 28-7921095, may be operated with the spinner dome removed.

NOTE 12 Maximum baggage may be increased to 200 lb. at (+117) by the installation of Piper Kit 756 962 and Sensenich propeller M74DM-58 or 74DM6-0-58. Maximum baggage may be increased to 300 lb. (200 lb. at +117 and 100 lb. at +133) by the installation of Piper Kit 756 962, Sensenich propeller M74DM-58 or 74DM6-0-58 and when modified in accordance with Piper Drawing 66671.

NOTE 13 Maximum baggage may be increased to 300 lb. (200 lb. at +117 and 100 lb. at +133) when modified in accordance with Piper Drawing 66671.

NOTE 14 The Model PA-28-235; S/N 28-10001 through 28-11378, and 28-7110001 through 28-7210023, may be operated with the spinner dome removed, or with the spinner dome and rear bulkhead removed on the constant speed propeller installation only.

NOTE 15 The Model PA-28-180, S/N 28-671 through 28-5859, may be operated to the expanded C.G. envelope:

(a) For S/N 28-671 through 28-3072 by the installation of P/N 65280-00 tube - Landing Gear Strut Piston in accordance with Piper Service Letter 567 and in accordance with FAA approved Airplane Flight Manual Supplement No. 2, dated September 14, 1970, for Model PA-28-180 (Piper Report VB-261).

(b) For S/N 28-3073 through 28-5859 in accordance with FAA approved Airplane Flight Manual Supplement No. 2, dated September 14, 1970, for Model PA-28-180 (Piper Report VB-261).

NOTE 16 The Model PA-28-235, S/N 28-10001 through 28-11378, may be operated to the expanded C.G. envelope in accordance with FAA approved Airplane Flight Manual Supplement No. 1, dated September 14, 1970, for Model PA-28-235 (Piper Report VB-274).

NOTE 17 The following serial numbered aircraft are not eligible for import certification to the U.S.: 28-5035, 28-5047, 28-5178, 28-5262, 28-5397, 28-5435, 28-11077, 28-11101, 28-11140, 28-11180, 28-11200, 28-11212, 28-11227, 28-11254, 28-11255, 28-24660, 28-24701, 28R-30861, 28R-30952, 28R-30972, 28R-31043, and 28R-31091. These aircraft have identification plates stamped "Ensenblado en Colombia."

NOTE 18 Two propeller flange bushings must be replaced with Lycoming #72068S bushings at propeller blade positions corresponding to noncounterbored bolt holes in order to use the McCauley propeller.

NOTE 19 Two propeller flange bushings must be replaced with Lycoming #72060S index bushing and Lycoming #721061S bushing, at flange index mark and opposite, in order to use the McCauley propellers. A spacer, Piper P/N 79528-0, is also required between propeller and engine flange.

NOTE 20 The following model and serial number aircraft are not eligible for import certification to the U.S.:

PA-28-140:

28-24660, 28-24701, 28-7225490, 28-7225491, 28-7225492, 28-7225493, 28-7225494, 28-7225495, 28-7225496, 28-7225497, 28-7225498, 28-7225499, 28-7325238, 28-7325371, 28-7325372, 28-7325373, 28-7325374, 28-7325375, 28-7325376, 28-7325377, 28-7325378, 28-7325379, 28-7325508, 28-7325516, 28-7325525, 28-7325526, 28-7325555, 28-7325556, 28-7325557, 28-7325558, 28-7325580, 28-7325581, 28-7325599, 28-7325600, 28-7425217, 28-7425222, 28-7425224, 28-7425271, 28-7425272, 28-7425273, 28-7425274, 28-7425275, 28-7425276, 28-7425277, 28-7425278, 28-7425279, 28-7425304, 28-7425305, 28-7425306, 28-7425307, 28-7425344, 28-7425383, 28-7425384, 28-7525142, 28-7525144, 28-7525177, 28-7525180, 28-7525181, 28-7525182, 28-7525197, 28-7525201, 28-7525215, 28-7525216, 28-7525217, 28-7525218, 28-7525230, 28-7525238, 28-7525243, 28-7525244, 28-7525246, 28-7525247, 28-7625060, 28-7625061, 28-7625130, 28-7625144, 28-7625272, 28-7625273, 28-7625274, 28-7625275, 28-7725053, and 28-7725188.

PA-28-161:

28-7816330, 28-7916235, 28-8016266, 28-8116157, 28-8116158, 28-8316031, 28-8316032, 28-8616006, 28-8616007, 2816006, 2816020, 2816021, and 2816022.

PA-28-180:

28-5047, 28-5178, 28-5262, 28-5397, 28-5435, 28-7305315, 28-7305316, 28-7305499, 28-7405136, 28-7405137, 28-7405138, 28-7405139, 28-7405158, 28-7405160, 28-7405161, 28-7405167, 28-7405184, 28-7405185, 28-7405186, 28-7405187, 28-7405223, 28-7505138, 28-7505148, 28-7505159, 28-7505168, 28-7505169, 28-7505179, 28-7505189, and 28-7505260.

NOTE 20 (cont.)

PA-28-181:
28-7690362, 28-7790343, 28-7790344, 28-7790388, 28-7790533, 28-7790571, 28-7790605, 28-7890060, 28-7890185, 28-7890290, 28-7890351, 28-7890352, 28-7890406, 28-7890407, 28-7890463, 28-7890464, 28-7890465, 28-7890466, 28-7890480, 28-7890481, 28-7890507, 28-7890508, 28-7890509, 28-7890510, 28-7890534, 28-7890550, 28-7890551, 28-7990158, 28-7990251, 28-8090203, 28-8090243, 28-8090274, 28-8090349, 28-8190032, 28-8190098, 28-8190099, 28-8190174, 28-8190175, 28-8190200, 28-8190201, 28-8190261, 28-8190262, 28-8190317, 28-8190318, 28-8290020, 28-8290021, 28-8290022, 28-8290122, 28-8290123, 28-8290124, 28-8290125, 28-8290146, 28-8290147, 28-8290148, 28-8290149, 28-8390031, 28-8390032, 28-8390057, 28-8390058, 28-8390059, 28-8390060, 28-8690061, 28-8690062, 2890035, and 2890036.

PA-28-201T:
28-7921085

PA-28-235:
28-11077, 28-11101, 28-11140, 28-11180, 28-11200, 28-11212, 28-11227, 28-11254, 28-11255, 28-11370, 28-11371, 28-11372, 28-11373, 28-7310074, 28-7310152, 28-7310153, 28-7310172, 28-7410074, 28-7410078, 28-7410089, 28-7410090, 28-7510072, 28-7510073, 28-7610087, 28-7610168, 28-7710033, 28-7710068, and 28-7710089.

PA-28-236:
28-7911027, 28-7911028, 28-7911029, 28-7911030, 28-7911136, 28-7911219, 28-7911220, 28-7911221, 28-7911252, 28-8011020, 28-8011021, 28-8011062, 28-8011092, 28-8011093, 28-8011094, 28-8011107, 28-8111030, 28-8111038, 28-8111058, 28-8111068, 28-8111069, 28-8111070, 28-8111095, 28-8411021, 28-8411022, 28-8411023, 28-8411024, 28-8411026, 28-8411027, 28-8411028, and 28-8411029.

PA-28R-180:
28R-31091

PA-28R-200:
28R-7335201, 28R-7335202, 28R-7335326, 28R-7335328, 28R-7335377, 28R-7335387, 28R-7335395, 28R-7335397, 28-7435214, 28-7435229, 28-7435252, 28-7435253, 28R-7535146, 28R-7535149, 28R-7535167, 28R-7535168, 28R-7535214, 28R-7535217, and 28R-7635377.

PA-28R-201:
28R-7737119, 28R-7837076, 28R-7837148, 28R-7837149, 28R-7837188, 28R-7837189, 28R-7837225, 28R-7837226, 28R-7837248, 28R-7837249, 28R-7837273, 28R-7837274, 28R-7837294, 28R-7837316, and 28R-7837317.

PA-28R-201T:
28R-7703069, 28R-7703132, 28R-7703184, 28R-7703185, 28R-7703285, 28R-7703382, 28R-7803064, 28R-7803156, 28R-7803207, 28R-7803208, 28R-7803251, 28R-7803291, 28R-7803292, 28R-7803293, 28R-7803294, 28R-7803295, 28R-7803299, 28R-7803300, 28R-7803317, 28R-7803318, 28R-7803319, 28R-7803320, 28R-7803344, 28R-7803360, 28R-7803361, 28R-7803370, 28R-7803371, 28R-7803372, and 28R-7803373.

PA-28RT-201:
28R-8118029, 28R-8118054, 28R-8118078, 28R-8218015, and 28R-8218016.

PA-28RT-201T:
28R-7931122, 28R-7931205, 28R-7931206, 28R-7931262, 28R-7931296, 28R-7931297, 28R-8031062, 28R-8131029, 28R-8131083, and 28R-8131183.

In addition, aircraft having the following serial number are not eligible for import certification to the U.S.:

AR28-7325238, AR28-7325371, AR28-7325372, AR28-7325373, AR28-7325374, AR28-7325375, AR28-7325376, AR28-7325377, AR28-7325378, AR28-7325379, AR28-7305315, AR28-7305316, AR28-7335201, AR28-7335202, AR28-7325508, AR28-7325516, AR28-7325525, AR28-7325526, AR28-7310152, AR28-7310153, AR28-7325555, AR28-7325556, AR28-7325557, AR28-7325558, AR28-7305480, AR28-7305499, AR28-7335326, AR28-7335328, AR28-7325580, AR28-7325581, AR28-7325599, AR28-7325600, AR28-7335395, and AR28-7335397.

NOTE 21 Engines with serial numbers ending with "A" require the F-4-13 propeller governor assembly. Other engines require the F-4-3( ) propeller governor assembly.

NOTE 22 Hartzell Propeller HC-C2YK-1( )/7666A-2 or HC-C2YK-1( )F/F7666A-2 approved with IO-360-C1C engine only (S/N 28R-7235001 through S/N 28R-7635516).

NOTE 23 McCauley Propeller B2D34C213/90DHA-16 approved with IO-360-C1C6 engine only (S/N 28R-7635517 through 28R-7635545).

NOTE 24 On Models PA-28-161; S/N 28-7816001 through 28-8616057, and S/N 2816001 through 2816109, and PA-28-181; S/N 28-7890001 through 28-8690056, 28-8690061, 28-8690062, 2890001 through 2890231, and 2843001 and up, the wheel fairings but not the landing gear strut fairings may be removed. If the wheel fairings are removed, the rudder centering springs must also be removed. This NOTE does not apply to the PA-28-161 (Cadet); S/N 2841001 through 2841365, or the PA-28-161 (Warrior III); S/N 2816110 through 2816119, and 2842001 and up, which are not equipped with wheel fairings.

NOTE 25 On Models PA-28-201T; S/N 28-7921001 through 28-7921095, and PA-28-236; S/N 28-7911001 through 28-8611008, 2811001 through 2811050, and 2845001 and up, the wheel fairings alone or the wheel fairings but not the landing gear strut fairings may be removed.

NOTE 26 With installation of Piper Kit 88050, PA-28-161 2325 lb. Maximum Gross Weight Modification, the following weights apply:

Normal Category: Ramp - 2332 lb.
Takeoff - 2325 lb.

Utility Category: Ramp - 2027 lb.
Takeoff - 2020 lb.

(See POH VB-1180 Supplement dated October 5, 1985.)

NOTE 27 With installation of Piper Kit 88168, PA-28-161 Cadet 2202 lb. Maximum Gross Weight Modification, the following weights apply:

Normal Category: Ramp - 2209 lb.
Takeoff - 2202 lb.

Utility Category: Ramp - 2027 lb.
Takeoff - 2020 lb.

(See POH VB-1410 dated March 14, 1990.)

NOTE 28 POH Supplement VB-1546 is applicable to POH VB-1180. POH Supplement VB-1545 is applicable to POH VB-1360. Supplements VB-1545 and VB-1546 restrict maximum r.p.m. limitation to 2600 r.p.m. for foreign countries requiring reduced noise level operation (Piper Kit No. 766 277 for PA-28-161 (Cadet) and Piper Kit No. 766 278 for PA-28-161 (Warrior II)).

...END...

3A19

	3A19
	Revision 40
	CESSNA
150	150J
150A	150K
150B	A150K
150C	150L
150D	A150L
150E	150M
150F	A150M
150G	152
150H	A152
	November 15, 1997

TYPE CERTIFICATE DATA SHEET NO. 3A19

This data sheet which is a part of type certificate No. 3A19 prescribes conditions and limitations under which the product for which the type certificate was issued meets the airworthiness requirements of the Federal Aviation Regulations.

Type Certificate Holder — Cessna Aircraft Company
P.O. Box 7704
Wichita, Kansas 67277

I - Model 150, 2 PCLM (Utility Category), Approved July 10, 1958
Model 150A, 2 PCLM (Utility Category), Approved June 14, 1960
Model 150B, 2 PCLM (Utility Category), Approved June 20, 1961
Model 150C, 2 PCLM (Utility Category), Approved June 15, 1962

Engine — Continental O-200-A

*Fuel — 80/87 min. grade aviation gasoline

*Engine limits — For all operations, 2750 r.p.m. (100 hp.)

Propeller and propeller limits —

1. Sensenich 69CK-0-52 — 24 lb. (-32)
   Diameter: not over 69 in., not under 67.5 in.
   Static r.p.m. at maximum permissible throttle setting:
   not over 2470, not under 2320
   No additional tolerance permitted
2. McCauley 1A100/MCM 6950 — 21 lb. (-32)
   Diameter: not over 69 in., not under 67.5 in.
   Static r.p.m. at maximum permissible throttle setting:
   not over 2475, not under 2375
   No additional tolerance permitted

*Airspeed limits (CAS)

Never exceed	157 m.p.h.	(136 knots)
Maximum structural cruising	120 m.p.h.	(104 knots)
Maneuvering	106 m.p.h.	(92 knots)
Flaps extended	85 m.p.h.	(74 knots)

C.G. range — (+33.4) to (+36.0) at 1500 lb.
(+32.2) to (+36.0) at 1250 lb. or less
Straight line variation between points given

Empty weight C.G. range — None

Leveling means — Top edge of fuselage splice plate

Page No.	1	2	3	4	5	6	7	8	9	10	11	12	13	14	15	16	17	18
Rev No.	40	40	40	40	40	40	40	40	37	37	38	38	39	30	30	32	32	37

I - Model 150, Model 150A, Model 150B, Model 150C (cont'd)

*Maximum weight	1500 lb.
No. of seats	2 at (+39); (for child's optional jump seat refer to Equipment List)
Maximum baggage	80 lb. at (+65)
Fuel capacity	26 gal. (22.5 gal. usable, two 13 gal. tanks in wings at +42) See NOTE 1 for data on system fuel
Oil capacity	6 qt. (-13.5; unusable 2 qt.) See NOTE 1 for data on system oil

Control Surface Movements

Wing flaps	Retracted	0°
	1st notch	10°
	2nd notch	20°
	3rd notch	30°
	4th notch	40°
Ailerons	Up 20°	Down 15°
Elevator	Up 25°	Down 15°
Elevator tab	Up 10°	Down 20°
Rudder	Right 16°	Left 16°

Serial Nos. eligible

Model 150:	617, 17001 through 17999, 59001 through 59018
Model 150A:	628, 15059019 through 15059350
Model 150B:	15059351 through 15059700
Model 150C:	15059701 through 15060087

II - Model 150D, 2 PCLM (Utility Category), Approved July 19, 1963
Model 150E, 2 PCLM (Utility Category), Approved June 18, 1964
Model 150F, 2 PCLM (Utility Category), Approved May 27, 1965

Engine	Continental O-200-A
*Fuel	80/87 min. grade aviation gasoline
*Engine limits	For all operations, 2750 r.p.m. (100 hp.)

| Propeller and propeller limits

1. Sensenich 69CK-0-52 — 24 lb. (-32)
   Diameter: not over 69 in., not under 67.5 in.
   Static r.p.m. at maximum permissible throttle setting:
   not over 2470, not under 2320
   No additional tolerance permitted
2. McCauley 1A100/MCM 6950 — 21 lb. (-32)
   Diameter: not over 69 in., not under 67.5 in.
   Static r.p.m. at maximum permissible throttle setting:
   not over 2475, not under 2375
   No additional tolerance permitted

*Airspeed limits (CAS)

Never exceed	162 m.p.h.	(141 knots)
Maximum structural cruising	120 m.p.h.	(104 knots)
Maneuvering	109 m.p.h.	(95 knots)
Flaps extended	100 m.p.h.	(87 knots)

C.G. range	(+32.9) to (+37.5) at 1600 lb. (+31.5) to (+37.5) at 1280 lb. or less Straight line variation between points given
Empty weight C.G. range	None
Leveling means	Top of tailcone

II - Model 150D, Model 150E, Model 150F (cont'd)

*Maximum weight	1600 lb.
No. of seats	2 at (+39); (for child's optional jump seat refer to Equipment List)
Maximum baggage	120 lb. at (+65) (150D, 150E) 120 lb. - Reference weight and balance data (150F)
Fuel capacity	26 gal. (22.5 gal. usable, two 13 gal. tanks in wings at +42) See NOTE 1 for data on system fuel
Oil capacity	6 qt. (-13.5; unusable 2 qt.) See NOTE 1 for data on system oil

Control Surface Movements

Wing flaps (150D, 150E)	Retracted	0°	
	1st Notch	10°	
	2nd Notch	20°	
	3rd Notch	30°	
	4th Notch	40°	
Wing flaps (150°F)		Down 0°	-40° ±2°
Ailerons	Up 20°	Down 15°	
Elevator	Up 25°	Down 15°	
Elevator tab	Up 10°	Down 20°	
Rudder (150D, 150E)	Right 16°	Left 16°	
(150F)	Right 23°	Left 23°	
	(measured parallel to chord)		

Serial Nos. eligible

Model 150D:	15060088 through 15060772
Model 150E:	644, 15060773 through 15061532
Model 150F:	15061533 through 15064532

III - Model 150G, 2 PCLM (Utility Category), Approved May 5, 1966
2 PCSM (Utility Category), Approved August 12, 1966
Model 150H, 2 PCL-SM (Utility Category), Approved August 10, 1967
Model 150J, 2 PCL-SM (Utility Category), Approved May 2, 1968
Model 150K, 2 PCL-SM (Utility Category), Approved June 5, 1969

Engine	Continental O-200-A
*Fuel	80/87 min. grade aviation gasoline
*Engine limits	For all operations, 2750 r.p.m. (100 hp.)

Propeller and propeller limits

1. Sensenich 69CK-0-52 — 24 lb. (-32)
   Diameter: not over 69 in., not under 67.5 in.
   Static r.p.m. at maximum permissible throttle setting:
   not over 2470, not under 2320
   No additional tolerance permitted
2. McCauley 1A100/MCM 6950 — 21 lb. (-32)
   Diameter: not over 69 in., not under 67.5 in.
   Static r.p.m. at maximum permissible throttle setting:
   not over 2475, not under 2375
   No additional tolerance permitted
3. McCauley 1A90/CF 7535 or 7538 (seaplane only) — 24 lb. (-32)
   Diameter: not over 75 in., not under 73.5 in.
   Static r.p.m. at maximum permissible throttle setting:
   not over 2600, not under 2500
   No additional tolerance permitted
4. McCauley 1A101/DCM 6948 or 6950 — 21 lb. (-32)
   Diameter: not over 69 in., not under 67 in.
   Static r.p.m. at maximum permissible throttle setting:
   not over 2600, not under 2500
   No additional tolerance permitted

III - Model 150G, Model 150H, Model 150J, Model 150K (cont'd)

*Airspeed limits (CAS)	Never exceed	162 m.p.h.	(141 knots)
	Maximum structural cruising	120 m.p.h.	(104 knots)
	Maneuvering	109 m.p.h.	(95 knots)
	Flaps extended	100 m.p.h.	(87 knots)

C.G. range
Landplane
(+32.9) to (+37.5) at 1600 lb.
(+31.5) to (+37.5) at 1280 lb. or less
Seaplane
(+33.8) to (+36.5) at 1650 lb.
(+33.0) to (+36.5) at 1400 lb. or less
Straight line variation between points given

Empty weight C.G. range — None

Leveling means — Top of tailcone

*Maximum weight
Landplane - 1600 lb.
Seaplane - 1650 lb. (Edo 88A-1650 floats)

No. of seats — 2 at (+39); (for child's optional jump seat, refer to Equipment List)

Maximum baggage — 120 lb. - Reference weight and balance data

Fuel capacity
Landplane
26 gal. (22.5 gal. usable, two 13 gal. tanks in wings at +42.0)
Seaplane
26 gal. (21.5 gal. usable, two 13 gal. tanks in wings at +42.0)
See NOTE 1 for data on system fuel

Oil capacity
6 qt. (-13.5; unusable 2 qt.)
See NOTE 1 for data on system oil

Control surface movements

Wing flaps			Down	0° -40° ±2°
Ailerons	Up	20° +2°, -0°	Down	14° +2°, -0°
Elevator	Up	25° ±1°	Down	15° ±1°
Elevator tab	Up	10° ±1°	Down	20° ±1°
Rudder	Right	23° +0°, -2°	Left	23° +0°, -2°

(measured perpendicularly to hinge line)

Serial Nos. eligible

Model 150G:	15064533 through 15067198 (except 15064970)
Model 150H:	649, 15067199 through 15069308
Model 150J:	15069309 through 15071128
Model 150K:	15071129 through 15072003

IV - Model A150K, Aerobat, 2 PCLM (Acrobatic Category), Approved June 5, 1969

Engine — Continental O-200-A

*Fuel — 80/87 min. grade aviation gasoline

*Engine limits — For all operations, 2750 r.p.m. (100 hp.)

| Propeller and propeller limits
1. McCauley 1A101/DCM 6948 or 6950 — 21 lb. (-32)
Diameter: not over 69 in., not under 67 in.
Static r.p.m. at maximum permissible throttle setting:
not over 2600, not under 2500
No additional tolerance permitted

IV - Model A150K (cont'd)

*Airspeed limits (CAS)	Never exceed	193 m.p.h.	(168 knots)
	Maximum structural cruising	140 m.p.h.	(122 knots)
	Maneuvering	118 m.p.h.	(103 knots)
	Flaps extended	100 m.p.h.	(87 knots)

C.G. range — (+32.9) to (+37.5) at 1600 lb.
(+31.5) to (+37.5) at 1280 lb. or less

Empty weight C.G. range — None

Leveling means — Top of tailcone

*Maximum weight — 1600 lb.

No. of seats — 2 at (+39); (for child's optional jump seat, refer to Equipment List)

Maximum baggage — 120 lb. - (reference weight and balance data)

Fuel capacity — 26 gal. (22.5 gal. usable, two 13 gal. tanks in wings at +42.0)

Oil capacity — 6 qt. (-13.5; unusable 2 qt.)
See NOTE 1 for data on system oil.

Control surface movements

Wing flaps			Down	0° -40° ±2°
Ailerons	Up	20° +2°, -0°	Down	14° +2°, -0°
Elevator	Up	25° ±1°	Down	15° ±1°
Elevator tab	Up	10° ±1°	Down	20° ±1°
Rudder	Right	23° +0°, -2°	Left	23° +0°, -2°

(measured perpendicularly to hinge line)

Serial Nos. Eligible — Model A150K: A1500001 through A1500226

V - Model 150L, 2 PCLM (Utility Category), Approved June 8, 1970

Engine — Continental O-200-A

*Fuel — 80/87 min. grade aviation gasoline

*Engine limits — For all operations, 2750 r.p.m. (100 hp.)

Propeller and propeller limits

1. McCauley 1A101/GCM 6948 — 27.7 lb. (-34.5)
   (1971, 1972, 1973 models)
   Diameter: not over 69 in., not under 67 in.
   Static r.p.m. at maximum permissible throttle setting:
   not over 2600, not under 2500
   No additional tolerance permitted

2. McCauley 1A101/HCM 6948 — 27.7 lb. (-34.5)
   (1973, 1974 models)
   Diameter: not over 69 in., not under 67 in.
   Static r.p.m. at maximum permissible throttle setting:
   not over 2600, not under 2500
   No additional tolerance permitted

3. McCauley 1A101/PCM 6948 — 27.0 lb. (-34.5)
   (1974 model)
   Diameter: not over 69 in., not under 67 in.
   Static r.p.m. at maximum permissible throttle setting:
   not over 2600, not under 2500
   No additional tolerance permitted

| **V - Model 150L** (cont'd)	4. McCauley 1A102/OCM 6948 (1971 through 1974 models) Diameter: not over 69 in., not under 67.5 in. Static r.p.m. at maximum permissible throttle setting: not over 2560, not under 2460 No additional tolerance permitted	27.0 lb. (-34.5)

*Airspeed limits (CAS)	Never exceed	162 m.p.h.	(141 knots)
	Maximum structural cruising	120 m.p.h.	(104 knots)
	Maneuvering	109 m.p.h.	(95 knots)
	Flaps extended	100 m.p.h.	(87 knots)

C.G. range	(+32.9) to (+37.5) at 1600 lb. (+31.5) to (+37.5) at 1280 lb. or less Straight line variation between points given
Empty weight C.G. range	None
Leveling means	Jig located nut plates and screws at Stations +94.63 and +132.94 on left side of tailcone
*Maximum weight	1600 lb.
No. of seats	2 at (+39); (for child's optional jump seat refer to Equipment List)
Maximum baggage	120 lb. - (Reference weight and balance data)
Fuel capacity	26 gal. (22.5 gal. usable, two 13 gal. tanks in wings at +42.0) See NOTE 1 for data on system fuel
Oil capacity	6 qt. (-13.5; unusable 2 qt.) See NOTE 1 for data on undrainable oil

Control surface movements					
Wing flaps			Down	0° -40° ±2°
Ailerons	Up	20° +2°, -0°	Down	14° +2°, -0°
Elevator	Up	25° ±1°	Down	15° ±1°
Elevator tab	Up	10° ±1°	Down	20° ±1°
Rudder	Right	23° +0°, -2°	Left	23° +0°, -2°

(measured perpendicularly to hinge line)

Serial Nos. eligible	15072004 through 15072628 (1971 Model) 15072629 through 15073658 (1972 Model) 15073659 through 15074850 (1973 Model) 15074851 through 15075781 (1974 Model)

VI - Model A150L, Aerobat, 2 PCLM (Acrobatic Category), Approved June 8, 1970

Engine	Continental O-200-A	
*Fuel	80/87 min. grade aviation gasoline	
*Engine limits	For all operations, 2750 r.p.m. (100 hp.)	
| Propeller and propeller limits	1. McCauley 1A101/GCM 6948 (1971, 1972, 1973 models) Diameter: not over 69 in., not under 67 in. Static r.p.m. at maximum permissible throttle setting: not over 2600, not under 2500 No additional tolerance permitted	27.7 lb. (-34.5)

VI - Model A150L (cont'd)		
	2. McCauley 1A101/HCM 6948 (1971, 1972, 1973 models) Diameter: not over 69 in., not under 67 in. Static r.p.m. at maximum permissible throttle setting: not over 2600, not under 2500 No additional tolerance permitted	27.7 lb. (-34.5)
	3. McCauley 1A102/OCM 6948 (1974 model) Diameter: not over 69 in., not under 67.5 in. Static r.p.m. at maximum permissible throttle setting: not over 2560, not under 2460 No additional tolerance permitted	27.0 lb. (-34.5)

*Airspeed limits (CAS)

Never exceed	193 m.p.h.	(168 knots)
Maximum structural cruising	140 m.p.h.	(122 knots)
Maneuvering	118 m.p.h.	(103 knots)
Flaps extended	100 m.p.h.	(87 knots)

C.G. range: (+32.9) to (+37.5) at 1600 lb.
(+31.5) to (+37.5) at 1280 lb. or less

Empty weight C.G. range: None

Leveling means: Jig located nut plates and screws at Stations +94.63 and +132.94 on left side of tailcone

*Maximum weight: 1600 lb.

No. of seats: 2 at (+39); (for child's optional jump seat refer to Equipment List)

Maximum baggage: 120 lb. - (Reference weight and balance data)

Fuel capacity: 26 gal. (22.5 gal. usable, two 13 gal. tanks in wings at +42.0)
See NOTE 1 for data on unusable fuel

Oil capacity: 6 qt. (-13.5; unusable 2 qt.)
See NOTE 1 for data on undrainable oil

Control surface movements

Wing flaps			Down	0° -40° ±2°
Ailerons	Up	20° +2°, -0°	Down	14° +2°, -0°
Elevator	Up	25° ±1°	Down	15° ±1°
Elevator tab	Up	10° ±1°	Down	20° ±1°
Rudder	Right	23° +0°, -2°	Left	23° +0°, -2°

(measured perpendicularly to hinge line)

Serial Nos. eligible:
A1500227 through A1500276 (1971 Model)
A1500277 through A1500342 (1972 Model)
A1500343 through A1500429 (1973 Model)
A1500430 through A1500523 (1974 Model) (Except A1500433)

VII - Model 150M, 2 PCLM (Utility Category), Approved May 6, 1974

Engine: Continental O-200-A

*Fuel: 80/87 min. grade aviation gasoline

*Engine limits: For all operations, 2750 r.p.m. (100 hp.)

VII - Model 150M (cont'd)

Propeller and propeller limits	1. McCauley 1A102/OCM 6948 Diameter: not over 69 in., not under 67 in. Static rpm at maximum permissible throttle setting: not over 2560, not under 2460 No additional tolerance permitted	27.7 lb. (-34.5)

*Airspeed limits (CAS)

15075782 through 15077005

Never exceed	162 m.p.h.	(141 knots)
Maximum structural cruising	120 m.p.h.	(104 knots)
Maneuvering	109 m.p.h.	(95 knots)
Flaps extended	100 m.p.h.	(87 knots)

*Airspeed limits (IAS) (See Note 4 on use of (IAS)

15077006 through 15079405

Never exceed	141 knots
Maximum structural cruising	107 knots
Maneuvering	97 knots
Flaps extended	85 knots

C.G. range	(+32.9) to (+37.5) at 1600 lb. (+31.5) to (+37.5) at 1280 lb. or less Straight line variation between points given
Empty weight C.G. range	None
Leveling means	Jig located nut plates and screws at Stations +94.63 and +132.94 on left side of tailcone
*Maximum weight	1600 lb.
No. of seats	2 at (+39); (for child's optional jump seat, refer to Equipment List)
Maximum baggage	120 lb. (Reference weight and balance data)
Fuel capacity	26 gal. (22.5 gal. usable, two 13 gal. tanks in wings at +42.0) See NOTE 1 for data on unusable fuel
Oil capacity	6 qt. (-13.5; unusable 2 qt.) See NOTE 1 for data on undrainable oil

Control surface movements

Wing flaps			Down	0° -40° ±2°
Ailerons	Up	20° +2°, -0°	Down	14° +2°, -0°
Elevator	Up	25° ±1°, -0°	Down	15° ±1°
Elevator tab	Up	10° ±1°	Down	20° ±1°
Rudder	Right	23° +0°, -2°	Left	23° +0°, -2°

(measured perpendicularly to hinge line)

Serial Nos. eligible

15075782 through 15077005 (1975 Model)
15077006 through 15078505 (1976 Model)
15078506 through 15079405 (1977 Model)

VIII - Model A150M, Aerobat, 2 PCLM (Acrobatic Category), Approved May 6, 1974

Engine	Continental O-200-A
*Fuel	80/87 min. grade aviation gasoline
*Engine limits	For all operations, 2750 r.p.m. (100 hp.)

VIII - Model A150M (cont'd)

Propeller and propeller limits	1. McCauley 1A102/OCM 6948 Diameter: not over 69 in., not under 67.5 in. Static r.p.m. at maximum permissible throttle setting: not over 2560, not under 2460 No additional tolerance permitted	27.0 lb. (-34.5)

*Airspeed limits (CAS)	15064970, A1500524 through A1500609	
	Never exceed	193 m.p.h. (168 knots)
	Maximum structural cruising	140 m.p.h. (122 knots)
	Maneuvering	118 m.p.h. (103 knots)
	Flaps extended	100 m.p.h. (87 knots)
*Airspeed limits (IAS) (See NOTE 4 on Use of IAS)	A1500610 through A1500734	
	Never exceed	164 knots
	Maximum structural cruising	123 knots
	Maneuvering	105 knots
	Flaps extended	85 knots

C.G. range: (+32.9) to (+37.5) at 1600 lb.
(+31.5) to (+37.5) at 1280 lb. or less
Straight line variation between points given

Empty weight C.G. range: None

Leveling means: Jig located nut plates and screws at Stations +94.63 and +132.94 on left side of tailcone

*Maximum weight: 1600 lb.

No. of seats: 2 at (+39); (for child's optional jump seat, refer to Equipment List)

Maximum baggage: 120 lb. - (Reference weight and balance data)

Fuel capacity: 26 gal. (22.5 gal. usable, two 13 gal. tanks in wings at +42.0)
See NOTE 1 for data on unusable fuel

Oil capacity: 6 qt. (-13.5; unusable 2 qt.)
See NOTE 1 for data on undrainable oil

Control surface movements:

Wing flaps			Down	0° -40° ±2°
Ailerons	Up	20° +2°, -0°	Down	14° +2°, -0°
Elevator	Up	23° ±1°, -0°	Down	15° ±1°
Elevator tab	Up	10° ±1°	Down	20° ±1°
Rudder	Right	23° +0°, -2°	Left	23° +0°, -2°

(measured perpendicularly to hinge line)

Serial Nos. eligible: 15064970, A1500524 through A1500609 (1975 Model)
A1500610 through A1500684 (1976 Model)
A1500685 through A1500734 (1977 Model)

IX - Model 152, 2 PCLM (Utility Category), Approved March 16, 1977

Engine: S/N 15279406 through 15285594
Lycoming O-235-L2C

S/N 15285595 and on aircraft reworked per SK152-15 or SK152-16
Lycoming O-235-N2C

*Fuel: 100LL/100 min. grade aviation gasoline

IX - Model 152 (cont'd)		
*Engine limits	S/N 15279406 through 15285594 For all operations, 2550 r.p.m. (110 hp.) S/N 15285595 and on For all operations 2550 r.p.m. (108 hp.)	
Propeller and propeller limits	1. (a) McCauley 1A103/TCM6958 Diameter: not over 69 in., not under 67.5 in. Static rpm at full throttle (carburetor heat off and mixture leased to maximum r.p.m.) is 2280 to 2380 r.p.m. For allowable variations in static r.p.m. at non-standard temperatures, refer to the Service Manual. (b) Spinner: Dwg. 0450073	23.2 lb. (-36.5)
*Airspeed Limits (IAS) (See NOTE 4 on Use of IAS)	Never exceed 149 knots Maximum structural cruising 111 knots Maneuvering 104 knots Flaps extended 85 knots	
C.G. range	(+32.65) to (+36.5) at 1670 lb. (+31.0) to (+36.5) at 1350 lb. or less Straight line variation between points given	
Empty weight C.G. range	None	
Leveling means	Jig located nut plates and screws at Stations +94.63 and +132.94 on left side of tailcone	
*Maximum weight	1670 lb. 1675 lb. ramp weight (S/N 15282032 and on)	
No. of seats	2 at (+39); (for child's optional jump seat, refer to Equipment List)	
Maximum baggage	120 lb. (Reference weight and balance data)	
Fuel capacity	26 gal. (24.5 gal. usable, two 13 gal. tanks in wings at +42.0) See NOTE 1 for data on unusable fuel	
Oil capacity	6 qt. (-14.7; unusable 2 qt.) See NOTE 1 for data on undrainable oil	

Control surface movements

Wing flaps			Down	0° -30° ±2°
Ailerons (aileron travel measured from 1° ±.5° droop)	Up	20° ± 2°	Down	15° ± 1°
Elevator	Up	25° ±1°	Down	18° ±1°
Elevator tab	Up	10° ±1°	Down	20° ±1°
Rudder (measured perpendicularly to hinge line)	Right	23° +0°, -2°	Left	23° +0°, -2°

Serial Nos. eligible

15279406 through 15282031 (1978 Model)
15282032 through 15283591 (1979 Model)
15283592 through 15284541 (1980 Model)
15284542 through 15285161 (1981 Model)
15285162 through 18285594 (1982 Model)
15285595 through 15285833 (1983 Model)
15285834 through 15285939 (1984 Model)
15285940 through 15286033 (1985 Model)

X - Model A152, Aerobat, 2 PCLM (Acrobatic Category), Approved March 16, 1977

Engine	S/N A1500433, A1520735, 681 through A521014 Lycoming O-235-L2C S/N A1521015 and on aircraft reworked per SK152-15 or SK152-16 Lycoming O-235-N2C
*Fuel	100LL/100 min. grade aviation gasoline
*Engine limits	S/N A1500433, A1520735, 681 through A1521014 For all operations, 2550 r.p.m. (110 hp.) S/N A1521015 and on For all operations 2550 r.p.m. (108 hp.)
Propeller and propeller limits	1. (a) McCauley 1A103/TCM6958 — 23.2 lb. (-36.5) Diameter: not over 69 in., not under 67.5 in. Static rpm at full throttle (carburetor heat off and mixture leaned to maximum r.p.m.) is 2280 to 2380 r.p.m. For allowable variations in static r.p.m. at non-standard temperatures, refer to the Service Manual. (b) Spinner: Dwg. 0450073
*Airspeed Limits (IAS) (See NOTE 4 on Use of IAS)	Never exceed 172 knots Maximum structural cruising 125 knots Maneuvering 108 knots Flaps extended 85 knots
C.G. range	(+32.65) to (+36.5) at 1670 lb. (+31.0) to (+36.5) at 1350 lb. or less
Empty weight C.G. range	None
Leveling means	Jig located nut plates and screws at Stations +94.63 and +132.94 on left side of tailcone
*Maximum weight	1670 lb. 1675 lb. ramp weight (S/N 681, A1520809 and on)
No. of seats	2 at (+39); (for child's optional jump seat, refer to Equipment List)
Maximum baggage	120 lb. (Reference weight and balance data)
Fuel capacity	26 gal. (24.5 gal. usable, two 13 gal. tanks in wings at +42.0) See NOTE 1 for data on unusable fuel
Oil capacity	6 qt. (-14.7; unusable 2 qt.) See NOTE 1 for data on undrainable oil

Control surface movements

Wing flaps			Down	0° -30° ±2°
Ailerons	Up	20° ± 1°	Down	15° ± 1°
(aileron travel measured from 1° ±.5° droop)				
Elevator	Up	25° ±1°	Down	18° ±1°
Elevator tab	Up	10° ±1°	Down	20° ±1°
Rudder	Right	23° +0°, -2°	Left	23° +0°, -2°
(measured perpendicularly to hinge line)				

X - Model A152 (cont'd)

Serial Nos. eligible	A1500433, A1520735 through A1520808	(1978 Model)
	681, A1520809 through A1520878	(1979 Model)
	A1520879 through A1520943	(1980 Model)
	A1520944 through A1520983	(1981 Model)
	A1520984 through A1521014	(1982 Model)
	A1521015 through A1521025	(1983 Model)
	A1521026 through A1521027	(1984 Model)
	A1521028 through A1521049	(1985 Model)

Data Pertinent to All Models

Datum — Fuselage station 0.0 front face of firewall

Certification basis — Part 3 of the Civil Air Regulations dated May 15, 1956, as amended by 3-4. In addition, effective S/N 15282032 and on for 152 and S/N 681, A1520809 and on for A152, FAR 23.1559 effective March 1, 1978. FAR 36 dated December 1, 1969, plus Amendments 36-1 through 36-5 for 152 and A152 only. In addition, effective S/N 15285940 and on, and S/N A1521028 and on, FAR 23.1545(a), Amendment 23-23 dated December 1, 1978.

Application for Type Certificate dated August 13, 1956.

Type Certificate No. 3A19 issued July 10, 1958, obtained by the manufacturer under delegation option procedures.

Equivalent Safety Items	S/N 15077006 through 15079405 S/N 15279406 and on S/N A1500610 through A1500734 S/N 681, A1500433, A1520735 and on
Airspeed Indicator	CAR 3.757 (See NOTE 4) (S/N 15279406 through 15285939 and 681, A1500433, A1520735 through A1521027)
Operating Limitations	CAR 3.778(a)

Production basis — Production Certificate No. 4. Delegation Option Manufacturer No. CE-1 authorized to issue airworthiness certificates under delegation option provisions of Part 21 of the Federal Aviation Regulations.

Equipment: The basic required equipment as prescribed in the applicable airworthiness regulations (see Certification Basis) must be installed in the aircraft for certification. This equipment must include a current Airplane Flight Manual effective S/N 15282032 and on, S/N 681, and S/N A1520809 and on. In addition, the following item of equipment is required:

1. Stall warning indicator, audible, Cessna Dwg. 0511062 (Model 150 through 150E)

2. Stall warning indicator, audible, Cessna Dwg. 0413029 (Model 150F through 150M, 1977 Model) (A150K through A150M, 1977 Model) (152 and on, A152 and on)

NOTE 1. Current weight and balance report together with list of equipment included in certificated empty weight, and loading instructions, when necessary, must be provided for each aircraft at the time of original certification.

Serial Nos. 17001 through 17999, 59001 through 59018, 15059019 through 15077005 and A1500001 through A1500609
The certificated empty weight and corresponding center of gravity location must include unusable fuel of 21 lb. at (+40) for landplanes or 27 lb. at (+40) for seaplanes and an undrainable oil of (0) lb. at (-13.5) for both landplane and seaplane.

NOTE 1. (cont'd)

Serial Nos. 15077006 through 15079405 and A1500610 through A1500734
The certificated empty weight and corresponding center of gravity location must include unusable fuel of 21 lb. at (+40) and full oil of 11.3 lb. at (-13.5) for landplane.

Serial Nos. 15279406 and on, and 681, A1500433, A1520735 and on
The certificated empty weight and corresponding center of gravity location must include unusable fuel of 9 lb. at (+40) and full oil of 11.3 lb. at (-14.7) for landplane.

NOTE 2. The following information must be displayed in the form of composite or individual placards.

A. In full view of the pilot:

(1) "This airplane must be operated in compliance with the operating limitations stated in the form of placards, markings and manuals."

(2) (a) Model 150, 150A, 150B and 150C
"Acrobatic maneuvers are limited to the following:

Maneuver	Entry Speed
Chandelle	106 m.p.h. (92 knots)
Steep turns	106 m.p.h. (92 knots)
Lazy eights	106 m.p.h. (92 knots)
Stalls (except whip)	Use slow deceleration
Spins	Use slow deceleration

Spin recovery - opposite rudder-neutral elevator
Intentional spins with flaps extended prohibited
Design maneuvering speed 106 m.p.h. (92 knots)"

(b) Model 150D, 150E, 150F, 150G, 150H, 150J, 150K
"Acrobatic maneuvers are limited to the following:

Maneuver	Entry Speed
Chandelle	109 m.p.h. (95 knots)
Steep turns	109 m.p.h. (95 knots)
Lazy eights	109 m.p.h. (95 knots)
Stalls (except whip)	Use slow deceleration
Spins	Use slow deceleration

Intentional spins with flaps extended prohibited
Spin recovery - opposite rudder-forward elevator
Maximum design weight - Landplane 1600 lb.
Seaplane 1650 lb.
Maximum maneuvering speed 109 m.p.h. (95 knots)
Maximum flight maneuvering load factors
Flaps Up +4.4 -1.76
Flaps Down +3.5"

(3) Model A150K
"This airplane must be operated as an Acrobatic Category airplane in compliance with the operating limitations stated in the form of placards, markings and manuals.

Acrobatic Category
Maximum design weight 1600 lb.
Maximum maneuvering speed 118 m.p.h. (103 knots)
Refer to weight and balance data for loading instructions
Flight maneuvering load factors: Flaps up +6.0 -3.0 Flaps down: +3.5
Aerobatic maneuvers with flaps extended are prohibited.
Inverted flight is prohibited.
Child's seat and/or baggage compartment must not be occupied during aerobatic maneuvering. Spin recovery: Apply opposite rudder, followed by forward elevator for normal recovery.

(3) (cont'd)

The following aerobatic maneuvers are approved:

Maneuver	Entry Speed		Maneuver	Entry Speed	
Chandelle	120 m.p.h.	(104 knots)	Lazy eights	120 m.p.h.	(104 knots)
Steep turns	110 m.p.h.	(96 knots)	Spins	Slow deceleration	
Barrel rolls	130 m.p.h.	(113 knots)	Aileron rolls	130 m.p.h.	(113 knots)
Snap rolls	90 m.p.h.	(78 knots)	Immelmanns	145 m.p.h.	(126 knots)
Loops	130 m.p.h.	(113 knots)	Cuban eights	145 m.p.h.	(126 knots)
Vertical reversements	90 m.p.h.	(78 knots)	Stalls (except whip stalls)	Slow deceleration"	

(4) Model 150L and 150M (1971 Model through 1975 Model)

"This airplane is approved in the utility category and must be operated in compliance with the operating limitations as stated in the form of placards, markings, and manuals.

	Maximums	
Maneuvering speed		109 m.p.h. CAS (95 knots)
Gross weight		1600 lb.
Flight load factor	Flaps Up	+4-4, -1.76
	Flaps Down	+3.5

Maneuver	Max. Entry Speed	Maneuver	Max. Entry Speed
Chandelles	109 m.p.h. (95 knots)	Spins	Slow deceleration
Lazy eights	109 m.p.h. (95 knots)	Stalls (except whip stalls)	Slow deceleration
Steep turns	109 m.p.h. (95 knots)		

Spin Recovery: opposite rudder - forward elevator - neutralize controls. Intentional spins with flaps extended are prohibited. Known icing conditions to be avoided. This airplane is certified for the following flight operations as of date of original airworthiness certificate:

(DAY - NIGHT - VFR - IFR)" (AS APPLICABLE)

(5) Model A150L and A150M (1971 Model through 1975 Model)

"This airplane is approved in the utility category and must be operated in compliance with the operating limitations as stated in the form of placards, markings, and manuals.

	Maximums	
Maneuvering speed		118 m.p.h. (CAS (103 knots)
Gross weight		1600 lb.
Flight load factor	Flaps up	+6.0, -3.0
	Flaps Down	+3.5

Aerobatic maneuvers with flaps extended are prohibited.
Inverted flight is prohibited.
Child's seat and/or baggage compartment must not be occupied during aerobatics.

Maneuver	Max. Entry Speed		Maneuver	Max. Entry Speed	
Chandelle	120 m.p.h.	(104 knots)	Lazy eights	120 m.p.h.	(104 knots)
Steep turns	110 m.p.h.	(96 knots)	Spins	Slow deceleration	
Barrell rolls	130 m.p.h.	(113 knots)	Aileron rolls	130 m.p.h.	(113 knots)
Snap rolls	90 m.p.h.	(78 knots)	Immelmanns	145 m.p.h.	(126 knots)
Loops	130 m.p.h.	(113 knots)	Cuban eights	145 m.p.h.	(126 knots)
Vertical reversements	90 m.p.h.	(78 knots)	Stalls (except whip stalls)	Slow deceleration	

(5) (cont'd)

Spin Recovery: opposite rudder - forward elevator - neutralize controls.
Known icing conditions to be avoided. This airplane is certified for the following flight operations as of date of original airworthiness certificate:

(DAY - NIGHT - VFR - IFR)" (As Applicable)

(6) Model 150M (1976 and 1977 Model)
"This airplane is approved in the utility category and must be operated in compliance with the operating limitations as stated in the form of placards, markings, and manuals.

Maximums		
Maneuvering speed		97 knots
Gross weight		1600 lb.
Flight load factor	Flaps up	+4.4, -1.76
	Flaps Down	+3.5

NO ACROBATIC MANEUVERS APPROVED EXCEPT THOSE LISTED BELOW

Maneuver	Max. Entry Speed	Maneuver	Max. Entry Speed
Chandelles	95 knots	Spins	Slow deceleration
Lazy eights	95 knots	Stalls (except whip stalls)	Slow deceleration
Steep turns	95 knots		

Abrupt use of controls prohibited above 97 knots.
Spin Recovery: opposite rudder - forward elevator - neutralize controls.
Intentional spins with flaps extended are prohibited. Flight into known icing conditions prohibited. This airplane is certified for the following flight operations as of date of original airworthiness certificate:

(DAY - NIGHT - VFR - IFR)" (As applicable)

(7) A150M (1976 and 1977 Model)
"This airplane is approved in the acrobatic category and must be operated in compliance with the operating limitations as stated in the form of placards, markings, and manuals.

Maximums		
Maneuvering speed (IAS)		105 knots
Gross weight		1600 lb.
Flight load factor -	Flaps up	+6.0, -3.0
	Flaps Down	+3.5

Aerobatic maneuvers with flaps extended are prohibited. Inverted flight is prohibited.
Baggage compartment and/or child's seat must not be occupied during aerobatics.

THE FOLLOWING AEROBATIC MANEUVERS ARE APPROVED

Maneuver	Recm. Entry Speed	Maneuver	Recm. Entry Speed
Chandelles	105 knots	Lazy eights	105 knots
Steep turns	100 knots	Spins	Slow deceleration
Barrel rolls	115 knots	Aileron rolls	115 knots
Snap rolls	80 knots	Immelmanns	130 knots
Loops	115 knots	Cuban eights	130 knots
Vertical reversements	80 knots	Stalls (except whip stalls)	Slow deceleration

Abrupt use of controls prohibited above 105 knots.
Spin Recovery: opposite rudder - forward elevator - neutralize controls.
Flight into known icing conditions prohibited. This airplane is certified for the following flight operations as of date of original airworthiness certificate:

(DAY - NIGHT - VFR - IFR)" (As Applicable)

(8) Model 152 (1978 Model)
"This airplane is approved in the utility category and must be operated in compliance with the operating limitations as stated in the form of placards, markings, and manuals.

Maximums		
Maneuvering speed (IAS)		104 knots
Gross weight		1670 lbs.
Flight load factor	Flaps up	+4.4, -1.76
	Flaps Down	+3.5

NO ACROBATIC MANEUVERS APPROVED EXCEPT THOSE LISTED BELOW

Maneuver	Recm. Entry Speed	Maneuver	Recm. Entry Speed
Chandelles	95 knots	Spins	Slow deceleration
Lazy eights	95 knots	Stalls (except whip stalls)	Slow deceleration
Steep turns	95 knots		

Abrupt use of controls prohibited above 104 knots.
Intentional spins with flaps extended are prohibited. Altitude loss in a stall recovery -- 160 ft. Flight into known icing conditions prohibited. This airplane is certified for the following flight operations as of date of original airworthiness certificate:
(DAY - NIGHT - VFR - IFR)" (As applicable)

(9) Model A152 (1978 Model and A1500433)
"This airplane is approved in the acrobatic category and must be operated in compliance with the operating limitations as stated in the form of placards, markings, and manuals.

Maximums		
Maneuvering speed (IAS)		108 knots
Gross weight		1670 lb.
Flight load factor	Flaps Up	+6.0, -3.0
	Flaps Down	+3.5

Aerobatic maneuvers with flaps extended are prohibited. Inverted flight is prohibited. Baggage compartment and/or child's seat must not be occupied during aerobatics.

THE FOLLOWING AEROBATIC MANEUVERS ARE APPROVED

Maneuver	Recm. Entry Speed	Maneuver	Recm. Entry Speed
Chandelles	105 knots	Lazy eights	105 knots
Steep turns	100 knots	Spins	Slow deceleration
Barrel rolls	115 knots	Aileron rolls	115 knots
Snap rolls	80 knots	Immelmanns	130 knots
Loops	115 knots	Cuban eights	130 knots
Vertical reversements	80 knots	Stalls (except whip stalls)	Slow deceleration

Abrupt use of controls prohibited above 108 knots.
Altitude loss in a stall recovery -- 160 ft. Flight into known icing conditions prohibited. This airplane is certified for the following flight operations as of date of original airworthiness certificate:
(DAY - NIGHT - VFR - IFR)" (As Applicable)

B. On the flap handle
(1) Models 150, 150A, 150B, 150C
"Flaps - Pull to extend
Retracted 0°
Takeoff - 1st Notch 10°
2nd Notch 20°
3rd Notch 30°
Landing - 4th Notch 40°"

B. (2) Models 150D. 150E
"Flaps - Pull to extend
Takeoff - Retracted 0°
Landing - 0°-40°"

C. In the baggage compartment
(1) Models 150. 150A. 150B. 150C
"Baggage - 80 lb. maximum."
(2) Model 150D. 150E
"Baggage - 120 lb. maximum."
(3) S/N 15279406 through 15282031. A1500433. A15200735 through A1520808
"120 lb. maximum baggage and/or auxiliary seat passenger. For additional loading instructions see Weight and Balance Data."

D. On the instrument panel
(1) Models 150K. A150K; 1971 Models 150L. A150L
"Do not turn off alternator in flight except in emergency."

E. Near fuel shut-off valve
(1) Models 150 through 150M (1977 Model) and A150K through A150M (1977 Model)
"Fuel 22.5 gals. ON-OFF."
(2) S/N 15279406 through 15282031. A1500433. A15200735 through A1520808
"Fuel 24.5 gals. ON-OFF."

F. On front door posts
(1) S/N A15200735 through A1520808. A1500433
"Emergency door release
1. Unlatch door
2. Pull 'D' ring."

G. On door near window latch
(1) Model A150K through A150M (1975 Model)
"Do not open window above 165 m.p.h."
(2) Model A150M (1976 and 1977 Model) (1978 Model A152)
"Do not open window above 143 knots IAS."

H. On the instrument panel near overvoltage light (Model 150L through 150M, and A150L through A150M, 1978 Model 152 and A152, and A1500433)
(1) "High Voltage"

I. On left hand instrument panel
(1) S/N 15279406 through 5282031, A1500433, A1520735 through A1520808
"Spin Recovery
1. Verify ailerons are neutral and throttle is closed.
2. Apply full opposite rudder.
3. Move control wheel briskly forward to break stall."

J. S/N 15282032 and on. S/N 681. and S/N A1520809 and on
All placards required in the Pilot's Operating Handbook and FAA Approved Airplane Flight Manual must be installed in the appropriate locations.

NOTE 3. Reserved

NOTE 4. The markings of the airspeed indicator with IAS provides an equivalent level of safety to CAR 3.757 when the approved airspeed calibration data presented in Section V of the Pilot's Operating Handbooks listed below is available to the pilot:

150M,	Cessna P/N D1055-13	(S/N 15077006 through 15078505)
A150M,	Cessna P/N D1056-13	(S/N A1500610 through A1500684)
150M,	Cessna P/N D1080-13	(S/N 1507506 through 15079405)
A150M,	Cessna P/N D1081-13	(S/N A1500685 through A1500734)
152,	Cessna P/N D1107-13	(S/N 15279406 through 15282031)
A152,	Cessna P/N D1108-13	(S/N A1500433 through A1520735 through A1520808)
152,	Cessna P/N D1136-13PH	(S/N 15282032 through 15283591)
A152,	Cessna P/N D1137-13PH	(S/N 681, A1520809 through A1520878)
152,	Cessna P/N D1170-13PH	(S/N 15283592 through 15284541)
A152,	Cessna P/N D1171-13PH	(S/N A1520879 through A1520943)
152,	Cessna P/N D1190-13PH	(S/N 15284542 through 15285161)
A152,	Cessna P/N D1191-13PH	(S/N A1520944 through A1520983)
152,	Cessna P/N D1210-13PH	(S/N 15285162 through 15285594)
A152,	Cessna P/N D1211-13PH	(S/N A1520984 through A1521014)
152,	Cessna P/N D1229-13PH	(S/N 15285595 through 15285833)
A152,	Cessna P/N D1230-13PH	(S/N A1521015 through A1521025)
152,	Cessna P/N D1249-13PH	(S/N 15285834 through 15285939)
A152,	Cessna P/N D1250-13PH	(S/N A1521026 through A1521027)

NOTE 5. Near fuel tank filler

A. 150 series through S/N 15079405 and A150 series through S/N A1500734 except A1500433:

"FUEL
80/87 min. grade aviation gasoline
Cap. 13.0 U.S. Gal."

B. S/N 15279406 through 15282031, A1500433, A1520735 through A1520808

"FUEL
100LL/100 min. grade aviation gasoline
Cap. 13.0 U.S. Gal."

NOTE 6. 14-volt electrical system
(150 series through S/N 15079405 and A150 series through S/N A1500734 except A1500433)

28-volt electrical system
(S/N 15279406 and on, S/N 681, A1500433, A/N A1520735 and on)

In addition to the placards specified above the prescribed operating limitations indicated by an asterisk (*) under Sections I through X of this data sheet must also be displayed by permanent markings.

...END...

E-295
Revision 11
AVCO Lycoming
O-540-A1A, -A1A5, -A1B5, -A1C5, -A1D, -A1D5, -A2B, -A3D5, -A4A5, -A4B5, -A4C5, -A4D5,
O-540-B1A5, -B1B5, -B1D5, -B2A5, -B2B5, -B2C5, -B4A5, -B4A5, -B4B5,
O-540-D1A5,
O-540-E4A5, -E4B5, -E4C5,
O-540-F1A5, -F1B5,
O-540-G1A5, -G2A5,
O-540-H1A5, -H2A5, -H1A5D, -H2A5D, -H1B5D, -H2B5D,
O-540-J1A5D, -J2A5D, -J1B5D, -J2B5D, -J3A5D, -J1C5D, -J2C5D, -J1D5D, -J2D5D, -J3C5D, -L3C5D

March 20, 1986

TYPE CERTIFICATE DATA SHEET NO. E-295

Engines of models described herein conforming with this data sheet (which is a part of Type Certificate No. 295) and other approved data on file with the Federal Aviation Administration meet the minimum standards for use in certificate aircraft in accordance with pertinent aircraft data sheets and applicable portions of the Civil Air Regulations/Federal Aviation Regulations provided they are installed, operated and maintained as prescribed by the approved manufacturer's manuals and other approved instructions.

Manufacturer — AVCO Lycoming Williamsport Division
AVCO Corporation
Williamsport, Pennsylvania 17701

Model Lycoming O-540	-A1A, -A1A5, -A1B5, -A1C5, -A1D, -A1D5, -A2B, -A3D5, -A4A5, -A4B5, -A4C5, -A4D5, -D1A5	-B1A5, -B1B5, -B1D5, -B2A5, -B2B5, -B2C5, -B4A5, -B4B5	-E4A5, -E4B5, -E4G5, -G1A5, -G2A5, -H1A5, -H2A5, -H1A5D, -H2A5D, -H1B5D, -H2B5D	-F1A5, -F1B5
Type 6H0A	Direct Drive	- -	- -	- -
Rating				
Maximum continuous, hp., r.p.m, in. Hg., at:				
Critical pressure altitude (ft.)	—	—	—	235-2800-25.0.4000
Sea level pressure altitude	250-2575-F.T.-S.L.	235-2575-F.T.-S.L.	260-2700-F.T.-S.L.	235-2800-26.0-S.L.
Takeoff (5 min.), hp., r.p.m., in. Hg., at:				
Critical pressure altitude (ft.)	—	—	—	260-2800-27.5-800
Sea level pressure altitude	250-2575-F.T.-S.L.	235-2575-F.T.-S.L.	260-2700-F.T.-S.L.	260-2800-28.0-S.L.
Fuel (Minimum grade aviation gasoline)	See NOTE 8	- -	- -	- -

"- -" indicates "same as preceding model"
"—" indicates "does not apply"

Page No.	01	02	03	04	05	06	07	08
Rev. No.	11	11	11	11	11	11	11	11

Model Lycoming O-540	-A1A, -A1A5, -A1B5, -A1C5, -A1D, -A1D5, -A2B, -A3D5, -A4A5, -A4B5, -A4C5, -A4D5, -D1A5	-B1A5, -B1B5, -B1D5, -B2A5, -B2B5, -B2C5, -B4A5, -B4B5	-E4A5, -E4B5, -E4G5, -G1A5, -G2A5, -H1A5, -H2A5, -H1A5D, -H2A5D, -H1B5D, -H2B5D	-F1A5, -F1B5
Lubricating oil (lubricants which conform to the specifications as listed or to subsequent revision thereto.)	No. 301-F	--	--	--
Bore and stroke. in.	5.125 X 4.375	--	--	--
Displacement, cu. in.	541.5	--	--	--
Compression ratio	See NOTE 8	--	--	--
Weight (dry)	See NOTE 5	--	--	--
C.G. location (dry)	See NOTE 5	--	--	--
From front face of prop shaft flange, in	17.9	--	--	--
Off propeller shaft C.L., in.	1.21 below 0.15 left	--	--	--
Propeller shaft-AS-127	Type 2 flange modified	--	--	--
Carburetion	Marvel-Schebler MA-4-5	--	--	--
Ignition, dual	See NOTE 8	--	--	--
Timing, °BTC	25	--	--	--
Spark plugs	See NOTE 7	--	--	--
Oil sump capacity, qt.	12	--	--	--
Crankshaft dampers	See NOTE 5 & 6	--	--	--
Minimum safe oil quantity qts.				
20°nose up or down attitude	2-3/4	--	--	--
30°nose up attitude	4	--	--	--
NOTES - As applicable	1 through 8, 10, 11	--	--	1 through 11

Model Lycoming O-540	-J1A5D, -J2A5D, -J1B5D, -J2B5D, -J3A5D	-J1C5D, -J2C5D, -J3C5D, -J1D5D, -J2D5D	-L3C5D (See NOTE 12)
Type 6H0A	Direct Drive	--	--
Rating			
Maximum continuous, hp., r.p.m, in. Hg., at:			
Critical pressure altitude (ft.)	—	—	—
Sea level pressure altitude	235-2400-F.T.-S.L.	235-2400-F.T.-S.L.	235-2400-F.T.-S.L.
Takeoff (5 min.), hp., r.p.m., in. Hg., at:			
Critical pressure altitude (ft.)	—	--	--
Sea level pressure altitude	235-2400-F.T.-S.L.	235-2400-F.T.-S.L.	235-2400-F.T.-S.L.
Fuel (Minimum grade aviation gasoline)	See NOTE 8	--	--
Lubricating oil (lubricants which conform to the specifications as listed or to subsequent revision thereto.)	No. 301-F	--	--

"- -" indicates "same as preceding model"
"—" indicates "does not apply"

Model Lycoming O-540	-J1A5D, -J2A5D, -J1B5D, -J2B5D, -J3A5D	-J1C5D, -J2C5D, -J3C5D, -J1D5D, -J2D5D	-L3C5D (See NOTE 12)
Bore and stroke. in.	5.125 X 4.375	--	--
Displacement, cu. in.	541.5	--	--
Compression ratio	See NOTE 8	--	--
Weight (dry)	See NOTE 5	--	--
C.G. location (dry)	See NOTE 5	--	--
From front face of prop shaft flange, in	17.75	17.94	18.10
Off propeller shaft C.L., in.	0.75 below 0.19 left	0.69 below 0.19 left	0.59 below 0.34 left
Propeller shaft-AS-127	Type 2 flange modified	--	--
Carburetion	Marvel Schebler HA-6	--	--
Ignition dual	See NOTE 8	25	--
Timing, °BTC	23	--	--
Spark plugs	See NOTE 7	--	--
Oil sump capacity, qts.	12	--	--
Crankshaft dampers	See NOTE 5 & 6	--	--
Minimum safe oil quantity qts.	--	--	--
20°nose up or down attitude	2-3/4	--	--
30°nose up attitude	2	--	--
		--	
NOTES - As applicable	1 through 8, 10, 11		1 through 8, 10, 11, 12, 13

"- -" indicates "same as preceding model"
"—" indicates "does not apply"

Certification basis:

Regulations and Amendments	Model	Date of Application	Date Type Certificate No. E-295 Issued/Revised
CAR 13 Effective June 15, 1956	O-540-A1A	July 2, 1957	October 31, 1957
As Amended By 13-1 & 13-2	O-540-A1A5	June 3, 1958	June 18, 1958
	O-540-A2P	July 24, 1958	July 24, 1958
	O-540-D1A5	October 21, 1958	August 12, 1959
	O-540-A1B5	January 21, 1959	February 10, 1959
	O-540-A1C5	March 16, 1959	April 2, 1959
	O-540-F1A5, -F1B5	April 3, 1959	August 12, 1959
	O-540-A1D, -A1D5	January 21, 1960	March 17, 1960
13-3	O-540-A3D5	May 17, 1960	June 22, 1960
	O-540-B1A5, -B2A5	November 30, 1960	May 3, 1961
	O-540-B1B5	April 17, 1961	May 3, 1961
	O-540-B2B5	December 8, 1961	December 26, 1961
13-4	O-540-A4A5, -A4B5, -A4C5, -A4D5, -B4A5, -B4B5	October 3, 1963	October 9, 1963
	O-540-E4A5, -E4B5	April 1, 1964	May 4, 1964
	O-540-E4C5	March 3, 1966	March 23, 1966
	O-540-B1D5, -B2C5	November 23, 1966	December 2, 1966
	O-540-G2A5	March 31, 1967	April 4, 1967
	O-540-G1A5	October 6, 1967	October 9, 1967
	O-540-H1A5, -H2A5	January 16, 1970	January 22, 1970
	O-540-H1B5D, H2B5D	July 30, 1971	August 4, 1971
	O-540-H1A5D, -H2A5D	July 27, 1971	October 21, 1971

Certification basis: (cont'd)

Regulations and Amendments	Model	Date of Application	Date Type Certificate No. E-295 Issued/Revised
13-4	-J1B5D, -J2B5D		
	O-540-J1C5D, -J2C5D	August 25, 1976	October 4, 1976
	-J1D5D, -J2D5D		
	O-540-J3C5D	February 4, 1977	February 15, 1977
	O-540-J3A5D	November 23, 1977	November 30, 1977
	O-540-L3C5D	July 21, 1977	June 19, 1978

Production basis: Production Certificate No. 3

NOTE 1. Maximum permissible temperatures are as follows:

Cylinder Head (well type)	Cylinder Base	Oil Inlet
500°F	325°F	245°F

NOTE 2. Pressure limits - p.s.i.

	Minimum	Maximum
	0.5	30.0 (O-540-L3C5D: See NOTE No. 13)
Fuel	0.5	8.0
Oil (Normal operation)	55.0	95.0
(Idle)	25.0	—
(Starting and warm-up)	—	115.0

NOTE 3. The following accessory provisions are incorporated:

Accessory	-A1A, -A1A5, -A1B5, -A1C5, -A1D, -A1D5, -A4A5, -A4B5, -A4C5, -A4D5 -E4A5, -E4B5 -E4C5	A3D5	-A2B -B2A5 -B2B5 -B2C5	-B1A5 -B1B5 -B1D5, -B4A5 -B4B5, -G1A5	-D1A5	-G2A5	-H1A5 -H2A5	-F1A5 -F1B5
Starter	*	*	*	*	*	*	*	—
Starter	—	—	—	—	—	—	—	*
Generator	*	*	*	*	*	*	—	*
Generator	**	**	**	**	**	**	—	—
Alternator	**	**	**	**	—	**	*	—
Alternator	**	**	**	**	**	**	**	**
Vacuum Pump	*	*	*	*	*	*	*	*
Hydraulic Pump	*	*	*	*	*	*	*	*
Hydraulic Pump	—	—	—	—	—	—	—	—
Tachometer	*	*	*	*	*	*	*	*
Propeller Governor	*	*	—	*	*	—	*	—
Propeller Governor	—	—	—	—	—	—	—	—
Fuel Pump	**	**	**	**	**	**	**	**
Fuel Pump (plunger)	**	*	**	**	**	**	**	**

Accessory	-L3C5D	-H1A5D -H2A5D -H1B5D -H2B5D	-J2D5D -J2C5D	-J1A5D -J2A5D -J3A5D -J1B5D -J2B5D -J3C5D -J1D5D -J1C5D	All Models Rotation Facing Drive Pad	Speed Ratio to Crankshaft	Maximum Torque (in. -lb.) Cont.	Static	Max. Overhang Moment (in. -lb.)
Starter	*	*	*	*	CC	16.556:1	—	450	150
Starter	—	—	—	—	CC	13.556:1	—	450	150
Generator	—	—	—	—	C	1.010:1	60	120	175
Generator	—	—	—	—	C	2.500:1	60	120	175
Alternator	*	*	*	*	C	3.250:1	60	120	175
Alternator	**	**	**	—	C	3.630:1	60	120	175
Vacuum Pump	*	*	*	*	CC	1.300:1	70	450	25
Hydraulic Pump	—	—	—	—	C	1.385:1	100	800	40
Hydraulic Pump	*	*	*	*	C	1.300:1	100	800	40
Tachometer	*	*	*	*	C	1.500:1	7	50	5
Propeller Governor	—	—	—	—	C	0.895:1	125	1200	25
Propeller Governor	*	*	—	*	C	0.947:1	125	1200	25
Fuel Pump	—	**	—	—	CC	1.000:1	25	—	25
Fuel Pump (plunger)	*	**	**	**	—	0.500:1	—	—	10

"C" - Clockwise "CC" - Counter clockwise
* - Standard
** - Optional

NOTE 4. These engines incorporate provisions for absorbing propeller thrust in both tractor and pusher type installations.

NOTE 5. These models incorporate additional characteristics as follows:

O-540-Models	Wt. dry, lb.	Characteristics
-A1A	374	Basic model, direct drive, six cylinder, horizontally opposed, air cooled engine with one each S6LN-20 and -21 Magnetos and two 6th order dampers.
-A1A5	374	Same as -A1A except has one fifth and one sixth order dampers.
-A1B5	375	Same as -A1A5 except has propeller governor pad with short studs to accommodate AN type governor.
-A1C5	375	Same as -A1A5 except has two S6LN-21 impulse coupling magnetos.
-A1D	375	Similar to -A1B5 except has one each S6LN-200 and S6LN-204 magnetos and two sixth order crankshaft torsional dampers.
-A1D5	375	Similar to -A1D except has one fifth and one sixth order crankshaft torsional dampers.
-A2B	374	Same as -A1B5 except for crankshaft damper arrangement and propeller flange has propeller locating bushings displaced 60° clockwise, viewed facing propeller.
-A3D5	373	Similar to -A1D5 except has provisions for Goodrich propeller deicing equipment.
-A4A5	374	Similar to -A1A5 except has heavier fifth and sixth order crankshaft counterweights.
-A4B5	375	Similar to -A1B5 except has heavier fifth and sixth order crankshaft counterweights.
-A4C5	375	Similar to -A1C5 except has heavier fifth and sixth order crankshaft counterweights.
-A4D5	375	Similar to -A1D5 except has heavier fifth and sixth order crankshaft counterweights.
-B1A5	366	Same as -A1D5 except has lower compression ratio and performance.
-B1B5	366	Field conversion of -A1A5, -A1B5, or -A1C5 to lower compression ratio.
-B1D5	367	Same as -B1A5 except for incorporation of Bendix 1200 series magnetos.
-B2A5	366	Similar to -B1A5 except does not have provisions for controllable pitch propeller.
-B2B5	366	Same as -B2A5 except has S6LN-20 and S6LN-21 magnetos.

NOTE 5. These models incorporate additional characteristics as follows: cont.

-B2C5	368	Same as -B2B5 except for incorporation of Bendix 1200 series magnetos and does not include generator as part of the engine.
-B4A5	366	Similar to -B1A5 except has heavier fifth and sixth order crankshaft counterweights.
B4B5	366	Similar to -B1B5 except has heavier fifth and sixth order crankshaft counterweights.
-D1A5	369	Same as -A1A5 except has increased strength crankcase.
-F4A5	368	Similar to -A4D5 except has hybrid camshaft permitting higher 260 hp. @ 2700 r.p.m.
-E4B5	369	Similar to -A4D5 except for left magneto S6LN-21 and minor difference in weight and length.
-E4C5	370	Same as model -E4B5 except has S6LN-1227 and S6LN-1209 magnetos.
-F1A5	367	Same as -A1A5 except rated for helicopter application and incorporates prototype bed mounting.
-F1B5	369	Same as -D1A5 except rated for helicopter application and incorporates provisions for either bed or dynafocal type mounting.
-G1A5	386	Similar to -E4C5 except incorporates heavier crankshaft, different crankcase and -A1D5 counterweights.
-G2A5	386	Similar to -G1A5 except does not provide for use of constant speed propeller.
-H1A5	385	Similar to -G1A5 except has different magnetos and incorporates piston cooling oil jets.
-H2A5	385	Similar to -G2A5 except has different magnetos and incorporates piston cooling oil jets.
-H1A5D	381	Similar to -H1A5 except incorporates dual magneto (impulse coupling).
-H2A5D	381	Similar to -H1A5D except does not have provision for controllable propeller.
-H1B5D	381	Similar to -H1A5 except incorporates dual magneto (retard).
-H2B5D	381	Similar to -H1B5D except does not have provision for controllable propeller.
-J1A5D	356	Similar to -A1A5 except incorporates dual magneto (impulse coupling), less weight and rated at 235 h.p. @ 2400 r.p.m.
-J2A5D	356	Similar to -J1A5D except does not have provision for controllable propeller.
-J1B5D	356	Similar to -A1A5 except incorporates dual magneto (retard), less weight and rated at 235 h.p. @ 2400 r.p.m.
-J2B5D	356	Similar to -J1B5D except does not have provision for controllable propeller.
-J1C5D	356	Same as -J1A5D except has horizontal carburetor and induction housing.
-J2C5D	356	Same as -J1C5D except has no provision for controllable propeller.
-J1D5D	356	Same as -J1C5D but with D6LN-3230 retard breaker dual magneto.
-J2D5D	356	Same as -J1D5D except does not have provision for controllable propeller.
-J3C5D	357	Same as -J1C5D except has heavier counterweights for use with McCauley controllable propeller.
-J3A5D	357	Same as -J1A5D except has heavier counterweights (same as O-540-J3C5D).
-L3C5D	367	Same as -J3C5D except for features to make engine suitable for turbocharging.

NOTE 6. These engines incorporate crankshafts with two sixth order dampers unless a "5" is part of the model designation, i.e., -A1A5. Engines so designated have one fifth order damper and one sixth order damper instead of two sixth order dampers.

NOTE 7. Spark plugs approved for use on these engines are listed in the latest revision of AVCO Lycoming Service Instruction No. 1042.

NOTE 8. Fuel grade, compression and ignition:

O-540- Models	Fuel - Aviation Gasoline	Compression Ratio	Ignition, Dual Bendix Models
-A1A	100 or 100 LL	8.50:1	S6LN-20, S6LN-21
-A1A5	100 or 100 LL	8.50:1	S6LN-20, S6LN-21
-A1B5	100 or 100 LL	8.50:1	S6LN-21, S6LN-21
-A1C5	100 or 100 LL	8.50:1	S6LN-21, S6LN-21
-A1D	100 or 100 LL	8.50:1	S6LN-204, S6LN-200
-A1D5	100 or 100 LL	8.50:1	S6LN-204, S6LN-200
-A2B	100 or 100 LL	8.50:1	S6LN-20, S6LN-21
-A3D5	100 or 100 LL	8.50:1	S6LN-204, S6LN-200
-A4A5	100 or 100 LL	8.50:1	S6LN-20, S6LN-21
-A4B5	100 or 100 LL	8.50:1	S6LN-21, S6LN-21
-A4C5	100 or 100 LL	8.50:1	26LN-21, S6LN-21
-A4D5	100 or 100 LL	8.50:1	26LN-204, S6LN-200
-B1A5	100 or 100 LL	7.20:1	S6LN-204, S6LN-200
-B1B5	100 or 100 LL	7.20:1	S6LN-20, S6LN-21
-B1D5	100 or 100 LL	7.20:1	S6LN-1209, S6LN-1208
-B2A5	100 or 100 LL	7.20:1	S6LN-204, S6LN-200
-B2B5	100 or 100 LL	7.20:1	S6LN-20, S6LN-21
-B2C5	100 or 100 LL	7.20:1	S6LN-1209, S6LN-1227
-B4A5	100 or 100 LL	7.20:1	S6LN-204, S6LN-200
-B4B5	100 or 100 LL	7.20:1	S6LN-20, S6LN-21
-D1A5	100 or 100 LL	8.50:1	S6LN-20, S6LN-21
-E4A5	100 or 100 LL	8.50:1	S6LN-204, S6LN-200
-E4B5	100 or 100 LL	8.50:1	S6LN-204, S6LN-200
-E4C5	100 or 100 LL	8.50:1	S6LN-204, S6LN-200
-F1A5	100 or 100 LL	8.50:1	S6LN-20, S6LN-21
-F1B5	100 or 100 LL	8.50:1	S6LN-204, S6LN-200
-G1A5	100 or 100 LL	8.50:1	S6LN-1227, S6LN-1209
-G2A5	100 or 100 LL	8.50:1	S6LN-1227, S6LN-1209
-H1A5	100 or 100 LL	8.50:1	S6LN-20, S6LN-21
-H2A5	100 or 100 LL	8.50:1	S6LN-20, S6LN-21
-H1A5D	100 or 100 LL	8.50:1	D6LN-3031
-H2A5D	100 or 100 LL	8.50:1	D6LN-3031
-H1B5D	100 or 100 LL	8.50:1	D6LN-3230
-H2B5D	100 or 100 LL	8.50:1	D6LN-3230
-J1A5D	100 or 100 LL	8.50:1	D6LN-3031
-J2A5D	100 or 100 LL	8.50:1	D6LN-3031
-J1B5D	100 or 100 LL	8.50:1	D6LN-3230
-J2B5D	100 or 100 LL	8.50:1	D6LN-3230
-J1C5D	100 or 100 LL	8.50:1	D6LN-3031
-J2C5D	100 or 100 LL	8.50:1	D6LN-3031
-J1D5D	100 or 100 LL	8.50:1	D6LN-3230
-J2D5D	100 or 100 LL	8.50:1	D6LN-3230
-J3C5D	100 or 100 LL	8.50:1	D6LN-3031
-J3A5D	100 or 100 LL	8.50:1	D6LN-3031

All models equipped with one impulse coupling magneto may use two impulse coupling magnetos as optional equipment.

NOTE 9. Engine models O-540-F1A5 and -F1B5 are approved for helicopter application and operation in a horizontal installation.

NOTE 10. Models O-540-A4A5, -A4B5, -A4C5, -A4D5, -B4A5, -B4B5, -E4B5, -E4A5, and -E4C5 are equipped with fifth and sixth order crankshaft counterweights which are heavier than the usual fifth and sixth order counterweights employed in other O-540 engine models.

NOTE 11. Starters, generators, and alternators approved for use on these engines are listed in the latest revision of AVCO Lycoming Service Instruction No. 1154.

NOTE 12. When equipped in accordance with Cessna Dwg. 2250065, this engine is certified for operation at a maximum manifold pressure of 31.0 in. Hg at 2400 r.p.m.

NOTE 13. When complying with Lycoming Service Instruction No. 1398, the minimum permissible fuel pressure increase from 0.5 psi to 3 psi. Therefore, revised fuel pressure gage marking indicating a minimum red line of 3 psi is required.

.....END.....

	A7CE
	Revision 42
	CESSNA
401	411A
401A	414
401B	414A
402	421
402A	421A
402B	421B
402C	421C
411	425
	October 15, 1997

A7CE

TYPE CERTIFICATE DATA SHEET NO. A7CE

This data sheet which is part of Type Certificate No. A7CE prescribes conditions and limitations under which the product for which the type certificate was issued meets the airworthiness requirements of the Federal Aviation Regulations.

Type Certificate Holder — Cessna Aircraft Company
P. O. Box 7704
Wichita, Kansas 67277

I - Model 411 (Normal Category), Approved August 17, 1964
Model 411A (Normal Category), Approved January 26, 1967

Engines — Two Continental GTSIO-520-C, reduction gear ratio .750:1

Fuel — Grade 100 or 100LL aviation gasoline

Engine Limits — For all operations, 2400 propeller r.p.m. (340 hp.)
34.5 in. Hg. Mp. up to critical altitude of 16,000 ft. in standard atmosphere. Above 16,000 ft. the following maximum Mp. applies for maximum r.p.m.

Altitude (ft.)	Max. Allowable Mp. (in. Hg.)
16,000	34.5
18,000	31.2
20,000	29.0
22,000	26.4
24,000	24.3
26,000	22.2
28,000	20.2
30,000	18.5

Propeller and Propeller Limits

1. Model 411 only
Two Hartzell full-feathering 3-bladed propeller installations
(a) Hartzell Hub HC-A3VF-2D with V8833 blades
Diameter: not over 88.4 in., not under 86.4 in.
(no further reduction permitted)
Pitch settings at 30 in. station:
low 14.0°, +0°, -2°
feathered 84.0°, +2°, -0°
(b) Hydraulic Governor Woodward A210444, 210439, C210446 or B210529
(c) Propeller spinner and bulkhead assembly, Hartzell 835-20

Page No.	1	2	3	4	5	6	7	8	9	10	11	12	13	14	15	16	17	18	19	20
Rev. No.	42	41	42	41	42	40	40	41	41	41	40	41	41	40	42	41	41	40	41	40
Page No.	21	22	23	24																
Rev. No.	41	40	42	41																

Propeller and Propeller Limits	or	2. Models 411 and 411A Two McCauley full-feathered 3-bladed propeller installations (a) McCauley hub 3AF34C74 with 90LF-0 blades or McCauley hub 3AF37C510 with 90LFB blades Diameter: not over 90 in., not under 84.0 in. with 90LF-0 blades or not under 88.0 in. with 90LFB-0 blades. (no further reduction permitted) Pitch settings at 30 in. station: low 14.0°, ±0.2° feathering 84.5°, ±0.3° (b) Hydraulic governor Woodward A210444, 210439, C210446 or B210529 (c) Propeller spinner and bulkhead assembly, McCauley D-3574 or D-3732 for use with C74 Model Propeller, or McCauley D-7229 for use with C510 Model Propeller.

Airspeed Limits (CAS)		
	Maneuvering	180 m.p.h. (156 knots)
	Maximum structural cruising	230 m.p.h. (200 knots)
	Never exceed	266 m.p.h. (231 knots)
	Landing gear operating	160 m.p.h. (139 knots)
	Landing gear extended	160 m.p.h. (139 knots)
	Flaps extended 15°	180 m.p.h. (156 knots)
	Flaps extended 45°	160 m.p.h. (139 knots)
	Minimum control	103 m.p.h. (90 knots)

C.G. Range (Landing Gear Extended)	(+150.6) to (+155.5) at 6500 lb. (+155.7) at 6100 lb. or less (+144.3) at 5200 lb. or less Straight line variation between points given Landing gear retracted moment change: +837 in.-lb.
Empty Wt. C.G. Range	None
Leveling Means	External screw heads on right side of fuselage at stations +213.65 and +238.00 on W.L. +93.80
Maximum Weight	Landing 6500 lb., takeoff 6500 lb.
No. of Seats	6, 7 or 8 (2 at +137.0, 2 at +175.5, 2 at +215.5, 1 or 2 at +238.0) (See manufacturer's equipment list for optional seating arrangements)
Maximum Baggage	Model 411: 120 lb. (+58.0), 240 lb. (+186.0), 340 lb. (+246.5) Model 411A: 350 lb. (+71.0), 240 lb. (+186.0), 340 lb. (+246.5)
Fuel Capacity	175 gal. (2 wing tip tanks, 51 gal. ea., 50 gal. usable at +152.0 and 2 wing tanks, 36.5 gal. ea., 35 gal. usable at +164.0) See NOTE 1 for data on unusable fuel
Oil Capacity	26 qt. (13 qt. in ea. engine at +115.4; usable 7.0 qt. per engine) See NOTE 1 for undrainable oil

I - Model 411, Model 411A (cont'd)

Control Surface Movements	Wing flaps			Down	45°, +1°, -0°
	Main surfaces				
	Aileron	Up	20°, +1°, -0°	Down	20°, +1°, -0°
	Elevator	Up	25°, +1°, -0°	Down	15°, +1°, -0°
	Rudder	Right	32°, +1°, -0°	Left	32°, +1°, -0°
	(Read degrees normal to rudder hinge line)				
	Tab (main surface in neutral)				
	Aileron	Up	20°, +1°, -0°	Down	20°, +1°, -0°
	Elevator	Up	10°, +1°, -0°	Down	26°, +1°, -0°
	Rudder	Right	17°, +1°, -0°	Left	22°, +1°, -0°
	(Read degrees normal to rudder hinge line)				

Serial Nos. Eligible — Model 411: 411-0001 through 411-0250
Model 411A: 411-0251 through 411-0300

II - Model 401 (Normal Category), Approved September 20, 1966
Model 401A (Normal Category), Approved October 29, 1968
Model 401B (Normal Category), Approved November 12, 1969

Engines — Two Continental TSIO-520-E or TSIO-520-EB

Fuel — Grade 100 or 100LL aviation gasoline

Engine Limits — For all operations, 2700 r.p.m. (300 hp.) 34.5 in. Hg. Mp. up to critical altitude of 16,000 ft. in standard atmosphere. Above 16,000 ft. the following maximum Mp. applies for maximum r.p.m.

Altitude (ft.)	Max. Allowable Mp. (in. Hg.)
16,000	34.5
18,000	31.8
20,000	29.5
22,000	27.3
24,000	25.1
26,000	23.0
28,000	22.0
30,000	19.0

Propeller and Propeller Limits — Two McCauley full-feathered 3-bladed propeller installations

(a) McCauley hub 3AF32C87 with 82NC-5.5 blades or McCauley hub 3AF32C504 with 82NEA-5.5 blades
Diameter: not over 76.5 in., not under 74.0 in.
(no further reduction permitted)
Pitch settings at 30 in. station:
low 14.2°, ±0.2°
feathered 81.2°, ±0.3°

(b) Model 401: Hydraulic Governor Woodward B210444, C210439, B210446 or A210529F
Model 401A and 401B: Hydraulic Governor Woodward B210444, C210439, B210446, or A210529F; McCauley DCF290D1/T3, DCF290D2/T3, DCF290D7/T3, DCFU290D1/T3, DCFU290D2/T3, DCFU290D7/T3, DCFU290D13/T3, DCFS290D1/T3, DCFS290D2/T3, DCFS290D7/T3, DCFUS290D1/T3, DCFUS290D2/T3, DCFUS290D7/T3, DCFUS290D13/T3.

(c) Propeller spinner and bulkhead assembly, McCauley D-3534/D-3537, D-3534/D-3796, and D-5212/D5214.

II - Model 401, Model 401A, Model 401B (cont'd)

Airspeed Limits (CAS)	Maneuvering	180 m.p.h. (156 knots)
	Maximum structural cruising	230 m.p.h. (200 knots)
	Never exceed	266 m.p.h. (231 knots)
	Landing gear operating	160 m.p.h. (139 knots)
	Landing gear extended	160 m.p.h. (139 knots)
	Flaps extended 15°	180 m.p.h. (156 knots)
	Flaps extended 45°	160 m.p.h. (139 knots)
	Minimum control	95 m.p.h. (83 knots)

C.G. Range (Landing Gear Extended)
(+150.8) to (+158.1) at 6300 lb.
(+158.5) at 5900 lb. or less
(+147.5) at 5000 lb. or less
Straight line variation between points given
Landing gear retracted moment change: +837 in.-lb.

Empty Wt. C.G. Range — None

Leveling Means — External screw heads on right side of fuselage at stations +213.65 and +238.00 on W.L. +93.80

Maximum Weight — Landing 6200 lb., takeoff 6300 lb.

No. of Seats — 6, 7 or 8 (2 at +137.0, 2 at +175.6, 2 at +215.5, 1 or 2 at +238.0)
(See manufacturer's equipment list for optional seating arrangements)

Maximum Baggage — 350 lb. (+71.0), 240 lb. (+186.0), 340 lb. (+246.5)

Fuel Capacity — 102 gal. (2 wing tip tanks, 51 gal. ea., 50 gal. usable at +152.0)
See NOTE 1 for data on unusable fuel

Oil Capacity — 26 qt. (13 qt. in ea. engine at +113.5; usable 6.5 qt. per engine)
See NOTE 1 for data on undrainable oil

Control Surface Movements

Wing flaps			Down	45°, +1°, -0°
Main surfaces				
Aileron	Up	20°, +1°, -0°	Down	20°, +1°, -0°
Elevator	Up	25°, +1°, -0°	Down	15°, +1°, -0°
Rudder	Right	32°, +1°, -0°	Left	32°, +1°, -0°
(Read degrees normal to rudder hinge line)				
Tab (main surface in neutral)				
Aileron	Up	20°, +1°, -0°	Down	20°, +1°, -0°
Elevator	Up	5°, +1°, -0°	Down	30°, +1°, -0°
Rudder	Right	7°, +1°, -0°	Left	9°, +1°, -0°
(Read degrees normal to rudder hinge line)				

Serial Nos. Eligible
Model 401: 401-0001 through 401-0322
Model 401A: 401A0001 through 401A0132
Model 401B: 401B0001 through 401B0221

III - Model 402 (Normal Category), Approved September 20, 1966
Model 402A (Normal Category), Approved January 3, 1969
Model 402B (Normal Category), Approved November 12, 1969

Engines — Two Continental TSIO-520-E or TSIO-520-EB

Fuel — Grade 100 or 100LL aviation gasoline

III - Model 402, Model 402A, Model 402B (cont'd)

Engine Limits

For all operations, 2700 r.p.m. (300 hp.)
34.5 in. Hg. Mp. up to critical altitude of 16,000 ft. in standard atmosphere. Above 16,000 ft. the following maximum Mp. applies for maximum r.p.m.

Altitude (ft.)	Max. Allowable Mp. (in. Hg.)
16,000	34.5
18,000	31.8
20,000	29.5
22,000	27.3
24,000	25.1
26,000	23.0
28,000	22.0
30,000	19.0

Propeller and Propeller Limits

Two McCauley full-feathered 3-bladed propeller installations

(a) McCauley hub 3AF32C87 with 82NC-5.5 blades or McCauley hub 3AF32C504 with 82NEA-5.5 blades
Diameter: not over 76.5 in., not under 74.0 in.
(no further reduction permitted)
Pitch settings at 30 in. station:
low 14.2°, ±0.2°
feathering 81.2°, ±0.3°

(b) Model 402, 402A and 402B, S/N 402B0001 thru 402B1200
Hydraulic governor, Woodward B210444, C210439, B210446F or A210529H; McCauley DCF290D1/T3, DCF290D2/T3, DCFS290D1/T3, DCFS290D2/T3, DCFU290D1/T3, DCFU290D2/T3, DCFUS290D1/T3, DCFUS290D2/T3, DCF290D7/T3, DCFS290D7/T3,DCFU290D7/T3, DCFU290D13/T3, DCFUS290D7/T3, or DCFUS290D13/T3.

Model 402B, S/N 402B1201 through 402B1300
Hydraulic governor, Woodward B210444, C210439; McCauley DCF290D1/T3, DCF290D2/T3, DCFU290D1/T3, DCFU290D2/T3, DCFS290D4/T3, DCFUS290D4/T3, DCFS290D5/T3, DCFUS290D5/T3, DCF290D7/T3, DCFU290D7/T3, DCFS290D7/T3, DCFUS290D7/T3, DCFU290D13/T3, or DCFUS290D13/T3.

Model 402B, S/N 402B1301 and up
Hydraulic governor, Woodward B210444, C210439; McCauley DCF290D1/T3, DCF290D2/T3, DCFU290D1/T3, DCFU290D2/T3, DCFS290D4/T3, DCFUS290D4/T3, DCFS290D6/T3, DCFUS290D6/T3, DCF290D7/T3, DCFU290D7/T3, DCFS290D7/T3, DCFUS290D7/T3, DCFS290D8/T3, DCFUS290D8/T3, DCFU290D13/T3, DCFUS290D12/T3, or DCFUS290D13/T3.

(c) Propeller spinner and bulkhead assembly, McCauley D-3534/D-3537, D-3534/D-3796, or D-5212/D5214.

A7CE

III - Model 402, Model 402A, Model 402B (cont'd)

Airspeed Limits (CAS)	Model 402, S/N 402-0001 and up	
	Model 402A, S/N 402A0001 and up	
	Model 402B, S/N 402B0001 through 402B0500	
	Maneuvering	180 m.p.h. (156 knots)
	Maximum structural cruising	230 m.p.h. (200 knots)
	Never exceed266 m.p.h. (231 knots)	
	Landing gear operating	160 m.p.h. (139 knots)
	Landing gear extended	160 m.p.h. (139 knots)
Airspeed Limits (Cont.) (CAS)	Flaps extended 15°	180 m.p.h. (156 knots)
	Flaps extended 45°	160 m.p.h. (139 knots)
	Minimum control	95 m.p.h. (83 knots)
	Model 402B, S/N 402B0501 through 402B1000	
	Maneuvering	156 KCAS (180 m.p.h.)
	Maximum structural cruising	200 KCAS (230 m.p.h.)
	Never exceed231 KCAS (266 m.p.h.)	
	Landing gear operating	140 KCAS (161 m.p.h.)
	Landing gear extended	140 KCAS (161 m.p.h.)
	Flaps extended 15°	160 KCAS (184 m.p.h.)
	Flaps extended 45°	140 KCAS (161 m.p.h.)
	Minimum control	83 KCAS (95 m.p.h.)
(IAS)	Model 402B, S/N 402B1001 and up	
	Maneuvering	156 KIAS (180 m.p.h.)
	Maximum structural cruising	199 KIAS (229 m.p.h.)
	Never exceed230 KIAS (265 m.p.h.)	
	Landing gear operating	140 KIAS (161 m.p.h.)
	Landing gear extended	140 KIAS (161 m.p.h.)
	Flaps extended 15°	160 KIAS (184 m.p.h.)
	Flaps extended 45°	140 KIAS (161 m.p.h.)
	Minimum control	82 KIAS (94 m.p.h.)

C.G. Range (Landing Gear Extended)
(+150.8) to (+159.7) at 6300 lb.
(+160.2) at 5900 lb. or less
(+147.5) at 5000 lb. or less
Straight line variation between points given
Landing gear retracted moment change: +837 in.-lb.

Empty Wt. C.G. Range
None

Leveling Means
External screw heads on right side of fuselage at stations +213.65 and +238.00 on W.L. +93.80

Maximum Weight
Models 402, 402A, 402B, S/N 402B0001 through 402B1300
Landing 6200 lb., takeoff 6300 lb.

Model 402B, S/N 402B1301 and up
Landing 6200 lb., rammp 6335 lb., takeoff 6300 lb.

No. of Seats
Model 402
9 (2 at +137.0, 2 at +166.0, 2 at +193.0, 2 at +220.0, 1 at +247.0)

Model 402A and 402B, S/N 402B0001 through 402B0300
9 or 10 (2 at +137.0, 2 at +166.0, 2 at +193.0, 2 at +220.0, 1 or 2 at +247.0)

III - Model 402, Model 402A, Model 402B (cont'd)

	Model 402B, S/N 402B0301 and up 6, 7 or 8 (2 at +137.0, 2 at +175.0, 2 at +218.0, 1 or 2 at +261.0) 9 (with photographic provisions option) (2 at +137.0, 2 at +162.0, 2 at +190.0, 2 at +218.0, 1 at +246.0) 10 (2 at +137.0, 2 at +162.0, 2 at +190.0, 2 at +218.0, 2 at +246.0) (See manufacturer's equipment list for optional seating arrangements)
Maximum Baggage	Models 402, 402A and 402B, S/N 402B0001 through 402B0300 350 lb. (+71.0), 240 lb. (+186.0), 170 lb. (+247.0) Model 402B, S/N 402B0301 and up 250 lb. (+32.0), 350 lb. (+71.0), 240 lb. (+186.0), 400 lb. (+266.0), 100 lb. (+282.0)
Fuel Capacity	102 gal. (2 wing tip tanks, 51 gal. ea., 50 gal. usable at +152.0) See NOTE 1 for data on unusable fuel
Oil Capacity	26 qt. (13 qt. in ea. engine at +113.5; usable 6.5 qt. per engine) See NOTE 1 for data on undrainable oil

Control Surface Movements

Wing flaps			Down	45°, +1°, -0°
Main surfaces				
Aileron	Up	20°, +1°, -0°	Down	20°, +1°, -0°
Elevator	Up	25°, +1°, -0°	Down	15°, +1°, -0°
Rudder	Right	32°, +1°, -0°	Left	32°, +1°, -0°
(Read degrees normal to rudder hinge line)				
Tab (main surface in neutral)				
Aileron	Up	20°, +1°, -0°	Down	20°, +1°, -0°
Elevator	Up	5°, +1°, -0°	Down	30°, +1°, -0°
Rudder	Right	7°, +1°, -0°	Left	9°, +1°, -0°
(Read degrees normal to rudder hinge line)				

Serial Nos. Eligible

Model 402:	402-0001 through 402-0322
Model 402A:	402A0001 through 402A0129
Model 402B:	402B0001 through 402B1384

IV - Model 421 (Normal Category), Approved May 1, 1967
Model 421A (Normal Category), Approved November 19, 1968

Engines	Two Continental GTSIO-520-D, reduction gear ratio .667:1
Fuel	Grade 100 or 100LL aviation gasoline
Engine Limits	For all operations, 2275 propeller r.p.m. (375 hp.) 39.5 in. Hg. Mp. up to critical altitude of 16,000 ft. in standard atmosphere. Above 16,000 ft. the following maximum Mp. applies for maximum r.p.m.

Model 421		Model 421A	
Altitude (ft.)	Max. Allowable Mp. (in. Hg.)	Altitude (ft.)	Max. Allowable Mp. (in. Hg.)
16,000	39.5	16,000	39.5
18,000	32.5	18,000	37.5
20,000	32.5	20,000	35.5
22,000	30.0	22,500	32.5
24,000	27.0	24,000	30.5
26,000	24.5	26,000	28.0
28,000	22.0	28,000	25.5
30,000	20.0	30,000	23.0

IV - Model 421, Model 421A (cont'd)

Propeller and Propeller Limits	Two McCauley full-feathered 3-bladed propeller installations (a) McCauley hub 3AF34C92 with 90LF-0 blades or McCauley hub 3AF37C516 with 90LFB-0 blades. Diameter: not over 90.0 in., not under 88.0 in. (no further reduction permitted) Pitch settings at 30 in. station: low 16.9°, ±0.2° feathering 84.5°, ±0.3°, (b) Hydraulic Governor Woodward 210594, 210595, 210596, or 210597. (c) Propeller spinner and bulkhead assembly, McCauley D-3573/D-3576, for use with C92 Model propeller, or McCauley D-7229 spinner and bulkhead assembly for use with C516 Model propeller.

Airspeed Limits (CAS)

Maneuvering	184 m.p.h. (160 knots)
Maximum structural cruising	230 m.p.h. (200 knots)
Never exceed272 m.p.h. (236 knots)	
Landing gear operating	165 m.p.h. (143 knots)
Landing gear extended	165 m.p.h. (143 knots)
Flaps extended 15°	180 m.p.h. (156 knots)
Flaps extended 45°	165 m.p.h. (143 knots)
Minimum control	106.5 m.p.h. (93 knots)

C.G. Range (Landing Gear Extended)

Model 421	Model 421A
(+151.9) to (+155.5) at 6800 lb.	(+152.1) to (+155.5) at 6840 lb.
(+155.7) at 6400 lb. or less	(+155.7) at 6500 lb. or less
(+144.3) at 5200 lb. or less	(+144.3) at 5200 lb. or less

Straight line variation between points given
Landing gear retracted moment change: +889 in.-lb.

Empty Wt. C.G. Range	None
Leveling Means	External screw heads on right side of fuselage at stations +213.29 and +238.55 on W.L. +93.80
Maximum Weight	Model 421 Landing 6500 lb., takeoff 6800 lb. (See NOTE 4 for takeoff 6840 lb.) Model 421A Landing 6500 lb., takeoff 6840 lb.
No. of Seats	Model 421 6 (2 at +137.0, 2 at +175.5, 2 at +215.5) Model 421A 6 or 7 (2 at +137.0, 2 at +175.5, 2 at +215.5, 1 at +246.5) (See manufacturer's equipment list for optional seating arrangement)
Maximum Baggage	350 lb. (+71.0), 240 lb. (+186.0), 340 lb. (+246.5)
Fuel Capacity	175 gal. (2 wing tip tanks, 51 gal. ea., 50 gal. usable at +152.0 and 2 wing tanks, 36.5 gal. ea., 35 gal. usable at +164.0) See NOTE 1 for data on unusable fuel
Oil Capacity	26 qt. (13 qt. in ea. engine at +115.4; usable 7.0 qt. per engine) See NOTE 1 for data on undrainable oil

IV - Model 421, Model 421A (cont'd)

Control Surface Movements	Wing flaps			Down	45°, +1°, -0°
	Main surfaces				
	Aileron	Up	20°, +1°, -0°	Down	20°, +1°, -0°
	Elevator	Up	25°, +1°, -0°	Down	15°, +1°, -0°
	Rudder	Right	25°, +1°, -0°	Left	25°, +1°, -0°
	(Read degrees normal to rudder hinge line)				
	Tab (main surface in neutral)				
	Aileron	Up	20°, +1°, -0°	Down	20°, +1°, -0°
	Elevator	Up	10°, +1°, -0°	Down	26°, +1°, -0°
	Rudder	Right	11°, +1°, -0°	Left	16°, +1°, -0°
	(Read degrees normal to rudder hinge line)				

Serial Nos. Eligible

Model 421: 421-0001 through 421-0200
Model 421A: 421A0001 through 421A0158

V - Model 414 (Normal Category), Approved September 24, 1969

Engines

Two Continental TSIO-520-J or TSIO-520-JB
(S/N 414-0001 through 414-0800)

Two Continental TSIO-520-N or TSIO-520-NB
(S/N 414-0801 and up)

Fuel

Grade 100 or 100LL aviation gasoline

Engine Limits

For all operations, 2700 r.p.m. (310 hp.)
36.0 in. Hg. Mp. (S/N 414-0001 through 414-0800) 38.0 in. Hg. Mp.
(S/N 414-0801 and up) up to critical altitude of 20,000 ft. in standard atmosphere.
Above 20,000 ft. the following maximum Mp. applies for maximum r.p.m.

S/N 414-0001 through 414-0800

Altitude (ft.)	Max. Allowable Mp. (in. Hg.)
20,000	36.0
22,000	33.6
24,000	31.2
26,000	28.8
28,000	26.4
30,000	24.0

S/N 414-0801 and up

Altitude (ft.)	Max. Allowable Mp. (in. Hg.)
20,000	38.0
22,000	35.2
24,000	32.3
26,000	29.8
28,000	27.4
30,000	25.0

Propeller and Propeller Limits

Two McCauley full-feathered 3-bladed propeller installations
(a) McCauley hub 3AF32C93 with 82NC-5.5 blades or McCauley hub 3AF32C505 with 82NEA-5.5 blades
Diameter: not over 76.5 in., not under 74.5 in. (S/N 414-0001 through S/N 414-0800), not under 75.0 in. (S/N 414-0801 and up)
(no further reduction permitted)
Pitch settings at 30 in. station:
low 14.9°, ±0.2°, feathering 81.2°, ±0.3°

V - Model 414 (Normal Category), Approved September 24, 1969

Propeller and Propeller Limits	(b) Model 414 S/N 414-0001 thru 414-0800 Hydraulic governor, Woodward B210444, C210439, B210446F, or A210529H McCauley DCF290D1/T3,DFC290D2/T3, DCF290D7/T3, DCFU290D1/T3, DCFS290D1/T3, DCFUS290D1/T3, DCFS290D2/T3, DCFU290D2/T3, DCFU290D7/T3, DCFU290D13/T3, DCFS290D7/T3, DCFUS290D2/T3, DCFUS290D7/T3 or DCFUS290D13/T3 Model 414 S/N 414-0801 and up McCauley DCFS290D4/T3, DCFUS290D4/T3, DCFS290D5/T3, DCFUS290D5/T3, DCFS290D7/T3, or DCFUS290D7/T3, DCFS290D8/T3, DCFUS290D8/T3, DCFUS290D12/T3, or DCFUS290D13/T3 (c) Propeller spinner and bulkhead assembly, McCauley D-3534/D-3537, D-3534/D-3796, or D-5212/D-5214.

Airspeed Limits (CAS)	S/N 414-0001 through 414-0450	
	Maneuvering	180 m.p.h. (156 knots)
	Maximum structural cruising	230 m.p.h. (200 knots)
	Never exceed266 m.p.h. (231 knots)	
	Flaps extended 15°	180 m.p.h. (157 knots)
	Flaps extended 45°	160 m.p.h. (139 knots)
	Landing gear operating	160 m.p.h. (139 knots)
	Landing gear extended	160 m.p.h. (139 knots)
	Minimum control	97 m.p.h. (84 knots)
	S/N 414-0451 through 414-0800	
	Maneuvering	156 KCAS (180 m.p.h.)
	Maximum structural cruising	200 KCAS (230 m.p.h.)
	Never exceed231 KCAS (266 m.p.h.)	
	Flaps extended 15°	160 KCAS (184 m.p.h.)
	Flaps extended 45°	140 KCAS (161 m.p.h.)
	Landing gear operating	140 KCAS (161 m.p.h.)
	Landing gear extended	140 KCAS (161 m.p.h.)
	Minimum control	84 KCAS (97 m.p.h.)
(IAS)	S/N 414-0801 and up	
	Maneuvering	160 KIAS (184 m.p.h.)
	Maximum structural cruising	205 KIAS (236 m.p.h.)
	Never exceed236 KIAS (272 m.p.h.)	
	Flaps extended 15°	164 KIAS (189 m.p.h.)
	Flaps extended 45°	147 KIAS (169 m.p.h.)
	Landing gear operating	143 KIAS (165 m.p.h.)
	Landing gear extended	143 KIAS (165 m.p.h.)
	Minimum control	82 KIAS (94 m.p.h.)

C.G. Range (Landing Gear Extended)	(+150.9) to (+159.7) at 6350 lb. (+160.2) at 5950 lb. or less (+147.5) at 5000 lb. or less Straight line variation between points given Landing gear retracted moment change: +837 in.-lb.
Empty Wt. C.G. Range	None
Leveling Means	External screw heads on right side of fuselage at stations +213.29 and +238.55 on W.L. +93.80
Maximum Weight	Landing 6200 lb., takeoff 6350 lb.

V - Model 414 (cont'd)

No. of Seats	S/N 414-0001 through 414-0350 6 or 7 (2 at +137.0, 2 at +175.5, 2 at +215.5, 1 at +246.5) S/N 414-0351 and up 6 (2 at +137.0, 2 at +175.0, 2 at +218.0) 7 (with toilet option) (2 at +137.0, 2 at +175.0, 2 at +218.0, 1 at +250.0) (See manufacturer's equipment list for optional seating arrangements)
Maximum Baggage	S/N 414-0001 through 414-0350 350 lb. (+71.0), 240 lb. (+186.0), 340 lb. (+246.5) S/N 414-0351 and up 350 lb. (+71.0), 240 lb. (+186.0), 400 lb. (+266.0), 100 lb. (+282.0)
Fuel Capacity	102 gal. (2 wing tip tanks, 51 gal. ea., 50 gal. usable at +152.0) See NOTE 1 for data on unusable fuel
Oil Capacity	26 qt. (13 qt. in ea. engine at +113.5; usable 6.5 qt. per engine) See NOTE 1 for data on undrainable oil

Control Surface Movements

Wing flaps			Down	45°, +1°, -0°
Main surfaces				
Aileron	Up	20°, +1°, -0°	Down	20°, +1°, -0°
Elevator	Up	25°, +1°, -0°	Down	15°, +1°, -0°
Rudder	Right	32°, +1°, -0°	Left	32°, +1°, -0°
(Read degrees normal to rudder hinge line)				
Tab (main surface in neutral)				
Aileron	Up	20°, +1°, -0°	Down	20°, +1°, -0°
Elevator	Up	5°, +1°, -0°	Down	30°, +1°, -0°
Rudder	Right	11°, +1°, -0°	Left	16°, +1°, -0°
(Read degrees normal to rudder hinge line)				

Serial Nos. Eligible — 414-0001 through 414-0965

VI - Model 421B, Golden Eagle, (Normal Category), Approved April 28, 1970

Engines	Two Continental GTSIO-520-H reduction gear ratio .667:1
Fuel	Grade 100 or 100LL aviation gasoline
Engine Limits	For all operations, 2275 propeller r.p.m. (375 hp.) 39.5 in. Hg. Mp. up to critical altitude of 18,000 ft. in standard atmosphere. Above 18,000 ft. the following maximum Mp. applies for maximum r.p.m.:

Altitude (ft.)	Max. Allowable Mp. (in. Hg.)
18,000	39.5
20,000	37.5
22,000	35.5
24,000	33.5
25,000	32.5
26,000	31.3
28,000	28.5
30,000	25.5

VI - Model 421B (cont'd)

Propeller and Propeller Limits

Two McCauley full-feathered 3-bladed propeller installations
(a) McCauley hub 3AF34C92 with 90LF-0 blades or
McCauley hub 3AF37C516 with 90LFB-0 blades
Diameter: not over 90.0 in., not under 88.0 in.
(no further reduction permitted)
Pitch settings at 30 in. station:
low 16.9°, ±0.2°
feathering 84.5°, ±0.3°
(b) Model 421B S/N 421B0001 thru 421B0500
Hydraulic governor Woodward 210594, 210595, 210596 or 210597
Model 421B S/N 421B0501 and up
McCauley DCF290D2/T4, DFC7290D2/T4, DCFS290D2/T4, DCFUS290D2/T4, DCF290D7/T4, DCFU290D7/T4, DCFS290D7/T4, DCFUS290D7/T4, DCFU290D13/T4 or DCFUS290D13/T4.
(c) Propeller spinner and bulkhead assembly, McCauley D-3534/D-3796.

Airspeed Limits (CAS)

Model 421B: S/N 421B0001 through 421B0500 (except as noted)

Maneuvering	175 m.p.h. (152 knots)
Maximum structural cruising	230 m.p.h. (200 knots)
Never exceed274 m.p.h. (238 knots)	
Landing gear operating	165 m.p.h. (143 knots)
Landing gear extended	165 m.p.h. (143 knots)
Flaps extended 15°	180 m.p.h. (156 knots) (S/N 421B0001 through 421B0200)
Flaps extended 15°	200 m.p.h. (174 knots) (S/N 421B0201 through 421B0500)
Flaps extended 45°	165 m.p.h. (143 knots)
Minimum control	100 m.p.h. (87 knots) (S/N 421B0001 through 421B0800)
	94 m.p.h. (82 knots) (S/N 421B0801 and up)

Model 421B: S/N 421B0501 and up

Maneuvering	152 KCAS (175 m.p.h.)
Maximum structural cruising	200 KCAS (230 m.p.h.)
Never exceed238 KCAS (274 m.p.h.)	
Landing gear operating	145 KCAS (167 m.p.h.)
Landing gear extended	145 KCAS (167 m.p.h.)
Flaps extended 15°	175 KCAS (202 m.p.h.)
Flaps extended 45°	145 KCAS (167 m.p.h.)
Minimum control	87 KCAS (100 m.p.h.)

C.G. Range (Landing Gear Extended)

S/N 421B0001 through 421B0200

6, 7, or 8 Place	10 Place
(+151.8) to (+156.4) at 7250 lb.	(+151.8) to (+157.7) at 7250 lb.
(+156.7) at 6850 lb. or less	(+158.0) at 6850 lb. or less
(+147.1) at 6100 lb. or less	(+147.1) at 6100 lb. or less

S/N 421B0201 and up

(+152.6) to (+156.5) at 7450 lb.	(+152.6) to (+157.8) at 7450 lb.
(+156.7) at 7050 lb. or less	(+158.0) at 7050 lb. or less
(+147.1) at 6100 lb. or less	(+147.1) at 6100 lb. or less

Straight line variation between points given
Landing gear retracted moment change: +889 in.-lb.

VI - Model 421B (cont'd)

Empty Wt. C.G. Range	None
Leveling Means	External screw heads on right side of fuselage at stations +213.9 and +238.55 on W.L. +93.80
Maximum Weight	Landing 7200 lb., takeoff 7250 lb. (S/N 421B0001 through 421B0200) Landing 7200 lb., takeoff 7450 lb. (S/N 421B0201 and up)
No. of Seats	S/N 421B0001 through 421B0300 6, 7, or 8 (2 at +137.0, 2 at +175.5, 2 at +215.5, 2 at +245.7) or 10 (2 at +137.0, 2 at +161.0, 2 at +190.0, 2 at +218.0, 2 at +249.0) S/N 421B0301 and up 6, 7, or 8 (2 at +137.0, 2 at +175.0, 2 at +218.0, 2 at +261.0) or 10 (2 at +137.0, 2 at +162.0, 2 at +190.0, 2 at +218.0, 2 at +246.0) (See manufacturer's equipment list for optional seating arrangements)
Maximum Baggage	S/N 421B0001 through 421B0300 250 lb. (+32.0), 350 lb. (+71.0), 400 lb. (+186.0), 340 lb. (+246.5) S/N 421B0301 and up 250 lb. (+32.0), 350 lb. (+71.0), 400 lb. (+186.0), 400 lb. (+266.0), 100 lb. (+282.0)
Fuel Capacity	175 gal. (2 wing tip tanks, 51 gal. ea., 50 gal. usable at +152.0 and 2 wing tanks, 36.5 gal. ea., 35 gal. usable at +164.0) See NOTE 1 for data on unusable fuel
Oil Capacity	26 qt. (13 qt. in ea. engine at +115.4; usable 7.0 qt. per engine) See NOTE 1 for data on undrainable oil

Control Surface Movements

Wing flaps			Down	45°, +1°, -0°
Main surfaces				
Aileron	Up	20°, +1°, -0°	Down	20°, +1°, -0°
Elevator	Up	25°, +1°, -0°	Down	15°, +1°, -0°
Rudder	Right	25°, +1°, -0°	Left	25°, +1°, -0° (S/N 421B0001 through 421B0800)
	Right	32°, +1°, -0°	Left	32°, +1°, -0° (S/N 421B0801 and up)
(Read degrees normal to rudder hinge line)				
Tab (main surface in neutral)				
Aileron	Up	20°, +1°, -0°	Down	20°, +1°, -0°
Elevator	Up	12°, +1°, -0°	Down	20°, +1°, -0°
Rudder	Right	11°, +1°, -0°	Left	16°, +1°, -0°
(Read degrees normal to rudder hinge line)				

Serial Nos. Eligible	421B0001 through 421B0970

VII - Model 421C, Golden Eagle, (Normal Category), Approved October 28, 1975

Engines	Two Continental GTSIO-520-L reduction gear ratio .667:1 (S/N 421C0001 through 421C1000) Two Continental GTSIO-520-N reduction gear ratio .667:1 (S/N 421C1001 and up)
Fuel	Grade 100 or 100LL aviation gasoline

VII - Model 421C (cont'd)

Engine Limits

For all operations, 2235 propeller r.p.m. (375 hp.)
39.0 in. Hg. Mp. up to critical altitude of 20,000 ft. in standard atmosphere. Above 20,000 ft. the following maximum Mp. applies for maximum r.p.m.:

Altitude (ft.)	Max. Allowable Mp. (in. Hg.)
20,000	39.0
22,000	36.5
24,000	34.0
25,000	32.5
26,000	31.0
28,000	28.0
30,000	25.0

Propeller and Propeller Limits

Two McCauley full-feathering 3-bladed propeller installations
(a) McCauley hub 3FF32C501 with 90UMB-0 blades
Diameter: not over 90.0 in., not under 88.0 in.
(no further reduction permitted)
Pitch settings at 30 in. station:
low 16.6°, ±0.2°, feathering 84.6°, ±0.3°
(b) S/N 421C0001 through 421C0800
Hydraulic Governor McCauley DCF290D2/T6, DCFU290D2/T6, DCFS290D2/T6, DCFUS290D2/T6, DCF290D7/T6, DCFU290D7/T6 or DCFU290D13/T6, DCFS290D7/T6, DCFUS290D7/T6 or DCFUS290D13/T6
S/N 421C0801 and up
Hydraulic Governor McCauley DCF290D7/T6, DCFU290D7/T6 or DCFU290D13/T6, DCFS290D9/T6, DCFUS290D9/T6
(c) Propeller spinner and bulkhead assembly, McCauley D-3534/D-4506 or McCauley D-5212/D-5217

Airspeed Limits (IAS)

Maneuvering	151 KIAS (174 m.p.h.)
Maximum structural cruising	201 KIAS (231 m.p.h.)
Never exceed240 KIAS (276 m.p.h.)	
Landing gear operating	176 KIAS (203 m.p.h.)
Landing gear extended	176 KIAS (203 m.p.h.)
Flaps extended 15°	176 KIAS (203 m.p.h.)
Flaps extended 45°	146 KIAS (168 m.p.h.)
Minimum control	80 KIAS (92 m.p.h.)

C.G. Range (Landing Gear Extended)

6, 7, 8, 9 or 10 Place
(+152.6) to (+158.0) at 7450 lb.
(+147.1) at 6100 lb. or less
Straight line variation between points given
Landing gear retracted moment change:
+917 in.-lb. (S/N 421C0001 through 421C0800)
+1318 in.-lb. (S/N 421C0801 and up)

Empty Wt. C.G. Range

None

Leveling Means

External screw heads on right side of fuselage at stations +213.9 and +238.55 on W.L. +93.80

Maximum Weight

S/N 421C0001 through 421C0400
Landing 7200 lb., takeoff 7450 lb.

S/N 421C0401 and up
Landing 7200 lb., takeoff 7450 lb., ramp 7500 lb.

VII - Model 421C (cont'd)

No. of Seats	6, 7 or 8 (2 at +137.0, 2 at +175.0, 2 at +218.0, 1 at +261.0) or 10 (2 at +137.0, 2 at +162.0, 2 at +190.0, 2 at +218.0, 2 at +246.0) (See manufacturer's equipment list for optional seating arrangements)
Maximum Baggage	250 lb. (+32.0), 350 lb. (+71.0), 400 lb. (+186.0), 400 lb. (+266.0), 100 lb. (+282.0)
Fuel Capacity	213.4 gal. (2 wing tanks, 106.7 gal. ea., 103.0 gal. usable at +161.0) See NOTE 1 for data on unusable fuel
Oil Capacity	26 qt. (13 qt. in ea. engine at +115.4; usable 7.0 qt. per engine) See NOTE 1 for data on undrainable oil

Control Surface Movements

Wing flaps			Down	45°, +1°, -0°
Main surfaces				
Aileron	Up	20°, +1°, -0°	Down	20°, +1°, -0°
Elevator	Up	25°, +1°, -0°	Down	15°, +1°, -0°
Rudder	Right	32°, +1°, -0°	Left	32°, +1°, -0°
(Read degrees normal to rudder hinge line)				
Tab (main surface in neutral)				
Aileron	Up	20°, +1°, -0°	Down	20°, +1°, -0°
Elevator	Up	12°, +1°, -0°	Down	20°, +1°, -0°
Rudder	Right	11°, +1°, -0°	Left	16°, +1°, -0°
(Read degrees normal to rudder hinge line)				

Serial Nos. Eligible 421C0001 through 421C1807

VIII - Model 414A, Chancellor, (Normal Category), Approved September 30, 1977

Engines
Two Continental TSIO-520-N or TSIO-520-NB (S/N 414A0001 through 414A0200)
Two Continental TSIO-520-NB (S/N 414A0201 and up)

| Fuel
Grade 100 or 100LL Aviation Gasoline

Engine Limits
For all operations, 2700 r.p.m., 310 hp., 38.0 in. Hg. Mp. up to critical altitude of 20,000 ft. in standard atmosphere.
Above 20,000 ft. the following maximum Mp. applies for maximum r.p.m.:

Altitude (ft.)	Max. Allowable Mp. (in. Hg.)
20,000	38.0
22,000	35.2
24,000	32.3
26,000	29.8
28,000	27.4
30,000	25.0

Propeller and Propeller Limits
Two McCauley full-feathering three-bladed propeller installations
(a) McCauley hub 3AF32C93 with 82NC-5.5 blades or McCauley hub 3AF32C505 with 82NEA-5.5 blades
Diameter: not over 76.5 in., not under 75.0 in.
(no further reduction permitted)
Pitch settings at 30 in. station:
low 14.9°, ±0.2°, feathering 81.2°, ±0.3°
or (b) McCauley hub 3AF32C93 with 82NC-5.5 blades or McCauley hub 3AF32C505 with 82NEA-5.5 blades
Diameter: not over 75.5 in., not under 75 in.
Pitch settings at 30 in. station:
low 15.2°, ±0.2°
feathered 81.2°, ±0.3°

VIII - Model 414A (cont'd)

Propeller and Propeller Limits	(c) S/N 414A0001 through 414A0801 Hydraulic governor McCauley DCF290D2/T3, DCFU290D2/T3, DCFS290D4/T3, DCFUS290D4/T3, DCFS290D6/T3, DCFUS290D6/T3, DCF290D7/T3, DCFU290D7/T3, DCFU290D13/T3, DCFS290D7/T3, DCFUS290D7/T3, DCFUS290D13/T3, DCFS290D8/T3, DCFUS290D8/T3 or DCFUS290D12/T13 S/N 414AC0801 and up Hydraulic governor McCauley DCF290D2/T3, DCFU290D2/T3, DCF290D7/T3, DCFU290D7/T3 or DCFU290D13/T3, DCFS290D9/T3, DCFUS290D9/T3 (d) Propeller spinner and bulkhead assembly, McCauley D-3534/D-3796, or McCauley D-5212/D-5214

Airspeed Limits (IAS)		
	Maneuvering	145 KIAS (167 m.p.h.)
	Max. structural cruising	203 KIAS (234 m.p.h.)
	Never exceed	237 KIAS (273 m.p.h.)
	Landing gear operating	177 KIAS (204 m.p.h.)
	Landing gear extended	177 KIAS (204 m.p.h.)
	Flaps extended 15°	177 KIAS (204 m.p.h.)
	Flaps extended 45°	146 KIAS (168 m.p.h.)
	Minimum control	79 KIAS (91 m.p.h.)

C.G. Range (Landing Gear Extended)	(+151.3) to (+160.0) at 6750 lb. (+147.8) at 5800 lb. or less Straight line variation between points given Landing gear retracted moment change: +917 in.-lb.
Empty Wt. C.G. Range	None
Leveling Means	External screw heads on right side of fuselage at stations +213.29 and +238.55 on W.L. +93.80
Maximum Weight	Ramp 6785 lb., takeoff and landing 6750 lb.
No. of Seats	6, 7 or 8 (2 at +137.0, 2 at +175.0, 2 at +218.0, Optional: 1 or 2 at +261.0 or with toilet option, 1 at +250.0) (See manufacturer's equipment list for optional seating arrangements)
Maximum Baggage	250 lb. (+32.0), 350 lb. (+71.0), 400 lb. (+186.0), 400 lb. (+266.0), 100 lb. (+282.0)
Fuel Capacity	S/N 414A0001 through 414A0200 213.4 gal. (2 wing tanks, 106.7 gal. ea., 103.0 gal. usable at +161.0) See NOTE 1 for data on unusable fuel S/N 414A0201 through 414A0400 213.4 gal. (2 wing tanks, 106.7 gal. ea., 102.0 gal. usable at +161.0) See NOTE 1 for data on unusable fuel S/N 414A0401 and up 213.4 gal. (2 wing tanks, 106.7 gal. ea., 103.0 gal. usable at +161.0) See NOTE 1 for data on unusable fuel
Oil Capacity	26 qt. (13 qt. in ea. engine at +110.9; usable 6.5 qt. per engine) See NOTE 1 for data on undrainable oil

VIII - Model 414A (cont'd)

Control Surface Movements	Wing flaps			Down	45°, +1°, -0°
	Main surfaces				
	Aileron	Up	20°, +1°, -0°	Down	20°, +1°, -0°
	Elevator	Up	25°, +1°, -0°	Down	15°, +1°, -0°
	Rudder	Right	32°, +1°, -0°	Left	32°, +1°, -0°
	(Read degrees normal to rudder hinge line)				
	Tab (main surface in neutral)				
	Aileron	Up	20°, +1°, -0°	Down	20°, +1°, -0°
	Elevator	Up	12°, +1°, -0°	Down	20°, +1°, -0°
	Rudder	Right	11°, +1°, -0°	Left	16°, +1°, -0°
	(Read degrees normal to rudder hinge line)				

Serial Nos. Eligible: 414A0001 through 414A1212

IX - Model 402C, Businessliner/Utililiner, (Normal Category), Approved September 25, 1978

Engines: Two Continental TSIO-520-VB rated at 325 hp.

Fuel: Grade 100 or 100LL aviation gasoline

Engine Limits: Takeoff and engine inoperative, 2700 r.p.m., 39.0 in. Hg. Mp. up to 12,000 ft. Above 12,000 ft. the following maximum Mp. applies for maximum r.p.m.

Altitude (ft.)	Max. Allowable Mp. (in. Hg.)
S.L. to 12,000	39.0
14,000	37.2
16,000	37.2
18,000	32.0
20,000	29.5
22,000	27.0
24,000	25.0
26,000	23.0
28,000	21.0
30,000	19.0

Propeller and Propeller Limits: Two McCauley full-feathering three-bladed propeller installations

(a) McCauley hub 3AF32C93 with 82NC-5.5 blades or McCauley hub 3AF32C505 with 82NEA-5.5 blades
Diameter: not over 76.5 in., not under 75.0 in.
(no further reduction permitted)
Pitch settings at 30 in. station:
low 14.9°, ±0.2°, feathering 82.2°, ±0.3°

or (b) McCauley hub 3AF32C93 with 82NC-6.5 blades or McCauley hub 3AF32C505 with 82NEA-6.5 blades
Diameter: not over 75.5 in., not under 75.0 in.
Pitch settings at 30 in. station:
low 15.2°, ±0.2°, feathering 82.2°, ±0.3°

(c) S/N 402C0001 through 402C0600
Hydraulic governor, Woodward B210444, C210439; McCauley DCF290D7/T3, DCFUS290D7/T3, DCFU290D13/T3, DCFS290D7/T3, DCFUS290D7/T3, DCFUS290D13/T3, DCFUS290D8/T3, or DCFUS290D12/T3
S/N 689, and 402C0601 and up
Hydraulic governor, Woodward B210444, C210439; McCauley DCF290D7/T3, DCFU290D7/T3 or DCFU290D13/T3, DCFS290D9/T3, DCFUS290D9/T3

(d) Propeller spinner and bulkhead assembly; McCauley D-3534/D-3537, D-3534/D-3796, or D-5212/D-5214

IX - Model 402C (cont'd)

Airspeed Limits (IAS)	Maneuvering	150 KIAS (173 m.p.h.)
	Max. structural cruising	205 KIAS (236 m.p.h.)
	Never exceed	235 KIAS (270 m.p.h.)
	Landing gear operating	180 KIAS (207 m.p.h.)
	Landing gear extended	180 KIAS (207 m.p.h.)
	Flaps extended 15°	180 KIAS (207 m.p.h.)
	Flaps extended 45°	149 KIAS (172 m.p.h.)
	Minimum control	80 KIAS (92 m.p.h.)

C.G. Range (Landing Gear Extended)
(+151.58) to (+160.67) at 6850 lb.
(+149.08) at 5800 lbs. or less
Straight line variation between points given
Landing gear retracted moment change: +917 in.-lb.

Empty Wt. C.G. Range
None

Leveling Means
External screw heads on right side of fuselage at stations +213.65 and +238.00 on W.L. +93.80

Maximum Weight
Ramp, 6885 lbs., takeoff and landing 6850 lbs.

No. of Seats
6, 7 or 8 (2 at +137.0, 2 at +175.0, 2 at +218.0, 1 or 2 at +261.0)
9 (with photographic provisions option) (2 at +137.0, 2 at +162.0, 2 at +190.0, 2 at +218.0, 1 at +246.0)
10 (2 at +137.0, 2 at +162.0, 2 at +190.0, 2 at +218.0, 2 at +246.0)
(See manufacturer's equipment list for optional seating arrangements)

Maximum Baggage
250 lbs. (+32.0), 350 lbs. (+71.0), 400 lbs. (+186.0), 400 lbs. (+266.0), 100 lbs. (+282.0)

Fuel Capacity
S/N 402C0001 through 402C0200
213.4 gal. (2 wing tanks, 106.7 gal. ea., 102 gal. usable at +161.0)
See NOTE 1 for data on unusable fuel

S/N 689, and 402C0201 and up
213.4 gal. (2 wing tanks, 106.7 gal. ea., 103 gal. usable at +161.0)
See NOTE 1 for data on unusable fuel

Oil Capacity
26 qt. (13 qt. in ea. engine at +110.9; usable 6.5 qt. per engine)
See NOTE 1 for data on undrainable oil

Control Surface Movements

Wing flaps			Down	45°, +1°, -0°
Main surfaces				
Aileron	Up	20°, +1°, -0°	Down	20°, +1°, -0°
Elevator	Up	25°, +1°, -0°	Down	15°, +1°, -0°
Rudder	Right	32°, +1°, -0°	Left	32°, +1°, -0°
(Read degrees normal to rudder hinge line)				
Tab (main surface in neutral)				
Aileron	Up	20°, +1°, -0°	Down	20°, +1°, -0°
Elevator	Up	12°, +1°, -0°	Down	20°, +1°, -0°
Rudder	Right	11°, +1°, -0°	Left	16°, +1°, -0°
(Read degrees normal to rudder hinge line)				

Serial Nos. Eligible
689, 402C0001 through 402C1020

X - Model 425, Corsair or Conquest I (See NOTE 7), (Normal Category), Approved July 1, 1980

Engines — Two Pratt & Whitney Aircraft of Canada, Ltd., PT6A-112 turboprop

Fuel — Aviation turbine fuel Jet A, Jet A-1, or Jet B, JP-4, JP-5 or JP-8. For required use of anti-icing additives and emergency use of aviation gasoline, refer to the Pilot's Operating Handbook and FAA Approved Airplane Flight Manual.

Engine Limits

	Operating Limits				
	Shaft Horsepower Power	Ng Gas Generator Speed (% rpm)	Indicated Torque (ft.-lbs.)	Prop. Shaft Speed (rpm)	Maximum Permissible Interturbine Temp. (°C.)
Takeoff static & max. continuous	450*	101.6	1244	1900	725
Starting (2 sec.)	- -	- -	- -	- -	1090
Maximum reverse	430	101.6	1244	1815	725

***Flat Rated:**
The engines may produce more power than that for which the airplane has been certificated. Under these conditions, the placarded torquemeter, ITT, or Ng limitations shall not be exceeded.

Propeller and Propeller Limits

(1) Two Hartzell three-bladed, full-feathered, reversible
Hub: HC-B3TN-3C
Blade: T10178B-8R
Diameter: Not over 93-3/8 in., not under 91 inches; no further reduction permitted
Pitch at 30-inch station:
Low pitch 20.2°
Feathered 86.7°
Reverse -10.9°

(2) Two McCauley three-bladed, full-feathered, reversible
Hub: 3GFR34C701
Blade: 93KB-0
Diameter: Not over 93 inches, not under 90-5/8 inches; no further reduction permitted
Pitch at 30-inch station:
Low pitch 18.5°
Feathered 85.5°
Reverse -13.5°

Propellers may be interchanged in any combination.

Airspeed Limits (IAS)

V_{MO} (Max Operating) Sea level to 21,800 ft.	230 knots	265 m.p.h.
M_{MO} Above 21,800 ft.	.52 mach	
V_A (Maneuvering) at 8200 lbs.	154 knots	177 m.p.h.
V_A (Maneuvering) at 8600 lbs.	157 knots	181 m.p.h.
V_{FE} (Flaps extended)		
45° (Landing)	145 knots	169 m.p.h.
15° (Takeoff & Approach)	175 knots	201 m.p.h.
V_{MCA} (Min. control speed) Air at 8200 lbs.	90 knots	104 m.p.h.
V_{MCA} (Min. control speed) Air at 8600 lbs.	92 knots	106 m.p.h.
V_{LE} (Landing gear extended)	175 knots	201 m.p.h.

X - Model 425 (cont'd)

C.G. Range (Landing Gear Extended)

S/N 425-0001 through 425-0176 (See NOTE 7)
(155.66) to (160.04) at 8200 lbs.
(150.65) to (160.04) at 6478 lbs. or less

S/N 425-0177 and up
(156.81) to (160.04) at 8600 lbs.
(150.65) to (160.04) at 6478 lbs. or less

Straight line variation between points given
Moment change due to retracting landing gear (+1448 in.-lb.)

Empty Wt. C.G. Range — None

Leveling Means — External screw heads on right side of fuselage at stations +213.9 and +238.55 on W.L. +93.80

Maximum Weight

	S/N 425-0001 through 425-0176 (See NOTE 7)	S/N 425-0177 and up
Takeoff	8200 lbs.	8600 lbs.
Landing	8000 lbs.	8000 lbs.
Zero fuel	6740 lbs.	7000 lbs.
Ramp	8275 lbs.	8675 lbs.

No. of Seats — 6, 7 or 8 (2 at +137.0, 2 at +175.0, 2 at +218.0, 2 at +261.0)
See manufacturer's equipment list for optional seating arrangements

Maximum Baggage — 250 lb. (+32.0), 350 lb. (+71.0), 400 lb. (+266.0), 100 lb. (+282.0)

Fuel Capacity — 2497.8 lb. (372.8 gal.) total in two wing tanks, 1248.9 lb. (186.4 gal.) each; 2452.2 lb. (366.0 gal.) usable total, 1226.1 lb. (133 gal.) in each tank at +163.3. Fuel weight based on 6.70 lb./gal. See NOTE 1 for data on unusable fuel.

Oil Capacity — 5.28 gal. total, 5.28 gal. usable (2.3 gal. in each engine-mounted tank at +125.3). See NOTE 1 for data on undrainable oil.

Maximum Operating Altitude — 30,000 ft.

Control Surface Movements

Wing flaps			Down	45°, +1°, -0°
Main surfaces				
Aileron	Up	20°, +1°, -0°	Down	20°, +1°, -0°
Elevator	Up	19°, +1°, -0°	Down	15°, +1°, -0°
Rudder	Right	32°, +1°, -0°	Left	32°, +1°, -0°
(Read degrees normal to rudder hinge line)				
Tab (main surface in neutral)				
Aileron	Up	20°, +1°, -0°	Down	20°, +1°, -0°
Elevator	Up	6°, +1°, -0°	Down	15°, +1°, -0°
Rudder	Right	11°, +1°, -0°	Left	16°, +1°, -0°
(Read degrees normal to rudder hinge line)				

Serial Nos. Eligible — 425-0001 through 425-0236

Data Pertinent to All Models

Datum — 100.00 in. forward face of fuselage bulkhead forward of rudder pedals.

Certification Basis

Models 401, 401A, 401B, 402, 402A, 402B, 411, 411A, 414, 421, 421A:
Part 3 of the Civil Air Regulations dated May 15, 1956, as amended by 3-1 through 3-5 and 3-8.

Model 421B:
Part 3 of the Civil Air Regulations dated May 15, 1956, except Subpart B, as amended by 3-1 through 3-5 and 3-8; Subpart B, paragraphs 23.25 through 23.253 of the Federal Aviation Regulations dated February 1, 1965, as amended by 23-1 through 23-7.

Models 414A and 421C:
Part 3 of the Civil Air Regulations dated May 15, 1956, as amended by 3-1 through 3-5 and 3-8, excluding the following portions:
Subpart B and paragraphs 3.356, 3.357, 3.358, 3.359, 3.411, 3.429, 3.433, 3.434, 3.435, 3.436, 3.437, 3.445, 3.581, 3.582, 3.583, 3.584, 3.585, 3.587, 3.628, 3.666, 3.672, 3.673, 3.674, 3.675, 3.700(c), 3.728, 3.767(a) and 3.767(b). Include the following portions of FAR 23 dated February 1, 1965, as amended by 23-1 through 23-14;
Subpart B and paragraphs 23.729, 23.901, 23.909, 23.951, 23.954, 23.955, 23.959, 23.973, 23.1041, 23.1043, 23.1047, 23.1143, 23.1305, 23.1387(e), 23.1435 and 23.1557(c); as amended by 23-1 through 23-21, paragraph 23.1385(c); as amended by 23-1 through 23-23, paragraph 23.1327. Add paragraph 23.1559(b) for Model 414A only. Findings of Equivalent Level of Safety were made for CAR 3.637, 3.757, and 3.778(a).

Model 402C:
Part 3 of the Civil Air Regulations dated May 15, 1956, as amended by 3-1 through 3-5 and 3-8, excluding the following portions: Subpart B and paragraphs 3.356, 3.357, 3.358, 3.359, 3.411, 3.429, 3.433, 3.434, 3.435, 3.436, 3.437, 3.445, 3.581, 3.582, 3.583, 3.584, 3.585, 3.587, 3.628, 3.666, 3.672, 3.673, 3.674, 3.675, 3.700(c), 3.728, 3.767(a) and 3.767(b). Include the following portions of FAR 23 dated February 1, 1965, as amended by 23-1 through 23-14: Subpart B and paragraphs 23.729, 23.901, 23.909, 23.951, 23.954, 23.955, 23.959, 23.973, 23.1041, 23.1043, 23.1047, 23.1143, 23.1305, 23.1387(e), 23.1435, 23.1557(c), and 23.1559(b); as amended by 23-1 through 23-21, paragraph 23.1385(c); as amended by 23-1 through 23-23, paragraph 23.1327. Part 36 of the Federal Aviation Regulations dated December 1, 1969, as amended by 36-1 through 36-7. Findings of Equivalent Level of Safety were made for CAR 3.637, 3.757, and 3.778(a).

Model 425:
Part 3 of the Civil Air Regulations dated May 15, 1956, as amended by 3-1 through 3-6 and 3-8 as follows: Paragraphs 3.0 through 3.20, 3.291 through 3.307, 3.317 through 3.347, 3.371 through 3.401, 3.651, 3.652, 3.655(c) and (d), 3.661, 3.662, 3.668, 3.686 through 3.699, 3.711 through 3.728, 3.749, 3.791, and 3.792; the following portions of FAR 23 dated February 1, 1965, as amended by 23-1 through 23-21: Paragraphs 23.21 through 23.33, 23.45(a) through (d), 23.49 through 23.179, 23.181(a), 23.201 through 23.572, 23.629, 23.723 through 23.735, 23.865, 23.867, 23.901 through 23.1017, 23.1019(a)(1) and (2), 23.1019(a)(4) and (5), 23.1019(b), 23.1021 through 23.1203, 23.1303(a) through (d), 23.1305(a) through (u) and (w), 23.1323, 23.1325, 23.1327, 23.1329, 23.1335, 23.1337, 23.1351 through 23.1357, 23.1385 through 23.1401, 23.1441 through 23.1449, 23.1501 through 23.1521, 23.1524, 23.1525, 23.1527(b), and 23.1529 through 23.1589; Paragraph 25.831(d) of FAR 25 dated February 1, 1965, as amended by 25-1 through 25-43; FAR 36 dated December 1, 1969, as amended by 36-1 through 36-10; SFAR No. 27, Fuel Venting and Exhaust Emission Requirements for Turbine Engine Powered Airplanes, effective February 1, 1974, as amended by SFAR's 27-1, 27-2, and 27-3; plus Special Conditions 23-93-CE-12 as amended by Amendment No. 1 dated June 25, 1980. (See NOTE 3.)

Model 414A (S/N 414A0401 and up, Model 421C (S/N 421C0801 and up)
In addition to the above certification basis, compliance with FAR 36, dated December 1, 1969, as amended by 36-1 through 36-10 (414A only) and 36-1 through 36-4 (421C only) has been demonstrated.

Model 402B, S/N 402B0501 and up
Model 402C
Model 414, S/N 414-0451 and up
Model 414A
Model 421B, S/N 421B0501 and up
Model 421C
Model 425
Markings, placards and manuals are primarily in knots instead of m.p.h. as required by CAR 3, but permitted by FAR 23, Amendment 23-7.

Model 402B, S/N 402B1001 and up
Model 414, S/N 414-0801 and up
Findings of equivalent level of safety were made for CAR 3.757 and 3.778(a).

Model 402B, S/N 402B0801 and up
Model 402C
Model 414, S/N 414-0601 and up
Model 414A
Model 421B, S/N 421B0801 and up
Model 421C
Model 425
In addition to the above certification basis, compliance with ice protection has been demonstrated in accordance with FAR 23.1419 of Amendment 23-14 effective December 20, 1973, when ice protection equipment is installed in accordance with Cessna Drawing 5914105 for 425, 5114400 for all other models, Factory Kit (FK) No. 194, Pilot's Operating Handbook and/or FAA Approved Airplane Flight Manual. Aircraft which have been modified in compliance with Accessory Kit (AK) No. 421-106 are considered to be equivalent to those with Factory Kit (FK) No. 194.

Application for Type Certificate dated September 18, 1961. Type Certificate No. A7CE issued August 17, 1964, obtained by the manufacturer under delegation option procedures.

Production Basis

Production Certificate No. 312 issued and Delegation Option Manufacturer No. CE-3 authorized to issue airworthiness certificates under delegation option provisions of Part 21 of the Federal Aviation Regulations. Effective February 15, 1985, and on, Production Certificate No. 4 is applicable to all spares production. See NOTE 8 for specific effectivity of P.C. 4 on new airplane serials.

Equipment: The basic required equipment as prescribed in the applicable airworthiness regulations (see Certification Basis) must be installed in the aircraft for certification. In addition, the following item of equipment is required.

1. Stall warning indicator, Cessna dwg. 5018100 (401, 402, 411, 411A)
   Stall warning indicator, Cessna dwg. 5118000 (421)
   Stall warning indicator, Cessna dwg. 5618002 (414)
   Stall warning indicator, Cessna dwg. 5218016 (401A, 402A, 401B, 402B0001 through 402B0300)
   Stall warning indicator, Cessna dwg. 5118310 (421A)
   Stall warning indicator, Cessna dwg. 5118402 (421B0001 through 421B0300)
   Stall warning indicator, Cessna dwg. 5618021 (414-0351 and up, 421B0301 and up)
   Stall warning indicator, Cessna dwg. 5218031 (402B0301 and up)
   Stall warning indicator, Cessna dwg. 5118627 (421C)
   Stall warning indicator, Cessna dwg. 5618041 (402C, 414A, 425)

or Angle of Attack Indicator System, Cessna Dwg. 0800302, Model 402B, 402C, 414, 414A, 421B, 421C.

NOTE 1. Current weight and balance report together with list of equipment included in certificated empty weight and loading instructions when necessary must be provided for each aircraft at the time of original certification.

The certificated empty weight and corresponding center of gravity location must include undrainable oil (not included in oil capacity) and unusable fuel as follows:

(a) Fuel. 12 lb. (tip) at (+152.0) (401, 401A, 401B, 402, 402A, 402B, 411, 411A, 414, 421, 421A, 421B)
18 lb. (wing, standard 73 gal. at +164.0) (411, 411A, 421, 421A, 421B)
24 lb. (wing, optional 100 gal. at +164.0) (411, 411A, 421, 421A, 421B, 402A, 402B, 414)
6 lb. (wing, optional 63 gal. at +164.0) (402B0301 and up and 414-0351 and up)
44 lb. (wing, 7.4 gal. at +165.2) (402C, S/N 689, and 402C0201 and up; 414A, S/N 414A0401 and up; 421C)
68 lb. (wing, 11.4 gal. at +165.2) (414A, S/N 414A0001 through S/N 414A0200)
56 lb. (wing, 9.4 gal. at +165.0) (402C, S/N 402C0001 through 402C0200; 414A, S/N 414A0201 through 414A0400)
45.6 lb. (wing, 6.8 gal. at +166.2) (425)

(b) If optional wing locker transfer tanks are installed 3.0 lb. (each 26 gal. tank) at (+176.0) (411, 411A, 421, 421A, 421B)
3.0 lb. (each 20 gal. tank) at (+175.0) (401, 401A, 401B, 402, 402A, 402B, 414)
2.0 lb. (each 28 gal. tank) at (+176.0) (421C0001 and up)

(c) Oil - 0.0 lb.

NOTE 2. The placards specified in the FAA Approved Airplane Flight Manual must be displayed.

NOTE 3. Service information
The appropriate airplane service manual contains structural retirement lives, which may not be changed without FAA Engineering approval, for the following components:

	Part Number	Hours	Model
Windshield	5111604-1 & -2	13,200	414, 414A, 421A, 421B, 421C, 425
Windshield, heated	9910013-1	13,200	421, 421A (S/N 421A0001 through 421A0117)
Windshield, heated	9910071-1	13,200	414, 421A, 421B (S/N 414-0001 through 414-0600, 421A0118 through 421B0800)
Windshield, heated	9910214-1 & -2	13,200	414, 414A, 421B, 421C (S/N 414-0601 and up, 421B0801 through 421C0800)
Windshield, heated	9910460-1 & -200	13,200	421C (S/N 421C0801 and up), 425
Upper cabin door latch pins	5111545-3	8,000	421 (S/N 421-0001 through 421-0079)
Upper cabin door latch pins	5111545-6	8,000	421 (S/N 421-0080 and up) 421A
Wing	5922125	13,000	425
Wing carry-thru	5911004, 5111225	30,000	425

Model 425 Special Conditions 23-93-CE-12, required, in part, that Cessna establish mandatory inspections of the Horizontal Tail Assembly in order to maintain continued structural integrity. Therefore, inspections are required for the horizontal stabilizer, elevators, elevator tab and tab actuator system. In order to comply with these requirements, airplanes must be inspected in accordance with inspection Item Codes A273002, A273101, A273102, B273109 and A551001 as contained in Model 425 Maintenance Manual, Part Number D2535-3-13, Revision 3 (or later revision). These inspection criteria are contained in Chapter 5, Subsection 5-10-01, and are applicable to Zones 331 and 332. All approved airplane inspection programs must include these mandatory inspections.

NOTE 4. Model 421, Serial Nos. 421-0001 and up, approved for 6840 lb. takeoff weight with C.G. range as follows when appropriate airplane flight manual, pilot's check list, weight and balance form, and other documents are provided as specified in Cessna Service Kit SK421-12.

C.G. Range (Landing Gear Extended)	(+152.1) to (+155.5) at 6840 lb.
	(+155.7) at 6500 lb.
	(+144.3) to (+155.7) at 5500 lb.

Straight line variation between points given

NOTE 5. McCauley propellers with 3AF32C87 and 3AF32C504 hubs may be interchanged in any combination. This also applies to propellers with 3AF32C93 and 3AF32C505m hubs; 3AF34C92 and 3AF37C516 hubs; 3AF34C74 and 3AF37C510 hubs.

NOTE 6. Model 425 aircraft in compliance with Cessna Drawing 5700018 are eligible for certification in The Netherlands.

NOTE 7. Model 425 S/N 425-0001 through 425-0176 (Corsair) are eligible for the maximum weights and C.G. range applicable to S/N 425-0177 and up (Conquest I), when modified in accordance with Cessna Service Kit SK425-17, and will be renamed Conquest I.

NOTE 8. Production Certificate No. 4 effective at Serials 402C1005 and on, 414A1208 and on, 421C1801 and on, and 425-0228 and on.

.....END.....

	3A13
	Revision 54
	CESSNA
182	182K
182A	182L
182B	182M
182C	182N
182D	182P
182E	182Q
182F	182R
182G	R182
182H	T182
182J	TR182
182S	
	December 15, 1997

TYPE CERTIFICATE DATA SHEET NO. 3A13

This data sheet which is part of Type Certificate No. 3A13 prescribes conditions and limitations under which the product for which the type certificate was issued meets the airworthiness requirements of the Federal Aviation Regulations.

Type Certificate Holder: Cessna Aircraft Company
P. O. Box 7704
Wichita, Kansas 67277

I - Model 182, Skylane, 4 PCLM (Normal Category), Approved March 2, 1956

Engine: Continental O-470-L

*Fuel: 80 minimum grade aviation gasoline

*Engine Limits: For all operations, 2600 r.p.m. (230 hp.)

Propeller and Propeller Limits:

1. Hartzell constant speed
   - (a) Hub HC82XF-1 or HCA2XF-1 or BHCA2XF-1 with 8433-2 blades
     Diameter: not over 82 in., not under 80 in.
     Pitch settings at 30 in. sta.:
     low 12°, high 24°
   - (b) Cessna spinner 0752006
   - (c) Woodward governor 210065, 210105, 210155 or 210340
2. McCauley constant speed
   - (a) Hub 2A36C with blades 90M-8
     Diameter: not over 82 in., not under 80 in.
     Pitch settings at 36 in. sta.:
     low 10.5°, high 22°
   - (b) Cessna spinner 0752004
   - (c) Woodward governor 210065, 210105, 210155, 210345 or 210452, or McCauley C290D2/T1 or C290D3/T1
3. Hartzell constant speed
   - (a) Hub BHC-C2YF-1 with 8468-2 blades
     Diameter: not over 82 in., not under 80 in.
     Pitch settings at 30 in. sta.:
     low 13°, high 24°

Page No..	1	2	3	4	5	6	7	8	9	10	11	12	13	14	15	16	17	18
Rev. No.	52	39	39	39	39	39	39	39	39	39	39	39	49	51	47	51	51	50
Page No..	19	20	21	22	23	24	25	26	27	28	29	30	31	32	33	34		
Rev. No.	51	51	51	51	51	51	43	51	43	51	51	51	51	54	54	52		

I - Model 182 (Cont'd)

Propeller and Propeller Limits (cont'd)	(b) Cessna spinner 0752619 (c) Woodward governor 210105AF, 210340 or 210451 4. McCauley constant speed (a) Hub 2A34C with 90A-8 or 90AT-8 blades Diameter: not over 82 in., not under 80 in. Pitch settings at 36 in. sta.: low 10.5°, high 21.5° (b) Cessna spinner 0752004 (c) Woodward governor 210065, 210105, 210155, 210345 or 210452 or McCauley C290D2/T1 or C290D3/T1
*Airspeed Limits (CAS)	Maneuvering 122 m.p.h. (106 knots) Maximum structural cruising 160 m.p.h. (139 knots) Never exceed 184 m.p.h. (160 knots) Flaps extended 100 m.p.h. (87 knots)
C.G. Range	(+39.5) to (+45.8) at 2550 lb. (+35.0) to (+45.8) at 2050 lb. or less Straight line variation between points given
Empty Wt. C.G. Range	None
*Maximum Weight	2550 lb.
No. of Seats	4 (2 at +36, 2 at +70)
Maximum Baggage	120 lb. (+95)
Fuel Capacity	60 gal. (55 gal. usable); two 30 gal. tanks in wings at +48. See NOTE 1 for data on unusable fuel
Oil Capacity	12 qt. (-15) (6 qt. usable) See NOTE 1 for data on undrainable oil

Control Surface Movements

Wing flaps	Takeoff		Retracted	0°
			1st notch	10°
			2nd notch	20°
	Landing		3rd notch	30°
			4th notch	40°
Ailerons	Up	20° ± 2°	Down	14° ± 2°
Adj. stabilizer	Up	1° 50' ± 15	Down	8° 20' ± 15'
Elevator	Up	25° ± 1°	Down	22° 50' ± 1°
(With stabilizer full down)				
Rudder	Right	24° ± 1°	Left	24° ± 1°

Serial Nos. Eligible	Model 182: 613 and 33000 through 33842 (1956 Model)

II - Model 182A, Skylane, 4 PCLM (Normal Category), Approved December 7, 1956

Engine	Continental O-470-L
*Fuel	80 minimum grade aviation gasoline
*Engine Limits	For all operations, 2600 r.p.m. (230 hp.)

II - Model 182A (cont'd)

Propeller and Propeller Limits	1. Hartzell constant speed (a) Hub HC82XF-1 or HCA2XF-1 or BHCA2XF-1 with 8433-2 blades Diameter: not over 82 in., not under 80 in. Pitch settings at 30 in. sta.: low 12°, high 24° (b) Cessna spinner 0752006 (c) Woodward governor 210065, 210105, 210155 or 210340 2. McCauley constant speed (a) Hub 2A36C with 90M-8 blades Diameter: not over 82 in., not under 80 in. Pitch settings at 36 in. sta.: low 10.5°, high 22° (b) Cessna spinner 0752004 (c) Woodward governor 210065, 210105, 210155 or 210452, or McCauley C290D2/T1 or C290D3/T1 3. Hartzell constant speed (a) Hub BHC-C2YF-1 with 8468-2 blades Diameter: not over 82 in., not under 80 in. Pitch settings at 30 in. sta.: low 13°, high 24° (b) Cessna spinner 0752619 (c) Woodward governor 210105AF, 210340 or 210451 4. McCauley constant speed (a) Hub 2A34C with 90A-8 or 90AT-8 blades Diameter: not over 82 in., not under 80 in. Pitch settings at 36 in. sta.: low 10.5°, high 21.5° (b) Cessna spinner 0752004 (c) Woodward governor 210065, 210105, 210155, 210345, 210452, or McCauley C290D2/T1 or C290D3/T1
*Airspeed Limits (CAS)	Maneuvering 122 m.p.h. (106 knots) Maximum structural cruising 160 m.p.h. (139 knots) Never exceed 184 m.p.h. (160 knots) Flaps extended 100 m.p.h. (87 knots)
C.G. Range	(+40.0) to (+45.8) at 2650 lb. (+33.5) to (+45.8) at 2100 lb. or less Straight line variation between points given
Empty Wt. C.G. Range	None
*Maximum Weight	2650 lb.
No. of Seats	4 (2 at +36, 2 at +70)
Maximum Baggage	120 lb. (+95)
Fuel Capacity	65 gal. (55 gal. usable); two 32.5 gal. tanks in wings at +48 See NOTE 1 for data on unusable fuel
Oil Capacity	12 qt. (-15) (6 qt. usable) See NOTE 1 for data on undrainable oil

II - Model 182A (cont'd)

Control Surface Movements	Wing flaps	Takeoff	Retracted	0°
			1st notch	10°
			2nd notch	20°
		Landing	3rd notch	30°
			4th notch	40°
	Ailerons	Up 20° ± 2°	Down	14° ± 2°
	Adj. stabilizer	Up 1° 50' ± 15'	Down	8° 20' ± 15'
	Elevator (With stabilizer full down)	Up 25° ± 1°	Down	22° 50' ± 1°
	Rudder	Right 24° ± 1°	Left	24° ± 1°

Serial Nos. Eligible

Model 182A: 33843 through 34753 (1957 Model)
Model 182A: 34755 through 34999 and 51001 through 51556 (1958 Model)

III - Model 182B, Skylane, 4 PCLM (Normal Category), Approved August 22, 1958

Engine — Continental O-470-L

*Fuel — 80 minimum octane aviation gasoline

*Engine Limits — For all operations, 2600 r.p.m. (230 hp.)

Propeller and Propeller Limits

1. Hartzell constant speed
   (a) Hub HC82XF-1 or HCA2XF-1 or BHCA2XF-1 with 8433-2 blades
   Diameter: not over 82 in., not under 80 in.
   Pitch settings at 30 in. sta.:
   low 12°, high 24°
   (b) Cessna spinner 0752006
   (c) Woodward governor 210065, 210105, 210155, or 210340
2. McCauley constant speed
   (a) Hub 2A36C with 90M-8 blades
   Diameter: not over 82 in., not under 80 in.
   Pitch settings at 36 in. sta.:
   low 10.5°, high 22°
   (b) Cessna spinner 0752004
   (c) Woodward governor 210065, 210105, 210155, 210345, 210452, or McCauley C290D2/T1 or C290D3/T1
3. Hartzell constant speed
   (a) Hub BHC-C2YF-1 with 8468-2 blades
   Diameter: not over 82 in., not under 80 in.
   Pitch settings at 30 in. sta.: low 13°, high 24°
   (b) Cessna spinner 0752619
   (c) Woodward governor 210105AF, 210340, or 210451
4. McCauley constant speed
   (a) Hub 2A34C with 90A-8 or 90AT-8 blades
   Diameter: not over 82 in., not under 80 in.
   Pitch settings at 36 in. sta.: low 10.5°, high 21.5°
   (b) Cessna spinner 0752004
   (c) Woodward governor 210065, 210105, 210155, 210345, 210452, or McCauley C290D2/T1 or C290D3/T1

*Airspeed Limits (CAS)	Maneuvering	122 m.p.h. (106 knots)
	Maximum structural cruising	160 m.p.h. (139 knots)
	Never exceed	184 m.p.h. (160 knots)
	Flaps extended	100 m.p.h. (87 knots)

III - Model 182B, Skylane (Cont'd)

C.G. Range	(+40.0) to (+45.8) at 2650 lb. (+33.5) to (+45.8) at 2100 lb. or less Straight line variation between points given
Empty Wt. C.G. Range	None
*Maximum Weight	2650 lb.
No. of Seats	4 (2 at +36, 2 at +70)
Maximum Baggage	120 lb. (+95)
Fuel Capacity	65 gal. (55 gal. usable); two 32.5 gal. tanks in wings at +48 See NOTE 1 for data on unusable fuel
Oil Capacity	12 qt. (-15) (6 qt. usable) See NOTE 1 for data on undrainable oil

Control Surface Movements

Wing flaps	Takeoff	Retracted	0°
		1st notch	10°
		2nd notch	20°
	Landing	3rd notch	30°
		4th notch	40°
Ailerons	Up 20° ±2°	Down	14° ±2°
Adj. stabilizer	Up 1° 50' ±15'	Down	8° 20' ±15'
Elevator	Up 25° ±1°	Down	22° 50' ±1°
(With stabilizer full down)			
Rudder	Right 24° ±1°	Left	24° ±1°

Serial Nos. Eligible: Model 182B: 34754, 51557 through 52358 except 51623 (1959 Model)

IV - Model 182C, Skylane, 4 PCLM (Normal Category), Approved July 8, 1959
Model 182D, Skylane, 4 PCLM (Normal Category), Approved June 14, 1960

Engine	Continental O-470-L
*Fuel	80 minimum octane aviation gasoline
*Engine Limits	For all operations, 2600 r.p.m. (230 hp.)

Propeller and Propeller Limits

1. Hartzell constant speed
   (a) Hub HC82XF-1 or HCA2XF-1 or BHCA2XF-1 with 8433-2 blades
   Diameter: not over 82 in., not under 80 in.
   Pitch settings at 30 in. sta.:
   low 12°, high 24°
   (b) Cessna spinner 0752006
   (c) Woodward governor 210065, 210105, 210155, or 210340
2. McCauley constant speed
   (a) Hub 2A36C with 90M-8 blades
   Diameter: not over 82 in., not under 80 in.
   Pitch settings at 36 in. sta.:
   low 10.5°, high 22°
   (b) Cessna spinner 0752004
   (c) Woodward governor 210065, 210105, 210155, 210345, 210452, or McCauley C290D2/T1 or C290D3/T1

IV - Model 182C, Model 182D (cont'd)

	3. Hartzell constant speed (a) Hub BHC-C2YF-1 with 8468-2 blades Diameter: not over 82 in., not under 80 in. Pitch settings at 30 in. sta.: low 13°, high 24° (b) Cessna spinner 0752619 (c) Woodward governor 210105AF, 210340, or 210451 4. McCauley constant speed (a) Hub 2A34C with 90A-8 or 90AT-8 blades Diameter: not over 82 in., not under 80 in. Pitch settings at 36 in. sta.: low 10.5°, high 21.5° (b) Cessna spinner 0752004 (c) Woodward governor 210065, 210105, 210155, 210345, 210452, or McCauley C290D2/T1 or C290D3/T1
*Airspeed Limits (CAS)	Maneuvering 122 m.p.h. (106 knots) Maximum structural cruising 160 m.p.h. (139 knots) Never exceed 184 m.p.h. (160 knots) Flaps extended 100 m.p.h. (87 knots)
C.G. Range	(+40.0) to (+45.8) at 2650 lb. (+33.5) to (+45.8) at 2100 lb. or less Straight line variation between points given
Empty Wt. C.G. Range	None
*Maximum Weight	2650 lb.
No. of Seats	4 (2 at +36, 2 at +70)
Maximum Baggage	120 lb. (+95)
Fuel Capacity	65 gal. (55 gal. usable); two 32.5 gal. tanks in wings at +48 See NOTE 1 for data on unusable fuel
Oil Capacity	12 qt. (-15) (6 qt. usable) See NOTE 1 for data on undrainable oil
Control Surface Movements	Wing flaps — Takeoff 0°, 10°, 20°; Landing 30°, 40° Ailerons Up 20° ±2° Down 14° ±2° Adj. stabilizer Up 0° 45' ±15' Down 8° 45' ±15' Elevator Up 25° ±1° Down 22° 50' ±1° (With stabilizer full down) Rudder Right 24° ±1° Left 24° ±1° (measured parallel to 0.0.W.L.)
Serial Nos. Eligible	Model 182C: 631, 52359 through 53007 (1960 Model) Model 182D: 51623, 18253008 through 18253598 (1961 Model)

V - Model 182E, Skylane, 4 PCLM (Normal Category), Approved June 27, 1961
Model 182F, Skylane, 4 PCLM (Normal Category), Approved August 1, 1962
Model 182G, Skylane, 4 PCLM (Normal Category), Approved July 19, 1963

Engine	Continental O-470-L or 0-470-R
*Fuel	80/87 minimum grade aviation gasoline
*Engine Limits	For all operations, 2600 r.p.m. (230 hp.)

Propeller and Propeller Limits

1. Hartzell constant speed
   (a) Hub HC82XF-1 or HCA2XF-1 or BHCA2XF-1 with 8433-2 blades
   Diameter: not over 82 in., not under 80 in.
   Pitch settings at 30 in. sta.:
   low 12°, high 24°
   (b) Cessna spinner 0752006
   (c) Woodward governor 210065, 210105, 210155, or 210340
   (Not eligible on O-470-R engine installation)
2. McCauley constant speed
   (a) Hub 2A36C with 90M-8 blades
   Diameter: not over 82 in., not under 80 in.
   Pitch settings at 36 in. sta.:
   low 10.5°, high 22°
   (b) Cessna spinner 0752004
   (c) Woodward governor 210065, 210105, 210155, 210345, or 210452, or McCauley C290D2/T1 or C290D3/T1
3. Hartzell constant speed
   (a) Hub BHC-C2YF-1 with 8468-2 blades
   Diameter: not over 82 in., not under 80 in.
   Pitch settings at 30 in. sta.: low 13°, high 24°
   (b) Cessna spinner 0752619
   (c) Woodward governor 210105AF, 210340, or 210451
4. McCauley constant speed
   (a) Hub 2A34C with 90A-8 or 90AT-8 blades
   Diameter: not over 82 in., not under 80 in.
   Pitch settings at 36 in. sta.: low 10.5°, high 21.5°
   (b) Cessna spinner 0752004
   (c) Woodward governor 210065, 210105, 210155, 210345, or 210452, or Garwin 34-828-01, or McCauley C290D2/T1 or C290D3/T1

*Airspeed Limits (CAS)		
	Maneuvering	128 m.p.h. (111 knots)
	Maximum structural cruising	160 m.p.h. (139 knots)
	Never exceed	193 m.p.h. (168 knots)
	Flaps extended	110 m.p.h. (96 knots)

C.G. Range
(+38.4) to (+47.4) at 2800 lb.
(+33.0) to (+47.4) at 2250 lb. or less
Straight line variation between points given

Empty Wt. C.G. Range	None
*Maximum Weight	2800 lb.
No. of Seats	4 (2 at +36, 2 at +71)
Maximum Baggage	120 lb. (+97)

V - Model 182E, Model 182F, Model 182G (cont'd)

Fuel Capacity	65 gal. (60 gal. usable); two 32.5 gal. tanks in wings at +48 See NOTE 1 for data on unusable fuel
Oil Capacity	12 qt. (-15) (6 qt. usable) See NOTE 1 for data on undrainable oil

Control Surface Movements					
Wing flaps					40° +1°, -2°
Elevator tab	Up	25° ±2°	Down	15° ±1°
Ailerons	Up	20° ±2°	Down	15° ±2°
Elevator (relative to stabilizer)	Up	26° ±1°	Down	17° ±1°
Rudder	Right	24° ±1°	Left	24° ±1°

Serial Nos. Eligible

Model 182E: 18253599 through 18254423 (1962 Model)
Model 182F: 18254424 through 18255058 (1963 Model)
Model 182G: 18255059 through 18255844 (1964 Model)

VI - Model 182H, Skylane, 4 PCLM (Normal Category), Approved September 17, 1964
Model 182J, Skylane, 4 PCLM (Normal Category), Approved October 20, 1965
Model 182K, Skylane, 4 PCLM (Normal Category), Approved August 3, 1966
Model 182L, Skylane, 4 PCLM (Normal Category), Approved July 28, 1967

Engine	Continental O-470-R
*Fuel	80/87 minimum grade aviation gasoline
*Engine Limits	For all operations, 2600 r.p.m. (230 hp.)
Propeller and Propeller Limits	1. McCauley constant speed (a) Hub 2A34C66/90AT-8 blades Diameter: not over 82 in., not under 80 in. Pitch settings at 36 in. sta.: low 10.5°, high 22° (b) Cessna spinner 0752637 (c) Woodward governor 210065, 210105, 210155, 210345, or 210452, or Garwin 34-828-01, or McCauley C290D2/T1 or C290D3/T1

*Airspeed Limits (CAS)		
Maneuvering	128 m.p.h.	(111 knots)
Maximum structural cruising	160 m.p.h.	(139 knots)
Never exceed	193 m.p.h.	(168 knots)
Flaps extended	110 m.p.h.	(96 knots)

C.G. Range	(+38.4) to (+47.4) at 2800 lb. (+33.0) to (+47.4) at 2250 lb. or less Straight line variation between points given
Empty Wt. C.G. Range	None
*Maximum Weight	2800 lb.
No. of Seats	4 (2 at +36, 2 at +71)
Maximum Baggage	120 lb. (+97)
Fuel Capacity	65 gal. (60 gal. usable); two 32.5 gal. tanks in wings at +48 See NOTE 1 for data on unusable fuel

VI - Model 182H, Model 182J, Model 182K, Model 182L (cont'd)

Oil Capacity	12 qt. (-15) (6 qt. usable) See NOTE 1 for data on undrainable oil

Control Surface Movements

Wing flaps				40° +1°, -2°
Elevator tab	Up	25° ±2°	Down	15° ±1°
Ailerons	Up	20° ±2°	Down	15° ±2°
Elevator(relative to stabilizer)	Up	26° ±1°	Down	17° ±1°
Rudder	Right	24° ±1°	Left	24° ±1°

Serial Nos. Eligible

Model 182H:	634, 18255846 through 18256684 (1965 Model)
Model 182J:	18256685 through 18257625 (1966 Model)
Model 182K:	18255845, 18257626 through 18257698, 18257700 through 18258505 (1967 Model)
Model 182L:	18258506 through 18259305 (1968 Model)

3A13

VII - Model 182M, Skylane, 4 PCLM (Normal Category), Approved September 19, 1968

Engine — Continental O-470-R

*Fuel — 80/87 minimum grade aviation gasoline

*Engine Limits — For all operations, 2600 r.p.m. (230 hp.)

Propeller and Propeller Limits

1. McCauley constant speed
   (a) Hub 2A34C66/90AT-8 blades
   Diameter: not over 82 in., not under 80 in.
   Pitch settings at 36 in. sta.:
   low 10.5°, high 22°
   (b) Cessna spinner 0752637
   (c) Woodward governor 210065, 210105, 210155, 210345, or 210452, or Garwin 34-828-01, or McCauley C290D2/T1 or C290D3/T1
2. McCauley constant speed
   (a) Hub 2A34C201/90DA-8 blades
   Diameter: not over 82 in., not under 80 in.
   Pitch settings at 30 in. sta.:
   low 13°, high 24.5°
   (b) Cessna spinner 0752637
   (c) Woodward governor 210065, 210105, 210155, 210345, or 210452, or Garwin 34-828-01, or McCauley C290D2/T1 or C290D3/T1
3. McCauley constant speed
   (a) Hub 2A34C203/90DCA-8 blades
   Diameter: not over 82 in., not under 80.5 in.
   Pitch settings at 30 in. sta.:
   low 12.5°, high 25°
   (b) Cessna spinner 0752637
   (c) Woodward governor 210065, 210105, 210155, 210345, or 210452, or Garwin 34-828-01, or McCauley C290D2/T1 or C290D3/T1

*Airspeed Limits (CAS)

Maneuvering	128 m.p.h. (111 knots)
Maximum structural cruising	160 m.p.h. (139 knots)
Never exceed	193 m.p.h. (168 knots)
Flaps extended	110 m.p.h. (96 knots)

C.G. Range

(+38.4) to (+47.4) at 2800 lb.
(+33.0) to (+47.4) at 2250 lb. or less
Straight line variation between points given

VII - Model 182M (cont'd)

Empty Wt. C.G. Range	None
*Maximum Weight	2800 lb.
No. of Seats	4 (2 at +36, 2 at +71)
Maximum Baggage	120 lb. (+97)
Fuel Capacity	65 gal. (60 gal. usable); two 32.5 gal. tanks in wings at +48 See NOTE 1 for data on unusable fuel
Oil Capacity	12 qt. (-15) (6 qt. usable) See NOTE 1 for data on undrainable oil

Control Surface Movements

Wing flaps				40° +1°, -2°
Elevator tab	Up	25° ±2°	Down	15° ±1°
Ailerons	Up	20° ±2°	Down	15° ±2°
Elevator(relative to stabilizer)	Up	26° ±1°	Down	17° ±1°
Rudder	Right	24° ±1°	Left	24° ±1°

Serial Nos. Eligible — Model 182M: 18257699, 18259306 through 18260055 (1969 Model)

VIII - Model 182N, Skylane, 4 PCLM (Normal Category), Approved September 17, 1969

Engine	Continental O-470-R Continental O-470-S (See NOTE 4)
*Fuel	80/87 minimum grade aviation gasoline
*Engine Limits	For all operations, 2600 r.p.m. (230 hp.)

Propeller and Propeller Limits

1. McCauley constant speed
   (a) Hub 2A34C201/90DA-8 blades
   Diameter: not over 82 in., not under 80 in.
   Pitch settings at 30 in. sta.: low 13°, high 24.5°
   (b) Cessna spinner 0752637
   (c) Woodward governor 210065, 210105, 210155, 210345, or A210452, or Garwin 34-828-01-2A, or McCauley C290D2/T1 or C290D3/T1
2. McCauley constant speed
   (a) Hub 2A34C66/90AT-8 blades
   Diameter: not over 82 in., not under 80 in.
   Pitch settings at 36 in. sta.: low 10.5°, high 22°
2. (b) Cessna spinner 0752637
   (c) Woodward governor 210065, 210105, 210155, 210345, or 210452, or Garwin 34-828-01, or McCauley C290D2/T1 or C290D3/T1
3. McCauley constant speed
   (a) Hub 2A34C203/90DCA-8 blades
   Diameter: not over 82 in., not under 80.5 in.
   Pitch settings at 30 in. sta.: low 12.5°, high 25°
   (b) Cessna spinner 0752637
   (c) Woodward governor 210065, 210105, 210155, 210345, or 210452, or Garwin 34-828-01, or McCauley C290D2/T1 or C290D3/T1

*Airspeed Limits (CAS)

Maneuvering	131 m.p.h. (114 knots)
Maximum structural cruising	160 m.p.h. (139 knots)
Never exceed	198 m.p.h. (172 knots)
Flaps extended	110 m.p.h. (96 knots)

VIII - Model 182N (cont'd)

C.G. Range	(+39.9) to (+47.4) at 2950 lb. (+38.4) to (+47.4) at 2800 lb. (+33.0) to (+47.4) at 2250 lb. or less Straight line variation between points given
Empty Wt. C.G. Range	None
*Maximum Weight	2950 lb. takeoff only, 2800 lb. landing
No. of Seats	4 Front standard (2 at +36 to +49) Optional (2 at +32 to +44) Rear (2 at +74)
Maximum Baggage	120 lb. (+97) (S/N 18260056 through 18260445) 120 lb. (+97) and 80 lb. (+117) (S/N 18260446 and up)
Fuel Capacity	65 gal. (60 gal. usable); two 32.5 gal. tanks in wings at +48 See NOTE 1 for data on unusable fuel
Oil Capacity	12 qt. (-15) (6 qt. usable) See NOTE 1 for data on undrainable oil

Control Surface Movements

Wing flaps			Down	40° +1°, -2°
Elevator tab	Up	25° ±2°	Down	15° ±1°
Ailerons	Up	20° ±2°	Down	15° ±2°
Elevator(rel. to stabilizer)	Up	26° ±1°	Down	17° ±1°
Rudder (parallel to 0.00 W.L.)	Right	24° ±1°	Left	24° ±1°
(Perpendicular to hinge line)	Right	27° 13' ±1°	Left	27° 13' ±1°

Serial Nos. Eligible — Model 182N: 18260056 through 18260445 (1970 Model) 18260446 through 18260825 (1971 Model)

IX - Model 182P, Skylane, 4 PCLM (Normal Category), Approved October 8, 1971

Engine	Continental O-470-R, Aircraft S/N 18260826 through 18263475 Continental O-470-S, Aircraft S/N 18260826 and up (See NOTE 4)
*Fuel	80/87 minimum grade aviation gasoline
*Engine Limits	For all operations, 2600 r.p.m. (230 hp.)

Propeller and Propeller Limits

1. McCauley constant speed
   (a) Hub 2A34C201/90DA-8 blades
   Diameter: not over 82 in., not under 80 in.
   Pitch settings at 30 in. sta.: low 13°, high 24.5°
   (b) Cessna spinner 0752637
   (c) Woodward governor 210065, 210105, 210155, 210345, or A210452, or Garwin 34-828-01-2A, or McCauley C290D2/T1 or C290D3/T1
2. McCauley constant speed
   (a) Hub 2A34C66/90AT-8 blades
   Diameter: not over 82 in., not under 80 in.
   Pitch settings at 36 in. sta.: low 10.5°, high 22°
   (b) Cessna spinner 0752637
   (c) Woodward governor 210065, 210105, 210155, 210345, or 210452, or Garwin 34-828-01, or McCauley C290D2/T1 or C290D3/T1

IX - Model 182P, Skylane (Cont'd)

3. McCauley constant speed
   (a) Hub 2A34C203/90DCA-8 blades
   Diameter: not over 82 in., not under 80.5 in.
   Pitch settings at 30 in. sta.: low 12.5°, high 25°
   (b) Cessna spinner 0752637
   (c) Woodward governor 210065, 210105, 210155, 210345, or 210452, or Garwin 34-828-01, or McCauley C290D2/T1 or C290D3/T1

*Airspeed Limits (CAS)	(S/N 675, 18260826 through 18264295)	
	Maneuvering	126 m.p.h. (109 knots)
	Maximum structural cruising	160 m.p.h. (139 knots)
	Never exceed	198 m.p.h. (172 knots)
	Flaps extended	110 m.p.h. (96 knots)
*Airspeed Limits (IAS) (See NOTE 5 on use of IAS)	(S/N 18264296 through 18265175)	
	Maneuvering	110 knots
	Maximum structural cruising	141 knots
	Never exceed	176 knots
	Flaps extended	95 knots

C.G. Range: (+39.5) to (+48.5) at 2950 lb.
(+33.0) to (+48.5) at 2250 lb. or less
Straight line variation between points given

Empty Wt. C.G. Range: None

*Maximum Weight: 2950 lb.

No. of Seats: 4 (2 front at +32.0 to +50.0)
(2 rear at +74)

Maximum Baggage: Serial Numbers 18260826 through 18263475
200 lb. (120 lb. at + 82.0 to +108.0)
(80 lb. at +108.0 to +124.0)
Serial Numbers 675 and 18263476 through 18265175
200 lb. (120 lb. at + 82.0 to +108.0)
(80 lb. at +108.0 to +136.0)

Fuel Capacity: (S/N 675, 18260826 through 18262250)
Standard Range Tanks:
65 gal. (60 gal. usable); two 32.5 gal. tanks in wings at +48
Long Range Tanks:
84 gal. (79 gal. usable); two 42.0 gal. tanks in wings at +48
(S/N 18262251 through 18265175)
Standard Range Tanks:
61 gal. (56 gal. usable); two 30.5 gal. tanks in wings at +48
Long Range Tanks:
80 gal. (75 gal. usable); two 40.0 gal. tanks in wings at +48

See NOTE 1 for data on unusable fuel

Oil Capacity: 12 qt. (-15) (6 qt. usable)
See NOTE 1 for data on undrainable oil

IX - Model 182P, Skylane (Cont'd)

Control Surface Movements	Wing flaps			Down	40° +1°, -2°
	Elevator tab	Up	25° ±2°	Down	15° ±1°
	Ailerons	Up	20° ±2°	Down	15° ±2°
	Elevator (rel. to stabilizer)	Up	26° ±1°	Down	17° ±1°
	Rudder(parallel to 0.00 W.L.)	Right	24° ±1°	Left	24° ±1°
	(perpendicular to hinge line)	Right	27° 13' ±1°	Left	27° 13' ±1°

Serial Nos. Eligible — Model 182P: 18260826 through 18261425 (1972 Model)
18261426 through 18262465 (1973 Model)
18262466 through 18263475 (1974 Model)
675, 18263476 through 18264295 except 18263479 (1975 Model)
18264296 through 18265175 (1976 Model)

X - Model 182Q, Skylane, 4 PCLM (Normal Category), Approved July 28, 1976

Engine — Continental O-470-U

*Fuel — 100/130 minimum aviation grade gasoline (S/N 18265176 through 18265965)
100LL/100 aviation grade gasoline (S/N 18265966 through 18267715)

*Engine Limits — For all operations, 2400 r.p.m. (230 hp.)

Propeller and Propeller Limits — McCauley constant speed
(a) Hub C2A34C204/90DCB-8 blades
Diameter: not over 82 in., not under 80.5 in.
Pitch settings at 30 in. sta.:
low 15°, high 29.4°
(b) Cessna spinner 0752637
(c) McCauley governor C290D3/T14

*Airspeed Limits (IAS) (See NOTE 5 on use of IAS)

Maneuvering	111 knots
Maximum structural cruising	143 knots
Never exceed	179 knots
Flaps extended	95 knots

C.G. Range — (+39.5) to (+48.5) at 2950 lb.
(+33.0) to (+48.5) at 2250 lb. or less
Straight line variation between points given

Empty Wt. C.G. Range — None

*Maximum Weight — 2950 lb.

No. of Seats — 4 (2 front at +32.0 to +50.0)
(2 rear at +74)

Maximum Baggage — 200 lb. (120 lb. at +82.0 to +108.0)
(80 lb. at +108.0 to +136.0)

Fuel Capacity — Standard Range Tanks:
61 gal. (56 gal. usable); two 30.5 gal. tanks in wings at +48
(S/N 18263479, 18265176 through 18266590)

Long Range Tanks:
80 gal. (75 gal. usable); two 40.0 gal. tanks in wings at +48
(S/N 18263479, 18265176 through 18266590)

X - Model 182Q (cont'd)

Fuel Capacity (Cont'd)	92 gal. (88 gal. usable); two 46.0 gal. integral tanks in wings at +46.5 (S/N 18266591 through 18267715) See NOTE 1 for data on unusable fuel
Oil Capacity	12 qt. (-15.0) (6 qt. usable) See NOTE 1 for data on undrainable oil

Control Surface Movements

Wing flaps			Down	40° +1°, -2°
Elevator tab	Up	25° ±2°	Down	15° ±1°
Ailerons	Up	20° ±2°	Down	15° ±2°
Elevator (rel. to stabilizer)	Up	26° ±1°	Down	17° ±1°
Rudder (parallel to 0.00 W.L.)	Right	24° ±1°	Left	24° ±1°
(perpendicular to hinge line)	Right	27° 13' ±1°	Left	27° 13' ±1°

Serial Nos. Eligible

Model 182Q:	18265176 through 18265965	(1977 Model)
	18263479, 18265966 through 18266590	(1978 Model)
	18266591 through 18267300	(1979 Model)
	18267301 through 18267715, except 18267302	(1980 Model)

XI - Model R182, Skylane RG, 4 PCLM (Normal Category), Approved July 7, 1977
Model TR182, Turbo Skylane RG, 4 PCLM (Normal Category), Approved September 12, 1978

Model R182

Engine	Lycoming O-540-J3C5D, rated at 235 hp.
*Fuel	100LL/100 aviation grade gasoline
*Engine Limits	Full throttle for all operations, 2400 r.p.m.

Propeller and Propeller Limits

1. McCauley constant speed (S/N R18200002 through R18201313)
   (a) Hub B2D34C214/90DHB-8 blades
       Diameter: not over 82 in., not under 80.5 in.
       Pitch settings at 30 in. sta.:
       low 15.8°, high 29.4°
   (b) Cessna spinner 2250003
   (c) McCauley governor C290D3/T16
2. McCauley constant speed (S/N R18201314 through R18201628)
   (a) Hub B2D34C218/90DHB-8 blades
       Diameter: not over 82 in., not under 80.5 in.
       Pitch settings at 30 in. sta.:
       low 15.8°, high 29.4°
   (b) Cessna spinner 2250124
   (c) McCauley governor C290D3/T22
3. McCauley constant speed (S/N R18201629 through R18202041 and aircraft reworked per SK182-71)
   (a) Hub B3D32C407/82NDA-3 blades
       Diameter: not over 79 in., not under 78 in.
       Pitch settings at 30 in. sta.: low 16.0°, high 31.7°
   (b) Cessna spinner 2252076
   (c) McCauley governor C290D3/T22

XI - Model R182, Model TR182, Turbo Skylane RG (cont'd)

Model TR182

Engine	Lycoming O-540-L3C5D, rated at 235 hp. (Turbocharged in accordance with Cessna Drawing No. 2250065)
*Fuel	100LL/100 aviation grade gasoline
*Engine Limits	For all operations, 2400 r.p.m., 31 in. hg. mp.
Propeller and Propeller Limits	1. McCauley constant speed (S/N R18200001, R18200584 through R18201313) (a) Hub B2D34C217/90DHB-8 Diameter: not over 82 in., not under 80.5 in. Pitch settings at 30 in. sta.: low 15.8°, high 31.9° (b) Cessna spinner 2250003 (c) McCauley governor C290D3/T21 2. McCauley constant speed (S/N R18201314 and up) (a) Hub B2D34C219/90DHB-8 Diameter: not over 82 in., not under 80.5 in. Pitch settings at 30 in. sta.: low 15.8°, high 31.9° (b) Cessna spinner 2250124 (c) McCauley governor C290D3/T22 3. McCauley constant speed (S/N R18201315, R18201629 and up and aircraft reworked per SK182-71 or SK182-72) (a) Hub B3D32C407/82NDA-3 Diameter: not over 79 in., not under 78 in. Pitch settings at 30 in. sta.: low 16.0°, high 31.7° (b) Cessna spinner 2252076 (c) McCauley governor C290D3/T22

Models R182, TR182

*Airspeed Limits (IAS) (See NOTE 5 on use of IAS)	1978 Model R182	Maneuvering	112 knots
		Maximum structural cruising	143 knots
		Never exceed	182 knots
		Flaps extended	95 knots
		Landing gear extension	140 knots
	1979 Model R182	Maneuvering	112 knots
		Maximum structural cruising	160 knots
		Never exceed	182 knots
		Flaps extended	95 knots
		Landing gear extension	140 knots
	Model TR182	Maneuvering	112 knots
		Maximum structural cruising	157 knots
		Never exceed	179 knots
		Flaps extended	95 knots
		Landing gear extension	140 knots
	1980 and up Model R182	Maneuvering	112 knots
		Maximum structural cruising	159 knots
		Never exceed	181 knots
		Flaps extended	95 knots
		Landing gear extension	140 knots
	Model TR182	Maneuvering	112 knots
		Maximum structural cruising	157 knots
		Never exceed	178 knots
		Flaps extended	95 knots
		Landing gear extension	140 knots

3A13

XI - Model R182, Model TR182 (cont'd)

C.G. Range

(a) S/N R18200001 through R18201628 except R18200975 & R18201315
(+40.9) to (+47.0) at 3100 lb.
(+35.5) to (+47.0) at 2700 lb.
(+33.0) to (+47.0) at 2250 lb. or less
Straight line variation between points given
Moment change due to retracting gear (+3052 in.-lb.)

(b) S/N R18200975, R18201315, R18201629 through R18202041
(+40.9) to (+46.0) at 3100 lb.
(+35.5) to (+46.0) at 2700 lb.
(+33.0) to (+46.0) at 2250 lb. or less
Straight line variation between points given
Moment change due to retracting gear (+3052 in.-lb.)

Empty Wt. C.G. Range — None

*Maximum Weight — 3100 lb.

No. of Seats — 4 (2 front at +32.0 to +50.0)
(2 rear at +74.0)

Maximum Baggage — 200 lb. (120 lb. at +82.0 to +110.0)
(80 lb. at +110.0 to +134.0)

Fuel Capacity

(a) S/N R18200002 through R18200583
Standard Range Tanks:
61 gal. (56 gal. usable); two 30.5 gal. tanks in wings at +48
Long Range Tanks:
80 gal. (75 gal. usable); two 40.0 gal. tanks in wings at +48

(b) S/N R18200001, R18200584 through R18202041
92 gal. (88 gal. usable); two 46.0 gal. integral tanks
in wings at +46.5

See NOTE 1 for data on unusable fuel

Oil Capacity — 9 qt. (-14.8)
See NOTE 1 for data on oil

Control Surface Movements

(a) S/N R18200001 through R18201628 except R18200975 & R18201315

Wing flaps			Down	40° +1°, -2°
Elevator tab	Up	25° ±2°	Down	15° ±1°
Ailerons	Up	20° ±2°	Down	15° ±2°
Elevator (rel. to stabilizer)	Up	28° ±1°	Down	17° ±1°
Rudder (parallel to 0.00 W.L.)	Right	24° ±1°	Left	24° ±1°
(Perpendicular to hinge line)	Right	27° 13' ±1°	Left	27° 13' ±1°

(b) S/N R18200975, R18201629 through R18201798

Wing flaps			Down	40° +1°, -2°
Elevator tab	Up	24° ±2°	Down	15° ±1°
Ailerons	Up	20° ±2°	Down	15° ±2°
Elevator (rel. to stabilizer)	Up	28° ±1°	Down	21° ±1°
Rudder (parallel to 0.00 W.L.)	Right	24° +0°, -1°	Left	24° +0°, -1°
(Perpendicular to hinge line)	Right	27° 13' +0°, -1°	Left	27° 13' +0°, -1°

XI - Model R182, Model TR182 (cont'd)

(c) S/N R18201315, R18201799 through R18202041

Wing flaps			Down	38° +0°, -1°
Elevator tab	Up	24° ±2°	Down	15° ±1°
Ailerons	Up	20° ±1°	Down	15° ±2°
Elevator (rel. to stabilizer)	Up	28° ±1°	Down	21° ±1°
Rudder (parallel to 0.00 W.L.)	Right	24° +0°, -1°	Left	24° +0°, -1°
(Perpendicular to hinge line)	Right	27° 13' +0°, -1°	Left	27° 13' +0°, -1°

Serial Nos. Eligible

Model R182:	R18200002 through R18200583	(1978 Model)
Model R182/TR182:	R18200001, R18200584 through R18201313	(1979 Model)
Model R182/TR182:	R18201314 through R18201628 except R18201315	(1980 Model)
Model R182/TR182:	R18201629 through R18201798	(1981 Model)
Model R182/TR182:	R18201799 through R18201928	(1982 Model)
Model R182/TR182:	R18201929 through R18201973	(1983 Model)
Model R182/TR182:	R18201974 through R18201999	(1984 Model)
Model R182/TR182:	R18201315, R18202000 through R18202031	(1985 Model)
Model R182/TR182:	R18202032 through R18202041	(1986 Model)

3A13

XII - Model 182R, 4 PCLM (Normal Category), Approved August 29, 1980
Model T182, 4 PCLM (Normal Category), Approved August 15, 1980

Model 182R

Engine — Continental O-470-U

*Fuel — 100LL/100 aviation grade gasoline

*Engine Limits — For all operations, 2400 r.p.m. (230 hp.)

Propeller and Propeller Limits — McCauley constant speed
(a) Hub C2A34C204/90DCB-8
Diameter: not over 82 in., not under 80.5 in.
Pitch settings at 30 in. sta.:
low 15°, high 29.4°
(b) Cessna spinner 0752637
(c) McCauley governor C290D3/T14

Model T182

Engine — Lycoming 0-540-L3C5D, rated at 235 hp.
(Turbocharged in accordance with Cessna Drawing No. 2250065)

*Fuel — 100LL/100 aviation grade gasoline

*Engine Limits — For all operations, 2400 r.p.m., 31 in. Hg. mp.

Propeller and Propeller Limits —
1. McCauley constant speed
(a) Hub B2D34C219/90DHB-8
Diameter: not over 82 in., not under 80.5 in.
Pitch settings at 30 in. sta.:
low 15.8°, high 31.9°
(b) Cessna spinner 2250124
(c) McCauley governor C290D3/T22

XII - Model 182R, Model T182 (Cont'd)

Models 182R/T182

Propeller and Propeller Limits (Cont'd)	2. McCauley constant speed (a) Hub B3D32C407/82NDA-3 Diameter: not over 79 in., not under 78 in. Pitch settings at 30 in. sta.: low 16.0°, high 31.7° (b) Cessna spinner 2252076 (c) McCauley governor C290D3/T22		
*Airspeed Limits (IAS) (See NOTE 5 on Use of IAS)	Model 182R	Maneuvering	111 knots
		Maximum structural cruising	143 knots
		Never exceed	179 knots
		Flaps extended	95 knots
	Model T182	Maneuvering	111 knots
		Maximum structural cruising	140 knots
		Never exceed	178 knots
		Flaps extended	95 knots

C.G. Range

Model 182R
(+40.9) to (+46.0) at 3100 lb.
(+33.0) to (+46.0) at 2250 lb. or less
Straight line variation between points given

Model T182
(+40.9) to (+46.0) at 3100 lb.
(+35.5) to (+46.0) at 2700 lb.
(+33.0) to (+46.0) at 2250 lb. or less
Straight line variation between points given

Empty Wt. C.G. Range: None

*Maximum Weight
3100 lb. takeoff/flight
2950 lb. landing

No. of Seats
4 (2 front at +32.0 to +50.0)
(2 rear at +74.0)

Maximum Baggage
200 lb. (120 lb. at +92.0 to +108.0)
(80 lb. at +108.0 to +136.0)

Fuel Capacity
92 gal. (88 gal. usable); two 46 gal. integral tanks in wings at +46.5
See NOTE 1 for data on unusable fuel

Oil Capacity

Model 182R	Model T182
12 qt. (-15.0) (6 qt. usable) (through S/N 18268055) 12 qt. (-14.1) (6 qt. usable) (S/N 18268056 and on) See NOTE 1 for data on oil	9 qt (-14.8) (6 qt. usable) See NOTE 1 for data on oil

XII - Model 182R, Model T182 (cont'd)

Control Surface Movements

(a) S/N 18267716 through 18268055

Wing flaps			Down	40° +1°, -2°
Elevator tab	Up	24° ±2°	Down	15° ±1°
Ailerons	Up	20° ±2°	Down	15° ±2°
Elevator (rel. to stabilizer)	Up	28° ±1°	Down	21° ±1°
Rudder (parallel to 0.00 W.L.)	Right	24° +0°, -1°	Left	24° +1°, -0°
(Perpendicular to hinge line)	Right	27° 13' +0°, -1°	Left	27° 13' +0°, -1°

(b) S/N 18268056 through 18268586

Wing flaps			Down	38° +0°, -1°
Elevator tab	Up	24° ±2°	Down	15° ±1°
Ailerons	Up	20° ±2°	Down	15° ±2°
Elevator (rel. to stabilizer)	Up	28° ±1°	Down	21° ±1°
Rudder (parallel to 0.00 W.L.)	Right	24° +0°, -1°	Left	24° +0°, -1°
(Perpendicular to hinge line)	Right	27° 13' +0°, -1°	Left	27° 13' +0°, -1°

Serial Nos. Eligible

Model	182R/T182:	18267302, 18267716 through 18268055	(1981 Model)
Model	182R/T182:	18268056 through 18268293	(1982 Model)
Model	182R/T182:	18268294 through 18268368	(1983 Model)
Model	182R/T182:	18268369 through 18268434	(1984 Model)
Model	182R/T182:	18268435 through 18268541	(1985 Model)
Model	182R:	18268542 through 18268586	(1986 Model)

(1986 Model)

Data Pertinent to Model Items I through XII

Datum — Front face of firewall

Leveling Means — Upper door sill. Top surface centerline of tailcone (S/N 18253599 through 18265965) Jig located nutplates and screws on left of tailcone (S/N 18263479, 18265966 through 18268586) (S/N R18200001 through 18202041)

Certification Basis

182 Series
Part 3 of the Civil Air Regulations dated November 1, 1949, as amended by 3-1 through 3-12 and Paragraph 3.112 as amended October 1, 1959, for the Model 182E and on. In addition, effective S/N 18266591 through 18268586, FAR 23.1559 effective March 1, 1978. FAR 36 dated December 1, 1969, plus Amendments 36-1 through 36-6 for Model 182Q and on. In addition, effective S/N 18268435 through 18268586, FAR 23.1545(a) Amendment 23-23 dated December 1, 1978.

Model T182
Part 3 of the Civil Air Regulations dated November 1, 1949, as amended by 3-1 through 3-12 and Paragraph 3.112 as amended October 1, 1959; and Sections 23.901, 23.909, 23.1041, 23.1043, 23.1143, and 23.1305 of the Federal Aviation Regulations dated February 1, 1965, as amended February 14, 1975; FAR 23.1559 effective March 1, 1978; FAR 36 dated December 1, 1969, plus Amendments 36-1 through 36-10. In addition, effective S/N 18268435 through 18268541, FAR 23.1545(a) Amendment 23-23 dated December 1, 1978.

Data Pertinent to Model Items I through XII, continued

Model R182
Part 3 of the Civil Air Regulations dated November 1, 1949, as amended by 3-1 through 3-12 and Paragraph 3.112 as amended October 1, 1959; and Sections 23.729, 23.777(e), 23.781, 23.1555(e)(1) and (2), and 23.1563 of the Federal Aviation Regulations dated February 1, 1965, as amended February 14, 1975. In addition, effective S/N R18200001, R18200584 and up, FAR 23.1559 effective March 1, 1978. FAR 36 dated December 1, 1969, plus Amendments 36-1 through 36-6. In addition, effective S/N R18202000 through R18202041, FAR 23.1545(a) Amendment 23-23 dated December 1, 1978.

Model TR182
Part 3 of the Civil Air Regulations dated November 1, 1949, as amended by 3-1 through 3-12 and Paragraph 3.112 as amended October 1, 1969; and Sections 23.729, 23.777(e), 23.781, 23.901, 23.909, 23.1041, 23.1043, 23.1143, 23.1305, 23.1555(e)(1) and (2), and 23.1563 of the Federal Aviation Regulations dated February 1, 1965, as amended February 14, 1975; FAR 23.1559 effective March 1, 1978; FAR 36 dated December 1, 1969, plus Amendments 36-1 through 36-9. In addition, effective S/N R18202000 through R18202041, FAR 23.1545(a) Amendment 23-23 dated December 1, 1978.

Application for Type Certificate dated July 11, 1955.

Type Certificate No. 3A13 issued March 2, 1956, obtained by the manufacturer under delegation option procedures.

Equivalent Safety Items:

S/N 18263479, 18264296 through 18267715

Airspeed Indicator	CAR 3.757 (See NOTE 5 on use of IAS)
Operating Limitations	CAR 3.778(a)

S/N 18267716 through 18268586

Airspeed Indicator	CAR 3.757 (See NOTE 5 on use of IAS) (S/N 18267716 through 18268434)
Operating Limitations	CAR 3.778(a)
Fuel System	CAR 3.430

S/N R18200001 through R18202041

Airspeed Indicator	CAR 3.757 (See NOTE 5 on use of IAS) (S/N R18200001 through R18201999)
Operating Limitations	CAR 3.778(a)
Fuel System	CAR 3.430

Production Basis

Production Certificate No. 4. Delegation Option Manufacturer No. CE-1 authorized to issue airworthiness certificates under delegation option provisions of Part 21 of the Federal Aviation Regulations.

Equipment:

The basic required equipment as prescribed in the applicable airworthiness requirements (see Certification Basis) must be installed in the aircraft for certification. This equipment must include a current Airplane Flight Manual effective S/N 18266591 through 18268586 and R18200584 through R18202041. In addition, the following item of equipment is required:

1. Stall warning indicator, Cessna Dwg. S1672-5.

The equipment portion of Aircraft Specification 3A13, Revision 15, or Cessna Publication TS3000-13 should be used for equipment references on all aircraft prior to the Model 182G. Refer to the applicable Equipment List for the Model 182G and subsequent models.

Data Pertinent to Model Items I through XII, continued

NOTE 1. Current weight and balance report including list of equipment included in certificated empty weight, and loading instructions when necessary must be provided for each aircraft at the time of original certification.

Serial Numbers 613 and 33000 through 34999
631 and 51001 through 53007
18253008 through 18264295 except 18263479

The certificated empty weight and corresponding center of gravity location must include unusable fuel of 30 lb. (+46) on Models 182, 182E, 182F, 182G, 182H, 182J, 182K, 182L, 182M, 182N and 182P through 18264295 and 60 lb. (+46) on Models 182A, 182B, 182C and 182D and undrainable oil of 0 lb.

Serial Numbers 18263479, 18264296 through 18266590
The certificated empty weight and corresponding center of gravity location must include unusable fuel of 30 lb. (+46) and full oil of 22.5 lb. at (-15.0).

Serial Numbers 18266591 through 18268055
The certificated empty weight and corresponding center of gravity location must include unusable fuel of 24 lb. at (+48) and full oil of 22.5 lb. at (-15.0) for the 182Q, 182R Model, and include oil of 16.9 lb. at (-14.8) for the T182 Model.

Serial Numbers 18268056 through 18268586
The certificated empty weight and corresponding center of gravity location must include unusable fuel of 24 lb. at (+48) and full oil of 24.4 lb. at (-14.1) for the 182R, and include oil of 16.9 lb. at (-14.8) for the T182.

Serial Numbers R18200002 through R18200583
The certificated empty weight and corresponding center of gravity location must include unusable fuel of 30 lb. (+46) and include oil of 16.9 lb. (-15.7).

Serial Numbers R18200001, R18200584 through R18202041
The certificated empty weight and corresponding center of gravity location must include unusable fuel of 24 lb. (+48) and include oil of 16.9 lb. (-14.8).

NOTE 2. The following placards must be displayed in locations as indicated:

A. Applicable to Model 182 only:

(1) In full view of the pilot:

(a) "This airplane must be operated as a normal category airplane in compliance with operating limitations stated in the form of placards, markings and manuals. No acrobatic maneuvers including spins approved.

Flight Maneuvering Load Factors

Flaps Up	+3.8	-1.52
Flaps Down	+3.5	

Maximum design weight 2550 lb.
Reference weight and balance data for loading instructions."

(b) "Both tanks on for takeoff and landing."

(c) "Flaps - Pull to extend

Takeoff	Retracted	0°
	1st Notch	10°
	2nd Notch	20°
Landing	3rd Notch	30°
	4th Notch	40°

(2) In baggage compartment
"Maximum baggage 120 lb. For additional loading instructions see weight and balance data."

3A13

Data Pertinent to Model Items I through XII, continued

B. Applicable to Models 182A, 182B, 182C and 182D

(1) In full view of the pilot:

(a) "This airplane must be operated as a normal category airplane in compliance with operating limitations stated in the form of placards, markings and manuals. No acrobatic maneuvers including spins approved.

Flight Maneuvering Load Factors

Flaps Up +3.8 -1.52

Flaps Down +3.5

Maximum design weight 2650 lb.

Reference weight and balance data for loading instructions."

(b) "Both tanks on for takeoff and landing."

(c) "Flaps - Pull to extend

Takeoff	Retracted	0°
	1st Notch	10°
	2nd Notch	20°
Landing	3rd Notch	30°
	4th Notch	40°"

(2) In baggage compartment

"Maximum baggage 120 lb. For additional loading instructions see weight and balance data."

C. Applicable to Models 182E, 182F, 182G, 182H, 182J, 182K, 182L, 182M

(1) In full view of the pilot:

(a) "This airplane must be operated as a normal category airplane in compliance with operating limitations stated in the form of placards, markings and manuals. No acrobatic maneuvers including spins approved.

Flight Maneuvering Load Factors

Flaps Up +3.8 -1.52

Flaps Down +3.5

Maximum design weight 2800 lb.

Reference weight and balance data for loading instructions."

(2) On the fuel selector valve plate:

"Both off. Left tank level flight only 31 gal. Both on for landing and takeoff all flight attitudes 60 gal. Right tank level flight only 31 gal."

(3) On the control lock:

"Control lock - Remove before starting engine."

(4) On the baggage door:

"120 lb. maximum baggage and/or auxiliary seat passengers. For additional loading instructions, see weight and balance data."

D. Applicable to Models 182N:

(1) In full view of the pilot:

(a) Serial Numbers 18260056 through 18260445

"This airplane must be operated as a normal category airplane in compliance with the operating limitations as stated in the form of placards, markings and manuals.

No acrobatic maneuvers, including spins, approved

	Maximums	
Design weight	2950 lb. takeoff	Alt. loss in stall recovery-160 ft.
	2800 lb. landing	Flight Maneuvering Load Factors
Maneuvering speed	131 m.p.h.-CAS	Flaps up +3.8, -1.52, Flaps down +3.5

Reference weight and balance data for loading instructions"

(b) Serial Numbers 182670446 through 18260825

"This airplane must be operated as a normal category airplane in compliance with the operating limitations as stated in the form of placards, markings and manuals.

Data Pertinent to Model Items I through XII, continued

D. Applicable to Models 182N, continued:

Maximums

Maneuvering speed	131 m.p.h. CAS (114 knots)	
Gross weight	Takeoff 2950 lb.	
	Landing 2800 lb.	
Flight load factor	Flaps up	+3.8, -1.52
	Flaps down	+3.5

No acrobatic maneuvers, including spins, approved. Altitude loss in a stall recovery 160 ft. Known icing conditions to be avoided. This airplane is certified for the following flight operations as of date of original airworthiness certificate: DAY-NIGHT-VFR-IFR" (as applicable)

(2) On the fuel selector valve plate:
"Both off. Left tank level flight only 31 gal. Both on for landing and takeoff all flight attitudes, 60 gal. Right tank level flight only 31 gal."

(3) On the control lock:
"Control lock - Remove before starting engine."

(4) On the baggage door:
(a) "120 lb. maximum baggage and/or auxiliary seat passengers. For additional loading instructions, see weight and balance data."
Applicable to Models 182N, S/N 18260056 through 18260445.
(b) "120 lb. maximum baggage and/or auxiliary passenger forward of baggage door latch, and 80 pounds maximum baggage aft of baggage door latch. Maximum 200 lb. combined. For additional loading instructions see weight and balance data." Applicable to Models 182N, S/N 18260446 and up.

(5) On flap control indicator:
(a) "0° to 20° - T.O."
(b) "10° - 20° - Full.
(Indices at these positions with blue color code and 160 m.p.h. callout, and white color code with 110 m.p.h. callout; mechanical detent at 10° and 20°)"

E. Applicable to Models 182P:
(1) In full view of the pilot:
(S/N 675, 18260826 through 18264295)
(a) "This airplane must be operated as a normal category airplane in compliance with the operating limitations as stated in the form of placards, markings and manuals.

Maximums

Maneuvering speed	126 m.p.h. CAS (109 knots)
Gross weight	2950 lb.
Flight load factor	Flaps up +3.8, -1.52
	Flaps down +2.0

No acrobatic maneuvers, including spins, approved. Altitude loss in a stall recovery 160 ft. Known icing conditions to be avoided. This airplane is certified for the following flight operations as of date of original airworthiness certificate: DAY-NIGHT-VFR-IFR." (as applicable)

(S/N 18264296 through 18265175)
(b) "This airplane must be operated as a normal category airplane in compliance with the operating limitations as stated in the form of placards, markings and manuals.

Data Pertinent to Model Items I through XII, continued

E. Applicable to Models 182P, continued:

	Maximums
Maneuvering speed (IAS)	110 knots
Gross weight	2950 lb.
Flight load factor	Flaps up +3.8, -1.52 Flaps down +2.0

No acrobatic maneuvers, including spins, approved. Altitude loss in a stall recovery 160 ft. Flight into known icing conditions prohibited. This airplane is certified for the following flight operations as of date of original airworthiness certificate: DAY-NIGHT-VFR-IFR" (as applicable)

(2) On the fuel selector valve plate: (S/N 675, 18260826 through 18262250)

Standard range tanks: "Off. Left tank level flight only 31 gal. Both on for landing and takeoff all flight attitudes, 60 gal. Right tank level flight only 31 gal."

Long range tanks: "Off. Left tank level flight only 39 gal. Both on for landing and takeoff all flight attitudes, 79 gal. Right tank level flight only 39 gal."

On the fuel selector valve plate: (S/N 182622251 through 18265175)

Standard range tanks: "Off. Left tank level flight only 29 gal. Both on for landing and takeoff all flight attitudes, 56 gal. Right tank level flight only 29 gal."

Long range tanks: "Off. Left tank level flight only 37 gal. Both on for landing and takeoff all flight attitudes, 75 gal. Right tank level flight only 37 gal."

(3) On the control lock: "Control lock - remove before starting engine."

(4) On the baggage door: (S/N 18260826 through 18263475)
"120 lb. maximum baggage and/or auxiliary passenger forward of baggage door latch, and 80 lb. maximum baggage aft of baggage door latch. Maximum 200 lb. combined. For additional loading instructions, see weight and balance data."

On the baggage door: (S/N 675, 18263476 through 18265175)
"Forward of baggage door latch, 120 lb. maximum baggage and/or auxiliary passenger. Aft of baggage door latch, 80 lb. maximum baggage including 25 lb. maximum in baggage wall hat shelf. Maximum 200 lb. combined. For additional loading instructions see weight and balance data."

(5) On flap control indicator: (S/N 675, 18260826 through 18264295)
"(a) 0° to 10° - (Blue color code and 160 m.p.h. callout; also, mechanical detent at 10°)
(b) 10° to 20°- Full (Indices at these positions with white color code and 110 m.p.h. callout; also, mechanical detent at 10° and 20°)"

On flap control indicator (S/N 18264296 through 18265175)
"(a) 0° to 10° - (Blue color code and 140 KTS callout; also, mechanical detent at 10°)
(b) 10° to 20°- Full (Indices at these positions with white color code and 95 KTS callout; also, mechanical detent at 10° and 20°)"

(6) Forward of the filler cap on the wing surface: (S/N 675, 18260826 through 18262250)

Standard range tanks: "Service this airplane with 80/87 minimum aviation grade gasoline. Capacity 32.5 gal."

Long range tanks: "Service this airplane with 80/87 minimum aviation grade gasoline. Capacity 42.0 gal."

Data Pertinent to Model Items I through XII, continued

E. **Applicable to Models 182P, continued:**

Forward of the filler cap on the wing surface: (S/N 18262251 through 18265175)

Standard range tanks: "Service this airplane with 80/87 minimum aviation grade gasoline. Capacity 30.5 gal."

Long range tanks: "Service this airplane with 80/87 minimum aviation grade gasoline. Capacity 40.0 gal."

(7) On aft panel of baggage compartment:
"Oxygen refill." (All models with oxygen)

(8) Adjacent to overvoltage light:
"High voltage."

(9) Above the left fuel gauge:
"Do not turn off alternator in flight except in emergency."
(Model 182P, S/N 18260826 through 18261425)

F. **Applicable to Models 182Q:**

(1) In full view of the pilot:

(a) S/N 18263479, 18265176 through 18266590

"This airplane must be operated as a normal category airplane in compliance with the operating limitations as stated in the form of placards, markings and manuals.

Maximums	
Maneuvering speed (IAS)	111 knots
Gross weight	2950 lb.
Flight load factor	Flaps up +3.8, -1.52
	Flaps down +2.0

No acrobatic maneuvers, including spins, approved. Altitude loss in a stall recovery 160 ft. Flight into known icing conditions prohibited. This airplane is certified for the following flight operations as of date of original airworthiness certificate: DAY-NIGHT-VFR-IFR." (as applicable)

S/N 18266591 through 18267715

"The markings and placards installed in this airplane contain operating limitations which must be complied with when operating this airplane in the Normal Category. Other operating limitations which must be complied with when operating this airplane in this category are contained in the Pilot's Operating Handbook and FAA Approved Airplane Flight Manual.

No acrobatic maneuvers, including spins, approved. Flight into known icing conditions prohibited. This airplane is certified for the following flight operations as of date of original airworthiness certificate: DAY-NIGHT-VFR-IFR." (as applicable)

(b) Near airspeed indicator:

S/N 18266591 through 18267715

"Maneuver Speed
111 KIAS"

Data Pertinent to Model Items I through XII, continued

F. **Applicable to Models 182Q, continued:**

(2) On the fuel selector valve plate:
S/N 18263479, 18265176 through 18266590

Standard range tanks: "Off.
Left - 29 gal. Level flight only.
Both - 56 gal. All flight attitudes.
Both on for takeoff and landing.
Right - 29 gal. Level flight only."

Long range tanks: "Off.
Left - 37 gal. Level flight only.
Both - 75 gal. All flight attitudes.
Both on for takeoff and landing.
Right - 37 gal. Level flight only."

S/N 18266591 through 18267715

"Take Off - Both - Landing,
All Flight - 88.0 Gal. - Attitudes
Left - 44.0 Gal. Level Flight Only
Right - 44.0 Gal. Level Flight Only
Off."

(3) On the control lock: "Control lock - remove before starting engine."

(4) On the baggage door: "Forward of baggage door latch, 120 pounds maximum baggage and/or auxiliary passenger. Aft of baggage door latch, 80 pounds maximum baggage including 25 pounds maximum in baggage wall hat shelf. Maximum 200 pounds combined. For additional loading instructions, see weight and balance data."

(5) On flap control indicator:
"0° to 10° - (Blue color code and 140 KTS callout; also, mechanical detent at 10°)"
"0° to 20° - Full (Indices at these positions with white color code and 95 KTS callout; also, mechanical detent at 10° and 20°)"

(6) Forward of the filler cap on the wing surface:
S/N 18265176 through 18265965

Standard range tanks: "Service this airplane with 100/130 minimum aviation grade gasoline. Capacity 30.5 gal."

Long range tanks: "Service this airplane with 100/130 minimum aviation grade gasoline. Capacity 40.0 gal."

S/N 18263479, 18265966 through 18266590

Standard range tanks: "Service this airplane with 100LL/100 aviation grade gasoline. Capacity 30.5 gal."

Long range tanks: "Service this airplane with 100LL/100 aviation grade gasoline. Capacity 40.0 gal."

S/N 18266591 through 18267715

"Fuel 100LL/100 minimum grade aviation gasoline.
Capacity 46 U.S. gal. Capacity 34.5 U.S. gal.
to bottom of filler collar."

Data Pertinent to Model Items I through XII, continued

F. Applicable to Models 182Q, continued:

(7) On aft panel of baggage compartment:
"Oxygen refill." (All models with oxygen)

(8) Adjacent to overvoltage light:
S/N 18263479, 18265176 through 18266590
"High Voltage"

S/N 18266591 through 18267715
"Low Voltage"

G. Applicable to Models R182 and TR182, S/N R18200001 through R18201928:

(1) In full view of the pilot:

(a) S/N R18200002 through R18200583
"This airplane must be operated as a normal category airplane in compliance with the operating limitations as stated in the form of placards, markings and manuals.

	Maximums
Gross weight	3100 lb.
Flight load factor	Flaps up +3.8, -1.52
	Flaps down +2.0

No acrobatic maneuvers, including spins, approved. Altitude loss in a stall recovery 240 ft. Flight into known icing conditions prohibited. This airplane is certified for the following flight operations as of date of original airworthiness certificate: DAY-NIGHT-VFR-IFR." (as applicable)

(b) S/N R18200001, R18200584 through R18202041
"The markings and placards installed in this airplane contain operating limitations which must be complied with when operating this airplane in the Normal Category. Other operating limitations which must be complied with when operating this airplane in this category are contained in the Pilot's Operating Handbook and FAA Approved Airplane Flight Manual.

No acrobatic maneuvers, including spins, approved. Flight into known icing conditions prohibited. This airplane is certified for the following flight operations as of date of original airworthiness certificate: DAY-NIGHT-VFR-IFR." (as applicable)

(c) Near Airspeed Indicator:
"MAX SPEED - KIAS

Maneuver	112
Gear Oper	140
Gear Down	140"

(2) On the fuel selector valve plate:

(a) S/N R18200002 through R18200583

Standard range tanks:	"Off Left - 29 gal. Level flight only. Both - 56 gal. All flight attitudes. Both on for takeoff and landing. Right - 29 gal. Level flight only."
Long range tanks:	"Off Left - 37 gal. Level flight only. Both - 75 gal. All flight attitudes. Both on for takeoff and landing. Right - 37 gal. Level flight only."

Data Pertinent to Model Items I through XII, continued

G. Applicable to Models R182 and TR182, S/N R18200001 through R18201928, continued:

(b) S/N R18200001, R18200584 through R18201798
"Take Off - Both - Landing,
All Flight - 88.0 Gal. - Attitudes
Left - 44.0 Gal. Level Flight Only
Right - 44.0 Gal. Level Flight Only
Off."

(c) S/N R18201799 through R18202041
"Both - 88.0 Gal. - Take Off - Landing - All Flight
Attitudes; Left - 44.0 Gal. - Level Flight Only
Right - 44.0 Gal. - Level Flight Only
Off - Off."

(3) On the control lock:
(a) S/N R18200001 through R18201798
"Control lock - Remove before starting engine."
(b) S/N R18201799 through R18202041
"Caution! Control Lock - Remove before starting engine."

(4) On the baggage door: "120 Pounds Maximum
Baggage And/Or Auxiliary Passenger
Forward of Baggage Door Latch And
80 Pounds Maximum
Baggage Aft of Baggage Door Latch
Maximum 200 Pounds Combined
For Additional Loading Instructions See Weight and Balance Data"

(5) On the flap control indicator:
"0° to 10° - (Blue color code and 140 KTS callout; also, mechanical detent at 10°)"

"0° to 20° - Full (Indices at these positions with white color code and 95 KTS callout; also, mechanical detent at 10° and 20°)"

(6) Forward of the filler cap on the wing surface:
(a) S/N R18200002 through R18200583
Standard range tanks: "Service this airplane with 100LL/100 aviation grade gasoline. Capacity 30.5 gal."

Long range tanks: "Service this airplane with 100LL/100 aviation grade gasoline. Capacity 40.0 gal."

(b) S/N R18200001, R18200584 through R18202041
Fuel 100LL/100 minimum grade aviation gasoline.
Capacity 46 U.S. gal. Capacity 34.5 U.S. gal. to
bottom of filler collar."

(7) Adjacent to overvoltage light:
(a) S/N R18200002 through R18200583
"High Voltage"

(b) S/N R18200001, R18200584 through R18202041
"Low Voltage"

Data Pertinent to Model Items I through XII, continued

G. **Applicable to Models R182 and TR182, S/N R18200001 through R18201928, continued:**

(8) Near gear hand pump:
"Manual Gear Extension
1. Select Gear Down
2. Pull Handle Fwd.
3. Pump Vertically
CAUTION
Do Not Pump With Gear
Up Selected"

(9) Forward of each fuel filler cap:
"Fuel Cap Forward - Arrow Alignment, Cap Must Not Rotate During Closing."

H. **Applicable to Models 182R and T182, S/N 18267302, 18267716 through 18268293:**

(1) In full view of the pilot:

(a) "The markings and placards installed in this airplane contain operating limitations which must be complied with when operating this airplane in the Normal Category. Other operating limitations which must be complied with when operating this airplane in this category are contained in the Pilot's Operating Handbook and FAA Approved Airplane Flight Manual.

No acrobatic maneuvers, including spins, approved. Flight into known icing conditions prohibited. This airplane is certified for the following flight operations as of date of original airworthiness certificate: DAY-NIGHT-VFR-IFR." (as applicable).

(b) Near airspeed indicator:
"Maneuver Speed
111 KIAS"

(2) On the fuel selector valve plate:

(a) S/N 18267716 through 18268055
"Take Off - Both - Landing, -
All Flight - 88.0 Gal. - Attitudes
Left - 44.0 Gal. Level Flight Only
Right - 44.0 Gal. Level Flight Only
Off."

(b) S/N 18268056 through 18268586
"Both - 88.0 Gal. - Takeoff - Landing - All Flight Attitudes
Left - 44.0 Gal. - Level Flight Only
Right - 44.0 Gal. - Level Flight Only
Off - Off."

(3) On the control lock:

(a) S/N 18267716 through 18268055
"Control Lock - Remove before starting engine."

(b) S/N 18268056 through 18268586
"Caution! Control Lock - Remove before starting engine."

Data Pertinent to Model Items I through XII, continued

H. Applicable to Models 182R and T182, S/N 18267302, 18267716 through 18268293, continued:

(4) On baggage door:
"120 Pounds Maximum
Baggage And/Or Auxiliary Passenger
Forward of Baggage Door Latch and
80 Pounds Maximum
Baggage Aft of Baggage Door Latch
Maximum 200 Pounds Combined

For Additional Loading Instructions
See Weight and Balance Data"

(5) On flap control indicator:
"0° to 10° - (Blue color code and 140 KTS callout; also, mechanical detent at 10°)"

"0° to 20° - Full (Indices at these positions with white color code and 95 KTS calout; also mechanical detent at 10° and 20°)"

(6) Forward of the filler cap on the wing surface:
"Fuel 100LL/100 minimum grade aviation gasoline. Capacity 46 U.S. gal. Capacity 34.5 U.S. gal. to bottom of filler collar."

(7) Forward of each fuel filler cap:
"Fuel cap fwd - arrow alignment, cap must not rotate during closing."

(8) Adjacent to overvoltage light:
"Low Voltage"

I. Applicable to Models R182 and TR182, S/N R18201929 through R18202041:
All placards required in the Pilot's Operating Handbook and FAA Approved Airplane Flight Manual must be installed in the appropriate locations.

J. Applicable to Models 182R and T182, S/N 18268294 through 18268586:
All placards required in the Pilot's Operating Handbook and FAA Approved Airplane Flight Manual must be installed in the appropriate locations.

NOTE 3. The cylinder head thermistors must be installed as follows:

Model	Engine and Cylinder Head Number				
	O-470-R	O-470-S	O-470-U	O-540-J	O-540-L
182N (1970 and 1971 Model)	3	3	N/A	N/A	N/A
182P (1972 and 1973 Model)	2	3	N/A	N/A	N/A
182P (1974 Model)	1	3	N/A	N/A	N/A
182P (1975 and 1976 Model)	N/A	3	N/A	N/A	N/A
182Q (1977 through 1980 Model)	N/A	N/A	3	N/A	N/A
182R (1981 Model through 18268160)	N/A	N/A	5	N/A	N/A
182R (18268161 through 18268586)	N/A	N/A	3	N/A	N/A
T182 (1981 Model through 1985 Model)	N/A	N/A	N/A	N/A	1
R182 (1978 and 1979 Model)	N/A	N/A	N/A	5	N/A
R182 (1980 Model through 1986 Model)	N/A	N/A	N/A	4	N/A
TR182 (1979 Model)	N/A	N/A	N/A	N/A	3
TR182 (1980 Model through 1986 Model)	N/A	N/A	N/A	N/A	5

NOTE 4. The installation of the 0-470-S engine in Model 182N and Model 182P (1970 through 1974) will require a change of the oil temperature gauge. Reference Cessna Service Letter SE75-2 for information and instructions for this change.

Data Pertinent to Model Items I through XII, continued

NOTE 5. The marking of the airspeed indicator with IAS provides an equivalent level of safety to CAR 3.757 when the approved airspeed calibration data presented in Section V of the Pilot's Operating Handbooks listed below is available to the pilot:

182P, Cessna P/N D1062-13	(S/N 18264296 through 18265175)
182Q, Cessna P/N D1087-13	(S/N 18265176 through 18265965)
182Q, Cessna P/N D1114-13	(S/N 18263479, 18265966 through 18266590)
182Q, Cessna P/N D1141-13PH	(S/N 18266591 through 18267300)
182Q, Cessna P/N D1176-13PH	(S/N 18267301 through 18267715)
182R, Cessna P/N D1196-13PH	(S/N 18267716 through 18268055)
182R, Cessna P/N D1215-13PH	(S/N 18268056 through 18268293)
182R, Cessna P/N D1233-13PH	(S/N 18268294 through 18268368)
182R, Cessna P/N D1254-13PH	(S/N 18268369 through 18268434)
T182, Cessna P/N D1197-13PH	(S/N 18267302, 18267716 through 18268055)
T182, Cessna P/N D1216-13PH	(S/N 18268056 through 18268293)
T182, Cessna P/N D1234-13PH	(S/N 18268294 through 18268368)
T182, Cessna P/N D1234R1-13PH	(Special) (S/N 18268365)
T182, Cessna P/N D1255-13PH	(S/N 18268369 through 18268434)
R182, Cessna P/N D1115-13	(S/N R18200002 through R18200583)
R182, Cessna P/N D1142-13PH	(S/N R18200584 through R18201313)
R182, Cessna P/N D1177-13PH	(S/N R18201314 through R18201628)
R182, Cessna P/N D1198-13PH	(S/N R18201629 through R18201798)
R182, Cessna P/N D1217-13PH	(S/N R18201799 through R18201928)
R182, Cessna P/N D1235-13PH	(S/N R18201929 through R18201973)
R182, Cessna P/N D1256-13PH	(S/N R18201974 through R18201999)
R182, Cessna P/N D1277-13PH	(S/N R18202000 through R18202031)
R182, Cessna P/N D1299-13PH	(S/N R18202032 through R18202041)
TR182, Cessna P/N D1143-13PH	(S/N R18200001, R18200584 through R18201313 except R18200975)
TR182, Cessna P/N D1143-2-13PH	(Special) (S/N R18200975)
TR182, Cessna P/N D1178-13PH	(S/N R18201314 through R18201628 except R18201315)
TR182, Cessna P/N D1199-13PH	(S/N R18201629 through R18201798)
TR182, Cessna P/N D1218-13PH	(S/N R18201799 through R18201928)
TR182, Cessna P/N D1236-13PH	(S/N R18201929 through R18201973)
TR182, Cessna P/N D1257-13PH	(S/N R18201974 through R18201999)
TR182, Cessna P/N D1278-13PH	(S/N R18201315, R18202000 through R18202031)
TR182, Cessna P/N D1300-13PH	(S/N R18202032 through R18202041)

NOTE 6. 14-volt electrical system
(182 series through S/N 18265965 except 18263479)
28-volt electrical system
(182 series S/N 18263479, 18265966 through 18268586)
(R182 and TR182 series S/N R18200001 through R18202041)

In addition to the above specified placards, the prescribed operating limitations indicated by an asterisk (*) under Sections I through XII must also be displayed by permanent markings.

XIII - Model 182S, Skylane, 4 PCLM (Normal Category), Approved October 3, 1996

Engine: Lycoming IO-540-AB1A5, rated 230 BHP

Fuel: 100/100LL minimum grade aviation gasoline

Engine Limits: For all operations, 2400 r.p.m.

Propeller and Propeller Limits:

(1) McCauley Constant Speed
- (a) Propeller: B2D34C235/90DKB-8 (2 blades)
  Diameter: not over 82 in., not under 80.5 in.
  Pitch settings at 30 in. sta.: Low 17.0°, High 31.8°
- (b) McCauley Spinner: D-7267-2
- (c) McCauley Governor: DC290D1/T8

(2) McCauley Constant Speed
- (a) Propeller: B3D36C431/80VSA-1 (3 blades)
  Diameter: not over 79 in., not under 77.5 in.
  Pitch settings at 30 in. sta.: Low 14.9°, High 31.7°
- (b) McCauley Spinner: D-7262-2
- (c) McCauley Governor: DC290D1/T8

Airspeed Limits:

Maneuvering:	110 Knots IAS	(108 Knots CAS)
Max. Structural Cruising:	140 Knots IAS	(138 Knots CAS)
Never Exceed:	175 Knots IAS	(170 Knots CAS)
Flaps Extended:	100 Knots IAS	(99 Knots CAS)

CG Range:

Normal Category:

(1) Aft Limits: 46.0 inches aft of datum at 3100 lbs. or less.

(2) Forward Limits: Linear variation from 40.9 inches aft of datum at 3100 pounds to 33.0 inches aft of datum at 2250 lbs.; 33.0 inches aft of datum at 2250 lbs. or less.

Empty Wt. C.G. Range: None

Reference Datum: Front Face of Firewall

MAC: 58.8 inches; Leading edge of MAC 25.98 inches aft of datum

Leveling Means: Left side of Tailcone at 139.65 inches and 171.65 inches aft of datum

Maximum Weights:

Maximum Ramp:	3110 lbs.
Maximum Takeoff:	3100 lbs.
Maximum Landing:	2950 lbs.

No. of Seats: 4 (2 at 32.0 to 50.0 inches aft of datum, 2 at 74.0 inches aft of datum)

Maximum Baggage: 200 lbs. (120 lbs. at 92.0 to 108.0 inches aft of datum)
(80 lbs. at 108.0 to 136.0 inches aft of datum)

Fuel Capacity (Gal.): 92 gal. total; 88 gal. usable
(Two 46 gal. integral tanks in wings at 46.5 inches aft of datum)
See NOTE 1 for data on usable fuel.

Oil Capacity (Qts.): 12.0 qts. at 14.8 inches forward of datum; 5 qts. usable

XIII - Model 182S, Skylane, continued

Control Surface Movements:

Wing Flaps:		Down 38° +0°, -1°
Elevator Tab:	Up 24° ± 2°	Down 15° ± 1°
Ailerons:	Up 20° ± 2°	Down 15° ± 2°
Elevator: (Relative to stabilizer)	Up 28° ± 1°	Down 21° ± 1°
Rudder:	Right: 24° +0°, -1° (Parallel to 0.00 W.L.)	Left: 24° +0°, -1°
	Right: 27°13' +0°, -1° (Perpendicular to hinge line)	Left: 27°13' +0°, -1°

Serial Nos. Eligible 18280001 and On

Data Pertinent to Model 182S:

Certification Basis (Model 182S)

Part 23 of the Federal Aviation Regulations effective February 1, 1965, as amended by 23-1 through 23-6, except as follows:

FAR 23.423; 23.611; 23.619; 23.623; 23.689; 23.775; 23.871; 23.1323; and 23.1563 as amended by Amendment 23-7. FAR 23.807 and 23.1524 as amended by Amendment 23-10. FAR 23.507; 23.771; 23.853(a),(b) and (c); and 23.1365 as amended by Amendment 23-14. FAR 23.951 as amended by Amendment 23-15. FAR 23.607; 23.675; 23.685; 23.733; 23.787; 23.1309 and 23.1322 as amended by Amendment 23-17. FAR 23.1301 as amended by Amendment 23-20. FAR 23.1353; and 23.1559 as amended by Amendment 23-21. FAR 23.603; 23.605; 23.613; 23.1329 and 23.1545 as amended by Amendment 23-23. FAR 23.441 and 23.1549 as amended by Amendment 23-28. FAR 23.779 and 23.781 as amended by Amendment 23-33. FAR 23.1; 23.51 and 23.561 as amended by Amendment 23-34. FAR 23.301; 23.331; 23.351; 23.427; 23.677; 23.701; 23.735; and 23.831 as amended by Amendment 23-42. FAR 23.961; 23.1093; 23.1143(g); 23.1147(b); 23.1303; 23.1357; 23.1361 and 23.1385 as amended by Amendment 23-43. FAR 23.562(a), 23.562(b)2, 23.562(c)1, 23.562(c)2, 23.562(c)3, and 23.562(c)4 as amended by Amendment 23-44. FAR 23.33; 23.53; 23.305; 23.321; 23.485; 23.621; 23.655 and 23.731 as amended by Amendment 23-45.

FAR 36 dated December 1, 1969, as amended by Amendments 36-1 through 36-21.

Equivalent Safety Items:

(1)	Induction System Icing Protection	FAR § 23.1093.
(2)	Throttle Control	FAR § 23.1143(g)
(3)	Mixture Control	FAR § 23.1147(b)

Date of Application for Amended Type Certificate was January 22, 1996.
Type Certificate No. 3A13 was amended October 3, 1996.

Production Basis (Model 182S)

None. Prior to original certification of each aircraft, an FAA representative must perform a detailed inspection for workmanship, materials, conformity with the approved data, and a check of the flight characteristics.

XIII - Model 182S, Skylane (Cont'd)

Equipment

The basic required equipment as prescribed in the applicable airworthiness regulations (see Certification Basis) must be installed in the airplane for certification.

NOTE 1: Weight and Balance:

Serial Nos. 18280001 and On: (Model 182S)

The certificated basic empty weight and corresponding center of gravity location must include unusable fuel of 24 lbs. at 48 inches aft of datum, and full oil of 16.2 lb. at 14.8 inches forward of datum.

NOTE 2: FAA Approved Airplane Flight Manual (AFM): Part number 182SPHUS00 or later FAA approved revisions are applicable to the Model 182S. The Airplane must be operated according to the appropriate AFM. Required placards are included in the AFM.

NOTE 3: The CHT probe must be installed on Head #1.

.....END....

A7SO
Revision 13
PIPER
PA-34-200
PA-34-200T
PA-34-220T

December 18, 1996

TYPE CERTIFICATE DATA SHEET NO. A7SO

This data sheet which is a part of type certificate No. A7SO, prescribes conditions and limitations under which the product for which the type certificate was issued meets the airworthiness requirements of the Federal Aviation Regulations.

Type Certificate Holder — The New Piper Aircraft, Inc.
2926 Piper Drive
Vero Beach, Florida 32960

I. - Model PA-34-200 (Seneca), 7 PCLM (Normal Category), Approved 7 May 1971.

Engines

S/N 34-E4, 34-7250001 through 34-7250214:
1 Lycoming LIO-360-C1E6 with fuel injector,
Lycoming P/N LW-10409 or LW-12586 (right side); and
1 Lycoming IO-360-C1E6 with fuel injector,
Lycoming P/N LW-10409 or LW 12586 (left side).

S/N 34-7250215 through 34-7450220:
1 Lycoming LIO-360-C1E6 with fuel injector,
Lycoming P/N LW-12586 (right side); and
1 Lycoming IO-360-C1E6 with fuel injector,
Lycoming P/N LW-12586 (left side).

Fuel — 100/130 minimum grade aviation gasoline

Engine Limits — For all operations, 2700 r.p.m. (200 hp)

Propeller and Propeller Limits

Left Engine
1 Hartzell, Hub Model HC-C2YK-2 ( ) E, Blade Model C7666A-0;
1 Hartzell, Hub Model HC-C2YK-2 ( ) EU, Blade Model C7666A-0;
1 Hartzell, Hub Model HC-C2YK-2 ( ) EF, Blade Model FC7666A-0;
1 Hartzell, Hub Model HC-C2YK-2 ( ) EFU, Blade Model FC7666A-0;
1 Hartzell, Hub Model HC-C2YK-2CG (F), Blade Model (F) C7666A
(This model includes the Hartzell damper); or
1 Hartzell, Hub Model HC-C2YK-2CGU (F), Blade Model (F) C7666A
(This model includes the Hartzell damper).

Page No.	1	2	3	4	5	6	7	8	9	10	11	12	13	14	15
Rev No.	13	12	12	12	12	12	13	11	13	13	13	13	13	12	13

Propeller and Propeller Limits (continued)	Right Engine 1 Hartzell, Hub Model HC-C2YK-2 ( ) LE, Blade Model JC7666A-0; 1 Hartzell, Hub Model HC-C2YK-2 ( ) LEU, Blade Model JC7666A-0; 1 Hartzell, Hub Model HC-C2YK-2 ( ) LEF, Blade Model FJC7666A-0; 1 Hartzell, Hub Model HC-C2YK-2 ( ) LEFU, Blade Model FJC7666A-0; 1 Hartzell, Hub Model HC-C2YK-2CLG (F), Blade Model (F) JC7666A (This model includes the Hartzell damper); or 1 Hartzell, Hub Model HC-C2YK-2CLGU (F), Blade Model (F) JC7666A (This model includes the Hartzell damper). Pitch setting: High 79° to 81°, Low 13.5° at 30" station. Diameter: Not over 76", not under 74". No further reduction permitted. Spinner: Piper P/N 96388 Spinner Assembly and P/N 96836 Cap Assembly, or P/N 78359-0 Spinner Assembly and P/N 96836-2 Cap Assembly (See NOTE 4) Governor Assembly: 1 Hartzell hydraulic governor, Model F-6-18AL (Right); 1 Hartzell hydraulic governor, Model F-6-18A (Left). Avoid continuous operation between 2200 and 2400 r.p.m. unless aircraft is equipped with Hartzell propellers which incorporates Hartzell damper on both left and right engine as noted above.

Airspeed Limits

V_{NE} (Never exceed)	217 m.p.h.	(188 knots)
V_{NO} (Maximum structural cruise)	190 m.p.h	(165 knots)
V_A (Maneuvering, 4200 lb.)	146 m.p.h.	(127 knots)
V_A (Maneuvering, 4000 lb.)	146 m.p.h.	(127 knots)
V_A (Maneuvering, 2743 lb.)	133 m.p.h	(115 knots)
V_{FE} (Flaps extended)	125 m.p.h	(109 knots)
V_{LO} (Landing gear operating)		
Extension	150 m.p.h.	(130 knots)
Retract	125 m.p.h.	(109 knots)
V_{LE} (Landing gear extended)	150 m.p.h	(130 knots)
V_{MC} (Minimum control speed)	80 m.p.h.	(69 knots)

C.G. Range (Gear Extended)	S/N 34-E4, 34-7250001 through 34-7250214 (See NOTE 3): (+86.4) to (+94.6) at 4000 lb. (+82.0) to (+94.6) at 3400 lb. (+80.7) to (+94.6) at 2780 lb. S/N 34-7250215 through 34-7450220: (+87.9) to (+94.6) at 4200 lb. (+82.0) to (+94.6) at 3400 lb. (+80.7) to (+94.6) at 2780 lb. Straight line variation between points given. Moment change due to gear retracting landing gear (-32 in.-lb.)
Empty Weight C.G. Range	None
Maximum Weight	S/N 34-E4, 34-7250001 through 34-7250214: 4000 lb.- Takeoff 4000 lb. - Landing See NOTE 3. S/N 34-7250215 through 34-7450220: 4200 lb. - Takeoff 4000 lb. - Landing

No. of Seats	7 (2 at +85.5, 3 at +118.1, 2 at +155.7)
Maximum Baggage	200 lb. (100 lb. at +22.5, 100 lb. at +178.7)
Fuel Capacity	98 gallons (2 wing tanks) at (+93.6) (93 gallons usable) See NOTE 1 for data on system fuel.
Oil Capacity	8 qts. per engine (6 qts. per engine usable) See NOTE 1 for data on system oil.

Control Surface Movements

Ailerons	(±2°)	Up	30°	Down	15°
Stabilator		Up	12.5° (+0, −1°)	Down	7.5° (±1°)
Rudder	(±1°)	Left	35°	Right	35°
Stabilator Trim Tab (Stabilator neutral)	(±1°)	Down	10.5°	Up	6.5°
Wing Flaps	(±2°)	Up	0°	Down	40°
Rudder Trim Tab (Rudder neutral)	(±1°)	Left	17°	Right	22°
Nose Wheel Travel	S/N 34-E4, 34-7250001 through 34-7350353:				
	(±1°)	Left	21°	Right	21°
Nose Wheel Travel	S/N 34-7450001 through 34-7450220:				
	(±1°)	Left	27°	Right	27°

Manufacturer's Serial Number 34-E4, 34-7250001 through 34-7450220 (See NOTE 7).

II. - Model PA-34-200T (Seneca II), 7 PCLM (Normal Category), Approved July 18, 1974.
Same as Model PA-34-200 series except engine installation, maximum gross weight, and other minor changes.

Engines	1 Teledyne Continental TSIO-360-E or TSIO-360-EB (left engine), 1 Teledyne Continental LTSIO-360-E or LTSIO-360-EB (right engine).
Fuel	100/130 minimum grade aviation gasoline
Engine Limits	For all operations, 2575 r.p.m. and 40" Hg. Manifold pressure, 200 hp @ S.L. and 215 hp @ 12,000 ft.

Propeller and Propeller Limits

Left engine
1 Hartzell, Hub Model BHC-C2YF-2 ( )F (See NOTE 10)
or BHC-C2YF-2 ( )UF; Blade Model FC8459-8R or FC8459B-8R.

Right engine
1 Hartzell, Hub Model BHC-C2YF-2 ( )L ( )F (See NOTE 10)
or BHC-C2YF-2 ( )L ( )UF; Blade Model FJC8459-8R or FJC8459B-8R.

Pitch setting at 30" station:
Hub Serial Numbers prior to AN3943:
High 79.3° ± 2.0°, Low 14.4° ± 0.2° or High 80.0° to 81.5°, Low 14.4° ± 0.2°.
Hub Serial Numbers AN3943 and subsequent:
High 80.0° to 81.5°, Low 14.4° ± 0.2°.

Diameter: Not over 76", not under 75".
No further reduction permitted.

Spinner: Piper P/N 37138-0 Spinner Assembly (left hand),
Piper P/N 37138-1 Spinner Assembly (right hand) (See NOTE 4).

Propeller and Propeller Limits (continued)	Governor Assembly: 1 Woodward hydraulic governor, Model C210659 (left), 1 Woodward hydraulic governor, Model 210658 (right); or 1 Hartzell hydraulic governor, Model E-3 (left) and 1 Hartzell hydraulic governor, Model E-3L (right); or 1 Hartzell hydraulic governor, Model E-8L (right) (E-8L Governor used with Synchrophaser). Avoid continuous operation between 2000 and 2200 r.p.m. with engine manifold pressure above 32" Hg. Avoid continuous ground operation in cross and tail winds over 10 knots between 1700 and 2100 r.p.m.. S/N 34-7970001 through 34-8170092: Left Engine 1 McCauley, Hub Model 3AF34C502, Blade Model 80 HA-4 Right Engine 1 McCauley, Hub Model 3AF34C503, Blade Model L80 HA-4 Pitch setting: High 81.0° to 83.5°, Low 12.0° ± .2° at 30" station. Diameter: Not over 76", not under 75". No further reduction permitted. Spinner: Piper P/N PS50077-49 Spinner Assembly See NOTE 4. Governor Assembly: 1 Woodward hydraulic governor, Model C210659 (left), 1 Woodward hydraulic governor, Model 210658 (right); 1 Hartzell hydraulic governor, Model E-3 (left), 1 Hartzell hydraulic governor, Model E-3L (right); or 1 Hartzell hydraulic governor, Model E-8L (right) (E-8L Governor used with Synchrophasers). Synchrophaser for S/N 34-7970001 through 34-8170092: Piper Drawing No. 36890 Synchrophaser Installation

Airspeed Limits

V_{NE} (Never exceed)	224 m.p.h.	(195 knots)
V_{NO} (Maximum structural cruise)	190 m.p.h.	(165 knots)
V_A (Maneuvering)	140 m.p.h.	(122 knots)
V_{FE} (Flaps extended)	125 m.p.h.	(109 knots)
V_{LO} (Landing gear operating)		
Extension	150 m.p.h.	(130 knots)
Retract	125 m.p.h.	(109 knots)
V_{LE} (Landing gear extended)	150 m.p.h.	(130 knots)
V_{MC} (Minimum control speed)	80 m.p.h.	(69 knots)

C.G. Range (Gear Extended)	(+90.6) to (+94.6) at 4570 lb. (+82.0) to (+94.6) at 3400 lb. Straight line variation between points given. Moment change due to retracting landing gear (-32 in.-lb.).
Empty Weight C.G. Range	None
Maximum Weight	4570 lb. - Takeoff 4342 lb. - Landing (All weight in excess of 4000 lb. must be fuel) Zero fuel weight may be increased up to a maximum of 4077.7 lb. when approved wing options are installed. See NOTE 11 for optional weights.

No. of Seats	7 (2 at +85.5, 3 at +118.1, 2 at +155.7) 7 (2 at +85.5, 3 at +118.1, 2 at +157.6) 6 (2 at +85.5, *2 at +119.1, 2 at +157.6) * - Optional Club Seats
Maximum Baggage	200 lb. (100 lb. at +22.5, 100 lb. at +178)
Fuel Capacity	98 gallons (2 wing tanks) at (+93.6) (93 gallons usable) * 128 gallons (2 wing tanks) at (+93.6) (123 gallons usable) * - Optional for S/N 34-7570001, 34-7670114 through 34-8170092. See NOTE 1 for data on system fuel.
Oil Capacity	8 qts. per engine (5 qts. per engine usable) See NOTE 1 for data on system oil.
Maximum Operating Altitude	25,000 feet

Control Surface Movements

Ailerons	(±2°)	Up	35°	Down	20°
Stabilator		Up	12.5° (+0°, −1°)	Down	7.5° (±1°)
Rudder	(±1°)	Left	35°	Right	35°
Stabilator Trim Tab (Stabilator neutral)	(±1°)	Down	10.5°	Up	6.5°
Wing Flaps	(±2°)	Up	0°	Down	40°
Rudder Trim Tab (Rudder neutral)	(±1°)	Left	25°	Right	25°
Nose Wheel Travel	(±1°)	Left	27°	Right	27°

Manufacturer's Serial Number — 34-7570001 through 34-8170092 (See NOTE 7).

IIIA. - Model PA-34-220T (Seneca III), 7 PCLM (Normal Category), Approved December 17, 1980.

Same as model PA-34-200T series except engines, windshield, instrument panel, landing gear, maximum gross weight and other minor changes.

Engines	1 Teledyne Continental TSIO-360-KB (left engine), 1 Teledyne Continental LTSIO-360-KB (right engine).
Fuel	100/100LL minimum grade aviation gasoline
Engine Limits	Takeoff, 5 minutes, 2800 r.p.m. and 40" Hg. manifold pressure (220 hp) Max. Continuous, 2600 r.p.m. and 40" Hg. manifold pressure (200 hp)
Propeller and Propeller Limits	Left Engine 1 Hartzell, Hub Model BHC-C2YF-2 ( ) UF, Blade Model FC8459-8R. Right Engine 1 Hartzell, Hub Model BHC-C2YF-2 ( )L ( )UF, Blade Model FJC8459-8R. Pitch setting: High 80.0° to 81.5°, Low 12.6° ± 0.2° at 30" station. Diameter: Not over 76", not under 75". No further reduction permitted. Spinner: Piper P/N 37138-0 assembly (left hand), Piper P/N 37138-1 assembly (right hand). See NOTE 4.

Propeller and Propellr Limits (cont'd)

Governor Assembly:
1 Hartzell hydraulic governor; Model E-3-7 (left),
1 Hartzell hydraulic governor; Model E-3-7L (right); or
1 Hartzell hydraulic governor; Model E-8-7L (14V) or E-8-8L (28V) (right) with Synchrophaser Installation, Piper Drawing 36890 or 87719.

Avoid continuous ground operation in cross and tail winds of over 10 knots between 1700 and 2100 r.p.m.

Avoid continuous operation between 2000 and 2200 r.p.m. with manifold pressure above 32" Hg.

Left Engine
1 McCauley, Hub Model 3AF32C508, Blade Model 82NFA-6,
Right Engine
1 McCauley, Hub Model 3AF32C509, Blade Model L82NFA-6.

Pitch setting: High 81.0° to 83.5°, Low 11.0° ± 0.2° at 30" station.
Diameter: Not over 76", not under 75".
No further reduction permitted.

Spinner: Piper P/N PS50077-49 or P/N PS50077-78 Assembly
See NOTE 4.

Governor Assembly:
1 Hartzell hydraulic governor; Model E-3-7 (left),
1 Hartzell hydraulic governor; Model E-3-7L (right); or
1 Hartzell hydraulic governor; Model E-8-7L (14V) or E-8-8L (28V) (right) with Synchrophaser Installation, Piper Drawing No. 36890 or 87719.

Airspeed Limits (IAS)

V_{NE} (Never exceed)	205 knots
V_{NO} (Maximum structural cruise)	166 knots
V_A (Maneuvering) at 4750 lb.	140 knots
V_{FE} (Flaps extended)	115 knots
V_{LO} (Landing gear retracting)	108 knots
V_{LO} (Landing gear extending)	130 knots
V_{LE} (Landing gear extended)	130 knots
V_{MC} (Minimum control speed)	66 knots

C.G. Range (Gear Extended)

(+90.6) to (+94.6) at 4750 lb.
(+86.7) to (+94.6) at 4250 lb.
(+82.0) to (+94.6) at 3400 lb.
Straight line variation between points given.
Moment change due to retracting landing gear (-32 in.-lb.)

Empty Weight C.G. Range

None

Maximum Weight

4773 lb. - Ramp
4750 lb. - Takeoff
4513 lb. - Landing
4470 lb. - Zero Fuel
See NOTE 12 for optional weights.

No. of Seats

7 (2 at +85.5, 3 at +118.1, 2 at +157.6)
6 (2 at +85.5, *2 at +119.1, 2 at +157.6)
* - Optional Club Seats

Maximum Baggage

200 lb. (100 lb. at +22.5, 100 lb. at +178.7)

Fuel Capacity	98 gallons (2 wing tanks) at (+93.6) (93 gallons usable) * 128 gallons (2 wing tanks) at (+93.6) (123 gallons usable) * - Optional installation See NOTE 1 for data on system fuel.
Oil Capacity	8 qts. per engine (5 qts. per engine usable) See NOTE 1 for data on system oil.
Maximum Operating Altitude	25,000 feet

Control Surface Movements

Ailerons	(±2°)	Up	35°	Down	20°
Stabilator		Up	12.5° (+0°, −1°)	Down	7.5° (±1°)
Rudder	(±1°)	Left	35°	Right	35°
Stabilator Trim Tab (Stabilator neutral)	(±1°)	Down	10.5°	Up	6.5°
Wing Flaps	(±2°)	Up	0°	Down	40°
Rudder Trim Tab (Rudder neutral)	(±1°)	Left	25°	Right	25°
Nose Wheel Travel	(±1°)	Left	27°	Right	27°

Manufacturer's Serial Number	34-8133001 through 34-8633031 (14V); 3433001 through 3433064; and 3448001 through 3448037 (28V) (See NOTE 7).

IIIB. - Model PA-34-220T (Seneca IV), 6 PCLM (Normal Category), Approved November 17, 1993.
Same as Model PA-34-220T (Seneca III) except nose bowl assembly, instrument panel, interior and other minor changes.

Engines	1 Teledyne Continental TSIO-360-KB (left engine), 1 Teledyne Continental LTSIO-360-KB (right engine).
Fuel	100/100LL minimum grade aviation gasoline
Engine Limits	Takeoff, 5 minutes, 2800 r.p.m. and 40" Hg. manifold pressure (220 hp) Max. Continuous, 2600 r.p.m. and 40" Hg. manifold pressure (200 hp)

Propeller and Propeller Limits

Left Engine
1 Hartzell, Hub Model BHC-C2YF-2 ( ) UF, Blade Model FC8459-8R.
Right Engine
1 Hartzell, Hub Model BHC-C2YF-2 ( )L ( )UF, Blade Model FJC8459-8R.

Pitch setting: High 80.0° to 81.5°, Low 12.6° ± 0.2° at 30 " station.
Diameter: Not over 76", not under 75".
No further reduction permitted.
Spinner: Piper P/N 37138-0 Assembly (left hand),
Piper P/N 37138-1 Assembly (right hand).
Governor Assembly:
1 Hartzell hydraulic governor; Model E-3-7 (left),
1 Hartzell hydraulic governor; Model E-3-7L (right); or
1 Hartzell hydraulic governor; Model E-8-8L (right) with Synchrophaser Installation, Piper Drawing No. 87719.

Avoid continuous ground operation in cross and tail winds between 1700 and 2100 r.p.m..
Avoid continuous operation between 2000 and 2200 r.p.m. with manifold pressure above 32" Hg.

Propeller and Propeller Limits (cont'd)	Left Engine 1 McCauley, Hub Model 3AF32C508, Blade Model 82NFA-6. Right Engine 1 McCauley, Hub Model 3AF32C509, Blade Model L82NFA-6. Pitch setting: High 81.0° to 83.5°, Low 11.0° ± 0.2° at 30" station. Diameter: Not over 76", not under 75". No further reduction permitted. Spinner: Piper P/N PS50077-78 Assembly Governor Assembly: 1 Hartzell hydraulic governor; Model E-3-7 (left), 1 Hartzell hydraulic governor; Model E-3-7L (right); or 1 Hartzell hydraulic governor; Model E-8-8L (right) with Synchrophaser Installation, Piper Drawing No. 87719.
Airspeed Limits (IAS)	V_{NE} (Never exceed) 205 knots V_{NO} (Maximum structural cruise) 166 knots V_A (Maneuvering) at 4750 lb. 140 knots V_{FE} (Flaps extended) 115 knots V_{LO} (Landing gear retracting) 108 knots V_{LO} (Landing gear extending) 130 knots V_{LE} (Landing gear extended) 130 knots V_{MC} (Minimum control speed) 66 knots
C.G. Range (Gear Extended)	(+90.6) to (+94.6) at 4750 lb. (+86.7) to (+94.6) at 4250 lb. (+82.0) to (+94.6) at 3400 lb. Straight line variation between points given. Moment change due to retracting landing gear (-32 in.-lb.)
Empty Weight C.G. Range	None
Maximum Weight	4773 lb. - Ramp 4750 lb. - Takeoff 4513 lb. - Landing 4470 lb. - Zero Fuel See NOTE 12 for optional weights.
No. of Seats	6 (2 at +85.5, 2 at +119.1, 2 at +157.6)
Maximum Baggage	200 lb. (100 lb. at +22.5, 100 lb. at +178.7)
Fuel Capacity	128 gallons (2 wing tanks) at (+93.6) (123 gallons usable) See NOTE 1 for data on system fuel.
Oil Capacity	8 qts. per engine (5 qts. per engine usable) See NOTE 1 for data on system oil.
Maximum Operating Altitude	25,000 feet

Control Surface Movements

Ailerons	(±2°)	Up	35°	Down	20°
Stabilator		Up	12.5° (+0°, −1°)	Down	7.5° (±1°)
Rudder	(±1°)	Left	35°	Right	35°
Stabilator Trim Tab (Stabilator neutral)	(±1°)	Down	10.5°	Up	6.5°
Wing Flaps	(±2°)	Up	0°	Down	40°
Rudder Trim Tab (Rudder neutral)	(±1°)	Left	25°	Right	25°
Nose Wheel Travel	(±1°)	Left	27°	Right	27°

Manufacturer's Serial Number: 3448038 through 3448079, and 3447001 through 3447029.

IIIC. - Model PA-34-220T (Seneca V), 6 PCLM (Normal Category), Approved December 11, 1996.
Same as Model PA-34-220T (Seneca IV) except engine installation, instrument panel, interior and other minor changes.

Engines: 1 Teledyne Continental TSIO-360-RB (left engine),
1 Teledyne Continental LTSIO-360-RB (right engine).

Fuel: 100/100LL minimum grade aviation gasoline

Engine Limits: Takeoff and Maximum Continuous Operation, 2600 r.p.m. and 38" Hg. manifold pressure (220 hp)

Propeller and Propeller Limits:

Left Engine
1 Hartzell, Hub Model BHC-J2YF-2CUF, Blade Model FC8459(B)-8R.
Right Engine
1 Hartzell, Hub Model BHC-J2YF-2CLUF, Blade Model FJC8459(B)-8R.

Pitch setting: High 80.0° to 81.5°, Low 14.6° ± 0.2° at 30" station.
Diameter: Not over 76", not under 75".
No further reduction permitted.

Spinner: Piper P/N 37138-6 Assembly (left hand),
Piper P/N 37138-7 Assembly (right hand).

Governor Assembly:
1 Hartzell hydraulic governor; Model E-3-9 (left),
1 Hartzell hydraulic governor; Model E-3-9L (right); or
1 Hartzell hydraulic governor; Model E-8-9L (right) with Synchrophaser Installation.

Avoid continuous ground operation in cross and tail winds between 1600 and 2100 r.p.m..

Avoid continuous operation between 1900 and 2100 r.p.m. with manifold pressure above 32" Hg.

Propeller and Propeller Limits (continued)	Left Engine 1 McCauley, Hub Model 3AF32C522, Blade Model 82NJA-6. Right Engine 1 McCauley, Hub Model 3AF32C523, Blade Model L82NJA-6. Pitch setting: Feather 82.1° ± 0.5°, Low 12.6° ± 0.2° at 30" station. Diameter: Not over 76", not under 75". No further reduction permitted. Spinner: Piper P/N 100738-2 Assembly Governor Assembly: 1 Hartzell hydraulic governor; Model E-3-9 (left), 1 Hartzell hydraulic governor; Model E-3-9L (right); or 1 Hartzell hydraulic governor; Model E-8-9L (right) with Synchrophaser Installation.
Airspeed Limits (IAS)	V_{NE} (Never exceed) 204 knots V_{NO} (Maximum structural cruise) 164 knots V_A (Maneuvering) at 4750 lb. 139 knots V_{FE} (Flaps extended) 113 knots V_{LO} (Landing gear retracting) 107 knots V_{LO} (Landing gear extending) 128 knots V_{LE} (Landing gear extended) 128 knots V_{MC} (Minimum control speed) 66 knots
C.G. Range (Gear Extended)	(+90.6) to (+94.6) at 4750 lb. (+86.7) to (+94.6) at 4250 lb. (+82.0) to (+94.6) at 3400 lb. Straight line variation between points given. Moment change due to retracting landing gear (-32 in.-lb.)
Empty Weight C.G. Range	None
Maximum Weight	4773 lb. - Ramp 4750 lb. - Takeoff 4513 lb. - Landing 4479 lb. - Zero Fuel See NOTE 12 for optional weights.
No. of Seats	6 (2 at +85.5, 2 at +119.1, 2 at +157.6)
Maximum Baggage	185 lb. (100 lb. at +22.5, 85 lb. at +178.7)
Fuel Capacity	128 gallons (2 wing tanks) at (+93.6) (122 gallons usable) See NOTE 1 for data on system fuel.
Oil Capacity	8 qts. per engine (5 qts. per engine usable) See NOTE 1 for data on system oil.
Maximum Operating Altitude	25,000 feet

Control Surface Movements						
	Ailerons	(±2°)	Up	35°	Down	20°
	Stabilator		Up	12.5° (+0°, −1°)	Down	7.5° (±1°)
	Rudder	(±1°)	Left	35°	Right	35°
	Stabilator Trim Tab (Stabilator neutral)	(±1°)	Down	10.5°	Up	6.5°
	Wing Flaps		Up	0° (±1°)	Down	40° (±2°)
	Rudder Trim Tab (Rudder neutral)	(±1°)	Left	26°	Right	26°
	Nose Wheel Travel	(±1°)	Left	27°	Right	27°

Manufacturer's Serial Number — 3449001 and up.

DATA PERTINENT TO ALL MODELS

Datum — 78.4" forward of wing leading edge from the inboard edge of the inboard fuel tank.

Leveling Means — Two screws left side fuselage below window.

Certification Basis — Type Certificate No. A7SO issued May 7, 1971, obtained by the manufacturer under the delegation option authorization.
Date of Type Certificate application July 23, 1968.

Model PA-34-200:
FAR 23 as amended by Amendment 23-6 effective August 1, 1967; FAR 23.959 as amended by Amendment 23-7 effective September 14, 1969; and FAR 23.1557(c)(1) as amended by Amendment 23-18 effective May 2, 1977. Compliance with FAR 23.1419 as amended by Amendment 23-14 effective December 20, 1973, has been established with optional ice protection provisions.

Model PA-34-200T:
FAR 23 as amended by Amendment 23-6 effective August 1, 1967; FAR 23.901, 23.909, 23.959, 23.1041, 23.1043, 23.1047, 23.1143, 23.1305(b)(c)(h)(p) and 23.1527(b) as amended by Amendment 23-7 effective September 14, 1969; and FAR 23.1557(c)(1) as amended by Amendment 23-18 effective May 2, 1977.

Model PA-34-220T (Seneca III and IV):
FAR 23 as amended by Amendment 23-6 effective August 1, 1967; FAR 23.207, 23.901, 23.909, 23.959, 23.1041, 23.1043, 23.1047, 23.1143, 23.1305(b)(c)(h)(p) and 23.1527 as amended by Amendment 23-7 effective September 14, 1969; FAR 23.201 and 23.203 as amended by Amendment 23-14 effective December 20, 1973; FAR 23.1557(c)(1) as amended by Amendment 23-18 effective May 2, 1977; FAR 23.175(a) and 23.1581(b)(2) as amended by Amendment 23-21 effective March 1, 1978; FAR 23.1545(a) as amended by Amendment 23-23 effective December 1, 1978; and FAR 36 through Amendment 36-9 effective January 15, 1979.

Certification Basis (continued)

Model PA-34-220T (Seneca V):
FAR 23 as amended by Amendment 23-6 effective August 1, 1967; FAR 23.901, 23.909, 23.1041, 23.1043, 23.1047, 23.1143, 23.1305(b)(c)(h)(p) and 23.1527 as amended by Amendment 23-7 effective September 14, 1969; FAR 23.959 as amended by Amendment 23-18 effective May 2, 1977; FAR 23.175(a), 23.201, 23.203, 23.1557(c)(1) and 23.1581 as amended by Amendment 23-21 effective March 1, 1978; FAR 23.1545(a) as amended by Amendment 23-23 effective December 1, 1978; FAR 23.1529 as amended by Amendment 23-26 effective October 14, 1980: FAR 23.1322 as amended by Amendment 23-43 effective May 10, 1993; FAR 23.207 as amended by Amendment 23-45 effective September 7, 1993; Removal of FAR 23.205 per Amendment 23-50 effective March 11, 1996; FAR 23.1305(b)(4)(ii) as amended by Amendment 23-52 effective July 25, 1996; and FAR 36, Appendix G through Amendment 36-16 effective December 18, 1988.

Compliance with the requirements of FAR 23.1419 as amended by Amendment 23-14 effective December 20, 1973, and FAR 23.1441 as amended by Amendment 23-9 effective June 17, 1970, has been established with optional ice protection provisions and optional supplemental oxygen equipment, respectively.

Production Basis

Production Certificate No. 206.
Production Limitation Record issued and the manufacturer is authorized to issue an airworthiness certificate under the delegation option provisions of FAR 21.

Equipment

The basic required equipment as prescribed in the applicable airworthiness regulations (see Certification Basis) must be installed in the aircraft for certification. In addition, the following items of equipment are required:

MODEL	*AFM/POH*	*REPORT NO.*	*APPROVED*	*SERIAL EFFECTIVITY*
PA-34-200 (Seneca)	AFM	VB-353	7/2/71	34-E4, 34-7250001 through 34-7250214
	AFM	VB-423	5/20/72	34-7250001 through 34-7250189 when Piper Kit 760-607 is installed; 34-7250190 through 34-7250214 when Piper Kit 760-611 is installed; and 34-7250215 through 34-7350353
	AFM	VB-563	5/14/73	34-7450001 through 34-7450220
	AFM Supp.	VB-588	7/20/73	34-7250001 through 34-7450039 when propeller with dampers are installed
	AFM Supp.	VB-601	11/9/73	34-7250001 through 34-745017 when ice protection system is installed
PA-34-200T (Seneca II)	AFM	VB-628	7/18/74	34-7570001 through 34-7670371
	POH	VB-850	8/23/76	34-7770001 through 34-8170092
	POH	VB-1140	6/30/80	34-7770001 through 34-8170092 when Piper Kit 764-048V is installed
	AFM	VB-1245	3/9/84	34-7570001 through 34-7670371 when Piper Kit 765-110 is installed

MODEL	AFM/POH	REPORT NO.	APPROVED	SERIAL EFFECTIVITY
PA-34-220T (Seneca III)	POH	VB-1110	1/8/81	34-8133001 through 34-8633031, and 3433001 through 3433064
	POH	VB-1150	2/20/81	34-8133001 through 34-8633031, and 3433001 through 3433064 when Piper Kit 764-099V is installed
	POH	VB-1257	10/20/89	3448001 through 3448037
	POH	VB-1259	11/20/89	3448001 through 3448037 when Piper Kit 766-203 is installed
PA-34-220T	POH	VB-1556	11/5/93	3448038 through 3448079
(Seneca IV)	POH	VB-1558	12/6/93	3448038 through 3448079 when Piper Kit 766-283 is installed
	POH	VB-1615	7/12/95	3447001 through 3447029
	POH	VB-1620	7/12/95	3447001 through 3447029 when Piper Kit 766-608 is installed
PA-34-220T	POH	VB-1638	12/6/96	3449001 and up
(Seneca V)	POH	VB-1649	1/23/97	3449001 and up when Piper Kit 766-632 is installed

NOTE 1 Current Weight and Balance Report, including list of equipment included in certificated empty weight, and loading instructions when necessary, must be provided for each aircraft at the time of original certification.

The certificated empty weight and corresponding center of gravity locations must include undrainable system oil (not included in oil capacity) and unusable fuel as noted below:

Fuel: 30.0 lb. at (+103.0) for PA-34 series, except Model PA-34-220T (Seneca V), S/N 3449001 and up
Fuel: 36.0 lb. at (+103.0) for Model PA-34-220T (Seneca V), S/N 3449001 and up
Oil: 6.2 lb. at (+ 39.6) for Model PA-34-200
Oil: 12.0 lb. at (+ 43.7) for Models PA-34-200T and PA-34-220T

NOTE 2 All placards required in the approved Airplane Flight Manual or Pilot's Operating Handbook and approved Airplane Flight Manual of Pilot's Operating Handbook supplements must be installed in the appropriate location.

NOTE 3 The Model PA-34-200; S/N 34-E4, 34-7250001 through 34-7250189, may be operated at a maximum takeoff weight of 4200 lb. when Piper Kit 760-607 is installed. S/N 34-7250190 through 34-7250214 may be operated at a maximum takeoff weight of 4200 lb. when Piper Kit 760-611 is installed.

NOTE 4 The Model PA-34-200; S/N 34-E4, 34-7250001 through 34-7250189, may be operated without spinner domes or without spinner domes and rear bulkheads when Piper Kit 760-607 has been installed. S/N 34-7250190 through 34-7250214 may be operated without spinner domes or without spinner domes and rear bulkhead when Piper Kit 760-611 has been installed. The Model PA-34-200; S/N 34-7250215 through 34-7450220, and the Model PA-34-200T; S/N 34-7570001 through 34-8170092, may be operated without spinner domes or without spinner domes and rear bulkheads.

The Model PA-34-200T; S/N 34-7970001 through 34-8170092, equipped with McCauley three-bladed propellers, may be operated with spinner dome and rear bulkhead removed.

The Model PA-34-220T; S/N 34-8133001 through 34-8633031, 3433001 through 3433064, and 3448001 through 3448037, with two-bladed Hartzell propellers may be operated without spinner domes or without spinner domes and rear bulkheads. With three-bladed McCauley propellers, this model may be operated without spinner dome and rear bulkhead.

NOTE 5 The Model PA-34-200 may be operated in known icing conditions when equipped with spinner assembly and the following kits:
(a) S/N 34-E4, 34-7250001 through 34-7250189: Piper Kit 760-781V and Piper Kit 760-607
(See NOTE 3).
(b) S/N 34-7250190 through 34-7250214: Piper Kit 760-781V and Piper Kit 760-611
(See NOTE 3).
(c) S/N 34-7250215 through 34-7450220: Piper Kit 760-781V.

NOTE 6 Model PA-34-200T; S/N 34-7570001 through 34-8170092, may be operated in known icing conditions when equipped with deicing equipment installed per Piper Drawing No. 37700 and spinner assembly.

NOTE 7 The following serial numbers are not eligible for import certification to the U.S.:
PA-34-200:
34-7350283, 34-7350299, 34-7350300, and 34-7450187.

PA-34-200T:
34-7570074, 34-7570136, 34-7570193, 34-7570292, 34-7670045, 34-7670071, 34-7670072, 34-7670168, 34-7670261, 34-7670312, 34-7770037, 34-7770137, 34-7770206, 34-7770288, 34-7770316, 34-7770357, 34-7770367, 34-7770368, 34-7770406, 34-7870069, 34-7870098, 34-7870133, 34-7870157, 34-7870171, 34-7870172, 34-7870173, 34-7870174, 34-7870212, 34-7870213, 34-7870214, 34-7870215, 34-7870216, 34-7870217, 34-7870252, 34-7870257, 34-7870258, 34-7870313, 34-7870314, 34-7870367, 34-7870368, 34-7870369, 34-7870410, 34-7870411, 34-7870443, 34-7870444, 34-7870445, 34-7870446, 34-7870473, 34-7870474, 34-7970021, 34-7970051, 34-7970052, 34-7970087, 34-7970088, 34-7970131, 34-7970132, 34-7970133, 34-7970205, 34-7970206, 34-7970207, 34-7970374, 34-7970375, 34-7970376, 34-7970472, 34-7970473, 34-7970474, 34-7970475, 34-7970512, 34-7970513, 34-7970514, 34-8070045, 34-8070096, 34-8070097, 34-8070098, 34-8070099, 34-8070132, 34-8070202, 34-8070203, 34-8070204, 34-8070205, 34-8070276, 34-8070277, 34-8070278, 34-8070279, 34-8070280, 34-8070298, 34-8070299, 34-8070300, 34-8070301, 34-8170012, 34-8170013, 34-8170014, and 34-8170015.

PA-34-220T:
34-8133039, 34-8133083, 34-8133125, 34-8133126, 34-8133127, 34-8133128, 34-8133129, 34-8133169, 34-8133208, 34-8133209, 34-8133210, 34-8133211, 34-8133212, 34-8133240, 34-8133241, 34-8133242, 34-8133243, 34-8133244, 34-8133261, 34-8133262, 34-8133263, 34-8133264, 34-8233129, 34-8233130, 34-8233131, 34-8233132, 34-8233158, 34-8233159, 34-8233160, 34-8233161, 34-8233196, 34-8233197, 34-8233198, 34-8233199, 34-8333014, 34-8333015, 34-8333016, 34-8333017, 34-8333034, 34-8333035, 34-8333036, 34-8333037, 34-8333081, 34-8333082, 34-8333083, 34-8333084, 34-8333121, 34-8333122, 34-8333123, 34-8333124, 34-8433010, 34-8433011, 34-8433012, 34-8433013, 34-8433042, 34-8433043, 34-8433044, 34-8433045, 34-8433084, 34-8433088, 34-8533014, 34-8533015, 34-8533016, 34-8533017, 34-8633018, 3433013, 3433014, 3433015, 3433026, 3433027, 3433028, 3433039, 3433040, 3433053, 3433054, 3433055, and 3433056.

NOTE 8 Model PA-34-200; S/N 34-E4, S/N 34-7250001 through 34-7450220, and Model PA-34-200T; S/N 34-7570001 through 34-8170092, and Model PA-34-220T may be operated subject to the limitations listed in the Airplane Flight Manual or Pilot's Operating Handbook with rear cabin and cargo door removed.

NOTE 9 In the following serial numbered aircraft, rear seat location is farther aft as shown and the center seats may be removed and replaced by CLUB SEAT INSTALLATION, which has a more aft C.G. location as shown in "No. of Seats," above:

PA-34-200T: S/N 34-7770001 through 34-8170092.

NOTE 10 These propellers are eligible on Teledyne Continental L/TSIO-360-E only.

NOTE 11 With Piper Kit 764-048V installed weights are as follows:
4407 lb. - Takeoff
4342 lb. - Landing (All weight in excess of 4000 lb. must be fuel)
Zero fuel weight may be increased to a maximum of 4077.7 lb. when approved wing options are installed (See POH VB-1140).

NOTE 12 With Piper Kit 764-099V installed, weights are as follows:
4430 lb. - Ramp
4407 lb. - Takeoff, Landing, and Zero Fuel (See POH VB-1150).

NOTE 13 With Piper Kit 766-203 installed, weights are as follows:
4430 lb. - Ramp
4407 lb. - Takeoff, Landing and Zero Fuel (See POH VB-1259).

NOTE 14 With Piper Kit 766-283 installed, weights are as follows:
4430 lb. - Ramp
4407 lb. - Takeoff, Landing and Zero Fuel (See POH VB-1558).

NOTE 15 With Piper Kit 766-608 installed, weights are as follows:
4430 lb. - Ramp
4407 lb. - Takeoff, Landing and Zero Fuel (See POH VB-1620).

NOTE 16 With Piper Kit 766-632 installed, weights are as follows:
4430 lb. - Ramp
4407 lb. - Takeoff, Landing and Zero Fuel (See POH VB-1649).

NOTE 17 The bolt and stack-up that connect the upper drag link to the nose gear trunnion are required to be replaced every 500 hours time-in-service. The part numbers are as follows:

1. Piper P/N 400 274 (AN7-35) bolt;
2. Piper P/N 407 591 (AN960-716L) washer, as applicable;
3. Piper P/N 407 568 (AN 960-716) washer, as applicable;
4. Piper P/N 404 396 (AN 320-7) nut; and
5. Piper P/N 424 085 cotter pin.

---END---

A11EA
Revision 9
American General
Aircraft Holding Co.
AA-1
AA-1A
AA-1B
AA-1C

June 7, 1995

TYPE CERTIFICATE DATA SHEET NO. A11EA

This data sheet, which is a part of Type Certificate No. A11EA, prescribes conditions and limitations under which the product for which the Type Certificate was issued meets the airworthiness requirements of the Federal Aviation Regulations.

Type Certificate Holder — American General Aircraft Holding Co., Inc.
2900 One Liberty Place
1650 Market. St.
Philadelphia, PA 19103

I. - Model AA-1, Yankee, 2 PCLM, Utility Category, Approved August 29, 1967, Normal Category Approved July 16, 1968.

Engine — Lycoming 0-235-C2C (Carburetor Setting 10-4953 or 10-3103-1)

Fuel — 80/87 minimum grade aviation gasoline

Engine limits — For all operations 2600 r.p.m. (108 h.p.)

Propeller and propeller limits

1. McCauley Model 1A105/SCM-7157 fixed pitch propeller. Static r.p.m. at maximum permissible throttle setting; not over 2300; not under 2150. Diameters: not over 71 inches, not under 69.5 inches.

2. McCauley Model 1A105/SCM-7153 and 1A105/SCM-7154 fixed pitch propellers. Static r.p.m. at maximum permissible throttle setting; not over 2400; not under 2250. Diameter: not over 71 inches, not under 69.5 inches.

3. McCauley Model 1A106/NCM-7157 fixed pitch propellers. Static r.p.m. at maximum permissible throttle setting; not over 2400; not under 2300. Diameter: not over 71 inches, not under 69.5 inches.

4. McCauley Model 1A106/NCM-7153 hub and fixed pitch propellers. Static r.p.m. at maximum permissible throttle setting; not over 2475; not under 2375. Diameter: not over 71 inches, not under 69.5 inches.

Airspeed limits (CAS)

V_{ne}	Never exceed	195 m.p.h. (169 knots)
V_{no}	Maximum structural cruising	144 m.p.h. (125 knots)
V_a	Maneuvering (Utility Category)	132 m.p.h. (115 knots)
V_a	Maneuvering (Normal Category)	125 m.p.h. (109 knots)
V_{fe}	Flaps extended	100 m.p.h. (87 knots)
	Canopy half open	130 m.p.h. (113 knots)

Page No.	1	2	3	4	5	6	7	8
Rev. No.	9	5	6	5	6	6	8	9

I. Model AA-1 (cont'd)

Center of gravity (C.G) range	(+78.5) to (+81.0) at 1500 lb. (+77.5) to (+81.0) at 1430 lb. (+75.0) to (+81.0) at 1245 lb. Straight line variation between points given.

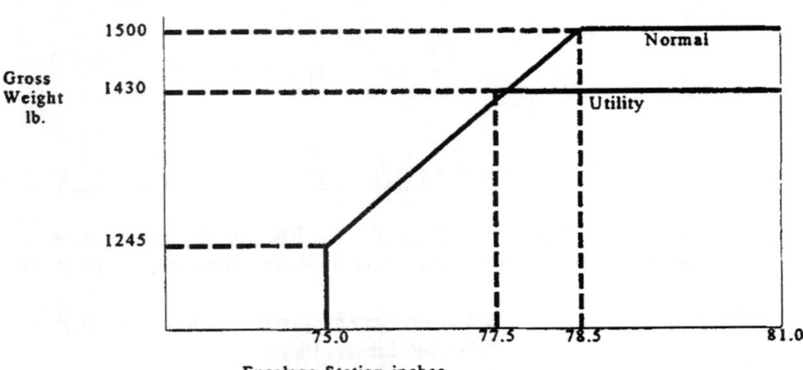

Empty weight C.G. range	None				
Maximum weight	1430 lb. (Utility Category) 1500 lb. (Normal Category)				
Number of seats	2 at (+92.5) (For optional child's seat refer to Equipment List.)				
Maximum baggage	100 lb. at (+120)				
Fuel capacity	24 gal. (2 wing tanks) at (+84.5) (See Note 1 for unusable fuel)				
Oil capacity	6 qt. at (+39) (2 qt. minimum)				
Control surface movements	Elevator	25° ± 2°	up	15° ± 2°	down
	Rudder25° ± 2°	left	25° ± 2°	right	
	Ailerons	25° ± 2°	up	20° ± 2°	down
	Flaps			30° ± 2°	down
	Elevator tab trim	21.5° ± 2°	up	11° ± 2°	down
Serial numbers eligible	AA1-0001 and up (Normal and Utility Category)				
Production basis	None. Prior to original certification of each aircraft manufactured subsequent to September 17, 1976, an FAA representative must perform a detailed inspection for workmanship, materials and conformity with the approved technical data and a check of the flight characteristics.				

II - Model AA-1A, Trainer, 2 PCLM, Utility Category, Approved January 14, 1971, Normal Category Approved January 14, 1971.

Engine	Lycoming 0-235-C2C (Carburetor Setting 10-4953 or 10-3103-1)
Fuel	80/87 minimum grade aviation gasoline
Engine limits	For all operations 2600 r.p.m. (108 h.p.)
Propeller and propeller limits	1. McCauley Model 1A105/SCM-7157 fixed pitch propeller. Static r.p.m. at maximum permissible throttle setting; not over 2300; not under 215C. Diameter: not over 71 inches, not under 69.5 inches.

II. Model AA-1A (cont'd)

2. McCauley Model 1A105/SCM-7153 and 1A105/SCM-7154 fixed pitch propellers. Static r.p.m. at maximum permissible throttle setting; not over 2400; not under 2250. Diameter: not over 71 inches, not under 69.5 inches.

3. McCauley Model 1A106/NCM-7157 fixed pitch propellers. Static r.p.m. at maximum permissible throttle setting; not over 2400; not under 2300. Diameter: not over 71 inches, not under 69.5 inches.

4. McCauley Model 1A106/NCM-7153 hub and fixed pitch propellers. Static r.p.m. at maximum permissible throttle setting; not over 2475; not under 2375. Diameter: not over 71 inches, not under 69.5 inches.

Airspeed limits (CAS)	V_{ne}	Never exceed	195 m.p.h. (169 knots)
	V_{no}	Maximum structural cruising	144 m.p.h. (125 knots)
	V_a	Maneuvering (Utility Category)	127 m.p.h. (110 knots)
	V_a	Maneuvering (Normal Category)	120 m.p.h. (104 knots)
	V_{fe}	Flaps extended	115 m.p.h. (100 knots)
		Canopy half open	130 m.p.h. (113 knots)

Center of gravity (C.G) range

(+78.5) to (+80.0) at 1500 lb.
(+77.5) to (+80.0) at 1430 lb.
(+75.0) to (+80.0) at 1245 lb.
Straight line variation between points given.

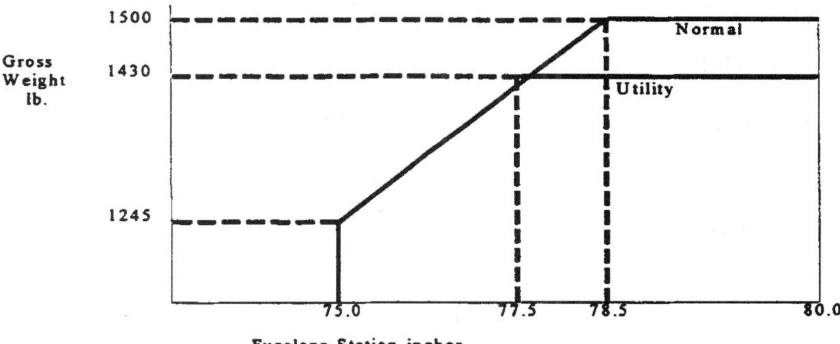

Empty weight C.G. range — None

Maximum weight — 1430 lb. (Utility Category)
1500 lb. (Normal Category)

Number of seats — 2 at (+92.5) (For optional child's seat refer to Equipment List.)

Maximum baggage — 100 lb. at (+120)

Fuel capacity — 24 gal. (2 wing tanks) at (+84.5) (See Note 1 for unusable fuel)

Oil capacity — 6 qt. at (+39) (2 qt. minimum)

Control surface movements				
	Elevator	25° ± 2°	up	15° ± 2° down
	Rudder25° ± 2°	left	25° ± 2°	right
	Ailerons	25° ± 2°	up	20° ± 2° down
	Flaps			30° ± 2° down
	Elevator tab trim	14.5° ± 2°	up	18° ± 2° down

Serial numbers eligible — AA1A-0001 and up (Normal and Utility Category)

Production basis	None. Prior to original certification of each aircraft manufactured subsequent to September 17, 1976, an FAA representative must perform a detailed inspection for workmanship, materials and conformity with the approved technical data and a check of the flight characteristics.

III - Model AA-1B, Trainer/TR-2, 2 PCLM, Utility Category, Approved June 30, 1972

Engine	Lycoming 0-235-C2C (Carburetor Setting 10-4953 or 10-3103-1)
Fuel	80/87 minimum grade aviation gasoline
Engine limits	For all operations 2600 r.p.m. (108 h.p.)
Propeller and propeller limits	1. McCauley Model 1A105 with 1A105/SCM hub and 7157 blades. Static r.p.m. at maximum permissible throttle setting; not over 2300; not under 2150. Diameter: not over 71 inches, not under 69.5 inches. 2. McCauley Model 1A105/SCM-7153 and 1A105/SCM-7154 fixed pitch propellers. Static r.p.m. at maximum permissible throttle setting; not over 2400; not under 2250. Diameter: not over 71 inches, not under 69.5 inches. 3. McCauley Model 1A106/NCM-7153 fixed pitch propellers. Static r.p.m. at maximum permissible throttle setting; not over 2400; not under 2300. Diameter: not over 71 inches, not under 69.5 inches. 4. McCauley Model 1A106/NCM-7157 fixed pitch propellers. Static r.p.m. at maximum permissible throttle setting; not over 2475; not under 2375. Diameter: not over 71 inches, not under 69.5 inches.

Airspeed limits (CAS)

V_{ne}	Never exceed	195 m.p.h. (169 knots)
V_{no}	Maximum structural cruising	144 m.p.h. (125 knots)
V_a	Maneuvering	135 m.p.h. (117 knots)
V_{fe}	Flaps extended	115 m.p.h. (100 knots)
	Canopy half open	130 m.p.h. (113 knots)

Center of gravity (C.G) range: (+78.25) to (+80.0) at 1560 lb.
(+75.0) to (+80.0) at 1300 lb.

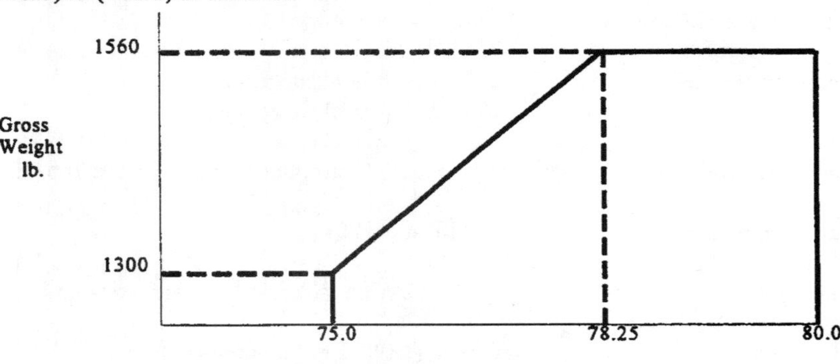

Empty weight C.G. range	None
Maximum weight	1560 lb.
Number of seats	2 at (+92.5) (For optional child's seat refer to Equipment List.)
Maximum baggage	100 lb. at (+120)

III - Model AA-1B (cont'd)

Fuel capacity	24 gal. (2 wing tanks) at (+84.5) (See Note 1 for unusable fuel)
Oil capacity	6 qt. at (+39) (2 qt. minimum)

Control surface movements

Elevator	25° ± 2°	up	15° ± 2°	down
Rudder25° ± 2°	left	25° ± 2°	right	
Ailerons	25° ± 2°	up	20° ± 2°	down
Flaps			30° ± 2°	down
Elevator tab trim	14.5° ± 2°	up	18° ± 2°	down

Serial numbers eligible	AA1B-0001 and up (Utility Category)
Production basis	Production Certificate No. 112/Production Certificate No. 3SO

IV - Model AA-1C, T-Cat/Lynx, 2 PCLM, Utility Category, Approved December 21, 1976. (Same as AA-1B except for engine, propeller, engine mount/baffles, and AA-5 elevator).

Engine	Lycoming 0-235-L2C (Carburetor Setting 10-4953 or 10-3103-1)
Fuel	100/130 minimum grade aviation gasoline
Engine limits	For all operations 2700 r.p.m. (115 h.p.)

Propeller and propeller limits

1. Sensenich Model 72CK-0-56 fixed pitch propeller. Static r.p.m. at maximum permissible throttle setting; not over 2275; not under 2125. No additional tolerance permitted. Diameter: not over 72 inches, not under 70.5 inches.

2. Sensenich Model 72CK-0-52 fixed pitch propellers. Static r.p.m. at maximum permissible throttle setting; not over 2475; not under 2325. No additional tolerance permitted. Diameter: not over 72 inches, not under 70.5 inches.

Airspeed limits (CAS)

V_{ne}	Never exceed	195 m.p.h. (169 knots)
V_{no}	Maximum structural cruising	144 m.p.h. (125 knots)
V_a	Maneuvering	135 m.p.h. (117 knots)
V_{fe}	Flaps extended	115 m.p.h. (100 knots)
	Canopy half open	130 m.p.h. (113 knots)

Center of gravity (C.G) range

(+78.00) to (+81.0) at 1600 lb.
(+75.5) to (+81.0) at 1385 lb.

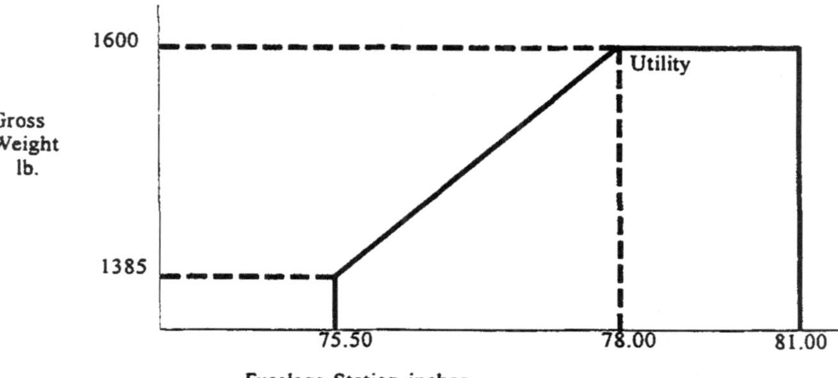

Empty weight C.G. range	None
Maximum weight	1600 lb.
Number of seats	2 at (+92.5) (For optional child's seat refer to Equipment List.)

IV - Model AA-1C (cont'd)

Maximum baggage	100 lb. at (+120)
Fuel capacity	24 gal. (2 wing tanks) at (+84.5) (See Note 1 for unusable fuel)
Oil capacity	6 qt. at (+39) (2 qt. minimum)

Control surface movements

Elevator	12° ± 1° up		28° ± 2°	down
Rudder25° ± 2°	left	25° ± 2°	right	
Ailerons	25° ± 2° up		20° ± 2°	down
Flaps			30° ± 2°	down
Elevator tab trim	15° ± 4° up		15° ± 2°	down

Serial numbers eligible	AA1B-0601 and AA1C-0001 and up (Utility Category)
Production basis	Production Certificate No. 3SO

DATA PERTINENT TO ALL MODELS:

Datum: 50.0 inches forward of front face of firewall (wing chord 48 inches for Model AA-1 and 49.32 inches for Models AA-1A, AA-1B, and AA-1C).

Leveling means: Top of fuselage canopy slide rail.

Certification basis: FAR 23 effective February 1, 1965, and amendments 23-1 and 23-2; and FAR 36 amended through 36-4 for the Model AA-1C.

Type Certificate No. A11EA issued August 29, 1967. Data of Application for Type Certificate October 22, 1965.

Equipment: The basic required equipment prescribed in the applicable airworthiness regulations (see Certification Basis) must be installed in the airplane for certification. In addition, equipment for the particular operation must be installed.

NOTE 1. Current weight and balance report including a list of equipment included in the certificated empty weight, and loading instructions when necessary must be provided for each aircraft at the time of original certification.

The certificated empty weight and corresponding center of gravity location must include 12 lb. (2 gal.) at (+84.5) of unusable fuel.

NOTE 2. The following placards must be installed in full view of the pilot:

(a) Models AA-1 and AA-1A:

"THIS AIRPLANE MUST BE OPERATED AS A NORMAL OR UTILITY CATEGORY AIRPLANE IN COMPLIANCE WITH THE OPERATING LIMITATIONS STATED IN THE FORM OF PLACARDS, MARKINGS, AND MANUALS."

NORMAL CATEGORY	AA-1	AA-1A
Maximum Design Weight	1500 lb.	1500 lb.
Design Maneuvering Speed, V_a	125 mph CAS	120 mph CAS
Flight Load Factors:		
Flaps Up	+3.8, -1.52	+3.8, -1.52
Flaps Down	+2.0	+3.5

NO ACROBATIC MANEUVERS INCLUDING SPINS APPROVED (AA-1 and AA-1A)

UTILITY CATEGORY	AA-1	AA-1A
Maximum Design Weight	1430 lb.	1430 lb.
Design Maneuvering Speed, V_a	130 mph CAS	127 mph CAS
Flight Load Factors:		
Flaps Up	+4.4, -1.76	+4.4, -1.76
Flaps Down	+2.0	+3.5

ACROBATIC MANEUVERS ARE LIMITED TO THE FOLLOWING:

MANEUVER	ENTRY SPEED (MPH, CAS)	
	AA-1	AA-1A
Chandelles	132	127
Lazy Eights	132	127
Steep Turns	132	127
Stalls (Except Whip Stalls)	Slow Deceleration	Slow Deceleration

Models AA-1B and AA-1C:

"THIS AIRPLANE MUST BE OPERATED AS A UTILITY CATEGORY AIRPLANE IN COMPLIANCE WITH THE OPERATING LIMITATIONS STATED IN THE FORM OF PLACARDS, MARKINGS, AND MANUALS."

		AA-1B	AA-1C
Maximum Design Weight		1560 Lb.	1600 Lb.
Design Maneuvering Speed, V_a	135 Mph Cas	117 Knots Cas	
Flight Load Factors:			
Flaps Up		+4.4, -1.76	+4.4, -1.76
Flaps Down		+3.5	+3.5

ACROBATIC MANEUVERS ARE LIMITED TO THE FOLLOWING:

MANEUVER	ENTRY SPEED (MPH, CAS)	ENTRY SPEED (KNOTS, CAS)
	AA-1B	AA-1C
Chandelles	135	117
Lazy Eights	135	117
Steep Turns	135	117
Stalls (Except Whip Stalls)	Slow Deceleration	Slow Deceleration

Maximum Altitude Loss In Stalls
300 Feet (AA-1)
250 Feet (AA-1A)
300 Feet (AA-1B)
200 Feet (AA-1C)

Demonstrated Crosswind Velocity
15 Mph (AA-1)
13 mph (AA-1A)
18 mph (AA-1B)
16 knots (AA-1C)

KNOWN ICING CONDITIONS TO BE AVOIDED. (Models AA-1, AA-1A, and AA-1B)

THIS AIRPLANE NOT APPROVED FOR FLIGHT IN ICING CONDITIONS. (Model AA-1C)

All Models:

THIS AIRPLANE IS CERTIFICATED FOR THE FOLLOWING OPERATIONS AS OF DATE OF ORIGINAL AIRWORTHINESS CERTIFICATE: IFR, VFR, DAY, NIGHT. (When properly equipped per FAR 91)

REFER TO WEIGHT AND BALANCE DATA FOR LOADING INSTRUCTIONS.

READ FUEL GAGES IN LEVEL FLIGHT ONLY.

FOR NORMAL OPERATION, MAINTAIN FUEL BALANCE.

DEMONSTRATED FUEL UNBALANCE 7 GAL.

(b) On left side of cabin:

"130 MPH MAX WITH CANOPY OPEN TO HERE. NO FLIGHT WITH CANOPY OPEN BEYOND THIS POINT." Placard Part No. 5803007-22 or equivalent. (Models AA-1, AA-1A, AA-1B)

"113 KNOTS MAX WITH CANOPY OPEN TO HERE. NO FLIGHT WITH CANOPY OPEN BEYOND THIS POINT." Placard Part No. 5803007-51 or equivalent. (Model AA-1C).

(c) In baggage compartment (All Models):

"BAGGAGE CAPACITY 100 LBS. MAX." Placard Part No. 803007-40 or equivalent.

(d) On instrument panel in full view of pilot (All Models):

"SPINS PROHIBITED." Placard Part No. 803007-56 or equivalent.

(e) On instrument panel near the airspeed indicator stall speed vs. bank angle placard.

Placard Part No. 803007-53 (Model AA-1), 803007-54 (Model AA-1A), 803007-55 (Model AA-1B), 803007-67 (Model AA-1C).

NOTE 3. The FAA Atlanta Aircraft Certification Office retains oversight responsibility for American General. By virtue of licensing agreement, product support and parts availability reside with Fletchair Inc., 9000 Randolph St., Houston, TX 77061, (713)-649-8700 or (800)-329-4647.

....END....

1A6
Revision 32
PIPER

PA-22
PA-22-108
PA-22-135
PA-22S-135
PA-22-150
PA-22S-150
PA-22-160
PA-22S-160

August 21, 1995

AIRCRAFT SPECIFICATION NO. 1A6

Type Certificate Holder — The New Piper Aircraft, Inc.
2926 Piper Drive
Vero Beach, Florida 32960

I - Model PA-22, 4 PCLM (Normal Category Only), Approved December 20, 1950

Engine — Lycoming O-290-D

Fuel — 80/87 minimum grade aviation gasoline

Engine Limits — For all operations, 2600 rpm (125 hp)

Airspeed Limits CAS

V_{ne}	(never exceed)	158 mph	(137 knots)
V_{no}	(maximum structural cruising)	126 mph	(110 knots)
V_p	(maneuvering)	106 mph	(92 knots)
V_{fe}	(flaps extended)	80 mph	(70 knots)

C. G. Range

(+17.5) to (+24.0) at 1800 lb.
(+10.0) to (+24.0) at 1380 lb. or less
Straight line variation between points given.

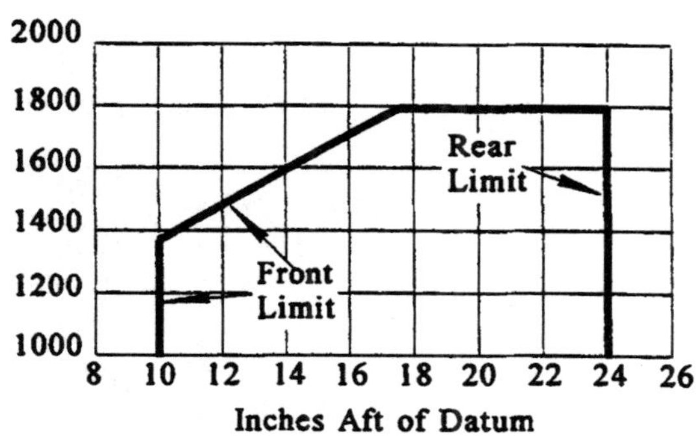

Empty Weight C. G. Range — None

Maximum Weight — 1800 lb.

Page No.	1	2	3	4	5	6	7	8	9	10	11	12	13	14	15
Rev. No.	32	31	31	31	31	31	31	31	32	31	31	32	32	31	31

I - Model PA-22 (cont'd)

Number Seats	4 (2 at +19.5 and 2 at +49)
Maximum Baggage	50 lb. (+67)
Fuel Capacity	36 gallons (2 Wing tanks at +24)
Oil Capacity	2 gallons (-29)

Control Surface Movements

Stabilizer	1°	Up	6½°	Down
Elevator	24°	Up	12°	Down
Aileron	15°	Up	15°	Down
Rudder	16°	Right	16°	Left
Flap	40°	Down		

Serial Numbers Eligible: 22-1 and up.

Required Equipment: In addition to the pertinent required basic equipment specified in CAR 3, the following items of equipment must be installed:
Items 1, 101, 201(a), 202, 205(a), 206, and 401(a).

II. Model PA-22-135, 4 PCLM (Normal Category), Approved May 5, 1952

Engine: Lycoming O-290-D2

Fuel: 80/87 minimum grade aviation gasoline

Engine Limits: For all operations, 2600 rpm (135 hp)

Airspeed Limits (CAS)

V_{ne}	(never exceed)	158 mph	(137 knots)
V_{no}	(maximum structural cruising)	126 mph	(110 knots)
V_p	(maneuvering)	106 mph	(92 knots)
V_{fe}	(flaps extended)	80 mph	(70 knots)

C. G. Range

(+17.5) to (+24.0) at 1950 lb.
(+10.0) to (+24.0) at 1380 lb. or less
Straight line variation between points given.

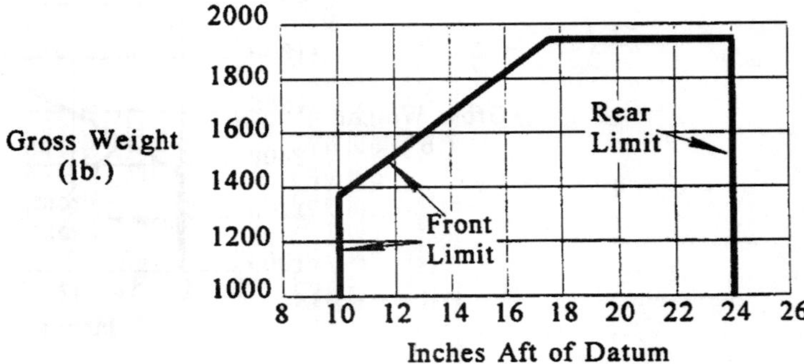

Empty Weight C. G. Range: None

Maximum Weight: 1950 lb.

Number of Seats: 4 (2 at +21 and 2 at +49)

II. Model PA-22-135 (cont'd)

Maximum Baggage	50 lb. (+67) May be increased to 100 lb. provided: (a) Baggage compartment placard is changed to "Maximum Baggage 100 Pounds." (b) Airplane Flight Manual, Item 401(c), is available in the airplane.
Fuel Capacity	36 gallons (2 wing tanks at +24). See Item 104 for reserve tank.
Oil Capacity	2 gallons (-29)
Control Surface Movements	Stabilizer 1° Up 6½° Down Elevator 24° Up 12° Down Aileron 15° Up 15° Down Rudder 16° Right 16° Left Flap 40° Down
Serial Numbers Eligible	22-534 and up.
Required Equipment	In addition to the pertinent required basic equipment specified in CAR 3, the following Items of equipment must be installed: Items 1, 103, 201(a), 202, 205(a), 206, and 401(b).

III - Model PA-22S-135, 3 PCSM (Normal Category), Approved May 14, 1954

Engine	Lycoming O-290-D2
Fuel	80/87 minimum grade aviation gasoline
Engine Limits	For all operations, 2600 r.p.m. (135 hp)
Airspeed Limits CAS	V_{ne} (never exceed) 140 mph (122 knots) V_{no} (maximum structural cruising) 117 mph (102 knots) V_p (maneuvering) 105 mph (91 knots) V_{fe} (flaps extended) 80 mph (70 knots)
C. G. Range	(+14.0) to (+20.0) at 1850 lb. (+10.0) to (+20.0) at 1300 lb. or less Straight line variation between points given.

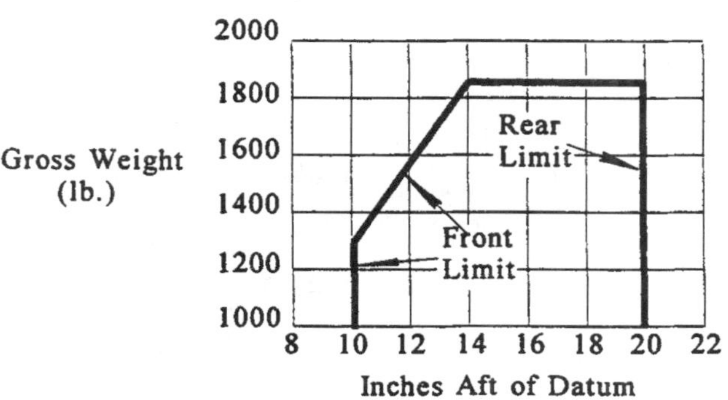

Empty Weight C. G. Range	None
Maximum Weight	1850 lb.
Number of Seats	4 (2 at +21 and 2 at +49)
Maximum Baggage	50 lb. (+67)

III - Model PA-22S-135 (cont'd)

Fuel Capacity	36 gallons (2 wing tanks at +24). See Item 104 for reserve tank.
Oil Capacity	2 gallons (-29)

Control Surface Movements

Stabilizer	1° Up	6½° Down
Elevator	24° Up	12° Down
Aileron	15° Up	15° Down
Rudder	16° Right	16° Left
Flap	40° Down	

Serial Numbers Eligible: 22-534 and up.

Required Equipment: In addition to the pertinent required basic equipment specified in CAR 3, the following Items of equipment must be installed:
Items 2, 103, 209, and 401(g).

IV - Model PA-22-150, 4 PCLM (Normal Category), Approved September 3, 1954.
Model PA-22-150, 2 PCLM (Utility Category), Approved May 24, 1957 (See NOTE 3 for limitations)

Engine: Lycoming O-320-A2A or O-320-A2B (Carburetor setting #10-3678-11, #10-3678-12 or #10-3678-32) (See Item 106 for optional engines)

Fuel: 80/87 minimum grade aviation gasoline

Engine Limits: For all operations, 2700 r.p.m. (150 hp)

Airspeed Limits CAS

V_{ne}	(never exceed)	170 mph	(148 knots)
V_{no}	(maximum structural cruising)	135 mph	(117 knots)
V_p	(maneuvering)	112 mph	(97 knots)
V_{fe}	(flaps extended)	95 mph	(82 knots)

C. G. Range

Normal Category:	(+17.5)	to	(+23.0)	at	2000 lb.
	(+12.0)	to	(+23.0)	at	1800 lb.
	(+9.5)	to	(+23.0)	at	1400 lb. or less
Utility Category:			(+13.5)	at	1680 lb.
	(+12.0)	to	(+13.5)	at	1665 lb.
	(+9.5)	to	(+13.5)	at	1400 lb. or less

Straight line variation between points given.

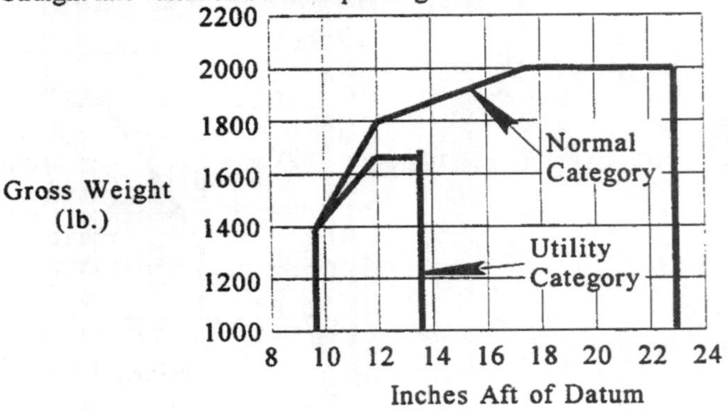

Empty Weight C. G. Range: None

Maximum Weight: Normal Category: 2000 lb.
Utility Category: 1680 lb.

Number of Seats: 4 (2 at +21 and 2 at +49)
Rear seats not to be used when operating in the Utility Category.

IV - Model PA-22-150, 2 PLCM (cont'd).

Maximum Baggage	100 lb. (+67) (No baggage allowed when operating in the Utility Category)
Fuel Capacity	36 gallons (2 wing tanks at +24) See Item 104 for reserve tank.
Oil Capacity	2 gallons (-29)
Control Surface Movements	Stabilizer 1° Up, 6½° Down Elevator 24° Up, 12° Down Aileron 15° Up, 15° Down Rudder 16° Right, 16° Left Flap 40° Down
Serial Numbers Eligible	22-2378, 22-2425 and up (Normal Category). See NOTE 3 for Utility Category.
Required Equipment	In addition to the pertinent required basic equipment specified in CAR 3, the following Items of equipment must be installed: Normal Category: Items 5, 103, 201(a), 202, 205(a), 206, and 401(h). Normal and Utility Category: Items 5, 103, 201(a), 202, 205(a), 206, 401(h), 401(r), and 407.

V. - Model PA-22S-150, 3 PCSM (Normal Category), Approved September 3, 1954

Engine	O-320-A2A Lycoming (Carburetor setting #10-3678-11, #10-3678-12) or O-320-A2B (Carburetor setting #10-3678-32) (See Item 106 for optional engines)
Fuel	80/87 minimum grade aviation gasoline
Engine Limits	For all operations, 2700 r.p.m. (150 hp)
Airspeed Limits CAS	V_{ne} (never exceed) 158 mph (137 knots) V_{no} (maximum structural cruising) 126 mph (109 knots) V_p (maneuvering) 111 mph (96 knots) V_{fe} (flaps extended) 80 mph (70 knots)
C. G. Range	(+14.0) to (+20.0) at 1950 lb. (+12.0) to (+20.0) at 1800 lb. (+10.0) to (+20.0) at 1500 lb. or less Straight line variation between points given.

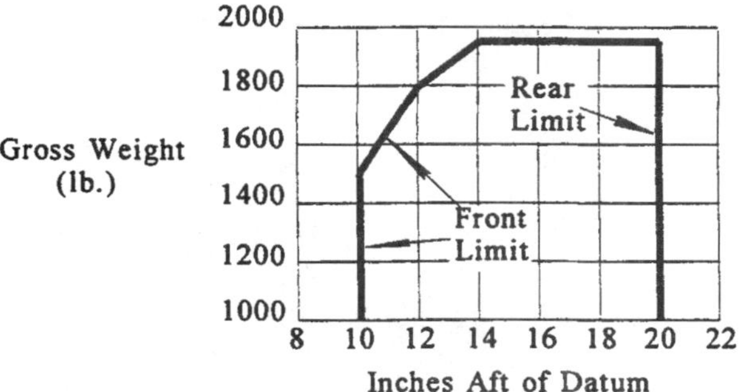

Empty Weight C. G. Range	None
Maximum Weight	1950 lb.
Number Seats	4 (2 at +21 and 2 at +49)

V. - Model PA-22S-150, 3 PCSM (cont'd)

Maximum Baggage	100 lb. (+67)
Fuel Capacity	36 gallons (2 wing tanks at +24). See Item 104 for reserve tank.
Oil Capacity	2 gallons (-29)

Control Surface Movements

Stabilizer	1°	Up	6½°	Down
Elevator	24°	Up	12°	Down
Aileron	15°	Up	15°	Down
Rudder	16°	Right	16°	Left
Flap	40°	Down		

Serial Numbers Eligible: 22-2378, 22-2425 and up.

Required Equipment: In addition to the pertinent required basic equipment specified in CAR 3, the following Items of equipment must be installed:
Items 5, 103, 209 and 401(i).

VI - Model PA-22-160, 4 PCLM (Normal Category), Approved August 27, 1957
Model PA-22-160, 2 PCLM (Utility Category), Approved August 27, 1957 (See NOTE 3)

Engine: Lycoming O-320-B2A or O-320-B2B (Carburetor setting #10-3678-11, #10-3678-12 or #10-3678-32) (See Item 106 for optional engines).

Fuel: 91/96 minimum grade aviation gasoline

Engine Limits: For all operations, 2700 r.p.m. (160 hp)

Airspeed Limits (CAS)

V_{ne}	(never exceed)	170 mph	(148 knots)
V_{no}	(maximum structural cruising)	135 mph	(117 knots)
V_p	(maneuvering)	112 mph	(97 knots)
V_{fe}	(flaps extended)	95 mph	(82 knots)

C. G. Range

Normal Category:	(+17.5)	to	(+23.0)	at	2000 lb.
	(+12.0)	to	(+23.0)	at	1800 lb.
	(+9.5)	to	(+23.0)	at	1400 lb. or less
Utility Category:			(+13.5)	at	1680 lb.
	(+12.0)	to	(+13.5)	at	1665 lb.
	(+9.5)	to	(+13.5)	at	1400 lb. or less

Straight line variation between points given.

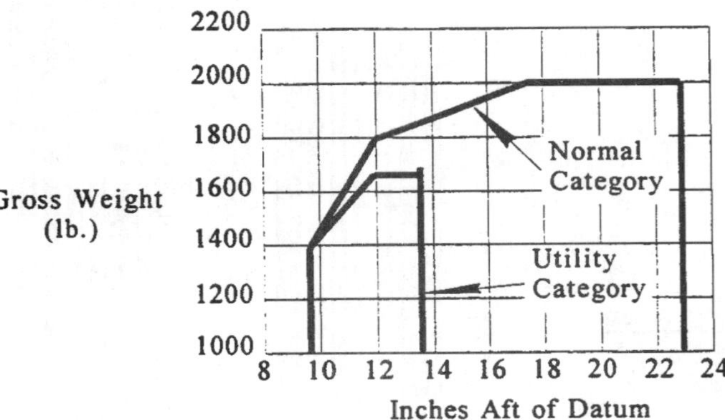

Empty Weight C. G. Range: None

VI - Model PA-22-160, 4 PCLM; Model PA-22-160, 2 PCLM (cont'd)

Maximum Weight	Normal Category: 2000 lb. Utility Category: 1680 lb.
Number of Seats	4 (2 at +21 and 2 at +49) Rear seats not to be used when operating in the Utility Category.
Maximum Baggage	100 lb. (+67) No baggage allowed when operating in the Utility Category.
Fuel Capacity	36 gallons (2 wing tanks at +24). See Item 104 for reserve tank.
Oil Capacity	2 gallons (-29)
Control Surface Movements	Stabilizer 1° Up 6½ Down Elevator 24° Up 12° Down Aileron 15° Up 15° Down Rudder 16° Right 16° Left Flap 40° Down
Serial Numbers Eligible	22-2378, 22-2425 and up (Normal Category). See NOTE 3 for Utility Category.
Required Equipment	In addition to the pertinent required basic equipment specified in CAR 3, the following Items of equipment must be installed: Normal Category: Items 7, 103, 201(a), 202, 205(a), 206, and 401(s). Normal and Utility Category: Items 7, 103, 201(a), 202, 205(a), 206, 401(s), 401(t), and 407.

VII - Model PA-22S-160, 3 PCSM (Normal Category), Approved October 25, 1957

Engine	Lycoming O-320-B2A (Carburetor setting #10-3678-11, #10-3678-12) or O-320-B2B (Carburetor setting #10-3678-32) (See Item 106 for optional engines).
Fuel	91/96 minimum grade aviation gasoline
Engine Limits	For all operations, 2700 r.p.m. (160 hp)
Airspeed Limits	V_{ne} (never exceed) 158 mph (137 knots) V_{no} (maximum structural cruising) 126 mph (109 knots) V_p (maneuvering) 111 mph (96 knots) V_{fe} (flaps extended) 80 mph (70 knots)
C. G. Range	(+14.0) to (+20.0) at 1950 lb. (+12.0) to (+20.0) at 1800 lb. (+10.0) to (+20.0) at 1500 lb. or less Straight line variation between points given.

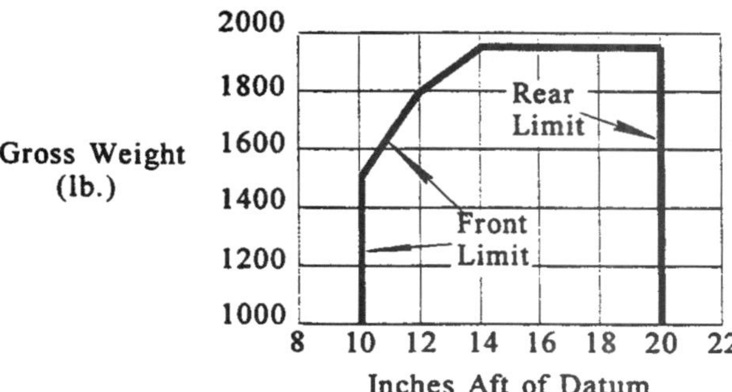

Empty Weight C. G. Range	None

VII - Model PA-22S-160, 3 PCSM (cont'd)

Maximum Weight	1950 lb.
Number of Seats	4 (2 at +21 and 2 at +49)
Maximum Baggage	100 lb. (+67)
Fuel Capacity	36 gallons (2 wing tanks at +24). See Item 104 for reserve tank.
Oil Capacity	2 gallons (-29)

Control Surface Movements

Stabilizer	1°	Up	6½°	Down
Elevator	24°	Up	12°	Down
Aileron	15°	Up	15°	Down
Rudder	16°	Right	16°	Left
Flap	40°	Down		

Serial Numbers Eligible	22-2378, 22-2425 and up.
Required Equipment	In addition to the pertinent required basic equipment specified in CAR 3, the following Items of equipment must be installed: Items 7, 103, 209, and 401(v).

VIII - Model PA-22-108, 2 PCLM (Normal and Utility Category), Approved October 21, 1960

Engine	Lycoming O-235-C1 or O-235-C1B (Carburetor setting #10-3103-1)
Fuel	80/87 minimum grade aviation gasoline
Engine Limits	For all operations, 2600 r.p.m. (108 hp)

Airspeed Limits (CAS)

V_{ne}	(never exceed)	138 mph	(120 knots)
V_{no}	(maximum structural cruising)	110 mph	(96 knots)
V_p	(maneuvering)	104 mph	(90 knots)

C. G. Range

Normal Category:	(+12.0)	to	(+16.25)	at	1650 lb.
	(+9.5)	to	(+16.25)	at	1300 lb. or less
Utility Category:	(+10.9)	to	(+14.00)	at	1500 lb.
	(+9.5)	to	(+14.00)	at	1300 lb. or less

Straight line variation between points given.

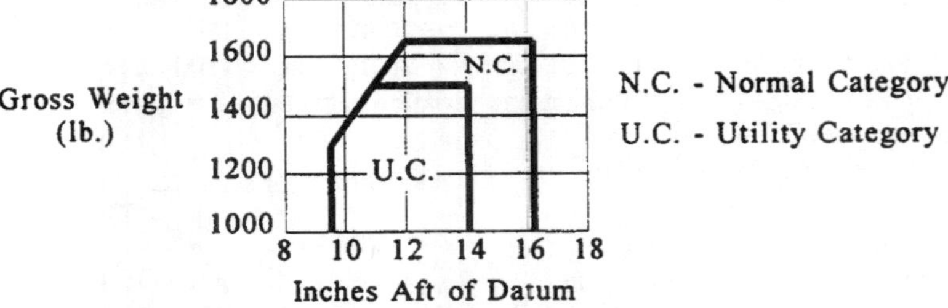

Empty Weight C. G. Range	None
Maximum Weight	Normal Category: 1650 lb. Utility Category: 1500 lb.
Number of Seats	2 at (+21)
Maximum Baggage	100 lb. (+45) (Normal category only)

VIII - Model PA-22-108, 2 PCLM (cont'd)

Fuel Capacity	18 gallons (+24) (See Item 108 for auxiliary tank)				
Oil Capacity	1.5 gallons (-29)				
Control Surface Movements	Stabilizer	1°	Up	6½°	Down
	Elevator	24°	Up	12°	Down
	Aileron	15°	Up	15°	Down
	Rudder	16°	Right	16°	Left
Serial Numbers Eligible	22-8000 and up.				
Required Equipment	In addition to the pertinent required basic equipment specified in CAR 3, the following Items of equipment must be installed: Items 8, 201(a) or 211(a), 202, 205(a), 206, and 401(y).				

Specifications Pertinent to All Models

Datum	Wing leading edge
Leveling Means	Plumb from hole in upper channel of front door to center punch mark on front seat cross tube.
Certification Basis	CAR 3, effective November 1, 1949, and Amendments 3-1 through 3-6, effective June 4, 1951. Type Certificate No. 1A6 issued December 20, 1950. Date of Application for Type Certificate September 13, 1950.
Production Basis	Approved for manufacture of spare parts only under Production Certificate No. 206.

1A6

Equipment

A plus (+) or minus (-) sign preceding the weight of an Item of equipment indicates net weight change when that Item is installed.

Approval for the installation of all Items of equipment listed herein has been obtained by the aircraft manufacturer except those Items preceded by an asterisk (*). The asterisk denotes that approval has been obtained by someone other than the aircraft manufacturer. An Item marked with an asterisk may not have been manufactured under an FAA monitored or approved quality control system, and therefore conformity must be determined if the Item is not identified by a Form FAA-186, PMA or other evidence or FAA production approval.

Propeller and Propeller Accessories

The following propellers are eligible at the limits shown for diameter and static r.p.m. at maximum permissible throttle setting, no additional tolerance permitted:

1. Propeller (with Lycoming O-290D or O-290-D2 engine)
   Sensenich 74FM59 or any other fixed pitch wood propeller which is rated for the engine power and speed: +11 lb. (-50)
   Static r.p.m.: Not over 2400, not under 2200.
   Diameter: Not over 74 inches, not under 70.5 inches
2. Propeller (with Lycoming O-290D or O-290-D2 engine) - fixed pitch metal
   (a) Sensenich M76AM-2 or +25 lb. (-50)
   (b) Sensenich M74DM +30 lb. (-50)
   Airplane Flight Manual shall be revised to reflect the subject propeller and limits.
   Landplane:
   Static r.p.m.: Not over 2450, not under 2150
   Diameter: Not over 74 inches, not under 72.5 inches
   Seaplane:
   Static r.p.m.: Not over 2450, not under 2350
   Diameter: Not over 74 inches, not under 72.5 inches

3. Propeller (with Lycoming O-290D or O-290-D2 engine)
Koppers Aeromatic, F200-H/00-74E +34 lb. (-50)
Parts List Assembly No. 4394H-1. Installation and operation must be accomplished in accordance with Koppers "Adjustment Instructions and Operation Limitations No. 58."
Low pitch setting 14° at 24 in sta.
Static r.p.m.: Not over 2600, not under 2550.
Diameter: Not over 74 inches, not under 72.5 inches
4. Propeller (with Lycoming O-290D or O-290-D2 engine)
Sensenich hub CS3FM-4, blades PC374A7 or C374E, two position controllable. +34 lb. (-50)
Propeller control installation required as per Sensenich Dwg. D-3028, Revision E.
Blade pitch setting at 3/4 radius (27.75 in. station):
Low 13°, high 16.6°
Diameter: Not over 74 inches, not under 72.5 inches
5. Propeller (with Lycoming O-320-A2A or O-320-A2B engine) - Fixed pitch metal
Sensenich M74DM +30 lb. (-50)
Landplane:
Static r.p.m.: Not over 2480, not under 2250.
Diameter: Not over 74 inches, not under 72.5 inches
Seaplane:
Static r.p.m.: Not over 2500, not under 2400
Diameter: Not over 74 inches, not under 72.5 inches
6. Propeller (with Lycoming O-320-A1A or O-320-A1B engine) - constant speed controllable
Hartzell hub HC82XG-6, blades 7636D-4 +54 lb. (-50)
Installed per Piper Dwg. No. 14747 when Item 105 (vacuum pump) is installed, or per Piper Dwg. No. 14792, without vacuum pump.
Not eligible when Item 407 is installed.
Note 2(f) placard required.
Blade pitch settings at 30 in. sta.: Low 12°, high 26°.
Diameter: Not over 72 inches, not under 70 inches
Eligible only on Models PA-22-150 and PA-22S-150, Serial Nos. 22-3218, 22-3387 and up.
When this propeller is used on Model PA-22S-150, the engine side cowls shall be installed per Piper Dwg. No. 14450.
7. Propeller (with Lycoming O-320-B2A or O-320-B2B engine) - fixed pitch metal
Sensenich M74DM +34 lb. (-50)
Landplane:
Static r.p.m.: Not over 2450, not under 2250
Diameter: Not over 74 inches, not under 72 inches
Seaplane:
Static r.p.m.: Not over 2500, not under 2400
Diameter: Not over 74 inches, not under 72 inches
Applicable Airplane Flight Manual shall be revised by the Modifier and approved by the applicable FAA Aircraft Certification Office to reflect this installation change.
8. Propeller (with Lycoming O-235-C1 or O-235-C1B engine) - fixed pitch metal
Sensenich M76AM-2 +25 lb. (-50)
Static r.p.m.: Not over 2450, not under 2200
Diameter: Not over 74 inches, not under 72.5 inches

Engines and Engine Accessories - Fuel and Oil Systems

101. Oil cooler - Harrison No. AP06CJ04-02 or AP06CU04-2 and Piper Air Duct +3 lb. (-18)
102. Oil filter, Fram PB-5, Kit No. K-520, Fram Dwg. No. 62832 and Instruction Sheet No. 62831 (weight includes 1 quart oil) +5 lb. (-18.5)
103. Oil Cooler Harrison No. AP13SJ03-01 or AP12CU03-01 installed in accordance with Piper Dwg. 13724 or 14368 +6 lb. (-46)
104. Reserve 8 gallons fuel tank with electric transfer fuel pump installed in accordance with Piper Dwg. 14454. When installed on Models PA-22S-135, PA-22S-150 or PA-22S-160, fuselage reinforcement channel, Part No. 14725, also required. +12 lb. (+46)
NOTE 2(e) placard required.
Airplane Flight Manual Supplement required:
Item 401(j), Model PA-22-150
Item 401(k) Model PA-22-135 (Serial Nos. 22-534 and up eligible),
Item 401(p) Model PA-22S-135 (Serial Nos. 22-807 and up eligible),
Item 401(q) Model PA-22S-150 (Serial Nos. 22-2378, 22-2425 and up eligible),
Item 401(u) Model PA-22-160 (Serial Nos. 22-2378, 22-2425 and up eligible),
or Item 401(w) Model PA-22S-160 (Serial Nos. 22-2378, 22-2425 and up eligible).
105. Vacuum pump
(a) Pesco Model 3P-194-F, Type B-11 +4 lb. (-25)
(b) Airborne Mechanisms Model 113A1 installed in accordance with Piper Dwg. 15163. (PA-22-108 only). +4 lb. (-25)
(c) Airborne Mechanisms Model 113A5 installed in accordance with Piper Dwg. 15163 or 15208. (PA-22-108 only). +4 lb. (-25)
106. Engines (Lycoming) O-320 Series
A. Model PA-22-150
(1) O-320
(2) O-320-A1A
(3) O-320-A1B
B. Model PA-22S-150
(1) O-320
(2) O-320-A1A
(3) O-320-A1B
107. Starter, Delco Remy Model 1109657 (12 v.) +17 lb. (-40)
108. Auxiliary 18 gallons fuel tank installed in accordance with Piper Dwg. 15147 (PA-22-108 only). NOTE 2(j) placard required. +25 lb. (+24)

Landing Gear

201. Two main wheel-brake assemblies, 6.00-6, Type III +14 lb. (+31.5)
(a) Cleveland Aircraft Products Model 6:00 DHB-3
Wheel Assembly No. C-38500H
Brake Assembly No. C-2000H
202. Two main 4-ply rating tires, 6.00-6, Type III, with regular tubes +17 lb. (+31.5)
205. One nose wheel, 6.00-6, Type III +5 lb. (-36)
(a) Cleveland Aircraft Products Wheel Assembly No. C-38500H (less brake-drum)
(b) Cleveland Aircraft Products Wheel Assembly No. 38501
206. One nose wheel 4-ply rating, tire, 6.00-6, Type III, with regular tube +9 lb. (-36)
*207. Nose wheel centering kit installed according to Javelin Aircraft Company (Wichita, Kansas) Dwg. 723 and Installation Instructions dated April 15, 1953. +2. lb. (-29)
208. Skis: Use Actual Weight Change
*(a) Federal A-2000A main skis and NA-1200A nose ski, per Federal Dwg. 11R951, Change E.
*(b) Federal AWB-2100 main skis and AWN-1200 nose ski, per Federal Dwg. 11R1117.
The following placard is required with this installation:
"Do not extend or retract skis while in motion on the ground."

209. Edo Model 89-2000 floats with water rudder installed in accordance with Edo Dwg. No. 16270.
Piper modifications must be made and installed in accordance with Piper Dwg. 14375 (Model PA-22S-135, Serial Nos. 22-534 to 22-2377, 22-2379 to 22-2424, inclusive) and Piper Dwg. 14450 (Model PA-22S-150 and PA-22S-160, Serial Nos. 22-2378, 22-2425 and up.) Serial Nos. 22-534 to 22-806, inclusive, require a fuselage reinforcement brace, Piper Part No. 12480.

210.	(a) Doyn Fiberglass wheel fairings installed in accordance with Doyn Dwg. No. 1300 and Doyn Process Specification for Fiberglass Part No. PS-100	Nose Fairing Main Fairing	+5.5 lb. +15.0 lb.	(-36) (+31.5)
or	(b) Piper wheel fairings installed in accordance with Piper Dwg. 15054 and 15058	Nose Fairing Main Fairing	+5.5 lb. +15.0 lb.	(-36) (+31.5)
or	(c) Piper wheel fairings installed in accordance with Piper Dwg. 15083	Nose Fairing Main Fairing	+5.5 lb. +15.0 lb.	(-36) (+31.5)

211. Two Main Wheel-Brake Assemblies, 6.00-6, Type III
(a) Cleveland Aircraft Products, Model 20-6 (Model PA-22-108 only) + 14.5 lb. (+31.5)
Wheel Assembly No. 40-28
Brake Assembly No. 30-18

Electrical Equipment

301. Battery - Reading S24-12V +25 lb. (+21)
302. Landing lights in wing leading edge per Piper Dwg. No. 12534 (Serial Nos. 22-534 to 22-2377, 22-2379 to 22-2424, inclusive) Piper Dwg. No. 14442 (Serial Nos. 22-2378, 22-2425 and up). +4 lb. (+5)
303. Battery - Reading R33-12V +28 lb. (+21)
Serial Nos. 22-267, 22-340, 22-349, 22-350, 22-351, 22-354 through 22-7999.

Interior Equipment

401. (a) CAA (FAA) approved Airplane Flight Manual dated December 20, 1950, for airplanes equipped with Lycoming O-290-D engines. (Required with 100 lb. baggage allowance.)
(b) FAA-DOA approved Airplane Flight Manual dated May 5, 1952, for airplanes equipped with Lycoming O-290-D2 engines.
(c) FAA-DOA approved Airplane Flight Manual dated October 23, 1952, for airplanes equipped with Lycoming O-290-D2 engines.
*(d) Supplement to Airplane Flight Manual dated January 17, 1952. (Required with Item 402(a) without altitude controller.)
*(e) Revised Supplement to Airplane Flight Manual dated January 19, 1953. (Required with Item 402(a) without altitude controller.)
*(f) Revised Supplement to Airplane Flight Manual dated November 18, 1953. (Required with Item 402(b) with approach coupler.)
(g) FAA-DOA approved Airplane Flight Manual dated May 14, 1954, for Model PA-22S-135 seaplanes equipped with Edo Model 89-2000 floats.
(h) FAA-DOA approved Airplane Flight Manual dated September 3, 1954, for Model PA-22-150.
(i) FAA-DOA approved Airplane Flight Manual dated September 3, 1954, for Model PA-22S-150 seaplanes equipped with Edo Model 89-2000 floats.
(j) FAA-DOA approved Supplement No. 1 to Airplane Flight Manual dated September 3, 1954, (Required with Item 104 Auxiliary Fuel System) for Model PA-22-150.
(k) FAA-DOA approved Supplement No. 1 to Airplane Flight Manual dated October 23, 1952, (Required with Item 104 Auxiliary Fuel System) for Model PA-22-135, Serial No. 22-534 and up.
*(l) Supplement to Airplane Flight Manual dated November 17, 1954. (Required with Item 404).
*(m) Supplement to Airplane Flight Manual dated April 20, 1955. (Required with Item 405).
(n) FAA-DOA approved Supplement to Airplane Flight Manual dated September 3, 1954, for Model PA-22-150 (Required with Item 6).

(o) FAA-DOA approved Supplement to Airplane Flight Manual dated September 3, 1954, for Model PA-22S-150 (Required with Item 6).
(p) FAA-DOA approved Supplement No. 1 to Airplane Flight Manual dated October 23, 1952, (Required with Item 104 Auxiliary Fuel System) for Model PA-22S-135.
(q) FAA-DOA approved Supplement No. 1 to Airplane Flight Manual dated September 3, 1954, (Required with Item 104 Auxiliary Fuel System) for Model PA-22S-150.
(r) FAA-DOA approved Supplement No. 3 to Airplane Flight Manual dated September 3, 1954, for Model PA-22-150 (Required with Item 407.).
(s) FAA-DOA approved Airplane Flight Manual dated August 27, 1957, for airplanes equipped with Lycoming O-320-B2A or O-320-B2B engines.
(t) FAA-DOA approved Supplement No. 1 to Airplane Flight Manual dated August 27, 1957, for Model PA-22-160 (Required with Item 407).
(u) FAA-DOA approved Supplement No. 2 to Airplane Flight Manual dated August 27, 1957, for Model PA-22-160 (Required with Item 104 Auxiliary Fuel System).
(v) FAA-DOA approved Airplane Flight Manual dated October 25, 1957, for Model PA-22S-160 seaplanes equipped with Edo Model 89-2000 floats.
(w) FAA-DOA approved Supplement No. 1 to Airplane Flight Manual dated October 25, 1957, for Model PA-22S-160 (Required with Item 104 Auxiliary Fuel System).
(x) FAA-DOA approved Supplement No. 3 to Airplane Flight Manual dated August 27, 1957 (Model PA-22-160); or FAA-DOA approved Supplement No. 4 to Airplane Flight Manual dated September 3, 1954 (Model PA-22-150) (Required with Item 408 Piper AutoControl, Mitchell Model AKO-64, Automatic Pilot) for Models PA-22-150 and PA-22-160, Serial No. 22-6328, 22-6344, 22-6352 and up.
(y) FAA-DOA approved Airplane Flight Manual dated October 21, 1960, revised November 22, 1960, for Model PA-22-108.
(z) FAA-DOA approved Supplement No. 1 to Airplane Flight Manual dated October 21, 1960, (Required with Item 409 Piper AutoControl, Mitchell Model AKO-64, Automatic Pilot) for Model PA-22-108, Serial No. 22-8000 and up.
(aa) FAA-DOA approved Supplement to Airplane Flight Manual dated December 20, 1950, for Model PA-22 (Required when rear door removed under provisions of NOTE 4).
(ab) FAA-DOA approved Supplement No. 3 to Airplane Flight Manual dated October 23, 1952, for Model PA-22-135 (Required when rear door removed under provisions of NOTE 4).
(ac) FAA-DOA approved Supplement No. 5 to Airplane Flight Manual dated September 3, 1954, for Model PA-22-150 (Required when rear door removed under provisions of NOTE 4).
(ad) FAA-DOA approved Supplement No. 4 to Airplane Flight Manual dated August 27, 1957 for Model PA-22-160 (Required when rear door removed under provisions of NOTE 4).

*402. Lear L-2B Automatic Pilot:
(An approved vacuum system to operate automatic pilot gyros and a 35 ampere generator meeting requirements of Aircraft Engine Specification E-229 are required. Servo pitch drum diameter for all three axes 1.375 inches.)
(a) Automatic pilot and altitude controller (optional equipment) installed in accordance with Lear Dwg. 95650. +51 lb. (+63)
Servo slip clutch stall torque, +0, -5 in.-lb. tolerance:

Aileron	40 in.-lb.
Elevator	25 in.-lb.
Rudder	50 in.-lb.

Items 401(d) or 401(e) and the following placard, installed in clear view of pilot, are required with this installation:
"Do not use Autopilot in normal operation below 75 feet above terrain including take-off, approach and landing."

1A6

(b) Automatic pilot and approach coupler (optional equipment) and altitude control (optional equipment) installed in accordance with Lear Dwg. 95650, Revision D. Servo slip clutch stall torque + 0, - 5 in.-lb tolerance: +7 lb. (+74)

Aileron 40 in.-lb.
Elevator 40 in.-lb.
Rudder 50 in.-lb.

Item 401(f) and the following placards, installed in clear view of the pilot, are required with this installation:

"Do no use Autopilot in normal operation below 300 feet above terrain except during take-off, approach and landing."

"During take-off, approach and landing, do not use Autopilot below 75 feet above terrain."

"Do not use transmitter #1 during an automatic approach."

*403. Javelin A2 single axis automatic pilot installed in accordance with Javelin Dwg. 721 and Instructions dated June 15, 1954. Item 207 required with this installation. +18 lb. (+94)

*404. Lear Arcon (Automatic rudder control) installed in accordance with Lear Dwg. 701944. Item 401(1) required with this installation. Model PA-22-135 only. +12 lb. (+65)

*405. Ross Control System Conversion Kit Model 10 installed in accordance with Ross (F. W. Ross, 755 Kalamath Drive, Del Mar, California) Dwgs. 10R100 through 9A114 on Drawing List dated November 5, 1955, and Installation Instructions dated November 5, 1955. Placard required on instrument panel: Use Actual Weight and Balance Change

"Equipped with Ross Control System - See Flight Manual Supplement."

Item 401(m) required with this installation.

*406. Deleted - November 26, 1957. Now covered by Supplemental Type Certificate No. SA1-108

407. Control modification kit (eliminating rudder and aileron interconnection) per Piper Dwg. No. 14926. Item 401(r) or 401(t) and NOTE 2(g) placard required. See limitations in NOTE 3.

408. Piper AutoControl (Mitchell Model AKO-64) Automatic Pilot installed in accordance with Piper Dwg. No. 14970. Item 105 and 401(x), and NOTE 2(h) placard required. (Models PA-22-150 and PA-22-160) +5 lb. (-10)

409. Piper Autocontrol (Mitchell Model AKO-64) Automatic Pilot installed in accordance with Piper Dwg. No. 14970. Item 105(b) or 105(c), and 401(z), and NOTE 2(h) placards required. (Model PA-22-108) +5 lb. (-10)

NOTE 1. Current weight and balance report including list of equipment included in certificated empty weight, and loading instructions when necessary, must be provided for each aircraft at the time of original certification.

NOTE 2. The following placards must be displayed:

(a) On the instrument panel in full view of the pilot (For all Models except PA-22-108):
    (1) "Operate in Normal Category in compliance with approved Flight Manual. Acrobatics (including spins) prohibited."
(b) On the baggage compartment (Serial Nos. 22-534 to 22-2377, 22-2379 to 22-2424):
    (1) "Maximum Baggage 50 Pounds." or
    (2) "Maximum Baggage 100 Pounds." (For Model PA-22-135 when Airplane Flight Manual, Item 401(c), is available in the airplane.)
(c) On the baggage compartment (Serial Nos. 22-2378, 22-2425 and up):
    (1) "Maximum Baggage 100 Pounds."
(d) Deleted, December 30, 1955.
(e) Adjacent to reserve tank selector valve when Item 104 is installed in aircraft:
    (1) "Reserve fuel
    pull on
    transfer fuel level flight only
    operate only in accordance with flight manual."
(f) Adjacent to the propeller pitch control when Item 6 is installed:
    (1) "Propeller-Push Increase R.P.M."
(g) On the instrument panel in full view of the pilot when Item 407 is installed:
    (1) "Operate in Normal or Utility Category in compliance with the approved Flight Manual. Airplane marked for Normal Category. Acrobatics (including spins) prohibited in Normal Category."

(h) When Item 408 or 409 is installed:
  (1) On left side of circuit breaker panel:
  "Piper Autocontrol
  Push to Engage
  Disengage During Take-off and Landing."
  (2) Between Directional Gyro and Gyro Horizon:
  "Turn Control
  Pull For Direction Control
  On 0° Heading Only"
  (3) On left side window channel in full view of the pilot:
  "Piper Autocontrol
  To Engage: Push turn control at D. G. in and center knobs then push in engaging control, rocking heel if necessary.
  To Turn: Move turn control in desired direction.
  For Heading Lock: Set D. G. at 0° pull put turn control knob, use trim knob to maintain exact 0° heading."

(i) On the instrument panel in full view of the pilot (For Model PA-22-108 only):
"This airplane must be operated as a normal or utility category airplane in compliance with approved Airplane Flight Manual. All markings and placards on this airplane apply to its operation as a normal category airplane. For utility category operation, refer to the Airplane Flight Manual. No acrobatics maneuvers (including spins) are approved for normal category operation."

(j) On the instrument panel in full view of the pilot (When Item 108 is installed):
"Right tank level flight only."

(k) On right fuel quantity gauge (Serial Nos. 22-1 to 22-7642)
"No take-off on right tank with less than 1/3 tank."

NOTE 3. Serial Nos. 22-3218, 22-3387 and up, of Model PA-22-150 or PA-22-160, are eligible to be operated as a Normal or Utility Category Airplane in compliance with the approved Airplane Flight Manual provided Item 407 (Control modification kit) is installed. Propeller Item 6 is not eligible when Item 407 is installed.

NOTE 4. Serial Nos. 22-1 through 22-7999 of Models PA-22, PA-22-135, PA-22-150, and PA-22-160, are eligible to be operated in the Normal Category with the rear door removed in compliance with the pertinent approved Flight Manual. Item 401(aa) for the PA-22; Item 401(ab) for the PA-22-135; Item 401(ac) for the PA-22-150; or Item 401(ad) for the PA-22-160, must be in each aircraft operated in this configuration.

(a) Airspeed Limits (CAS)

V_{ne}	(never exceed)	128 mph	(111 knots)
V_{no}	(max. structural cruising)	100 mph	(87 knots)
V_p	(maneuvering)	100 mph	(87 knots)
V_{fe}	(flaps extended)	80 mph	(70 knots)

(b) When the rear door is removed the following placards must be displayed in full view of the pilot:
  (1) "Airplane maneuvers are limited to normal take-offs, climbs, banks not to exceed 30°, glides and landings at speeds not in excess of 128 mph."
  (2) "No smoking permitted."

(c) No baggage may be carried when the aircraft is flown with the rear door removed.

.....END.....

E-273
Revision 36

CONTINENTAL

O-470-A, -B, -E, -G, -H, -J, -K, -L, -M, -N, -P, -R, -S, -T, -U
O-470-B-CI, -G-CI, K-CI, L-CI, M-CI (NOTE 6)
IO-470-A, -C

September 29, 1995

TYPE CERTIFICATE DATA SHEET NO. E-273

Engines of models described herein conforming with this data sheet (which is part of type certificate No. 273) and other approved data on file with the Federal Aviation Administration, meet the minimum standards for use in certificated aircraft in accordance with pertinent aircraft data sheets and applicable portions of the Civil Air Regulations provided they are installed, operated and maintained as prescribed by the approved manufacturer's manuals and other approved instructions.

Type Certificate Holder — Teledyne Continental Motors
P.O. Box 90
Mobile, Alabama 36601

Model	O-470-A	O-470-E	O-470-J	O-470-K, -L, -R, -S	O-470-B, -M, -N
Type	6HOA	---	---	---	---
Rating, ICAO or ARDC standard atmosphere					
Max. continuous hp, rpm, at sea level pressure altitude	225-2600	225-2600	225-2550	230-2600	240-2600
Takeoff hp, 5 min., rpm, full throttle at sea level pressure altitude	225-2600	225-2600	225-2550	230-2600	240-2600
Fuel, (aviation gasoline, minimum grade)	80/87	---	---	---	91/96
Lubricating oil, ambient air	See NOTE 9	---	---	---	---
temperature: Above 40° F.	Oil Grade SAE 50	---	---	---	---
Below 40° F.	Oil Grade SAE 30	---	---	---	---
Bore and stroke, in.	5.00 x 4.00	---	---	---	---
Displacement, cu. in.	471	---	---	---	---
Compression ratio	7:1	---	---	---	---
Weight (dry), lb.	378	390	378	404 (-K, -L) 401 (-R, -S)	410
C.G. location (basic engine)					
Fwd. of rear face, engine Accessory case, in.	12.8	---	---	12.0	11.3
Below crankshaft center line, in.	0.1	---	---	0.3	0.5
Beside crankshaft center line, toward 1-3-5 side, in.	═	═	═	═	0.2

Page No.	1	2	3	4	5	6
Rev. No.	35	36	36	36	36	36

Model	O-470-A	O-470-E	O-470-J	O-470-K, -L, -R, -S	O-470-B, -M, -N
Propeller Shaft	Special integral flange 4 7/8 in. o.d. with six ½ in. bolt holes in 4 in. diameter circle	---	---	---	---
Carburetion or Fuel Injection	Marvel-Schebler MA-4-5 (TCM #535207 or 538872)	Bendix-Stromberg PSD-5C (TCM #536911)	Marvel-Schebler MA-4-5 (TCM #535207 or 538872)	Marvel-Schebler M-4-5 (TCM #539883) (-L, -K) 641139 (-S, -R)	Bendix-Stromberg PSD-5C (TCM #535503)
Ignition, dual magnetos	NOTE 13	---	---	---	---
Timing, ° BTC	26	---	20	22	24
Spark plugs	See NOTE 11	---	---	---	---
Oil sump capacity, qt.	12; 6 usable at 15° noseup and nosedown attitudes; 7 usable at 10° noseup and nosedown attitudes	---	---	---	---
NOTES	1, 2, 3, 4, 9, 10, 11	1, 2, 3, 4, 5, 9, 10, 11	1, 2, 3, 4, 5, 9, 10, 11	1, 2 ,3, 4, 5, 6, 9, 10, 11	1, 2, 3, 4, 5, 6, 7, 8, 9, 10, 11

Model	O-470-H	O-470-G, -P	IO-470-A	IO-470-C	O-470-T, -U
Type	6HOA	---	---	---	---
Rating, ICAO or ARDC standard atmosphere					
Max. continuous hp, rpm, at sea level pressure altitude	240-2600	240-2600	240-2600	250-2600	230-2400
Takeoff hp, 5 min., rpm, full throttle at sea level pressure altitude	240-2600	240-2600	240-2600	250-2600	230-2400
Fuel, (aviation gasoline, minimum grade)	91/96	---	---	---	100, 100LL or B95/130 CIS
Lubricating oil, ambient air	See NOTE 9	---	---	---	---
temperature: Above 40° F.	Oil Grade SAE 50	---	---	---	---
Below 40° F.	Oil Grade SAE 30	---	---	---	---
Bore and stroke, in.	5.00 x 4.00	---	---	---	---
Displacement, cu. in.	471	---	---	---	---
Compression ratio	8:1	---	---	---	8.6:1
Weight (dry), lb.	495	432	410	432	410 (-T) 412 (-U)
C.G. location (basic engine)					
Fwd. of rear face, engine Accessory case, in.	14.2	12.0	11.3	12.0	11.76 (U-T) 12.07 (-U)
Below crankshaft center line, in.	1.0	1.2	0.5	1.2	.88 (-T) .31(-U)
Beside crankshaft center line, toward 1-3-5 side, in.	0.2	0.5	0.2	0.5	.35 (-T), .11 (-U)
Propeller Shaft	SAE 20 Spline Extension	Special integral flange 4 7/8 in. o.d. with six ½ in. bolt holes in 4 in. diameter circle	---	---	---

Model	O-470-H	O-470-G, -P	IO-470-A	IO-470-C	O-470-T, -U
Carburetion or Fuel Injection	Berndix-Stromberg PSD-5C (TCM#535503)	Bendix-Stromberg PSH-5BO (TCM#625203)	TCM Injector Eq #5580	TCM Injector Eq. #5620 or 5827	Marvel-Schebler MA-4-5 (TCM #641860)
Ignition, dual magnetos	NOTE 13	- - -	- - -	- - -	- - -
Timing, ° BTC	24	- - -	- - -	26	24
Spark plugs	See NOTE 11	- - -	- - -	- - -	- - -
Oil sump capacity, qt.	12; 6 usable at 15° noseup and nosedown attitudes; 7 usable at 10° noseup and nosedown attitudes	12; 10 usable at 18° noseup and 14° nosedown attitudes	12; 6 usable at 15° noseup and nosedown attitudes; 7 usable at 10° noseup and nosedown attitudes	12; 9 usable at 34° noseup and 27° nosedown attitudes; 10 usable at 28° noseup and nosedown attitudes; 11 usable at 16° noseup and nosedown attitudes	12; 6 usable at 15° noseup and nosedown attitudes
NOTES	1, 2, 3, 5, 9, 10, 11	1, 2, 3, 5, 6, 9, 10, 11	1, 2, 3, 5, 9, 10, 11	1, 2, 3, 5, 9, 10, 11	1, 2, 3, 4, 5, 9, 10, 11

"- - -" indicates "same as preceding model."
"═══" indicates "does not apply."

Certification Basis — CAR 13
Type Certificate No. 273 issued December 4, 1952.

Production Basis — P.C. 508

NOTE 1. Maximum permissible temperatures:

Cylinder head
(Spark plug gasket)

All engines except	O-470-G, -N	525° F.
	O-470-G, -N	500° F.
	O-470-A, -E, -J, -N	450° F.
	O-470-B, -H, -IO-470-A	475° F.
	O-470-G, -K, -L, -P, -R, -S, -R, -U; IO-470-C	460° F.
Cylinder barrel		290° F.
Oil inlet		225° F., 240° F. (-S, -T, -U)

NOTE 2. Fuel inlet and oil pressure limits:

Model		Minimum		Maximum
-A, -J, -K, -L		0.5 p.s.i.		6.0 p.s.i
-B, -E, -G, -H, -M, -N		9.0 p.s.i.		15.0 p.s.i.
IO-470-A, O-470-B-CI, -M-CI	minus	0.75 p.s.i	plus	1.50 p.s.i.
-G-CI	minus	2.25 p.s.i	plus	10.0 p.s.i.
-K-CI, -L-CI	minus	1.0 p.s.i.	plus	12.0 p.s.i.
IO-470-C	minus	2.0 p.s.i.	plus	10.0 p.s.i.
O-470-R,-S		15.5 in.	gasoline	6.0 p.s.i.
O-470-T, -U		14.0 in.	gasoline	6.0 p.s.i.

Oil pressure limits: 2-4-6 side (normal) 30 to 60 p.s.i. (idle 10 p.s.i. min.)

NOTE 3. The following accessory drive or mounting provisions are available:

Original Accessory	**Direction of Rotation	Speed Ratio to Crankshaft	Max. Torque Continuous	(in.-lb.) Static	Maximum Overhang Moment (in.-lb.)
Governor	C	1.0:1	29	825	50
****Tachometer	CC	.5:1			25
Optional (2)					
Left & Right Hand	C	1.5:1	***100	800	40
Generator (Belt driven)	CC	2:1	100	800	100
Alternator (Gear driven)	CCW	3:1	150	800	150
*Fuel pump	C	1.0:1	25	680	60
Oil cooler	==	==	==	==	65
Starter:	CC	32:1	200	400	60

O-470-B, -B-CI engines eligible with TCM P/N 537241.
All others eligible with TCM P/N 535856, 539910, 626960, 627842, 628482, or 637847.

* Special equipment on O-470-A, -J, -K, and -L models.
** "C" indicates clockwise viewing drive pad; "CC" counter clockwise.
*** One drive eligible at 160 in.-lb. continuous torque load provided the other drive does not exceed 100 in.-lb. continuous torque load.
**** O-470-G clockwise; O-470-V and -VO optional rotation.

NOTE 4. Crankshaft damper configuration: O-470-A, S/N 41000 and up, and -E, -J, -R, -S, and -T engines are equipped with one 5th and one 6th order damper.
O-470-B, -H, and -N have two 6-½ order dampers.
O-470-K, -L, -M, -P and IO-470-A and -C have four 6th order dampers.
O-470-G has one 6-½ and one 9th order damper.
O-470-A, S/N 40001 through 40655, and -P, have two 6th order dampers.
O-470-U has two 6th, one 5th, and one 4½ order dampers.

NOTE 5. The following similarities and differences exist between the various models:
O-470-B is similar to O-470-A except for increased power rating, different damper configuration, incorporation of inclined valve cylinders, downdraft pressure carburetor and related induction system changes.
O-470-E is same as O-470-A except for incorporation of downdraft pressure carburetor and related induction system changes.
O-470-G is similar to O-470-M except for crankshaft damper configuration, revised oil sump integral cast intake air passage and mounting brackets.
O-470-J is same as O-470-A except for reduced rated speed and minor changes in induction system risers, manifold and balance tube.
O-470-K is similar to O-470-J except for ratings, crankshaft damper configuration and incorporation of shell-molded cylinder heads and revised mounting brackets.
O-470-L is same as O-470-K except for relocated carburetor and revised intake manifold oil sump.
O-470-M is same as O-470-B except for crankshaft damper configuration and incorporation of shell-molded cylinder heads.
O-470-N is same as O-470-M except for crankshaft damper configuration.
O-470-P is identical to O-470-G except for crankshaft damper configuration.
IO-470-A is same as O-470-M except incorporates CMC continuous flow fuel injection system instead of Bendix carburetor.
IO-470-C is same as O-470-G except for crankshaft damper configuration and incorporation of CMC continuous flow fuel injection system instead of Bendix carburetor.
O-470-H is same as O-470-B except incorporates extension propeller shaft and is approved for pusher operation.
O-470-R is same as O-470-L except for crankshaft damper configuration.
O-470-S is same as O-470-R except for piston oil cooling and semi-keystone piston rings.
O-470-T is similar to the O-470-S except for crankcase design and rating.
O-470-U is similar to the O-470-S except for rating and crankshaft damper configuration.

NOTE 6. O-470-B, -G, -K, -L, and -M engines are eligible for incorporation of TCM continuous flow fuel injection system (Eq. No. 5580 for -B, -M; Eq. No. 5701 or 5702 for -G; Eq. No. 5613 for -K, -L) replacing carburetion system with no change in weight. When this modification is accomplished the engines will be designated as O-470-B-CI, O-470-G-CI, O-470-K-CI, O-470-L-CI and O-470-M-CI and the nameplate changed accordingly.

NOTE 7. O-470-B engine mounting brackets are eligible for use with O-470-M engines.

NOTE 8. O-470-M engines with S/N's suffixed with the letter "P" are approved for pusher type installation.

NOTE 9. Straight mineral or ashless disperant oil meeting TCM Spec. MHS #24 is approved for use in engines, except the O-470-S, -T, and -U which must use ashless disperant oil conforming to MHS-24. TCM instructions should be followed when changing types of oil.

NOTE 10. A full flow oil filter may be used with these engines if the installation incorporates a filter bypass valve which opens between 12 and 16 p.s.i. Oil sump housing is eligible for direct mounting of oil filter having a maximum weight of 6 lb. and overhang moment of 25 in.-lb.

NOTE 11. The following spark plugs are approved on these engines:

Models O-470-A, -E, -J, -K, -L, -R, -S

AC	HSR83IR, SR83IR, HSR83P, SR83P, HSR87, SR87, A88, S88, HSR88, HS88, SR88, S88D, SR88D
Auto Lite	SH2M, SH15, SH15R, SH20, SH20A, SH200A, SH150
BG	RB485S, 706S, RB919SR, 919SR5, RB955S
Champion	RC26S, C27S, REM38P, RHM38P, RED39N, RHD39N, REM39N, RHM39N, REM40E, RHM40E, D41N, ED41N, EM41N, EM42E
Red Seal	SE190, SE230, SJ190, SJ230

Models O-470-B, -G, -H, -M, -N, -P; IO-470-A

AC	SR83IR, HSR83IR, HSR83P, SR83P, S86R, SR86, HSR86, SR87, HSR87
Auto Lite	SH20A, SH200A, SH26, SH260, PH26, PH260
BG	RB485S, RB955S
Champion	RC26S, REM38E, REM38P, RHM38E, RHM38P, RED39N, REM39N, RHD39N, RHM39N, REM40E, RHM40E
Red Seal	SE230, SJ230, SE270, SJ270

Model IO-470-C

AC	SR83IR, HSR83IR, HSR83P, SR83P, SR86, HSR86, S86R, HSR87, SR87
Auto Lite	SH26, SH260, PH26, PH260
Champion	R25S, RC26CS, RED37N, REM37N, RHD37N, REM38E, REM38P, RHM38E, RHM39P, RED39N, RHD39N, RHM39N, REM40E, RHM40E, RHM37N, REM39N
Red Seal	SE270, SJ270

Model O-470-T, -U

AC	SR86L, HSR86L, HSR87LIR HSR87LP, 171, 181, 271, 273, 281, 281IR, 283, 283IR
Auto Lite	SL350
Champion	RHA32N, RHB32N, RHB32E, RHB33E, RHB36P, RHB37E, REA37N, REB37N, RHA37N, RHB37N, RHB38E, R115
Red Seal	LE310, LJ8310

NOTE 12. Teledyne Crittenden Alternator P/N 642056 and Drive Coupling P/N 642362 eligible for use with Model O-470-T engine. Alternator compatibility with aircraft must be accomplished by installer.

E-273
Page 6

NOTE 13.	The following magnetos equipped with an appropriate harness are eligible on these engines at the Indicated Weight Changes:	
	Two TCM/Bendix S6RN-25	None
	One Ea. TCM/Bendix S6RN-201 & S6RN-205	-2 lb.
	Two Bendix Scintilla 1225	-1 lb.
	Two TCM S6RSC-25	None
	One Ea. TCM S6RSC-201(L) & S6RSC-205(R)	None
	Two Slick Electro 662	None
	Two Slick Electro 680	None
	Two Slick Electro 6210	-5 lb.
	Two Slick model 6310	-5 lb.

.....END.....

P57GL
REVISION 9
McCAULEY
3AF32C(5--)
3AF34C(5--)
3AF36C(5--)
3AF37C(5--)
August 16, 1996

TYPE CERTIFICATE DATA SHEET NO. P57GL

Propellers of models described herein conforming with this data sheet, which is part of Type Certificate No. P57GL and other approved data on file with the Federal Aviation Administration, meet the minimum standards for use in certificated aircraft in accordance with the pertinent aircraft data sheets and applicable portions of the Federal Aviation Regulations provided they are installed, operated, and maintained as prescribed by the approved manufacturer's manuals and other approved instructions.

Type Certificate Holder:	McCauley Propeller Systems 3535 McCauley Drive Vandalia, Ohio 45377
Type	Constant speed; hydraulic (see Notes 3 and 4)
Engine Shaft	Special flange 4.00 inch B.C.
Hub Material	Aluminum Alloy
Blade Material	Aluminum Alloy
No. of Blades	Three
Hubs Eligible	3AF34C502, 3AF34C503, 3AF32C504, 3AF32C505, 3AF32C506, 3AF32C507, 3AF32C508, 3AF32C509, 3AF37C510, 3AF32C511, 3AF32C512, 3AF36C514, 3AF32C515, 3AF37C516, 3AF32C521, 3AF32C522 and 3AF32C523

Blades (See Note 2)	Maximum Continuous HP	Maximum Continuous RPM	Take-Off HP	Take-Off RPM	Diameter Limits (See Note 2)	Approx. Max. Weight Complete (For Ref. Only)
			Hub Model 3AF34C502			
80H[X]-0 to 80H[X]-8	215	2575	215	2575	80" - 72" (-0 to -8)	76.0 Lbs.
			Hub Model 3AF34C503			
L80H[X]-0 to L80H[X]-8	215	2575	215	2575	80" - 72" (-0 to -8)	76.0 Lbs.
			Hub Models 3AF32C504, 3AF32C505, 3AF32C511, and 3AF32C512			
82NE[X]-2 to 82NE[X]-8	325	2700	325	2700	80" - 74" (-2 to -8)	70.0 Lbs. 75.8 Lbs. *

Page No.	1	2	3	4	5
Rev. No.	9	9	9	8	9

Blades (See Note 2)	Maximum Continuous HP	Maximum Continuous RPM	Take-Off HP	Take-Off RPM	Diameter Limits (See Note 2)	Approx. Max. Weight Complete (For Ref. Only)
			Hub Model 3AF32C506			
82NE[X]-2 to 82NE[X]-10	250	2400	250	2400	80" - 72" (-2 to -10)	71.5 Lbs.
			Hub Model 3AF32C507			
L82NE[X]-2 to L82NE[X]-10	250	2400	250	2400	80" - 72" (-2 to -10)	71.5 Lbs.
			Hub Model 3AF32C508			
82NF[X]-2 to 82NF[X]-8	220	2800	220	2800	80" - 74" (-2 to -8)	69.5 Lbs.
			Hub Model 3AF32C509			
L82NF[X]-2 to L82NF[X]-8	220	2800	220	2800	80" - 74" (-2 to -8)	69.5 Lbs.
			Hub Model 3AF37C510			
90LF[X]-0 to 90LF[X]-10	375	2400	375	2400	90" - 80" (-0 to -10)	86.9 Lbs.
			Hub Model 3AF36C514			
80VMF[X]-0 to 80VMF[X]-6	350	2700	350	2700	80" - 74" (-0 to -6)	75.8 Lbs.
			Hub Model 3AF32C515			
82NL[X]-2 to 82NL[X]-8	350	2700	350	2700	80" - 74" (-2 to -8)	74.0 Lbs.
			Hub Model 3AF37C516			
90LF[X]-0 to 90LF[X]-6	375	2275	375	2275	90" - 84" (-0 to -6)	86.9 Lbs.
			Hub Model 3AF32C521			
82NL[X]-4 to 82NL[X]-10	350	2700	350	2700	78" - 72" (-4 to -10)	80.5 Lbs.
			Hub Model 3AF32C522			
82NJ[X]-2 to 82NJ[X]-8	220	2800	220	2800	80" - 74" (-2 to -8)	69.5 Lbs.
			Hub Model 3AF32C523			
L82NJ[X]-2 to L82NJ[X]-8	220	2800	220	2800	80" - 74" (-2 to -8)	69.5 Lbs.

* Higher Weight applies to -C511 model only.

Certification Basis	Type Certificate No. P57GL issued July 17, 1978, under Delegation Option Authorization Provisions of Part 21, Subpart J, of the Federal Aviation Regulations. Date of application for Type Certificate, July 12, 1978. Models 3AF34C502, 3AF34C503, 3AF32C504, 3AF32C505, 3AF32C508, 3AF32C509: Federal Aviation Regulations Part 35 including Amendments 35-1 through 35-4 (May 2, 1977) thereto. Models 3AF32C506, 3AF32C507, 3AF32C511, 3AF32C512, 3AF36C514: Federal Aviation Regulations Part 35 including Amendments 35-1 through 35-5 (October 14, 1980) thereto. Models 3AF37C510, 3AF32C515, 3AF37C516, 3AF32C521, 3AF32C522, 3AF32C523: Federal Aviation Regulations Part 35 including Amendments 35-1 through 35-6 (August 18, 1990) thereto.
Production Basis	Production Certificate No. 3

NOTE 1. Hub Model Designation.

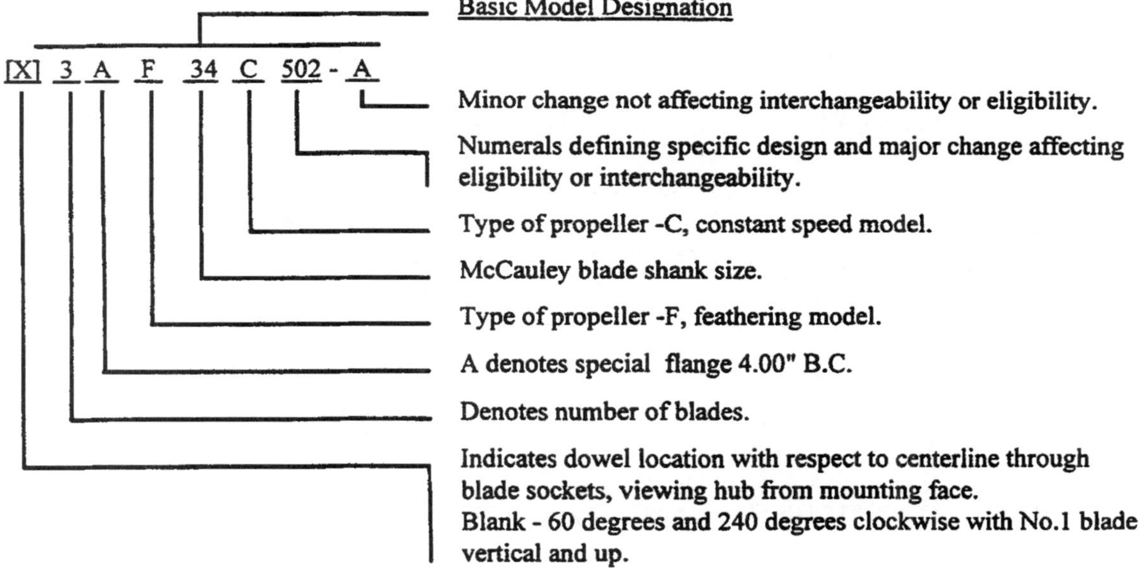

NOTE 2. Blade Model Designation.

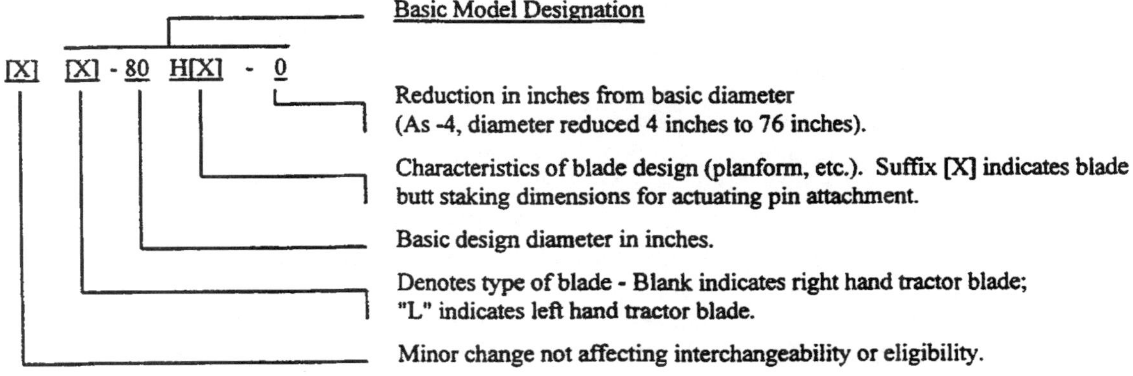

P57GL

NOTE 3. Pitch Control. With the following governors:

McCauley Model DCF290D[X]/T[X]	Wt. 3.0 lbs.
McCauley Model DCFU290D[X]/T[X]	Wt. 3.0 lbs.
McCauley Model DCFS290D[X]/T[X]	Wt. 3.0 lbs.
McCauley Model DCFUS290D[X]/T[X]	Wt. 3.0 lbs.
Hartzell Model E-[X]-[X]	Wt. 4.5 lbs.
Hartzell Model E-[X]-[X]L	Wt. 4.5 lbs.
Hartzell Model U-[X]-[X]	Wt. 4.5 lbs.
Hartzell Model U-[X]-[X]L	Wt. 4.5 lbs.
Woodward Model [X]2106[X][X]	Wt. 3.5 lbs.

NOTE 4. Feathering. With full feathering control installed in accordance with the propeller manufacturer's instructions. Controls may include unfeathering, synchronizing or synchrophasing features.

NOTE 5. Not applicable.

NOTE 6. Not applicable.

NOTE 7. Accessories

a. Propeller Anti-icing/Deicing

(1) Model 80HA, L80HA, 82NFA, and L82NFA blades per Goodrich installation drawing 7E1391.

(2) Model -C504/82NEA and -C505/82NEA blades per McCauley assembly drawing E-5186.

(3) Model -C511/82NEA and -C512/82NEA blades per McCauley assembly drawing E5358.

(4) Model 82NEB or L82NEB blades per McCauley assembly drawing E-5203.

(5) Model 80VMF blades per McCauley assembly drawing E-6312, and deice installation drawing D-40486.

(6) Model 3AF32C515/82NLA per McCauley assembly drawing E-5186 and deice installation drawing C-40219.

(7) Model 3AF37C516/90LFB per McCauley assembly drawing E-7110.

(8) Model 3AF37C510/90LFB per McCauley assembly drawing E-7272.

b. Propeller Spinners

(1) Model 3AF34C502/80HA or 3AF34C503/L80HA with plain or electric deice spinner; reference D-4986 Dome, D-4984 Bulkhead and D-4987 Installation.

(2) Model 3AF32C504/82NEA or 3AF32C505/NEA with plain or electric deice spinner; reference D-3651 Dome, D-3925 Bulkhead and D-4042 Installation.

(3) Model 3AF32C506/82NEB or 3AF32C507/L82NEB with plain or electric deice spinner; reference D-5285 Dome, D-5274 Bulkhead and D-5275 Installation.

(4) Model 3AF32C508/82NFA or 3AF32C509/L82NFA with plain or electric deice spinner; reference D-4986 Dome, D-4984 Bulkhead and D-4987 Installation.

(5) Model 3AF32C511/82NEA with plain or electric deice spinner; reference D5370 Dome, D5371-2 Bulkhead and D-5311 Installation.

(6) Model 3AF32C512/82NEA with plain or electric deice or liquid anti-ice spinner; reference D-5370 Dome, D-5499-1 and -3 Bulkhead and D-5309 and D-5310 Installation.

(7) Model 3AF36C514/80VMFA with plain or electric deicing spinner; reference E-6190 Dome, E-6178 Bulkhead and D-6176 Installation.

(8) Model 3AF32C515/82NLA with electric deice spinner; reference D-5215 Installation.

(9) Model 3AF37C516/90LFB per assembly drawing E-7110.

(10) Model 3AF37C510/90LFB per assembly drawing E-7272.

(11) Model 3AF32C522/82NJA per assembly drawing E-7315.

(12) Model 3AF32C523/L82NJA per assembly drawing E-7316.

NOTE 8. Not applicable.

NOTE 9. Not applicable.

NOTE 10. Special Notes. Aircraft installation must be approved as part of the aircraft type certificate upon compliance with the applicable aircraft airworthiness requirements.

... END ...

DEPARTMENT OF TRANSPORTATION

FEDERAL AVIATION ADMINISTRATION

P-920
Revision 21
HARTZELL
HC-C2Y
BHC-C2Y
CHC-C2Y
DHC-C2Y
April 25, 1996

TYPE CERTIFICATE DATA SHEET NO. P920

Propellers of models described herein conforming with this data sheet, (which is part of Type Certificate No. P-920) and other approved data on file with the Federal Aviation Administration, meet the minimum standards for use in certificated aircraft in accordance with pertinent aircraft data sheets and applicable portions of the Federal Aviation Regulations provided they are installed, operated, and maintained as prescribed by the approved manufacturer's manuals and other approved instructions.

Type Certificate Holder	Hartzell Propeller, Inc. Piqua, Ohio 45356
Type	Constant speed; hydraulic (See NOTES 3 and 4)
Engine shaft	SAE #2 flange, special flange 4" B.C.
Hub material	Aluminum alloy
Blade material	Aluminum alloy
Number of blades	Two
Hub models	HC-C2YF-1, -2, -4; BHC-C2YF-1, -2, -4; CHC-C2YF-1, -2, -4; DHC-C2YF-1, -2, -4; HC-C2YK-1, -2, -4; CHC-C2YK-1, -2, -4; HC-C2YL-1, -2, -4; HC-C2YR-1, -2, -4; CHC-C2YR-1, -2, -4 (See NOTES 1 and 4)

Blades (See NOTES 2 & 6)	Maximum Continuous HP	Maximum Continuous RPM	Takeoff HP	Takeoff RPM	Diameter Limits	*Hub and Blades Approx. Wt. (See NOTES 3 and 7)
			Non-Counterweighted Blades - Hub Models: All -1			
7068-0 to 7068-10	300	2700	300	2700	70" - 60" (-0 to -10)	55.5 lb
7280+1/2 to 7280-7	250	2700	250	2700	72-1/2" - 65" (+1/2 to -7)	51 lb.
7663-0 to 7663-8	210	2800	210	2800	76" - 68" (-0 to -8)	46 lb.
7666-0 to 7666-8	180 or 250	2900 or 2700	180 or 250	2900 or 2700	76" - 68" (-0 to -8)	51 lb.
7681-0 to 7681-8	250	2700	250	2700	76" - 68" (-0 to -8)	51 lb.
7692-0 to 7692-8	180 or 250	2900 or 2700	180 or 250	2900 or 2700	76" - 68" (-0 to -8)	46 lb.
8052-0 to 8052-8	310	2600	310	2600	80" - 72" (-0 to -8)	50.5 lb.
8459-0 to 8459-18	260	2800	260	2800	84" - 66" (-0 to -18)	48 lb.

Page No.	1	2	3	4	5	6	7	8	9	10
Rev. No.	21	21	20	20	21	20	20	20	21	-

P-920

Blades (See NOTES 2 & 6)	Maximum Continuous HP	Maximum Continuous RPM	Takeoff HP	Takeoff RPM	Diameter Limits	*Hub and Blades Approx. Weight (See NOTES 3 and 7)
8465-0 to 8465-14	315	2575	2575	315	84" - 70" (-0 to -14)	50 lb.
8467-0 to 8467-12	285	2700	285	2700	84" - 72" (-0 to -12)	52 lb.
8468-0 to 8468-12	260	2700	260	2700	84" - 72" (-0 to -12)	50 lb.
8470-0 to 8470-8	260	2700	250	2700	84" - 76" (-0 to -8)	49 lb.
8475+2 to 8475-4	310	2700	310	2700	86" - 80" (+2 to -4)	52 lb.
8475-4 to 8475-6	350	2700	350	2700	80" - 78" (-4 to -6)	51 lb.
8475-6 to 8475-14	310	2700	310 or 300	2700 or 2850	78" - 70" (-6 to -14)	50 lb.
8477-0 to 8477-4	310 or 260	2575 or 2700	310 or 260	2575 or 2700	84" - 80" (-6 to -4)	54 lb.
8477-4 to 8477-6	350	2700	350	2700	80" - 78" (-4 to -6)	53 lb.
8477-6 to 8477-14	310	2700	310 or 300	2700 or 2850	78" - 70" (-6 to -14)	52 lb.
9587-0 to 9587-2	320	2200	320	2200	95" - 93" (-0 to -2)	50 lb.
9587-2 to 9587-20	320 or 300	2200 or 2400	320 or 300	2200 or 2400	93" - 75" (-2 to -20)	50 lb.
			<u>Counterweighted Blades - Hub Models: All -2 and -4</u>			
C7663-0 to C7663-8	210	2800	210	2800	76" - 68" (-0 to -8)	50 lb.
C7666-0 to C7666-8	180 or 250	2850 or 2700	180 or 250	2850 or 2700	76" - 68" (-0 to -8)	55 lb.
C7681-0 to C7681-8	250	2700	250	2700	76" - 68" (-0 to -8)	55 lb.
C7692-0 to C7692-8	180 or 250	2900 or 2700	180 or 250	2900 or 2700	76" - 68" (-0 to -8)	50 lb.

Blades (See NOTES 2 & 6)	Maximum Continuous HP	Maximum Continuous RPM	Takeoff HP	Takeoff RPM	Diameter Limits	*Hub and Blades Approx. Weight (See NOTES 3 and 7)
C8052-0 to C8052-8	310	2600	310	2600	80" - 72" (-0 to -8)	54.4 lb.
C8459-0 to C3459-12	260	2800	260	2800	84" - 72" (-0 to -12)	52 lb.
C8465-0 to C8465-14	315	2575	315	2575	84" - 70" (-0 to -14)	54 lb.
C8465-6 to C8465-14	260	2700	250	2700	78" - 70" (-6 to -14)	53 lb.
C8467-0 to C8467-12	285	2700	285	2700	84" - 72" (-0 to -12)	56 lb.
C8468-0 to C8468-12	260	2700	260	2700	84" - 72" (-0 to -12)	54 lb.
C8470-0 to C8470-8	260	2700	260	2700	84" - 76" (-0 to -8)	53 lb.
C8475+2 to C8475-4	310	2700	310	2700	86" - 80" (+2 to -4)	56 lb.
C8475-4 to C8475-6	350	2700	350	2700	80" - 78" (-4 to -6)	55 lb.
C8475-6 to C8475-14	310	2700	310 or 300	2700 or 2850	78" - 70" (-6 to -14)	54 lb.
C8477-0 to C8477-4	310 or 260	2575 or 2700	310 or 260	2575 or 2700	84- 80" (-0 to -4)	58 lb.
C8477-4 to C8477-6	350	2700	350	2700	80" - 78" (-4 to -6)	57 lb.
C8477-6 to C8477-14	310	2700	310 or 300	2700 or 2850	78" - 70" (-6 to -14)	56 lb.
C9587-0 to C9587-2	320	2200	320	2200	95" - 93" (-0 to -2)	54 lb.
C9587-2 to C9587-20	320 or 300	2200 or 2400	320 or 300	2200 or 2400	93" - 75" (-2 to -20)	54 lb.

*Weights apply to -1 constant speed hub with "F' flange. Add 1.2 lb. for "L", "K", and "R": flange, 3.0 lb. for feathering -2 hubs, and 5.5 lb. for feathering -2R hubs. Add 4 lb. for -4 model.

P-920

Certification basis

Civil Air Regulations Part 14 effective December 25, 1956
Type Certificate No. P-920 issued July 24, 1961
Date of Application for Type Certificate March 24, 1959
Models listed are approved under Delegation Option
Authorization Provisions of FAR 21 Subpart J with corresponding approval dates indicated below.
HC-C2Y( )-1/7681 & HC-C2Y( )-2/C7681 approved May 23, 1967
HC-C2YF-4 & HC-C2YK-4 approved June 16, 1970
HC-C2YL-4 approved February 17, 1971

Production basis

Production Certificate No. 10

NOTE 1. Hub Model Designation.

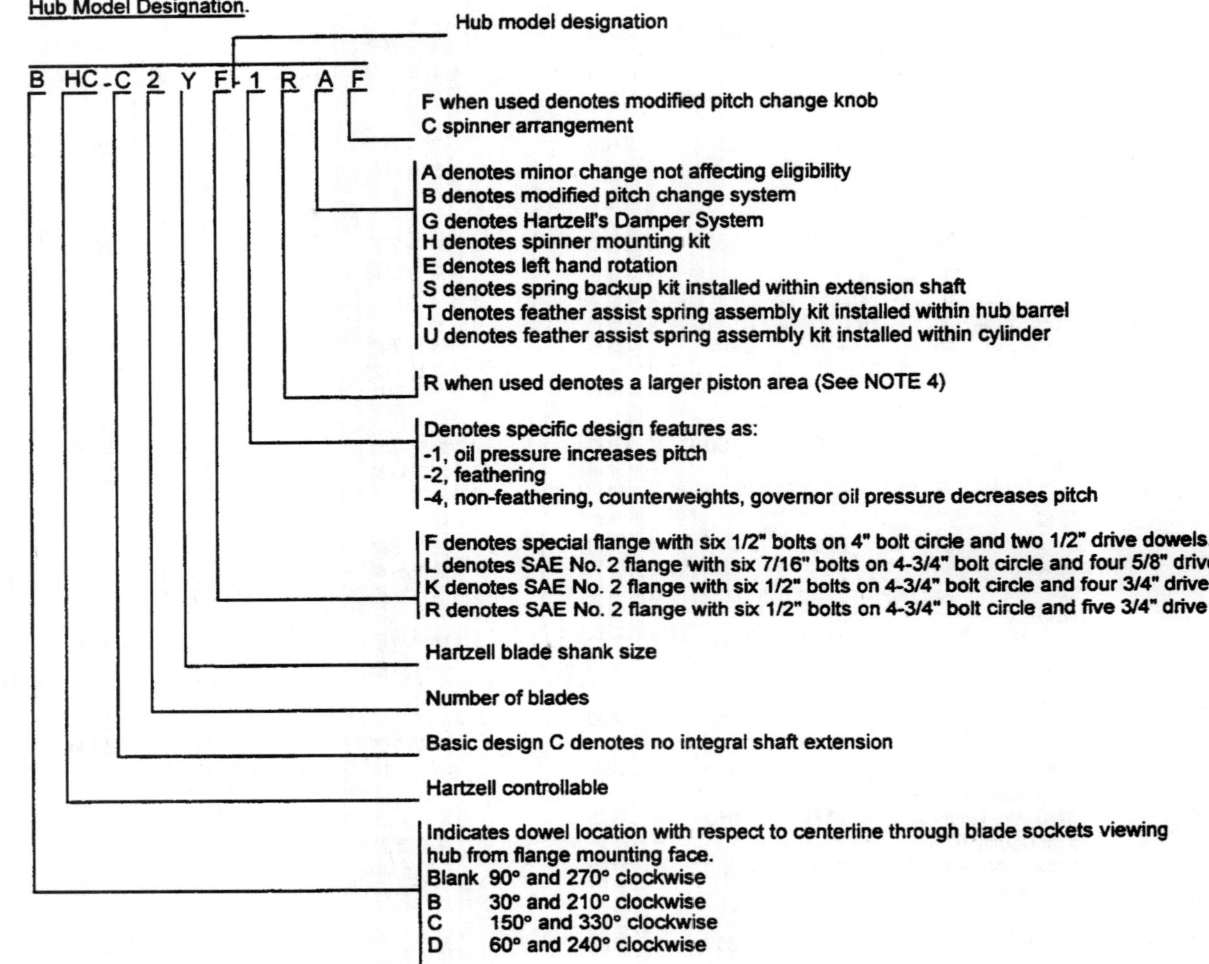

NOTE 2. Blade Model Designation.

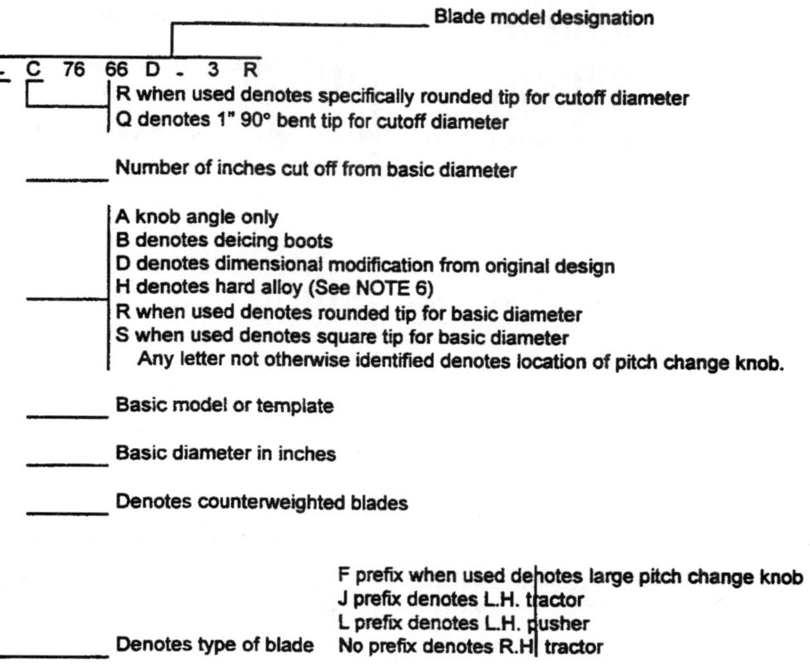

NOTE 3: Pitch Control. (See Note 10).
Approved with Hartzell governors per drawing list C-4770 or C-4772.
Governor model designation sample:

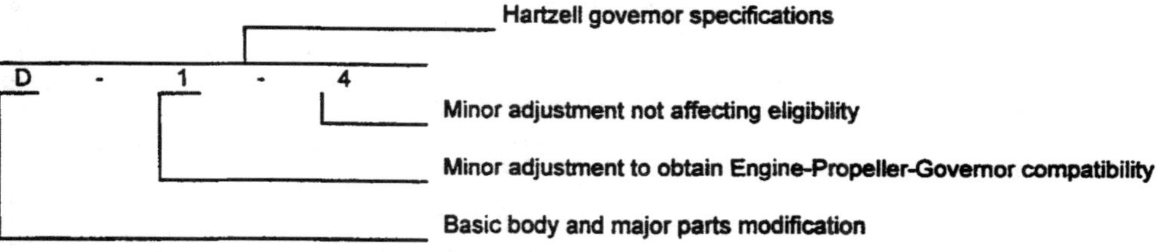

Approved with the following governors:
*Woodward Model X210XXX or X210X-XXX
*Edo-Aire Model 34-828-xxx Wt. 3.0 lb.
**McCauley Model C290D3-X/TXX Wt. 2.8 lb.
* Not approved with air feathering propellers without counterweights.
**Not approved with counterweights or feathering propeller.

NOTE 4. (a) Feathering. The -1 and -4 models do not feather. The -2 models incorporate feathering and unfeathering features.
(b) Reversing. Not applicable.
(c) Piston size. The -2R model differs from the -2 model in that the -2R has a piston area of 20.2 sq. in. and the -2 has a piston area of 16.25 sq. in.

NOTE 5. Left-Hand Models. The left-hand version of a model propeller is approved at the same rating and diameter as listed for right-hand model. See NOTES 1 and 2.

NOTE 6. Interchangeability.
(a) Blades.
(1) Blades with counterweights (having "C" prefix) can replace non-counterweighted blades on feathering propellers (Hub Model Suffix -2 or -2R) only, providing the air charge is reduced to 80 p.s.i. @ 70° F. Attached decal specifying air charge must be changed accordingly.

(2) Hard and soft alloy blades of the same model designation are interchangeable but only on seaplanes and amphibious aircraft.

(b) Propellers.
(1) "F" type propellers with larger pitch change knobs are interchangeable with corresponding propellers with standard pitch change system (See NOTES 1 and 2 above).

NOTE 7. Accessories. (See NOTE 10)

(a) Propeller Anti-Icing

(1) Fluid feed shoes or Icex boots installed in accordance with Hartzell Special Instructions No. 59A.

(2) Hartzell fluid feed equipment on propeller models for which the equipment is available.

(b) Propeller Deicing.

(1) Goodyear Ice Guards (electrical propeller deicer) when installed in accordance with instructions outlined in Goodyear Report No. AP-147 dated October 23, 1961.

(2) Goodrich De-icing Kit 77-xxx, 67-xxx, or 65-xxx when installed in accordance with Goodrich Report No. 59-728 ( ).

(c) Propeller Spinner

(1) Approved with Hartzell spinners (weight of spinner extra).

(d) Propeller Damper C-1576

(1) Approved for Hartzell Propeller Model HC-C2Y.

NOTE 8. Shank Fairings. Not applicable.

NOTE 9. Special Limits.

Table of Propeller-Engine Combinations
Approved Vibrationwise for Use on Normal Category Single-Reciprocating Engine Tractor Aircraft

The maximum and minimum propeller diameters that can be used from a vibrationwise standpoint are shown below. No reduction below the minimum diameter listed is permissible, since this figure includes the diameter reduction allowable for repair purposes.

Hub Model	Blade Model	Engine Model	Max. Dia. (Inches)	Min. Dia. (Inches)	Placards
HC-C2YR	F7068-( )	Lycoming IO-360-B1A, -B1B, -B1C, -B1D, -B1E, -B1F, -E1A, -F1A O-360-A1A, -A1AD, -A1C, -A1D, -A1F, -A1G, -A1H, -A1LD	68	67	"Stabilized operation is prohibited above 25" manifold pressure between 2300-2350 RPM and below 15" manifold pressure above 2600 RPM"
BHC-C2YF	7663	Continental O-300-E	72	70	None
HC-C2YK	7663	Continental IO-346-B	76	76	None
BHC-C2YF	7663	Continental IO-346-A, -B, -C, -D, -E	76	72	None
BHC-C2YF	F7663( )	Continental IO-360-H, -HB	76	72	None
HC-C2YL	7663	Lycoming O-290-D2A	72	70	None
HC-C2YL	7663	Lycoming O-320-A3A, -A3B, -A3C, -B3A, -B3B, -B3C, -C3A, -C3B, -C3C, -D1A, -D1B, -D1C, -E1A, -E1B, -E1C, -E1F	72	70	None
HC-C2YL	7663	Lycoming IO-320-A1A, -A2A, -B1A, -B1B, -B1C, -B1D, -B2A, -C1A, -D1A, -D1B, -E1A	72	70	None
HC-C2YL	7663	Lycoming IO-320-E1A	72	70	None
HC-C2YK HC-C2YR	7666 F7666	Lycoming O-360-A1A, -A1AD, -A1C, -A1D, -A1F, -A1G, -A1LD, -B1A, -B1B, -C1A, -C1C, -C1F, -D1A	76	72	"Avoid continuous operation between 2000 and 2250 r.p.m."
HC-C2Y(KR)	F7666A-2Q	Lycoming O-360-A1A, -A1C, -A1D, -A1F, -A1G, -B1A, -B1B, -C1A, -C1C, -C1F, -D1A	74	74	"Avoid continuous operation between 2000 and 2250 r.p.m."
HC-C2YK CHC-C2YK	7666 C7666	Lycoming IO-360-A1A, -A1B, -A1C, -B1A, -B1C, -C1A, -C1B, -C1C, -D1A	74	72	"Avoid continuous operation between 2000 and 2350 r.p.m."

Hub Model	Blade Model	Engine Model	Max. Dia. (Inches)	Min. Dia. (Inches)	Placards
HC-C2YR HC-C2YK	F7666( )-3Q	Lycoming IO-360-A3B6D	73	73	None
HC-C2YK HC-C2YR	FC7666 F7666A FJC7666	Lycoming O-360-E1A6D Lycoming LO-360-E1A6D	74	72	None
HC-C2YK HC-C2YR	F7666A-2	Lycoming O-360-A1F6D	74	73	None
HC-C2YR	FJC7666A-( )R FC7666A-( )R	Lycoming LTO-360-E1A6D TO-360-E1A6D	74	72	None
( )HC-C2YK ( )HC-C2YR	( ) 7666 ( ) -4Q	Lycoming IO-360-B1A, -B1B, -B1D, -B1E, -B1F, -E1A, -F1A	72	72	"Avoid continuous operation between 2000 and 2250 r.p.m."
HC-C2YK HC-C2YR	F( )7666A -4Q	Lycoming O-360-A1A, -A1C, -A1D, -A1F, -A1G, -B1A, -B1B, -C1A, -C1C, -D1A	72	72	"Avoid continuous operation between 2000 and 2250 r.p.m."
HC-C2YK	F7666A -4Q	Lycoming IO-360-A1B6	72	72	None
HC-C2YK HC-C2YR	7666 -4Q	Lycoming IO-360-A1A, -A1B, -A1C, -C1A, -C1B, -C1C, -D1A	72	72	"Avoid continuous operation between 2000 and 2250 r.p.m."
HC-C2YK	7666	Lycoming IO-360-B1A, -B1B, -B1C, -B1D, -B1E, -B1F, -E1A, -F1A	74	72	"Avoid continuous operation between 2000 and 2250 r.p.m."
HC-C2YK	7666	Lycoming IO-360-B1A, -B1B, -B1D, -B1E, -B1F, -E1A, -F1A	76	74 1/2	"Avoid continuous operation between 2000 and 2250 r.p.m."
HC-C2YK HC-C2YR	7666 C7666	Lycoming IO-360-A1B6, -A1D6, -C1E6, -C1C6	76	76	"None - which contain inertia damper Hartzell P/N C1576"
HC-C2YK HC-C2YR	7666 C7666	Lycoming IO-360-A1B6, -A1D6, -C1E6, -C1C6	76	76	"Avoid continuous operation between 2000 and 2400 r.p.m."
HC-C2YK HC-C2YR	7666	Lycoming O-360-F1A6	74	72	None
HC-C2YK HC-C2YR	( )7666	Lycoming IO-360-A1B6D	74	72	None
HC-C2YK HC-C2YR	JC7666	Lycoming IO-360-C1E6	76	76	"None - which contain inertia damper Hartzell P/N C1576"
HC-C2YK HC-C2YR	JC7666	Lycoming LIO-360-C1E6	76	76	"Avoid continuous operation between 2200 and 2400 r.p.m."
HC-C2YK	7666	Lycoming AIO-360-A1A, -A1B, -B1B	74	72	"Avoid continuous operation between 2000 and 2350 r.p.m."
HC-C2YK HC-C2YR	FC7666A F7666A	Lycoming TO-360-C1A6D	76	75	"Do not operate about 36 inches of manifold pressure at engine speeds below 2400 r.p.m."

P-920

Hub Model	Blade Model	Engine Model	Max. Dia. (Inches)	Min. Dia. (Inches)	Placards
HC-C2YR	7666	Lycoming AEIO-360-H1A	74	72	"Avoid continuous operation between 2000 and 2350 r.p.m."
HC-C2YK HC-C2YR	F7666	Lycoming IO-360-A1B6, -A1D6, -C1E6, -C1C6	74	72	None
BHC-C2YF	8052	Continental TSIO-520-BE	80	78	None
BHC-C2YF	FC8459 FJC8459 F8459-( )R	Continental TSIO-360-E, -EB, -KB Continental LTSIO-360-E, -EB, -KB Continental TSIO-360-F, -FB, -G	76	75	"Avoid continuous operation between 2000 and 2200 r.p.m. with engine manifold pressure above 32 in. Hg. Avoid continuous ground operation in cross and tail winds of over 10 knots between 1700 and 2100 r.p.m.
BHC-C2YF	F8459( )-( )R	Continental IO-360-ES	76	75	"Avoid continuous ground operation between 1700 and 2100 RPM in cross and tail winds of over 10 knots."
HC-C2YF	8459 C8459	Franklin 6A-350-C1, -C2	80	76	None
HC-C2YL	8459	Lycoming O-320-A3A, -A3B, -A3C, -B3A, -B3B, -B3C, -C3A, -C3B, -C3C, -D1A, -D1B, -D1C, -E1A, -E1B, -E1C, -E1F	66	66	None
HC-C2YL	8459	Lycoming IO-320-A1A, -A2A, -B1A, -B2A, -B1B, -B1C, -B1D, -C1A, -E2A	66	64	None
BHC-C2YK CHC-C2YK DHC-C2YK	8465	Continental IO-470-L	78	76	None
HC-C2YK HC-C2YR	8467	Lycoming IO-540-D4A5	77	75	"Avoid continuous operation between 2500 and 2600 r.p.m. above 25" Hg. manifold pressure."
HC-C2YK HC-C2YR	F8467	Lycoming IO-540-R1A5 with Ray Jay turbosuper charger (up to 29" Hg manifold pressure absolute)	77	75	None
HC-C2YK HC-C2YR	8467-( )R	Lycoming O-540-B4A5, -B4B5, -E4A5, -E4B5, -E4C5	77	75	"Avoid continuous operation between 2500 and 2600 r.p.m. above 25" Hg. manifold pressure."
HC-C2YK HC-C2YR	8467-( )R	Lycoming IO-540-T4A5D	77	75	None
HC-C2YF BHC-C2YF	8468	Continental O-470-R	84	80	None
HC-C2YF BHC-C2YF	8468	Continental IO-470-E	84	83	"Avoid continuous operation between 2100 and 2225 r.p.m."
HC-C2YF	8468	Continental IO-470-D, -E, -F, -G, -H, -M, -N, -R, -S	82	80	None

Hub Model	Blade Model	Engine Model	Max. Dia. (Inches)	Min. Dia. (Inches)	Placards
HC-C2YF	8468 C8468	Continental IO-470-D, -E, -F, -G, -H, -M, -N, -R, -S	78	78	"Do not exceed 23" Hg. manifold pressure below 2300 r.p.m."
HC-C2YF	8468R	Continental IO-520-BA	84	84	None
BHC-C2YF	F8468R F8468AR	Continental IO-520-BB	84	84	None
HC-C2YL	8468	Lycoming O-320-A3A,-A3B, -A3C, -B3A, -B3B, -B3C, -C3A, -C3B, -C3C, -D1A, -D1B, -D1C, -E1A, -E1B, -E1C, -E1F	80	74	None
HC-C2YK	8468-10R C8468-10R	Lycoming TIO-360-A1A, -A1B	74	74	"Avoid continuous operation between 1975 and 2200 r.p.m."
HC-C2YK HC-C2YR	8468	Lycoming O-540-B4A5, -B4B5	84	77	None
HC-C2YR	F8468A-3R	Lycoming O-540-J1A5D	81	77	None
HC-C2YF	8475 C8475	Continental IO-520-A, -J Continental TSIO-520-A, -C, -H	80	77	None
HC-C2YF	8475 C8475	Continental IO-520-D, -E, -F, -K, -L Continental TSIO-520-G	78	77	None
BHC-C2YF	8475 C8475	Continental IO-520-B & -C Continental TSIO-520-B, -D	80	77	None
BHC-C2YF	8475 C8475	Continental TSIO-520-E	78	77	None
HC-C2YK HC-C2YR	8475R	Lycoming IO-540-K1B5, -K1C5, -L1A5, -M1A5	84	84	None
HC-C2YK HC-C2YR	8475R	Lycoming IO-540-K1A5	84	78	None
HC-C2YK HC-C2YR	8475	Lycoming IO-540-K1A5, -K1B5, -K1C5, -K1D5, -L1A5, -M1A5	86	86	"Do not exceed 24 in. Hg. manifold pressure between 2300 and 2475 r.p.m."
HC-C2YK HC-C2YR	8475D	Lycoming IO-540-K1G5, -K1G5D, -K1A5D	83	78	None
HC-C2YK HC-C2YR	8475	Lycoming IO-540-K1A5, -K1B5, -K1C5, -L1A5, -M1A5	83	78	None
HC-C2YK HC-C2YR	8475	Lycoming TIO-541-A1A	80	80	None
HC-C2YK HC-C2YR	F8477-6Q or FC8477-6Q	Lycoming IO-540-D4A5, -D4B5, -D4C5	78	78	None
HC-C2YK HC-C2YR	8477	Lycoming O-540-A4A5, -A4B5, -A4C5, -A4D5, -E4A5, -E4B5, -E4C5	84	76	None
HC-C2YK HC-C2YR	C8477	Lycoming O-540-A4A5, -A4B5, -A4C5, -A4D5, -E4A5, -E4B5, -E4C5	80	78	None

P-920

Hub Model	Blade Model	Engine Model	Max. Dia. (Inches)	Min. Dia. (Inches)	Placards
HC-C2YK HC-C2YR	8477-8R	Lycoming O-540-A4A5, -A4B5, -A4C5, -A4D5, -E4A5, -E4B5, -E4C5	76	76	None
HC-C2YK HC-C2YR	8477	Lycoming O-540-G1A5	84	83	None
HC-C2YK HC-C2YR	8477 C8477	Lycoming IO-540-C4B5, -C4C5, -D4A5	84	76	None
HC-C2YK HC-C2YR	C8477	Lycoming IO-540-K1A5, -K1B5, -K1C5, -K1D5, -L1A5, -M1A5	80	80	"Do not exceed 23" Hg. manifold pressure below 2200 r.p.m."
HC-C2YK HC-C2YR	FC8477A	Lycoming IO-540-K1D5	80	78	"Do not exceed 23" Hg. manifold pressure below 2200 r.p.m."
HC-C2YF	9587A	Continental 6-285-B, 6-285-C	95	93	"Avoid continuous operation on the ground between 1900 and 2300 engine r.p.m. in winds above 15 m.p.h."

NOTE 10. Special Notes. Aircraft installation must be approved as part of the aircraft type certificate and demonstrate compliance with the applicable aircraft airworthiness requirements.

...END...

2A4
Revision 45
Twin Commander

560-F	681
680	690
680E	685
680F	690A
720	690B
680FL	690C
680FL(P)	690D
680T	695
680V	695A
680W	695B

January 1, 1990

TYPE CERTIFICATE DATA SHEET NO.2A4

This data sheet, which is a part of Type Certificate No. 2A4 prescribes conditions and limitations under which the product for which the Type Certificate was issued meets the airworthiness requirements of the Civil Air Regulations.

Type Certificate Holder: Twin Commander Aircraft Corporation
19003 - 59th Drive N.E.
Arlington, Washington 98223

I - Model 680, 7 PCLM (Normal Category), Approved October 14, 1955 (See NOTE 3 for RL-26-D (See NOTE 7 for conversion to Model 680E)

Engines: 2 Lycoming GSO-480-A1A6, Carburetor Bendix PS-7BD, Part Listing No. 391663-3, -4, -5, -6, or -7, or GSO-480-B1A6 (See NOTE 4).

Fuel: 100/130 minimum grade aviation gasoline.

Engine Limits: (Straight line manifold pressure variation with altitudes shown)

	HP.	R.P.M.	M.P.	ALT.
Takeoff	340	3400	48.0	S.L.
Takeoff	340	3400	44.5	8000
Maximum continuous	320	3200	45.0	S.L.
Maximum continuous	320	3200	43.0	8000

Propeller and Propeller Limits: 2 Hartzell 3-Bladed feathering propellers

a. H.C.-83x20-2 Hubs with 9333c blades
   Pitch settings at 30 in. Station: Low 17°, Feather 83°
   Diameter: 93 in., no cutoff permitted
   NOTE: Letters appearing after the dash numbers of the above listed hub model do not affect eligibility; however, for best synchronization, hubs with different numbers should not be combined on the same aircraft.

b. Spinner: 2 Hartzell, Dome C-888-3, Bulkhead C-807-3 or 2 Hartzell 835-10 assemblies or 2 Hartzell 836-7A assemblies (installed with alcohol anti-icing system per P/N 5890047).

c. Governor: 2 Woodward 210075

Page No.	1	2	3	4	5	6	7	8	9	10	11	12	13
Rev. No.	45	45	45	45	45	45	45	45	45	45	45	45	45
Page No.	14	15	16	17	18	19	20	21	22	23	24	25	26
Rev.No.	45	45	45	45	45	45	45	45	45	45	45	45	45
Page No.	27	28	29	30	31	32	33	34					
Rev.No.	45	45	45	45	45	45	45	45					

I - Model 680 (cont'd)

Airport Limits	Maneuvering	160 m.p.h.	(139K) True Ind.
	Max. Struc. cruising	210 m.p.h.	(182K) True Ind.
	Never exceed	270 m.p.h.	(235K) True Ind.
	Flaps extended - half	150 m.p.h.	(230K) True Ind.
	Flaps extended - full	130 m.p.h.	(113K) True Ind.
	Landing gear extended	180 m.p.h.	(156K) True Ind.

C.G. range — (+166.4) to +175.8) (Gear extended)
Effect of retracting landing gear +6655 in.-lb.

Empty Weight C.G. range — None

Datum — 152 in. forward of wing landing edge at center section.

Leveling means — Longitudinal - Top of fuselage on centerline aft of wing trailing edge.
Lateral - Transverse beams a: front or rear of baggage compartment floor.

Maximum weight — 7000 lb.

No. of seats — 7 (2 at +95, 2 at +128, and 3 at +168)

Maximum baggage — 350 lb. (+200)

Fuel capacity — Center tank 158.5 gal. (+187), usable fuel 156 gal. Outboard tanks 33.5 gal. each (+178), usable fuel 33.5 gal. each.
Total capacity 225.5 gal., usable fuel 223 gal.
(See NOTE 1 for system fuel)

Oil capacity — 8.5 gal. total (4.25 gal. each tank) (+191)
8.5 gal. usable (See NOTE 1 for system oil)

Control surface

Elevator	Up	$20° {}^{+1}_{-0}$	Down	$10° {}^{+2}_{-0}$
Elevator tab	Up	$20° {}^{+2}_{-0}$	Down	$20° {}^{+2}_{-0}$
Rudder	Right	$20° {}^{+2}_{-0}$	Left	$20° {}^{+2}_{-0}$
Rudder tab	Right	$26° {}^{+2}_{-0}$	Left	$26° {}^{+2}_{-0}$
Aileron	Up	23° ± 2	Down	15° ± 2
Flap outboard			Down	40° ± 2
Flap inboard			Down	40° ± 2

Serial Nos. eligible — Under the delegation option, provisions of Part 21 of the Federal Aviation Regulations, Delegation Option Manufacturer No. SW-2 is authorized to approve design and production charges on airplane serial numbers 680-244-2 to 680-658-255. (See NOTES 15 and 22)

II. - Model 680-E, 7 PCLM (Normal Category) Approved June 19, 1958
(Same as Model 680 except for extended wing and increased maximum weight)

Engines — 2 Lycoming GSO-480-B1A6, Carburetor Bendix PA-7 BD, Part Listing No. 391663-3, -4, -5, -6, and -7.

Fuel — 100/130 minimum grade aviation gasoline.

II. - Model 680-E (cont'd)

Engine limits	(Straight line manifold pressure variation with altitudes shown)				
		HP.	R.P.M.	M.P.	ALT.
	Takeoff	340	3400	48.0	S.L.
	Takeoff	340	3400	44.5	8000
	Maximum continuous	320	3200	45.0	S.L.
	Maximum continuous	320	3200	43.0	8000

Propeller and Propeller Limits

2 Hartzell 3-Bladed feathering propellers

a. HC-83x20-2 or HC-A3x20-2 Hubs with 9333c blades.
   Pitch settings at 30 in. Station: Low 17°, Feather 83°
   Diameter: 93 in., no cutoff permitted
   NOTE: Letters appearing after the dash numbers of the above listed hub model do not affect eligibility; however, for best synchronization hubs with different numbers should not be combined on the same aircraft.

b. Spinner: 2 Hartzell, Dome C-888-3, Bulkhead C-807-3 or 2 Hartzell 835-10 assemblies or 2 Hartzell 836-7A assemblies (installed with alcohol anti-icing system per P/N 5890047) or 2 Hartzell 836-22S assemblies (installed with alcohol anti-icing system per P/N 5890047).

c. Governor: 2 Woodward 210075

Airspeed Limits

Maneuvering	160 m.p.h.	(139K) True Ind.
Max. Struc. cruising	210 m.p.h.	(182K) True Ind.
Never exceed	270 m.p.h.	(235K) True Ind.
Flaps extended - half	150 m.p.h.	(130K) True Ind.
Flaps extended - full	135 m.p.h.	(117K) True Ind.
Landing gear extended	180 m.p.h.	(156K) True Ind.

C.G. range

(+166.0) to (+175.1) (Gear extended)
Effect of retracting landing gear +6655 in.-lb.

Empty Weight C.G. Range

None

Datum

152 in. forward of wing leading edge at center section.

Leveling means

Longitudinal - Top of fuselage on centerline aft of wing trailing edge.
Lateral: Transverse beams at front or rear of baggage compartment floor.

Maximum Weight

7500 lb.

No. of seats

7 (2 at +94, 2 at +128, and 3 at +168)

Maximum baggage

350 lb. (+200)

Fuel capacity

Center tank 158.5 gal. (+187), usable fuel 156 gal.
Outboard tanks 33.5 gal. each (+178), usable fuel 33.5 gal. ea.
total capacity 225.5 gal., usable fuel 223 gal.
(See NOTE 1 for system fuel)

Oil capacity

8.5 gal. total (4.25 gal. each tank) (+191)
8.5 gal. usable (See NOTE 1 for system oil)

II. - Model 680-E (cont'd)

Control surface movements	Elevator	Up 30° ± 1 0	Down 10° ± 2 0
	Elevator Tab	Up2 1/2° ± 2 1/2	Down 20° ± 2 0
	Rudder	Right 20° ± 2 0	Left 20° + 2 0
	Rudder tab	Right 26° ± 2 0	Left 26° + 2 0
	Aileron	Up 23° ± 2	Down 15° ± 2
	Flap outboard		Down 40° ± 2
	Flap inboard		Down 40° ± 2

Serial Nos. eligible

Under the delegation option, provisions of Part 21 of the Federal Aviation Regulations, Delegation Option Manufacturer No. SW-2 is authorized to approve design and production changes on airplane serial numbers 680-E-242-102, 680-E-623-1 to 680-E-892-100. (See NOTES 15 and 22.)

III. - Model 720, 6 PCLM (Normal Category), Approved December 5, 1958

(Same as Model 680 except for pressurized cabin, structural modifications to the fuselage, extended wing and increased maximum weight)

Engines

2 Lycoming GSO-480-B1A6, AMC Carburetor Bendix PS-7BD, Part Listing Nos. 391714-1, -2, -3, and -4.

Fuel

100/130 minimum grade aviation gasoline.

Engine limits

(Straight line manifold pressure variation with altitudes shown)

	HP	R.P.M.	M.P.	ALT.
Takeoff	340	3400	48.0	S.L.
Takeoff	340	3400	44.5	8000
Maximum continuous	320	3200	45.0	S.L.
Maximum continuous	320	3200	43.0	8000

Propeller and Propeller Limits

2 Hartzell 3-Bladed feathering propellers

a. HC-83x20-2 Hubs with 9333c blades
   Pitch settings at 30 in. Station: Low 17°, Feather 83°
   Diameter: 93 in., no cutoff permitted
   NOTE: Letters appearing after the dash numbers of the above listed hub model do not affect eligibility; however, for best synchronization hubs with different numbers should not be combined on the same aircraft.
b. Spinner: 2 Hartzell, Dome C-888-3, Bulkhead C-807-3 or 2 Hartzell 835-10 assemblies or 2 Hartzell 836-7A assemblies (installed with alcohol anti-icing system per P/N 5890047).
c. Governor: 2 Woodward 210075

Airspeed Limits

Maneuvering	160 m.p.h.	(139K) True Ind.
Max. Struc. cruising	210 m.p.h.	(182K) True Ind.
Never exceed	270 m.p.h.	(235K) True Ind.
Flaps extended - half	150 m.p.h.	(130K) True Ind.
Flaps extended - full	135 m.p.h.	(117K) True Ind.
Landing gear extended	180 m.p.h.	(156K) True Ind.

C.G. Range

(+166.0) to (+175.1) (Gear extended)
Effect of retracting landing gear +6655 in.-lb.

Empty Weight C.G. Range

None

Datum

152 in. forward of wing leading edge at center section.

III. - Model 720 (cont'd)

Leveling means	Longitudinal - top of fuselage on centerline aft of wing trailing edge. Lateral - Transverse beams at front or rear of baggage compartment floor.
Maximum weight	7500 lb.
No. of seats	6 (2 at +94, 2 at +128, and 3 at +168)
Maximum baggage	175 lb. (+200)
Fuel capacity	Center tank 158.5 gal. (+187), usable fuel 156 gal. Outboard tanks 33.5 gal. each (+178), usable fuel 33.5 gal. ea. Total capacity 225.5 gal., usable fuel 223 gal. (See NOTE 1 for system fuel)
Oil capacity	8.5 gal. total (4.25 gal. each tank) (+191) 8.5 gal. usable (See NOTE 1 for system oil)

Control surface movements

Surface				
Elevator	Up	30° ± 1 0	Down	10° ± 2 0
Elevator tab	Up	2 1/2° ± 2 1/2	Down	20° ± 2 0
Rudder	Right	20° ± 2 0	Left	20° ± 2 0
Rudder tab	Right	26° ± 2 0	Left	26° ± 2 0
Aileron	Up	23° ± 2	Down	15° ± 2
Flap outboard			Down	40° ± 2
Flap inboard			Down	40° ± 2

Serial Nos. eligible — Under the delegation option, provisions of Part 21 of the Federal Aviation Regulations, Delegation Option Manufacturer No. SW-2 is authorized to approve design and production changes on airplane serial numbers 720-501-1 to 720-850. (See NOTES 15 and 22).

IV - MODEL 680-F, 7 PCLM (Normal Category), Approved August 23, 1960

(Same as 680-E, except for fuel injection engine, new nacelles, new main gear and increased maximum weight.)
(See NOTE 5 for pressurized version).

Engines	2 Lycoming IGSO-540-B1A or IGSO-540-B1C, fuel injector Simmonds Model 580, Parts Listing No. 580056-B or Model 582 Parts Listing No. 582025 or Model 582, Parts Listing No. 582026.
Fuel	100/130 minimum grade aviation gasoline.

Engine limits — (Straight line manifold pressure variation with altitudes shown)

	HP	R.P.M	M.P.	ALT.
Takeoff (2 min.limit)	380	3400	47.0	S.L.
Takeoff (2 min. limit)	380	3400	43.5	12,000
Maximum continuous	360	3200	45.0	S.L.
Maximum continuous	360	3200	40.5	11,500

Propeller and Propeller Limits — 2 Hartzell 3-Bladed feathering propellers

a. HC-B3Z-30-2 Hubs with 9349 or 9349-6.5 propellers
Pitch settings at 30 in. Station: Low 18°, Feather 86°
Diameter: (For 9349) 93.5 in.
(For 9349-6.5) 87.0 in., no cutoff permitted
NOTE: Letters appearing after the dash numbers of the above listed hub model do not affect eligibility; however, for best synchronization hubs with different numbers should not be combined on the same aircraft.

IV - MODEL 680-F (cont'd)	b. Spinner: 2 Hartzell C2504 assemblies or 2 Hartzell C2535 assemblies (installed with alcohol anti-icing system per P/N 5890047). c. Governor: 2 Woodward B210310 or 2 Woodward B210410 (when propeller unfeathering system, Drawing 5640030, is installed). NOTE: Prefix B on part number or type number denotes based orientation only and may or may not be stamped on the nameplate. Governor part numbers may differ from governor type numbers. For best synchronization, governors with different part numbers should not be combined on the same aircraft.
Airspeed Limits	Maneuvering 157 m.p.h. (137K) True Ind. Max. Struc. cruising 230 m.p.h. (200K) True Ind. Never exceed 288 m.p.h. (250K) True Ind. Flaps extended - half 150 m.p.h. (130K) True Ind. Flaps extended - full 136 m.p.h. (118K) True Ind. Landing gear extended 180 m.p.h. (156K) True Ind.
C.G. Range	(+167.4) to (+174.4) (Gear extended) Effect of retracting landing gear +10,073 in.-lb.
Empty Weight C.G. Range	None
Datum	152 in. forward of wing leading edge at center section.
Leveling means	Longitudinal - Top of fuselage on centerline aft of wing trailing edge. Lateral - Transverse beams at front or rear of baggage compartment floor.
Maximum weight	8000 lb.
No. of seats	7 (2 at +94, 2 at +128, and 3 at +168)
Maximum baggage	350 lb. (+200)
Fuel capacity	Center tank 158.5 gal. (+187), usable fuel 156 gal. Outboard tanks 33.5 gal. each (+187), usable fuel 33.5 gal. ea. Total capacity 225.5 gal., usable fuel 223 gal. (See NOTE 1 for system fuel)
Oil capacity	10 gal. total (5.00 gal. each tank) (+191) 9.12 gal. usable (See NOTE 1 for system oil)

Control surface

Surface				
Elevator	Up	$30° {}^{+1}_{0}$	Down	$10° {}^{+2}_{0}$
Elevator tab	Up	$2\ 1/2° {}^{+2}_{0}$	Down	$20° {}^{+2}_{0}$
Rudder	Right	$20° {}^{+2}_{0}$	Left	$20° {}^{+2}_{0}$
Rudder tab	Right	$26° {}^{+2}_{0}$	Left	$26° {}^{+2}_{0}$
Aileron	Up	23° ± 2	Down	15° ± 2
Flap outboard			Down	40° ± 2
Flap inboard			Down	40° ± 2
*Elevator tab 680-F-971 and up			Down	26° ± 2 / 0

Serial Nos. eligible	Under the delegation option, provisions of Part 21 of the Federal Aviation Regulations, Delegation Option Manufacturer No. SW-2 is authorized to approve design and production changes on airplane serial numbers 680-F-871-1, 680-F-820-2 to 680-F-1447-152. (See NOTES 15 and 22.)

V - Model 560-F, 7 PCLM (Normal Category), Approved February 8, 1961

(Same as Model 680-F except unsupercharged engine and reduced gross weight)

Engine: 2 Lycoming IGO-B1A or 2 Lycoming IGO-540 B1C with Aero Commander Vapor Separator 4630193 installed, fuel injector Bendix Model RS10ED2, Parts Lifting No. 391825-1 (or any combination of these installations).

Fuel: 100/130 minimum grade aviation gasoline.

Engine limits

	HP.	R.P.M.
Takeoff (2 min.)	350	3400
Minimum continuous	325	3000

Propeller and Propeller Limits

1. 2 Hartzell 3-Bladed feathering propellers
   a. HC-B3Z-20-2 Hubs with 9349 blades
      Pitch settings at 30 in. Station: Low 15°, Feather 87°
      Diameter: 93.5 in., no cutoff permitted
      NOTE: Letters appearing after the dash numbers of the above listed hub model do not affect eligibility; however, for best synchronization hubs with different numbers should not be combined on the same aircraft.
   b. Spinner: 2 Hartzell C2504 assemblies or 2 Hartzell C2535 assemblies (installed with alcohol anti-icing system per P/N 5890047).
   c. Governor: 2 Woodward B210310 or 2 Woodward B210410 (when propeller unfeathering system, Drawing 5640030, is installed). NOTE: Prefix B on part number or type number denotes based orientation only and may or may not be stamped on the nameplate. Governor part numbers may differ from governor type numbers. For best synchronization, governors with different part numbers should not be combined on the same aircraft.
2. 2 Hartzell 3-Bladed feathering propellers
   a. HC-B3Z-30-2 Hubs with 9349-6.5 blades
      Pitch settings at 30 in. Station: Low 18°, Feather 86°
      Diameter: 87.0 in., no cutoff permitted
      NOTE: Letters appearing after the dash numbers of the above listed hub model do not affect eligibility; however, for best synchronization hubs with different numbers should not be combined on the same aircraft.
   b. Spinner: 2 Hartzell C2504 assemblies or 2 Hartzell C2535 assemblies (installed with alcohol anti-icing system per P/N 5890047).
   c. Governor: 2 Woodward B210310 or 2 Woodward B210410 (when propeller unfeathering system, Drawing 5640030, is installed). NOTE: Prefix B on part number or type number denotes based orientation only and may or may not be stamped on the nameplate. Governor part numbers may differ from governor type numbers. For best synchronization, governors with different part numbers should not be combined on the same aircraft.

Airspeed Limits

Maneuvering	155 m.p.h.	(135K) True Ind.
Max. Struc. cruising	230 m.p.h.	(200K) True Ind.
Never exceed	288 m.p.h.	(250K) True Ind.
Flaps extended - half	150 m.p.h.	(130K) True Ind.
Flaps extended - full	136 m.p.h.	(118K) True Ind.
Landing gear extended	180 m.p.h.	(156K) True Ind.

C.G. Range: (+167.4) to (+174.4) (Gear extended)
Effect of retracting landing gear +10,073 in.-lb.

Empty Weight C.G. Range: None

Datum: 152 in. forward of wing leading edge at center section.

V - Model 560-F (cont'd)

Leveling means	Longitudinal - Top of fuselage on centerline aft of wing trailing edge. Lateral - Transverse beams at front or rear of baggage compartment floor.
Maximum weight	7500 lb.
No. of seats	7 (2 at +94, 2 at +128, and 3 at +168)
Maximum baggage	350 lb. (+200)
Fuel capacity	Center tank 158.5 gal. (+187), usable fuel 156 gal. Outboard tanks 33.5 gal. each (+178), usable fuel 33.5 gal. ea. Total capacity 225.5 gal., usable fuel 223 gal. (See NOTE 1 for system fuel)
Oil capacity	10 gal. total (5.0 gal. each tank) (+191) 9.12 gal. usable (See NOTE 1 for system oil)

Control surface movements

Elevator	Up	30° ± 1 0	Down	10° ± 2 0
Elevator tab	Up	2 1/2° ± 2 1/2	Down	26° ± 2 0
Rudder	Right	20° ± 2 0	Left	20° ± 2 0
Rudder tab	Right	26° ± 2 0	Left	26° ± 2 0
Aileron	Up	23° ± 2	Down	15° ± 2
Flap outboard			Down	40° ± 2
Flap inboard			Down	40° ± 2

Serial Nos. eligible — Under the delegation option, provisions of Part 21 of the Federal Aviation Regulations, Delegation Option Manufacturer No. SW-2 is authorized to approve design and production changes on airplane serial numbers 560-F-951-1 to 560-F-1496-73. (See NOTES 15 and 22).

VI - MODEL 680-FL, 11 PCLM (Normal Category), Approved May 24, 1963

(Same as 680-F, except extended fuselage)

Engines	2 Lycoming IGSO-540-B1A or IGSO-540-B1C, fuel injector Simmonds Model 580, Parts Listing No. 580056-B or Model 582 Parts Listing No. 582025 or Model 582 Parts Listing No. 582026. (582026 required for 8500 lb. aircraft.)
Fuel	100/130 minimum grade aviation gasoline
Engine limits	(Straight line manifold pressure variation with altitudes shown)

	HP	R.P.M.	M.P.	ALT.
Takeoff (2 min. limit)	380	3400	47.0	S.L.
Takeoff (2 min. limit)	380	3400	43.5	12,000
Maximum continuous	360	3200	45.0	S.L.
Maximum continuous	360	3200	40.5	11,500

Propeller and Propeller Limits

2 Hartzell 3-Bladed feathering propellers

a. HC-B3Z-30-2 Hubs with 9349 or 9349-6.5 propellers
   Pitch settings at 30 in. Station: Low 28°, Feather 86°
   Diameter: (For 9349) 93.5 in.
   (For 9349-6.5) 87.0 in., no cutoff permitted
   NOTE: Letters appearing after the dash numbers of the above listed hub model do not affect eligibility; however, for best synchronization hubs with different numbers should not be combined on the same aircraft.

V - Model 560-F (cont'd)

b. Spinner: 2 Hartzell C2504 assemblies
c. Governor: 2 Woodward B210310 or 2 Woodward B210410 (when propeller unfeathering system, Drawing 5640030, is installed). NOTE: Prefix B on part number or type number denotes based orientation only and may or may not be stamped on the nameplate. Governor part numbers may differ from governor type numbers. For best synchronization, governors with different part numbers should not be combined on the same aircraft.

Airspeed Limits

Maneuvering	157 m.p.h. (137K) True Ind. @ 8000 lb.
	161 m.p.h. (140K) True Ind. @ 8500 lb.
Max. Struc. cruising	230 m.p.h. (200K) True Ind. @ 8000 lb. and 8500 lb.
Never exceed	288 m.p.h. (250K) True Ind. @ 8000 lb. and 8500 lb.
Flaps extended - half	150 m.p.h. (130K) True Ind. @ 8000 lb. and 8500 lb.
Flaps extended - full	136 m.p.h. (118K) True Ind. @ 8000 lb.
	146 m.p.h. (127K) True Ind. @ 8500 lb.
Landing gear extended	180 m.p.h. (156K) True Ind. @ 8000 lb. and 8500 lb.

C.G.Range (Gear extended)

Weight	Fwd.		Aft.	
lb.	Sta.(in)	% MAC	Sta.(in)	% MAC
Up to 7000	203.0	10	218.4	32
8000	206.5	15	218.4	32
8500	208.3	17.5	218.4	32

Straight line variation between points given
Effect of retracting landing gear +10,073 in.-lb.

Empty Weight C.G. Range — None

Datum — 196 in. forward of wing leading edge at center section.

Leveling means — Longitudinal - Top of fuselage on centerline aft of wing trailing edge.
Lateral - Transverse beams at front or rear of baggage compartment floor.

Maximum weight — (See NOTE 6)

No. of seats — 11 (Pilot + 10 passengers; pilot, co-pilot + 9 passengers)

Maximum baggage (std) — 400 lb. (+258)

Maximum baggage (with extended baggage compartment) — 600 lbs. (+258)

Fuel capacity — Center tank 158.5 gal. (+231), usable fuel 156 gal.
Outboard tanks 33.5 gal. each (+222), usable fuel 33.5 gal. ea.
Total capacity 225.5 gal., usable fuel 223 gal. (See NOTE 1 for system fuel)

Oil capacity — 10 gal. total (5.00 gal. each tank) (+235)
9.12 gal. usable (See NOTE 1 for system oil)

Control surface movements

ElevatorUp	Up	30° ± 1 / 0	Down	10° ± 2 / 0
Elevator tab	Up	2 1/2° ± 2 / 1/2	Down	26° ± 2 / 0
Rudder	Right	20° ± 2 / 0	Left	20° ± 2 / 0
Rudder tab	Right	26° ± 2 / 0	Left	26° ± 2 / 0
Aileron	Up	23° ± 2	Down	15° ± 2
Flap outboard			Down	40° ± 2
Flap inboard			Down	40° ± 2

V - Model 560-F (cont'd)

Serial Nos. eligible: (See NOTE 6). Under the delegation option, provisions of Part 21 of the Federal Aviation Regulations, Delegation Option Manufacturer No. SW-2 is authorized to: Issue Airworthiness Certificates for airplane serial numbers 680-FL-1553-107 and up; and approve design and production changes on airplane serial numbers 680-FL-1261 through 1853-157. (See NOTES 15 and 22).

VII - MODEL 680-FL(P), 11 PCLM (Normal Category), approved October 8, 1964

(Same as 680-FL, S/N 1461 and up, except pressurization)

Engines: 2 Lycoming IGSO-540-B1A or IGSO-540-B1C, fuel injector Simmonds Model 582, Parts Listing No. 582026.

Fuel: 100/130 minimum grade aviation gasoline.

Engine limits: (Straight line manifold pressure variation with altitudes shown)

	HP.	R.P.M	M.P.	ALT.
Takeoff (2 min. limit)	380	3400	47.0	S.L.
Takeoff (2 min. limit)	380	3400	43.5	12,000
Maximum continuous	360	3200	45.0	S.L.
Maximum continuous	360	3200	40.5	11,500

Propeller and Propeller limits: 2 Hartzell 3-Bladed feathering propellers

a. HC-B3Z-30-2 Hubs with 9349-6.5 blades
   Pitch settings at 30 in. Station: Low 18°, Feather 86°
   Diameter: 87.0 in., no cutoff permitted
   NOTE: Letters appearing after the dash numbers of the above listed hub model do not affect eligibility; however, for best synchronization hubs with different numbers should not be combined on the same aircraft.
b. Spinner: 2 Hartzell C2504 assemblies
c. Governor: 2 Woodward B210310 or 2 Woodward B210410 (when propeller unfeathering system, Drawing 5640030, is installed). NOTE: Prefix B on part number or type number denotes based orientation only and may or may not be stamped on the nameplate. Governor part numbers may differ from governor type numbers. For best synchronization, governors with different part numbers should not be combined on the same aircraft.

Airspeed Limits:

Maneuvering	161 m.p.h. (140K) True Ind.
Max. Struc. cruising	230 m.p.h. (200K) True Ind.
Never exceed	288 m.p.h. (250K) True Ind.
Flaps extended - half	150 m.p.h. (130K) True Ind.
Flaps extended - full	146 m.p.h. (127K) True Ind.
Landing gear extended	180 m.p.h. (156K) True Ind.

C.G. Range (Gear extended):

Weight	Fwd.		Aft.	
lb.	Sta.(in)	% MAC	Sta.(in)	% MAC
Up to 7000	203.0	10	218.4	32
8500	208.3	17.5	218.4	32

Straight line variation between points given.
Effect of retracting landing gear +10,073 in.-lb.

Empty Weight C.G. Range: None

Datum: 196 in. forward of wing leading edge at center section

Leveling means: Longitudinal - Top of fuselage on centerline aft of wing trailing edge.
Lateral - Transverse beams at front or rear of baggage compartment floor.

Maximum weight: 8500 lb.

VII - MODEL 680-FL(P) (cont'd)

Maximum No. of seats	11 (Pilot - 10 passengers; pilot, co-pilot +9 passengers)
Maximum baggage	400 lb. (+258)
Fuel capacity	Center tank 158.5 gal. (+231), usable fuel 156 gal. Outboard tanks 33.5 gal. each (+222), usable fuel 33.5 gal. ea. Total capacity 225.5 gal. usable fuel 223 gal. (See NOTE 1 for system fuel)
Oil capacity	10 gal. total (5.00 gal. each tank) (+235) 9.12 gal. usable (See NOTE 1 for system oil)

Control surface movements

Elevator	Up	30° ± 1 0	Down	10° ± 2 0
Elevator tab	Up	6 1/2° ± 1	Down	24° ± 1
Rudder	Right	20° ± 2 0	Left	20° ± 2 0
Rudder tab	Right	26° ± 2 0	Left	26° ± 2 0
Aileron	Up	23° ± 2	Down	15° ± 2
Flap outboard			Down	40° ± 2
Flap inboard			Down	40° ± 2

Serial No. eligible — Under the delegation option, provisions of Part 21 of the Federal Aviation Regulations, Delegation Option Manufacturer No. SW-2 is authorized to: Issue Airworthiness Certificates for airplane serial numbers 680-FLP-1559-25 and up; and approve design and production changes on airplane serial numbers 680-FLP-1471-2 through 1854-38. (See NOTES 15 and 22)

VIII - MODEL 680-T - 11 PCLM (Normal Category), approved September 15, 1965
(See NOTE 9 conversion to Model 680V)

Engines	2 AiResearch Model TPE-331-43 Turboprop engines (Rockwell P/N 6610400-501) or TPE-331-43A (Rockwell P/N 6610400-505) (See NOTE 11 for requirements)
Fuel	Aviation turbine fuels ASTM designation D1655-63T, Types Jet A, Jet B, and Jet A-1; and MIL-J-5624G(1), Grades JP-4 & JP-5 and MIL-F-5516-1, JP-1 (See Aerocom Serv. Ltr. 170)
Oil	BRACO 880F (MIL-L-7808D) and Sinclair Turbo S Oil 15 (MIL-L-7808D&E) (See Aerocom Service Letter 170)

Engine limits

	HP.	R.P.M.	EGT
Takeoff	575	100%	576°C
Maximum continuous	500	100%	550°C

Propeller and Propeller Limits — 2 Hamilton Standard 3-bladed feathering and reversing propellers Rockwell Assembly No. 640050.

a. 33LF-325 Hubs with 1033A-O Blades
   Pitch settings at 30 in. Station: Flt. Idle 9.0° ± 0.2°,
   Feather 86.5° ± 0.5°, Reverse -9.5° ± 1.5°
   Diameter: 90 in., no cutoff permitted.
   NOTE: Use AiResearch oil transfer tube No. 866678-2.

b. Spinner: 2 Rockwell 2640050-7

c. Governor: 2 AiResearch 865423-4 or 865423-5-1

VIII - MODEL 680-T (cont'd)

Airspeed Limits

Maneuvering	164 m.p.h. ((143K) CAS
Maximum Operating	250 m.p.h. (217K) CAS
Flaps extended - half	150 m.p.h. (130K) CAS
Flaps extended - full	146 m.p.h. (127K) CAS
Landing gear extended	180 m.p.h. (156K) CAS

C.G. range

Rear:	217.78 (30.19%)	8950 lbs. (Gear down
	216.94 (29.02%)	5300 lbs. (Gear down)
Fwd:	208.14 (16.83%)	8950 lbs. (Gear down)
	203.50 (10.40%)	7500 lbs. (Gear down)

Straight line variation between points given.
Effect of retracting landing gear +10,073 in.-lb.

Datum

196 in. forward of wing landing edge at center section

Leveling means

Longitudinal - Top of fuselage on centerline aft of wing trailing edge.
Lateral - Transverse beams at front of rear baggage compartment floor.

Maximum weight

Maximum takeoff 8950 lbs. (ramp weight 9000 lbs.)
Maximum landing 8500 lbs.

Maximum operating altitude

25,000 feet

Maximum No. of seats

11 (Pilot + 10 passengers; pilot, co-pilot + 9 passengers)

Maximum baggage

400 lb. (+258)

Fuel capacity

Center tank 221.5 gal. (+231), usable fuel 219.5 gal.
Outboard tanks 33.5 gal. each (+222), usable fuel 33.5 gal. ea.
Total capacity 288.5 gal., usable fuel 286.5 gal.
(See NOTE 1 for system fuel) (See NOTE 12 for auxiliary fuel)

Oil capacity

15.0 qts. total (7.5 qts. each tank) (+188)
11.8 qts. usable (See NOTE 1 for system oil)

Control surface movements

Elevator	Up	30° ± 1 0	Down	10° ± 2 0
Elevator tab	Up	6 1/2° ± 1	Down	24° ± 1
Rudder	Right	20° ± 2 0	Left	20° ± 2 0
Rudder tab	Right	26° ± 2 0	Left	26° ± 2 0
Aileron	Up	23° ± 2	Down	15° ± 2
Flap outboard			Down	40° ± 2
Flap inboard			Down	40° ± 2

Serial Nos. eligible

Under the delegation option, provisions of Part 21 of the Federal Aviation Regulations, Delegation Option Manufacturer No. SW-2 is authorized to: Issue Airworthiness Certificates for airplane serial numbers 680-T-1473, 680-T-1519, 680-T-1532, 680-T-1536, and 680-T-1550-11 and up; and approve design and production changes on airplane serial numbers 680-T-1473 through 1720. (See NOTES 15 and 22).

IX - MODEL 680-V, 11 PCLM (Normal Category), Approved June 13, 1967

Engines	2 AiResearch Model TPE-331-43 Turboprop engines (Rockwell P/N 6610400-501) or TPE-331-43A (Rockwell P/N 6610400-505) (See NOTE 11 for requirements).
Fuel	Aviation turbine fuels ASTM designation D1655-63T, Types Jet A, Jet B, and Jet A-1; and MIL-J-5624G(1), Grades JP-4 & JP-5 and MIL-F-5616-1, JP-1. (See Aerocom Serv. Ltr. 170)
Oil	BRACO 880F (MIL-L-7808D) and Sinclair Turbo S Oil 15 (MIL-L-7808D&E) (See Aerocom Service Letter 170)

Engine Limits

	HP.	R.P.M.	EGT
Takeoff	575	100%	576°C
Maximum continuous	500	100%	550°C

Propeller and Propeller Limits

2 Hamilton Standard 3-bladed feathering and reversing propellers.
Rockwell Assembly No. 640050.

a. 33LF-325 Hubs with 1033A-0 blades
   Pitch settings at 30 in. Station: Flt. Idle 9.0° ± 0.2°
   Feather 86.5° ± 0.5°, Reverse -9.5° ± 1.5°
   Diameter: 90 in., no cutoff permitted
   NOTE: Use AiResearch oil transfer tube No. 866678-2.
b. Spinner: 2 Rockwell 2640050-7
c. Governor: 2 AiResearch 865423-4 or 865423-5-1

Airspeed Limits

Maneuvering	164 m.p.h. (143K) CAS
Maximum Operating	250 m.p.h. (217K) CAS
Flaps extended - half	150 m.p.h. (130K) CAS
Flaps extended - full	146 m.p.h. (127K) CAS
Landing gear extended	180 m.p.h. (156K) CAS

C.G. Range

Rear: 215.68 (27.28%) 9450 lbs. (Gear down)
216.73 (28.73%) 9400 lbs. (Gear down)
217.87 (30.31%) 9346 lbs. (Gear down)
216.94 (29.02%) 5300 lbs. (Gear down)
Fwd: 209.74 (19.04%) 9450 lbs. (Gear down)
209.60 (18.83%) 9400 lbs. (Gear down)
203.50 (10.40%) 7500 lbs. (Gea down)
Straight line variation between points given.
Effect of retracting landing gear +10,073 in.-lb.

Datum	196 in. forward of wing leading edge at center section.
Leveling means	Longitudinal - Top of fuselage on centerline aft of wing trailing edge. Lateral - Transverse beams at front of rear baggage compartment floor.
Maximum weight	Maximum takeoff 9400 lbs. (ramp weight 9450 lbs.) Maximum landing 9000 lbs. Zero fuel 8000 lbs.
Maximum operating altitude	25,000 feet
Maximum No. of seats	11 (Pilot + 10 passengers; pilot, co-pilot + 9 passengers)
Maximum baggage	500 lb. (+258)
Fuel capacity	Center tank 221.5 gal. (+231), usable fuel 219.5 gal. Outboard tanks 33.5 gal. each (+222), usable fuel 33.5 gal. ea. Total capacity 288.5 gal., usable fuel 286.5 gal. (See NOTE 1 for system fuel) (See NOTE 12 for auxiliary fuel).

IX - MODEL 680-V, 11 PCLM (Normal Category), Approved June 13, 1967

Oil capacity	15.0 qts. total (7.5 qts. each tank) (+188) 11.8 qts. usable (See NOTE 1 for system oil)

Control surface movements

Elevator	Up	30° + 1 / 0	Down	10° + 2 / 0
Elevator tab	Up	6 1/2° ± 1	Down	24° ± 1
Rudder	Right	20° + 2 / 0	Left	20° + 2 / 0
Rudder tab	Right	26° + 2 / 0	Left	26° + 2 / 0
Aileron	Up	23° ± 2	Down	15° ± 2
Flap outboard			Down	40° ± 2
Flap inboard			Down	40° ± 2

Serial Nos. eligible — Under the delegation option, provisions of Part 21 of the Federal Aviation Regulations, Delegation Option Manufacturer No. SW-2 is authorized to: Issue Airworthiness Certificates for airplane serial numbers 680-V-1550 through 680-V-1725; and approve design and production changes on airplane serial numbers 680-V-1473 through 1720. (See NOTES 15 and 22).

X - MODEL 680-W, 11 PCLM (Normal Category), approved February 5, 1968

Engines — 2 AiResearch Model TPE-331-43BL Turboprop engines (Rockwell P/N 6610400-503)

Fuel — Aviation turbine fuels ASTM designation D1655-63T, Types A, Jet B, and Jet A-1; and MIL-J-5624G(1), Grades JP-4 & JP-5; and MIL-F-5616-1, JP-1, (See Aerocom Serv. Ltr. 170)

Oil — BRACO 880F (MIL-L-7808D) and Sinclair Turbo S Oil 15 (MIL-L-7808D&E) (See Aerocom Service Letter 170)

Engine limits

	HP.	R.P.M.	EGT
Takeoff	575	100%	576°C
Maximum continuous	500	100%	550°C

Propeller and Propeller Limits — 2 Hamilton Standard 3-bladed feathering and reversing propellers. Rockwell Assembly No. 640050.

a. 33LF-325 Hubs with 1033A-0 Blades
   Pitch settings at 30 in. Station: Flt. Idle 9.0° ± 0.2°, Feather 86.5° ± 0.5°, Reverse -9.5° ± 1.5°.
   Diameter: 90 in., no cutoff permitted.
   NOTE: Use AiResearch oil transfer tube No. 866678-2.
b. Spinner: 2 Rockwell 2640050-7
c. Governor: 2 AiResearch 869132-2-1

Airspeed Limits

Maneuvering	164 m.p.h. (143K) CAS
Maximum Operating	250 m.p.h. (217K) CAS
Flaps extended - half	150 m.p.h. (130K) CAS
Flaps extended - full	146 m.p.h. (127K) CAS
landing gear extended	180 m.p.h. (156K) CAS

C.G. Range

Rear: 215.68 (27.28%) 9450 lbs. (Gear down)
216.73 (28.73%) 9400 lbs. (Gear down)
217.87 (30.31%) 9346 lbs. (Gear down)
216.94 (29.02%) 5300 lbs. (Gear down)
Fwd.: 209.74 (19.04%) 9450 lbs. (Gear down)
209.60 (18.83%) 9400 lbs. (Gear down)
203.50 (10.40%) 7500 lbs. (Gear down)
Straight line variation between points given.
Effect of retracting landing gear +10,073 in.-lb.

X - MODEL 680-W (cont'd)

Datum	196 in. forward of wing leading edge at center section.
Leveling means	Longitudinal - Top of fuselage on centerline aft of wing trailing edge. Lateral - Transverse beams at front of each baggage compartment floor.
Maximum weight	Maximum takeoff 9400 lbs. (ramp weight 9450 lbs.) Maximum landing 9000 lbs. Zero fuel 8000 lbs.
Maximum operating altitude	25,000 feet
Maximum No. of seats	11 (Pilot + 10 passengers; pilot, co-pilot + 9 passengers)
Maximum baggage	500 lb. (+258) Serial numbers eligible for Model 680-W-1721 through 1850.
Fuel capacity	Center tank 221.5 gal. (+231), usable fuel 219.5 gal. Outboard tanks 33.5 gal. each (+222), usable fuel 33.5 gal. ea. Total capacity 288.5 gal., usable fuel 286.5 gal. (See NOTE 1 for system fuel.) (See NOTE 12 for auxiliary fuel.)
Oil capacity	15.0 qts. total (7.5 qts. each tank) (+188) 11.8 qts. usable (See NOTE 1 for system oil)

Control surface movements

Elevator	Up	30° ± 1 0	Down	10° ± 2 0
Elevator tab	Up	6 1/2° ± 1	Down	24° ± 1
Rudder	Right	20° ± 2 0	Left	20° ± 2 0
Rudder tab	Right	26° ± 2 0	Left	26° ± 2 0
Aileron	Up	23° ± 2	Down	15° ± 2
Flap outboard			Down	40° ± 2
Flap inboard			Down	40° ± 2

Serial Nos. eligible	Under the delegation option, provisions of Part 21 of the Federal Aviation Regulations, Delegation Option Manufacturer No. SW-2 is authorized to: Issue Airworthiness Certificates and approve design and production changes on airplane serial numbers 680-W-1721 through 1850, (See NOTES 15 and 22).

XI - MODEL 681, 11 PCLM (Normal Category), Approved March 20, 1969

Engines	2 AiResearch Model TPE-331-43BL Turboprop engines (Rockwell P/N 6610400-507)
Fuel	Aviation turbine fuels ASTM designation D1655-64T, Types Jet A, Jet B, and Jet A-1; and MIL J-5624G(1), Grades JP-4 & JP-5. (See Aerocom Service Letter 170)
Oil	BRACO 880F (MIL-L-7808D) and Sinclair Turbo S Oil 15 (MIL-L-7808D&E) (See Aerocom Service Letter 170)

Engine limits

	HP.	R.P.M.	EGT
Takeoff	575	100%	576°C
Maximum continuous	500	100%	550°C

XI - MODEL 681 (cont'd)

Propeller and Propeller limits

2 Hamilton Standard 3-bladed feathering and reversing propellers
Rockwell Assembly No. 640050.

a. 33LF-325 Hubs with 1033 A-0 Blades
   Pitch settings at 30 in. Station: Flt. Idle 9.0° ± 0.2°
   Feather 86.5° ± 0.5°, Reverse -9.5 ° ± 1.5°
   Diameter: 90 in., no cutoff permitted.
   NOTE: Use AiResearch oil transfer tube No. 866678-2.
b. Spinner: 2 Rockwell 2640050-7
c. Governor: 2 AiResearch 869132-2-1

Airspeed Limits

Maneuvering	164 m.p.h. (143K) CAS
Maximum Operating	250 m.p.h. (217K) CAS
Flaps extended - half	150 m.p.h. (130K) CAS
Flaps extended - full	149 m.p.h. (129K) CAS
Landing gear extended	180 m.p.h. (156K) CAS

C.G. Range

Rear: 215.68 (27.28%) 9450 lbs. (Gear down)
216.73 (28.73%) 9400 lbs. (Gear down)
217.87 (30.31%) 9346 lbs. (Gear down)
216.94 (29.02%) 5300 lbs. (Gear down)
Fwd.: 209.74 (19.04%) 9450 lbs. (Gear down)
209.60 (18.83%) 9400 lbs. (Gear down)
203.50 (10.40%) 7500 lbs. (Gear down)
Straight line variation between points given.
Effect of retracting landing gear +10,073 in.-lb.

Datum

196 in. forward of wing leading edge at center section

Leveling means

Longitudinal - Top of fuselage on centerline aft of wing trailing edge.
Lateral - Transverse beams at front of rear baggage compartment floor.

Maximum weight

Maximum takeoff 9400 lbs. (ramp weight 9450 lbs.)
Maximum landing 9000 lbs.
Zero fuel 8500 lbs.

Maximum operating altitude

25,000 feet

Maximum No. of seats

11 (Pilot + 10 passengers; pilot, co-pilot + 9 passengers)

Maximum baggage

500 lb. (+258)

Fuel capacity

Center tank 221.5 gal. (+231), usable fuel 219.5 gal.
Outboard tanks 33.5 gal. each (+222), usable fuel 33.5 gal. ea.
Total capacity 288.5 gal., usable fuel 286.5 gal.
(See NOTE 1 for system fuel) (See NOTE 12 for auxiliary fuel)

Oil capacity

15.0 qts. total (7.5 qts. each tank) (+188)
11.8 qts. usable (See NOTE 1 for system oil)

Control surface movements

Elevator	Up	30° ± 1	Down	10° ± 2
		0		0
Elevator tab	Up	6 1/2° ± 1	Down	24° ± 1
Rudder	Right	20° ± 2	Left	20° ± 2
		0		0
Rudder tab	Right	26° ± 2	Left	26° ± 2
		0		0
Aileron	Up	23° ± 2	Down	15° ± 2
Flap outboard			Down	40° ± 2
Flap inboard			Down	40° ± 2

XI - MODEL 681 (cont'd)

Serial Nos. eligible	Under the delegation option, provisions of Part 21 of the Federal Aviation Regulations, Delegation Option Manufacturer No. SW-2 is authorized to: Issue Airworthiness Certificates and approve design and production changes on airplane serial numbers 681-6001 through 6072. (See NOTES 15 and 22).

XII - MODEL 690, 11 PCLM (Normal Category), approved July 19, 1971

Engines — 2 AiResearch Model TPE-331-5-251K Turboprop engines (Rockwell P/N 610495)

Fuel — Aviation turbine fuels ASTM designation D1655-68, Types Jet A, Jet B, and Jet A-1; and MIL-T-5624G(1), Grades JP-4 & JP-5. (See Rockwell Service Letter 170H)

Oil — MIL-L-23699A and MIL-L-7808G. (See Rockwell Service Letter 170H)

Engine limits

	HP.	R.P.M.	I.T.T.
Takeoff	717.5	101%	923°C
Maximum continuous	717.5	101%	923°C

Propeller and Propeller limits

2 Hartzell 3-bladed feathering and reversing propellers.
Rockwell Assembly No. 640053.

a. HC-B3TN-5FL Hubs with LT10282H-4 or LT10282H(B)+4 or LT10282+4 or LT10282(B)+4 or LT10282A+4 or LT10282AB+4 blades

OR HC-B3TN-5DL or HC-B3TN-5NL hubs with LT10282A+4 or LT10282AB+4 blades.
Pitch settings at 30 in. Station: Low 13.5° ± 0.2°,
Feather 90.0° ± 0.5°, Reverse -8.0° ± 0.5°,
Start Locks +2.5° ± 0.2°
Diameter: 106 in, 1/2 in. reduction per blade allowed.
NOTE: Use AiResearch oil transfer tube No. 866533-3.
See NOTE 16.

b. Spinner: 2 Hartzell 836-57

c. Governor: 2 AiResearch 895490-1 or 895490-3.

Airspeed Limits

Maneuvering	167 m.p.h. (145K) CAS
Maximum Operating	280 m.p.h. (234K) CAS
Flaps extended - half	180 m.p.h. (156K) CAS
Flaps extended - full	157 m.p.h. (136K) CAS
Landing gear extended	230 m.p.h. (200K) CAS

C.G. Range

Forward
212.93 inches aft of datum (22.72% MAC) at 10,250 lbs.
203.75 inches aft of datum (10.40% MAC) at 7,500 lbs.
203.75 inches aft of datum (10.40% MAC) at 5,750 lbs.
Straight line variation between points.

Aft
218.70 inches aft of datum (30.47% MAC) at 10,250 lbs.
217.81 inches aft of datum (29.28% MAC) at 5,750 lbs.
Variation between points:
Inches aft of datum = 219.84 - (11653/Weight)

Datum — 196 in. forward of wing leading edge at center section.

Leveling means — Longitudinal - top of fuselage on centerline aft of wing trailing edge.
Lateral - Transverse beam at front of rear baggage compartment floor.

2A4

XII - MODEL 690 (cont'd)

Maximum weight	Maximum takeoff 10,250 lbs. (ramp weight 10,300 lbs.) Maximum loading 9600 lbs. Zero fuel 8750 lbs.
Maximum operating altitude	25,000 feet
Maximum No. of seats	11 (Pilot + 10 passengers; pilot, co-pilot + 9 passengers)
Maximum baggage	600 lbs. (+260)
Fuel capacity	Total capacity 389.0 gal., usable fuel 384.0 gal. (see NOTE 1 for system fuel)
Oil capacity	Oil capacity per engine @ +188 AiResearch Tank No. 896062-1 6.25 qt. total 5.25 qt. usable AiResearch Tank No. 896417-1 6.00 qt. total 5.00 qt. usable (See NOTE 1 for system oil)

Control surface movements

Elevator	Up	30° + 1	Down	10° + 2
		0		0
Elevator tab	Up	1/2° ± 1	Down	4° ± 1
Rudder	Right	0° ± 2	Left	0° ± 2
		0		0
Rudder tab	Right	26° ± 2	Left	26° ± 2
		0		0
Aileron	Up	23° ± 2	Down	15° ± 2
Flaps			Down	40° ± 2
Aileron tab	Up	17° ± 2.5°	Down	17° ± 2.5°

Serial Nos. eligible — Under the delegation option, provisions of Part 21 of the Federal Aviation Regulations, Delegation Option Manufacturer No. SW-2 is authorized to: Issue Airworthiness Certificate and approve design and production changes on airplane serial numbers 690-11001 through 11099. (See NOTES 15 and 22).

XIII - MODEL 685, 9 PCLM (Normal Category), Approved September 17, 1971

Engines	2 Continental Model GTSIO-520-F or GTSIO-520-K Turbosupercharged engines (See NOTE 14) (Rockwell P/N 610503)
Fuel	Aviation gasoline, 100/130 octane.
Oil	Teledyne Continental Specification MHS-24A.

Engine limits

	HP	R.P.M.	M.A.P.
Takeoff	435	3400	44.5 In.Hg
Maximum continuous	435	3400	44.5 In.Hg

Propeller and Propeller limits

2 Hartzell 3-bladed feathering propellers
Rockwell Drawing No. 610505

a. HC-H3YN-2 or HC-H3YN-2F Hubs with C8475+2, FC8475+2, or FC8475B+2 blades.
   Pitch settings at 30 in. Station: Low 18.1° ± 1.0°
   Feathered 83.5° ± 1.0°
   Diameter: 88 in., 1/2 in. reduction per blade allowed.
b. Spinner: 2 Hartzell D-3273-1
c. Governor: 2 Rockwell 610445-1, 610445-501, or 610445-503

XIII - MODEL 685 (cont'd)

Airspeed Limits	Maneuvering	156 m.p.h. (136K) CAS
	Never exceed	290 m.p.h. (252K) CAS
	Never exceed Mach	0.554
	Flaps extended - half	180 m.p.h. (156K) CAS
	Flaps extended - full	149 m.p.h. (130K) CAS
	Landing gear extended	230 m.p.h. (200K) CAS
	Max structural cruise	258 m.p.h. (224K)
	Max. structural cruise Mach	0.493

C.G.Range

Rear: 216.88 (28.0%) 9,000 lbs. (Gear down)
216.18 (27.1%) 5, 850 lbs. (Gear down)
Variation between points: inches = 218.15 - (11653/Weight)
Fwd: 208.67 (17.0%) 9,000 lbs. (Gear down)
203.45 (10.0%) 7,500 lbs. (Gear down)
203.45 (10.0%) 5,850 lbs. (Gear down)
Straight line variation between points given. Effect of retracting landing gear +11,653 in.-lb.

Datum

196 in. forward of wing leading edge at center section.

Leveling means

Longitudinal - Top of fuselage on centerline aft of wing trailing edge.
Lateral - Transverse beams at front of rear baggage compartment floor.

Maximum weight

Maximum takeoff 9000 lbs. (ramp weight 9050 lbs.)
Maximum landing 9000 lbs.

Maximum operating altitude

25,000 feet

Maximum number of seats

9 (Pilot + 8 passengers; pilot, co-pilot + 7 passengers)

Maximum baggage

600 lb. (+260)

Fuel capacity

Total capacity 261.0 gal., usable fuel 256.0 gal.
Auxiliary (option) 66.0 gal. total usable 322.0 gal.
Total undrainable 10.7 lbs. (without auxiliary option)
total undrainable 13.0 lbs. (with auxiliary option)

Oil capacity

24.0 qts. total (12.0 qts. each engine, 9.0 qts. usable - (See NOTE 1 for system oil) (+188)
Auxiliary with optional fuel 27.2 qts. total (13.6 qts. each engine, 10.6 qts. usable) (+188)

Control surface movements

Elevator	Up	30° ± 1 0	Down	10° ± 2 0
Elevator tab	Up	6 1/2° ± 1	Down	24° ± 1
Rudder	Right	20° ± 2 0	Left	20° ± 2 0
Rudder tab	Right	26° ± 2 0	Left	26° ± 2 0
Aileron	Up	23° ± 2	Down	15° ± 2
Flaps			Down	40° ± 2
Aileron tab	Up	17° ± 2.5°	Down	17° ± 2.5°

Serial Nos. eligible

Under the delegation option, provisions of Part 21 of the Federal Aviation Regulations, Delegation Option Manufacturer No. SW-2 is authorized to: Issue Airworthiness Certificates and approve design and production changes on airplane serial numbers 685-12000 through 12066. (See NOTES 15 and 22).

2A4

XIV - MODEL 690A, 11 PCLM (Normal Category), Approved April 25, 1973

Engines	2 AiResearch Model TPE-331-5-251K Turboprop engines (Rockwell P/N 610495)
Fuel	Aviation turbine fuels ASTM designation D1655-68, Types Jet A, Jet B, and Jet A-1; and MIL-T-5624G(1), Grades JP-4 & JP-5. (See Rockwell Service Letter 170H) (See Mfg. Data Part V Approved F/M for List of Approved Fuels)
Oil	MIL-L-23699A and MIL-L-7808G. (See Mfg. Data Part V Approved F/M for List of Approved Lubricants)

Engine limits

	HP.	R.P.M.	I.T.T.
Takeoff	717.5	101%	923°C
Maximum continuous	717.5	101%	923°C

Propeller and Propeller Limits

2 Hartzell 3-bladed feathering and reversing propellers.
Rockwell Assembly No. 640053.

a. HC-B3TN-5FL Hubs with LT10282H-4 or LT10282H(B)+4 or LT10282+4 or LT10282(B)+4 or LT10282+4 or LT10282AB+4 blades

OR HC-B3TN-5DL or HC-B3TN-5NL hubs with LT10282A+4 or LT10282AB+4 blades.
Pitch settings at 30 in. Station: Low 13.5° ± 0.2°
Feather 90.0° ± 0.5°, Reverse -8.0° ± 0.5°
Start Locks +2.5° ± 0.2°
Diameter: 106 in, 1/2 in. reduction per blade allowed.
NOTE: Use AiResearch oil transfer tube No. 866533-3.
(See NOTE 16)

b. Spinner: 2 Hartzell 836-57P

c. Governor: 2 AiResearch 895490-1 or 895490-3

Airspeed Limits

Maneuvering	167 m.p.h. (145K) CAS
Maximum Operating	280 m.p.h. (243K) CAS .52 MACH
Flaps extended - half	207 m.p.h. (180K) CAS
Flaps extended - full	161 m.p.h. (140K) CAS
Landing gear extended	230 m.p.h. (200K) CAS

C.G. Range

Forward
212.93 inches aft of datum (22.72% MAC) at 10,250 lbs.
203.75 inches aft of datum (10.40% MAC) at 7,500 lbs.
203.75 inches aft of datum (10.40% MAC) at 6,749 lbs.
214.58 inches aft of datum (24.93% MAC) at 6,000 lbs.
Straight line variation between points
Aft
218.70 inches aft of datum (30.47% MAC) at 10,250 lbs.
217.98 inches aft of datum (29.50% MAC) at 6,278 lbs.
Variation between points:
Inches aft of datum = 219.84 - (11653/Weight)

Datum	196 in forward of wing leading edge at center section
Leveling means	Longitudinal - top of fuselage on centerline aft of wing trailing edge. Lateral - Transverse beams at front of rear baggage compartment floor.
Maximum weight	Maximum takeoff 10,250 lbs. (ramp weight 10,300 lbs.) Maximum landing 9600 lbs. Zero fuel 8750 lbs.
Maximum operating altitude	31,000 feet
Maximum No. of seats	11 (Pilot +10 passengers; pilot, co-pilot + 9 passengers)

XIV - MODEL 690A (cont'd)

Maximum baggage	600 lb. (+260)
Fuel capacity	Total capacity 389.0 gal., usable fuel 384.0 gal. (See NOTE 1 for system fuel)
Oil capacity	12.0 qts. total (6.0 qts. total each tank) (+188) 10.0 qts. usable (See NOTE 1 for system oil)

Control surface movements

Elevator	Up	30° ± 1 0	Down	10° ± 2 0
Elevator tab	Up	6 1/2° ± 1	Down	24° ± 1
Rudder	Right	20° ± 2 0	Left	20° ± 2 0
Rudder tab	Right	26° ± 2 0	Left	26° ± 2 0
Aileron	Up	23° ± 2	Down	15° ± 2
Flaps			Down	40° ± 2
Aileron tab	Up	17° ± 2.5°	Down	17° ± 2.5°

Serial Nos. eligible: Under the delegation option, provisions of Part 21 of the Federal Aviation Regulations, Delegation Option Manufacturer No. SW-2 is authorized to: Issue Airworthiness Certificates and approve design and production changes on airplane serial numbers 690A-11100 through 11349. (See NOTES 15 and 22).

XV - MODEL 690B, 10 PCLM (Normal Category), Approved October 5, 1976

Engines: 2 AiResearch Model TPE-331-5-251K Turboprop engines (Rockwell P/N 610495), S/N 11350 through 11542
2 AiResearch Model TPE-331-5-252K Turboprop engines (Rockwell P/N 610495), S/N 11431, S/N 11543 and subs.

Fuel: Aviation turbine fuels ASTM designation D1655-68, Types Jet A, Jet B, and Jet A-1; and MIL-T-5624G(1), Grades JP-4 & JP-5. (See Rockwell Services Letter 170H) (See Mfg. Data Part V Approved F/M for List of Approved Fuels).

Oil: MIL-L-23699A and MIL-L-7808G (See Mfg. Data Part V Approved F/M for List of Approved Lubricants).

Engine limits

	HP.	R.P.M.	I.T.T.
Takeoff	717.5	101%	923°C
Maximum continuous	717.5	101%	923°C

Propeller and Propeller Limits: 2 Hartzell 3-bladed feathering and reversing propellers.
Rockwell Assembly No. 640053.

a. HC-B3TN-5FL Hubs with LT10282H-4 or LT10282H(B)+4 or LT10282+4 or LT10282(B)+4 or LT10282+4 or LT10282AB+4 blades

OR HC-B3TN-5DL or HC-B3TN-5NL hubs with LT10282A+4 or LT10282AB+4 blades.

Pitch settings at 30 in. Station: Low 13.5° ± 0.2°
Feather 90.0° ± 0.5°, Reverse -8.0° ± 0.5°
Start Locks +2.5° ± 0.2°
Diameter: 106 in, 1/2 in. reduction per blade allowed.
NOTE: Use AiResearch oil transfer tube No. 866533-3.
(See NOTE 16)

b. Spinner: 2 Hartzell 836-57P

c. Governor: 2 AiResearch 895490-1 or 895490-3 (for aircraft with TPE 331-5-251K engines)
2 AiResearch 895490-5 (for aircraft with TPE 331-5-252K engines)

XV - MODEL 690B (cont'd)

Airspeed Limits	Maneuvering	171 m.p.h. (149K) CAS
	Maximum Operating	280 m.p.h. (243K) CAS .52 MACH
	Flaps extended - half	207 m.p.h. (180K) CAS
	Flaps extended - full	161 m.p.h. (140K) CAS
	Landing gear extended	230 m.p.h. (200K) CAS

C.G. Range

Forward
213.14 inches aft of datum (23.00% MAC) at 10,325 lbs.
203.75 inches aft of datum (10,40% MAC) at 7,500 lbs.
203.75 inches aft of datum (10.40% MAC) at 6,749 lbs.
214.58 inches aft of datum (24.93% MAC) at 6,000 lbs.
Straight line variation between points.
Aft
218.64 inches aft of datum (30.39% MAC) at 10,325 lbs.
217.85 inches aft of datum (29.33% MAC) at 6,267 lbs.
Variation between points.
Inches aft of datum = 219.84 - (12444/Weight)

Datum

196 in. forward of wing leading edge at center section.

Leveling means

Longitudinal - Top of fuselage on centerline aft of wing trailing edge.
Lateral - Transverse beams at front of rear baggage compartment floor.

Maximum weight

Maximum takeoff 10,325 lbs. (ramp weight 10,375 lbs.)
Maximum landing 9675 lbs.
Zero fuel 8750 lbs.

Maximum operating altitude

31,000 feet

Maximum No. of seats

10 (Pilot + 9 passengers; pilot, co-pilot + 8 passengers)

Maximum baggage

600 lb. (+260)

Fuel capacity

Total capacity 389.0 gal., usable fuel 384.0 gal.
(See NOTE 1 for systems fuel)

Oil capacity

12.0 qts. total (6.0 qts. total each tank) (+188)
10.0 qts. usable (See NOTE 1 for system oil)

Control surface movements

Elevator	Up	$30° \pm 1$ 0	Down	$10° \pm 2$ 0
Elevator tab	Up	$6\ 1/2° \pm 1$	Down	$24° \pm 1$
Rudder	Right	$20° \pm 2$ 0	Left	$20° \pm 2$ 0
Rudder tab	Right	$26° \pm 2$ 0	Left	$26° \pm 2$ 0
Aileron	Up	$23° \pm 2$	Down	$15° \pm 2$
Flaps			Down	$40° \pm 2$
Aileron tab	Up	$17° \pm 2.5°$	Down	$17° \pm 2.5°$

Serial Nos. eligible

Under the delegation option, provisions of Part 21 of the Federal Aviation Regulations, Delegation Option Manufacturer No. SW-2 is authorized to: Issue Airworthiness Certificates and approve design and production changes on airplane serial numbers 690B-11350 through 11566. (See NOTES 15 and 22).

XVI - MODEL 690C, 11 PCLM (Normal Category), Approved September 7, 1979

Engines	2 AiResearch Model TPE-331-5-254K Turboprop engines (Rockwell P/N 610495).
Fuel	Aviation turbine fuels ASTM designation D1655-68, Types Jet A, Jet A-1, and Jet B; MIL-T-5624G-1, Grades JP-4 and JP-5; MIL-T-83133, Grade JP-8 and MIL-F-46005A(MR)-1, Types I and II.
Oil	MIL-L-23699B Type II, MIL-L-7808G Type I (See Mfg. Data Part VIII Approved POH for List of Approved Lubricants).

Engine limits

	HP.	R.P.M.	I.T.T.
Takeoff	717.5	101%	923°C
Maximum continuous	717.5	101%	923°C

Propeller and Propeller Limits

2 Dowty-Rotol Ltd. 3-bladed feathering and reversing propellers.
Rockwell Assembly No. 640080.

a. Dowty-Rotol Ltd. Type No. (C) R306/3-82-F/7-(c) VP2926 includes B. F. Goodrich propeller de-icing kit No. 65-330-1 or Dowty Rotol Ltd. Type No. (C) R306/3-82-F/7-(c) VP 3027 includes Dowty Rotol Drice Boots 660709275 as B. F. Goodrich De-Ice Boots 4E 2598-10. See NOTE 17.
   Dowty-Rotol Propeller Blade Assembly P/N 660706330-XX
   Pitch settings at .7 radius station:
   Feather 83 10' ± 20", Reverse -13.75° ± 1.0°
   Start Locks -1.25° ± 1.0°, Flight Idle 6.0° ± 0.5°
   Diameter: 106 in., 1/2 in. reduction per blade allowed.
   NOTE: Use AiResearch oil transfer tube No. 897458-2.
   NOTE: All engine ground running for maintenance test purposes with the airplane stationary, must be done with the airplane headed into the wind.
b. Spinner: 2 Dowty-Rotol Ltd. Type No. (C)SB7/3/1
c. Governor: 2 AiResearch P/N 895490-5, 897410-2B, or 897410-4

Airspeed Limits

Maneuvering	158 m.p.h. (137K) CAS
Maximum Operating	280 m.p.h. (234K) CAS .52 MACH
Flaps extended - half	207 m.p.h. (180K) CAS (S/N 11600-11729)
	230 m.p.h. (200K) CAS (S/N 11730-11999)
Flaps extended - full	161 m.p.h. (140K) CAS (S/N 11600-11729)
	184 m.p.h. (160K) CAS (S/N 11730-11999)
Landing gear extended	230 m.p.h. (200K) CAS

C.G. Range

Forward
210.51 inches aft of datum (20.06% MAC) at 10,325 lbs.
204.70 inches aft of datum (12.03% MAC) at 7,500 lbs.
204.70 inches aft of datum (12.03% MAC) at 6,798 lbs.
215.10 inches aft of datum (26.42% MAC) at 6,240 lbs.
Straight line variation between points.
Aft
218.67 inches aft of datum (31.35% MAC) at 10,325 lbs.
217.88 inches aft of datum (30.25% MAC) at 6,332 lbs.
Variation between points:
Inches aft of datum = 219.93 - (13029/Weight)

Datum	196 in. forward of wing leading edge at center section
Leveling means	Longitudinal - Top of fuselage on centerline aft of wing trailing edge. Lateral - Transverse beams at front of rear baggage compartment floor.
Maximum weight	Maximum takeoff 10,325 lbs. (ramp weight 10,375 lbs.) Maximum landing 9675 lbs. Zero fuel 8800 lbs.

XVI - MODEL 690C (cont'd)

Maximum operating altitude	31,000 feet
Maximum No. of seats	11 (Pilot + 10 passengers; pilot, co-pilot + 9 passengers)
Maximum baggage	600 lb. (+260)
Fuel capacity	Total standard capacity 430 gal., usable 425 gal. Total capacity with optional system 482 gal., usable 474 gal. (See NOTE 1 for systems fuel.)
Oil capacity	12.0 qts. total (6.0 qts. total each tank) (+188) 10.0 qts. usable (See NOTE 1 for system oil)

Control surface movements

Elevator	Up	30° ± 1 0	Down	10° ± 2 0
Elevator tab	Up	3° ± 1	Down	24° ± 1
Rudder	Right	20° ± 2 0	Left	20° ± 2 0
Rudder tab	Right	20° ± 2 0	Left	20° ± 2 0
Aileron	Up	23° ± 2	Down	15° ± 2
Flaps			Down	40° ± 2
Aileron tab	Up	17° ± 2.5°	Down	17° ± 2.5°

Serial Nos. eligible — Under the delegation option, provisions of Part 21 of the Federal Aviation Regulations, Delegation Option Manufacturer No. SW-2 is authorized to: Issue Airworthiness Certifictes and approve design and production changes on airplane serial numbers 11600 through 11735. (See NOTE 22)

XVII - MODEL 695, 11 PCLM (Normal Category), Approved November 1, 1979

Engines — 2 AiResearch Model TPE-331-10-501K Turboprop Engines (Rockwell P/N 610653) or 2 Garrett Model TPE-331-10-511K Turboprop Engines (Gulfstream P/N 610653) See NOTE 19.

Fuel — Aviation turbine fuel ASTM designation D1655-68, Types Jet A and Jet A-1, and Jet B; MIL-T-5624G-1, Grades JP-4 and JP-5; MIL-T-83133, Grade JP-8, MIL-F-46005A(MR)-1, Types I and II.

Oil — MIL-L-23699B Type II, MIL-L-7808G Type I (See Mfg. Data Part VIII Approved POH for List of Approved Lubricants).

Engine Limits

	HP	R.P.M.	E.G.T.
Takeoff	733	101%	650°C
Maximum continuous	733	101%	650°C

Propeller and Propeller Limits — 2 Dowty-Rotol Ltd. 3-bladed feathering and reversing propellers. Rockwell Assembly No. 640080.

a. Dowty-Rotol Ltd. Type No. (C) R306/3-82-F/7-(c) VP2926 includes B. F. Goodrich propeller de-icing kit No. 65-330-1 or Dowty Rotol Ltd. Type No. (C) R306/3-82-F/7-(c) VP 3027 includes Dowty Rotol Deice Boots 660709275 as B. F. Goodrich De-Ice Boots 4E 2598-10. See NOTE 17.
Dowty-Rotol Propeller Blade Assembly P/N 660706330-XX
Pitch settings at .7 radius station:
Feather 83 10' ± 20', Reverse -13.75° ± 1.0°,
Start Locks -1.25° ± 1.0°, Flight Idle 6.0° ± 0.5°.
Diameter: 106 in., 1/2 in. reduction per blade allowed.
NOTE: Use AiResearch oil transfer tube Part No. 897458-2.
NOTE: Downwind ground operation above taxi power is prohibited when airplane is stationary.

XVII - MODEL 695 (cont'd)

Propeller and Propeller Limits (cont'd)	b. Spinner: 2 Dowty-Rotol Ltd. Type No. (C)SB7/3/1 c. Governor: 2 AiResearch P/N 897410-2B or 897410-4.
Airspeed Limits	Maneuvering 158 m.p.h. (137K) CAS Maximum Operating 280 m.p.h. (143K) CAS .52 MACH Flaps extended - half 207 m.p.h. (180K) CAS Flaps extended - full 161 m.p.h. (140k) CAS Landing gear extended 230 m.p.h. (200K) CAS
C.G. Range	Forward 210.51 inches aft of datum (20.06% MAC) at 10,325 lbs. 204.70 inches aft of datum (12.03% MAC) at 7,500 lbs. 204.70 inches aft of datum (12.03% MAC) at 6,798 lbs. 215.10 inches aft of datum (26.42% MAC) at 6,240 lbs. Straight line variation between points. Aft 218.67 inches aft of datum (31.35% MAC) at 10,325 lbs. 217.88 inches aft of datum (30.25% MAC) at 6,332 lbs. Variation between points: Inches aft of datum = 219.93 - (13029/Weight)
Datum	196 in. forward of wing leading edge at center section
Leveling means	Longitudinal - Top of fuselage on centerline aft of wing trailing edge. Lateral - Transverse beams at front of rear baggage compartment floor.
Maximum weight	Maximum takeoff 10,325 lbs. (ramp weight 10,375 lbs.) Maximum landing 9,675 lbs. Zero fuel 8,800 lbs.
Maximum operating altitude	31,000 feet
Maximum No. of seats	11 (Pilot + 10 passengers; pilot, co-pilot + 9 passengers)
Maximum baggage	600 lb. (+260)
Fuel capacity	Total standard capacity 430 gal., usable 425 gal. (S/N 95000 thru 95040). Total standard capacity 482 gal., usable 474 gal. (S/N 95041 thru 95999). (See NOTE 1 for systems fuel.)
Oil capacity	12.0 qts. total (6.0 qts. total each tank) (+188) 10.0 qts. usable (See NOTE 1 for system oil).

Control Surface movements

Elevator	Up	30° ± 1 0	Down	10° ± 2 0
Elevator tab	Up	3° ± 1	Down	24° ± 1
Rudder	Right	20° ± 2 0	Left	20° ± 2 0
Rudder tab	Right	20° ± 2 0	Left	20° ± 2 0
Aileron	Up	23° ± 2	Down	15° ± 2
Flaps	Down	40° ± 2		
Aileron tab	Up	17° ± 2.5°	Down	17° ± 2.5°

Serial Nos. eligible

Under the delegation option, provisions of Part 21 of the Federal Aviation Regulations, Delegation Option Manufacturer No. SW-2 is authorized to: Issue Airworthiness Certificates and approve design and production changes on airplane serial numbers 95000 through 95084. (See NOTE 22.)

XVIII - MODEL 695A, 11 PCLM (Normal Category), Approved April 30, 1981

Engines	2 AiResearch Model TPE-331-10-501K Turboprop Engines (Rockwell P/N 610653) or 2 Garrett Model TPE-331-10-511K Turboprop Engines (Gulfstream P/N 610653) See NOTE 19.
Fuel	Aviation turbine fuels ASTM designation D1655-68, Types Jet A, Jet A-1, and Jet B; MIL-T-5624G-1, Grades JP-4 and JP-5; MIL-T-83133, Grade JP-8, and MIL-F-46005A(MR)-1, Types I and II.
Oil	MIL-L-23699B Type II (See Mfg. Data Part VIII Approved POH for List of Approved Lubricants).

Engine Limits

	Torque	RPM	EGT
Takeoff and Maximum continuous	102.5%(820)	101.0%	650°C

Propeller and Propeller Limits

2 Dowty-Rotol Ltd. 3-bladed feathering and reversing propellers.
Rockwell Assembly No. 640080.

a. Dowty-Rotol Ltd. Type No. (C) R306/3-82-F/7-(c) VP2926 includes B. F. Goodrich propeller de-icing kit No. 65-330-1 or Dowty Rotol Ltd. Type No. (C) R306/3-82-F/7-(c) VP 3027 includes Dowty Rotol Deice Boots 660709275 as B. F. Goodrich De-Ice Boots 4E 2598-10. See NOTE 17.
   Dowty-Rotol Propeller Blade Assembly P/N 660706330-XX
   Pitch settings at .7 radius station:
   Feather 83 10' ± 20', Reverse -13.75° ± 1.0°,
   Start Locks -1.25° ± 1.0°, Flight Idle 6.0° ± 0.5°.
   Diameter: 106 in., 1/2in. reduction per blade allowed.
   NOTE: Use AiResearch oil transfer tube Part No. 897458-2.
   NOTE: Downwind ground operation above taxi power is prohibited when airplane is stationary, must be done with the airplane headed into the wind.
b. Spinner: 2 Dowty-Rotol Ltd. Type No. (C)SB7/3/1
c. Governor: 2 AiResearch P/N 897410-2B or 897410-4.

Airspeed Limits

Maneuvering	162 m.p.h. (141K) CAS
Maximum Operating	290 m.p.h. (252K) CAS .60 MACH
Flaps extended - half	207 m.p.h. (180K) CAS (S/N 96000-96055)
	230 m.p.h. (200K) CAS (S/N 96056-96999)
Flaps extended- full	161 m.p.h. (140K) CAS (S/N 96000-96055)
	184 m.p.h. (160K) CAS (S/N 96056-96999)
Landing gear extended	230 m.p.h. (200K) CAS

C.G. Range

Forward
209.78 inches aft of datum (19.1% MAC) at 11,200 lbs.
204.34 inches aft of datum (11.5% MAC) at 8,500 lbs.
204.34 inches aft of datum (11.5% MAC) at 7,010 lbs.
214.18 inches aft of datum (25.1% MAC) at 6,466 lbs.
Straight line variation between points
Aft
218.77 inches aft of datum (31.5% MAC) at 11,200 lbs.
217.95 inches aft of datum (30.4% MAC) at 6,582 lbs.
Variation between points:
   Inches aft of datum = 219.93 - (13029/Weight)

Datum	196 in forwad of wing leading edge at center section.
Leveling means	Longitudinal - Top of fuselage on centerline aft of wing trailing edge. Lateral - Transverse beams at front of rear baggage compartment floor.

XVIII - MODEL 695A (cont'd)

Maximum weight	Maximum takeoff	11,200 lbs. (ramp weight 11,250 lbs.)
	Maximum landing	10,550 lbs.
	Zero fuel	9,500 lbs.

Maximum operating altitude — 35,000 feet

Maximum No. of seats — 11 (Pilot + 10 passengers; pilot, co-pilot + 9 passengers)

Maximum baggage — 600 lb. (+290) Non pressurized compartment (See NOTE 18)
100 lb. (+245) Pressurized compartment

Fuel capacity — Total standard capacity 482 gal., usable 474 gal.
(See Note 1 for systems fuel)

Oil capacity — 12.0 qts. total (6.0 qts. total each tank) (+188)
10.0 qts. usable (See Note 1 for system oil)

Control Surface movements

Surface				
Elevator	Up	30° ± 1 0	Down	10° ± 2 0
Elevator tab	Up	3° ± 1	Down	24° ± 1
Rudder	Right	20° ± 2 0	Left	20° ± 2 0
Rudder tab	Right	20° ± 2 0	Left	20° ± 2 0
Aileron	Up	23° ± 2	Down	15° ± 2
Flaps			Down	40° ± 2
Aileron tab	Up	17° ± 2.5°	Down	17° ± 2.5°

Serial Nos. eligible — Under the delegation option, provisions of Part 21 of the Federal Aviation Regulations, Delegation Option Manufacturer No. SW-2 is authorized to: Issue Airworthiness Certificates and approve design and production changes on airplane serial numbers 96000 through 96100. (See Note 21 and 22)

XIX - MODEL 690D, 11 PCLM (Normal Category) Approved December 2, 1981

Engines — 2 AiResearch Model TPE 331-5-254K Turboprop Engines (Gulfstream P/N 610495).

Fuel — Aviation turbine fuels ASTM designation D1655-68, types Jet A, Jet A-1 and Jet B; MIL-T-5624G-1, Grades JP-4 and JP-5; MIL-T-83133, Grade JP-8 and MIL-F-46005A(MR)-1, Types I and II.

Oil — MIL-L-23699B Type II or MIL-L-7808G type I (See Mfg. Data Part VIII Approved POH for List of Approved Lubricants).

Engine limits

	HP	R.P.M.	ITT
Takeoff and Maximum continuous	748	101.0%	923°

Propeller and Propeller Limits — 2 Dowty-Rotol Ltd. 3-bladed feathering and reversing propellers.
Rockwell Assembly No. 640080.

a. Dowty-Rotol Ltd. Type No. (C) R306/3-82-F/7-(c) VP2926 includes B. F. Goodrich propeller de-icing kit No. 65-330-1 or Dowty Rotol Ltd. Type No. (C) R306/3-82-F/7-(c) VP 3027 includes Dowty Rotol Deice Boots 660709275 as B. F. Goodrich De-Ice Boots 4E 2598-10. See NOTE 17.
Dowty-Rotol Propeller Blade Assembly P/N 660706330-XX
Pitch settings at .7 radius station:

<u>XIX - MODEL 690D</u> (cont'd)

Propeller and Propeller Limits (cont'd)	Feather 83° 10' ± 20', Reverse -13.75° ± 1.0° Start Locks -1.25° ± 1.0°, Flight Idle 6.0° ± 0.5°. Diameter: 106 in., 1/2in. reduction per blade allowed. <u>NOTE</u>: Use AiResearch oil transfer tube Part No. 897458-2. <u>NOTE</u>: All engine ground running for maintenance test purposes, with the airplane stationary, must be done with the airplane head into the wind. b. Spinner: 2 Dowty-Rotol Ltd. Type No. (C)SB7/3/1 c. Governor: 2 AiResearch P/N 897410-2B or -4.
Airspeed Limits	Maneuvering 160 m.p.h. (139K) CAS Maximum Operating 290 m.p.h. (252K) CAS .60 MACH Flaps extended - half 207 m.p.h. (180K) CAS (S/N 15000-15024) 230 m.p.h. (200K) CAS (S/N 15025-15999) Flaps extended - full 161 m.p.h. (140K) CAS (S/N 15000-15024) 184 m.p.h. (160K) CAS (S/N 15025-15999) landing gear extended 230 m.p.h. (200K) CAS
C.G. Range	Forward 208.77 inches aft of datum (17.7% MAC) at 10,700 lbs. 204.34 inches aft of datum (11.5% MAC) at 8,500 lbs. 204.34 inches aft of datum (11.5% MAC) at 7,010 lbs. 214.18 inches aft of datum (25.1% MAC) at 6,466 lbs. Straight line variation between points. Aft 218.72 inches aft of datum (31.4% MAC) at 10,700 lbs. 217.94 inches aft of datum (30.4% MAC) at 6,582 lbs. Variation between points Inches aft of datum = 219.93 - (13029/Weight)
Datum	196 in. forward of wing leading edge at center section
Leveling means	Longitudinal - Top of fuselage on centerline aft of wing trailing edge. Lateral - Transverse beams at front of rear baggage compartment floor.
Maximum weight	Maximum takeoff 10,700 lbs. (ramp weight 10,775 lbs.) Maximum landing 10,550 lbs. Zero fuel 9,500 lbs.
Maximum operating altitude	31,000 feet
Maximum No. of seats	11 (Pilot + 10 passengers; pilot, co-pilot + 9 passengers)
Maximum baggage	600 lb. (+290) Non pressurized compartment 100 lb. (+245) Pressurized compartment
Fuel capacity	Total standard capacity 430 gal., usable fuel 425.0 gal. Total capacity with optional system 482 gal., usable 474 gal. (See NOTE 1 for systems fuel).
Oil capacity	12.0 qts. total (6.0 qts. total each tank) (+188) 10.0 qts. usable (See NOTE 1 for system oil)

XIX - MODEL 690D (cont'd)

Control Surface movements

Elevator	Up	$30^\circ {}^{+1}_{0}$	Down	$10^\circ {}^{+2}_{0}$
Elevator tab	Up	3° ± 1	Down	24° ± 1
Rudder	Right	$20^\circ {}^{+2}_{0}$	Left	$20^\circ {}^{+2}_{0}$
Rudder tab	Right	$20^\circ {}^{+2}_{0}$	Left	$20^\circ {}^{+2}_{0}$
Aileron	Up	23° ± 2	Down	15° ± 2
Flaps			Down	40° ± 2
Aileron tab	Up	17° ± 2.5°	Down	17° ± 2.5°

Serial Nos. eligible

Under the delegation option, provisions of Part 21 of the Federal Aviation Regulations, Delegation Option Manufacturer No. SW-2 is authorized to: Issue Airworthiness Certificates and approve design and production changes on airplane serial numbers 15000 through 15042. (See Notes 21 and 22.)

XX - MODEL 695B, 11 PCLM (Normal Category), Approved February 15, 1984

Engines

2 Garrett Model TPE 331-10-511K Turboprop engines (Gulfstream P/N 610653).

Fuel

Aviation Turbine fuels ASTM designation D1655-68, types Jet A, Jet A-1, and Jet B; MIL-T-5624G-1, Grades JP-4 and JP-5; MIL-T-83133, Grade JP-8; and MIL-F-46005A(MR)-1, Types I and II; British D.ENG.R.D. 2486 Issue 2; British D.ENG.R.D. 2494 Issue 4; and NATO Equivalents.

Oil

MIL-L-23699B type II (See Mfg. Data Part VIII Approved POH for List of approved lubricants.)

Engine Limits

	Torque (HP)	RPM	EGT
Takeoff and Maximum Continuous	102.5% (820)	101.0%	650°C

Propeller and Propeller Limits

2 Dowty-Rotol Ltd. 3-bladed feathering and reversing propellers.
Gulfstream Assembly No. 640080

a. Dowty-Rotol Ltd. Type No. (C) R306/3-82-F/7-(c) VP 3027 includes
Dowty-Rotol Deice Boots 660709275 or B. F. Goodrich De-ice Boots 4E2498-10.
Dowty-Rotol Propeller Blade Assembly P/N 660706330-XX
Pitch settings at .7 radius stations:
Feather 83° 10'± 20', Reverse -13.75° ± 1.0°
Start Locks -1.25° ± 1.0°, Flight Idle 6.0° ± 0.5°
Diameter: 106 In., 1/2 in. reduction per blade allowed.
NOTE: Use Garrett oil transfer tube Part No. 897458-2.
NOTE: All engine ground running for maintenance and test purposes, with the airplane stationary, must be done with the airplane headed into the wind.

b. Spinner: 2 Dowty-Rotol Ltd. Type No. (C) SB7/3/1

c. Governor: 2 Garrett P/N 897410-4

Airspeed Limits

Maneuvering	182 m.p.h. (158K) CAS	
Maximum Operating	290 m.p.h. (252K) CAS	.60 MACH
Flaps extended - half	230 m.p.h. (200K) CAS	
Flaps extended - full	184 m.p.h. (160K) CAS	
Landing gear extemded	230 m.p.h. (200K) CAS	

2A4

XX - MODEL 695B (cont'd)

C.G. Range

Forward
210.91 inches aft of datum (20.6% MAC) at 11,750 lbs.
204.34 inches aft of datum (11.5% MAC) at 8,500 lbs.
204.34 inches aft of datum (11.5% MAC) at 6,836 lbs.
211.56 inches aft of datum (21.5% MAC) at 6,410 lbs.
Straight line variation between points.
Aft
217.03 inches aft of datum (29.1% MAC) at 11,750 lbs.
218.71 inches aft of datum (31.4% MAC) at 11,628 lbs.
217.85 inches aft of datum (30.2% MAC) at 6,639 lbs.
Straight line variation except between 11,628 lbs. and 6,639 lbs.
Inches aft of datum = 219.87 - (13402/weight)

Datum

196 In. forward of wing leading edge at center section

Leveling means

Longitudinal - Top of fuselage on centerline aft of wing trailing edge.
Lateral - Transverse beams at front of rear baggage compartment floor.

Maximum Weight

Maximum takeoff 11,750 lbs. (Maximum Ramp 11,800 lbs.)
Maximum landing 11,000 lbs.
Zero Fuel 9,800 lbs.

Maximum operating altitude

35,000 feet

Maximum No. of seats

11 (Pilot + 10 passengers; pilot, co-pilot + 9 passengers)

Maximum baggage

750 lb. (+290) Nonpressurized compartment
100 lb. (+245) Pressurized compartment

Fuel capacity

Total standard capacity 482 gal., usable 474 gal.
(See NOTE 1 for systems fuel).

Oil capacity

12.0 qts. total (6.0 qts. total each tank) (+188)
10.0 qts. usable (See NOTE 1 for system oil).

Control Surface movements

Elevator	Up	30° ± 1	Down	10° ± 2
		0		0
Elevator tab	Up	3° ± 1	Down	24° ± 1
Rudder	Right	20° ± 2	Left	20° ± 2
		0		0
Rudder tab	Right	20° ± 2	Left	20° ± 2
		0		0
Aileron	Up	23° ± 2	Down	15° ± 2
Flaps			Down	40° ± 2
Aileron tab	Up	17° ± 2.5°	Down	17° ± 2.5°

Serial Nos. eligible

Under the Delegation Option Provisions of Part 21 of the Federal Aviation Regulations, Delegaton Option Manufacturer No. SW-2 is authorized to: Issue Airworthiness Certificates and approve design and production changes on airplane Serial Numbers 96201 thru 96208 (See NOTES 20 and 22).

Specifications Pertinent to All Models

Certification basis — Type Certificate No. 2A4

Models 680, 680E:	**CAR 3** effective Nov. 1, 1949, through Amdt. 3-12 dated May 18, 1954.
Model 720:	**CAR 3** effective Nov. 1, 1949, through Amdt. 3-12 dated May 18, 1954, and 3.197, 3.395, 3.396 of Amdt. 3-2 dated August 12, 1957.
Models 560, 680F, 680FL:	**CAR 3** effective May 15, 1956, including Amdts. 3-3 dated May 17, 1958, and 3-4 dated October 6, 1958.
Models 680F (Pressurized) 680 FL (Pressurized):	**CAR 3** effective May 15, 1956, including 3.197, 3.395, 3.396 of Amd. 3-2 dated Aug.12, 1957, and Amdt. 3-3 dated May 17, 1958, and 3-4 dated October 6, 1958.
Model 680T:	**CAR 3** effective May 15, 1956, including 3.197, 3.395, 3.396 of Amdt. 3-2 dated August 12, 1957, and Amdts. 3-3 dated May 17, 1958, 3-4 dated Oct. 6, 1958, Amdt. 3-6 dated Sept.13, 1961, plus Special Conditions dated April 1, 1965.
Models 680V, 680W, 681:	**CAR 3** effective May 15, 1956, including 3.197, 3.270, 3.395, 3.396 of Amdt. 3-2 dated August 12, 1957, and Amdts. 3-3 dated May 17, 1958, 3-4 dated Oct.6, 1958, Amdt. 3-6 dated Sept.13, 1961, plus Special Conditions dated April 1, 1965.
Models 690, 690A, 690B	**CAR 3** dated May 15, 1956, including Pars. 3.197, 3.270, 3.395, and 3.396 of Amdt. 3-2 dated Aug.12, 1957, and Amdt. 3-3 dated May 17, 1958, 3-4 dated Oct.6, 1958, 3-6 dated Sept.13, 1961, Par. 23.473, 23.479, 23.481, and 23.483 of FAR 23, Amdt. 23-7 dated Sept.14, 1969, plus Special Conditions dated April 1, 1965, and August 12, 1970; Docket #10506
Model 685:	**CAR 3** dated May 15, 1956, including Pars. 3.197, 3.270, 3.395, and 3.396 of Amdt. 3-2 dated August 12, 1957, and Amdt. 3-3 dated May 17, 1958, 3-4 dated Oct.6, 1958, 3-6 dated Sept. 13, 1961.
Models 690C, 695	**CAR 3** dated May 15, 1956, including Pars. 3.197, 3.270, 3.395, and 3.396 of Amdt. 3-2 dated August 12, 1957, and Amdt. 3-3 dated May 17, 1958, 3-4 dated Oct.6, 1958, 3-6 dated Sept. 13, 1961, Pars. 23.473, 23.479, 23.481, and 23.483 of FAR 23, Amdt. 23-7 dated Sept. 14, 1969, plus Special Conditions dated April 1, 1965, and Aug.12, 1970; Docket #10506, and FAR 36 dated Dec.1, 1969, through Amdt. 36-6 dated Jan.24, 1977.
Model 695A, 690D	**CAR 3** dated May 15, 1956, including Pars. 3.197, 3.270, 3.395, and 3.396 of Amdt. 3-2 dated August 12, 1957, and Amdt. 3-3 dated May 1, 1958, 3-4 dated Oct. 6, 1958, 3-6 dated Sept. 13, 1961, Pars. 23.253, 23.335(b)(4), 23.473, 23.479, 23.481, 23.483, 23.571(a), 23.572(a)(1), and 23.1505(c) of FAR 23, Amdt. 23-7 dated Sept. 14, 1969, FAR 23.1303(e)(2) of Amdt. 23-17 dated Feb. 1, 1977, plus special Conditions dated April 1, 1965, and August 12, 1970, Docket No. 10506, and FAR 36 dated December 1, 1969, through Amdt. 36-6 dated Jan.24, 19'
Model 695B	**CAR 3** dated May 15, 1956, including Pars. 3.395 and 3.396 of Amdt.3-2 dated August 12, 1957, and Amdt. 3-3 dated May 1, 1958, 3-4 dated Oct. 6, 1958, 3-6 dated Sept.13, 1961, except for Subpart C, plus Pars. 23.253, 23.1303(e)(2), and 23.1505(c), and Subpart C of FAR 23 as amended thru Change 17 dated Sept.13, 1982, plus Special Conditions dated April 1, 1965, and Aug.12, 1970, Docket No. 10506 and FAR 36 dated December 1, 1969, through Amdt. 36-6 dated January 24, 1977.

Production basis — Production Certificate No. 203

2A4

Equipment

The basic required equipment as prescribed in the applicable airworthiness regulations (see Certification Basis) must be installed in the aircraft for certification. This equipment must include a current Airplane Flight Manual except for Models 690B, 690C, 690D, 695, 695A, and 695B which require a current Pilot's Operating Handbook.

In addition, the following item(s) are required:

1. Stall warning system:
   Models 560F, 680F, 680F(P), 680FL, 680FLP, 680T, 680W, 681, 690, 685, 690A (throuogh S/N 11268 except 11249) - Gulfstream Dwgs. 850016 and 850195.
   Models 690A (11249, 11269 through 11349), 690B - Gulfstream Dwgs. 850016 and 8000644
   Model 690C, 690D and 695 - Gulfstream Dwgs. 200036 and 800644.
   Model 695A and 695B - Gulfstream Dwgs. 200036, 800644 and 800746.

2. Outside Air Temperature Thermometer
   Models 680T, 680V, 680W, 681 - Gulfstream Dwg. 850295
   Models 690, 690A, 690B, 690C, 690D, 695, 695A, and 695B - Gulfstream Dwg. 850478.

3. EGT System
   Model 685 (with Service Letter 300 installed) Gulfstream Dwg. 890412.

NOTE 1:

Current weight and balance report, including list of equipment, included in certificated empty weight and loading instructions must be in each aircraft at the time of original airworthiness certification and at all times thereafter (except in the case of air carrier operators having an approved weight control system.)

The certificated empty weight and corresponding center of gravity location must include unusable fuel (included in total fuel capacity and undrainable oil (included in total oil capacity) as follows:

Model	680	680-E	720	680-F & 680-F Press	680-FL & 680-FL(P)
Fuel	15.5 lb.(+187)	15.5 lb.(+187)	15.5 lb.(+187)	15.5 lb.(+187)	15.5 lb.(+231)
Oil	15.0 lb.(+191)	15.0 lb.(+191)	15.0 lb.(+191)	17.4 lb.(+150)	17.4 lb.(+194)

Model	560-F	680W, 681 680T, 680V	690, 690A 690B	685
Fuel	15.5 lb.(+187)	13 lb. (+231)	31 lb.(+231)	27 lb.(+231)
Oil	17.4 lb.(+191)	6.5 lb.(+188)	4 lb.(+188)	0 lb.(+188)

Model	690C, 690D		695		695A, 695B
	Standard	Std + Optional	(SN 95000-95040)	(95041-95999)	
Fuel	33.5 lb.(+230)	53.6 lb.(+230)	33.6 lb.(+230)	53.6 lb.(+230)	53.6 lb.(+230)
Oil	4.0 lb.(+188)	4.0 lb.(+188)	4.0 lb.(+188)	4.0 lb.(+188)	4.0 lb.(+188)

NOTE 2:

The placards specified in the Airplane Flight Manual must be displayed in front of and in clear view of the pilots.

NOTE 3:

Serial Numbers 466, 471, 529, and 530 of Military RL-26-D as defined by Aero Commander Dwg. 6100012-A are eligible as Model 680 airplanes.

NOTE 4:

When Lycoming GSO-480-B1A6 engines are installed, the following pertains: The oil cooler outlet gills must be relocated in accordance with Service Letter No. 62 and oil temperature gage markings changed per Service Letter No. 63. Engines must be operated in accordance with Airplane Flight Manual.

NOTE 5:

An optional pressurized version of the Model 680-F designated "680-F (Pressurized)" was approved June 29, 1962. This model is a standard 680-F incorporating a factory modification per Aero Commander Dwg. 610021. Note the special required equipment list and the special equipment column for this modified 680-F in Revision No. 24 or Service Information SI-118.

NOTE 6: Model 680FL S/N 1471 and up are manufactured as 8500 lb. gross weight aircraft. Serial Numbers 1261 through 1470 are manufactured as 8000 lb. gross weight aircraft and become 8500 lb. aircraft when modified per Aero Commander Dwg. 6100028. Serial Number 1441through 1470 were modified per Rockwell Dwg. 6100028 at the factory.

NOTE 7: The Model 680 is eligible as a Model 680E when modified in accordance with Aero Commander Report G10-163.

NOTE 8: All Model 680T aircraft are to be modified or manufactured per Aero Commander Report G10-227 and are to be 8950 lb. gross weight aircraft.

NOTE 9: The Model 680T is eligible as a Model 680V when modified in accordance with Aero Commander Dwg. 6100034.

NOTE 10: Icing Approval:

a. The Models 680T, 680V, 680W, and 681 may be flown through known icing conditions when equipped in accordance with Aero Caommander Service Letter No. 196.
b. The Model 690 may be flown through known icing conditions when equipped in accordance with Aero Commander Service Letter No. 241A or Drawing 890338. Flight Manual Supplement 4 dated 6/10/71 is required.
c. Models 690A and 690B are fully equipped and approved for flight into known icing. See Flight Manual (Pilots Operating Handbook) for list of required operable equipment. Safe Flight P/N C-01426 and C-01427 required to provide stall warning.
d. Model 690C Serial Numbers 11600 thru 11619 approved for flight into known icing after compliance with Rockwell Service Letter No. 329. Serial Numbers 11620 and Subs are fully equipped for flight into known icing. See Pilots Operating Handbook for list of required operable equipment.
e. Model 695, 695A, 695B and690D are fully equipped for flight into known icing. See Pilots Operating Handbook for list of required operable equipment.

NOTE 11: The Models 680T and 680V may have the AiResearch engines TPE-331-43A installed as a product improvement item and in accordance with Aero Commander Service Letter No. 208.

NOTE 12: The Models 680T, 680V, 680W, and 681 may have auxiliary fuel tanks installed in accordance with Aero Commander Drawing 890326. These provide 25.5 usable gals. each side. (51 gal. total) Unusable added is negligible.

NOTE 13: The Model 685 may be approved for flight into known icing coorditions when equipped in accordance with Aero Commander Service Letter No. 241 or Drawing No. 890338. Flight Manual Supplement 5 dated April 15, 1972, is required.

NOTE 14: With GTS10-520-K engine installed, 2 Alcor turbine inlet temperature indicators must be installed per Rockwell Service Letter 300. Flight Manual Revision No. 5.

NOTE 15: In some cases, the serial number contains the basic number plus a dash followed by a second set of numbers. This second number is a model unit number and the basic serial number applies with or without the second number. Example as follows: 680FL-1779-148 can be referred to as S/N 1779-148 or by S/N 1779.

NOTE 16: If blades LT10673 or LT10673B are installed per STC SA546GL, propeller blade angles at the 42 inch station are: Reverse 14.0° ± .5°, Start Locks -8.7° ± .5°; Low 6.0° ± .5°, and Feather 77.9° ± .5°.

NOTE 17: Airframe electrical modifications per 800 788 required when installing Dowty Rotol boots 660709275 or B. F. Goodrich boots 4E2498-10 in place of previously installed B. F. Goodrich de-ice Kit 65-330-1.

NOTE 18: Maximum Baggage Weight increased to 750 pounds for Model 695A Serial Numbers 96063, 96069, 96075, 96078, and 96085.

NOTE 19: TPE 331-10-501K effective on Models 695 S/N 95000 through 95084, 695A S/N 96001through 96071 except those complying with Service Information Letter 189. TPE 331-1Q-511K effective on Models 695 S/N 95087 and Subs. 695A S/N 96000, 96072 and Subs. plus those complying with Service Information Letter 189. It is acceptable to have one each -501K and -511K engine installed.

NOTE 20: **Model 695A Serial Numbers 96062, 96063, 96069, 96075, and 96078, and 96085 are eligible as a Model 695B when modified in accordance with Gulfstream Aerospace Drawing 100062 Rework EO No. 3 except that the maximum value of zero fuel weights is limited to 9500 pounds.**

NOTE 21: **Model 690D airplanes, Serial Numbers 15000 through 15042, are eligible for conversion to Model 695A when modified in accordance with Gulfstream Drawing 100068.**

NOTE 22: Delegation Option Authorization No. SW-2 expired July 17, 1986.

...END...

E-284
Revision 9
Textron Lycoming
GSO-480-A1A6, -A1C6, -A2A6
GSO-480-B1A6, -B1B6 (O-480-1), -B1C6, -B1E6
-B1F6, -B1G6, -B1J6, -B2C6, -B2D6,
-B1B3, -B2G6, -B2H6
IGSO-480-A1A6 (0-480-3), -A1B6, -A1C6, -A1D6,
-A1E6, -A1F3-A1F6, -A1G6

May 15, 1988

TYPE CERTIFICATE DATA SHEET NO. E-284

Engines of models described herein conforming with this data sheet (which is a part of type certificate No. 284) and other approved data on file with the Federal Aviation Administration, meet the minimum standards for use in certificated aircraft in accordance with pertinent aircraft data sheets and applicable portions of the Civil Air Regulations/Federal Air Regulations provided they are installed, operated and maintained as prescribed by the approved manufacturer's manuals and other approved instructions.

Type Certificate Holder — Textron Lycoming/Subsidiary of Textron, Inc.
Williamsport Plant
Williamsport, Pennsylvania 17701

Model Lycoming	GSO-480-A1A6, -A1C6, -A2A6, -B1A6, -B1B6, -B1C6, -B1E6, -B1F6, -B1G6, -B1J6, -B2C6, -B2D6, -B2G6, -B2H6, -B1B3	IGSO-480-A1A6, -A1B6, -A1C6, -A1D6, -A1E6, -A1F6, -A1G6, -A1F3
Type 6HOA-Reduction Gear Ratio	77:120	- -
Rating		
Max. continuous, hp, r.p.m., in Hg., at:		
Rated pressure alt. (ft.)	320-3200-43.3-8000	320-3200-41.3-11,000
Sea level pressure alt. (ft.)	320-3200-45.0-S.L.	320-3200-45.0-S.L.
Takeoff (5 min.), hp, r.p.m. in. Hg., at:		
Rated pressure alt. (ft.)	340-3400-45.8-8000	340-3400-44.0-11,000
Sea level pressure alt. (ft.)	340-3400-48.0-S.L.	340-3400-48.0-S.L.
Fuel (min. grade aviation gasoline)*	100/130	- -
Lubricating Oil		
(lubricant should conform to the specifications as listed or to subsequent revisions thereto)	Lycoming Spec. No. 301-F and Service Instruction No. 1014	- - - -
Bore and stroke, in.	5.125 x 3.875	- -
Displacement, cu. in.	479.7	- -
Supercharging ratio	11.27:1	- -
Compression ratio	7.3:1	- -
Weight (dry) lb.	See NOTE No. 8	- -
C.G. location (dry)	See NOTE No. 8	- -
Propeller shaft, SAE No.	See NOTE No. 8	- -
Carburetion	See NOTE No. 8	- -
Ignition, dual	See NOTE No. 8	- -
Timing °BTC	25	- -
Spark Plugs	See NOTE No. 9	- -
Oil Sump - capacity	Dry Sump	- -
Notes 1 through 9 as applicable	1,2,3,4,5,6,7,8,9	- -

"- -" indicates "same as preceding model"
"#" indicates "does not apply"
"" See latest revision of Lycoming Service Instruction No. 1070 for alternate fuel grades.*

Page No.	1	2	3	4
Rev. No.	9	9	9	9

Reformatted 1/95

E-284

Certification basis:

Regulations & Amendments	Model	Date of Application	Date of Type Certificate No. 284 Issued/Revised
CAR 13 Effective March 5, 1952			
As Amended by 13-1 and 13-2	GS0-480-A1A6	December 13, 1954	June 30, 1955
CAR 13 Effective June 15, 1956	0-480-1	November 27, 1956	December 5, 1956
	GSO-480-B1A6	April 26, 1957	May 9, 1957
	GSO-480-B1B6	April 26, 1957	May 9, 1957
	GSO-480-B1C6	April 26, 1957	May 9, 1957
	GSO-480-A1C6	June 18, 1957	June 27, 1957
As Amended by 13-1	IGSO-480-A1A6	January 10, 1958	May 14, 1958
	GSO-480-B2D6	February 21, 1958	March 6, 1958
CAR 13 Effective June 15, 1956			
As Amended by 13-1, 13-2, 13-3	GSO-480-A2A6	April 13, 1960	May 3, 1960
	IGSO-480-A1B6	June 11, 1960	August 25, 1960
	GSO-480-B1E6	May 26, 1961	June 19, 1961
	GSO-480-B1F6	May 26, 1961	June 19, 1961
	GSO-480-B1G6	May 26, 1961	June 19, 1961
	GSO-480-B2H6	May 26, 1961	June 19, 1961
	GSO-480-B2C6	June 1, 1961	June 19, 1961
	GSO-480-B2G6	June 1, 1961	June 19, 1961
	O-480-3	June 26, 1961	June 14, 1961
	IGSO-480-A1C6	September 13, 1961	October 17, 1961
	IGSO-480-A1D6	May 2, 1962	May 6, 1963
And 13-4	IGSO-480-A1F6	July 6, 1962	August 16, 1962
	IGSO-480-A1E6	August 27, 1964	October 23, 1964
	IGSO-480-A1G6	August 16, 1966	August 26, 1966
	IGSO-480-B1J6	January 5, 1967	January 21, 1967
	GSO-480-B1B3	June 21, 1971	July 7, 1971
	IGSO-480-A1F3	January 28, 1980	February 21, 1980

Production basis: Production Certificate No. 3

NOTE 1. Maximum permissible temperatures:

Cylinder Head Well type	Cylinder Base*	Oil Inlet
500°F	350°F	225°F - GSO-480-A1A6, -A2A6, -A1C6
		245°F - All others

*This parameter dispensed with where pistons are internally cooled by oil jets.

NOTE 2.

	Minimum	Maximum
Fuel Pressure Limits:	9 p.s.i.	15 p.s.i. (17 p.s.i. min., 65 p.s.i. max. for IGSO-480-A1E6, -A1D6, -A1G6)
Oil Pressure Limits:		
(Normal Operations)	55 p.s.i.	85 p.s.i.
(Idling)	25 p.s.i. (35 p.s.i. for IGSO-480-A1A6, -A1B6, -A1C6, -A1F6, -A1F3)	

NOTE 3. The following accessory provisions are made:

Accessory	Rotation Facing Drive Pad	Speed Ratio to Crankshaft	Maximum Torque (in. -lb.) Continuous	Static	Maximum Overhang Moment (in. -lb.)
Starter	C	1.000:1	#	12000	300
Generator	C	2.600:1	500	2200	400
Fuel Pump	CC	.803:1	25	450	25
Vacuum Pump	C	1.219:1	200	800	25
Hydraulic Pump	C	1.083:1	400	1650	175
Tachometer	CC	.500:1	7	50	#
Propeller Governor	C	.801:1	125	1200	25

"C" - Clockwise, "CC" - Counter-Clockwise
"#" Indicates "does not apply"

NOTE 4. The "6" in the engine model designation indicates the crankshaft has five 3rd order and one 6th order torsional vibration dampers. The IGSO-480-A1F3 and GSO-480-B1B3 have four heavy 3rd order and two 6th order torsional vibration dampers.

NOTE 5. All engines incorporate provisions for absorbing propeller thrust in both tractor and pusher type installations.

NOTE 6. Military Models 0-480-1 and -3 are identical to the corresponding civil designated engines except for ignition, which are the Scintilla S6LN-22 and S6RN-23 with AN 3105, primary ground terminal. When installed in certificate aircraft, the corresponding commercial model designations and type certificate number should be added to the engine data plate.

NOTE 7. The above models incorporate additional characteristics as follows:

Models	Characteristics
GSO-480-A1A6	Basic model. Geared drive, six cylinder, horizontally opposed, supercharged, dry sump, aircooled engine with side mounted accessory drives and accessories.
GSO-480-A1C6	Similar to GSO-480-A1A6 except has provisions for a supercharger bearing thermocouple.
GSO-480-A2A6	Similar to GSO-480-A1A6 except has flange type propeller shaft with 2-way oil for reversible propeller.
GSO-480-B1A6	Similar to GSO-480-A1C6 except incorporates crankcase oil jets for increased piston cooling, provisions for supercharger inlet and an updraft carburetor.
GSO-480-B1B6	Similar to GSO-480-B1A6 except has a horizontal elbow and carburetor under the engine.
GSO-480-B1B3	Same as GSO-480-B1B6 except that the torsional damper system has been modified. (SEE NOTE 4)
GSO-480-B1C6	Similar to GSO-480-B1A6 except has a horizontal carburetor mounted directly on a straight-through air inlet supercharger housing.
GSO-480-B1E6	Similar to GSO-480-B1A6 excepting magnetos.
GSO-480-B1F6	Similar to GSO-480-B1B6 excepting magnetos.
GSO-480-B1G6	Similar to GSO-480-B1C6 excepting magnetos.
GSO-480-B1J6	Same as GSO-480-B1A6 except incorporates 1200 series Bendix magnetos.
GSO-480-B2C6	Similar to GSO-480-B1C6 except has flanged propeller shaft and provision for reversible propeller.
GSO-480-B2D6	Similar to GSO-480-A2A6 except has internal piston cooling, special supercharger inlet for down-draft carburetor and is also similar to the -B1 series engines except incorporates a flange type propeller shaft.
GSO-480-B2G6	Similar to GSO-480-B2C6 excepting magnetos.
GSO-480-B2H6	Similar to GS-470-B2D6 excepting magnetos.
IGSO-480-A1A6	Basic fuel injection model.
IGSO-480-A1B6	Similar to IGSO-480-A1A6 except has retard breaker magnetos.
IGSO-480-A1C6	Similar to IGSO-480-A1A6 except has horizontal air inlet housing and throttle.
IGSO-480-A1D6	Similar to GSO-480-B1A6, except for incorporation of service kit which included Bendix RS10-FB1 fuel injector and supercharger air inlet housing assembly, P/N 74323.
IGSO-480-A1E6	Similar to IGSO-480-A1D6 except for different configuration of supercharger air inlet housing and incorporation of retard breaker magnetos.
IGSO-480-A1F3	Similar to IGSO-480-A1F6 except that it has two 6th and four heavy 3rd order dynamic counterweights.
IGSO-480-A1F6	Similar to IGSO-480-A1C6 except has retard breaker magnetos in place of impulse type magnetos.
IGSO-480-A1G6	Same as IGSO-A1E6 with 1200 series magnetos but without the Bendix modulator unit.

E-284

NOTE 8. For all models - weights, carburetion, ignition, C.G. location and propeller shaft SAE designations.

Models	Weight (dry) lb.	Carburetion	Ignition, dual	C.G. Location, Dry: From front face of thrust nut, in.	C.G. Location, Dry: Off propeller shaft C.L. in. lateral	C.G. Location, Dry: Off propeller shaft C.L. in. vertical	Propeller shaft, SAE No.
GSO-480-A1A6	498	Bendix PS-7BD Bendix S6	Bendix S6LN-20, S6RN-21	21.74	0.22 left	0.59 above	20 spline
-A1C6	498	Bendix PS-7BD	Bendix S6LN-20, S6RN-21	21.74	0.22 left	0.59 above	20 spline
-A2A6	498	Bendix PS-7BD	Bendix S6LN-20, S6RN-21	21.74	0.22 left	0.59 above	flange, ARP 502
-B1A6	513	Bendix PS-7BD	Bendix S6LN-20, S6RN-21	22.32	0.18 left	0.22 above	20 spline
-B1B6	515	Bendix PSH-7BD	Bendix S6LN-20, S6RN-21	22.18	0.18 left	0.01 below	20 spline
*O-480-1							
-B1B3	517	Bendix PSH-7BD	Bendix S6LN-20, S6RN-21	22.18	0.18 left	0.01 below	20 spline
GSO-480-B1C6	512	Bendix PSH-7BD	Bendix S6LN-20, S6RN-21	22.54	0.16 left	0.59 above	20 spline
-B1E6	513	Bendix PS-7BD	Bendix S6LN-204, S6RN-200 or S6LN-604, S6RN-600	22.32	0.18 left	0.22 above	20 spline
-B1F6	515	Bendix PSH-7BD	Bendix S6LN-204, S6RN-200 or S6LN-604, S6RN-600	22.18	0.18 left	0.01 below	20 spline
-B1G6	512	Bendix PSH-7BD	Bendix S6LN-204, S6RN-200 or S6LN-604, S6RN-600	22.54	0.16 left	0.59 above	20 spline
-B1J6	515	Bendix PS-7BD	Bendix S6LN-1209, S6RN-1227	22.29	0.18 left	0.22 above	20 spline
-B2C6	512	Bendix PSH-7BD	Bendix S6LN-20, S6RN-21	22.54	0.16 left	0.59 above	flange, ARP 502
GSO-480-B2D6	513	Bendix PSD-7BD	Bendix S6LN-20, S6RN-21	22.39	0.25 left	0.71 above	flange, ARP 502
-B2G6	512	Bendix PSH-7BD	Bendix S6LN-20, S6RN-21, S6LN-204, S6RN-200, S6LN-604, S6RN-600	22.54	0.16 left	0.59 above	flange, ARP 502
-B2H6	513	Bendix PSD-7BD	Bendix S6LN-204, S6RN-200, S6LN-604, S6RN-600	22.39	0.25 left	0.71 above	flange, ARP 502
IGSO-480-A1A6	512	Fuel Injector Simmonds Type 570	Bendix S6LN-20, S6RN-21	22.00	0.34 left	0.71 above	20 spline
*O-480-3 IGSO-480-A1B6	512	Simmonds Type 570	Bendix S6LN-204, S6RN-200, S6LN-604, S6RN-600	22.00	0.34 left	0.71 above	20 spline
-A1C6	513	Simmonds Type 570	Bendix S6LN-20, S6RN-21	22.00	0.34 left	0.71 above	20 spline
-A1D6	514	Bendix RS10-FB1	Bendix S6LN-20, S6RN-21	22.29	0.21 left	0.35 above	20 spline
-A1E6	514	Bendix RS10-FB1	Bendix S6LN-204, S6RN-200	22.29	0.21 left	0.35 above	20 spline
-A1F6	513	Simmonds Type 570	Bendix S6LN-204, S6RN-200, S6LN-604, S6RN-600	22.00	0.34 left	0.71 above	20 spline
-A1G6	515	Bendix RS10-FB1	Bendix S6LN-1209, S6RN-1208	22.29	0.21 left	0.35 above	20 spline
-A1F3	517	Simmonds Type 570	Bendix S6LN-204 S6RN-200	22.00	0.34 left	0.71 above	20 spline

* See NOTE No. 6.

NOTE 9. Spark Plugs: See latest revision of Lycoming Service Instruction No. 1042 for approved equipment.

.....END.....

DEPARTMENT OF TRANSPORTATION
FEDERAL AVIATION ADMINISTRATION

A9CE
Revision 27
CESSNA
188 A188A
188A A188B
188B T188C
A188
March 31, 2003

TYPE CERTIFICATE DATA SHEET NO. A9CE

This data sheet which is part of Type Certificate A9CE prescribes conditions and limitations under which the product for which the type certificate was issued meets the airworthiness requirements of the Federal Aviation Regulations.

Type Certificate Holder — Cessna Aircraft Company
P O Box 7704
Wichita KS 67277

I. Model 188, AGwagon 230, 1 PCLM (Normal and Restricted Category), approved February 14, 1966

Engine — Continental O-470-R

*Fuel — 80/87 minimum grade aviation gasoline

*Engine limits — For all operations, 2600 rpm (230 hp)

Propeller and propeller limits

1. (a) McCauley 1A200/AOM fixed pitch
   Static rpm at max. permissible throttle setting:
   not over 2300, not under 2200
   No additional tolerance permitted
   Diameter: not over 90 in., not under 88 in.
2. (a) McCauley constant speed, 2A34C50 hub with 90A-2 blades
   Diameter: not over 88 in., not under 86 in.
   Pitch settings at 36 in. sta.: low 8°, high 22°
   (b) Governor: Garwin 34-828-01, McCauley C290D2/T1 or C290D3/T1, or Woodward A210452
3. (a) McCauley constant speed, 2A34C66 hub with 90AT-2 blades
   Diameter: not over 88 in., not under 86 in.
   Pitch settings at 36 in. sta.: low 8°, high 22°
   (b) Governor: Garwin 34-828-01, McCauley C290D2/T1 or C290D3/T1, or Woodward A210452
4. (a) McCauley constant speed, 2A34C201 hub with 90DA-2 blades
   Diameter: not over 88 in., not under 86.5 in.
   Pitch settings at 30 in. sta.: low 10.5°, high 24.5°
   (b) Governor: Garwin 34-828-01, McCauley C290D2/T1 or C290D3/T1, or Woodward A210452
5. (a) McCauley constant speed, 2A34C203 hub with 90DCA-2 blades
   Diameter: not over 88 in., not under 86.5 in.
   Pitch settings at 30 in. sta.:
   Low 10.0°, high 24.5°
   (b) Governor: Garwin 34-828-01, McCauley C290D2/T1 or C290D3/T1, or Woodward A210452

*Airspeed Limits (CAS) (Normal Category)

Never exceed181 mph (157 knots)
Maximum structural cruising 144 mph (125 knots)
Maneuvering 127 mph (110 knots)
Flaps extended 110 mph (96 knots)
(See Additional Limitation for Restricted Category.)

Page No.	1	2	3	4	5	6	7	8	9	10	11	12	13	14	15	16	17
Rev. No.	27	21	21	21	19	21	21	25	25	22	26	26	26	25	26	26	27

A9CE

I. Model 188, AGwagon 230 (cont'd)

C.G. Range (Normal Category)	(+39.0) to (+45.5) at 2300 lb. or less (+41.0) to (+45.5) at 3300 lbs. Straight line variation between points given.
Empty weight C.G. range	None
*Maximum weight	3300 lb. (Normal Category)
Number of Seats (Max.)	1 (at +91 to +95)
Maximum Baggage	100 lb. (+12.0) (optional)
Fuel Capacity	37 gal. (+11.0; 36.5 gal. usable) *See Note 1 for data on unusable fuel.*
Oil Capacity	12 qt. (-17.0; includes 9 lb. unusable) *See Note 1 for data on undrainable oil.*

Control surface movements

Wing flaps (S/N 188-0001 through 188-0293)			0° -	28° ± 2°
Wing flaps (S/N 188-0294 and on)			0° -	20° ± 1°
Ailerons (from neutral)	Up	18° ± 1°	Down	10° ± 1°
Elevators	Up	26° 30' ± 1°	Down	21° ± 1°
Elevator tab	Up	12° ± 1°	Down	27° ± 1°
Rudder	Right	24° + 0°, -1°	Left	24° + 0°, -1°

(Neutral aileron is rigged with trailing edge 3° ± 30' below trailing edge of wing.)

Additional Limitations for Restricted Category

*Airspeed limits (CAS)	Maximum operating speed in agricultural operations 120 mph (104 knots)
*C.G. Range	(+39.0) to (+45.5) at 2300 lbs. or less (+42.0) to (+45.5) at 3800 lbs.
*Maximum Weight	3800 lb. *(See Note 3.)*
Serial numbers eligible	653, 188-0001 through 188-0572

II. Model A188, AGwagon 300, 1 PCLM (Normal and Restricted Category), approved February 14, 1966

Engine	Continental IO-520-D
*Fuel	100/130 minimum grade aviation gasoline
*Engine limits	Takeoff (5 min.) at 2850 rpm (300 hp) For all other operations, 2700 rpm (285 hp)

Propeller and propeller limits

1. (a) McCauley D2A34C58 hub or D2A34C58-0 (oil filled) hub with 90AT-4 blades
   Diameter: not over 86 in., not under 84 in.
   Pitch settings at 36 in. sta.:
   Low 8°, high 25°
   (b) Governor: Garwin 34-828-01 or McCauley C290D2/T9 or C290D3/T9, or Woodward A210462
   (c) Spinner, Cessna 0752040 (optional)
2. (a) McCauley F2A34C58 hub with 90AT-4 blades
   Diameter: not over 86 in., not under 84 in.
   Pitch settings at 36 in. sta.:
   Low 8°, high 25°

II. Model A188, AGwagon 300 (cont'd)

(b) Governor: Garwin 34-828-01 or McCauley C290D2/T9 or C290D3/T9, or Woodward A210462

3. (a) McCauley D2A34C58/90AT-8 or D2A34C58-0/90AT-8 (oil filled)
Diameter: not over 82 in., not under 80 in.
Pitch settings at 36 in. sta.:
Low 8.8°, high 25.8°

(b) Governor: Garwin 34-828-01, McCauley C290D2/T9 or C290D3/T9, or Woodward A210462

4. (a) McCauley D2A34C98/90AT-8 or D2A34C98-0/90AT-8 (oil filled)
Diameter: not over 82 in., not under 80 in.
Pitch settings at 36 in. sta.:
Low 8°, high 25°

(b) Governor: Garwin 34-828-01, McCauley C290D2/T9 or C290D3/T9, or Woodward A210462

(c) Spinner, Cessna 0752040 (optional)

*Airspeed Limits (CAS) (Normal Category)	Never exceed	181 mph (157 knots)
	Maximum structural cruising	144 mph (125 knots)
	Maneuvering	127 mph (110 knots)
	Flaps extended	110 mph (96 knots)
	(See Additional Limitation for Restricted Category.)	

C.G. Range (Normal Category): (+39.0) to (+45.0) at 2300 lbs. or less
(+41.0) to (+45.5) at 3300 lbs.
Straight line variation between points given.

Empty weight C.G. range: None

*Maximum weight: 3300 lbs. (normal category)

Number of seats (maximum): 1 (at +91 to +95)

Maximum baggage: 100 lb. (+12.0) (optional)

Fuel capacity: 37 gal. (+11.0; 36.5 gal. usable)
See Note 1 for data on unusable fuel.

Oil capacity: 12 qt. (-17.0; includes 9 lb. usable)
See Note 1 for data on undrainable oil.

Control surface movements:

Wing flaps (S/N 188-0001 through 188-0293)				0° - 28° ± 2°
Wing flaps (S/N 188-0294 and on)				0° - 20° ± 1°
Ailerons (from neutral)	Up	18° ± 1°	Down	10° ± 1°
Elevators	Up	26° 30' ± 1°	Down	21° ± 1°
Elevator tab	Up	12° ± 1°	Down	27° ± 1°
Rudder	Right	24° + 0°, -1°	Left	24° + 0°, -1°

(Neutral aileron is rigged with trailing edge 3° ± 30' below trailing edge of wing.)

Additional Limitations for Restricted Category

*Airspeed limits (CAS): Maximum operating speed in agricultural operations 120 mph (104 knots)

C.G. range: (+39.0) to (+45.5) at 2300 lbs. or less
(+42.4) to (+45.5) at 4000 lbs.

*Maximum weight: 4000 lbs. (*See Note 3.*)

Serial numbers eligible: 653, 188-0001 through 188-0572

A9CE

III. Model 188A, AGwagon "A" & "B", 1 PCLM (Normal and Restricted Category), approved September 26, 1969

Engine	Continental O-470-R
*Fuel	80/87 minimum grade aviation gasoline
*Engine limits	For all operations, 2600 rpm (230 hp)
Propeller and propeller limits	1. (a) McCauley 1A200/AOM fixed pitch Static rpm at maximum permissible throttle setting: Not over 2300, not under 2200 No additional tolerance permitted. Diameter: not over 90 in., not under 88 in. 2. (a) McCauley constant speed, 2A34C50 hub with 90A-2 blades Diameter: not over 88 in., not under 86 in. Pitch settings at 36 in. sta.: low 8°, high 22° (b) Governor: Woodward A210452, Garwin 34-828-01, McCauley C290D2/T1 or C290D3/T1 3. (a) McCauley constant speed, 2A34C66 hub with 90AT-2 blades Diameter: not over 88 in., not under 86 in. Pitch settings at 36 in. sta.: low 8°, high 22° (b) Governor: Woodward A210452, Garwin 34-828-01, McCauley C290D2/T1 or C290D3/T1 4. (a) McCauley constant speed, 2A34C201 hub with 90DA-2 blades Diameter: not over 88 in., not under 86.5 in. Pitch settings at 30 in. sta.: low 10.5°, high 24.5° (b) Governor: Woodward A210452, Garwin 34-828-01, McCauley C290D2/T1 or C290D3/T1 5. (a) McCauley constant speed 2A34C203 hub with 90 DCA-2 blades Diameter: not over 88 in., not under 86.5 in. Pitch settings at 30 in. sta.: low 10.0°, high 24.5° (b) Governor: Woodward A210452, Garwin 34-828-01 McCauley C290D2/T1 or C290D3/T1
*Airspeed Limits (CAS)	Never exceed 181 mph (157 knots) Maximum structural cruising 144 mph (125 knots) Maneuvering 127 mph (110 knots) Flaps extended 110 mph (96 knots) *(See Additional Limitation for Restricted Category.)*
C.G. range (normal category)	(+39.0) to (+45.5) at 2300 lbs. or less (+41.0) to (+45.5) at 3300 lbs. Straight line variation between points given.
Empty weight C.G. range	None
*Maximum weight	3300 lbs. (normal category)
Number of seats (max.)	1 (at +91 to 95)
Maximum baggage	100 lb. (+12.0) (optional)
Fuel capacity	37 gal. (+11.0; 36.5 usable) *See Note 1 for data on unusable fuel.*

III. Model 188A, AGwagon "A" & "B" (cont'd)

Oil capacity	12 qt. (-17.0; includes 9 lb. unusable) *See Note 1 for data on undrainable oil.*

Control surface movements

Wing flaps			Down	20° ± 1°
Ailerons (from neutral)	Up	18° ± 1°	Down	10° ± 1°
Elevators	Up	26° ± 1°	Down	21° ± 1°
Elevator tab	Up	12° ± 1°	Down	27° ± 1°
Rudder	Right	24° + 0°, -1°	Left	24° + 0°, -1°

(Neutral aileron is rigged with trailing edge 3° ± 30' below trailing edge of wing.)

Additional Limitations for Restricted Category

*Airspeed limits (CAS)	Maximum operating speed in agricultural operations 120 mph (104 knots)
C.G. range	(+39.0) to (+45.5) at 2300 lbs. or less (+42.0) to (+45.5) at 3800 lbs. Straight line variation between points given.
*Maximum weight	*See Note 3.*
Serial numbers eligible	18800573 through 18800832

IV. Model A188A, AGwagon "A" & "B", 1 PCLM (Normal and Restricted Category), approved September 26, 1969

Engine	Continental IO-520-D
*Fuel	100/130 minimum grade aviation gasoline
*Engine limits	Takeoff (5 min.) at 2850 rpm (300 hp) For all other operations, 2700 rpm (285 hp)

1. (a) McCauley D2A34C58 hub or D2A34C58-0 (oil filled) hub with 90AT-4 blades
   Diameter: not over 86 in., not under 84 in.
   Pitch settings at 36 in. sta.:
   Low 8°, high 25°
   (b) Governor: Garwin 34-828-01, McCauley C290D2/T9 or C290D3/T9, or Woodward A210462
2. (a) McCauley F2A34C58 hub with 90AT-4 blades
   Diameter: not over 86 in., not under 84 in.
   Pitch settings at 36 in. sta.:
   Low 8°, high 25°
   (b) Governor: Garwin 34-828-01, McCauley C290D2/T9 or C290D3/T9, or Woodward A210462
3. (a) McCauley D2A34C58/90AT-8 or D2A34C58-0/90AT-8 (oil filled)
   Diameter: not over 82 in., not under 80 in.
   Pitch settings at 36 in. sta.:
   Low 8.8°, high 25.8°
   (b) Governor: Garwin 34-828-01, McCauley C290D2/T9 or C290D3/T9, or Woodward A210462
4. (a) McCauley D2A34C98/90AT-4 or D2A34C98-0/90AT-4 (oil filled)
   Diameter: not over 86 in., not under 84 in.
   Pitch settings at 36 in. sta.:
   Low 8°, high 25°
   (b) Governor: Garwin 34-828-01, McCauley C290D2/T9 or C290D3/T9

IV. Model A188A, AGwagon "A" & "B" (cont'd)

5. (a) McCauley D2A34C98/90AT-8 or D2A34C98-0/90AT-8 (oil filled)
Diameter: not over 82 in., not under 80 in.
Pitch settings at 36 in. sta.:
Low 8.8°, high 25.8°
(b) Governor: Garwin 34-828-01, McCauley C290D2/T9 or C290D3/T9

*Airspeed Limits (CAS)	Never exceed	181 mph (157 knots)
	Maximum structural cruising	144 mph (125 knots)
	Maneuvering	127 mph (110 knots)
	Flaps extended	110 mph (96 knots)
	(See Additional Limitation for Restricted Category.)	

C.G. Range (Normal Category): (+39.0) to (+45.5) at 2300 lbs. or less
(+41.0) to (+45.5) at 3300 lbs.
Straight line variation between points given.

Empty weight C.G. Range: None

*Maximum weight: 3300 lbs. (normal category)

Number of seats (max.): 1 (at +91 to +95)

Maximum baggage: 100 lb. (+12.0) (Optional)

Fuel capacity: 37 gal. (+11.0; 36.5 gal. usable)
See Note 1 for data on unusable fuel.

Oil capacity: 12 qt. (-17.0; includes 9 lbs. unusable)
See Note 1 for data on undrainable oil.

Control surface movements:

Wing flaps			Down	20° ± 1°
Ailerons (from neutral)	Up	18° ± 1°	Down	10° ± 1°
Elevators	Up	26° ± 1°	Down	21° ± 1°
Elevator tab	Up	12° ± 1°	Down	27° ± 1°
Rudder	Right	24° + 0°, -1°	Left	24° + 0°, -1°

(Neutral aileron is rigged with trailing edge 3° ± 30' below trailing edge of wing.)

Additional Limitations for Restricted Category

*Airspeed Limits (CAS): Maximum operating speed in agricultural operations 120 mph (104 knots)

C.G. Range: (+39.0) to (+47.5) at 2300 lbs. or less
(+39.4) to (+47.5) at 2500 lbs.
(+42.4) to (+45.5) at 4000 lbs.
Straight line variation between points given.

*Maximum weight: *See Note 3.*

Serial numbers eligible: 18800573 through 18800832

**V. Model 188B, AGpickup, 1 PCLM (Restricted Category), approved December 20, 1971
Model 188B, AGpickup, 1 PCLM (Normal Category) (See required equipment, item 2), approved December 20, 1971**

Engine: Continental O-470-R (S/N 18800833 through 18801824)
Continental O-470-S (S/N 18801825 and up) *(See Note 6.)*

*Fuel: 80/87 minimum grade aviation gasoline

V. Model 188B, AGpickup (cont'd)

*Engine limits	For all operations, 2600 rpm (230 hp)
Propeller and propeller limits	1. (a) McCauley 1A200/AOM Fixed Pitch Static rpm at max. permissible throttle setting: Not over 2300, not under 2200 No additional tolerance permitted. Diameter: not over 90 in., not under 88 in. 2. (a) McCauley Constant Speed, 2A34C50 hub with 90A-2 blades Diameter: not over 88 in., not under 86 in. Pitch settings at 36 in. sta.: Low 8°, high 22° (b) Governor: Woodward A210452, Edo-Aire 34-828-01 or McCauley C290D2/T1 or C290D3/T1 3. (a) McCauley constant speed, 2A34C66 hub with 90AT-2 blades Diameter: not over 88 in., not under 86 in. Pitch settings at 36 in. sta.: Low 8°, high 22° (b) Governor: Woodward A210452, Edo-Aire 34-828-01 or McCauley C290D2/T1 or C290D3/T1 4. (a) McCauley constant speed, 2A34C201 hub with 90DA-2 blades Diameter: not over 88 in., not under 86.5 in. Pitch settings at 30 in. sta.: Low 10.5°, high 24.5° (b) Governor: Woodward 4210452, Edo-Aire 34-828-01 or McCauley C290D2/T1 or C290D3/T1 5. (a) McCauley constant speed, 2A34C203 hub with 90DCA-2 blades Diameter: not over 88 in., not under 86.5 in. Pitch settings at 30 in. sta.: Low 10.0°, high 24.5° (b) Governor: Woodward A210452, Edo-Aire 34-828-01, McCauley C290D2/T1 or C290D3/T1
*Airspeed Limits (CAS)	Never exceed 181 mph (157 knots) Maximum structural cruising 144 mph (125 knots) Maneuvering 116 mph (101 knots) Flaps extended (5°) 120 mph (104 knots) (10° - 20°) 110 mph (96 knots)
C.G. Range (normal category)	(+39.0) to (+45.5) at 2300 lbs. or less (+41.0) to (+45.5) at 3300 lbs. Straight line variation between points given.
Empty weight C.G. range	None
*Maximum weight	3300 lbs. (normal category)
Number of seats (max.)	1 (at +91 to +95)
Maximum cargo	26.7 cubic feet within operational gross weight
Fuel capacity	37 gal. (+11.0, 36.5 usable) *See Note 1 for data on unusable fuel.*
Oil capacity	12 qt. (-17.0; includes 9 lb. unusable) *See Note 1 for data on undrainable oil.*

A9CE

V. Model 188B, AGpickup (cont'd)

Control surface movements	Wing flaps			Down	20° ± 1°
	Ailerons (from neutral)	Up	18° ± 1°	Down	10° ± 1°
	Elevators	Up	26° ± 1°	Down	21° ± 1°
	Elevator tab	Up	12° ± 1°	Down	27° ± 1°
	Rudder	Right	24° + 0°, -1°	Left	24° + 0°, -1°
	(Neutral aileron is rigged with trailing edge 3° ± 30' below trailing edge of wing.)				

Additional Limitations for Restricted Category

*Airspeed limits (CAS) — Maximum operating speed in agricultural operations 120 mph (104 knots)

C.G. Range — (+39.0) to (+45.5) at 2300 lbs. or less
(+42.0) to (+45.5) at 3800 lbs.
Straight line variation between points given.

*Maximum Weight — *See Note 3.*

Serial numbers eligible — 18800833 through 18802348

VI. Model A188B, AGwagon"C" and AGtruck, 1 PCLM (Restricted Category), approved December 20, 1971, Model A188B, Agwagon "C" and AGtruck, 1 PCLM (Normal Category), (see required equipment, Item 2), approved December 20, 1971

Engine — Continental IO-520-D

*Fuel — 100/130 minimum grade aviation gasoline (S/N 18800833 through 18803046)
100LL/130 minimum grade aviation gasoline (S/N 678T, 18803047 and on)

*Engine limits — Takeoff (5 min.) at 2850 rpm (300 hp)
For all other operations, 2700 rpm (285 hp)

Propeller and propeller limits

1. S/N 678T, 18800833 through 18803721
   (a) McCauley D2A34C58/90AT-8 or D2A34C98/90AT-8 or D2A34C58-0/90AT-8 (oil filled) or D2A34C98-0/90AT-8 (oil filled)
   Diameter: not over 82 in., not under 80 in.
   Pitch setting at 36 in. sta.:
   Low 8.8°, high 25.8°
   (b) Governor: Edo-Aire 34-828-01-1, McCauley C290D2/T9 or C290D3/T9, or Woodward A210462
2. S/N 678T, 18800833 through 18803721
   (a) McCauley D2A34C58/90AT-4 or D2A34C98/90AT-4 or D2A34C58-0/90AT-4 (oil filled) or D2A34C98-0/90AT-4 (oil filled)
   Diameter: not over 86 in., not under 84 in.
   Pitch settings at 36 in. sta.:
   Low 8°, high 25°
   (b) Governor: Edo-Aire 34-828-01-1, McCauley C290D2/T9 or C290D3/T9, or Woodward A210462
3. S/N 678T, 18802002 through 18803721 and those aircraft reworked per SE75-4
   (a) McCauley D3A32C90/82NC-2 or D3A32C90-N/82NC-2 (oil filled)
   Diameter: not over 80 in., not under 78.5 in.
   Pitch setting at 30 in. sta.:
   Low 10.4°, high 28.1°
   (b) Governor: McCauley C290D2/T9 or C290D3/T9, Edo-Aire 34-828-01-1 or Woodward A210462

VI. Model A188B (cont'd)	4. S/N 18803722 and on and those aircraft reworked per Cessna Service Kit SK188-76 or SK188-77 (a) McCauley B2A34C205/90DHA-4 Diameter: not over 86 in., not under 84.5 in. Pitch setting at 30 in. sta.: Low 9.7°, high 28.5° (b) Governor: McCauley C290D3/T9 5. S/N 18803722 and on (a) McCauley D3A32C408/82NDA-2 Diameter: not over 80 in., not under 78.5 in. Pitch setting at 30 in. sta.: Low 10.4°, high 28.1° (b) Governor: McCauley C290D3/T9
*Airspeed limits (CAS)	(S/N 18800833 through 18802348) Never exceed 181 mph (157 knots) Maximum structural cruising 144 mph (125 knots) Maneuvering 116 mph (101 knots) Flaps extended (5° 120 mph (104 knots) (10° - 20°) 110 mph (96 knots)
(IAS) *(See Note 7 on use of IAS)*	(S/N 678T, 18802349 through 18803721) Never exceed 182 mph (158 knots) Maximum structural cruising 146 mph (126 knots) Maneuvering 118 mph (103 knots) Flaps extended (5°) 121 mph (105 knots) (10° - 20°) 109 mph (95 knots)
(IAS) *(See Note 7 on use of IAS)*	(S/N 18803722 and on) Never exceed 179 mph (156 knots) Maximum structural cruising 144 mph (125 knots) Maneuvering 118 mph (102 knots) Flaps extended (5°) 122 mph (106 knots) (10° - 20°) 112 mph (97 knots)
C.G. Range (Normal Category)	(+39.0) to (+45.5) at 2300 lbs. or less (+41.0) to (+45.5) at 3300 lbs. Straight line variation between points given.
Empty weight C.G. Range	None
*Maximum weight	3300 lbs. (Normal Category)
Number of seats (maximum)	1 at (+91) to (+95)
Maximum cargo	1670 lb. at +43.0 sta. *(see Note 5)*
Fuel capacity	37 gal. (+11.0); (36.5 gal. usable) fuselage tank (through S/N 18802745) 56 gal. (+48.0); (54 gal. usable) wing tanks (through S/N 18801346) 54 gal. (+48.0); (52 gal. usable) wing tanks (S/N 678T, 18801347 and on) *See Note 1 for data on unusable fuel.*
Oil capacity	12 qt. (-17.0; includes 9 lb. unusable through S/N 18803856) 13 qt. (-15.9) (9 lb. unusable) (S/N 18803857T and on) *See Note 1 for data on undrainable oil.*

VI. Model A188B (cont'd)

Control surface movements	Wing flaps			Down	20° ± 1°
	Ailerons (from neutral)	Up	18° ± 1°	Down	10° ± 1°
	Elevators	Up	26° ± 1°	Down	21° ± 1°
	Elevator tab	Up	12° ± 1°	Down	27° ± 1°
	Rudder	Right	24° + 0°, -1°	Left	24° + 0°, -1°
	(Neutral aileron is rigged with trailing edge 3° ± 30' below trailing edge of wing.)				

Additional Limitations for Restricted Category

*Airspeed Limits (CAS)	Max. operation speed in agricultural operations (S/N 18800833 through 18802348)	120 mph (104 knots)
	Max. operation speed in agricultural operations (S/N 678T, 18802349 through 18803721)	121 mph (105 knots)
	Max. operation speed in agricultural operations (S/N 18803722 and on)	130 mph (113 knots)

C.G. Range

(+39.0) to (+47.5) at 2300 lbs. or less
(+39.4) to (+47.5) at 2500 lbs.
(+41.0) to (+46.4) at 3300 lbs.
(+39.3) to (+45.2) at 4200 lbs. (see Note 3)
Straight line variation between points given.

*Maximum Weight — *See Note 3.*

Serial numbers eligible — 678T, 18800833 through 18803973 *(See Note 5.)*

VII. Model T188C, Aghusky, 1 PCLM (Restricted Category), approved September 8, 1978

Engine — Continental TSIO-520-T

*Fuel — 100LL/100 minimum grade aviation gasoline

*Engine limits — 310 hp at 2700 rpm and 39.5 in. Hg. for all operations

Propeller and propeller limits

1. (a) McCauley D3A34C402/90DFA-10
   Diameter: not over 80 in., not under 78.5 in.
   Pitch settings at 30 in. sta.:
   Low 12.4°, high 28.5°
   Avoid continuous operation between 2000 and 2250 rpm above 27 in. mp.
   (b) Cessna spinner 0750286
   (c) McCauley hydraulic governor C161031-0110

*Airspeed limits (IAS) *(See Note 7 on use of IAS.)*	Maximum operational speed in agricultural operations	130 mph (113 knots)
	Flaps extended (5°)	121 mph (105 knots)
	(10° - 20°)	109 mph (95 knots)

C.G. Range (Normal Category)

(+39.0) to (+45.9) at 2300 lbs. or less
(+39.7) to (+45.9) at 3300 lbs.
(+40.0) to (+45.5) at 3300 lbs.
(+39.2) to (+44.0) at 4400 lbs. *(See Note 3.)*
Straight line variation between points given

Empty weight C.G. Range — None

*Maximum weight — 3300 lbs. *(See Note 3.)*

VII. Model T188C (cont'd)

Number of seats (Maximum)	1 at (+91) to (+95)
Maximum cargo	*See Note 5.*
Fuel capacity	54 gal. (+48.0); 52 gal. usable *See Note 1 for data on unusable fuel.*
Oil capacity	13 qt. (-18.7; includes 9 lb. unusable) *See Note 1.*
Maximum operating altitude	14,000 MSL

Control surface movements

Wing flaps			Down	20° ± 1°
Ailerons (from neutral)	Up	18° ± 1°	Down	10° ± 1°
Elevators	Up	26° ± 1°	Down	21° ± 1°
Elevator tab	Up	12° ± 1°	Down	27° ± 1°
Rudder	Right	24° + 0°, -1°	Left	24° + 0°, -1°

(Neutral aileron is rigged with trailing edge 3° ± 30' below trailing edge of wing.)

Serial numbers eligible: T18802839T, T18803307T, T18803308T, T18803325T through T18803974T

Data Pertinent to All Models

Datum: Fuselage station 0.0 (front face of firewall)

Leveling means: Two jig located nutplates and screws on left of tailcone

Certification basis

Part 21 of the Federal Aviation Regulations dated February 1, 1965, for Restricted Category.

Part 23 of the Federal Aviation Regulations dated February 1, 1965, for Normal Category.

In addition, (S/N 18803297 and on) FAR 23.1559 effective March 1, 1978, for Normal Category.

For the T188C only, Part 21 of the Federal Aviation Regulations dated February 1, 1965, and Part 23 of the Federal Aviation Regulations dated February 1, 1965, with exception to 23.221 per 21.25(a)(1). In addition, FAR 23.1559 effective March 1, 1978.

Application for Type Certificate dated April 7, 1965.
Type Certificate NO. A9CE issued February 14, 1966, obtained by the manufacturer under delegation option procedures.

Equivalent Safety Items	S/N 678T, 18802349 and on S/N T18802839T, T18803307T, T18803308T, T18803325T and on
Airspeed Indicator	FAR 23.1545 *(See Note 7 on use of IAS.)*
Airspeed Limitations	FAR 23.1583(a)(1)

Production Basis: Production Certificate No. 4. Delegation Option Manufacturer No. CE-1 authorized to issue airworthiness certification under delegation option provisions of Part 21 of the Federal Aviation Regulations.

Equipment: The basic required equipment as specified in the applicable airworthiness regulations (see Certification Basis) must be installed in the aircraft for certification. This equipment must include a current Airplane Flight Manual effective S/N 678T, 18803297 and on and T18802839T and T18803307T, T18803308T, and T18803325T and on. In addition, the following items of equipment are required:

(1) Stall Warning Indicator, Cessna Dwg. 1670056.
(2) Model 188B and A188B eligible for normal category certification when Cessna spring 1660206-3 replaces 1660206-2.

NOTE 1. Current weight and balance report together with list of equipment included in the certificated empty weight, and loading instructions when necessary, must be provided for each aircraft at the time of original certification.

The certificated empty weight and corresponding center of gravity location must include unusable fuel of 3 lbs. at +6.0 with the fuselage tank, or 42 lbs. at +48.0 Serials 188-0446 through 188-0572 (or 12 lbs. at +37.3 Serials 18800573 and on) when wing tanks are installed, and undrainable oil of 0.0 lb. at -17.0 through S/N 18802348, or full oil of 22.5 lb. at -17.5 S/N 678T, 18802349 through S/N 18803856; 24.4 lb. at -15.9 S/N 18803857T and on; 24.4 lb. at -18.7 S/N T18802389T, T18803307T, T18803308T, T18803325T and on.

NOTE 2. The following information must be displayed in the form of composite or individual placards.

(a) In full view of the pilot: (S/N 188-0001 through 188-0572 and 18800573 through 18800832)

(1) "This airplane must be operated as a normal category airplane in compliance with the operating limitations as stated in the form of placards, markings, and manuals. For restricted category operations, refer to additional placards and limitations."

(2) "No acrobatic maneuvers including spins approved."

(3) "Maximum design weight - 3300 lb. (Reference weight and balance data for loading instructions)."

(4) "Maximum maneuvering speed - 127 mph, CAS."

(5) "Maximum altitude loss in stall recovery - 200 ft."

(6) "Maximum flight maneuvering load factors:

Flaps Up	+3.8, -1.52
Flaps Down	+3.0"

(7) Maximum flap extension speed - 110 mph, CAS."

(8) "Airplane controllable in 15 knot crosswind."

(9) "VFR - DAY" or

(10) "VFR - DAY - NIGHT."

(b) (1) In full view of the pilot: (S/N 18800833 through 18802348)

"This airplane must be operated as a normal category airplane in compliance with the operating limitations as stated in the form of placards, markings, and manuals. For restricted category operations refer to additional placards and limitations.

MAXIMUMS

Maneuvering speed		116 mph CAS (101 knots)
Gross weight (normal category)		3300 lb.
Altitude loss in stall recovery		140 ft.
Demonstrated crosswind		15 knots
Flight load factor	Flaps Up	+3.8, -1.52
	Flaps Down 5°	+2.5
	Flaps Down 10° - 20°	+2.0

Reference weight and balance data for loading instructions. No acrobatic maneuvers, including spins, approved. Known icing conditions to be avoided. This airplane is certified for the following flight operations as of date of original airworthiness certificate.

VFR - DAY - NIGHT" (as applicable)

(2) In full view of the pilot: (S/N 18802349 through S/N 18803296)

"This airplane must be operated as a normal category airplane in compliance with the operating limitations as stated in the form of placards, markings, and manuals. For restricted category operations refer to additional placards and limitations.

MAXIMUMS

Maneuvering speed		118 mph IAS
Gross weight (normal category)		3300 lb.
Altitude loss in stall recovery		140 ft.
Demonstrated crosswind		15 knots
Flight load factor	Flaps Up	+3.8, -1.52
	Flaps Down 5°	+2.5
	Flaps Down 10° - 20°	+2.0

Reference weight and balance data for loading instructions. No acrobatic maneuvers, including spins, approved. Known icing conditions to be avoided. This airplane is certified for the following flight operations as of date of original airworthiness certificate.

VFR - DAY - NIGHT" (as applicable)

(3) In full view of the pilot: (S/N 678T, 18803297 and on)
"The markings and placards installed in this airplane contain operating limitations which must be complied with when operating this airplane in the Normal Category. Other operating limitations which must be complied with when operating this airplane in this category or in the Restricted Category are contained in the Airplane Flight Manual.

Refer to weight and balance data for loading instructions.
No acrobatic maneuvers, including spins, approved.
Flight into known icing conditions prohibited.

This airplane is certified for the following flight operations as of date of original airworthiness certificate.

DAY - NIGHT - VFR" (as applicable)

(4) In full view of the pilot: (S/N T18802839T, T18803307T, T18803308T, T18803325T and on)
"The markings and placards installed in this airplane contain operating limitations which must be complied with when operating this airplane in the Restricted Category. Other operating limitations which must be complied with when operating this airplane in this category are contained in the Airplane Flight Manual. Reference weight and balance data for loading instructions. No acrobatic maneuvers, including spins, approved. Flight into known icing conditions prohibited. This airplane is certified for the following flight operations as of date of original airworthiness certificate.

VFR - DAY - NIGHT" (as applicable)

(c) (1) On crash pad: (S/N 188-0001 through 18802348)
Flaps 5° 120 mph
Flaps 10° and 20° 110 mph
(2) On crash pad: (S/N 18802349 through 18803296)
Flaps 5° 121 mph IAS
Flaps 10° and 20° 109 mph IAS
(3) On crash pad: (effective S/N 678T, 18803297 through 18803721)

MAXIMUM AIRSPEEDS	
Maneuver	118 MIAS
Flaps 5°	121 MIAS
Flaps 10° and 20°	109 MIAS
Agricultural operations	121 MIAS

(4) On crash pad: (effective S/N 18803722 and on)

MAXIMUM AIRSPEEDS - MIAS	
Maneuver (3300 lbs.)	118
Flaps 5°	122
Flaps 10° and 20°	112
Agricultural operations	130

A9CE

(d) (1) On flap handle: (S/N 188-0001 through 188-0293)
"FLAPS - WARNING Avoid slips with flaps extended."
"FLAPS - PULL TO EXTEND

Takeoff	Retracted
	1st Notch 10°
	2nd Notch 20°
Landing	0 to 3rd Notch 30° "

(2) On flap handle: (S/N 188-0294 through 188-0572 and 18800573 through 18800832)
"FLAPS - PULL TO EXTEND

Takeoff and Landing	Retracted	0°
	1st Notch	10°
	2nd Notch	20°

(3) On flap handle: (S/N 678T, 18800833 and on)
"FLAPS - PULL TO EXTEND

Takeoff	Retracted	0°
	1st Notch	5°
and	2nd Notch	10°
Landing	3rd Notch	20° "

(e) (1) Adjacent to the fuel valve control:
"Fuel Valve Push-on; 36.5 gals. usable." (through S/N 18802745)
(2) Adjacent to the fuel valve control for models equipped with wing fuel tanks:
"Fuel Valve Push-on; 49 gals. usable." (S/N 188-0446 through 188-0572)
"Fuel Valve Push-on; 54 gals. usable." (S/N 18800573 through 18801346)
"Fuel Valve Push-on; 52 gals. usable." (S/N 678T, 18801347 and on)

(f) On Doors:
"Do not open doors in flight."

(g) On Baggage Door: (S/N 188-0001 through 188-0572 and S/N 18800573 through 18800832)
"Maximum baggage capacity 100 lb., articles stowed in this compartment to be securely tied down." Refer to Owner's Manual for details.

(h) On Instrument Panel:
"No Smoking." (Except with optional ash tray installation)

(i) On Hopper Lid:
(1) "Hopper capacity 200 U.S. Gal."
Serial 188-0001 through 18801040
"Maximum allowable hopper load - 1670 lb. See Weight and Balance Data."
Serial 18801041 and on
(2) "Max. allowable hopper load - 1800 lb. See Weight and Balance Data."
(On aircraft serials with "T" suffix)
(3) "Max. allowable hopper load - 1900 lb. See Weight and Balance Data."
(On aircraft serials with prefix and suffix "T")

(j) Adjacent to the master switch: (S/N 18800573 through 18801040)
(1) "Do not turn off alternator in flight except in emergency."

(k) Below the fuel flow gauge: (A188, A188A, and A188B through S/N 18802745)
"Fuel Flows at Full Throttle

	2850 rpm	2700 rpm
S.L.	24	23
4000 ft.	22	21
8000 ft.	20	19"

A188B (S/N 678T, 18802746 through 18803296)
"Max. Power Settings and Fuel Flow Takeoff (5 min. only) 2850 rpm
Max. Continuous Power 2700 rpm

Fuel Flows at Full Throttle

	2700 rpm	2850 rpm
S.L.	23 gph	24 gph
4,000 ft.	21 gph	22 gph
8,000 ft.	19 gph	20 gph"

A188B (S/N 18803297 and on)
"Min. Fuel Flows at Full Throttle

RPM	S.L.	4000	8000	12,000
2700	23 GPH	21 GPH	19 GPH	17 GPH
2850	24 GPH	22 GPH	20 GPH	18 GPH"

T188C (S/N T18802839T, T18803307T, T18803308T, T18803325T and on)
"Maximum Allowable Manifold Pressure

Press Alt.	MP. in. Hg.
S.L.	39.5
2500	38.8
5000	38.1
7500	37.3"

(l) (1) Adjacent to or on the fuel filler cap as applicable (fuselage tank)
"80/87 Octane 37 U.S. Gal. Cap." (O-470 engine)
"100/130 Octane 37 U.S. Gal. Cap." (IO-520 engine)
(2) Adjacent to or on the fuel filler caps (wing tanks)
"100/130 Octane 28 U.S. Gal. Cap." (through S/N 18801346)
"100/130 Octane 27 U.S. Gal. Cap." (S/N 18801347 through 18803046)
"Service this airplane with 100LL/100 Min.
Aviation Grade Gasoline - Capacity 27.0 Ga." (S/N 678T, 18803047 and on)

(m) Near tailwheel lock control: (S/N 678T, 18800833 and on) (except for serials with "T" prefix) "Lock for flight."

(n) On outside of cockpit doors:
"For emergency door removal pull out hinge pins."

(o) Below each door sill on inside of cockpit:
"Pull - Emergency Door Release."

(p) On Control Lock:
"Control Lock - Unlock before starting engine."

(q) On Crash Pad (T18802839T, T18803307T, T18803308T, T18803325T and on)
"Avoid Continuous Operation above 27 in. M.P. between 2000 and 2250 rpm."

NOTE 3. When operating in restricted category, operators may approve higher maximum weights as permitted by FAA Advisory Circular No. 20-33B and Civil Aeronautics Manual 8. With respect to this action, these aircraft have demonstrated satisfactory operation in the restricted category envelope given at 1500 ft. altitude and standard day at the following restricted gross weights:

188 Series		3800 lb.
A188 Series	(Serials 188-0001 and on)	4000 lb.
	(Serials 18800967T through 18801374T)	4000 lb.
	(Serials 678T, 18801375T and on)	4200 lb.
T188C Series	(Serials T18802839T, T18803307T, T18803308T, T18803325T and on)	4400 lb.

The following additional information must be displayed in the form of placards when operating in the Restricted Category:

(a) On Instrument Panel in full view of the pilot:
(1) "Maximum operating speed in agricultural operations - 120 mph (104 knots)"
(S/N 188-0001 through 18802348)
(2) "Maximum operating speed in agricultural operations - 121 mph IAS. (105 knots IAS)."
(S/N 18802349 through 18803296)
(3) T188C (Serials T18802839T, T18803307T, T18803308T, T18803325T and on)

MAXIMUM AIRSPEEDS

Maneuver (3300 lbs.)	117 MIAS
Flaps 5°	121 MIAS
Flaps 10° to 20°	109 MIAS
Agricultural Operation	130 MIAS"

(4) "Hopper Dump - Pull"
(S/N 188-0001 through 18801374) (Airplanes with Transland dump plate assembly)
"Hopper Dump - - - - - - -→"
(S/N 188-0390 and on) (on dump handle) (Airplanes with Transland or Cessna gate box assembly)
"Dump"
(S/N 18802311 and on) (Airplanes with Transland P/N 21767 Australian dump plate assembly)

(b) On canopy, side, window or fuselage side panel:
"RESTRICTED"

NOTE 4. Cylinder head probe location No. 1 cylinder through S/N 18803046; S/N 18803722 and on. No. 5 cylinder S/N 678T, S/N 18803047 through S/N 18803721. No. 2 cylinder S/N T18802839T, T18803307T, T18803308T, T18803325T and on.

NOTE 5. The letter "T" suffix after the serial number indicates an A188 series aircraft with an 1800 lb. maximum capacity hopper (Ex: 18800967T). Serial numbers with prefix "T" and suffix "T" indicate T188C aircraft with 1900 lb. maximum capacity hopper. (Ex: T18803329T)

NOTE 6. The installation of the O-470-S engine in Model 188B (1972 through 1974) will require a change of the oil temperature gauge. Reference Cessna Service Letter SE 75-2 for this change.

NOTE 7. (a) The marking of the airspeed indicator with IAS provides an equivalent level of safety to FAR 23.1545 when the approved airspeed calibration data presented in Section VI of the Owner's Manual listed below is available to the pilot:

A188B	Cessna P/N D1064-13	(S/N 18802349 through S/N 18802745)
A188B	Cessna P/N D1089-13	(S/N 18802746 through S/N 18803046)
A188B	Cessna P/N D1117-13	(S/N 18803047 through S/N 18803296)

(b) The marking of the airspeed indicator with IAS provides an equivalent level of safety to FAR 23.1545 when the approved airspeed calibration data presented in the FAA approved Airplane Flight Manual listed below is available to the pilot:

A188B	Cessna P/N D1166-13	(S/N 678T, 18803297 through S/N 18803521)
T188C	Cessna P/N D1168-13	(S/N T18803307T, T18803308T, T18803325T through S/N T18803521T)
A188B	Cessna P/N D1180-13FM	(S/N 18803522 through S/N 18803721)
T188C	Cessna P/N D1181-13FM	(S/N T18803522T through T18803721T)
A188B	Cessna P/N D1201-13FM	(S/N 18803722 through 18803856)
T188C	Cessna P/N D1202-13FM	(S/N T18803722T through T18803856T)
A188B	Cessna P/N D1220-13FM	(S/N 18803857T through 18803926T)
T188C	Cessna P/N D1221-13FM	(S/N T18803857T through T18803926T
A188B	Cessna P/N D1238-13FM	(S/N 18803927T through 18803973T)
T188C	Cessna P/N D1239-13FM	(S/N T18802839T, T18803927T through T18803974T)

NOTE 8. 14 volt electrical system
188/A188 series through Serial 18803046

28 volt electrical system
A188 Series, Serial 678T, 18803047 and on
T188 Series, Serial T18803307T, T18803308T, T18803325T and on

In addition to the placards specified above, the prescribed operating limitations indicated by an asterisk (*) under Sections I through VII of this data sheet must also be displayed by permanent markings.

Note: For 188, A188, and T188:

"WARNING": Use of alcohol-based fuels can cause serious performance degradation and fuel system component damage, and is therefore prohibited on Cessna airplanes."

....END....

DEPARTMENT OF TRANSPORTATION
FEDERAL AVIATION ADMINISTRATION

	3A12
	Revision 72
	CESSNA
172	172I
172A	172K
172B	172L
172C	172M
172D	172N
172E	172P
172F (USAF T-41A)	172Q
172G	172R
172H (USAF T-41A	172S
	January 5, 2006

"**WARNING**: Use of alcohol-based fuels can cause serious performance degradation and fuel system component damage, and is therefore prohibited on Cessna airplanes."

TYPE CERTIFICATE DATA SHEET NO. 3A12

This data sheet which is part of Type Certificate No. 3A12 prescribes conditions and limitations under which the product for which the type certificate was issued meets the airworthiness requirements of the Federal Aviation Regulations.

Type Certificate Holder — Cessna Aircraft Company
P.O. Box 7704
Wichita, Kansas 67277

I. Model 172, 4 PCLM (Normal Category), approved November 4, 1955; 2 PCLM (Utility Category), approved December 14, 1956

Engine — Continental O-300-A or O-300-B

*Fuel — 80/87 minimum grade aviation gasoline

*Engine limits — For all operations, 2700 rpm (145 hp)

Propeller and propeller limits

1. Propeller
   (a) McCauley 1A170
       Static rpm at maximum permissible throttle setting:
           Not over 2360, not under 2230
           No additional tolerance permitted
       Diameter: not over 76 in., not under 74.5 in.
   (b) Spinner, Dwg. 0550162
2. Propeller
   (a) Sensenich M74DR or 74DR
       Static rpm at maximum permissible throttle setting:
           Not over 2430, not under 2300
           No additional tolerance permitted
       Diameter: not over 74 in., not under 72.0 in.
   (b) Spinner, Dwg. 0550162

Page No.	1	2	3	4	5	6	7	8	9	10	11	12	13	14	15	16	17	18	19	20
Rev. No.	72	60	60	60	60	57	64	60	64	60	51	60	67	60	59	59	50	50	55	55
Page No.	21	22	23	24	25	26	27	28	29	30	31									
Rev. No.	50	72	59	65	65	70	70	71	70	70	70									

I. Model 172, 4 PCLM (Normal Category) (cont'd)

	3. Propeller (a) McCauley 1C172/MDM 7652, 53, or 55 30 lb. (-39.0) Static rpm at maximum permissible throttle setting: Not over 2350, not under 2250 No additional tolerance permitted Diameter: not over 76 in., not under 74.5 in. (b) Spinner, Dwg. 0550216
*Airspeed Limits (CAS)	Maneuvering 115 mph (100 knots) Maximum structural cruising 140 mph (122 knots) Never exceed160 mph (139 knots) Flaps extended 100 mph (87 knots)
C.G. range	Normal (+40.8) to (+46.4) at 2200 lbs. (+36.4) to (+46.4) at 1733 lbs. Utility category (+38.4) to (+40.3) at 1950 lbs. (+36.4) to (+40.3) at 1733 lbs. or less Straight line variation between points given.
Empty weight C.G. range	None
*Maximum Weight	Normal category 2200 lbs. Utility category 1950 lbs.
Number of seats	4 (2 at +36, 2 at +70) (For child's optional jump seat, refer to Equipment List.)
Maximum baggage	120 lbs. (+95)
Fuel capacity	42 gal. total, 37 gal. usable (two 21 gal. tanks in wings at +48) *See Note 1 for weight of unusable fuel and oil.*
Oil capacity	2 gal. (-20), includes 1 gal. unusable

Control surface movements

Wing flaps	Takeoff		Retracted	0°
			1st notch	10°
	Landing		2nd notch	20°
			3rd notch	30°
			4th notch	40°
Ailerons	Up	20°	Down	14°
Elevator tab	Up	28°	Down	13°
Elevator	Up	28°	Down	26°
Rudder	Right	16°	Left	16°

Serial numbers eligible: 610, 612, 615, 28000 through 29999, 36000 through 36999 and 46001 through 46754

II. Model 172A, 4 PCL-SM (Normal Category), 2 PCLM (Utility Category), approved July 16, 1959; Model 172B, Skyhawk, 4 PCL-SM (Normal Category), 2 PCLM (Utility Category), approved June 14, 1960

Engine	Continental O-300-C or O-300-D
*Fuel	80/87 minimum grade aviation gasoline
*Engine limits	For all operations, 2700 rpm (145 hp)
Propeller and propeller limits	1. Propeller (a) McCauley 1C172/EM 7652, 53, or 55 Static rpm at maximum permissible throttle setting: Not over 2350, not under 2230 No additional tolerance permitted Diameter: not over 76 in., not under 74.5 in. (b) Spinner, Dwg. 0550216, 0550221 or 0550228

II. Model 172A, 4 PCL-SM (Normal Category), 2 PCLM (Utility Category) (cont'd)
Model 172B, Skyhawk, 4 PCL-SM (Normal Category), 2 PCLM (Utility Category) (cont'd)

2. Propeller (seaplane only)
   (a) McCauley 1A175/SFC 8040
   Static rpm at maximum permissible throttle setting:
   Not over 2480, not under 2380
   No additional tolerance permitted
   Diameter: not over 80 in., not under 78.4 in.
   (b) Spinner, Dwg. 0550216 or 0550221
3. Propeller
   (a) Sensenich 74DC-0-56
   Static rpm at maximum permissible throttle setting:
   Not over 2420, not under 2300
   No additional tolerance permitted
   Diameter: not over 74 in., not under 72.5 in.

*Airspeed Limits (CAS)	Maneuvering	115 mph (100 knots)
	Maximum structural cruising	140 mph (122 knots)
	Never exceed	160 mph (139 knots)
	Flaps extended	100 mph (87 knots)
C.G. range	Landplane (Model 172A):	
	Normal category	(+40.8) to (+46.4) at 2200 lbs.
		(+36.4) to (+46.4) at 1733 lbs. or less
	Utility category	(+38.4) to (+40.3) at 1950 lbs.
		(+36.4) to (+40.3) at 1733 lbs. or less
	Straight line variation between points given.	
	Landplane (Model 172B):	
	Normal category	(+40.4) to (+46.4) at 2200 lbs.
		(+36.4) to (+46.4) at 1850 lbs. or less
	Utility category	(+37.4) to (+40.3) at 1950 lbs.
		(+36.4) to (+40.3) at 1850 lbs. or less
	Seaplane (Models 172A and 172B):	
	Normal category	(+39.8) to (+45.5) at 2220 lbs.
		(+36.4) to (+45.5) at 1825 lbs. or less
	Straight line variation between points given.	
Empty weight C.G. range	None	
*Maximum weight	Landplane:	
	Normal category	2200 lb.
	Utility category	1950 lb.
	Seaplane:	
	Normal category	2220 lb.
Number of seats	4 (2 at +36, 2 at +70) (For child's optional jump seat, refer to Equipment List.)	
Maximum baggage	120 lb. (+95)	
Fuel capacity	42 gal. total, 37 gal. usable (172A); 39 gal. usable (172B) (two 21 gal. tanks in wings at +48) *See Note 1 for weight of unusable fuel and oil.*	
Oil capacity	2 gal. (-20), 1 gal. usable	

II. Model 172A, 4 PCL-SM (Normal Category), 2 PCLM (Utility Category) (cont'd)
Model 172B, Skyhawk, 4 PCL-SM (Normal Category), 2 PCLM (Utility Category) (cont'd)

Control surface movements	Wing flaps	Takeoff		Retracted	0°
				1st notch	10°
		Landing		2nd notch	20°
				3rd notch	30°
				4th notch	40°
	Ailerons	Up	20°	Down	15°
	Elevator tab	Up	28°	Down	13°
	Elevator	Up	28°	Down	26°
	Rudder (landplane)	Right	16°	Left	16°
	(seaplane)	Right	19°	Left	15°
	(Measured parallel to W.L.)				

Serial numbers eligible — Model 172A: 622, 625, 46755 through 47746
Model 172B: 630, 17247747 through 17248734

III. Model 172C, 4 PCL-SM (Normal Category), 2 PCLM (Utility Category), approved July 18, 1961

Engine — Continental O-300-C or O-300-D

*Fuel — 80/87 minimum grade aviation gasoline

*Engine limits — For all operations, 2700 rpm (145 hp)

Propeller and propeller limits

1. Propeller
   (a) McCauley 1C172/EM 7652, 53, or 55
   Static rpm, at maximum permissible throttle setting:
   Not over 2350, not under 2230
   No additional tolerance permitted
   Diameter: not over 76 in., not under 74.5 in.
   (b) Spinner, Dwg. 0550216, 0550221 or 0550228
2. Propeller (seaplane only)
   (a) McCauley 1A175/SFC 8040
   Static rpm, at maximum permissible throttle setting:
   Not over 2480, not under 2380
   No additional tolerance permitted
   Diameter: not over 80 in., not under 78.4 in.
   (b) Spinner, Dwg. 0550216 or 0550221
3. Propeller
   (a) Sensenich 74DC-0-56
   Static rpm at maximum permissible throttle setting:
   Not over 2420, not under 2300
   No additional tolerance permitted
   Diameter: not over 74 in., not under 72.5 in.

*Airspeed limits (CAS)

Maneuvering	115 mph (100 knots)
Maximum structural cruising	140 mph (122 knots)
Never exceed	160 mph (139 knots)
Flaps extended	100 mph (87 knots)

C.G. range

Landplane		
Normal category	(+40.5) to (+46.4)	at 2250 lbs.
	(+36.4) to (+46.4)	at 1850 lbs. or less
Utility category	(+37.4) to (+40.3)	at 1950 lbs.
	(+36.4) to (+40.3)	at 1850 lbs. or less
Seaplane		
Normal category	(+39.8) to (+45.5)	at 2220 lbs.
	(+36.4) to (+45.5)	at 1825 lbs. or less

Straight line variation between points given.

III. Model 172C, 4 PCL-SM (Normal Category), 2 PCLM (Utility Category) (cont'd)

Empty weight C.G. range	None
*Maximum weight	Landplane Normal category 2250 lbs. Utility category 1950 lbs. Seaplane Normal category 2220 lbs.
Number of seats	4 (2 at +36, 2 at +70) (For child's optional jump seat, refer to Equipment List.)
Maximum baggage	120 lbs. (+95)
Fuel capacity	39 gal. total, 36 gal. usable (two 19.5 gal. tanks in wings at +48) *See Note 1 for weight of unusable fuel and oil.*
Oil capacity	2 gal. (-20), includes 1 gal. unusable

Control surface movements

Wing flaps	Takeoff		Retracted	0°
			1st notch	10°
	Landing		2nd notch	20°
			3rd notch	30°
			4th notch	40°
Ailerons	Up	20°	Down	15°
Elevator tab	Up	28°	Down	13°
Elevator	Up	28°	Down	26°
Rudder (Landplane)	Right	16°	Left	16°
(Seaplane)	Right	19°	Left	15°

(Measured parallel to W.L.)

Serial numbers eligible 17248735 through 17249544

IV. Model 172D, 4 PCL-SM (Normal Category), 2 PCLM (Utility Category), approved June 19, 1962
Model 172E, 4 PCL-SM (Normal Category), 2 PCLM (Utility Category), approved June 27, 1963
Model 172F (USAF T-41A), 4 PCL-SM (Normal Category), 2 PCLM (Utility Category), approved April 21, 1964
Model 172G, 4 PCL-SM (Normal Category), 2 PCLM (Utility Category), approved June 15, 1965
Model 172H (USAF) T-41A), 4 PCL-SM (Normal Category), 2 PCLM (Utility Category), approved June 7, 1966

Engine — Continental O-300-C or O-300-D

*Fuel — 80/87 minimum octane aviation gasoline

*Engine limits — For all operations, 2700 rpm (145 hp)

Propeller and propeller limits

1. Propeller
   (a) McCauley 1C172/EM 7652, 53
   Static rpm at maximum permissible throttle setting:
   Not over 2420, not under 2230
   No additional tolerance permitted
   Diameter: not over 76 in., not under 74.5 in.
   (b) Spinner
   Model 172D, E, F, Dwg. 0550216, 0550221 or 0550228
   Model 172G, H, Dwg. 0550236
2. Propeller (Seaplane only)
   (a) McCauley 1A175/SFC 8040
   Static rpm at maximum permissible throttle setting:
   Not over 2480, not under 2380
   No additional tolerance permitted
   Diameter: not over 80 in., not under 78.4 in.

3A12

IV. Model 172D, Model 172E, Model 172F, Model 172G, Model 172H (cont'd)

(b) Spinner
Model 172D, E, F, Dwg. 0550216, 0550221
Model 172G, H, Dwg. 0550236

*Airspeed limits (CAS)	Maneuvering	122 mph (106 knots)
	Maximum structural cruising	142 mph (122 knots)
	Never exceed	174 mph (151 knots)
	Flaps extended	100 mph (87 knots)
C.G. range	Landplane	
	Normal category	(+38.5) to (+47.3) at 2300 lbs.
		(+35.0) to (+47.3) at 1950 lbs. or less
	Utility category	(+35.5) to (+40.5) at 2000 lbs.
		(+35.0) to (+40.5) at 1950 lbs. or less
	Seaplane	
	Normal category	(+39.8) to (+45.5) at 2220 lbs.
		(+36.4) to (+45.5) at 1825 lbs. or less
	Straight line variation between points given.	
Empty weight C.G. range	None	
*Maximum Weight	Landplane:	
	Normal category	2300 lbs.
	Utility category	2000 lbs.
	Seaplane:	
	Normal category	2220 lbs.
Number of seats	4 (2 at +36, 2 at +70) (For child's optional jump seat, refer to Equipment List.)	
Maximum Baggage	120 lbs. (+95)	
Fuel Capacity	39 gal. total, 36 gal. usable (two 19.5 gal. tanks in wings at +48) *See Note 1 for weight of unusable fuel and oil.*	
Oil capacity	2 gal. (-20), 1 gal. usable	

Control surface movements

Wing flaps	Takeoff		Retracted	0°
			1st notch	10°
	Landing		0°	40°
Ailerons	Up	20°	Down	15°
Elevator tab	Up	28°	Down	13°
Elevator	Up	28°	Down	23°

(Neutral position is with bottom of balance area flush with bottom of stabilizer.)

Rudder (landplane)	Right	16°	Left	16°
(seaplane)	Right	19°	Left	15°

Serial numbers eligible

Model 172D:	17249545 through 17250572
Model 172E:	639, 17250573 through 17251822
Model 172F:	17251823 through 17253392
Model 172G:	17253393 through 17254892
Model 172H:	638, 17254893 through 17256512 (except 17256493)

V. Model 172I, 4 PCL-SM (Normal Category), 2 PCLM (Utility Category), approved December 15, 1967
Model 172K, 4 PCL-SM (Normal Category), 2 PCLM (Utility Category), approved May 9, 1968

Engine — Lycoming O-320-E2D

*Fuel — 80/87 minimum grade aviation gasoline

*Engine limits — For all operations, 2700 rpm (150 hp)

Propeller and propeller limits

1. Propeller
   (a) McCauley 1C172/MTM 7653
       Static rpm at maximum permissible throttle setting:
           Not over 2360, not under 2260
           No additional tolerance permitted (see Note 3)
       Diameter: not over 76 in., not under 74 in.
   (b) Spinner, Dwg. 0550320
2. Propeller (seaplane only)
   (a) McCauley 1A175/ATM 8042
       Static rpm at maximum permissible throttle setting:
           Not over 2480, not under 2380
           No additional tolerance permitted (see Note 3)
       Diameter: not over 80 in., not under 78.4 in.
   (b) Spinner, Dwg. 0550320
3. Propeller
   (a) McCauley 1C160/CTM 7553
       Static rpm at maximum permissible throttle setting:
           Not over 2370, not under 2270
           No additional tolerance permitted (see Note 3)
       Diameter: not over 75 in., not under 74 in.
   (b) Spinner, Dwg. 0550320
4. Propeller (seaplane only)
   (a) McCauley 1A175/ETM 8042
       Static rpm at maximum permissible throttle setting:
           Not over 2480, not under 2380
           No additional tolerance permitted (see Note 3)
       Diameter: not over 80 in., not under 78.4 in.
   (b) Spinner, Dwg. 0550321
5. Propeller
   (a) McCauley 1C160/DTM 7553
       Static rpm at maximum permissible throttle setting:
           Not over 2370, not under 2270
           No additional tolerance permitted (see Note 3)
       Diameter: not over 75 in., not under 74 in.
   (b) Spinner, Dwg. 0550320

*Airspeed Limits (CAS)

Maneuvering	122 mph (106 knots)
Maximum structural cruising	140 mph (122 knots)
Never exceed	174 mph (151 knots)
Flaps extended	100 mph (87 knots)

C.G. range

Landplane

Normal category	(+38.5) to (+47.3) at 2300 lbs.
	(+35.0) to (+47.3) at 1950 lbs. or less
Utility category	(+35.5) to (+40.5) at 2000 lbs.
	(+35.0) to (+40.5) at 1950 lbs. or less

Seaplane (Edo 89-2000 or 89A2000 floats)

Normal category	(+39.8) to (+45.5) at 2220 lbs.
	(+36.4) to (+45.5) at 1825 lbs. or less

Straight line variation between points given.

V. Model 172I, Model 172K (cont'd)

Empty weight C.G. range	None
*Maximum Weight	Landplane: Normal category 2300 lbs. Utility category 2000 lbs. Seaplane: Normal category 2220 lbs.
Number of seats	4 (2 at +34 to +46, 2 at +73) (Occupant on child's optional jump seat at +93)
Maximum baggage	120 lb. at +95
Fuel capacity	42 gal. total, 38 gal. usable (two 21 gal. tanks in wings at +48) *See Note 1 for weight of unusable fuel and oil.*
Oil capacity	2 gal. (-14.0), 1-1/2 gal. usable
Control surface movements	Wing flaps Takeoff 0° - 10° Landing 0° - 40° ±2° Ailerons Up 20° ±1° Down 15° ±1° Elevator tab Up 28° +1°, -0° Down 13° +1°, -0° Elevator Up 28° +1°, -0° Down 23° +1°, -0° (Neutral position is with bottom of balance area flush with bottom of stabilizer.) Rudder (landplane) Right 16° ±1° Left 16° ±1° (seaplane) Right 19° ±1° Left 15° ±1° (Measured parallel to W.L.)
Serial numbers eligible	Model 172I: 17256513 through 17257161 Model 172K: 17257162 through 17258486 (1969 model) 17258487 through 17259223 (1970 model)

VI. Model 172L, 4 PCL-SM (Normal Category), 2 PCLM (Utility Category), approved May 13, 1970

Engine	Lycoming O-320-E2D
*Fuel	80/87 minimum grade aviation gasoline
*Engine limits	For all operations, 2700 rpm (150 hp)
Propeller and propeller limits	1. Propeller (a) McCauley 1C172/MTM 7653 Static rpm at maximum permissible throttle setting: Not over 2360, not under 2260 No additional tolerance permitted (see Note 3) Diameter: not over 76 in., not under 74 in. (b) Spinner, Dwg. 0550320 2. Propeller (seaplane only) (a) McCauley 1A175/ATM 8042 Static rpm at maximum permissible throttle setting: Not over 2480, not under 2380 No additional tolerance permitted (see Note 3) Diameter: not over 80 in., not under 78.4 in. (b) Spinner, Dwg. 0550320

VI. Model 172L, 4 PCL-SM (Normal Category), 2 PCLM (Utility Category) (cont'd)

3. Propeller
   (a) McCauley 1C160/CTM 7553
   Static rpm at maximum permissible throttle setting:
   Not over 2370, not under 2270
   No additional tolerance permitted (see Note 3)
   Diameter: not over 75 in., not under 74 in.
   (b) Spinner, Dwg. 0550320
4. Propeller
   (a) McCauley 1A160/DTM 7553
   Static rpm at maximum permissible throttle setting:
   Not over 2370, not under 2270
   No additional tolerance permitted (see Note 3)
   Diameter: not over 75 in., not under 74 in.
   (b) Spinner, Dwg. 0550320
5. Propeller (Seaplane only)
   (a) McCauley 1A175/ETM 8042
   Static rpm at maximum permissible throttle setting:
   Not over 2480, not under 2380
   No additional tolerance permitted (see Note 3)
   Diameter: not over 80 in., not under 78.4 in.
   (b) Spinner, Dwg. 0550321
6. Propeller
   (a) McCauley 1C160/DTM 7553
   Static rpm at maximum permissible throttle setting:
   Not over 2370, not under 2270
   No additional tolerance permitted (see Note 3)
   Diameter: not over 75 in., not under 74 in.
   (b) Spinner, Dwg. 0550320

*Airspeed Limits (CAS)	Maneuvering	122 mph (106 knots)
	Maximum structural cruising	140 mph (122 knots)
	Never exceed	174 mph (151 knots)
	Flaps extended	100 mph (87 knots)
C.G. range	Landplane	
	Normal category	(+38.5) to (+47.3) at 2300 lbs.
		(+35.0) to (+47.3) at 1950 lbs. or less
	Utility category	(+35.5) to (+40.5) at 2000 lbs.
		(+35.0) to (+40.5) at 1950 lbs. or less
	Straight line variation between points given.	
	Seaplane (Edo 89-2000 or 89A2000 floats)	
	Normal category	(+39.8) to (+45.5) at 2220 lbs.
		(+36.4) to (+45.5) at 1825 lbs. or less
	Straight line variation between points given.	
Empty weight C.G. range	None	
*Maximum Weight	Landplane:	
	Normal category	2300 lbs.
	Utility category	2000 lbs.
	Seaplane:	
	Normal category	2220 lbs.
Number of seats	4 (2 at +34 to +46, 2 at +73) (Occupant on child's optional jump seat at +96)	
Maximum baggage	120 lb. at +95	
Fuel capacity	42 gal. total, 38 gal. usable (two 21 gal. tanks in wings at +48) *See Note 1 for weight of unusable fuel.*	

3A12

VI. Model 172L, 4 PCL-SM (Normal Category), 2 PCLM (Utility Category) (cont'd)

Oil capacity	2 gal. (-14.0), 1-1/2 gal. usable *See Note 1 for data on undrainable oil.*
Control surface movements	Wing flaps — Takeoff 0° - 10°; Landing 0° - 40° ±2° Ailerons Up 20° ±1° Down 15° ±1° Elevator tab Up 28° +1°, -0° Down 13° +1°, -0° Elevator Up 28° +1°, -0° Down 23° +1°, -0° (Neutral position is with bottom of balance area flush with bottom of stabilizer.) Rudder (landplane) Right 16° ±1° Left 16° ±1° (seaplane) Right 19° ±1° Left 15° ±1° (Measured parallel to W.L.)
Serial numbers eligible	Model 172L: 17259224 through 17259903 (1971 model) Model 172L: 17259904 through 17260758 (1972 model)

VII. Model 172M, Skyhawk, 4 PCL-SM (Normal Category), 2 PCLM (Utility Category, approved May 12, 1972

Engine	Lycoming O-320-E2D
*Fuel	80/87 minimum grade aviation gasoline
*Engine limits	For all operations, 2700 rpm (150 hp)

Propeller and propeller limits

1. Propeller
   - (a) McCauley 1C160/CTM 7553
     - Static rpm at maximum permissible throttle setting:
       - Not over 2370, not under 2270
       - No additional tolerance permitted (see Note 3)
     - Diameter: not over 75 in., not under 74 in.
   - (b) Spinner: Dwg. 0550320
2. Propeller
   - (a) McCauley 1C160/DTM 7553
     - Static rpm at maximum permissible throttle setting:
       - Not over 2370, not under 2270
       - No additional tolerance permitted (see Note 3)
     - Diameter: not over 75 in., not under 74 in.
   - (b) Spinner, Dwg. 0550320
3. Propeller (seaplane only)
   - (a) McCauley 1A175/ATM 8042
     - Static rpm at maximum permissible throttle setting:
       - Not over 2545, not under 2445
       - No additional tolerance permitted (see Note 3)
     - Diameter: not over 80 in., not under 78.4 in.
   - (b) Spinner, Dwg. 0550320
4. Propeller (seaplane only)
   - (a) McCauley 1A175/ETM 8042
     - Static rpm at maximum permissible throttle setting:
       - Not over 2545, not under 2445
       - No additional tolerance permitted (see Note 3)
     - Diameter: not over 80 in., not under 78.4 in.
   - (b) Spinner, Dwg. 0550320

*Airspeed Limits (CAS)

17256493, 17260759 through 17265684

Maneuvering	112 mph (97 knots)
Maximum structural cruising	145 mph (126 knots)
Never exceed	182 mph (158 knots)
Flaps extended	100 mph (87 knots)

VII. Model 172M, Skyhawk, 4 PCL-SM (Normal Category), 2 PCLM (Utility Category) (cont'd)

*Airspeed Limits (CAS) (See Note 4 on use of CAS)	17265685 through 17267584 Maneuvering 97 knots Maximum structural cruising 128 knots Never exceed 160 knots Flaps extended 85 knots
C.G. range	Landplane: Normal category (+38.5) to (+47.3) at 2300 lbs. (+35.0) to (+47.3) at 1950 lbs. or less Utility category (+35.5) to (+40.5) at 2000 lbs. (+35.0) to (+40.5) at 1950 lbs. or less Seaplane: (Edo 89-2000 or 89A2000 floats) Normal category (+39.8) to (+45.5) at 2220 lbs. (+36.4) to (+45.5) at 1825 lbs. or less Straight line variation between points given.
Empty weight C.G. Range	None
*Maximum weight	Normal category: 2300 lb. (landplane); 2220 lb. (seaplane) Utility category: 2000 lb. (landplane)
Number of seats	4 (2 at +34 to +46, 2 at +73) (Occupant on child's optional jump seat at +96)
Maximum baggage	120 lb. at +95
Fuel capacity	42 gal. total, 38 gal. usable (two 21 gal. tanks in wings at +48) *See Note 1 for data on unusable fuel.*
Oil capacity	2 gal. (-14.0), 1-1/2 gal. usable *See Note 1 for data on undrainable oil.*
Control surface movements	Wing flaps Takeoff 0° - 10° (landplane) (seaplane) Landing 0° - 40° +0°, -2° (landplane) 0° - 30° ±2° (seaplane) Ailerons Up 20° ±1° Down 15° ±1° Elevator tab Up 28° +1°, -0° Down 13° +1°, -0° Elevator Up 28° +1°, -0° Down 23° +1°, -0° (Neutral position is with bottom of balance area flush with bottom of stabilizer.) Rudder (landplane) Right 16° ±1° Left 16° ±1° (landplane) (seaplane) Right 19° ±1° Left 15° ±1° (seaplane) (Measured parallel to W.L.)
Serial numbers eligible	17256493, 17260759 through 17261898 (1973 model) (except 17261445 and 17261578) 17261899 through 17263458 (1974 model) 17263459 through 17265684 (1975 model) 17265685 through 17267584 (1976 model)

VIII. Model 172N, Skyhawk, 4 PCL-SM (Normal Category), 2 PCLM (Utility Category), approved May 17, 1976

Engine	Lycoming O-320-H2AD
*Fuel	100/130 minimum grade aviation gasoline (S/N 17261445, 17267585 through 17269309)

VIII. Model 172N, Skyhawk, 4 PCL-SM (Normal Category), 2 PCLM (Utility Category) (cont'd)

	100LL/100 minimum grade aviation gasoline (S/N 17261578, 17269310 through 17274009)
*Engine limits	For all operations, 2700 rpm (160 hp)
Propeller and propeller limits	1. Propeller (a) McCauley 1C160/DTM 7557 Static rpm at maximum permissible throttle setting: Not over 2400, not under 2280 No additional tolerance permitted Diameter: not over 75 in., not under 74 in. (b) Spinner: Dwg. 0550320 2. Propeller (seaplane only) (a) McCauley 1A175/ETM 8042 Static rpm at maximum permissible throttle setting: Not over 2570, not under 2470 No additional tolerance permitted Diameter: not over 80 in., not under 78.5 in. (b) Spinner: Dwg. 0550320
*Airspeed limits (CAS) (See Note 4 on use of CAS)	1977 Model through 1979 Model Maneuvering 97 knots Maximum structural cruising 128 knots Never exceed 160 knots Flaps extended 85 knots 1980 Model Maneuvering 97 knots Maximum structural cruising 127 knots Never exceed 158 knots Flaps extended 85 knots
C.G. range	Landplane: Normal category (+38.5) to (+47.3) at 2300 lbs. (+35.0) to (+47.3) at 1950 lbs. or less Utility category (+35.5) to (+40.5) at 2000 lbs. (+35.0) to (+40.5) at 1950 lbs. or less Seaplane: (Edo 89-2000 or 89A2000 floats) Normal category (+39.8) to (+45.5) at 2220 lbs. (+36.4) to (+45.5) at 1825 lbs. or less Straight line variation between points given.
Empty weight C.G. Range	None
*Maximum weight	Normal category: 2300 lb. (landplane); 2220 lb. (seaplane) Utility category: 2000 lb. (landplane)
Number of seats	4 (2 at +34 to +46, 2 at +73) (Occupant on child's optional jump seat at +96)
Maximum baggage	120 lb. at +95
Fuel capacity	42 gal. total, 40 gal. usable (two 21.5 gal. tanks in wings at +48) *See Note 1 for data on unusable fuel.*
Oil capacity	1.5 gal. (-14.0), 1.0 gal. usable

VIII. Model 172N, Skyhawk, 4 PCL-SM (Normal Category), 2 PCLM (Utility Category) (cont'd)

Control surface movements	Wing flaps	Takeoff	0° - 10°	(landplane) (seaplane)	
		Landing	0° - 40° +0°, -2° (landplane)		
			0° - 30° ±2° (seaplane)		
	Ailerons	Up	20° ±1°	Down	15° ±1°
	Elevator tab	Up	28° +1°, -0°	Down	13° +1°, -0°
	Elevator	Up	28° +1°, -0°	Down	23° +1°, -0°
	(Neutral position is with bottom of balance area flush with bottom of stabilizer.)				
	Rudder (landplane)	Right	16° ±1°	Left	16° ±1° (landplane)
	(seaplane)	Right	19° ±1°	Left	15° ±1° (seaplane)
	(Measured parallel to W.L.)				

Serial numbers eligible

17261445, 17267585 through 17269309 (1977 model)
17261578, 17269310 through 17271034 (1978 model) (except 17270050)
17271035 through 17272884 (1979 model)
17270050, 17272885 through 17274009 (1980 model)

IX. Model 172P, Skyhawk, 4 PCL-SM (Normal Category), 2 PCLM (Utility Category), approved May 13, 1980

Engine — Lycoming O-320-D2J

*Fuel — 100LL/100 minimum grade aviation gasoline

*Engine limits — For all operations, 2700 rpm (160 hp)

Propeller and propeller limits

1. Propeller
   (a) McCauley 1C160/DTM 7557
   Static rpm at maximum permissible throttle setting:
   Not over 2420, not under 2300
   No additional tolerance permitted
   Diameter: not over 75 in., not under 74 in.
   (b) Spinner: Dwg. 0550320
2. Propeller (floatplane only)
   (a) McCauley 1A175/ETM 8043
   Static rpm at maximum permissible throttle setting:
   Not over 2570, not under 2470
   No additional tolerance permitted
   Diameter: not over 80 in., not under 78.5 in.
   (b) Spinner: Dwg. 0550320

*Airspeed limits (CAS) *(See Note 4 on use of CAS)*

Maneuvering	99 knots (landplane)
	96 knots (floatplane)
Maximum structural cruising	127 knots
Never exceed	158 knots
Flaps extended	85 knots

C.G. range

Landplane:

Normal category	(+39.5) to (+47.3) at 2400 lbs.
	(+35.0) to (+47.3) at 1950 lbs. or less
Utility category	(+36.5) to (+40.5) at 2100 lbs.
	(+35.0) to (+40.5) at 1950 lbs. or less

Seaplane: (Edo 89-2000 or 89A2000 floats)

Normal category	(+39.8) to (+45.5) at 2220 lbs.
	(+36.4) to (+45.5) at 1825 lbs. or less

Straight line variation between points given.

Empty weight C.G. Range — None

3A12

IX. Model 172P, Skyhawk, 4 PCL-SM (Normal Category), 2 PCLM (Utility Category) (cont'd)

*Maximum weight	Normal category: 2400 lb. (landplane); 2220 lb. (seaplane) Utility category: 2100 lb. (landplane)
Number of seats	4 (2 at +34 to +46, 2 at +73) (Occupant on child's optional jump seat at +96)
Maximum baggage	120 lb. at +95
Fuel capacity	42 gal. total, 40 gal. usable (two 21.5 gal. tanks in wings at +48) *See Note 1 for data on unusable fuel.*
Oil capacity	2 gal. (-13.1), 3.5 gal. usable

Control surface movements

Wing flaps			Takeoff	0° - 10°
			Landing	0° - 30° +0°, -2°
Ailerons	Up	20° ±1°	Down	15° ±1°
Elevator tab	Up	28° +1°, -0°	Down	13° +1°, -0° (floatplane)
	Up	22° +1°, -0°	Down	19° +1°, -0° (landplane)
Elevator	Up	28° +1°, -0°	Down	23° +1°, -0°

(Neutral position is with bottom of balance area flush with bottom of stabilizer.)

Rudder (landplane)	Right	16° ±1°	Left	16° ±1° (landplane)
(seaplane)	Right	19° ±1°	Left	15° ±1° (seaplane)

(Measured parallel to W.L.)

Serial numbers eligible

17274010 through 17275034 (1981 model)
17275035 through 17275759 (1982 model)
17275760 through 17276079 (1983 model)
17276080 through 17276259 (1984 model)
17276260 through 17276516 (1985 model)
17276517 through 17276654 (1986 model)

X. Model 172Q, Cutlass, 4 PCLM (Normal Category), approved October 15, 1982

Engine	Lycoming O-360-A4N
*Fuel	100LL/100 minimum grade aviation gasoline
*Engine limits	For all operations, 2700 rpm (180 hp)

Propeller and propeller limits

1. Propeller
   (a) McCauley 1A170E/JFA 7658
   Static rpm at maximum permissible throttle setting:
   Not over 2450, not under 2350
   No additional tolerance permitted
   Diameter: not over 76 in., not under 74.5 in.
   (b) Spinner: Dwg. 0509077

*Airspeed limits

Maneuvering	105 knots
Maximum structural cruising	127 knots
Never exceed	158 knots
Flaps extended	85 knots

C.G. range

Normal category (+41.0) to (+47.3) at 2550 lbs.
(+35.0) to (+47.3) at 1950 lbs. or less
Straight line variation between points given.

Empty weight C.G. range	None
*Maximum weight	Normal category: 2550 lb.

X. Model 172Q, Cutlass, 4 PCLM (Normal Category) (cont'd)

Number of seats: 4 (2 at +34 to +46, 2 at +73) (Occupant on optional child's seat at +96)

Maximum baggage: 120 lbs. at +95

Fuel capacity: 54 gal. total, 50 gal. usable (two 27 gal. tanks in wings at +48)
See Note 1 for data on unusable fuel.

Oil capacity: 9 qt. at -15.5, 2 qt. unusable

Control surface movements:

Wing flaps			Takeoff	0° - 10°
			Landing	0° - 30° +0°, -2°
Ailerons	Up	20° ±1°	Down	15° ±1°
Elevator tab	Up	22° +1°, -0°	Down	19° +1°, -0°
Elevator	Up	28° +1°, -0°	Down	23° +1°, -0°

(Neutral position is with bottom of balance area flush with bottom of stabilizer.)

Rudder	Right	16° ±1°	Left	16° ±1°

(Measured parallel to W.L.)

Serial numbers eligible: 17275869 through 17276054 (1983 model)
17276101 through 17276211 (1984 model)

DATA PERTINENT TO ALL MODELS 172 THROUGH 172Q

Datum: Front face of firewall (28000 through 47746)
Lower front face of firewall (17247747 through 17276654)

Leveling means: Upper doorsill

Certification basis:

Models 172 through 172P
Part 3 of the Civil Air Regulations effective November 1, 1949, as amended by 3-1 through 3-12. In addition, effective S/N 17271035 and on, FAR 23.1559 effective March 1, 1978. FAR 36 dated December 1, 1969, plus Amendments 36-1 through 36-5 for Model 172N; FAR 36 dated December 1, 1969, plus Amendments 36-1 through 36-12 for Model 172P through 172Q. In addition, effective S/N 17276260 and on, FAR 23.1545(a), Amendment 23-23 dated December 1, 1978.

Equivalent Safety Items	17261445, 17261578, 17265685
Airspeed Indicator	CAR 3.757 (see Note 4 on use of CAS) (17261445, 17261578, 17265685 through 17276259)
Operating Limitations	CAR 3.778(a)

Model 172Q
Part 3 of the Civil Air Regulations dated November 1, 1949, as amended by 3-1 through 3-12. In addition, FAR 23.1559 effective March 1, 1978; FAR 25.951(b)(2), Amendment 23-15 effective October 31, 1974; and FAR 23.1545(a), Amendment 23-23 effective December 1, 1978. FAR 36 dated December 1, 1969, plus amendments 36-1 through 36-12.

Application for Type Certificate dated July 11, 1955. Type Certificate No. 3A12 issued November 4, 1955, obtained by the manufacturer under Delegation Option Procedures.

Production basis: Production Certificate No. 4. Delegation Option Manufacturer No. CE-1 authorized to issue airworthiness certificates under delegation option provisions of Part 21 of the Federal Aviation Regulations.

DATA PERTINENT TO ALL MODELS 172 THROUGH 172Q (cont'd)

Equipment: The basic required equipment as prescribed in the applicable airworthiness requirements (see Certification Basis) must be installed in the aircraft for certification. This equipment must include a current Airplane Flight Manual effective S/N 17271035 and on.

1. Model 172 through 172G: Stall warning indicator, Dwg. 0511062.
2. Model 172H and on: Stall warning indictor, Dwg. 0523112.

The equipment portion of Aircraft Specification 3A12, Revision 17, or Cessna Publication TS1000-13 should be used for equipment references on all aircraft prior to the Model 172E. Refer to applicable equipment list for the Model 172E and subsequent models.

NOTE 1: Current weight and balance report including list of equipment included in certificated empty weight, and loading instructions when necessary must be provided for each aircraft at the time of original certification.

Serial Nos. 28000 through 29999, 36000 through 36999 and 46001 through 47746, 17247747 through 17265684
The certificated empty weight and corresponding center of gravity location must include unusable fuel of 30 lbs. at (+46) on Models 172 and 172A, or 18 lbs. at (+46) for Models 172B through 172H, or 24 lbs. at (+46) for Models 172I through 172M (17265684) and undrainable oil of (0) lb. at -20) for 172 through 172H and (0) lb. at (-14) for 172I through 172M (17265684).

Serial Nos. 17261578, 17261445, 17265685 through 17274009
The certificated empty weight and corresponding center of gravity location must include unusable fuel of 24 lbs. at (+46) through 172M (17267584) or 18 lbs. at (+46) 17267585 and on and full oil of 11.3 lb. at (-14).

Serial Nos. 17274010 through 17276654: (Model 172P)
The certificated empty weight and corresponding center of gravity location must include unusable fuel of 18 lb. at (+46) and full oil of 15 lb. at (-13.1).

Serial Nos. 17275869 through 17276211; (Model 172Q)
The certificated empty weight and corresponding center of gravity location must include unusable fuel of 24 lb. at (+46) and full oil of 16.88 lb. at (-15.5).

NOTE 2. The following placards must be displayed as indicated:

A. In full view of the pilot:
(1) Models 172, 172A and 172B
"This airplane must be operated in compliance with the operating limitations stated in the form of placards, markings, and manuals.

NORMAL CATEGORY

Maximum design weight	2200 lbs.		
Refer to weight and balance data for loading instructions.			
Flight maneuvering load factors	Flaps up	+3.8	-1.52
	Flaps down	+3.5	
No acrobatic maneuvers including spins approved.			

UTILITY CATEGORY

Maximum design weight	1950 lbs.		
Baggage compartment and rear seat must not be occupied			
Flight maneuvering load factors	Flaps up	+4.4	-1.76
	Flaps down	+3.5	

DATA PERTINENT TO ALL MODELS 172 THROUGH 172Q (cont'd)

(1) Models 172, 172A and 172B (cont'd)

No acrobatic maneuvers approved except those listed below.

Maneuver	Entry speed
Chandelles	115 mph (100 knots)
Lazy eights	115 mph (100 knots)
Steep turns	115 mph (100 knots)
Spins	Slow deceleration
Stalls (except whip stalls)	Slow deceleration"

(2) Model 172C

"This airplane must be operated in compliance with the operating limitations stated in the form of placards, markings, and manuals.

NORMAL CATEGORY

Maximum design weight 2250 lbs.

Refer to weight and balance data for loading instructions.

Flight maneuvering load factors	Flaps up	+3.8	-1.52
	Flaps down	+3.5	

No acrobatic maneuvers including spins approved.

UTILITY CATEGORY

Maximum design weight 1950 lbs.

Baggage compartment and rear seat must not be occupied.

Flight maneuvering load factors	Flaps up	+4.4	-1.76
	Flaps down	+3.5	

No acrobatic maneuvers approved except those listed below.

Maneuver	Entry speed
Chandelles	115 mph (100 knots)
Lazy eights	115 mph (100 knots)
Steep turns	115 mph (100 knots)
Spins	Slow deceleration
Stalls (except whip stalls)	Slow deceleration"

(3) Models 172D, 172E, 172F, 172G, 172H, 172I, and 172K

"This airplane must be operated in compliance with the operating limitations stated in the form of placards, markings, and manuals.

NORMAL CATEGORY

Maximum design weight 2300 lbs.

Refer to weight and balance data for loading instructions.

Flight maneuvering load factors	Flaps up	+3.8	-1.52
	Flaps down	+3.5	

No acrobatic maneuvers including spins approved.

UTILITY CATEGORY

Maximum design weight 2000 lbs.

Baggage compartment and rear seat must not be occupied.

Flight maneuvering load factors	Flaps up	+4.4	-1.76
	Flaps down	+3.5	

No acrobatic maneuvers except those listed below.

Maneuver	Max. Entry speed
Chandelles	122 mph (106 knots)
Lazy eights	122 mph (106 knots)
Steep turns	122 mph (106 knots)
Spins	Slow deceleration
Stalls (except whip stalls)	Slow deceleration"

DATA PERTINENT TO ALL MODELS 172 THROUGH 172Q (cont'd)

(4) Model 172L (1971 model)
"This airplane must be operated in compliance with the operating limitations stated in the form of placards, markings, and manuals.

	MAXIMUMS			
	Normal Category		Utility Category	
Maneuvering speed (CAS)	122 mph (106 knots)		122 mph (106 knots)	
Gross weight	2300 lbs.		2000 lbs.	
Flight load factor				
Flaps up	+3.8	-1.52	+4.4	-1.76
Flaps down	+3.5		+3.5	

Normal category - No acrobatic maneuvers including spins approved
Utility category - Baggage compartment and rear seat must not be occupied.

No acrobatic maneuvers approved except those listed below.

Maneuver	Entry speed
Chandelles	122 mph (106 knots)
Lazy eights	122 mph (106 knots)
Steep turns	122 mph (106 knots)
Spins	Slow deceleration
Stalls (except whip stalls)	Slow deceleration"

Spin recovery: opposite rudder - forward elevator - neutralize controls

Known icing conditions to be avoided. This airplane is certified for the following flight operations as of date of original airworthiness certificate:

(DAY NIGHT VFR IFR)" (as applicable)

(5) Model 172L (1972 model)
"This airplane must be operated in compliance with the operating limitations as stated in the form of placards, markings, and manuals:

	MAXIMUMS			
	Normal Category		Utility Category	
Maneuvering speed (CAS)	122 mph (106 knots)		122 mph (106 knots)	
Gross weight	2300 lbs.		2000 lbs.	
Flight load factor				
Flaps up	+3.8	-1.52	+4.4	-1.76
Flaps down	+3.5		+3.5	

Normal category - No acrobatic maneuvers including spins approved
Utility category - Baggage compartment and rear seat must not be occupied.

No acrobatic maneuvers approved except those listed below.

Maneuver	Max. Entry speed
Chandelles	122 mph (106 knots)
Lazy eights	122 mph (106 knots)
Steep turns	122 mph (106 knots)
Spins	Slow deceleration
Stalls (except whip stalls)	Slow deceleration"

Spin recovery: opposite rudder - forward elevator - neutralize controls. Intentional spins with flaps extended are prohibited. Known icing conditions to be avoided. This airplane is certified for the following flight operations as of date of original airworthiness certificate:

(DAY NIGHT VFR IFR)" (as applicable)

DATA PERTINENT TO ALL MODELS 172 THROUGH 172Q (cont'd)

(6) Model 172M (Landplane) 17256493, 17260759 through 17265684 except 17261445 and 17261578

"This airplane must be operated in compliance with the operating limitations as stated in the form of placards, markings, and manuals.

	MAXIMUMS			
	Normal Category		Utility Category	
Maneuvering speed (CAS)	112 mph (97 knots)		112 mph (97 knots)	
Gross weight	2300 lbs.		2000 lbs.	
Flight load factor				
Flaps up	+3.8	-1.52	+4.4	-1.76
Flaps down	+3.0		+3.0	

Normal category - No acrobatic maneuvers including spins approved
Utility category - Baggage compartment and rear seat must not be occupied.

No acrobatic maneuvers approved except those listed below.

Maneuver	Recommended Entry speed	Maneuver	Recommended Entry Speed
Chandelles	120 mph (104 knots)	Spins	Slow deceleration
Lazy eights	120 mph (104 knots)	Stalls (except whip stalls)	Slow deceleration
Steep turns	112 mph (97 knots)		

Altitude loss in stall recovery -- 180 feet.
Abrupt use of the controls prohibited above 112 mph
Spin recovery: opposite rudder -- forward elevator -- neutralize controls
Intentional spins with flaps extended are prohibited. Flight into known icing conditions prohibited. This airplane is certified for the following flight operations as of date of original airworthiness certificate:

(DAY - NIGHT - VFR - IFR)" (as applicable)

Model 172M (Floatplane) 17256493, 17260759 through 17265684 except 17261445 and 17261578

"This airplane must be operated as a normal category airplane in compliance with the operating limitations as stated in the form of placards, markings, and manuals.

MAXIMUMS		
Maneuvering speed	110 mph (96 knots) (CAS)	
Gross weight	2220 lbs.	
Flight load factor	Flaps up	+3.8, -1.52
	Flaps down	+3.0

WATER RUDDER: Extend for taxi; retract for takeoff, flight, and landing.

No acrobatic maneuvers, including spins approved. Altitude loss in a stall recovery - 200 ft. Flight into known icing conditions prohibited. This airplane is certified for the following flight operations as of date of original airworthiness certificate:

(DAY - NIGHT - VFR - IFR)" (as applicable)

(7) Model 172M and 172N (Landplane) (17261445, 17261578, 17265685 through 17271034 except 17270050)

"This airplane must be operated in compliance with the operating limitations stated in the form of placards, markings, and manuals.

DATA PERTINENT TO ALL MODELS 172 THROUGH 172Q (cont'd)

	MAXIMUMS			
	Normal Category		Utility Category	
Maneuvering speed (CAS)	97 knots		97 knots	
Gross weight	2300 lbs.		2000 lbs.	
Flight load factor				
Flaps up	+3.8	-1.52	+4.4	-1.76
Flaps down	+3.0		+3.0	

Normal category - No acrobatic maneuvers including spins approved.
Utility category - Baggage compartment and rear seat must not be occupied.

NO ACROBATIC MANEUVERS EXCEPT THOSE LISTED BELOW:

Maneuver	Recommended Entry speed	Maneuver	Recommended Entry Speed
Chandelles	105 knots	Spins	Slow deceleration
Lazy eights	105 knots	Stalls (except	Slow deceleration
Steep turns	95 knots)	whip stalls)	

Altitude loss in stall recovery - 180 feet.
Abrupt use of the controls prohibited above 97 knots

Spin recovery: opposite rudder - forward elevator - neutralize controls. Intentional spins with flaps extended are prohibited. Flight into known icing conditions prohibited. This airplane is certified for the following flight operations as of date of original airworthiness certificate.

(DAY - NIGHT - VFR - IFR)" (as applicable)

Model 172M and 172N (Floatplane) (17265685 through 17271034)

FLOATPLANE

"This airplane must be operated as a normal category airplane in compliance with the operating limitations as stated in the form of placards, markings, and manuals.

MAXIMUMS		
Maneuvering speed (CAS)	96 knots	
Gross weight	2220 lbs.	
Flight load factor	Flaps up	+3.8, -1.52
	Flaps down	+3.0

Water Rudder: Extend for taxi; retract for takeoff, flight and landing.

No acrobatic maneuvers, including spins approved. Altitude loss in a stall recovery - 200 ft. Flight into known icing conditions prohibited. This airplane is certified for the following flight operations as of date of original airworthiness certificate:

(DAY - NIGHT - VFR - IFR)" (as applicable)

B. Forward of fuel selector valve: (All models through S/N 17265684 except 17261445 and 17261578)

"Both tanks on for takeoff and landing."

C. On the fuel selector valve (at appropriate location)

(1) Model 172 and 172A

"Both - 37 gal.
Left - 18.5 gal.
Right - 18.5 gal.
Off"

DATA PERTINENT TO ALL MODELS 172 THROUGH 172Q (cont'd)

(2) Model 172B
"Both - 39 gal.
Left - 19.5 gal.
Right - 19.5 gal.
Off"

(3) Model 172C, 172D, 172E, 172F, 172G, and 172H
"Both - 36 gal.
Left - 18 gal.
Right - 18 gal.
Off"

(4) Model 172I through 172M (except 17261445 and 17261578)
"Both - 38 gal. (all flight attitudes)
Left - 19 gal. (level flight only)
Right - 19 gal. (level flight only)
Off"

(5) Model 172N (17261445, 17261578, 17267585 through 17271034, excluding 17270050)
"Both - 40 gal. (all flight altitudes) (Takeoff-landing)
Left - 20 gal. (level flight only)
Right - 20 gal. (level flight only)
Off"

D. On flap handle, Models 172 through 172E
(1) "Flaps - Pull to extend
Takeoff Retract 0°
1st notch 10°
Landing 0° - 40°
(2) "Avoid slips with flaps down."

E. Near flap indicator Models 172F (electric flaps) through 17271034, excluding 17270050)
"Avoid slips with flaps extended."

F. In baggage compartment:
(1) Models 172 through 172B
"Maximum baggage 120 lb. For additional loading instructions, see weight and balance data."
(2) Model 172C through 172M (1973 model)
"120 lb. maximum baggage and/or auxiliary seat passenger. For additional loading instructions see weight and balance data."
(3) 17261899 through 17271034, excluding 17270050
"120 lb. maximum baggage and/or auxiliary passenger forward of baggage door latch."
"50 lb. maximum baggage aft of baggage door latch maximum 120 lb. combined. For additional loading instructions see weight and balance data."

G. Near ammeter (Models 17258487 through 17259903)
"Do not turn off alternator in flight except in emergency."

H. Additional placards required in seaplane.
(1) Model 172A through 172I in full view of the pilot.
"Operate as normal category airplane except:
Maximum weight 2220 lbs.
Maximum altitude loss in stall recovery 120 ft.
Flaps - takeoff - 1st notch -10°
Water rudder - pull to extract
Retract - takeoff, flight and landing
Extend - taxi."

DATA PERTINENT TO ALL MODELS 172 THROUGH 172Q (cont'd)

H. (2) Model 172K in full view of the pilot:
"THIS AIRPLANE MUST BE OPERATED IN COMPLIANCE WITH THE OPERATING LIMITATIONS AS STATED IN THE FORM OF PLACARDS, MARKINGS, AND MANUALS

NORMAL CATEGORY - FLOATPLANE

Maximum weight	2220 lb.	
Refer to weight and balance data for loading instructions.		
Flight maneuvering load factors	Flaps up	+3.8, -1.52
	Flaps down	+3.5

No acrobatic maneuvers including spins approved.
Maximum altitude loss in stall recovery - 120 ft.
Flaps: Takeoff - 10° . . . Water rudder: Pull to retract . . .
Retract: Takeoff, flight and landing . . . . Extend: Taxi."

(3) Model 172F through 17271034, excluding 17270050, in full view of the pilot.
"Floatplane Max. Flaps - 30°."

(4) Model 172L in full view of the pilot:

"FLOATPLANE
THIS AIRPLANE MUST BE OPERATED AS A NORMAL CATEGORY AIRPLANE IN COMPLIANCE WITH THE OPERATING LIMITATIONS AS STATED IN THE FORM OF PLACARDS, MARKINGS, AND MANUALS.

"MAXIMUMS

Maneuvering speed		122 mph CAS (106 knots)
Gross weight		2220 lbs.
Flight load factor	Flaps up	+3.8, -1.52
	Flaps down	+3.5

WATER RUDDER: Extend for taxi; retract for takeoff, flight and landing.

FLAPS: 10° for takeoff

No acrobatic maneuvers, including spins, approved. Altitude loss in stall recovery - 120 ft. Known icing conditions to be avoided. This airplane is certified for the following flight operations as of date of original airworthiness certificate:

DAY NIGHT VFR IFR" (as applicable)

I. Near tachometer on Models 172I, 172K and 172L (with IC172/MTM propeller):
"Avoid continuous operation
1. Above 75 percent power in cruise
2. Above 2500 rpm in full throttle climb."

J. Near ammeter and adjacent to overvoltage light:
(1) Model 172L (1972) through Model 172N (1978)
"High Voltage"

K. Near fuel selector valve on models with serial numbers 28000 through 17258855, except those with Cessna Kit No. SK-172-31B or SK-172-32 installed:

"SWITCH TO SINGLE TANK OPERATION IMMEDIATELY UPON REACHING CRUISE ALTITUDES ABOVE 5000 FEET."

DATA PERTINENT TO ALL MODELS 172 THROUGH 172Q (cont'd)

L. Near fuel tank filler

(1) Model 172, 172A and 172B

"FUEL
80/87 min. grade aviation gasoline
Cap. 21 U.S. gal."

(2) Model 172C, 172D, 172E, 172F, 172G, and 172H

"FUEL
80/87 min. grade aviation gasoline
Cap. 19.5 U.S. gal."

(3) Model 172I through 172M (except 17261445 and 17261578)

"FUEL
80/87 min. grade aviation gasoline
Cap. 21 US. gal."

(4) Model 172N (17261445, 17267585 through 17269309)

"FUEL
100/130 min. grade aviation gasoline
Cap. 21.5 US. gal."

(5) Model 172N (17261578, 17269310 through 17271034, excluding 17270050)

"FUEL
100LL/100 min. grade aviation gasoline
Cap. 21.5 US. gal."

M. Effective 17270050, 17271035 through 17276654
All placards required in the Pilot's Operating Handbook and FAA Approved Airplane Flight Manual must be installed in the appropriate locations.

NOTE 3. Compliance with Service Letter SE74-16 - Carburetor Nozzle Replacement - allows rpm's as follows:

Landplane: not over 2420, not under 2300
Seaplane: not over 2570, not under 2445

NOTE 4. The marking of the airspeed indicator in CAS provides an equivalent level of safety to CAR 3.757 when approved airspeed calibration data presented in Section V of the Pilot's Operating Handbooks listed below is available to the pilot (TIAS is exactly equal to CAS):

172M, Cessna P/N D1057-13	(S/N 17265685 through 17267584)
172N, Cessna P/N D1082-13	(S/N 17261445, 17267585 through 17269309)
172N, Cessna P/N D1109-13	(S/N 17261578, 17269310 through 17271034 except 17270050)
172N, Cessna P/N D1138-13PH	(S/N 17271035 through 17272884)
172N, Cessna P/N D1172-13PH	(S/N 17270050, 17272885 through 17274009)
172P, Cessna P/N D1192-13PH	(S/N 17274010 through 17275034)
172P, Cessna P/N D1212-13PH	(S/N 17275035 through 17275759)
172P, Cessna P/N D1231-13PH	(S/N 17275760 through 17276079)
172P, Cessna P/N D1251-13PH	(S/N 17276080 through 17276259)

NOTE 5. 14-volt electrical system
(172 series through S/N 17269309, except 17258105 through 17258112 and 17261578)

28-volt electrical system
(S/N 17258105 through 17258112, 17261578 and 17269310 through 17276654)

NOTE 6: Special Ferry Flight Authorization. Flight Standards District Offices are authorized to issue Special overweight ferry flight authorizations. These airplanes are structurally satisfactory for ferry flight if maintained within the following limits: (1) Takeoff weight must not exceed 130% of the maximum weight for Normal Category; and (2) The Never Exceed Airspeed (V_{NE}) and Maximum Structural Cruising Speed (V_C) must be reduced by 30%; and (3) Forward and aft center of gravity limits may not be exceeded; and (4) Structural load factors of +2.5 g. to -1.0 g. may not be exceeded. Requirements for any additional oil should established in accordance with Advisory Circular AC23.1011-1. Increased stall speeds and reduced climb performance should be expected for the increased weights. Flight characteristics and performance at the increased weights have not been evaluated. Flight Permit for operations of overweight aircraft may be found in Advisory Circular AC21-4B

3A12

DATA PERTINENT TO ALL MODELS 172 THROUGH 172Q (cont'd)

In addition to the placards specified above, the prescribed operating limitations indicated by an asterisk (*) under Sections I through X of this data sheet must also be displayed by permanent markings.

XI - Model 172R, Skyhawk, 4 PCLM (Normal Category), 2 PCLM (Utility Category), Approved June 21, 1996

Engine	Lycoming IO-360-L2A, Rated 160 Horsepower When Modified by Cessna Modification Kit MK172-72-01 (See NOTE 4) Lycoming IO-360-L2A, Rated 180 Horsepower
Fuel	100/100LL minimum grade aviation gasoline
Engine Limits	For all operations, 2,400 RPM When Modified by Cessna Modification Kit MK172-72-01 (See NOTE 4) For all operations, 2,700 RPM
Propeller	(a) McCauley Model IC235/LFA7570 (b) Spinner: Drawing No. 0550236 When Modified by Cessna Modification Kit MK172-72-01 (See NOTE 4) (a) McCauley Model 1A170E/JHA7660 (b) Spinner: Drawing No. 0550236
Propeller limits	Static RPM at full throttle: Not over 2,165; Not Under 2,065 No Additional Tolerance Permitted Diameter: Not over 75 inches; not under 74 inches When Modified by Cessna Modification Kit MK172-72-01 (See NOTE 4) Static RPM at full throttle: Not over 2,400; Not Under 2,300 No Additional Tolerance Permitted Diameter: Not over 76 inches; not under 75 inches

Airspeed Limits

Maneuvering	99 Knots IAS	(97 Knots CAS)
Max Structural Cruising	129 Knots IAS	(126 Knots CAS)
Never Exceed	163 Knots IAS	(160 Knots CAS)
Flaps Extended	85 Knots IAS	(84 Knots CAS)

When Modified by Cessna Modification Kit MK172-72-01 (See NOTE 4)

Maneuvering	105 Knots IAS	(102 Knots CAS)
Max Structural Cruising	129 Knots IAS	(126 Knots CAS)
Never Exceed	163 Knots IAS	(160 Knots CAS)
Flaps Extended	85 Knots IAS	(84 Knots CAS)

C.G. Range

Normal Category

(1) Aft Limits	47.3 inches aft of datum at 2,450 pounds or less.
(2) Forward Limits	Linear variation from 40.0 inches aft of datum at 2,450 pounds to 35.0 inches aft of datum at 1,950 pounds; 35.0 inches aft of datum at 1,950 pounds or less.

Utility Category

(1) Aft Limits	40.5 inches aft of datum at 2,100 pounds or less.
(2) Forward Limits	Linear variation from 36.5 inches aft of datum at 2,100 pounds to 35.0 inches aft of datum at 1,950 pounds; 35.0 inches aft of datum at 1,950 pounds or less.

XI - Model 172R (cont'd)

C.G. Range	When Modified by Cessna Modification Kit MK172-72-01 (See NOTE 4)	
	Normal Category	
	(1) Aft Limits	47.3 inches aft of datum at 2,550 pounds or less.
	(2) Forward Limits	Linear variation from 41.0 inches aft of datum at 2,550 pounds to 35.0 inches aft of datum at 1,950 pounds; 35.0 inches aft of datum at 1,950 pounds or less.
	Utility Category	
	(1) Aft Limits	40.5 inches aft of datum at 2,200 pounds or less.
	(2) Forward Limits	Linear variation from 37.5 inches aft of datum at 2,200 pounds to 35.0 inches aft of datum at 1,950 pounds; 35.0 inches aft of datum at 1,950 pounds or less.
Empty Wt. C.G. Range	None	
Reference Datum	Lower portion of front face of firewall	
MAC	58.8 inches; Leading edge of MAC 25.9 inches aft of datum	
Leveling Means	Left side of Tailcone at 108.0 inches and 142.0 inches aft of datum	
Maximum Weights	Normal Category	
	Maximum Ramp	2,457 pounds
	Maximum Takeoff and Landing	2,450 pounds
	Utility Category	
	Maximum Ramp	2,107 pounds
	Maximum Takeoff and Landing	2,100 pounds
	When Modified by Cessna Modification Kit MK172-72-01 (See NOTE 4)	
	Normal Category	
	Maximum Ramp	2,558 pounds
	Maximum Takeoff and Landing	2,550 pounds
	Utility Category	
	Maximum Ramp	2,208 pounds
	Maximum Takeoff and Landing	2,200 pounds
No. of Seats	4 (2 at 34.0 to 46.0 inches aft of datum; 2 at 73.0 inches aft of datum)	
Maximum Baggage	120 pounds at 95.0 inches aft of datum	
	When Modified by Cessna Modification Kit MK172-72-01 (See NOTE 4) 120 pounds at 82.0 to 108.0 inches aft of datum 50 pounds at 108.0 to 142.0 inches aft of datum (Maximum combined weight capacity for baggage areas is 120 pounds.)	
Fuel Capacity (Gal.)	56 gallons total; 53 gallons usable (Two 28 gallon tanks in wings at 48.0 inches aft of datum) See NOTE 1 for data on usable fuel.	
Oil Capacity (Gal.)	2.0 gallons at 13.1 inches forward of datum 3.5 quarts usable	
	When Modified by Cessna Modification Kit MK172-72-01 (See NOTE 4) 2.0 gallons at 13.1 inches forward of datum 3.0 quarts usable	

3A12

XI - Model 172R (cont'd)

Control surface movements	Wing flaps	Takeoff	0° - 10°		
		Landing	0° - 30° +0°/-2°		
	Ailerons	Up	20° ± 1°	Down	15° ± 1°
	Elevator tab	Up	22° + 1°/-0°	Down	19° + 1°/-0°
	Elevator	Up	28° + 1°/-0°	Down	23° + 1°/-0°

(Neutral position is with bottom of balance area flush with bottom of stabilizer)

Rudder (Measured parallel to W.L.): Right 16° 10'± 1° Left 16° 10' ± 1°

Rudder (Measured perpendicular to Hinge: Right 17° 44' ± 1° Left 17° 44' ± 1°

Certification Basis

Part 23 of the Federal Aviation Regulations effective February 1, 1965, as amended by 23-1 through 23-6, except as follows:
FAR 23.423; 23.611; 23.619; 23.623; 23.689; 23.775; 23.871; 23.1323; and 23.1563 as amended by Amendment 23-7. FAR 23.807 and 23.1524 as amended by Amendment 23-10. FAR 23.507; 23.771; 23.853(a),(b) and (c); and 23.1365 as amended by Amendment 23-14. FAR 23.951 as amended by Amendment 23-15. FAR 23.607; 23.675; 23.685; 23.733; 23.787; 23.1309 and 23.1322 as amended by Amendment 23-17. FAR 23.1301 as amended by Amendment 23-20. FAR 23.1353; and 23.1559 as amended by Amendment 23-21. FAR 23.603; 23.605; 23.613; 23.1329 and 23.1545 as amended by Amendment 23-23. FAR 23.441 and 23.1549 as amended by Amendment 23-28. FAR 23.779 and 23.781 as amended by Amendment 23-33. FAR 23.1; 23.51 and 23.561 as amended by Amendment 23-34. FAR 23.301; 23.331; 23.351; 23.427; 23.677; 23.701; 23.735; and 23.831 as amended by Amendment 23-42. FAR 23.961; 23.1093; 23.1143(g); 23.1147(b); 23.1303; 23.1357; 23.1361 and 23.1385 as amended by Amendment 23-43. FAR 23.562(a), 23.562(b)2, 23.562(c)1, 23.562(c)2, 23.562(c)3, and 23.562(c)4 as amended by Amendment 23-44. FAR 23.33; 23.53; 23.305; 23.321; 23.485; 23.621; 23.655 and 23.731 as amended by Amendment 23-45.

FAR 36 dated December 1, 1969, as amended by Amendments 36-1 through 36-21.

Additions for the Garmin G1000 Integrated Cockpit System (ICS) Only:

14 CFR 23.303; 23.307; 23.601; 23.1163(a); 23.1367 and 23.1381 as amended by Amendment 23- N/C. 14 CFR 23.1589 as amended by Amendment 23-13. 14 CFR 23.771(a) as amended by Amendment 23-14. 14 CFR 23.607 and (Electrical System) 23.1309(a)(1)(2), (c) as amended by Amendment 23-17. 14 CFR 23.1301; 23.1327 and 23.1547(e) as amended by Amendment 23-20. 14 CFR 23.1501 and 23.1541(a)(1), (a)(2), (b)(1), (b)(2) as amended by Amendment 23-21. 14 CFR 23.603 and 23.605 as amended by Amendment 23-23. 14 CFR 23.1529 as amended by Amendment 23-26. 14 CFR 23.561(e); 23.1523; 23.1581(a)(2); and 23.1583(a), (c), (d), (f) as amended by Amendment 23-34. 14 CFR 23.301 as amended by Amendment 23-42. 14 CFR 23.1322; 23.1331 and 23.1357(a)(b)(c)(d) as amended by Amendment 23-43. 14 CFR 23.305; 23.773(a)(1), (a)(2); 23.1525and 23.1549 as amended by Amendment 23-45. 14 CFR 23.1303(a)(b)(c)(f); 23.1309(a)(1)(i), (a)(1)(ii), (a)(2), (b)(1), (b)(2)(i), (b)(2)(ii), (b)(3),(b)(4)(i), (b)(4)(ii), (b)(4)(iii), (b)(4)(iv), (c)(1), (c)(2)(iii), (c)(3), (d), (e), (f)(1); 23.1311; 23.1321 (a)(c)(d)(e); 23.1323(a), (b)(1), (b)(2), (c); 23.1329 (g)(h); 23.1351(a)(1), (a)(2)(i), (b)(1)(iii), (b)(2)(3), (c)(4), (d)(1); 23.1353(a)(b)(c)(d)(e); 23.1359(c); 23.1361; 23.1365(a)(b)(d)(e)(f) and 23.1431(a)(b)(d)(e) as amended by Amendment 23-49. 14 CFR 23.1325(a), (b)(1), (b)(2)(i), (b)(3), (c)(d)(e); 23.1543(b)(c); 23.1545(a), (b)(1), (b)(2), (b)(3), (b)(4); 23.1553; 23.1555(a)(b); 23.1563(a) and 23.1567(a) as amended by Amendment 23-50. 14 CFR 23.777(a)(b); 23.955(a)(2); 23.1337(a)(1), (a)(2), (b)(1), (c) as amended by Amendment 23-51. 14 CFR 23.1305(a)(1), (a)(2), (a)(3), (b)(2), (b)(3)(i), (b)(4)(i), (b)(5), (b)(6)(i) as amended by Amendment 23-52. 14 CFR 23.901(a)(b) as amended by Amendment 23-53.

Equivalent Safety Items

(1)	Induction System Icing Protection	FAR § 23.1093; Refer to FAA letter dated 5/3/96
(2)	Throttle Control	FAR § 23.1143(g); Refer to FAA letter dated 3/22/96
(3)	Mixture Control	FAR § 23.1147(b); Refer to FAA letter dated 3/22/96

Date of Application for Amended Type Certificate was September 25, 1995.
Type Certificate No. 3A12 was amended June 21, 1996.

Serial numbers eligible 17280001 and On

Special Conditions as follows:

No. 23-159-SC, "Special Conditions: Cessna Aircraft Company; Cessna Model 172R Airplane; Installation of Electronic Flight Instrument System and and the Protection of the System From High Intensity Radiated Fields (HIRF)."

Data Pertinent to Model 172R:

Production Basis
Production Certificate No. PC-4 issued March 28, 1997. Applies to airplane serial numbers 17280014, 17280015, 17280017, 17280021 through 17280029, and 17280031 and on. Airplane serial numbers not listed were produced under Type Certificate only. Cessna is authorized to issue airworthiness certificates under the delegation provisions of Delegation Option Authorization No. CE-1 in accordance with Part 21 of the Federal Aviation Regulations.

Equipment
The basic required equipment as prescribed in the applicable airworthiness regulations (see Certification Basis) must be installed in the airplane for certification.

NOTE 1: Weight and Balance:

Serial Nos. 17280001 and On
The certificated empty weight and corresponding center of gravity location must include unusable fuel of 18 pounds at 46.0 inches aft of datum, and full oil of 15.0 pounds at 13.1 inches forward of datum.

NOTE 2: The airplane must be operated according to the appropriate Pilot's Operating Handbook and FAA Approved Airplane Flight Manual (POH/AFM). POH/AFM part number 172RPHUS00 (or later approved revision) is applicable to Production Model 172R. POH/AFM part number 172R180PH00 (or later approved revision) is applicable to Production Model 172R airplanes when modified by Cessna Modification Kit MK172-72-01. All POH/AFM Supplements approved for part number 172RPHUS00, are also applicable to part number 172R180PH00, unless specifically noted otherwise in the Supplement. All FAA required placards are included in Section 2 of the applicable POH/AFM. Placards may also be found in the Maintenance Manual, part number 172RMM00 (or later revision) , Chapter Eleven (11), "Placards and Markings."

FAA Approved Airplane Flight Manual (AFM): Part Number 172RPHAUS-00 (or later FAA approved revisions) is applicable to the Model 172R equipped with Garmin G1000 Integrated Cockpit System. The airplane must be operated according to the appropriate AFM. Required placards are included in the AFM.

NOTE 3: Special Ferry Flight Authorization. Flight Standards District Offices are authorized to issue Special overweight ferry flight authorizations. This airplane is structurally satisfactory for ferry flight if maintained within the following limits: (1) Takeoff weight must not exceed 130% of the maximum weight for Normal Category; and (2) The Never Exceed Airspeed (V_{NE}) and Maximum Structural Cruising Speed (V_C) must be reduced by 30%; and (3) Forward and aft center of gravity limits may not be exceeded; and (4) Structural load factors of +2.5 g. to -1.0 g. may not be exceeded. Requirements for any additional oil should be established in accordance with Advisory Circular AC23.1011-1. Increased stall speeds and reduced climb performance should be expected for the increased weights. Flight characteristics and performance at the increased weights have not been evaluated. Flight Permit for operations of overweight aircraft may be found in Advisory Circular AC21-4B.

3A12

NOTE 4: Only certain Model 172R airplane serial numbers are eligible for modification by Cessna Modification Kit MK172-72-01. Applicable serial numbers are as follows:

17280159	17280242	17280251	17280253	17280257
17280262	17280281	17280292	17280301	17280305
17280426	17280488	17280606	17280607	17280608
17280609	17280610	17280613	17280614	17280616
17280621	17280622	17280623	17280624	17280631
17280632	17280633	17280634	17280638	17280639
17280640	17280646	17280648	17280652	17280659
17280660	17280661	17280662	17280664	17280667
17280668	17280669	17280670	17280672	17280673
	17280674	17280675	17280701	17280707

XII - Model 172S, Skyhawk SP, 4 PCLM (Normal Category), 2 PCLM (Utility Category), Approved May 1, 1998

Engine — Lycoming IO-360-L2A, Rated 180 Horsepower

Fuel — 100/100LL minimum grade aviation gasoline

Engine Limits — For all operations, 2,700 RPM

Propeller
(a) McCauley Model 1A170E/JHA7660
(b) Spinner: Drawing No. 0550236

Propeller limits
Static RPM at full throttle: Not over 2400; Not Under 2300
Diameter: Not over 76 inches; not under 75 inches

Airspeed Limits

Maneuvering	105 Knots IAS	(102 Knots CAS)
Max Structural Cruising	129 Knots IAS	(126 Knots CAS)
Never Exceed	163 Knots IAS	(160 Knots CAS)
Flaps Extended	85 Knots IAS	(85 Knots CAS)

C.G. Range

Normal Category
(1) Aft Limits — 47.3 inches aft of datum at 2,550 pounds or less.
(2) Forward Limits — Linear variation from 41.0 inches aft of datum at 2,550 pounds to 35.0 inches aft of datum at 1,950 pounds; 35.0 inches aft of datum at 1,950 pounds or less.

Utility Category
(1) Aft Limits — 40.5 inches aft of datum at 2,200 pounds or less.
(2) Forward Limits — Linear variation from 37.5 inches aft of datum at 2,200 pounds to 35.0 inches aft of datum at 1,950 pounds; 35.0 inches aft of datum at 1,950 pounds or less.

Empty Wt. C.G. Range — None

Reference Datum — Lower portion of front face of firewall

MAC — 58.8 inches; Leading edge of MAC 25.9 inches aft of datum

Leveling Means — Left side of Tailcone at 108.0 inches and 142.0 inches aft of datum

Maximum Weights

Normal Category

Maximum Ramp	2,558 pounds
Maximum Takeoff and Landing	2,550 pounds

	Utility Category Maximum Ramp 2,208 pounds Maximum Takeoff and Landing 2,200 pounds
No. of Seats	4 (2 at 34.0 to 46.0 inches aft of datum; 2 at 73.0 inches aft of datum)

XII - Model 172S (cont'd)

Maximum Baggage	120 pounds at 82.0 to 108.0 inches aft of datum 50 pounds at 108.0 to 142.0 inches aft of datum (Max. combined weight capacity for baggage areas is 120 pounds)
Fuel Capacity (Gal.)	56 gallons total; 53 gallons usable (Two 28 gallon tanks in wings at 48.0 inches aft of datum) See NOTE 1 for data on usable fuel.
Oil Capacity (Gal.)	8.0 quarts at 13.1 inches forward of datum 3.0 quarts usable
Control surface movements	Wing flaps Takeoff 0° - 10° Landing 0° - 30° +0°/-2° Ailerons Up 20° ± 1° Down 15° ± 1° Elevator tab Up 22° + 1°/-0° Down 19° + 1°/-0° Elevator Up 28° + 1°/-0° Down 23° + 1°/-0° (Neutral position is with bottom of balance area flush with bottom of stabilizer) Rudder (Measured parallel to W.L.): Right 16° 10'± 1° Left 16° 10' ± 1° Rudder (Measured perpendicular to Hinge: Right 17° 44' ± 1° Left 17° 44' ± 1°
Certification Basis	Part 23 of the Federal Aviation Regulations effective February 1, 1965, as amended by 23-1 through 23-6, except as follows:

FAR 23.423; 23.611; 23.619; 23.623; 23.689; 23.775; 23.871; 23.1323; and 23.1563 as amended by Amendment 23-7. FAR 23.807 and 23.1524 as amended by Amendment 23-10. FAR 23.507; 23.771; 23.853(a),(b) and (c); and 23.1365 as amended by Amendment 23-14. FAR 23.951 as amended by Amendment 23-15. FAR 23.607; 23.675; 23.685; 23.733; 23.787; 23.1309 and 23.1322 as amended by Amendment 23-17. FAR 23.1301 as amended by Amendment 23-20. FAR 23.1353; and 23.1559 as amended by Amendment 23-21. FAR 23.603; 23.605; 23.613; 23.1329 and 23.1545 as amended by Amendment 23-23. FAR 23.441 and 23.1549 as amended by Amendment 23-28. FAR 23.779 and 23.781 as amended by Amendment 23-33. FAR 23.1; 23.51 and 23.561 as amended by Amendment 23-34. FAR 23.301; 23.331; 23.351; 23.427; 23.677; 23.701; 23.735; and 23.831 as amended by Amendment 23-42. FAR 23.961; 23.1093; 23.1143(g); 23.1147(b); 23.1303; 23.1357; 23.1361 and 23.1385 as amended by Amendment 23-43. FAR 23.562(a), 23.562(b)2, 23.562(c)1, 23.562(c)2, 23.562(c)3, and 23.562(c)4 as amended by Amendment 23-44. FAR 23.33; 23.53; 23.305; 23.321; 23.485; 23.621; 23.655 and 23.731 as amended by Amendment 23-45.

FAR 36 dated December 1, 1969, as amended by Amendments 36-1 through 36-21.

Additions for the Garmin G1000 Integrated Cockpit System (ICS) Only: Additions for the Garmin G1000 Integrated Cockpit System (ICS) Only:

14 CFR 23.303; 23.307; 23.601; 23.1163(a); 23.1367 and 23.1381 as amended by Amendment 23- N/C. 14 CFR 23.1589 as amended by Amendment 23-13. 14 CFR 23.771(a) as amended by Amendment 23-14. 14 CFR 23.607 and (Electrical System) 23.1309(a)(1)(2), (c) as amended by Amendment 23-17. 14 CFR 23.1301; 23.1327 and

23.1547(e) as amended by Amendment 23-20. 14 CFR 23.1501 and 23.1541(a)(1), (a)(2), (b)(1), (b)(2) as amended by Amendment 23-21. 14 CFR 23.603 and 23.605 as amended by Amendment 23-23. 14 CFR 23.1529 as amended by Amendment 23-26. 14 CFR 23.561(e); 23.1523; 23.1581(a)(2); and 23.1583(a), (c), (d), (f) as amended by Amendment 23-34. 14 CFR 23.301 as amended by Amendment 23-42. 14 CFR 23.1322; 23.1331 and 23.1357(a)(b)(c)(d) as amended by Amendment 23-43. 14 CFR 23.305; 23.773(a)(1), (a)(2); 23.1525 and 23.1549 as amended by Amendment 23-45. 14 CFR 23.1303(a)(b)(c)(f); 23.1309(a)(1)(i), (a)(1)(ii), (a)(2), (b)(1), (b)(2)(i), (b)(2)(ii), (b)(3),(b)(4)(i), (b)(4)(ii), (b)(4)(iii), (b)(4)(iv), (c)(1), (c)(2)(iii), (c)(3), (d), (e), (f)(1); 23.1311; 23.1321 (a)(c)(d)(e); 23.1323(a), (b)(1), (b)(2), (c); 23.1329 (g)(h); 23.1351(a)(1), (a)(2)(i), (b)(1)(iii), (b)(2)(3), (c)(4), (d)(1); 23.1353(a)(b)(c)(d)(e); 23.1359(c); 23.1361; 23.1365(a)(b)(d)(e)(f) and 23.1431(a)(b)(d)(e) as amended by Amendment 23-49. 14 CFR 23.1325(a), (b)(1), (b)(2)(i), (b)(3), (c)(d)(e); 23.1543(b)(c); 23.1545(a), (b)(1), (b)(2), (b)(3), (b)(4); 23.1553; 23.1555(a)(b); 23.1563(a) and 23.1567(a) as amended by Amendment 23-50. 14 CFR 23.777(a)(b); 23.955(a)(2); 23.1337(a)(1), (a)(2), (b)(1), (c) as amended by Amendment 23-51. 14 CFR 23.1305(a)(1), (a)(2), (a)(3), (b)(2), (b)(3)(i), (b)(4)(i), (b)(5), (b)(6)(i) as amended by Amendment 23-52. 14 CFR 23.901(a)(b) as amended by Amendment 23-53.

Equivalent Safety Items

(1)	Induction System Icing Protection	FAR § 23.1093; Refer to FAA letter dated 5/1/98
(2)	Throttle Control	FAR § 23.1143(g); Refer to FAA letter dated 5/1/98
(3)	Mixture Control	FAR § 23.1147(b); Refer to FAA letter dated 5/1/98

Date of Application for Amended Type Certificate for the 172S was November 13, 1997.
Type Certificate No. 3A12 was amended May 1, 1998 for the Model 172S.

Serial numbers eligible 172S8001 and On

Special Conditions as follows:

No. 23-159-SC, "Special Conditions: Cessna Aircraft Company; Cessna Model 172S Airplane; Installation of Electronic Flight Instrument System and the Protection of the System From High Intensity Radiated Fields (HIRF)."

Data Pertinent to Model 172S:

Production Basis
Production Certificate No. PC-4 issued August 27, 1998. Applies to airplane serial numbers 172S80003 and on. Airplane serial numbers not listed were produced under Type Certificate only. Cessna is authorized to issue airworthiness certificates under the delegation provisions of Delegation Option Authorization No. CE-1 in accordance with Part 21 of the Federal Aviation Regulations.

Equipment
The basic required equipment as prescribed in the applicable airworthiness regulations (see Certification Basis) must be installed in the airplane for certification.

NOTE 1: Weight and Balance:

Serial Nos. 172S8001 and On
The certificated empty weight and corresponding center of gravity location must include unusable fuel of 18 pounds at 46.0 inches aft of datum, and full oil of 15.0 pounds at 13.1 inches forward of datum.

NOTE 2: Pilot's Operating Handbook and FAA Approved Airplane Flight Manual (POH/AFM): part number 172SPHUS00 (or later approved revision) is applicable to the Model 172S. The airplane must be operated according to the appropriate POH/AFM. All FAA required placards are included in Section 2 of the POH/AFM. Placards may also be found in the Maintenance Manual, part number 172RMM02 (or later revision) for the Model 172S, Chapter 11, Placards and Markings."

FAA Approved Airplane Flight Manual (AFM): Part Number 172SPHAUS-00 (or later FAA approved revisions) is applicable to Model 172S equipped with Garmin G1000 Integrated Cockpit System. The airplane must be operated according to the appropriate AFM. Required placards are included in the AFM.

NOTE 3: Special Ferry Flight Authorization. Flight Standards District Offices are authorized to issue Special overweight ferry flight authorizations. This airplane is structurally satisfactory for ferry flight if maintained within the following limits: (1) Takeoff weight must not exceed 130% of the maximum weight for Normal Category; and (2) The Never Exceed Airspeed (V_{NE}) and Maximum Structural Cruising Speed (V_C) must be reduced by 30%; and (3) Forward and aft center of gravity limits may not be exceeded; and (4) Structural load factors of +2.5 g. to -1.0 g. may not be exceeded. Requirements for any additional oil should be established in accordance with Advisory Circular AC23.1011-1. Increased stall speeds and reduced climb performance should be expected for the increased weights. Flight characteristics and performance at the increased weights have not been evaluated. Flight Permit for operations of overweight aircraft may be found in Advisory Circular AC21-4B

.....END....

DEPARTMENT OF TRANSPORTATION
FEDERAL AVIATION ADMINISTRATION

A16CE
CESSNA
Revision 21
207 T207
207A T207A

March 31, 2003

TYPE CERTIFICATE DATA SHEET NO. A16CE

This data sheet which is part of Type Certificate A16CE prescribes conditions and limitations under which the product for which the type certificate was issued meets the airworthiness requirements of the Federal Aviation Regulations.

Type Certificate Holder — Cessna Aircraft Company
P. O. Box 7704
Wichita, Kansas 67277

I - Model 207/T207, Skywagon/Turbo Skywagon, 7 PCLM (Normal Category), Approved December 31, 1968

Model 207

Engine — Continental IO-520-F

*Fuel — 100/130 minimum grade aviation gasoline

*Engine Limits — Takeoff (5 min.) at 2850 r.p.m. (300 hp.)
For all other operations, 2700 r.p.m. (285 hp.)

Propeller and Propeller Limits — Landplane

1. (a) McCauley D2A34C58/90AT-8 (C161004-0106)
   Diameter: not over 82 in., not under 80 in.
   Pitch settings at 36 in. sta.:
   low 9.5°, high 25.8°
   (b) Cessna spinner dome 1250909-3
   (c) Woodward hydraulic governor 210462
   (d) McCauley hydraulic governor C290D2/T4 or C290D4/T4
2. (a) McCauley D3A32C90/82NC-2 (C161006-0205)
   Diameter: not over 80 in., not under 78 in.
   Pitch settings at 30 in. sta.:
   low 11.5°, high 28.1°
   (b) Cessna spinner dome 1250909-8
   (c) Woodward hydraulic governor 210462
   (d) McCauley hydraulic governor C290D2/T4 or C290D4/T4

Model T207

Engine — Continental TSIO-520-G

*Fuel — 100/130 minimum grade aviation gasoline

*Engine Limits — Takeoff (5 min.) at 2700 r.p.m. (300 hp.)
For all other operations, 2600 r.p.m. (285 hp.)

A16CE

Page No.	1	2	3	4	5	6	7	8	9	10	11
Rev. No.	21	10	14	20	13	10	15	15	15	17	21

Propeller and Propeller Limits	Landplane 1. (a) McCauley D2A34C78/90AT-8.5 (C161004-0108) Diameter: not over 81.5 in., not under 80.5 in. Pitch settings at 36 in. sta.: low 11.8°, high 32.0° (b) Cessna spinner dome 1250909-3 (c) Woodward hydraulic governor G210452 (d) McCauley hydraulic governor C290D2/T2 or C290D4/T2 2. (a) McCauley D3A32C90/82NC-2 (C161006-0204) Diameter: not over 80 in., not under 79 in. Pitch settings at 30 in. sta.: low 14°, high 33° (b) Cessna spinner dome 1250909-8 (c) Woodward hydraulic governor G210452 (d) McCauley hydraulic governor C290D2/T2 or C290D4/T2

Models 207 & T207

*Airspeed Limits (CAS)	S/N 20700001 through 20700314	
	Never exceed	210 m.p.h. (182 knots)
	Maximum structural cruising	170 m.p.h. (148 knots)
	Maneuvering (3800 lb. landplane)	148 m.p.h. (129 knots)
	Flaps extended 0° - 10°	160 m.p.h. (139 knots)
	10° - 30°	110 m.p.h. (96 knots)
(IAS) (See NOTE 5 on Use of IAS)	S/N 20700315 and up	
	Never exceed	186 knots
	Maximum structural cruising	151 knots
	Maneuvering (3800 lb. landplane)	132 knots
	Flaps extended 0° - 10°	140 knots
	10° - 30°	100 knots

*C.G. Range	Landplane (+43.0) to (+50.5) at 3800 lb. (+31.0) to (+50.5) at 2600 lb. or less Straight line variation between points given
Empty Wt. C.G. Range	None
*Maximum Weight	Landplane 3800 lb.
No. of Seats	(S/N 20700001 through 20700148) 7 (2 at +35 to +47, 2 at +68 to +78, 2 at +99 to +109, 1 at +130) (S/N 20700149 and on) 7 (2 at +34 to +48, 2 at +69 to +79, 2 at +100 to +110, 1 at +124 to +130)
Maximum Baggage	Reference weight and balance data
Fuel Capacity	(S/N 20700001 through 20700225) 65 gal. (58 gal. usable), two 32.5 gal. tanks in wings at +48 (S/N 20700226 and on) 61 gal. (54 gal. usable), two 30.5 gal. tanks in wings at +48 See NOTE 1 for data on unusable fuel
Oil Capacity	12 qt. at -37.4 (6 qt. usable) See NOTE 1 for data on undrainable oil

Control Surface Movements					
	Wing flaps				30° +1° -2°
	Ailerons	Up	21° ±2°	Down	14° 30' ±2°
	Elevator	Up	21° ±1°	Down	19° ±1°
	Elevator tab	Up	25° +1° -0°	Down	5° +1° -0°
	Rudder (measured perpendicular to hinge line)	Right	27° 13' ±1°	Left	27° 13' ±1°
	(measured parallel to 0.0.W.L.)	Right	24° ±1°	Left	24° ±1°

Serial Nos. Eligible

20700001 through 20700148 1969 Model
20700149 through 20700190 1970 Model
20700191 through 20700205 1971 Model
20700206 through 20700215 1972 Model
20700216 through 20700227 1973 Model
20700228 through 20700267 1974 Model
20700268 through 20700314 1975 Model
20700315 through 20700362 1976 Model

II - Model 207A/T207A, Skywagon/Turbo Skywagon; Stationair/Turbo Stationair, 7 PCLM (Normal Category), Approved July 12, 1976; 8 PCLM (Normal Category), Approved September 11, 1979

Model 207A

Engine: Continental IO-520-F

*Fuel: 100/130 minimum grade aviation gasoline (S/N 20700363 through 20700414)
100LL/100 minimum aviation grade gasoline (S/N 20700415 and up)

*Engine Limits: Takeoff (5 min.) at 2850 r.p.m., 300 hp.
For all other operations, 2700 r.p.m., 285 hp.

Propeller and Propeller Limits:

1. (a) McCauley D3A32C90/82NC-2 (S/N 20700363 through 20700482)
Diameter: not over 80 in., not under 78 in.
Pitch settings at 30 in. sta.:
low 11.5°, high 28.1°
(b) Cessna spinner 1250909
(c) Woodward hydraulic governor 210462 or McCauley hydraulic governor C290D4/T4
2. (a) McCauley D3A34C404/80VA-0 (S/N 20700483 and up)
Diameter: not over 80 in., not under 78.5 in.
Pitch settings at 30 in. sta.:
low 11.0°, high 27.0°
(b) Cessna spinner 1250030
(c) McCauley hydraulic governor C290D4/T4

Model T207A

Engine: Continental TSIO-520-M

*Fuel: 100/130 minimum grade aviation gasoline (S/N 20700363 through 20700414)
100LL/100 minimum aviation grade gasoline (S/N 20700415 and up)

*Engine Limits: Takeoff (5 min.) at 2700 r.p.m., 36.5 in. Hg. mp., 310 hp.
For all other operations, 2600 r.p.m., 35 in. Hg. mp., 285 hp.

Propeller and Propeller Limits:

1. (a) McCauley D3A34C401/90DFA-10
Diameter: not over 80 in., not under 78.5 in.
Pitch settings at 30 in. sta.:
low 12.4°, high 28.5°
Avoid continuous operation between 1850 and 2150 r.p.m. above 24 in. mp.
(b) Cessna spinner 1250909
(c) McCauley hydraulic governor C290D4/T2

Models 207A & T207A

*Airspeed Limits (IAS) (See NOTE 5 on use of IAS)	S/N 20700363 through 20700482	
	Never exceed (207A)	186 knots
	(T207A)	182 knots
	Maximum structural cruising (207A)	151 knots
	(T207A)	148 knots
	Maneuvering	130 knots
	Flaps extended 0° - 10°	140 knots
	10° - 30°	100 knots
	S/N 20700483 and up	
	Never exceed	182 knots
	Maximum structural cruising	148 knots
	Maneuvering	130 knots
	Flaps extended 0° - 10°	140 knots
	10° - 30°	105 knots

*C.G. Range: (+43.0) to (+50.5) at 3800 lb.
(+31.0) to (+50.5) at 2600 lb. or less
Straight line variation between points given

Empty Wt. C.G. Range: None

*Maximum Weight: 3800 lb.

No. of Seats: 7 (2 at +34 to +48, 2 at +69 to +79, 2 at +100 to +110, 1 at +124 to +130)
S/N 20700363 through 20700562
8 (2 at +34 to +48, 2 at +69 to +79, 2 at +100 to +110, 2 at +124 to +130)
S/N 20700563 and up

Maximum Baggage: Reference weight and balance data

Fuel Capacity: Std.: 61 gal. (54 gal. usable), two 30.5 gal. tanks in wings at +48
Opt.: 80 gal. (73 gal. usable), two 40 gal. tanks in wings at +48
See NOTE 1 for data on unusable fuel

Oil Capacity: 12 qt. at -37.4 (6 qt. usable)
See NOTE 1 for data on undrainable oil

Control Surface Movements

Wing flaps				30° +1° -2°
Ailerons	Up	21° ±2°	Down	14° 30' ±2°
Elevator	Up	21° ±1°	Down	19° ±1°
Elevator tab	Up	25° +1° -0°	Down	5° +1° -0°
Rudder (measured perpendicular to hinge line)	Right	27° 13' ±1°	Left	27° 13' ±1°
(measured parallel to 0.0.W.L.)	Right	24° ±1°	Left	24° ±1°

Serial Nos. Eligible

20700363 through 20700414	1977 Model
20700415 through 20700482	1978 Model
20700483 through 20700562	1979 Model
20700563 through 20700654	1980 Model
20700655 through 20700729	1981 Model
20700730 through 20700762	1982 Model
20700763 through 20700767	1983 Model
20700768 through 20700788	1984 Model

Data Pertinent to All Models

Datum — Fuselage sta. 0.0 (front face of lower baggage bulkhead)

Leveling Means — Screws and nutplates located on the left hand side of the fuselage at 0.0.W.L. and sta. +25.57 and -1.00

Certification Basis — Part 23 of the Federal Aviation Regulations effective February 1, 1965, as amended by 23-1 through 23-6. In addition, effective S/N 20700483 and up, FAR 23.1559 effective March 1, 1978. FAR 36 dated December 1, 1969, plus Amendments 36-1 through 36-6 for S/N 20700363 and up.

Application for Type Certificate dated May 15, 1968.

Type Certificate No. A16CE issued December 31, 1968, obtained by the manufacturer under delegation option procedures.

Equivalent Safety Items	S/N 20700315 and on
Airspeed Indicator	FAR 23.1545 (See NOTE 5 on use of IAS)
Airspeed Limitations	FAR 23.1583(a)(1)

Production Basis — Production Certificate No. 4. Delegation Option Manufacturer No. CE-1 authorized to issue airworthiness certificates under delegation option provisions of Part 21 of the Federal Aviation Regulations.

Equipment: The basic required equipment as prescribed in the applicable airworthiness regulations (see Certification Basis) must be installed in the aircraft for certification. This equipment must include a current Airplane Flight Manual effective S/N 20700480 and on. In addition, the following item of equipment is required:

1. Stall Warning Indicator, Cessna Dwg. S1672-5

NOTE 1. Current weight and balance report including list of equipment included in the certificated empty weight and loading instructions when necessary, must be provided for each aircraft at the time of original certification. The certificated empty weight and corresponding center of gravity location must include unusable fuel of 42 lb. at +48 on the 207 and T207 Series, and undrainable oil of 0.0 at (-37.4) through S/N 20700314 and full oil of 22.5 lb. at (-37.4) for S/N 20700315 and on.

NOTE 2. The following placards must be displayed as indicated:

A. Applicable to Models 207 and T207 Landplane

(1) In full view of the pilot:

(a) S/N 20700001 through 20700314

"This airplane must be operated as a normal category airplane in compliance with the operating limitations as stated in the form of placards, markings and manuals. No acrobatic maneuvers including spins approved.

Maximums

Maneuvering speed 148 m.p.h. (CAS)
Gross weight 3800 lb.
Flight maneuvering load factors:
Flaps up +3.8; -1.52 Flaps down +2.40
Altitude loss in stall recovery 350 ft.
Flap extension speed 110 m.p.h. (CAS) 0° - 30°
160 m.p.h. (CAS) 0° - 10°
Airplane is controllable in 20 knot cross-winds.
Known icing conditions to be avoided.
This airplane is certified for the following flight operations as of date of original airworthiness certification:

VFR - IFR - DAY - NIGHT" (as applicable)

A16CE

(b) S/N 20700315 and up
"This airplane must be operated as a normal category airplane in compliance with the operating limitations as stated in the form of placards, markings, and manuals.

Maximums		
Maneuvering speed (IAS)		132 knots
Gross weight		3800 lb.
Flight load factor	Flaps Up	+3.8 -1.52
	Flaps Down	+2.4

No acrobatic maneuvers, including spins, approved. Altitude loss in a stall recovery -350 ft. Flight into known icing conditions prohibited. This airplane is certified for the following flight operations as of date of original airworthiness certificate:
DAY - NIGHT - VFR - IFR" (As applicable)

(2) On control lock:
"Control lock - remove before starting engine."

(3) On fuel selector plate: (S/N 20700001 through 20700221)
(Standard range tanks) "Off - Left tank 29.0 gal. Right tank 29.0 gal.
Use full rich mixture to switch tanks. Take off and land on fuller tank."
(Optional long range tanks)
"Off - Left tank 38.5 gal. Right tank 38.5 gal.
Use full rich mixture to switch tanks. Take off and land on fuller tank."

(S/N 20700222 through 20700225)
(Standard range tanks) "Off - Left tank 29.0 gal. Right tank 29.0 gal.
Take off and land on fuller tank."
(Optional long range tanks)
"Off - Left tank 38.5 gal. Right tank 38.5 gal.
Take off and land on fuller tank."

(S/N 20700226 and up)
(Standard range tanks) "Off - Left tank 27.0 gal. Right tank 27.0 gal.
Take off and land on fuller tank."
(Optional long range tanks)
"Off - Left tank 36.5 gal. Right tank 36.5 gal.
Take off and land on fuller tank."

(4) On fuel tank filler cap: (S/N 20700001 through 20700203)
(Standard range tanks) "Tank capacity 32.5 U.S. Gal., 100/130."
(Optional long range tanks)
"Tank capacity 42 U.S. Gal., 100/130."
Forward of fuel tank filler cap: (S/N 20700204 through 20700225)
(Standard range tanks) "Service this airplane with 100/130 min. aviation grade gasoline - capacity 32.5 gal."
(Optional long range tanks)
"Service this airplane with 100/130 min. aviation grade gasoline - capacity 42.0 gal."

Forward of fuel tank filler cap: (S/N 20700226 and on)
(Standard range tanks) "Service this airplane with 100/130 min. aviation grade gasoline - capacity 30.5 gal."
(Optional long range tanks)
"Service this airplane with 100/130 min. aviation grade gasoline - capacity 40.0 gal."

(5) Above selector valve: (S/N 20700001 through 20700227)
"When switching from dry tank turn pump on 'HI' momentarily."
(S/N 20700228 and up)
"When switching from dry tank turn auxiliary fuel pump 'on' momentarily."

(6) On cargo door: "Baggage net 180 lb. max. capacity. Refer to weight and balance data for baggage/cargo loading."

(7) On the following model(s) near manifold pressure gauge:

207

"Fuel flow at full throttle

	2850 rpm	2700 rpm
Sea level	24 gph	23 gph
4,000 ft.	22 gph	21 gph
8,000 ft.	20 gph	19 gph

T207

Maximum Power Settings and Fuel Flow

Takeoff (5 min. only	2700 rpm
35 In. Mp.	30 gph
Max. continuous power	2600 rpm

Alt. Ft.	Man. Press In. Hg.	Fuel Flow G.P.H.
S.L. to 17,000	35	28
18,000	34	27
20,000	32	25
22,000	30	23
24,000	28	21
26,000	26	19
28,000	24	18
30,000	22	17
75% Power Climb:		2500 rpm

28 In. MP., 20 GPH."

(8) On instrument panel above fuel pump switch (S/N 20700001 through 20700148)
"Use 'HI' for emergency only."

(9) On the baggage door:
"Max. baggage 120 lb. Refer to weight and balance data for baggage/cargo loading."

(10) Below oil temperature gauge: (S/N 20700216 and up)
"High voltage."

(11) On the flap control indicator for the following models:
(a) S/N 20700001 through 20700314
"(i) Up to 10° (Partial flap range with blue color code and 160 m.p.h. callout; also mechanical detent at 10°).
(ii) 10° to Full (Indices at these positions with white color code and 110 m.p.h. callout; also mechanical detent at 20°)."
(b) S/N 20700315 through 20700362
"(i) Up to 10° (Partial flap range with blue color code and 140 knot callout; also mechanical detent at 10°).
(ii) 10° to Full (Indices at these positions with white color code and 100 knot callout; also mechanical detent at 20°)."

(12) In full view of the pilot:
"MAJOR FUEL FLOW FLUCTUATIONS/POWER SURGES
1. AUX FUEL PUMP ON ADJUST MIXTURE
2. SELECT OPPOSITE TANK
3. WHEN FUEL FLOW STEADY, RESUME NORMAL OPERATIONS
SEE PROCEDURE CARD DL189-13 FOR EXPANDED INSTRUCTIONS."

B. Applicable to Models 207A and T207A

(1) In full view of the pilot:

(a) S/N 20700363 through 20700482

"This airplane must be operated as a normal category airplane in compliance with the operating limitations as stated in the form of placards, markings, and manuals.

Maximums		
Maneuvering speed (IAS)		130 knots
Gross weight		3800 lb.
Flight load factor	Flaps Up	+3.8 -1.52
	Flaps Down	+2.4

No acrobatic maneuvers, including spins, approved. Altitude loss in a stall recovery -350 ft. Flight into known icing conditions prohibited. This airplane is certified for the following flight operations as of date of original airworthiness certificate:

DAY - NIGHT - VFR - IFR" (As applicable)

(b) S/N 20700483 through 20700729

"The markings and placards installed in this airplane contain operating limitations which must be complied with when operating this airplane in the Normal Category. Other operating limitations which must be complied with when operating this airplane in this category are contained in the Pilot's Operating Handbook and FAA Approved Airplane Flight Manual.

No acrobatic maneuvers, including spins, approved. Flight into known icing conditions prohibited. This airplane is certified for the following flight operations as of date of original airworthiness certificate:

DAY - NIGHT - VFR - IFR" (As applicable)

(2) On control lock through 20700729:
"Control lock - remove before starting engine."

(3) On fuel selector plate through 20700729:

(Standard range tanks) "Off - Left on 27.0 gal. Right on 27.0 gal. Take off and land on fuller tank."

(Optional long range tanks) "Off - Left on 36.5 gal. Right on 36.5 gal. Take off and land on fuller tank."

(4) (a) Forward of fuel tank filler cap: (S/N 20700363 through 20700414)
(Standard range tanks) "Service this airplane with 100/130 min. aviation grade gasoline - capacity 30.5 gal."
(Optional long range tanks)
"Service this airplane with 100/130 min. aviation grade gasoline - capacity 40.0 gal."

(b) Forward of fuel tank filler cap: (S/N 20700415 through 20700729)
(Standard range tanks) "Service this airplane with 100LL/100 min. aviation grade gasoline - capacity 30.5 gal."
(Optional long range tanks)
"Service this airplane with 100LL/100 min. aviation grade gasoline - capacity 40.0 gal."

(5) Above selector valve through 20700729:
"When switching from dry tank turn auxiliary fuel pump 'on' momentarily."

(6) On cargo door through 20700729: "Baggage net 180 lb. max. capacity. Refer to weight and balance data for baggage/cargo loading."

(7) Near the manifold pressure gauge:

(a) Model 207A:

S/N 20700363 through 20700482

"Maximum power setting and fuel flow

Takeoff (5 min. only): 2850 r.p.m., maximum continuous pwr.: 2700 r.p.m.,

Fuel flow at full throttle

	2700 r.p.m.	2850 r.p.m.
S.L.	23 g.p.h.	24 g.p.h.
4000 ft.	21 g.p.h.	22 g.p.h.
8000 ft.	19 g.p.h.	20 g.p.h.
12000 ft.	17 g.p.h.	18 g.p.h."

S/N 20700483 through 20700729

"Min. fuel flows at full throttle

R.P.M.	S.L.	4000	8000	12000
2700	23 g.p.h.	21 g.p.h.	19 g.p.h.	17 g.p.h.
2850	24 g.p.h.	22 g.p.h.	20 g.p.h.	18 g.p.h."

(b) Model T207A

(1) S/N 20700363 through 20700482

"Maximum power setting and fuel flow

Takeoff (5 min. only): 2700 r.p.m., 36.5 in. mp., 31 g.p.h.

Maximum continuous power: 2600 r.p.m., 35.0 in. mp., 27 g.p.h.

Alt. Ft.	Man. Press In. Hg.	Fuel Flow G.P.H.
S.L. to 17,000	35	27
18,000	34	26
20,000	32	24
22,000	30	22
24,000	28	20
26,000	26	18
28,000	24	17
30,000	22	16

normal climb 2500 r.p.m. 30.0 in. mp., 22 g.p.h."

S/N 20700483 through 20700729

"MINIMUM FUEL FLOWS

TAKEOFF		Maximum Continuous Power: 2600 RPM							
2700 RPM	ALT - FT/1000	SL-17	18	20	22	24	26	28	30
36.5 In. Hp.	MP. In. Hg.	35	34	32	30	28	26	24	22
31 GPH	Fuel flow - GPH	27	26	24	22	20	18	17	16"

(2) S/N 20700363 through 20700729

"Avoid continuous operation between 1850 and 2150 r.p.m. above 24 in. mp."

(8) On the baggage door through 20700729:

"Max. baggage 120 lb. Refer to weight and balance data for baggage/cargo loading."

(9) Adjacent to the voltage light:

S/N 20700363 through 20700482

"High Voltage"

S/N 20700483 through 20700729

"Low Voltage"

(10) (a) S/N 20700363 through 20700482
On the flap control indicator
"Up to 10° (Partial flap range with blue color code and 140 knot callout; also mechanical detent at 10°).
10° to Full (Indices at these positions with white color code and 100 knot callout; also mechanical detent at 20°)."

(b) S/N 20700483 through 20700729
On the flap control indicator
"Up to 10° (Partial flap range with blue color code and 140 knot callout; also mechanical detent at 10°).
10° to Full (Indices at these positions with white color code and 105 knot callout; also mechanical detent at 20°)."

(11) Near airspeed indicator:
S/N 20700483 through 20700729
"Maneuver Speed
130 KIAS"

(12) In full view of the pilot:

(a) Model 207A and T207A, S/N 20700363 through 20700482
"MAJOR FUEL FLOW FLUCTUATIONS/POWER SURGES
1. AUX FUEL PUMP ON ADJUST MIXTURE
2. SELECT OPPOSITE TANK
3. WHEN FUEL FLOW STEADY, RESUME NORMAL OPERATIONS
SEE PROCEDURE CARD D1189-13 FOR EXPANDED INSTRUCTIONS."

(b) Model 207A, S/N 20700483 through 20700562
"MAJOR FUEL FLOW FLUCTUATIONS/POWER SURGES
1. AUX FUEL PUMP ON ADJUST MIXTURE
2. SELECT OPPOSITE TANK
3. WHEN FUEL FLOW STEADY, RESUME NORMAL OPERATIONS
SEE P.O.H. FOR EXPANDED INSTRUCTIONS."

(c) Model T207A, S/N 20700483 through 20700729
"MAJOR FUEL FLOW FLUCTUATIONS/POWER SURGES
1. AUX FUEL PUMP ON ADJUST MIXTURE
2. SELECT OPPOSITE TANK
3. WHEN FUEL FLOW STEADY, RESUME NORMAL OPERATIONS
SEE P.O.H. FOR EXPANDED INSTRUCTIONS."

(13) Effective 20700730 and up:
All placards required in the Pilot's Operating Handbook and FAA Approved Airplane Flight Manual must be installed in the appropriate locations."

In addition to the above placards, the prescribed operating limitations indicated by an asterisk (*) under Sections I and II of this data sheet must also be displayed by permanent markings.

NOTE 3. Reserved.

NOTE 4. The cylinder head thermistors must be installed as follows:

MODEL	CYLINDER HEAD NUMBER
207	3
T207	1
207A (1977 & 1978 Models)	3
207A (1979 Model and on)	6
T207A	6

NOTE 5. The marking of the airspeed indicator with IAS provides an equivalent level of safety to FAR 23.1545 when the approved airspeed calibration data presented in Section V of the Pilot's Operating Handbooks listed below is available to the pilot:

207	Cessna P/N D1068-13
T207	Cessna P/N D1067-13
207A (1977	Cessna P/N D1092-13
T207A (1977)	Cessna P/N D1093-13
207A (1978)	Cessna P/N D1120-13
T207A (1978)	Cessna P/N D1121-13
207A (1979)	Cessna P/N D1149-13PH
T207A (1979)	Cessna P/N D1150-13PH
207A (1980)	Cessna P/N D1184-13PH
T207A (1980)	Cessna P/N D1185-13PH
207A (1981)	Cessna P/N D1205-13PH
T207A (1981)	Cessna P/N D1206-13PH
207A (1982)	Cessna P/N D1224-13PH
T207A (1982)	Cessna P/N D1225-13PH
207A (1983)	Cessna P/N D1242-13PH
T207A (1983)	Cessna P/N D1243-13PH
207A (1984)	Cessna P/N D1263-13PH
T207A (1984)	Cessna P/N D1264-13PH

NOTE 6. 14-volt electrical system
(207 series through S/N 20700414)

28-volt electrical system
(207 series S/N 20700415 and up)

"WARNING: Use of alcohol-based fuels can cause serious performance degradation and fuel system component damage, and is therefore prohibited on Cessna airplanes."

.....END.....

A16CE

DEPARTMENT OF TRANSPORTATION
FEDERAL AVIATION ADMINISTRATION

	3A21
	Revision 46
	CESSNA
210	210K
210A	T210K
210B	210L
210C	T210L
210D	210M
210E	T210M
210F	210N
T210F	P210N
210G	T210N
T210G	210R
210H	P210R
T210H	T210R
210J	210-5 (205)
T210J	210-5A (205A)
	March 31, 2003

TYPE CERTIFICATE DATA SHEET NO. 3A21

This data sheet which is part of Type Certificate No.3A21 prescribes conditions and limitations under which the product for which the type certificate was issued meets the airworthiness requirements of the Federal Aviation Regulations.

Type Certificate Holder — Cessna Aircraft Company
P. O. Box 7704
Wichita, Kansas 67277

I - Model 210, 4 PCLM (Normal Category), Approved April 20, 1959

Engine — Continental IO-470-E

*Fuel — 100/130 minimum grade aviation gasoline

*Engine Limits — For all operations, 2625 r.p.m. (260 b.hp.)

Propeller and Propeller Limits —

1. (a) Hartzell HC-A2XF-1/8433-2
   Diameter: not over 82 in., not under 80 in.
   Pitch settings at 30 in. sta.:
   low 13.5°, high 28.0°
   (b) Cessna spinner 0752006

or 2. (a) McCauley D2A36C33/90M-8 or D2A34C49/90A-8 or D2A34C58/90AT-8
   Diameter: not over 82 in., not under 80 in.
   Pitch settings at 36 in. sta.:
   low 10.8°, high 25.8°
   (b) Cessna spinner 0752004

3. Woodward hydraulic governor 210270, 210280, 210340 or 210345

Page No.	*1*	*2*	*3*	*4*	*5*	*6*	*7*	*8*	*9*	*10*	*11*	*12*	*13*	*14*	*15*	*16*	*17*	*18*	*19*	*20*
Rev.No.	46	30	30	27	27	29	27	29	27	27	27	27	31	31	34	36	40	40	44	44
Page No.	*21*	*22*	*23*	*24*	*25*	*26*	*27*	*28*	*29*	*30*	*31*	*32*	*33*	*34*	*35*	*36*	*37*	*38*	*39*	*40*
Rev.No.	44	30	30	27	27	45	27	29	27	27	27	27	31	31	34	36	40	40	44	44
Page No.	*41*	*42*	*43*																	
Rev.No.	44	44	46																	

I - Model 210 (cont'd)

*Airspeed Limits (CAS)	Never exceed	200 m.p.h. (174 knots)
	Maximum structural cruising	175 m.p.h. (152 knots)
	Maneuvering	130 m.p.h. (113 knots)
	Flaps extended	110 m.p.h. (96 knots)
	Landing gear operating speed	160 m.p.h. (139 knots)
	Landing gear extension speed	160 m.p.h. (139 knots)

C.G. Range (Landing Gear Extended) — (+38.4) to (+46.5) at 2900 lb.
(+34.5) to (+46.5) at 2550 lb. or less
Straight line variation between points given.
Moment change due to retracting landing gear (+2456 in.-lb.)

Empty Wt. C.G. Range — None

*Maximum Weight — 2900 lb.

No. of Seats — 2 (2 at +36, 2 at +70)

Maximum Baggage — 120 lb. (+95)

Fuel Capacity — 65 gal. (55 gal. usable); two 32.5 gal. tanks in wings at +48.
See NOTE 1 for data on unusable fuel

Oil Capacity — 12 qt. (-19.4), 6 qt. usable
See NOTE 1 for data on undrainable oil

Control Surface Movements

Wing flaps	Up	0°	Down	38° +2°, -1°
Ailerons	Up	20° ±2°	Down	14° ±2°
Elevator	Up	26°30' ±1°	Down	22° ±1°
Elevator tab	Up	25° +1°, -0°	Down	15° +1°, -0°
Rudder	Right	24° ±1°	Left	24° ±1°

(measured parallel to 0.0 W.L.)

Serial Nos. Eligible — Model 210: 618, 57001 through 57575 (1960 Model)

II - Model 210A, 4 PCLM (Normal Category), Approved June 14, 1960

Engine — Continental IO-470-E

*Fuel — 100/130 minimum grade aviation gasoline

*Engine Limits — For all operations, 2625 r.p.m. (260 b.hp.)

Propeller and Propeller Limits

1. (a) Hartzell HC-A2XF-1/8433-2
   Diameter: not over 82 in., not under 80
   Pitch settings at 30 in. sta.:
   low 13.5°, high 28.0°
   (b) Cessna spinner 0752006

or 2. (a) McCauley D2A36C33/90M-8 or D2A34C49/90A-8 or D2A34C58/90AT-8
   Diameter: not over 82 in., not under 80 in.
   Pitch settings at 36 in. sta.:
   low 10.8°, high 25.8°
   (b) Cessna spinner 0752004

3. Woodward hydraulic governor 210270, 210280, 210340, 210345

II - Model 210A (cont'd)	
*Airspeed Limits (CAS)	Never exceed 200 m.p.h. (174 knots) Maximum structural cruising 175 m.p.h. (152 knots) Maneuvering 130 m.p.h. (113 knots) Flaps extended 110 m.p.h. (96 knots) Landing gear operating speed 160 m.p.h. (139 knots) Landing gear extended speed 160 m.p.h. (139 knots)
C.G. Range (Landing Gear Extended)	(+38.4) to (+44.4) at 2900 lb. (+33.7) to (+44.4) at 2250 lb. or less Straight line variation between points given. Moment change due to retracting landing gear (+2456 in.-lb.)
Empty Wt. C.G. Range	None
*Maximum Weight	2900 lb.
No. of Seats	4 (2 at +36, 2 at +70)
Maximum Baggage	120 lb. (+103)
Fuel Capacity	65 gal. (55 gal. usable); two 32.5 gal. tanks in wings at +48. See NOTE 1 for data on unusable fuel
Oil Capacity	12 qt. (-19.4), 6 qt. usable See NOTE 1 for data on undrainable oil
Control surface movements	Wing flaps Up 0° Down 38° +2°, -1° Ailerons Up 20° ±2° Down 14° ±2° Elevator Up 26°30' ±1° Down 22° ±1° Elevator tab Up 10° +2°, -0° Down 25° +2°, -0° Rudder Right 24° ±1° Left 24° ±1° (measured parallel to 0.0. W.L.)
Serial Nos. Eligible	Model 210A: 616, 21057576 through 21057840 (1961 Model)

III - Model 210B, 4 PCLM (Normal Category), Approved June 27, 1961
Model 210C, 4 PCLM (Normal Category), Approved June 14, 1962

Engine	Continental IO-470-S
*Fuel	100/130 minimum grade aviation gasoline
*Engine Limits	For all operations, 2625 r.p.m. (260 b.hp.)
Propeller and Propeller Limits	1. (a) Hartzell HC-A2XF-1/8433-2 Diameter: not over 82 in., not under 80 in. Pitch settings at 30 in. sta.: low 13.5°, high 28.0° (b) Cessna spinner 0752006 or 2. (a) McCauley D2A36C33/90M-8 or D2A34C49/90A-8 or D2A34C58/90AT-8 Diameter: not over 82 in., not under 80 in. Pitch settings at 36 in. sta.: low 10.8°, high 25.8° (b) Cessna spinner 0752004 3. Woodward hydraulic governor 210270, 210280, 210340, 210345, 210451, 210452

III - Model 210B, Model 210C (cont'd)

*Airspeed Limits (CAS)	Never exceed 225 m.p.h. (196 knots) Maximum structural cruising 190 m.p.h. (165 knots) Maneuvering 132 m.p.h. (115 knots) Flaps extended 110 m.p.h. (96 knots) Landing gear operating speed 160 m.p.h. (139 knots) Landing gear extended speed 160 m.p.h. (139 knots)
C.G. Range (Landing Gear Extended)	(+39.2) to (+45.0) at 3000 lb. (+33.0) to (+45.0) at 2250 lb. or less Straight line variation between points given. Moment change due to retracting landing gear (+2456 in.-lb.)
Empty Wt. C.G. Range	None
*Maximum Weight	3000 lb.
No. of Seats	4 (2 at +36, 2 at +70)
Maximum Baggage	120 lb. (+103)
Fuel Capacity	65 gal. (63.4 gal. usable); two 32.5 gal. tanks in wings at +48. See NOTE 1 for data on unusable fuel
Oil Capacity	12 qt. (-19.4), 6 qt. usable. See NOTE 1 for data on undrainable oil

Control Surface Movements

Wing flaps	Up	0°	Down	40° +1°, -2°
Ailerons	Up	20° ±2°	Down	14° ±2°
Elevator	Up	26°30' ±1°	Down	18° ±1°
Elevator tab	Up	20° +1°, -0°	Down	20° +1°, -0°
Rudder	Right	24° ±1°	Left	24° ±1°

(measured parallel to 0.0 W.L.)

Serial Nos. Eligible: Model 210B: 21057841 through 21058085 (1962 Model)
Model 210C: 21058086 through 21058139 and 21058141 through 21058220 (1963 Model)

IV - Model 210-5 (205), 6 PCLM (Normal Category), Approved June 14, 1962
Model 210-5A (205A), 6 PCLM (Normal Category), Approved July 19, 1963

Engine	Continental IO-470-S
*Fuel	100/130 minimum grade aviation gasoline
*Engine Limits	For all operations, 2625 r.p.m. (260 b.hp.)

Propeller and Propeller Limits

1. (a) Hartzell HC-A2XF-1A13.5/8433-2
   Diameter: not over 82 in., not under 80 in.
   Pitch settings at 30 in. sta.:
   low 13.5°, high 28.0°
   (b) Cessna spinner 0752614

or 2. (a) McCauley D2A36C33/90M-8 or D2A34C49/90A-8 or D2A34C58/90AT-8
   Diameter: not over 82 in., not under 80 in.
   Pitch settings at 36 in. sta.:
   low 10.8°, high 25.8°
   (b) Cessna spinner 0752614

3. Woodward hydraulic governor 210270, 210280, 210340, 210345, 210451, 210452

IV - Model 210-5 (205), Model 210-5A (205A) (cont'd)

*Airspeed Limits (CAS)	Never exceed 210 m.p.h (182 knots) Maximum structural cruising 170 m.p.h. (148 knots) Maneuvering 138 m.p.h. (120 knots) Flaps extended 110 m.p.h. (96 knots)
C.G. Range (Landing Gear Extended)	(+40.5) to (+47.4) at 3300 lb. (+33.0) to (+47.4) at 2250 lb. or less Straight line variation between points given.
Empty Wt. C.G. Range	None
*Maximum Weight	3300 lb.
No. of Seats	6 (2 at +36, 2 at +69, 2 at +100)
Maximum Baggage	Reference weight and balance data
Fuel Capacity	65 gal. (63.4 gal. usable); two 32.5 gal. tanks in wings at +48. See NOTE 1 for data on unusable fuel.
Oil Capacity	12 qt. (-19.4), 6 qt. usable. See NOTE 1 for data on undrainable oil.
Control Surface Movements	Wing flaps Up 0° Down 40° +1°, -2° Ailerons Up 20° ±2° Down 14° ±2° Elevator Up 26°30' ±1° Down 18° ±1° Elevator tab Up 20° +1°, -0° Down 20° +1°, -0° Rudder Right 24° ±1° Left 24° ±1° (measured parallel to 0.0. W.L.)
Serial Nos. Eligible	Model 210-5 (205) : 641, 205-0001 through 205-0480 (1963 Model) Model 210-5A (205A) : 205-0481 through 205-0577 (1964 Model)

V - Model 210D, 4 PCLM (Normal Category), Approved July 19, 1963

Engine	Continental IO-520-A
*Fuel	100/130 minimum grade aviation gasoline
*Engine Limits	For all operations, 2700 r.p.m. (285 b.hp.)
Propeller and propeller limits	1. (a) McCauley D2A34C58/90AT-8 Diameter: not over 82 in., not under 80 in. Pitch settings at 36 in. sta.: low 10.3°, high 25.8° (b) Cessna spinner 0752004 (c) Woodward hydraulic governor D210452
*Airspeed limits (CAS)	Never exceed 225 mph. (196 knots) Maximum structural cruising 190 mph. (165 knots) Maneuvering 134 mph. (116 knots) Flaps extended 110 mph. (96 knots) Landing gear operating speed 160 mph. (139 knots) Landing gear extended speed 160 mph. (139 knots)
C.G. range (landing gear extended)	(+39.2) to (+46.6) at 3100 lb. (+33.0) to (+46.6) at 2250 lb. or less Straight line variation between points given. Moment change due to retracting landing gear (+2456 in.-lb.)

3A21

V - Model 210D (cont'd)

Empty wt. C.G. range	None
*Maximum weight	3100 lb.
No. of seats	4 (2 at +36, 2 at +70)
Maximum baggage	Reference weight and balance data
Fuel capacity	65 gal. (63.4 gal. usable); two 32.5 gal. tanks in wings at +48. See Note 1 for data on unusable fuel.
Oil capacity	12 qt. (-19.4), 6 qt. usable. See Note 1 for data on undrainable oil.
Control surface movements	Wing flaps Up 0° Down 40° +1°, -2° Ailerons Up 21° ±2° Down 14°30' ±2° Elevator Up 26°30' ±1° Down 18° ±1° Elevator tab Up 20° +1°, -0° Down 10° +1°, -0° Rudder Right 24° ±1° Left 24° ±1° (measured parallel to 0.0. W.L.)
Serial Nos. eligible	Model 210D: 21058221 through 21058510 (1964 Model)

VI - Model 210E, 4 PCLM (Normal Category), Approved September 17, 1964

Engine	Continental IO-520-A
*Fuel	100/130 minimum grade aviation gasoline
*Engine limits	For all operations, 2700 rpm. (285 b.hp.)
Propeller and propeller limits	1. (a) McCauley E2A34C64/90AT-8 Diameter: not over 82 in., not under 80 in. Pitch settings at 36 in. sta.: low 10.3°, high 25.8° (b) Cessna spinner 1250411 (c) Woodward hydraulic governor D210452 2. (a) McCauley E2A34C73/90AT-8 Diameter: not over 82 in., not under 80 in. Pitch settings at 36 in. sta.: low 10.3°, high 25.8° (b) Cessna spinner 1250415 (c) Woodward hydraulic governor D210452
*Airspeed limits (CAS)	Never exceed 225 mph. (196 knots) Maximum structural cruising 190 mph. (165 knots) Maneuvering 134 mph. (116 knots) Flaps extended 110 mph. (96 knots) Landing gear operating speed 160 mph. (139 knots) Landing gear extended speed 160 mph. (139 knots)
C.G. range (landing gear extended)	(+39.2) to (+46.6) at 3100 lb. (+33.0) to (+46.6) at 2250 lb. or less Straight line variation between points given. Moment change due to retracting landing gear (+2456 in.-lb.)
Empty wt. C.G. range	None

VI - Model 210E (cont'd)

*Maximum weight	3100 lb.
No. of seats	4 (2 at +36, 2 at +70)
Maximum baggage	Reference weight and balance data
Fuel capacity	65 gal. (63.4 gal. usable); two 32.5 gal. tanks in wings at +48. See Note 1 for data on unusable fuel.
Oil capacity	12 qt. (-19.5), 6 qt. usable See Note 1 for data on undrainable oil.

Control surface movements

Wing flaps	Up	0°	Down	40° +1°, -2°
Ailerons	Up	21° ±2°	Down	14°30' ±2°
Elevator	Up	26°30' ±1°	Down	18° ±1°
Elevator tab	Up	20° +1°, -0°	Down	10° +1°, -0°
Rudder	Right	24° ±1°	Left	24° ±1°

(measured parallel to 0.0. W.L.)

Serial Nos. eligible — Model 210E: 21058511 through 21058715 (1965 Model)

VII - Model T210F, 4 PCLM (Normal Category), Approved August 3, 1965

Engine	Continental TSIO-520-C
*Fuel	100/130 minimum grade aviation gasoline
*Engine limits	For all operations, 2700 r.p.m., 32.5 in. Hg. mp. (285 b.hp.)

Propeller and propeller limits

1. (a) McCauley E2A34C70/90AT-8
   Diameter: not over 82 in., not under 80 in.
   Pitch settings at 36 in. sta.:
   low 11.8°, high 32.0°
   (b) Cessna spinner 1250415
   (c) Woodward hydraulic governor G210452
2. (a) McCauley D3A32C77/82NK-2
   Diameter: not over 80 in., not under 78 in.
   Pitch settings at 30 in. sta.:
   low 13.2°, high 32.5°
   (b) Cessna spinner 1250419-2
   (c) Woodward hydraulic governor G210452
3. (a) McCauley D3A32C88/82NC-2
   Diameter: not over 80 in., not under 78 in.
   Pitch settings at 30 in. sta.:
   low 14.0°, high 33.0°
   (b) Cessna spinner 1250419-2
   (c) Woodward hydraulic governor G210452

*Airspeed limits (CAS)

Never exceed	225 mph.	(196 knots)
Maximum structural cruising	190 mph.	(165 knots)
Maneuvering	131 mph.	(114 knots)
Flaps extended	110 mph	(96 knots)
Landing gear operating speed	160 mph.	139 knots)
Landing gear extended speed	160 mph.	(139 knots)

VII - Model T210F (cont'd)

C.G. range (landing gear extended)	(+39.0) to (+46.6) at 3300 lb. (+33.0) to (+46.6) at 2480 lb. or less Straight line variation between points given. Moment change due to retracting landing gear (+2456 in.-lb.)
Empty wt. C.G. range	None
*Maximum weight	3300 lb.
No. of seats	4 (2 at +36, 2 at +70)
Maximum baggage	Reference weight and balance data
Fuel capacity	65 gal. (63 gal. usable); two 32.5 gal. tanks in wings at +48. See Note 1 for data on unusable fuel.
Oil capacity	12 qt. (-19.4), 6 qt. usable. See Note 1 for data on undrainable oil.

Control surface movements

Wing flaps	Up	0°	Down	40° +1°, -2°
Ailerons	Up	21° ±2°	Down	14°30' ±2°
Elevator	Up	26°30' ±1°	Down	18° ±1°
Elevator tab	Up	20° ±1°	Down	20° ±1°
Rudder	Right	24° ±1°	Left	24° ±1°

(measured parallel to 0.0. W.L.)

Serial Nos. eligible — Model T210F: T210-0001 through T210-0197 (1966 Model)

VIII - Model 210F, 4 PCLM (Normal Category), Approved August 3, 1965

Engine	Continental IO-520-A
*Fuel	100/130 minimum grade aviation gasoline
*Engine limits	For all operations, 2700 rpm. (285 b.hp.)

Propeller and propeller limits

1. (a) McCauley E2A34C73/90AT-8
   Diameter: not over 82 in., not under 80 in.
   Pitch settings at 36 in. sta.:
   low 10.3°, high 25.8°
   (b) Cessna spinner 1250415
   (c) Woodward hydraulic governor D210452
2. (a) McCauley D3A32C77/82NK-2
   Diameter: not over 80 in., not under 78 in.
   Pitch settings at 30 in. sta.:
   low 11.3°, high 27.6°
   (b) Cessna spinner 1250419-2
   (c) Woodward hydraulic governor D210452
3. (a) McCauley D3A32C88/82NC-2
   Diameter: not over 80 in., not under 78 in.
   Pitch settings at 30 in. sta.:
   low 13.8°, high 28.1°
   (b) Cessna spinner 1250419-2
   (c) Woodward hydraulic governor D210452

VIII - Model 210F (cont'd)

*Airspeed limits (CAS)	Never exceed 225 mph.	(196 knots)	
	Maximum structural cruising	190 mph.	(165 knots)
	Maneuvering	131 mph	(114 knots)
	Flaps extended	110 mph	(96 knots)
	Landing gear operating speed	160 mph	(139 knots)
	Landing gear extended speed	160 mph.	(139 knots)

C.G. range (landing gear extended)
(+39.0) to (+46.6) at 3300 lb.
(+33.0) to (+46.6) at 2400 lb. or less
Straight line variation between points given.
Moment change due to retracting landing gear (+2456 in.-lb.)

Empty wt. C.G. range: None

*Maximum weight: 3300 lb.

No. of seats: 4 (2 at +36, 2 at +70)

Maximum baggage: Reference weight and balance data

Fuel capacity: 65 gal. (63 gal. usable), two 32.5 gal. tanks in wings at +48.
See Note 1 for data on unusable fuel.

Oil capacity: 12 qt. (-19.4), 6 qt. usable
See Note 1 for data on undrainable oil.

Control surface movements

Wing flaps	Up	0°	Down	40° +1°, -2°
Ailerons	Up	21° ±2	Down	14°30' ±2°
Elevator	Up	26°30' ±1°	Down	18° ±1°
Elevator tab	Up	20° ±1°	Down	20° ±1°
Rudder	Right	24° ±1°	Left	24° ±1°

(measured parallel to 0.0. W.L.)

Serial Nos. eligible: Model 210F: 21058716 through 21058818 (1966 Model)

IX - Model T210G, 4 PCLM (Normal Category), Approved August 23, 1966
Model T210H, 4 PCLM (Normal Category), Approved August 16, 1967

Engine: Continental TSIO-520-C

*Fuel: 100/130 minimum grade aviation gasoline

*Engine limits: For all operations, 2700 rpm., 32.5 in. Hg. mp. (285 b.hp.)

Propeller and propeller limits

1. (a) McCauley E2A34C70/90AT-8
   Diameter: not over 82 in., not under 80 in.
   Pitch settings at 36 in. sta.:
   low 11.8°, high 32.0°
   (b) Cessna spinner 1250415
   (c) Woodward hydraulic governor G210452
   (d) McCauley hydraulic governor C290D2/T2 or C290D4/T2
2. (a) McCauley D3A32C88/82NC-2
   Diameter: not over 80 in., not under 78 in.
   Pitch settings at 30 in. sta.:
   low 14.0°, high 33.0°
   (b) Cessna spinner 1250419-2
   (c) Woodward hydraulic governor G210452
   (d) McCauley hydraulic governor C219D2/T2 or C290D4/T2

IX - Model T210G, Model T210H (cont'd)

Propeller and propeller limits	3. (a) McCauley D3A32C77/82NK-2 (T-210G Only) Diameter: not over 80 in., not under 78 in. Pitch settings at 30 in. sta.: low 13.2°, high 32.5° (b) Cessna spinner 1250419-2 (c) Woodward hydraulic governor G210452
*Airspeed limits (CAS)	Never exceed 225 mph. (196 knots) Maximum structural cruising 190 mph (165 knots) Maneuvering 135 mph. (117 knots) Flaps extended 110 mph. (96 knots) Landing gear operating speed 160 mph. (139 knots) Landing gear extended speed 160 mph. (139 knots)
C.G. range (landing gear extended)	(+39.7) to (+47.8) at 3400 lb. (+35.5) to (+47.8) at 2800 lb. or less Straight line variation between points given. Moment change due to retracting landing gear (+2456 in.-lb.)
Empty wt. C.G. range	None
*Maximum weight	3400 lbs.
No. of seats	4 (2 at +36, 2 at +70)
Maximum baggage	Reference weight and balance data.
Fuel capacity	90 gal. (89 gal. usable), two 45.0 gal. tanks in wings at +43. See Note 1 for data on unusable fuel
Oil capacity	12 qt. (-19.4), 6 qt. usable. See Note 1 for data on undrainable oil
Control surface movements	Wing flaps Up 0° Down 30° Ailerons Up 20° ±2° Down 15° ±2° Elevator Up 23° ±1° Down 15° ±1° Elevator tab Up 20° ±1° Down 5° ±1° Rudder Right 24° ±1° Left 24° ±1° (measured parallel to 0.0. W.L.)
Serial Nos. eligible	Model T210G: T210-0198 through T210-0307 (1967 Model) Model T210H: T210-0308 through T210-0392 (1968 Model)

X - Model 210G, 4 PCLM (Normal Category), Approved August 23, 1966
Model 210H, 4 PCLM (Normal Category), Approved August 16, 1967

Engine	Continental IO-520-A
*Fuel	100/130 minimum grade aviation gasoline
*Engine limits	For all operations, 2700 rpm. (285 b.hp.)
Propeller and propeller limits	1. (a) McCauley E2A34C73/90AT-8 Diameter: not over 82 in., not under 80 in. Pitch settings at 36 in. sta.: low 10.3°, high 25.8° (b) Cessna spinner 1250415 (c) Woodward hydraulic governor D210452 (d) McCauley hydraulic governor C290D2/T5 or C290D3/T5

X - Model 210G, Model 210H (cont'd)	2. (a) McCauley D3A32C88/82NC-2 Diameter: not over 80 in., not under 78 in. Pitch settings at 30 in. sta.: low 13.8°, high 28.1° (b) Cessna spinner 1250419-2 (c) Woodward hydraulic governor D210452 (d) McCauley hydraulic governor C290D2/T5 or C290D3/T5
*Airspeed limits (CAS)	Never exceed 225 mph (196 knots) Maximum structural cruising 190 mph (165 knots) Maneuvering 135 mph. (117 knots) Flaps extended 110 mph. (96 knots) Landing gear operating speed 160 mph. (139 knots) Landing gear extended speed 160 mph. (139 knots)
C.G. range (landing gear extended)	(+39.7) to (+47.8) at 3400 lb. (+35.5) to (+47.8) at 2800 lb. or less Straight line variation between points given. Moment change due to retracting landing gear (+2456 in.-lb.)
Empty wt. C.G. range	None
*Maximum weight	3400 lb.
No. of seats	4 (2 at +36, 2 at +70)
Maximum baggage	Reference weight and balance data
Fuel capacity	90 gal. (89 gal. usable); two 45.0 gal. tanks in wings at +43. See Note 1 for data on unusable fuel.
Oil capacity	12 qt. (-19.4); 6 qt. usable See Note 1 for data on undrainable oil.
Control surface movements	Wing flaps Up 0° Down 30° Ailerons Up 20° ±2° Down 15° ±2° Elevator Up 23° ±1° Down 15° ±1° Elevator tab Up 20° ±1° Down 5° ±1° Rudder Right 24° ±1° Left 24° ±1° (measured parallel to 0.0. W.L.)
Serial Nos. eligible	Model 210G: 21058819 through 21058936 (1967 Model) Model 210H: 21058937 through 21059061 (1968 Model)

XI - Model T210J, 4 PCLM (Normal Category), Approved July 17, 1968

Engine	Continental TSIO-520-H
*Fuel	100/130 minimum grade aviation gasoline
*Engine limits	For all operations, 2700 rpm., 32.5 in. Hg. mp. (285 b.hp.)
Propeller and propeller limits	1. (a) McCauley E2A34C70/90AT-8 Diameter: not over 82 in., not under 80 in. Pitch settings at 36 in. sta.: low 11.8°, high 32.0° (b) Cessna spinner 1250415 (c) Woodward hydraulic governor G210452 (d) McCauley hydraulic governor C290D2/T2 or C290D4/T2

3A21

XI - Model T210J (cont'd)

	2. (a) McCauley D3A32C88/82NC-2 Diameter: not over 80 in., not under 78 in. Pitch settings at 30 in. sta.: low 14.0°, high 33.0° (b) Cessna spinner 1250419-2 (c) Woodward hydraulic governor G210452 (d) MCauley hydraulic governor C219D2/T2 or C290D4/T2
Airspeed limits (CAS)	Never exceed 225 mph. (196 knots) Maximum structural cruising 90 mph. (165 knots) Maneuvering 135 mph (117 knots) Flaps extended 110 mph. (96 knots) Landing gear operating speed 160 mph (139 knots) Landing gear extended speed 160 mph (139 knots)
C.G. range (landing gear extended)	(+39.7) to (+47.8) at 3400 lb. (+35.5) to (+47.8) at 2800 lb. or less Straight line variation between points given. Moment change due to retracting landing gear (+2456 in.-lb.)
Empty wt. C.G. range	None
*Maximum weight	3400 lb.
No. of seats	4 (2 at +36, 2 at +70)
Maximum baggage	Reference weight and balance data.
Fuel capacity	90 gal. (89 gal. usable), two 45.0 gal. tanks in wings at +43. See Note 1 for data on unusable fuel.
Oil capacity	10 qt. (-12.5), 8 qt. usable See Note 1 for data on undrainable oil.
Control surface movements	Wing flaps Up 0° Down 30° Ailerons Up 20° ±2° Down 15° ±2° Elevator Up 23° ±1° Down 15° ±1° Elevator tab Up 20° ±1° Down 5° ±1° Rudder Right 24° ±1° Left 24° ±1° (measured parallel to 0.0. W.L.)
Serial Nos. eligible	Model T210J: 21058140, T210-0393 through T210-0454 (1969 Model)

XII - Model 210J, 4 PCLM (Normal Category), Approved July 17, 1968

Engine	Continental IO-520-J
*Fuel	100/130 minimum grade aviation gasoline
*Engine limits	For all operations, 2700 rpm. (285 b.hp.)
Propeller and propeller limits	1. (a) McCauley E2A34C73/90AT-8 Diameter: not over 82 in., not under 80 in. Pitch settings at 36 in. sta.: low 10.3°, high 25.8° (b) Cessna spinner 1250415 (c) Woodward hydraulic governor D210452 (d) McCauley hydraulic governor C290D2/T5 or C290D3/T5

XII - Model 210J (cont'd)	2. (a) McCauley D3A32C88/82NC-2 Diameter: not over 80 in., not under 78 in. Pitch settings at 30 in. sta.: low 13.8°, high 28.1° (b) Cessna spinner 1250419-2 (c) Woodward hydraulic governor D210452 (d) McCauley hydraulic governor C290D2/T5 or C290D3/T5
*Airspeed limits (CAS)	Never exceed 225 mph. (196 knots) Maximum structural cruising 190 mph (165 knots) Maneuvering 135 mph. (117 knots) Flaps extended 110 mph. (96 knots) Landing gear operating speed 160 mph. (139 knots) Landing gear extended speed 160 mph. (139 knots)
C.G. range (landing gear extended)	(+39.7) to (+47.8) at 3400 lb. (+35.5) to (+47.8) at 2800 lb. or less Straight line variation between points given. Moment change due to retracting landing gear (+2456 in.-lb.)
Empty wt. C.G. range	None
*Maximum weight	3400 lb.
No. of seats	4 (2 at +36, 2 at +70)
Maximum baggage	Reference weight and balance data
Fuel capacity	90 gal. (89 gal. usable); two 45.0 gal. tanks in wings at +43. See Note 1 for data on unusable fuel.
Oil capacity	10 qt. (-12.5); 8 qt. usable See Note 1 for data on undrainable oil.
Control surface movements	Wing flaps Up 0° Down 30° Ailerons Up 20° ±2° Down 15° ±2° Elevator Up 23° ±1° Down 15° ±1° Elevator tab Up 20° ±1° Down 5° ±1° Rudder Right 24° ±1° Left 24° ±1° (measured parallel to 0.0. W.L.)
Serial Nos. eligible	Model 210J: 21059062 through 21059199 (1969 Model)

XIII - Model 210K/T210K, 6 PCLM (Normal Category), Approved September 26, 1969
Model 210L/T210L, 6 PCLM (Normal Category), Approved October 7, 1971

Model 210K/210L

Engine	Continental IO-520-L
*Fuel	100/130 minimum grade aviation gasoline
*Engine limits	Takeoff (5 min.) at 2850 rpm. (300 hp.) For all other operations, 2700 r.p.m. (285 hp.)

XIII Model 210K/T210K, Model 210L/T210L (cont'd)

Propeller and propeller limits	1. Model 210K/210L (S/N 21059200 through 21060539) (a) McCauley E2A34C73/90AT-8 Diameter: not over 82 in., not under 80 in. Pitch settings at 36 in. sta.: low 10.3°, high 25.8° (b) Cessna spinner 1250419 (c) Woodward hydraulic governor 2104562 (d) McCauley hydraulic governor C290D2/T4 or C290D4/T4 2. (a) McCauley D3A32C88/82NC-2 Diameter: not over 80 in., not under 78.5 in. Pitch settings at 30 in. sta.: low 11.5°, high 28.1° (b) Cessna spinner 1250419-2 (c) Woodward hydraulic governor 210462 (d) McCauley hydraulic governor C290D2/T4 or C290D4/T4

Model T210K/T210L

Engine	Continental TSIO-520-H
*Fuel	100/130 minimum grade aviation gasoline
*Engine limits	For all operations, 2700 rpm., 32.5 in. Hg. mp. (285 b.hp.)
Propeller and Propeller Limits	1. Model T210K/T210L (S/N 21059200 through 21060539) (a) McCauley E2A34C70/90AT-8 Diameter: not over 82 in., not under 80 in. Pitch settings at 36 in. sta.: low 11.8°, high 32.0° (b) Cessna spinner 1250415 (c) Woodward hydraulic governor G210452 (d) McCauley hydraulic governor C290D2/T2 or C290D4/T4 2. (a) McCauley D3A32C88/82NC-2 Diameter: not over 80 in., not under 78.5 in. Pitch settings at 30 in. sta.: low 14.0°, high 33.0° (b) Cessna spinner 1250419-2 (c) Woodward hydraulic governor G210452 (d) McCauley hydraulic governor C290D2/T2 or C290D4/T2

Models 210K/210L/T210K/T210L

*Airspeed Limits (CAS)

Model 210K/T210K, 210L/T210L (S/N 21059200 through 21061039)

Never exceed	225 m.p.h	(196 knots)
Maximum structural cruising	190 m.p.h	(165 knots)
Maneuvering	135 m.p.h	(117 knots)
Flaps extended (210K/T210K)	110 m.p.h	(96 knots)
Flaps extended (210L/T210L)	120 m.p.h	(104 knots)
Landing gear operating speed	160 m.p.h	(139 knots)
Landing gear extended speed	160 m.p.h	(139 knots)

(IAS)
(See NOTE 4 on use of IAS)

Model 210L/T210L (S/N 21061040 through 21061573)

Never exceed	199 knots
Maximum structural cruising	168 knots
Maneuvering	119 knots
Flaps extended	105 knots
Landing gear operating speed	140 knots
Landing gear extended speed	140 knots

Models 210K/210L/T210K/T210L (cont'd)

C.G. Range (Landing Gear Extended)	(+42.5) to (+53.0) at 3800 lb. (+37.0) to (+53.0) at 3000 lb. or less Straight line variation between points given. Moment change due to retracting landing gear (+3207 in.-lb.)
Empty Wt. C.G. Range	None
*Maximum Weight	3800 lb.
No. of Seats	Standard 6 (2 at +34 to +46, 2 at +61 to +77, 2 at +101) Optional 4 (2 at +34 to +46, 2 at +77) (210K/T210K)
Maximum Baggage	Reference weight and balance data
Fuel Capacity	90 gal. (89 gal. usable); two 45.0 gal. tanks in wings at +43 See NOTE 1 for data on unusable fuel.
Oil Capacity	10 qt. (-12.5); 8 qt. usable See NOTE 1 for data on undrainable oil.

Control Surface Movements

Wing flaps	Up	0°	Down	30° +1°, -2°
Ailerons	Up	20° ±2°	Down	15° ±2°
Elevator	Up	23° ±1°	Down	17° ±1°
Elevator tab	Up	25° ±1°	Down	10° ±1°
Rudder	Right	24° ±1°	Left	24° ±1°
(measured parallel to 0.0 W.L.)				
Rudder	Right	27°13' ±1°	Left	27°13' ±1°
(measured perpendicular to hinge line)				

Serial Nos. Eligible

Models 210K/T210K:	21059200 through 21059351	(1970 Model)
	21059352 through 21059502	(1971 Model)
Models 210L/T210L:	21059503 through 21059719	(1972 Model)
	21059720 through 21060089	(1973 Model)
	21060090 through 21060539 1974 Model)	
	21060540 through 21061039	1975 Model)
	21061040 through 21061041	1976 Model)
	21061043 through 21061573 (1976 Model)	

XIV - Model 210M/T210M, 6 PCLM (Normal Category), October 7, 1976

Model 210M

Engine	Continental IO-520-L
*Fuel	Model 210M (S/N 21061574 through 21062273) 100/130 minimum grade aviation gasoline Model 210M (S/N 21062274 through 21062953) 100LL/100 minimum grade aviation gasoline
*Engine Limits	Takeoff (5 min.) at 2850 r.p.m. (300 hp.) For all other operations, 2700 r.p.m. (285 hp.)

XIV - Model 210M/T210M (cont'd)

Propeller and Propeller Limits	1. Model 210M (S/N 21061574 through 21062273) (a) McCauley D3A32C88/82NC-2 Diameter: not over 80 in., not under 78.5 in. Pitch settings at 30 in. sta.: low 11.5°, high 28.1° (b) Cessna spinner 1250419-2 (c) Woodward hydraulic governor 210462 (d) McCauley hydraulic governor C290D4/T4 2. Model 210M (S/N 21062274 and up) (a) McCauley D3A34C404/80VA-0 Diameter: not over 80 in., not under 78.5 in. Pitch settings at 30 in. sta.: low 11.0°, high 27.0° (b) Cessna spinner 1250419 (c) McCauley hydraulic governor C290D4/T4
*Airspeed Limits (IAS) (See NOTE 4 on use of IAS)	1. Model 210M (S/N 21061574 through 21062273) Never exceed 199 knots Maximum structural cruising 168 knots Maneuvering 119 knots Flaps extended 105 knots Landing gear operating speed 140 knots Landing gear extended speed 140 knots 2. Model 210M (S/N 21062274 through 21062953) Never exceed 199 knots Maximum structural cruising 168 knots Maneuvering 119 knots Flaps extended 115 knots Landing gear operating speed 140 knots Landing gear extended speed 199 knots

Model T210M

Engine	Continental TSIO-520-R
*Fuel	Model T210M (S/N 21061574 through 21062273) 100/130 minimum grade aviation gasoline Model T210M (S/N 21062274 through 21062953) 100LL/100 minimum grade aviation gasoline
Engine Limits	Takeoff (5 min. at 2700 r.p.m., 36.5 in. Hg. mp. (310 hp.) For all other operations 2600 r.p.m., 35 in. Hg. mp. (285 hp.)
Propeller and Propeller Limits	1. (a) McCauley D3A34C402/90DFA-10 Diameter: not over 80 in., not under 78.5 in. Pitch settings at 30 in. sta.: low 12.4°, high 28.5° (b) Cessna spinner 1250419-10 (c) McCauley hydraulic governor C290D4/T2 (d) Woodward hydraulic governor G210452
*Airspeed Limits (IAS) (See NOTE 4 on use of IAS)	1. Model T210M (S/N 21061574 through 21062273) Never exceed 195 knots Maximum structural cruising 165 knots Maneuvering 119 knots Flaps extended 105 knots Landing gear operating speed 140 knots Landing gear extended speed 140 knots

2. Model T210M (S/N 21062274 through 21062953

Never exceed	195 knots
Maximum structural cruising	165 knots
Maneuvering	119 knots
Flaps extended	115 knots
Landing gear operating speed	140 knots
Landing gear extended speed	195 knots

Models 210M/T210M

C.G. Range (Landing Gear Extended): (+42.5) to (+53.0) at 3800 lb.
(+37.0) to (+53.0) at 3000 lb. or less
Straight line variation between points given
Moment change due to retracting landing gear (+3207 in.-lb.)

Empty Wt. C.G. Range: None

*Maximum Weight: 3800 lb.

No. of Seats: 6 (2 at +34 to +46, 2 at +61 to +77, 2 at +101)

Maximum Baggage: Reference weight and balance data

Fuel Capacity: 90 gal. (89 gal. usable), two 45.0 gal. tanks in wings at +43.
See NOTE 1 for data on unusable fuel

Oil Capacity: 10 qt. (-12.5), 8 qt. usable

Control Surface Movements:

Wing flaps	Up	0°	Down	30° +1°, -2°
Ailerons	Up	20° ±2°	Down	15° ±2°
Elevator	Up	23° ±1°	Down	17° ±1°
Elevator tab	Up	25° ±1°	Down	10° ±1°
Rudder	Right	24° ±1°	Left	24° ±1°
(measured parallel to 0.0 W.L.)				
Rudder	Right	27° 13' ±1°	Left	27° 13' ±1°
(measured perpendicular to hinge line)				

Serial Nos. Eligible: Models 210M/T210M: 21061574 through 21062273 (1977 Model)
21061042, 21062274 through 21062954 (1978 Model)

XV - Model P210N, Pressurized Centurion, 6 PCLM (Normal Category), Approved August 10, 1977

Engine: Model P210N (S/N P21000001 through P21000760: Continental TSIO-520-P
Model P210N (S/N P21000761 and up): Continental TSIO-520-AF

*Fuel: 100LL/100 minimum grade aviation gasoline

*Engine Limits: Model P210N (S/N P21000001 through P21000760)
Takeoff (5 min.) at 2700 r.p.m., 36.5 in. Hg. mp. (310 hp.)
For all other operations 2600 r.p.m., 33.5 in. Hg. mp. (285 hp.)
Model P210N (S/N P21000761 and up)
Takeoff (5 min.) at 2700 r.p.m., 35.5 in. Hg. mp. (310 hp.)
For all other operations, 2600 r.p.m., 34.5 in. Hg. mp. (285 hp.)

XV - Model P210N (cont'd)

Propeller and Propeller Limits	1. (a) McCauley D3A34C402/90DFA-10 Diameter: not over 80 in., not under 78.5 in. Pitch settings at 30 in. sta.: low 12.4°, high 28.5° Model P210N (S/N P21000001 through P21000760) Avoid continuous operation between 1850 and 2150 r.p.m. above 24 in. mp. Model P210N (S/N P21000761 and up) Avoid continuous operation between 1850 and 2150 r.p.m. above 23 in. mp. (b) Cessna spinner 1250419 (c) McCauley hydraulic governor C290D4/T2
*Airspeed Limits (IAS) (See NOTE 4 on use of IAS)	1. Model P210N (S/N P21000001 through P21000150) Never exceed 200 knots Maximum structural cruising 167 knots Maneuvering 130 knots Flaps extended 115 knots Landing gear operating speed 140 knots Landing gear extended speed 200 knots 2. Model P210N (S/N P21000151 and up) Never exceed 200 knots Maximum structural cruising 167 knots Maneuvering 130 knots Flaps extended 115 knots Landing gear operating speed 165 knots Landing gear extended speed 200 knots
C.G. Range (Landing Gear Extended)	(+43.9) to (+52.0) at 4000 lb. (+42.5) to (+52.0) at 3800 lb. (+37.0) to (+52.0) at 3000 lb. or less Straight line variation between points given Moment change due to retracting landing gear (+3207 in.-lb.) S/N P21000001 through P21000150 (+2907 in.-lb.) S/N P21000151 and up
Empty Wt. C.G. Range	None
*Maximum Weight	4000 lb. takeoff and flight 3800 lb. landing 4016 lb. ramp, S/N 21000151 and up
No. of Seats	6 (2 at +34 to +46, 2 at +61 to +77, 2 at +101)
Maximum Baggage	Reference weight and balance data
Fuel Capacity	90 gal. (89 gal. usable), S/N P21000001 through P21000760 90 gal. (87 gal. usable), S/N P21000761 and up two 45.0 gal. tanks in wings at +43 See NOTE 1 for data on unusable fuel.
Oil Capacity	10 qt. (-12.5); 8 qt. usable

Control Surface Movements

Wing flaps	Up	0°	Down	30° +1°, -2°
Ailerons	Up	20° ±2°	Down	15° ±2°
Elevator	Up	23° ±1°	Down	17° ±1°
Elevator tab	Up	25° ±1°	Down	10° ±1°
Rudder (measured parallel to 0.0 W.L.)	Right	24° ±1°	Left	24° ±1°
Rudder (measured perpendicular to hinge line)	Right	27° 13' ±1°	Left	27° 13' ±1°

XV - Model P210N (cont'd)

Serial Nos. Eligible	Model P210N: P21000001 through P21000150 (1978 Model) P21000151 through P21000385 (1979 Model) P21000386 through P21000590 (1980 Model) P21000591 through P21000760 (1981 Model) P21000761 through P21000811 (1982 Model) P21000812 through P21000834 (1983 Model)

XVI - Model 210N/T210N, Centurion/Turbo System Centurion, 6 PCLM (Normal Category), approved October 19, 1978

Model 210N

Engine	Continental IO-520-L
*Fuel	100LL/100 minimum grade aviation gasoline
*Engine Limits	Takeoff full throttle (5 min.) at 2850 r.p.m. (300 hp. rating) For all other operations, full throttle 2700 r.p.m. (285 hp. rating)
Propeller and Propeller Limits	1. (a) McCauley D3A34C404/80VA-0 Diameter: not over 80 in., not under 78.5 in. Pitch settings at 30 in. sta.: low 11.0°, high 27.0° (b) Cessna spinner 1250419 (c) McCauley hydraulic governor C290D4/T4
*Airspeed Limits (IAS) (See NOTE 4 on Use of IAS)	1. Model 210N (S/N 21062954 and up) Never exceed 200 knots Maximum structural cruising 165 knots Maneuvering 125 knots Flaps extended 115 knots Landing gear operating speed 165 knots Landing gear extended speed 200 knots
C.G. Range (Landing Gear Extended)	(+42.5) to (+53.0) at 3800 lb. (+37.0) to (+53.0) at 3000 lb. or less Straight line variation between points given Moment change due to retracting landing gear (+2907 in.-lb.)
Empty Wt. C.G. Range	None
*Maximum Weight	3800 lb. 3812 lb. ramp
No. of Seats	6 (2 at +34 to +46, 2 at +61 to +77, 2 at +101)
Maximum Baggage	Reference weight and balance data
Fuel Capacity	90 gal. (89 gal. usable), S/N 21062955 through 21064535 90 gal. (87 gal. usable), S/N 21064536 and up two 45.0 gal. tanks in wings at +43 See NOTE 1 for data on unusable fuel.
Oil Capacity	10 qt. (-12.5), 8 qt. usable

Model 210N (cont'd) Control Surface Movements	Wing flaps Up 0° Down 30° +1°, -2° Ailerons Up 20° ±2° Down 15° ±2° Elevator Up 23° ±1° Down 17° ±1° Elevator tab Up 25° ±1° Down 10° ±1° Rudder Right 24° ±1° Left 24° ±1° (measured parallel to 0.0 W.L.) Rudder Right 27° 13' ±1° Left 27° 13' ±1° (measured perpendicular to hinge line)
Serial Nos. Eligible	Model 210N: 21062955 through 21063640 (1979 Model) 21063641 through 21064135 (1980 Model) 21064136 through 21064535 (1981 Model) 21064536 through 21064772 (1982 Model) 21064773 through 21064822 (1983 Model) 21064823 through 21064897 (1984 Model)
Model T210N Engine	Continental TSIO-520-R
Fuel	100LL/100 minimum grade aviation gasoline
*Engine Limits	Takeoff (5 min.) at 2700 r.p.m., 36.5 in. Hg. mp. (310 hp. rating) For all other operations 2600 r.p.m., 35 in. Hg. mp. (285 hp. rating)
Propeller and Propeller Limits	1. (a) McCauley D3A34C402/90DFA-10 Diameter: not over 80 in., not under 78.5 in. Pitch settings at 30 in. sta.: low 12.4°, high 28.5° Avoid continuous operation between 1850 and 2150 r.p.m.. above 24 in. mp. (b) Cessna spinner 1250419 (c) McCauley hydraulic governor C290D4/T2 or Woodward hydraulic governor G210452
*Airspeed Limits (IAS) (See NOTE 4 on Use of IAS)	1. Model T210N (S/N 21062954 and up) Never exceed 203 knots Maximum structural cruising 168 knots Maneuvering 130 knots Flaps extended 115 knots Landing gear operating speed 165 knots Landing gear extended speed 203 knots
C.G. Range (Landing Gear Extended)	(+43.9) to (+52.0) at 4000 lbs. (+42.5) to (+53.0) at 3800 lbs. (+37.0) to (+53.0) at 3000 lbs. Straight line variation between points given Moment change due to retracting landing gear (+2907 in.-lb.)
Empty Wt. C.G. Range	None
*Maximum Weight	4000 lb. takeoff and flight 3800 lb. landing 4016 lb. ramp
No. of Seats	6 (2 at +34 to 46, 2 at +61 to +77, 2 at +101)
Maximum Baggage	Reference weight and balance data

Model T210N (cont'd)

Fuel Capacity	90 gal. (89 gal. usable), S/N 21062955 through 21064535 90 gal. (87 gal. usable), S/N 21064536 and up two 45.0 gal. tanks in wings at +43 See NOTE 1 for data on unusable fuel.
Oil Capacity	10 qt. (-12.5); 8 qt. usable

Control Surface Movements

Wing flaps	Up	0°	Down	30° +1°, -2°
Ailerons	Up	20° ±2°	Down	15° ±2°
Elevator	Up	23° ±1°	Down	17° ±1°
Elevator tab	Up	25° ±1°	Down	10° ±1°
Rudder	Right	24° ±1°	Left	24° ±1°
(measured parallel to 0.0 W.L.)				
Rudder	Right	27° 13' ±1	Left	17° 13' ±1°
(measured perpendicular to hinge line)				

Serial Nos. Eligible

Model T210N: 21062955 through 21063640 (1979 Model)
21063641 through 21064135 (1980 Model)
21064136 through 21064535 (1981 Model)
21064536 through 21064772 (1982 Model)
21064773 through 21064822 (1983 Model)
21064823 through 21064897 (1984 Model)

XVII - Model P210R, Pressurized Centurion, 6 PCLM (Normal Category), Approved September 24, 1984

Engine	Continental TSIO-520-CE
*Fuel	100LL/100 minimum grade aviation gasoline
*Engine Limits	For all operations 2700 r.p.m., 37 in. Hg. mp. (325 hp.)
Propeller and Propeller Limits	1. (a) McCauley D3A36C410/80VMB-0 Diameter: not over 80 in., not under 78.5 in. Pitch settings at 30 in. sta.: low 14.2°, high 36.5° (b) Cessna spinner 2150150 (c) McCauley hydraulic governor C290D4/T2
*Airspeed Limits (IAS)	Never exceed 200 knots Maximum structural cruising 167 knots Flaps extended 115 knots Maneuvering 130 knots Landing gear operating speed 165 knots Landing gear extended speed 200 knots
C.G. Range (Landing Gear Extended)	(+42.0) to (+52.0) at 4100 lb. (+37.0) to (+52.0) at 3350 lb. or less Straight line variation between points given Moment change due to retracting landing gear (+2907 in.-lb.)
Empty Wt. C.G. Range	None
*Maximum Weight	4100 lb. takeoff and flight 3900 lb. landing 4116 lb. ramp
No. of Seats	6 (2 at +34 to 46, 2 at +61 to +77, 2 at +101)
Maximum Baggage	Reference weight and balance data

XVII - Model P210R (cont'd)

Fuel Capacity	Std.: 90 gal. (87 gal. usable) Two 45.0 gal. tanks in wings at +42.5 Opt.: 120 gal. (115 gal. usable) Two 60.0 gal. tanks in wings at +42.5 See NOTE 1 for data on unusable fuel
Oil Capacity	10 qt. (-12.5), 8 qt. usable
Maximum Operating Altitude	25,000 ft.
Control Surface Movements	Wing flaps Up 0° Down 30° +1°, -2° Ailerons Up 20° ±2° Down 15° ±2° Elevator Up 25° ±1° Down 20° ±1° Elevator tab Up 20° ±1° Down 15° ±1° Rudder Right 24° ±1° Left 24° ±1° (measured parallel to 0.0 W.L.) Rudder Right 27° 13' ±1° Left 27° 13' ±1° (measured perpendicular to hinge line)
Serial Nos. Eligible	Model P210R: P21000835 through P21000866 (1985 Model) P21000867 through P21000874 (1986 Model)

XVIII - Model T210R, Turbo System Centurion, 6 PCLM (Normal Category), Approved December 4, 1984
Model 210R, Centurion, 6 PCLM (Normal Category), Approved December 20, 1984

Model 210R

Engine	Continental IO-520-L
*Fuel	100LL/100 minimum grade aviation gasoline
*Engine Limits	Takeoff full throttle (5 min.) at 2850 r.p.m. (300 hp. rating) For all other operations, full throttle 2700 r.p.m. (285 hp. rating)
Propeller and Propeller Limits	1. (a) McCauley D3A34C404/80VA-0 Diameter: not over 80 in., not under 78.5 in. Pitch settings at 30 in. sta.: low 11.0°, high 27.0° (b) Cessna spinner 1250419 (c) McCauley hydraulic governor C290D4/T4
*Airspeed Limits (IAS) (See NOTE 4 on use of IAS)	Never exceed 200 knots Maximum structural cruising 167 knots Maneuvering 125 knots Flaps extended 115 knots Landing gear operating speed 165 knots Landing gear extended speed 200 knots
C.G. Range (Landing) Gear Extended)	(+40.33) to (+52.0) at 3850 lb. (+37.0) to (+52.0) at 3350 lb. or less Straight line variation between points given Moment change due to retracting landing gear (+2907 in.-lb.)
Empty Wt. C.G. Range	None
*Maximum Weight	3850 lb. 3862 lb. ramp
No. of Seats	6 (2 at +34 to 46, 2 at +61 to +77, 2 at +101)

XVIII - Model T210R, 210R (cont'd)

Maximum Baggage	Reference weight and balance data
Fuel Capacity	Std.: 90 gal. (87 gal. usable) Two 45.0 gal. tanks in wings at +42.5 Opt: 120 gal. (115 gal. usable) Two 60 gal. tanks in wings at +42.5 See NOTE 1 for data on unusable fuel.
Oil Capacity	10 qt. (-12.5), 8 qt. usable
Control Surface Movements	Wing flaps Up 0° Down 30° +1°, -2° Ailerons Up 20° ±2° Down 15° ±2° Elevator Up 25° ±1° Down 20° ±1° Elevator tab Up 20° ±1° Down 15° ±1° Rudder Right 24° ±1° Left 24° ±1° (measured parallel to 0.0 W.L.) Rudder Right 27° 13' ±1° Left 27° 13' ±1° (measured perpendicular to hinge line)
Serial Nos. Eligible	Model 210R: 21064898 through 21064949 (1985 Model) 21064950 through 21065009 (1986 Model)

Model T210R

Engine	Continental TSIO-520-CE
*Fuel	100LL/100 minimum grade aviation gasoline
*Engine Limits	For all operations 2700 r.p.m., 37 in. Hg. mp. (325 hp.)
Propeller and Propeller Limits	1. (a) McCauley D3A36C410/80VMB-0 Diameter: not over 80 in., not under 78.5 in. Pitch settings at 30 in. sta.: low 14.2°, high 36.5° (b) Cessna spinner 2150150 (c) McCauley hydraulic governor C290D4/T2
*Airspeed Limits (IAS)	Never exceed 203 knots Maximum structural cruising 167 knots Maneuvering 130 knots Flaps extended 115 knots Landing gear operating speed 165 knots Landing gear extended speed 200 knots
C.G. Range (Landing Gear Extended)	(+42.0) to (+52.0) at 4100 lb. (+37.0) to (+52.0) at 3350 lb. Straight line variation between points given Moment change due to retracting landing gear (+2907 in.-lb.)
Empty Wt. C.G. Range	None
*Maximum Weight	4100 lb. takeoff and flight 3900 lb. landing 4116 lb. ramp
No. of Seats	6 (2 at +34 to 46, 2 at +61 to +77, 2 at +101)
Maximum Baggage	Reference weight and balance data

3A21

Model T210R (cont'd)

Fuel Capacity	Std.: 90 gal. (87 gal. usable) Two 45.0 gal. tanks in wings at +42.5 Opt: 120 gal. (115 gal. usable) Two 60 gal. tanks in wings at +42.5 See NOTE 1 for data on unusable fuel
Oil Capacity	10 qt. (-12.5), 8 qt. usable

Control Surface Movements

Wing flaps	Up	0°	Down	30° +1°, -2°
Ailerons	Up	20° ±2°	Down	15° ±2°
Elevator	Up	25° ±1°	Down	20° ±1°
Elevator tab	Up	20° ±1°	Down	15° ±1°
Rudder	Right	24° ±1°	Left	24° ±1°
(measured parallel to 0.0 W.L.)				
Rudder	Right	27° 13' ±1°	Left	27° 13' ±1°
(measured perpendicular to hinge line)				

Serial Nos. Eligible — Model T210R: 21064898 through 21064949 (1985 Model)
21064950 through 21065009 (1986 Model)

Data Pertinent to All Models

Datum — Fuselage station 0.0 (front face of firewall)

Leveling Means — Baggage compartment floor (except for 210-5(205) and 210-5A(205A)) -
Top of tailcone (except 210K/T210K/P210N and up, screws on left side tailcone)

Certification Basis — Models 210/210A: Part 3 of the Civil Air Regulations effective May 15, 1956, with no amendments.
Models 210B, 210C, 210D, 210E, 210F, T210F, 210G, T210G, 210H, T210H, 210J, T210J, 210K, T210K, 210L, T210L, 210M, T210M, 210N, T210N, 210R, 210-5(205), 210-5A(205A): Part 3 of the Civil Air Regulations effective May 15, 1956, and Paragraph 3.112 as amended October 1, 1959. FAR 36 dated December 1, 1969, plus Amendments 36-1 through 36-4 for Models 210M/T210M/210N/210R; Amendments 36-1 through 36-9 for the T210N. In addition, FAR 23.1559 effective March 1, 1978, for the Models 210N/T210N/210R.

Models P210N, P210R: Part 3 of the Civil Air Regulations dated May 15, 1956, Paragraph 3.112 as amended October 1, 1959, and 23.365, 23.571, 23.775, 23.841, 23.843, 23.901, 23.909, 23.1041, 23.1043, 23.1143, 23.1305, 23.1325, 23.1441 and 23.1527 of FAR 23 effective February 1, 1965, as amended to February 14, 1975. FAR 36 dated December 1, 1969, plus Amendments 36-1 through 36-6 for P210N; Amendments 36-1 through 36-12 for P210R. Also FAR 23.1559 effective March 1, 1978, for P21000151 and up. Also for P210R, FAR 23.1323 effective September 1, 1977, and FAR 23.1545 effective December 1, 1978.

Model T210R: Part 3 of the Civil Air Regulations dated May 15, 1956, Paragraph 3.112 as amended October 1, 1959, and 23.901, 23.909, 23.1041, 23.1043, 23.1143, 23.1305 of FAR 23 effective February 1, 1965, as amended to February 14, 1975; FAR 23.1323 effective September 1, 1977; FAR 23.1545 effective December 1, 1978; and FAR 23.1559 effective March 1, 1978; FAR 36 dated December 1, 1969, plus Amendments 36-1 through 36-12.

Compliance with ice protection has been demonstrated in accordance with FAR 23.1419, as amended through Amendment 23-14, when ice protection equipment is installed in accordance with the airplane equipment list (Models P210N, T210N, P210R, and T210R only).

Certification basis (cont'd)

Application for type certificate dated August 13, 1956.

Type Certificate No. 3A21 issued April 20, 1959, obtained by the manufacturer under delegation option procedures.

Equivalent Safety Items (S/N 21061040 through 21064897 (T210 only), and S/N P21000001 through P21000835)

Airspeed Indicator CAR 3.757 (See NOTE 4 for effectivity)
Operating Limitations CAR 3.778(a)
(210 S/N 21061040 through 21065009)
(T210 S/N 21061040 through 21064897)
(P210 S/N P21000001 through P21000834)

Airspeed Indicating System CAR 3.663
(210N, S/N 21062955 through 21064897)
(210R, S/N 21064898 through 21065009)

Production Basis

Production Certificate No. 4. Delegation Option Manufacturer No. CE-1 authorized to issue airworthiness certificates under delegation option provisions of Part 21 of the Federal Aviation Regulations.

Equipment

The basic required equipment as prescribed in the applicable airworthiness regulations (see Certification Basis) must be installed in the aircraft for certification. This equipment must include a current Airplane Flight Manual effective S/N 21062955 and up and P21000151 and up. In addition, the following item of equipment is required:

1. Stall warning indicator, Cessna Dwg. 0511062-4: S/N 21057001 through 21058818
S/N T210-0001 through T210-0197
Cessna Dwg. S-1672-1: S/N 21058819 and up
S/N T210-0198 through T210-0454
S/N P21000001 and up

NOTE 1.

Current weight and balance report including list of equipment included in certificated empty weight, and loading instructions when necessary must be provided for each aircraft at the time of original certification. The certificated empty weight and corresponding center of gravity location must include unusable fuel of 60 lb. at (+46) on Models 210 and 210A, 9 lb. at (+46) on the 210B, 210C, 210D, 210E, 210-5(205) 210-5A(205A); 12 lb. at (+46) on the 210F, T210F; and 6 lb. at (+23) on the 210G, T210G, 210H, T210H, 210J, T210J, 210K, T210K, 210L, T210L, 210M, T210M, 210N, T210N, P210N through S/N's 21064535 and P21000760; and 18 lb. at (+38) on S/N's 21064536 and up, and P21000761 and up; and undrainable oil of 0 lb. at (-19) through S/N 21061039 and full oil of 18.8 lb. at (-12.5) S/N 21061040 and up, and S/N P21000001 and up.

NOTE 2.

The following placards must be displayed in locations as indicated:

A. Applicable to Models 210/210A
 (1) In full view of the pilot:
  (i) "This airplane must be operated as a normal category airplane in compliance with the operating limitations as stated in the form of placards, markings and manuals. No acrobatic maneuvers, including spins, approved. Maximum maneuvering speed - 130 m.p.h. - CAS. Maximum design weight 2900 lb. Maximum flight maneuvering load factors - Flaps up +3.8, -1.52; Flaps down +3.5. Maximum gear extension speed 160 m.p.h. - CAS. Maximum flap extension speeds 10° flaps - 160 m.p.h. - CAS; 10°-40° flaps - 110 m.p.h. - CAS.

Before takeoff	Before landing
1. Set tabs	1. Gear down
2. Flaps 0°-20°	2. Flaps down
3. Check induction air-cold	3. Check induction air-cold
4. Mixture rich	4. Mixture rich
5. Propeller full in	5. Propeller full in
6. Check cowl flaps open	6. Check cowl flaps closed
7. Check fuel selector on fullest tank	7. Check fuel selector on fullest tank"

NOTE 2. (cont'd)

or

(i) "This airplane must be operated as a normal category airplane in compliance with the operating limitations as stated in the form of placards, markings and manuals. No acrobatic maneuvers, including spins, approved. Maximum maneuvering speed - 130 mph - CAS. Maximum design weight 2900 lb. Maximum flight maneuver load factors - Flaps up +3.8, -1.52; Flaps down +3.5. Maximum gear extension speed 160 mph - CAS. Maximum flap extension speeds 10° flaps - 160mph - CAS; 10° - 40° flaps - 110 mph - CAS.

Before takeoff	Before landing
1. Set tabs	1. Gear down
2. Fuel selector full tank	2. Fuel selector full tank
3. Cowl flaps open	3. Cowl flaps closed
4. Mixture rich	4. Mixture rich
5. Propeller full in	5. Propeller full in
6. Flaps 0° -20°	6. Flaps down"

(2) On the control lock: "Control lock - remove before starting engine."

(3) On the upper pack cover: "To extend gear manually, place gear handle in full down position, pull emergency handle and pump vertically."

(4) On fuel selector valve plate: "Both off. Left tank - 27.5 gal. Right tank 27.5 gal. Use full rich mixture to switch tanks. Take off and land on fullest tank."

(5) On the baggage door: "Maximum baggage 120 lb. For additional loading instructions see weight and balance data."

(6) On the fuel tank filler cap: "Tank capacity 32.5 U.S. gallons, 100/130."

(7) On the instrument panel directly below the fuel gauge indicators: "Avoid landing approaches in red arc and over 30 second slips under 1/2 tank. (Reference Owner's Manual)."

(8) In full view of the pilot:
"MAJOR FUEL FLOW FLUCTUATIONS/POWER SURGES
1. AUX FUEL PUMP ON ADJUST MIXTURE
2. SELECT OPPOSITE TANK
3. WHEN FUEL FLOW STEADY, RESUME NORMAL OPERATIONS
SEE PROCEDURE CARD D1189-13 FOR EXPANDED INSTRUCTIONS."

B. Applicable to Models 210B/210C

(1) In full view of the pilot:
"This airplane must be operated as a normal category airplane in compliance with the operating limitations as stated in the form of placards, markings and manuals. No acrobatic maneuvers including spins approved. Maximum maneuvering speed - 132 m.p.h. - CAS. Maximum design weight 3000 lb. Maximum flight maneuvering load factors - Flaps up +3.8, -1.52; Flaps down +3.5. Maximum gear extension speed 160 m.p.h. - CAS; Maximum flap extension speeds 10° flaps - 160 m.p.h. - CAS; 10°-40° flaps - 110 m.p.h. - CAS.

Before Takeoff	Before Landing
1. Set tabs	1. Gear down
2. Fuel selector	2. Fuel selector full tank
3. Cowl flaps open	3. Cowl flaps closed
4. Mixture rich	4. Mixture rich
5. Propeller full in	5. Propeller full in
6. Flaps 0°-20°	6. Flaps down."

(2) On the control lock: "Control lock - remove before starting engine."

(3) On the upper pack cover: "To extend gear manually, place gear handle in full down position, pull emergency handle and pump vertically."

(4) On fuel selector valve plate: "Both off. Left tank - 31.7 gal. Right tank - 31.7 gal. Use full rich mixture to switch tanks. Take off and land on fullest tank."

(5) On the baggage door: "Maximum baggage 120 lb. For additional loading instructions see weight and balance data."

(6) On the fuel tank filler cap: "Tank capacity 32.5 U.S. gallons, 100/130."

(7) In full view of the pilot:
"MAJOR FUEL FLOW FLUCTUATIONS/POWER SURGES
1. AUX FUEL PUMP ON ADJUST MIXTURE
2. SELECT OPPOSITE TANK
3. WHEN FUEL FLOW STEADY, RESUME NORMAL OPERATIONS
SEE PROCEDURE CARD D1189-13 FOR EXPANDED INSTRUCTIONS."

C. Applicable to Model 210-5(205) and 210-5A(205A)

(1) In full view of the pilot:
"This airplane must be operated as a normal category airplane in compliance with the operating limitations as stated in the form of placards, markings and manuals. No acrobatic maneuvers including spins approved. Maximum maneuvering speed - 138 m.p.h. - CAS. Maximum design weight 3300 lb. Maximum flight maneuvering load factors - Flaps up +3.8, -1.52; Flaps down +3.0; altitude load in stall recovery 200 ft.; Flap extension speed - 110 m.p.h. - CAS."

(2) On the control lock: "Control lock - remove before starting engine."

(3) On fuel selector valve plate: "Both off. Left tank - 31.7 gal.
Right tank - 31.7 gal. Use full rich mixture to switch tanks. Take off and land on fullest tank."

(4) On the fuel tank filler cap: "Tank capacity 32.5 U.S. gallons, 100/130."

(5) In full view of the pilot:
"MAJOR FUEL FLOW FLUCTUATIONS/POWER SURGES
1. AUX FUEL PUMP ON ADJUST MIXTURE
2. SELECT OPPOSITE TANK
3. WHEN FUEL FLOW STEADY, RESUME NORMAL OPERATIONS
SEE PROCEDURE CARD D1189013 FOR EXPANDED INSTRUCTIONS."

D. Applicable to Models 210D/210E

(1) In full view of the pilot:
"This airplane must be operated as a normal category airplane in compliance with the operating limitations as stated in the form of placards, markings and manuals. No acrobatic maneuvers including spins approved. Maximum maneuvering speed - 134 m.p.h. - CAS. Maximum design weight 3100 lb. Maximum flight maneuvering load factors - Flaps up +3.8, -1.52; Flaps down +3.5. Maximum gear extension speed 160 m.p.h. - CAS; Maximum flap extension speeds 10°, flaps - 160 m.p.h. - CAS; 10°-40° flaps - 110 m.p.h. - CAS; altitude loss in stall recovery 130 ft.

Before Takeoff	Before Landing
1. Set tabs	1. Gear down
2. Fuel selector full tank	2. Fuel selector full tank
3. Cowl flaps open	3. Cowl flaps closed
4. Mixture rich	4. Mixture rich
5. Propeller full in	5. Propeller full in
6. Flaps 0°-20°	6. Flaps down."

(2) On the control lock: "Control lock - remove before starting engine."

(3) On the upper pack cover: "To extend gear manually, place gear handle in full down position, pull emergency handle out and pump vertically."

NOTE 2.

(4) On fuel selector valve plate: "Both off. Left tank - 31.7 gal. Right tank - 31.7 gal. Use full rich mixture to switch tanks. Take off and land on fullest tank."

(5) On baggage door: "Maximum weight each child's seat, 140 lb. Refer to weight and balance data for baggage/cargo loading."

(6) On the fuel tank filler cap: "Tank capacity 32.5 U.S. gallons, 100/130."

(7) Above selector valve: "Turn pump on 'HI' when switching from a dry tank to a tank containing fuel."

(8) In full view of the pilot:
"MAJOR FUEL FLOW FLUCTUATIONS/POWER SURGES
1. AUX FUEL PUMP ON ADJUST MIXTURE
2 SELECT OPPOSITE TANK
3. WHEN FUEL FLOW STEADY, RESUME NORMAL OPERATIONS
SEE PROCEDURE CARD D1189-13 FOR EXPANDED INSTRUCTIONS."

E. Applicable to Models 210F/T210F

(1) In full view of the pilot:
"This airplane must be operated as a normal category airplane in compliance with the operating limitations as stated in the form of placards, markings and manuals. No acrobatic maneuvers including spins approved. Maximum maneuvering speed - 131.0 m.p.h. - CAS. Maximum design weight 3300 lb. Maximum flight maneuvering load factors - Flaps up +3.8, -1.52; Flaps down +3.0. Maximum gear extension speed 160 m.p.h. - CAS; Maximum flap extension speeds 10° flaps - 160 m.p.h. - CAS; 10°-40° flaps - 110 m.p.h. - CAS; Altitude loss in stall recovery 240 feet.

Before Takeoff	Before Landing
1. Set tabs	1. Gear down
2. Fuel selector full tank	2. Fuel selector full tank
3. Cowl flaps open	3. Cowl flaps closed
4. Mixture rich	4. Mixture rich
5. Propeller full in	5. Propeller full in
6. Flaps 0°-20°	6. Flaps down."

(2) On control lock: "Control lock - remove before starting engine."

(3) On the power pack cover: "To extend gear manually, place gear handle in full down position, pull emergency handle and pump vertically."

(4) On fuel selector valve plate: "Both off. Left tank - 31.5 gal. Right tank - 31.5 gal. Use full rich mixture to switch tanks. Take off and land on fullest tank."

(5) On baggage door: "Maximum weight each child's seat, 140 lb. Refer to weight and balance data for baggage/cargo loading."

(6) On the fuel tank filler cap: "Tank capacity 32.5 U.S. gallons, 100/130."

(7) Above selector valve: "Turn pump on 'HI' when switching from a dry tank to a tank containing fuel."

NOTE 2. (cont'd)

(8) Near the engine power instruments: (T210F only)

*Altitude in Feet Sea Level to:	Manifold Pressure in. Hg.	Fuel Flow Gal/Hr
19,000	32.5	28
20,000	31.5	26
22,000	29.5	24
24,000	27.5	22
26,000	25.5	20
28,000	23.5	19
30,000	21.5	18

75% power climb - 2500 r.p.m. - 27.5 manifold pressure - 20 g.p.h."

(9) On instrument panel above fuel boost pump switch:
"Use 'HI' for emergency only ↓."

(10) In full view of the pilot:
"MAJOR FUEL FLOW FLUCTUATIONS/POWER SURGES
1. AUX FUEL PUMP ON ADJUST MIXTURE
2. SELECT OPPOSITE TANK
3. WHEN FUEL FLOW STEADY, RESUME NORMAL OPERATIONS
SEE PROCEDURE CARD D1189-13 FOR EXPANDED INSTRUCTIONS."

F. Applicable to Models 210G, T210G, 210H, T210H, 210J, T210J

(1) In full view of the pilot:
"This airplane must be operated as a normal category airplane in compliance with the operating limitations as stated in the form of placards, markings and manuals. No acrobatic maneuvers, including spins, approved. Maximum maneuvering speed - 135 m.p.h. - (CAS). Maximum design weight 3400 lb. Maximum flight maneuvering load factors - Flaps up +3.8, -1.52; Flaps down +3.0. Maximum gear extension speed - 160 m.p.h. - (CAS); Maximum flap extension speeds 10° flaps - 160 m.p.h. - (CAS); 10°-30° flaps - 110 m.p.h. - (CAS); Altitude loss in stall recovery 250 feet.

Before Takeoff	Before Landing
1. Set tabs	1. Gear down
2. Fuel selector full tank	2. Fuel selector full tank
3. Cowl flaps open	3. Cowl flaps closed
4. Mixture rich	4. Mixture rich
5. Propeller full in	5. Propeller full in
6. Flaps 0°-20°	6. Flaps down."

(2) On control lock: "Control lock - remove before starting engine"

(3) On the power pack cover: "To extend gear manually, place gear handle in full down position, pull emergency handle out and pump vertically."

(4) On fuel selector valve plate: "Both off. Left-44.5 gal. Right-44.5 gal. Use full rich mixture to switch tanks. Take off and land on fullest tank."

(5) On baggage door: "Maximum weight each child's seat 140 lb. Refer to weight and balance data for baggage/cargo loading."

(6) Aft of the filler cap on the adapter plate: "Tank capacity 45.0 U.S. gallons. Service this airplane with 100/130 minimum grade aviation gasoline."

3A21

NOTE 2. (cont'd)

(7) Above selector valve: "Turn pump on 'HI' when switching from a dry tank to a tank containing fuel."

(8) Near the engine power instruments: (T210G/T210H/T210J)

*Altitude in Feet Sea Level to:	Manifold Pressure in. Hg.	Fuel Flow Gal/Hr
19,000	32.5	28
20,000	31.5	26
22,000	29.5	24
24,000	27.5	22
26,000	25.5	20
28,000	23.5	19
30,000	21.5	18

75% power climb - 2500 r.p.m. - 27.5 manifold pressure - 20 g.p.h."

(9) On instrument panel above fuel boost pump switch:
"Use 'HI' for emergency only ↓."

(10) In full view of the pilot:
"MAJOR FUEL FLOW FLUCTUATIONS/POWER SURGES
1. AUX FUEL PUMP ON ADJUST MIXTURE
2. SELECT OPPOSITE TANK
3. WHEN FUEL FLOW STEADY, RESUME NORMAL OPERATIONS
SEE PROCEDURE CARD D1189-13 FOR EXPANDED INSTRUCTIONS."

G. Applicable to Model 210K/T210K (S/N 21059200 through 21059351)

(1) In full view of the pilot:
"This airplane must be operated as a normal category airplane in compliance with the operating limitations as stated in the form of placards, markings and manuals. No acrobatic maneuvers, including spins, approved. Maximum maneuvering speed - 135 m.p.h.(CAS). Maximum design weight 3800 lb. Maximum flight maneuvering load factors - Flaps up +3.8, -1.52; Flaps down +2.0. Maximum gear extension speed - 160 m.p.h.- (CAS); Maximum flap extension speed 10° flaps - 160 m.p.h. - (CAS); 10°-30° flaps - 110 m.p.h. - (CAS); Altitude loss in stall recovery 300 feet.

Checklist Placard

Before Takeoff	Before Landing
1. Adjust trim controls	1. Fuel selector full tank
2. Fuel selector full tank	2. Gear down
3. Cowl flaps open	3. Cowl flaps closed
4. Mixture rich	4. Mixture rich
5. Propeller full in	5. Propeller full in
6. Flaps 0°-10°	6. Flaps down."

(2) On control lock: "Control lock - remove before starting engine."

(3) On the power pack cover: "To extend gear manually, place gear handle in full down position, pull emergency handle and pump vertically."

(4) On fuel selector valve plate: "Both off. Left on-44.5 gal. Right on -44.5 gal. Take off and land on fuller tank."

(5) On baggage door: "Maximum baggage 120 lb. Refer to weight and balance data for baggage/cargo loading."

(6) Aft of the filler cap on the adapter plate: "Tank capacity 45.0 U.S. gallons. Service this airplane with 100/130 minimum grade aviation gasoline."

NOTE 2. (cont'd) G. (7) Above selector valve: "When switching from a dry tank turn pump on 'HI' momentarily."

(8) Above fuel flow and manifold pressure indicator: (Model 210K)

"Fuel flow at Full Throttle

	2700 r.p.m.	2850 r.p.m.
Sea Level	23 gal/hr	24 gal/hr
4000 ft.	21 gal/hr	22 gal/hr
8000 ft.	19 gal/hr	20 gal/hr"

(9) Near the engine power instruments: (Model T210K)

*Altitude in Feet Sea Level to:	Manifold Pressure in. Hg.	Fuel Flow Gal/Hr
19,000	32.5	28
20,000	31.5	26
22,000	29.5	24
24,000	27.5	22
26,000	25.5	20
28,000	23.5	19
30,000	21.5	18

75% power climb - 2500 r.p.m. - 27.5 manifold pressure - 20 g.p.h."

(10) On flap control indicator:
"a. 0°-10° - T.O. (Takeoff range with blue color code and 160 m.p.h. callout; also mechanical detent at 10°)"
"b. 10°-20° - Full (Indices at these positions with white color code and 110 m.p.h. callout; also, mechanical detent at 20°."

(11) In plain view of the pilot:
"MAJOR FUEL FLOW FLUCTUATIONS/POWER SURGES
1. AUX FUEL PUMP ON ADJUST MIXTURE
2. SELECT OPPOSITE TANK
3. WHEN FUEL FLOW STEADY, RESUME NORMAL OPERATIONS
SEE PROCEDURE CARD D1189-13 FOR EXPANDED INSTRUCTIONS."

H. Applicable to Model 210K/T210K (S/N 21059352 through 21059502)
Applicable to Model 210L/T210L (S/N 21059503 through 21061039)

(1) In full view of the pilot:

(a) Applicable to Model 210K/T210K (S/N 21059352 through 21059502)
Applicable to Model 210L/T210L (S/N 21059503 through 21061039)
"This airplane must be operated as a normal category airplane in compliance with the operating limitations as stated in the form of placards, markings, and manuals.

MAXIMUMS

Maneuvering speed	135 m.p.h. CAS (117 knots)
Gear extension speed	160 m.p.h. CAS (139 knots)
Gross weight	3800 lbs.
Flight load factor	Flaps up +3.8, -1.52
	Flaps down +2.0

No acrobatic maneuvers, including spins, approved. Altitude loss in a stall recovery - 300 ft. Known icing conditions to be avoided. This airplane is certificated for the following flight operations as of date of original airworthiness certificate:

DAY - NIGHT - VFR - IFR" (As applicable)

NOTE 2. (cont'd) H. (1) (b) Applicable to Model 210L/T210L (S/N 21061040 and up)
"This airplane must be operated as a normal category airplane in accordance with the operating limitations as stated in the form of placards, markings, and manuals.

MAXIMUMS

Maneuvering speed (IAS)	119 knots
Gear extension speed (IAS)	140 knots
Gross weight	3800 lbs
Flight load factor	Flaps up +3.8, -1.52 Flaps down +2.0

No acrobatic maneuvers, including spins, approved. Altitude loss in a stall recovery - 300 ft. Flight into known icing conditions prohibited. This airplane is certified for the following flight operations as of date of original airworthiness certificate:
DAY - NIGHT - VFR - IFR" (As applicable)

Checklist Placard (Model 210K/T210K)(S/N 21059352 through 21059502)

"Checklist Placard

Before Takeoff	Before Landing
1. Adjust trim controls	1. Fuel selector full tank
2. Fuel selector full tank	2. Gear down
3. Cowl flaps open	3. Cowl flaps closed
4. Mixture rich	4. Mixture rich
5. Propeller full in	5. Propeller full in
6. Flaps 0°-10°	6. Flaps down."

Checklist (Model 210L/T210L)(S/N 21059503 through 21060539)
(Stowed - not required for flight)

"Cessna 210L & T210L or Centurion & Centurion II (as applicable)
Checklist

Before Takeoff	Before Landing
1. Controls - free and correct	1. Fuel selector - fullest tank
2. Elevator and rudder trim - set	2. Landing gear - DN 160 m.p.h. max
3. Fuel seelctor - fullest tank	3. Mixture - rich
4. Cowl flaps - open	4. Propeller - high r.p.m.
5. Propeller - high r.p.m.	5. Airspeed - 100 m.p.h. flaps up 90 m.p.h. flaps down"
6. Mixture - as required	
7. Flaps - 0° to 10°	
8. Instruments - check and set	
9. Seats and belts - secure	

(2) On control lock: "Control lock - remove before starting engine."

(3) On the power pack cover: (210K/T210K) (S/N 21059200 through 21059502)
To extend gear manually, place gear handle in full down position, pull emergency handle out and pump vertically."
On hand pump cover: (210L/T210L) (S/N 21059503 and up)
"Manual gear extension: 1. select gear down; 2. pull handle forward; 3. pump vertically."

(4) On fuel selector valve plate: "Off. Left on -44.5 gal.
Right on -44.5 gal. Takeoff and land on fuller tank."

(5) On baggage door: "Maximum baggage 120 lb. Refer to weight and balance data for baggage/cargo loading."

NOTE 2. (cont'd) H. (6) Aft of the filler cap on the adapter plate: "Service this airplane with 100/130 minimum aviation grade gasoline. Total capacity 45.0 gal."

(7) Above fuel selector valve: "When switching from dry tank, turn pump on 'HI' momentarily" (210L/T210L) (S/N 21059503 through 21060089)

Above fuel selector valve: "When switching from dry tank, turn Auxiliary fuel pump 'ON' momentarily" (210L/T210L) (S/N 21060090 and up).

(8) In front of pilot on lower instrument panel knee pad: "Alternate static air ↓ on."

(9) Above ammeter: "Do not turn off alternator in flight except in emergency." (Model 210K/T210K) (S/N 21059200 through 21059502)

(10) Adjacent to overvoltage light: "High voltage" (Models 210L/T210L) (S/N 21059503 and up)

(11) Above left fuel gauge: "Do not turn off alternator in flight except in emergency." (Models 210L/T210L) (S/N 21059503 through 21059719)

(12) Above fuel flow and manifold pressure indicator: (Model 210K/210L)

"Fuel flow at full throttle

	2700 r.p.m.	2850 r.p.m.
S.L.	138 lbs/hr	144 lbs/hr
400 ft.	126 lbs/hr	132 lbs/hr
8000 ft.	114 lbs/hr	120 lbs/hr"

(13) Near the engine power instruments (Models T210K/T210L)

"Max. allowable manifold press. & climb fuel flow

Alt.-ft/1000	SL-19	20	22	24	26	28	30
M.P.-In. Hg.	32.5	31.5	29.5	27.5	25.5	23.5	21.5
Fluel flow-lbs/hr	168	156	144	132	120	114	108

75% power climb - 2500 r.p.m., 27.5 in. M.P., 120 lbs/hr"

(14) On lower surface of right hand wing just outboard of fuselage: "Oxygen filler door." (All models with oxygen)

(15) On flap control indicator: (210K/T210K) (S/N 21059352 through 21059502)
"a. 0°-10° (Takeoff range with blue color code and 160 m.p.h. callout; also mechanical detent at 10°)"

b. 10°-20° Full (Indices at these positions with white color code and 110 m.p.h. callout; also mechanical detent at 20°)"

On flap control indicator: (210L/T210L) (S/N 21059503 through 21061039)
"a. 0°-10° (Takeoff range with blue color code and 160 m.p.h. callout; also mechanical detent at 10°)"

b. 10°-20° Full (Indices at these positions with white color code 120 m.p.h. callout; also mechanical detent at 20°)"

3A21

NOTE 2. (cont'd) H. (15) On flap control indicator: (210L/T210L) (S/N 21051040 and up)

"a. 0°-10° (Takeoff range with blue color code and 140 knots callout; also mechanical detent at 10°)"

b. 10°-20° - Full (Indices at these positions with white color code and 105 knots callout; also mechanical detent at 20°)"

(16) On inside nose wheel doors:
"WARNING - before working in wheel well area pull hydraulic pump circuit breaker off." (Model 210L/T210L) (S/N 21059503 and up)

(17) In full view of the pilot:
"MAJOR FUEL FLOW FLUCTUATIONS/POWER SURGES
1. AUX FUEL PUMP ON ADJUST MIXTURE
2. SELECT OPPOSITE TANK
3. WHEN FUEL FLOW STEADY, RESUME NORMAL OPERATIONS
SEE PROCEDURE CARD D1189-13 FOR EXPANDED INSTRUCTIONS."

J. Applicable to Model 210M/T210M, 210N/T210N, 210R/T210R

(1) In full view of the pilot:

(a) Applicable to Model 210M/T210M (S/N 21061574 through 21062273)
"This airplane must be operated as a normal category airplane in compliance with operating limitations as stated in the form of placards, markings and manuals.

MAXIMUMS

Maneuvering speed (IAS)	119 knots
Gear extension speed (IAS)	140 knots
Gross weight	3800 lbs.
Flight load factor	Flaps up +3.8, -1.52
	Flaps down +2.0

No acrobatic maneuvers, including spins, approved. Altitude loss in a stall recovery - 300 ft. Flight into known icing conditions prohibited. This airplane is certified for the following flight operations as of date of original airworthiness certificate:

DAY - NIGHT - VFR - IFR" (As applicable)

(b) Applicable to Model 210M/T210M (S/N 21061042, 21062274 through 21062954
"This airplane must be operated as a normal category airplane in compliance with the operating limitations as stated in the form of placards, markings and manuals.

MAXIMUMS

Maneuvering speed (IAS)		119 knots
Gross weight		3800 lbs.
Flight load factor	Flaps up	+3.8, -1.52
	Flaps down	+2.0

No acrobatic maneuvers, including spins, approved. Altitude loss in a stall recovery 300 ft. Flight into known icing conditions prohibited. This airplane is certified for the following flight operations as of date of original airworthiness certificate:
DAY - NIGHT - VFR - IFR" (As applicable)

(c) Applicable to Models 210N/T210N (S/N 21062955 through 21064535)
"The markings and placards installed in this airplane contain operating limitations which must be complied with when operating this airplane in the Normal Category. Other operating limitations which must be complied with when operating this airplane in this category are contained in the Pilot's Operating Handbook and FAA Approved Airplane Flight Manual.

No acrobatic maneuvers, including spins, approved.
Flight into known icing conditions prohibited.

NOTE 2. (cont'd) J. (1) (c) This airplane is certified for the following flight operations as of date of original airworthiness certificate:

DAY - NIGHT - VFR - IFR" (As applicable)

(2) On control lock through 21064535: "Control Lock - Remove Before Starting Engine."

(3) On the hand pump cover:
(S/N 21061574 through 21062273)
"Manual gear extension: 1. Select gear down; 2. pull handle forward; 3. pump vertically."

(S/N 21061042, 21062274 through 21064535)
"Manual gear extension: 1. Select gear down; 2. pull handle forward; 3. pump vertically.

CAUTION: Do not pump with gear up selected"

(4) On fuel selector valve plate through 21064535:
"Off. Left on - 44.5 gal. Right on - 44.5 gal.
Takeoff and land on fuller tank."

(5) 210M/T210M (S/N 21061042, 21061574 through 21062954)
On baggage door: "Maximum baggage 120 lb. Refer to weight and balance data for baggage/cargo loading."

210N/T210N (S/N 21062955 through 21064535)
On baggage door: "Maximum baggage 200 lbs. total. Refer to weight and balance data for baggage/cargo loading."

(6) Near the wing filler caps:
(S/N 21061574 through 21062273)
"Service this airplane with 100/130 minimum aviation grade gasoline. Total capacity 45.0 gal."

(S/N 21061042, 21062274 through 21064535)
"Service this airplane with 100LL/100 minimum aviation grade gasoline. Total capacity 45.0 gal."

(7) Near fuel selector valve through 21064535:
"When switching from dry tank, turn auxiliary fuel pump on momentarily."

(8) In front of pilot on lower instrument panel:
(S/N 21061574 through 21062273)
"Alternate static air ↓ pull on."

(S/N 21061042, 21062274 through 21064535)
"Alternate static air pull on."

(9) 210M/T210M (S/N 21061042 through 21062954)
Adjacent to overvoltage light: "High Voltage."

210N/T210N (S/N 21062955 through 21064535)
Adjacent to low voltage light: "Low Voltage"

(10) Near the engine power instruments (Model 210M, S/N 21061574 through 21062954):
"Fuel Flow at Full Throttle

	2700 r.p.m.	2850 r.p.m.
S.L.	138 lbs/hr	144 lbs/hr
400 ft.	126 lbs/hr	132 lbs/hr
8000 ft.	114 lbs/hr	120 lbs/hr"

"Max. power setting
Takeoff (5 min. only) 2850 r.p.m.
Max. continuous power 2700 r.p.m."

3A21

NOTE 2. (cont'd) J. (10)

Near the engine power instruments (Model 210N, S/N 21062955 through 21064535:
"Min. Fuel Flows at Full Throttle

	2700 r.p.m.	2850 r.p.m.
S.L.	138 lbs/hr	144 lbs/hr
4000 ft.	126 lbs/hr	132 lbs/hr
8000 ft.	114 lbs/hr	120 lbs/hr
12000 ft.	102 lbs/hr	108 lbs/hr"

(11) Near the engine power instruments (T210M):
(S/N 21061574 through 21062273)

"Maximum power setting & fuel flow
T.O. (5 min. only): 2700 r.p.m. Normal climb: 2500 r.p.m.
36.5 in. mp., 186 lbs/hr 30.0 in. mp., 126 lbs/hr

Max. continuous power: 2600 r.p.m.

Alt.-ft/1000	SL-17	18	20	22	24	26	28	30
M.P.-In. Hg.	35	34	32	30	28	26	24	22
Fluel flow-lbs/hr	162	156	144	132	120	108	102	96"

"Avoid continuous operation between 1850 and 2150 r.p.m. above 24 in. M.P."

(S/N 21061042, 21062274 through 21062953)
"Maximum power setting & fuel flow
T.O. (5 min. only): 2700 r.p.m. Normal climb: 2500 r.p.m.
36.5 in. mp., 186 lbs/hr 30.0 in. mp., 120 lbs/hr

Max. continuous power: 2600 r.p.m.

Alt.-ft/1000	SL-17	18	20	22	24	26	28	30
M.P.-In. Hg.	35	34	32	30	28	26	24	22
Fluel flow-lbs/hr	162	156	144	132	120	108	102	96"

"Avoid continuous operation between 1850 and 2150 r.p.m. above 24 in. M.P."

Near the engine power instruments (T210N, S/N 21062955 through 21064535):
"Minimum Fuel Flows
T.O.: 2700 r.p.m.
36.5 in. mp., 186 lbs/hr
Maximum continuous power: 2600 r.p.m.

Alt.-ft/1000	SL-17	18	20	22	24	26	28	30
M.P.-In. Hg.	35	34	32	30	28	26	24	22
Fluel flow-lbs/hr	162	156	144	132	129	108	102	96"

"Avoid continuous operation between 1850 and 2150 r.p.m. above 24 in. M.P."

(12) On lower surface of right hand wing just outboard of fuselage through 21064535:
"Oxygen filler door." (All models with oxygen.)

(13) On flap indicator:
(S/N 21061574 through 21062273)
a. "0° - 10° - (Partial flap range with blue color code and 140 knots callout; also, mechanical detent at 10°)"
b. "10°- 20° - Full - (Indices at these positions with white color code and 105 knots callout; also, mechanical detent at 20°)"

NOTE 2. (cont'd) J. (13) (S/N 21061042, 21062274 through 21063640)

a. "0° - 10° - (Partial flap range with blue color code and 150 knots callout; also, mechanical detent at 10°)"

b. "10°- 20° - Full - (Indices at these positions with white color code and 115 knots callout; also, mechanical detent at 20°)"

(S/N 21063641 through 21064535)

a. "0° - 10° - (Partial flap range with dark blue color code and 160 knot callout; also, mechanical detent at 10°)"

b. "10°- 20° - (Indices at these positions with light blue color code and 130 knot callout; also, mechanical detent at 10°)"

c. "20°- 30° - (Indices at these positions with white color code and 115 knot callout)"

(14) On inside nose wheel doors, strut doors and main wheel doors through 21062954 and on inside of nose wheel doors S/N 21064535: "Warning - Before working in the wheel well area pull hydraulic pump circuit breaker off."

(15) Applicable to the Model 210M: (S/N 21062274 through 21062954)
Near the gear selector handle:
"Maximum speed IAS
Gear oper. 140 knots
Gear down 199 knots"

(16) Applicable to the Model T210M: (S/N 21061042, 21062274 through 21062953)
Near the gear selector handle:
"Maximum speed IAS
Gear oper. 140 knots
Gear down 195 knots"

(17) Applicable to the Model 210N: (S/N 21062955 through 21064535)
Near the gear selector handle:
"Maximum speed IAS
Gear oper. 165 knots
Gear down 200 knots"

(18) Applicable to the Model T210N: (S/N 21062955 through 21064535)
Near the gear selector handle:
"Maximum speed IAS
Gear oper. 165 knots
Gear down 203 knots"

(19) Near the airspeed indicator

(a) Model 210N (S/N 21062955 through 21064535)
"Maneuver Speed 125 KIAS"

(b) Model T210N (S/N 21062955 through 21064535)
"Maneuver Speed 130 KIAS"

(20) Near the fuel cap
Models 210N/T210N (S/N 21062955 through 21063640)
"For 32 gal. fuel load fill to bottom of filler neck extension."

Models 210N/T210N (S/N 21063641 through 21064535)
"Capacity 33.5 gallons to bottom of filler neck extension."

NOTE 2. (cont'd) J. (21) Near the oil filler
Models 210N/T210N (S/N 21062955 through 21064135)
"Oil 10 qts."

(22) On the nose gear strut
Models 210N/T210N (S/N 21062955 through 21064135)
"WARNING
Release air and fluid pressure before removing any part of this assembly."

(23) In full view of the pilot:

(a) Models 210M/T210M (S/N 21061574 through 21062954)
"MAJOR FUEL FLOW FLUCTUATIONS/POWER SURGES
1. AUX FUEL PUMP ON ADJUST MIXTURE
2. SELECT OPPOSITE TANK
3. WHEN FUEL FLOW STEADY, RESUME NORMAL OPERATIONS
SEE PROCEDURE CARD D1189-13 FOR EXPANDED INSTRUCTIONS."

(b) Model 210N (S/N 21062955 through 21063640)
"MAJOR FUEL FLOW FLUCTUATIONS/POWER SURGES
1. AUX FUEL PUMP ON ADJUST MIXTURE
2 SELECT OPPOSITE TANK
3. WHEN FUEL FLOW STEADY, RESUME NORMAL OPERATIONS
SEE P.O.H. FOR EXPANDED INSTRUCTIONS."

(c) Model T210N (S/N 21062955 through 21064535)
"MAJOR FUEL FLOW FLUCTUATIONS/POWER SURGES
1. AUX FUEL PUMP ON, ADJUST MIXTURE
2. SELECT OPPOSITE TANK
3. WHEN FUEL FLOW STEADY, RESUME NORMAL OPERATIONS
SEE P.O.H. FOR EXPANDED INSTRUCTIONS."

(24) Effective S/N 21064536 and up:
"All placards required in the Pilot's Operating Handbook and FAA Approved Airplane Flight Manual must be installed in the appropriate locations."

K. <u>Applicable to Model P210N and P210R</u>

(1) In full view of the pilot:
Model P210N (S/N P21000001 through P21000150)
"This airplane must be operated as a normal category airplane in compliance with the operating limitations as stated in the form of placards, markings and manuals.

MAXIMUMS

Operating altitude		23,000 ft.
Maneuvering speed (IAS)		130 knots
Gross weight	Takeoff	4000 lbs.
	Landing	3800 lbs.
Flight load factor	Flaps up	+3.8, -1.52
	Flaps down	+2.0

No acrobatic maneuvers, including spins, approved. Landing with cabin pressurized is prohibited. Altitude loss in a stall recovery - 300 ft. Flight into known icing conditions prohibited. This airplane is certified for the following flight operations as of date of original airworthiness certificate:

DAY - NIGHT - VFR - IFR" (As applicable)

NOTE 2. (cont'd) K. (1) Model P210N (S/N P21000151 and up)
"The markings and placards installed in this airplane contain operating limitations which must be complied with when operating this airplane in the Normal Category. Other operating limitations which must be complied with when operating this airplane in this category are contained in the Pilot's Operating Handbook and FAA Approved Airplane Flight Manual.

No acrobatic maneuvers, including spins, apaproved.
Landing with cabin pressurized is prohibited.
Flight into known icing conditions prohibited.

This airplane is certified for the following flight operations as of date of original airworthiness certificate:

DAY - NIGHT - VFR - IFR" (As applicable)

(2) On control lock through P21000760: "Control Lock - Remove Before Starting Engine."

(3) On the hand pump cover through P21000760:
"Manual gear extension: 1. Select gear down; 2. pull handle forward;
3. pump vertically. CAUTION: Do Not Pump With Gear Up Selected."

(4) On fuel selector valve plate through P21000760: "Off. Left on - 44.5 gal., Right on - 44.5 gal., Takeoff and land on fuller tank"

(5) On baggage door through P21000760:
"Maximum baggage 200 lbs. total. Raised area aft of baggage door 80 lbs. maximum.
Refer to weight and balance data for baggage cargo loading."

(6) Near the wing filler caps through P21000760: "Service this airplane with 100LL/100 minimum aviation grade gasoline. Total capacity 45.0 gal."

(7) Near fuel selector valve through P21000760: "When switching from dry tank, turn auxiliary fuel pump on momentarily."

(8) P210N (S/N P21000001 through P21000150)
Adjacent to over voltage light: "HIGH VOLTAGE"

P210N (S/N P21000151 through P21000760)
Adjacent to low voltage light: "LOW VOLTAGE"

(9) Near the engine power instruments through P21000760:

"Minimum Fuel Flows

TAKEOFF		MAX. CONTINUOUS POWER: 2600 RPM						
2700 R.P.M.	ALT-FT/1000	SL-17	18	19	20	21	22	23
36.5 In.M.P	M.P. IN. HG.	35.5	34.5	33.5	32.5	31.5	30.5	29.5
180 LBS/HR	Fuel Flow - lbs/hr	162	156	150	144	138	132	126"

(10) On flap indicator:
P210N (S/N P21000001 through P21000385)

a. "0° - 10° - (Partial flap range with dark blue color code and 150 knots callout; also, mechanical detent at 10°)"
b. "10°- 20° - Full - (Indices at these positions with white color code and 115 knot callout; also, mechanical detent at 20°)"

P210N (S/N P21000386 through P21000760)

a. "0° - 10° - (Partial flap range with dark blue color code and 160 knot callout; also, mechanical detent at 10°)"
b. "10°- 20° - Full - (Indices at these positions with light blue color code and 130 knot callout; also, mechanical detent at 20°)"
c. "20°- 30° - (Indices at these positions with white color code and 115 knot callout)" (Full)

NOTE 2. (cont'd) K. (11) On inside nose wheel doors, strut doors and main wheel doors:
"Warning - Before working in wheel well area pull hydraulic pump circuit breaker off."

(12) Near the gear selector handle:
P210N (S/N P21000001 through P21000150)
"Maximum speed IAS
Gear oper. 140 knots
Gear down 200 knots"

P210N (S/N P21000151 through P21000760)
"Maximum speed IAS
Gear oper. 165 knots
Gear down 200 knots"

(13) Near the pilot's outside door handle through P21000760:
"Close ↶
Open ↷ "

(14) Near the emergency button to unlock the pilot's cabin door from the outside through P21000760:
"Emergency
Push to unlock"

(15) Near the secondary lock for the inside pilot's door handle through P21000760:
"Door Handle Safety Lock
Push Flush to Lock
Pull To Unlock"

(16) Near the pilot's inside door handle through P21000760:
"Close
Open ←→ Lock"

(17) Near the right exit handle through P21000760:
"Open ←→ Close ←→ Latch
Push Flush
to Lock
Close and Lock for Flight"

(18) Near the airspeed indicator:
P210 (S/N P21000151 through P21000760)
"Maneuver Speed - 130 KIAS"

(19) Near the oil filler:
P210N (S/N P21000151 through P21000760)
"Oil 10 qts"

(20) Near the fuel cap:
P210N (S/N P21000151 through P21000760)
"For 32 gal. fuel load fill to bottom of filler neck extension."

(21) On emergency exit through P21000760:
"Emergency Exit - To Open
1. Lift handle (Do not pull inward)
2. Rotate counter clockwise to 'OPEN' position
3. Push door outward"

NOTE 2. (cont'd) K. (22) On the main cabin door through P21000760:
"Door Handle Safety Lock
Push Flush To Lock
Pull to Unlock"

And

"To Open Door
1. Unlock safety lock (pull out)
2. Rotate handle to 'OPEN' position
3. Push door outward"

(23) In full view of the pilot:
S/N P21000001 through P21000150
"MAJOR FUEL FLOW FLUCTUATIONS/POWER SURGES
1. AUX FUEL PUMP ON ADJUST MIXTURE
2. SELECT OPPOSITE TANK
3. WHEN FUEL FLOW STEADY, RESUME NORMAL OPERATIONS
SEE PROCEDURE CARD D1189-13 FOR EXPANDED INSTRUCTIONS."

S/N P21000151 through P21000760:
"MAJOR FUEL FLOW FLUCTUATIONS/POWER SURGES
1. AUX FUEL PUMP ON ADJUST MIXTURE
2. SELECT OPPOSITE TANK
3. WHEN FUEL FLOW STEADY, RESUME NORMAL OPERATIONS
SEE P.O.H. FOR EXPANDED INSTRUCTIONS."

(24) When equipped with optional EGT gauge: - On the left forward side panel near instrument panel (S/N P21000001 through P21000150):

"EGT LIMITATION
USE OF EGT GAUGE IS PROHIBITED
AT ALL R.P.M. SETTINGS ABOVE 2500
R.P.M. AT ALL ALTITUDES"

(25) When equipped with optional EGT gauge: - On the left side panel near instrument panel (S/N P21000001 through P21000150):

"EGT LIMITATIONS
USE OF EGT GAUGE IS PROHIBITED AT ALL POWER SETTINGS
ABOVE 80% AT ALL ALTITUDES; OR ABOVE THE FOLLOWING
POWERS AT THE LISTED ALTITUDES WHEN OAT IS ABOVE STANDARD.
75% AT 17,000 FEET OR HIGHER
70% AT 20,000 FEET OR HIGHER
65% AT 22,000 FEET OR HIGHER
CONTINUOUS OPERATION LEANER THAN SHOWN IN THE TABLE IS PROHIBITED."

EXHAUST GAS TEMPERATURE (°F RICH OF PEAK)

POWER	2500 R.P.M.	2400 R.P.M.	2300 R.P.M.	2200 R.P.M.
76 to 80%	100%	75%	75%	50%
71 to 75%	75°	75°	50°	50°
66 to 70%	75°	50°	50°	25°
61 to 65%	50°	50°	25°	25°
56 to 60%	50°	25°	25°	Peak EGT
51 to 55%	25°	25°	Peak EGT	Peak EGT
46 to 50%	25°	Peak EGT	Peak EGT	Peak EGT
45% or less	Peak EGT	Peak EGT	Peak EGT	Peak EGT

2105030-1

3A21

NOTE 2. K. (26) Effective P21000761 and up:
"All placards required in the Pilot's Operating Handbook and FAA Approved Airplane Flight Manual must be installed in the appropriate locations."

NOTE 3. The cylinder head thermistors must be installed as follows:

Model		Cylinder Head Number
210, 210A	(1960-61 Model)	3
210B, 210C, 210D	(1962-63-64 Model)	1
210E,210F,210G,210H,210J	(1965-66-67-68-69 Model)	2
210F,T210G,T210H,T210J	(1966-67-68-69 Model)	5
210K	(1970-71 Model)	3
T210K	(1970-71 Model)	5
210L	(1972-73-74-75-76 Model)	3
T210L	(1972-73 Model)	5
T210L	(1974-75-76 Model)	1
210M	(1977 Model)	3
210M	(1978 Model)	1
T210M	(1977-78 Model)	1
P210N	(1978-81 Model)	5
210N	(1979-81 Model)	1
T210N	(1979 Model)	1
T210N	(1980-81 Model)(Non-Air Cond)	5 or 1
T210N	(1980-81 Model)(With Air Cond)	1
P210N	(1982-83 Model)	4
210N, 210R	(1982 Model and up)(Non Air Cond)	4
210N, 210R	(1982 Model and up)(With AirCond)	1
T210N	(1982 Model and up)	3
P210R, T210R	(1985 Model and up)	1

NOTE 4. The marking of the airspeed indicator with I.A.S. provides an equivalent level of safety to CAR 3.757 when the approved airspeed calibration data presented in Section V of the Pilot's Operating Handbooks listed below is available to the pilot:

210L	Cessna P/N D1069-13	(S/N 21061040 through 21061573)
T210L	Cessna P/N D1070-13	(S/N 21061040 through 21061573 except 21061042)
210M	Cessna P/N D1094-13	(S/N 21061574 through 21062273)
T210M	Cessna P/N D1095-13	(S/N 21061574 through 21062273)
210M	Cessna P/N D1122-13	(S/N 21062274 through 21063954)
T210M	Cessna P/N D1123-13	(S/N 21061042, 21062274 through 21062954)
P210N	Cessna P/N D1124-13	(S/N P21000001 through P21000150)
210N	Cessna P/N D1151-13PH	(S/N 21062955 through 21063640)
T210N	Cessna P/N D1152-13PH	(S/N 21062955 through 21063640)
P210N	Cessna P/N D1153-13PH	(S/N P21000151 through P21000385)
210N	Cessna P/N D1186-13PH	(S/N 21063641 through 21064135)
T210N	Cessna P/N D1187-13PH	(S/N 21063641 through 21064135)
P210N	Cessna P/N D1188-13PH	(S/N P21000386 through P21000590)
210N	Cessna P/N D1207-13PH	(S/N 21064136 through 21064535)
T210N	Cessna P/N D1208-13PH	(S/N 21064136 through 21064535)
P210N	Cessna P/N D1209-13PH	(S/N P21000591 through P21000760)
210N	Cessna P/N D1226-13PH	(S/N 21064536 through 21064772)
T210N	Cessna P/N D1227-13PH	(S/N 21064536 through 21064772)
P210N	Cessna P/N D1228-13PH	(S/N P21000761 through P21000811)
210N	Cessna P/N D1244-13PH	(S/N 21064773 through 21064822)
T210N	Cessna P/N D1245-13PH	(S/N 21064773 through 21064822)
P210N	Cessna P/N D1246-13PH	(S/N P21000812 through P21000834)
210N	Cessna P/N D1265-13PH	(S/N 21064823 through 21064897)
T210N	Cessna P/N D1266-13PH	(S/N 21064823 through 21064897)
210R	Cessna P/N D1288-13PH	(S/N 21064898 through 21065009)

NOTE 5. Service information applicable to Models P210N and P210R:

Components subject to the establishment of a retirement life as shown below with the corresponding retirement life hours:

Component Name	Retirement Hours
Windshield, rear cabin top windows Side windows, and ice detector light lens	13,000 hours

NOTE 6. 14-volt electrical system
(210/T210 series through S/N 21059502)
(205 series through S/N 205-0577)

28-volt electrical system
(210/T210 series effective S/N 21059503 and up)
(P210 series effective S/N P21000001 and up)

In addition to the placards specified above, the prescribed operating limitations indicated by an asterisk (*) under Sections I through XVIII of this data sheet must also be displayed by permanent markings.

"WARNING: Use of alcohol-based fuels can cause serious performance degradation and fuel system component damage, and is therefore prohibited on Cessna airplanes."

...END...

DEPARTMENT OF TRANSPORTATION
FEDERAL AVIATION ADMINISTRATION

	A3SO Revision 29 PIPER
PA-32-260	PA-32R-301 (SP)
PA-32-300	PA-32R-301 (HP)
PA-32S-300	PA-32R-301T
PA-32R-300	PA-32-301
PA-32RT-300	PA-32-301T
PA-32RT-300T	
PA-32-301FT	
PA-32-301XTC	October 28, 2005

TYPE CERTIFICATE DATA SHEET NO. A3SO

This data sheet which is a part of Type Certificate No. A3SO, prescribes conditions and limitations under which the product for which the Type Certificate was issued meets the airworthiness requirements of the Federal Aviation Regulations.

Type Certificate Holder — The New Piper Aircraft, Inc.
2926 Piper Drive
Vero Beach, Florida 32960

I. - Model PA-32-260 (Cherokee Six 260), 6 PCLM (Normal Category), Approved March 4, 1965; 7 PCLM (Normal Category), Approved November 15, 1966.

Engine — Lycoming O-540-E4B5 with carburetor setting 10-4404, 10-5042, or 10-5054
Oil cooler P/N 8529245 required with 10-5042 setting

Fuel — 100/130 minimum grade aviation gasoline

Engine Limits — For all operations, 2700 r.p.m. (260 hp)

Propeller and Propeller Limits — McCauley fixed pitch metal 1P235PFA82 (See NOTE 8)
Static r.p.m. at maximum permissible throttle setting, not over 2480 r.p.m., not under 2270 r.p.m.
Diameter: Not over 82 in., not under 80.5 in.
Spinner: P/N 63760-00 or 63760-03 (See NOTE 6)

Hartzell constant speed Model HC-C2YK-1( ) and Blade Model 8477-2, or
Hartzell constant speed Model HC-C2YK-1( )F and Blade Model F8477-2
Pitch: High 32° ± 2°, Low 12.0° ± .2° at 30 in. station
Diameter: Not over 82 in., not under 80.5 in.
Governor Assembly: Hartzell F-4-4( ) or F-4-11( ) (See NOTE 10)
Spinner: P/N 68713 or 66785 Spinner Tip and P/N 66786 Spinner Shell or P/N 67790-0 Spinner, P/N 67791-0 Bulkhead, P/N 67793-0 Bulkhead, P/N 99499-0 Plate, two each P/N 67794-0 Cuff or Kit 760-452V (See NOTE 6)

Page No.	1	2	3	4	5	6	7	8	9	10	11	12	13	14	15	16	17	18	19
Rev No.	29	25	27	25	27	25	27	27	24	27	27	24	29	27	25	27	27	24	27

Page No.	20	21	22	23	24	25	26	27	28	29	30
Rev No.	29	29	27	29	29	28	28	28	27	27	27

I. - Model PA-32-260 (cont'd)

Airspeed Limits

Never exceed	212 m.p.h. (184 knots) CAS
Maximum structural cruise	168 m.p.h. (146 knots) CAS
Maneuvering	149 m.p.h. (130 knots) CAS
Flaps extended	125 m.p.h. (109 knots) CAS

C.G. Range (gear extended)

(+91.4) to (+95.5) at 3400 lb.
(+90.2) to (+96.2) at 3300 lb.
(+81.4) to (+96.2) at 2600 lb.
(+78.0) to (+96.2) at 2060 lb. or less
Straight line variation between points given.

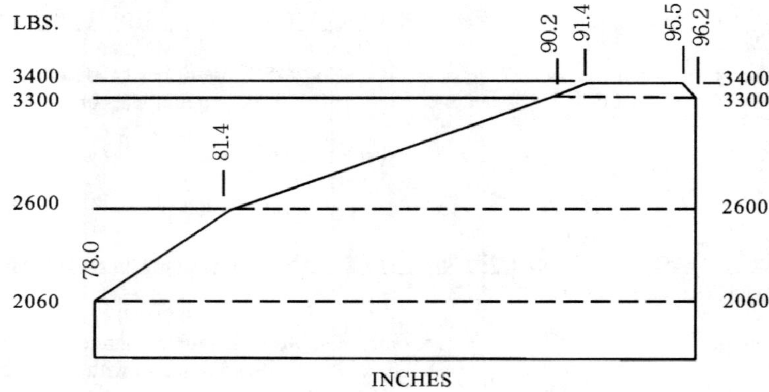

(S/N 32-1 through 32-1075)

(+91.4) to (+95.5) at 3400 lb.
(+89.0) to (+96.2) at 3300 lb.
(+80.0) to (+96.2) at 2900 lb.
(+76.0) to (+96.2) at 2400 lb. or less
Straight line variation between points given.

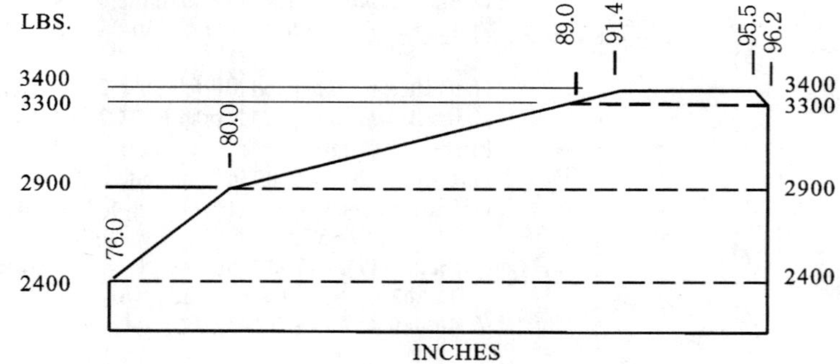

(S/N 32-1111 through 32-1297, and 32-7100001 through 32-7800008)

Empty Weight C.G. Range — None

Maximum Weight — 3400 lb.

I. - Model PA-32-260 (cont'd)

No. of Seats	6 (2 at +85.5, 2 at +118.1, 2 at +155.7) 7 (2 at +85.5, 3 at +118.1, 2 at +155.7) (See NOTE 3) 6 (2 at +85.5, 2 at +118.1, 2 at +157.6) 7 (2 at +85.5, 3 at +118.1, 2 at +157.6) (See NOTE 3) 6 (2 at +85.5, *2 at +119.1, 2 at +157.6) (See NOTE 11) * - Optional Club Seats
Maximum Baggage	200 lb. (100 lb. at +42.0, 100 lb. at +178.7)
Fuel Capacity	84 gallons at +95.0 (4 wing tanks) See NOTE 1 for data on system fuel
Oil Capacity	12 qt. at +16.6 (9-1/4 qt. usable) See NOTE 1 for data on system oil

Control Surface Movements

Wing Flaps	Up	0° (±2°)	Down	40° (±2°)
Ailerons	Up	30° (±2°)	Down	15° (±2°)
Rudder	Left	27° (±2°)	Right	27° (±2°)
Stabilator	Up	16° (±1°)	Down	2° (±1°)
Stabilator Tab	Up	5° (±1°)	Down	8° (±1°)

Nose Wheel Travel

S/N 32-1 through 32-1297, and 32-7100001 through 32-7300066:
Left 30° (±2°) Right 30° (±2°)
S/N 32-7400001 through 32-7800008:
Left 24° (±2°) Right 24° (±2°)

Manufacturer's Serial Nos. — 32-03, 32-04, 32-1 through 32-1297, and 32-7100001 through 32-7800008. The manufacturer is authorized to issue airworthiness certificates for airplane serial numbers 32-1034 through 32-1297, and 32-7100001 through 32-7800008 under the delegation option provisions of FAR 21.

II. - Model PA-32-300 (Cherokee Six 300), 6 PCLM (Normal Category), Approved May 27, 1966; 7 PCLM (Normal Category), Approved November 15, 1966.

Same as Model PA-32-260 except for engine installation and fuel system.

Engine	Lycoming IO-540-K1A5, Bendix injector type RSA-10ED1 Lycoming IO-540-K1G5 (See NOTE 12) Flow Setting No. 2524273
Fuel	100/130 minimum grade aviation gasoline
Engine Limits	For all operations, 2700 r.p.m. (300 hp)
Propeller and Propeller Limits	Hartzell constant speed Model HC-C2YK-1( ), Blade Models 8475-4 & 8475D-4, or Hartzell constant speed Model HC-C2YK-1( )F, Blade Models F8475D-4 Pitch: High 34° ± 1°, Low 13.5° ± .2° at 30 in. station Diameter: Not over 80 in., not under 78.5 in. Governor Assembly: Hartzell F-4-4( ) or F-4-11( ) (See NOTE 10) Spinner: P/N 68713 or P/N 66785 Spinner Tip and P/N 66786 Spinner Shell, or P/N 67790-0 Spinner, P/N 67791-0 Bulkhead, P/N 67793-0 Bulkhead, P/N 99499-0 Plate, two each P/N 67794-0 Cuff or Kit 760-452V (See NOTE 6)

A3SO

II. - Model PA-32-300 (cont'd)

Propeller and Propeller Limits (continued)	Hartzell constant speed Model HC-C2YK-1( ), Blade Model 8475R-0, or Hartzell constant speed Model HC-C2YK-1( )F, Blade Model F8475R-0 Pitch: High 29° ± 1°, Low 12.4° ± .2° at 30 in. station Diameter: Not over 84 in., not under 82.3 in. Governor Assembly: Hartzell F-4-4( ) or F-4-11( ) (See NOTE 10) Spinner: P/N 68713 or P/N 66785 Spinner Tip and P/N 66786 Spinner Shell or P/N 67790-0 Spinner, P/N 67791-0 Bulkhead, P/N 67793-0 Bulkhead, P/N 99499-0 Plate, two each P/N 67794-0 Cuff or Kit 760-452V (See NOTE 6)
Airspeed Limits	Never exceed 212 m.p.h. (184 knots) CAS Maximum structural cruise 168 m.p.h. (146 knots) CAS Maneuvering 149 m.p.h. (130 knots) CAS Flaps extended 125 m.p.h. (109 knots) CAS
C.G. Range (gear extended)	(+91.4) to (+95.5) at 3400 lb. (+90.2) to (+96.2) at 3300 lb. (+81.4) to (+96.2) at 2600 lb. (+78.0) to (+96.2) at 2060 lb. or less Straight line variation between points given.

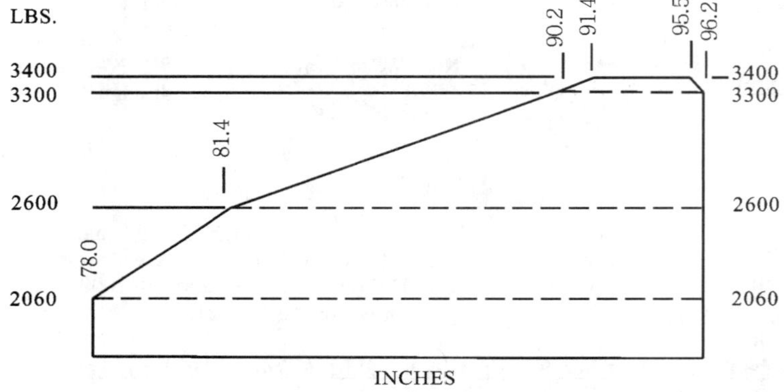

(S/N 32-40001 through 32-40565)

(+91.4) to (+95.5) at 3400 lb.
(+89.0) to (+96.2) at 3300 lb.
(+80.0) to (+96.2) at 2900 lb.
(+76.0) to (+96.2) at 2400 lb. or less
Straight line variation between points given.

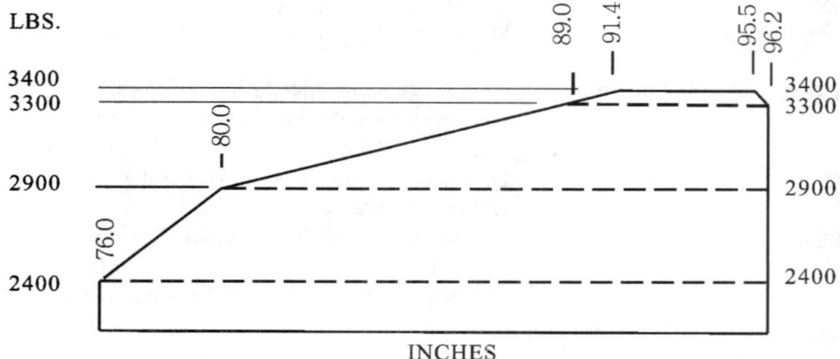

(S/N 32-40566 through 32-40974, and 32-7140001 through 32-7940290)

II. - Model PA-32-300 (cont'd)

Empty Weight C.G. Range	None
Maximum Weight	3400 lb.
No. of Seats	6 (2 at +85.5, 2 at +118.1, 2 at +155.7) 7 (2 at +85.5, 3 at +118.1, 2 at +155.7) (See NOTE 3) 6 (2 at +85.5, 2 at +118.1, 2 at +157.6) 7 (2 at +85.5, 3 at +118.1, 2 at +157.6) (See NOTE 3) 6 (2 at +85.5, *2 at +119.1, 2 at +157.6) (See NOTE 11) * - Optional Club Seats
Maximum Baggage	200 lb. (100 lb. at +42.0, 100 lb. at +178.7)
Fuel Capacity	S/N 32-15, 32-21, 32-40000 through 32-40974, and 32-7140001 through 32-7840202: 84 gallons at +95.0 (4 wing tanks) S/N 32-7940001 through 32-7940290: 98 gallons at +93.6 (2 wing tanks) (94 gallons usable) See NOTE 1 for data on system fuel
Oil Capacity	12 qt. at +16.6 (9-1/4 qt. usable) See NOTE 1 for data on system oil

Control Surface Movements

Wing Flaps	Up	0° (±2°)	Down	40° (±2°)
Ailerons	Up	30° (±2°)	Down	15° (±2°)
Rudder	Left	27° (±2°)	Right	27° (±2°)
Stabilator	Up	16° (±1°)	Down	2° (±1°)
Stabilator Tab	Up	5° (±1°)	Down	8° (±1°)

Nose Wheel Travel: S/N 32-40001 through 32-40974, and 32-7140001 through 32-7340191:
Left 30° (±2°) Right 30° (±2°)
S/N 32-7400001 through 32-7940290:
Left 24° (±2°) Right 24° (±2°)

Manufacturer's Serial Nos.: 32-15, 32-21, 32-40000 through 32-40974, and 32-7140001 through 32-7940290. The manufacturer is authorized to issue airworthiness certificates for airplane serial numbers 32-40382, 32-40385, 32-40403, 32-40465 through 32-40469, 32-40471 through 32-40974, and 32-7140001 through 32-7940290 under the delegation option provisions of FAR 21 (See NOTE 7 and 9).

III. - Model PA-32S-300 (Cherokee Six Seaplane), 7 PCSM (Normal Category), Approved February 14, 1967.
Same as Model PA-32-300 except for float installation.

Engine	Lycoming IO-540-K1A5 Flow Setting No. 2524273
Fuel	100/130 minimum grade aviation gasoline
Engine Limits	For all operations, 2700 r.p.m. (300 hp)

A3SO

III. - Model PA-32S-300 (cont'd)

Propeller and Propeller Limits

Hartzell constant speed Model HC-C2YK-1( ), Blade Models 8475-4 & 8475D-4, or
Hartzell constant speed Model HC-C2YK-1( )F, Blade Model F8475D-4
Pitch: High 34° ± 1°, Low 13.5° ± .2° at 30 in. station
Diameter: Not over 80 in., not under 78.5 in.
Governor Assembly: Hartzell F-4-4( ) or F-4-11( ) (See NOTE 10)
Spinner: P/N 68713 or P/N 66785 Spinner Tip and P/N 66786 Spinner Shell (See NOTE 6)

Hartzell constant speed Model HC-C2YK-1( ), Blade Model 8475R-0, or
Hartzell constant speed Model HC-C2YK-1( )F, Blade Model F8475R-0
Pitch: High 29° ± 1°, Low 12.4° ± .2° at 30 in. station
Diameter: Not over 84 in., not under 82.3 in.
Governor Assembly: Hartzell F-4-4( ) or F-4-11( ) (See NOTE 10)
Spinner: P/N 68713 or P/N 66785 Spinner Tip and P/N 66786 Spinner Shell or P/N 67790-0 Spinner, P/N 67791-0 Bulkhead, P/N 67793-0 Bulkhead, P/N 99499-0 Plate, two each P/N 67794-0 Cuff or Kit 760-452V (See NOTE 6)

Airspeed Limits

Never exceed	176 m.p.h. (153 knots) CAS
Maximum structural cruise	140 m.p.h. (122 knots) CAS
Maneuvering	140 m.p.h. (122 knots) CAS
Flaps extended	125 m.p.h. (109 knots) CAS

C.G. Range (gear extended)

(+87.6) to (+94.5) at 3400 lb.
(+82.6) to (+94.5) at 2940 lb.
(+79.8) to (+94.5) at 2400 lb.
Straight line variation between points given.
(See NOTE 4 for operation in landplane configuration)

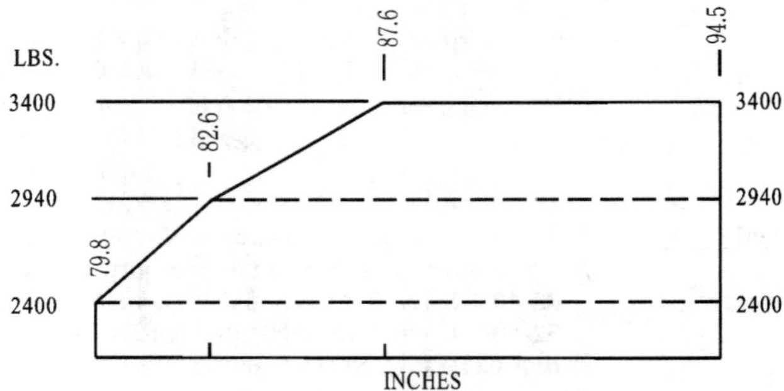

(S/N 32S-40001 through 32S-40974, and 32S-7140001 through 32S-7240137)

Empty Weight C. G. Range: None

Maximum Weight: 3400 lb.

No. of Seats: 7 (2 at +85.5, 2 at +118.1, 2 at +155.7)

Maximum Baggage: 200 lb. (100 lb. at +42.0, 100 lb. at +178.7)

III. - Model PA-32S-300 (cont'd)

Fuel Capacity	84 gallons at +95.0 (4 wing tanks) See NOTE 1 for data on system fuel
Oil Capacity	12 qt. at +16.6 (9-1/4 qt. usable) See NOTE 1 for data on system oil

Control Surface Movements

Wing Flaps	Up	0° (±2°)	Down	40° (±2°)
Ailerons	Up	30° (±2°)	Down	15° (±2°)
Rudder	Left	27° (±2°)	Right	27° (±2°)
Stabilator	Up	16° (±1°)	Down	2° (±1°)
Stabilator Tab	Up	5° (±1°)	Down	8° (±1°)

Manufacturer's Serial Nos. — 32S-15, 32S-40000 through 32S-40974, and 32S-7140001 through 32S-7240137. The manufacturer is authorized to issue airworthiness certificates for airplane serial numbers 32S-40382, 32S-40385, 32S-40403, 32S-40465 through 32S-40469, 32S-40471 through 32S-40974, and 32S-7140001 through 32S-7240137 under the delegation option provisions of FAR 21 (See NOTE 7 and 9).

IV. - Model PA-32R-300 (Lance), 7 PCLM (Normal Category), Approved February 25,1975.

Same as Model PA-32-300 except for redesigned wing and engine mount to accommodate retractable landing gear, gross weight increase, increased capability fuel system and other minor changes.

Engine	Lycoming IO-540-K1A5D Lycoming IO-540-K1G5D for S/N 32R-7680141 through 32R-7880068 (See NOTE 13) Flow Setting No. 2524273
Fuel	100/130 minimum grade aviation gasoline
Engine Limits	For all operations, 2700 r.p.m. (300 hp)
Propeller and Propeller Limits	Hartzell constant speed Model HC-C2YK-1( )F, Blade Model F8475D-4 Pitch: High 34° ± 1°, Low 13.5° ± .2° at 30 in. station Diameter: Not over 80 in., not under 78.5 in. Governor Assembly: Hartzell F-4-11B( ) Spinner: P/N 67790-0 Spinner, P/N 67791-0 Bulkhead, P/N 67793-0 Bulkhead, P/N 99499-0 Plate, and two each P/N 67794-0 Cuff (See NOTE 6)

Airspeed Limits

Never exceed	217 m.p.h. (188 knots) CAS
Maximum structural cruise	172 m.p.h. (149 knots) CAS
Maneuvering	125 m.p.h. (109 knots) CAS
Maximum flaps extended	125 m.p.h. (109 knots) CAS
Maximum gear extension	150 m.p.h. (130 knots) CAS
Maximum gear retraction	125 m.p.h. (109 knots) CAS

IV. - Model PA-32R-300 (cont'd)

C.G. Range (gear extended)	(+91.4) to (+95.0) at 3600 lb. (+80.0) to (+95.0) at 2900 lb. (+76.0) to (+95.0) at 2400 lb. or less Straight line variation between points given.

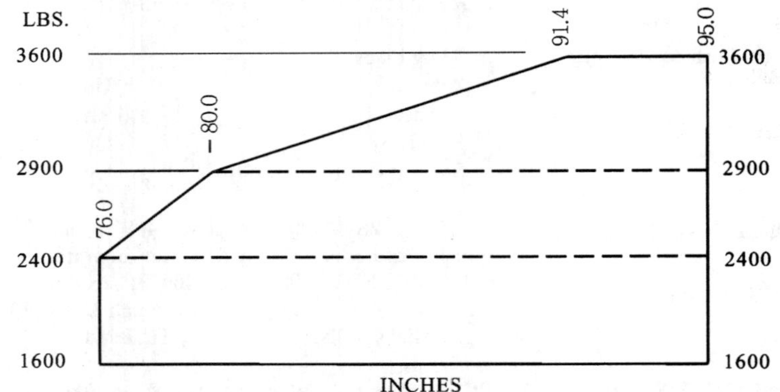

Empty Weight C.G. Range	None
Maximum Weight	3600 lb.
No. of Seats	7 (2 at +85.5, 3 at +118.1, 2 at +155.7) 7 (2 at +85.5, 3 at +118.1, 2 at +157.6) 6 (2 at +85.5, *2 at +119.1, 2 at +157.6) (See NOTE 11) * - Optional Club Seats
Maximum Baggage	200 lb. (100 lb. at +42.0, 100 lb. at +178.7)
Fuel Capacity	98 gallons at +93.6 (2 wing tanks) (94 gallons usable) See NOTE 1 for data on system fuel
Oil Capacity	12 qt. at +16.6 (9-1/4 qt. usable) See NOTE 1 for data on system oil

Control Surface Movements

Wing Flaps	Up	0° (±2°)	Down	40° (±2°)
Ailerons	Up	30° (±2°)	Down	15° (±2°)
Rudder	Left	27° (±2°)	Right	27° (±2°)
Stabilator	Up	16° (±1°)	Down	2° (±1°)
Stabilator Tab	Up	5° (±1°)	Down	8° (±1°)

Manufacturer's Serial Nos.	32R-7680001 through 32R-7880068. The manufacturer is authorized to issue airworthiness certificates for airplane serial numbers 32R-7680001 through 32R-7880068 under the delegation option provisions FAR 21 (See NOTE 7).

V. - Model PA-32RT-300 (Lance II), 7 PCLM (Normal Category), Approved December 13, 1977.
Same as Model PA-32R-300 except for redesigned tail surfaces in "T" configuration and other minor changes.

Engine	Lycoming IO-540-K1G5D Flow Setting No. 2524273
Fuel	100/130 minimum grade aviation gasoline
Engine Limits	For all operations, 2700 r.p.m. (300 hp)
Propeller and Propeller Limits	Hartzell constant speed Model HC-C2YK-1( )F, Blade Model F8475D-4 Pitch: High 34° ± 1°, Low 13.5° ± .2° at 30 in. station Diameter: Not over 80 in., not under 78.5 in. Governor Assembly: Hartzell F-4-11B( ) Spinner: P/N 99374 (See NOTE 6)

Airspeed Limits

Never exceed	217 m.p.h. (189 knots) CAS
Maximum structural cruise	173 m.p.h. (150 knots) CAS
Maneuvering (with 3600 lb. gross weight)	152 m.p.h. (132 knots) CAS
Maximum flaps extended	125 m.p.h. (109 knots) CAS
Maximum gear extension	150 m.p.h. (130 knots) CAS
Maximum gear retraction	125 m.p.h. (109 knots) CAS

C.G. Range (gear extended)

(+91.4) to (+96.0) at 3600 lb.
(+84.0) to (+96.0) at 3000 lb.
(+82.0) to (+96.0) at 2500 lb. or less
Straight line variation between points given.

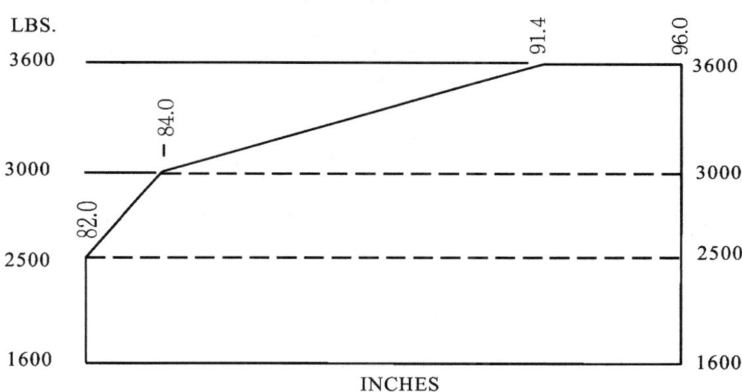

Empty Weight C.G. Range	None
Maximum Weight	3600 lb.
No. of Seats	7 (2 at +85.5, 3 at +118.1, 2 at +157.6) 6 (2 at +85.5, *2 at +119.1, 2 at +157.6) (See NOTE 11) * - Optional Club Seats
Maximum Baggage	200 lb. (100 lb. at +42.0, 100 lb. at +178.7)

V. - Model PA-32RT-300 (cont'd)

Fuel Capacity	98 gallons at +93.6 (2 wing tanks) (94 gallons usable) See NOTE 1 for data on system fuel
Oil Capacity	12 qt. at +16.6 (9-1/4 qt. usable) See NOTE 1 for data on system oil

Control Surface Movements

Wing Flaps	Up	0° (±2°)	Down	30° (±2°)
Ailerons	Up	30° (±2°)	Down	15° (±2°)
Rudder	Left	36° (±2°)	Right	36° (±2°)
Stabilator	Up	14.5° (±.5°)	Down	10° (±1°)
Stabilator Tab	Up	2.5° (±1°)	Down	10° (±.5°)

Manufacturer's Serial Nos. 32R-7885002 through 32R-7985106. The manufacturer is authorized to issue airworthiness certificates for airplane serial numbers 32R-7885002 through 32R-7985106 under the delegation option provisions FAR 21 (See NOTE 7).

VI. - Model PA-32RT-300T (Turbo Lance II), 7 PCLM (Normal Category), Approved April 20, 1978.
Same as Model PA-32RT-300 except for turbocharged engine installation and other minor changes.

Engine	Lycoming TIO-540-S1AD Bendix Injector Type RSA-10ED1 Flow Setting No. 2524693 for S/N 32R-7787001, 32R-7887002 through 32R-7887041 Bendix Injector Type RSA-10ED2 Flow Setting No. 2524791 for S/N 32R-7787001, 32R-7887002 through 32R-7987126
Fuel	100/130 minimum grade aviation gasoline
Engine Limits	For 5 minute takeoff, 2700 r.p.m. and 36.0" Hg MAP (300 hp) For maximum continuous operation, 2575 r.p.m. and 33.0" Hg MAP (270 hp)
Propeller and Propeller Limits	Hartzell constant speed Model HC-E2YR-1( )F, Blade Model F8477-4 Pitch: High 34° ± 1°, Low 15.6° ± .2° at 30 in. station Diameter: Not over 80 in., not under 78.5 in. Governor Assembly: Hartzell F-4-11B or F-4-11B( ) Spinner: Piper P/N 98708-2 or Hartzell P/N A-2298-2

Airspeed Limits

Never exceed	217 m.p.h. (189 knots) CAS
Maximum structural cruise	173 m.p.h. (150 knots) CAS
Maneuvering (with 3600 lb. gross weight)	152 m.p.h. (132 knots) CAS
Maximum flaps extended	125 m.p.h. (109 knots) CAS
Maximum gear extension	150 m.p.h. (130 knots) CAS
Maximum gear retraction	125 m.p.h. (109 knots) CAS

VI. - Model PA-32RT-300T (cont'd)

C.G. Range (gear extended) (+91.4) to (+95.0) at 3600 lb.
(+80.0) to (+95.0) at 2900 lb. or less
Straight line variation between points given.

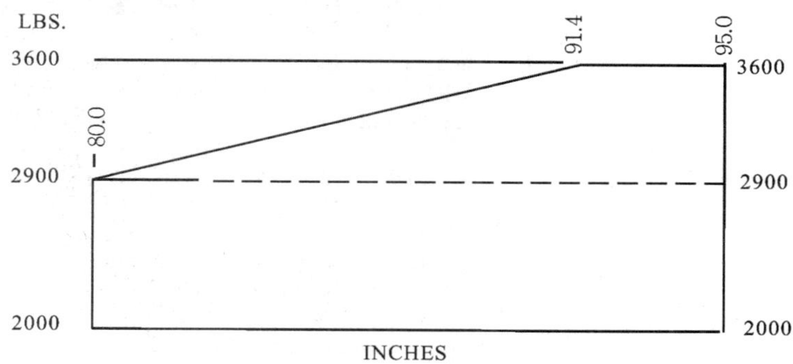

Empty Weight C.G. Range	None
Maximum Weight	3600 lb.
No. of Seats	7 (2 at +85.5, 3 at +118.1, 2 at +157.6) 6 (2 at +85.5, *2 at +119.1, 2 at +157.6) (See NOTE 11) * - Optional Club Seats
Maximum Baggage	200 lb. (100 lb. at +42.0, 100 lb. at +178.7)
Fuel Capacity	98 gallons at +93.6 (2 wing tanks) (94 gallons usable) See NOTE 1 for data on system fuel
Oil Capacity	12 qt. at +16.6 (9-1/4 qt. usable) See NOTE 1 for data on system oil

Control Surface Movements

Wing Flaps	Up	0° (±2°)	Down	40° (±2°)
Ailerons	Up	30° (±2°)	Down	15° (±2°)
Rudder	Left	36° (±2°)	Right	36° (±2°)
Stabilator	Up	14.5° (±.5°)	Down	10° (±1°)
Stabilator Tab	Up	1.0° (±1°)	Down	10° (±.5°)

Manufacturer's Serial Nos. 32R-7787001, 32R-7887002 through 32R-7987126. The manufacturer is authorized to issue airworthiness certificates for airplane serial numbers 32R-7787001, 32R-7887002 through 32R-7987126 under the delegation option provisions of FAR 21 (See NOTE 7).

VII. - Model PA-32R-301 (Saratoga SP), 7 PCLM (Normal Category), Approved November 7, 1979.

Same as Model PA-32R-300 except for tapered wings and other minor changes.

Engine	Lycoming IO-540-K1G5D Bendix Injector Type RSA-10ED1 Flow Setting No. 2524273
Fuel	100 or 100LL aviation grade fuel

A3SO

VII. - Model PA-32R-301 (cont'd)

Engine Limits

For airplanes equipped with standard Hartzell 2 blade propeller
HC-C2Y(K,R)-1( )F/F8475D-4:
For 5 minute takeoff, 2700 r.p.m. and full throttle (300 rated hp)
For maximum continuous operation, 2600 r.p.m. and full throttle (294 rated hp)

For airplanes equipped with optional Hartzell 3 blade propeller
HC-C3YR-1( )F/F7663R-0:
For all operations, 2700 r.p.m. and full throttle (300 rated hp)

Propeller and Propeller Limits

Hartzell constant speed Model HC-C2Y(K,R)-1( )F/F8475D-4 (standard 2 blade):
Pitch: High 34° ± 1°, Low 13.5° ± .2° at 30 in. station
Diameter: Not over 80 in., not under 78.5 in.
Governor Assembly: Hartzell F-4-11B or F-4-11B( )
Spinner: Piper P/N 98708-2 or Hartzell P/N A-2298-2

Hartzell constant speed Model HC-C3YR-1( )F/F7663R-0 (optional 3 blade):
Pitch: High 32.0° ± 1°, Low 12.4° ± .2° at 30 in. station
Diameter: Not over 78 in., not under 76 in.
Governor Assembly: Hartzell F-4-11B or F-4-11B( )
Spinner: Piper PS50077-56 or Hartzell P/N 835-47

Airspeed Limits (Indicated)

Never exceed	197 knots (226 m.p.h.)
Maximum structural cruise	154 knots (177 m.p.h.)
Maneuvering (with 3600 lb. gross weight)	134 knots (154 m.p.h.)
Maximum flaps extended	112 knots (129 m.p.h.)
Maximum gear extension	132 knots (151 m.p.h.)
Maximum gear retraction	110 knots (126 m.p.h.)
Maximum gear extended	132 knots (151 m.p.h.)

C.G. Range (gear extended)

(+91.4) to (+95.0) at 3600 lb.
(+83.5) to (+95.0) at 3200 lb.
(+78.0) to (+95.0) at 2400 lb.
Straight line variation between points given.

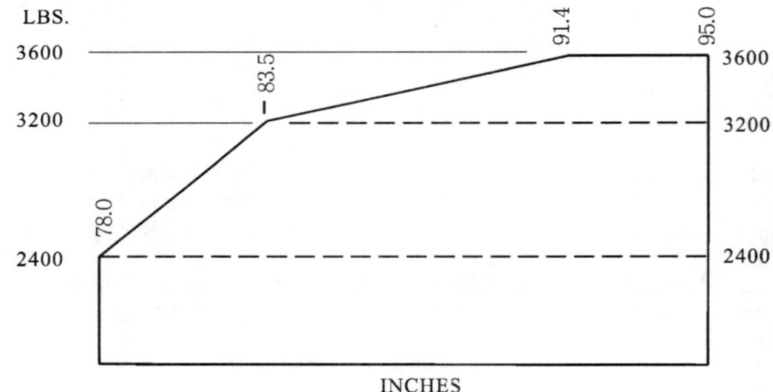

Empty Weight C.G. Range

None

Maximum Weight

Ramp: 3615 lb.
Takeoff: 3600 lb.
Landing: 3600 lb.

No. of Seats

7 (2 at +85.5, 3 at +118.1, 2 at +157.6)
6 (2 at +85.5, *2 at +119.1, 2 at +157.6) (See NOTE 11)
* - Optional Club Seats

VII. - Model PA-32R-301 (cont'd)

Maximum Baggage	200 lb. (100 lb. at +42.0, 100 lb. at +178.7)
Fuel Capacity	107 gallons at +94.0 (2 wing tanks) (102 gallons usable) See NOTE 1 for data on system fuel
Oil Capacity	12 qt. at +16.6 (9-1/4 qt. usable) See NOTE 1 for data on system oil

Control Surface Movements

Wing Flaps	Up	0° (±1°)	Down	40° (±2°)
Ailerons	Up	28° (±1°)	Down	22° (±1°)
Rudder	Left	28° (±1°)	Right	28° (±1°)
Stabilator	Up	14.5° (±.5°)	Down	5.5° (±.5°)
Stabilator Tab	Up	5° (±1°)	Down	8° (±1°)

Manufacturer's Serial Nos. 32R-8013001 through 32R-8613006, 3213001 through 3213028, and 3213030 through 3213041. The manufacturer is authorized to issue airworthiness certificates for airplane serial numbers 32R-8013001 through 32R-8613006, 3213001 through 3213028, and 3213030 through 3213041 under the delegation option provisions of FAR 21.

VIII. - Model PA-32R-301 (Saratoga II HP), 7 PCLM (Normal Category), Approved May 26, 1993.
Same as Model PA-32R-301, Saratoga SP, except for engine cowling, engine model designation and other minor changes.

Engine

Lycoming IO-540-K1G5
Precision Airmotive Injector, Type RSA-10ED1
Flow Setting No. 2524273 for S/N 3213042 through 3213103, and 3246001 and up

Lycoming IO-540-K1G5D for S/N 3213029 only

Fuel

100 or 100LL aviation grade fuel

Engine Limits

Equipped with Hartzell 3 blade propeller HC-I3YR-1RF/F7663DR:
For all operations, 2700 r.p.m. and full throttle (300 rated hp)

Propeller and Propeller Limits

Hartzell constant speed Model HC-I3YR-1RF/F7663DR (3 blade)
Hartzell constant speed Model HC-I3YR-1RF/F7663DRB
(3 blade with TKS Ice Protection System)
Pitch: High 32.0° ± 1°, Low 12.4° ± .2° at 30 in. station
Diameter: Not over 78 in., not under 77 in.
Governor: Hartzell V-5-4
Spinner Assy: Hartzell P/N C3575-1 (P)
Dome: Hartzell P/N C-3532-16P (with TKS Ice Protection System)
Do not exceed 23" manifold pressure below 2100 r.p.m.

Airspeed Limits (Indicated)

For S/N 3213029, 3213042 through 3213103, and 3246001 through 3246017:

Never exceed	193 knots (222 m.p.h.)
Maximum structural cruise	160 knots (184 m.p.h.)
Maneuvering (with 3600 lb. gross weight)	132 knots (152 m.p.h.)
Flaps extended	108 knots (124 m.p.h.)
Maximum gear extension	130 knots (150 m.p.h.)
Maximum gear retraction	108 knots (124 m.p.h.)
Maximum gear extended	130 knots (150 m.p.h.)

For S/N 3246018 and up:

Never exceed	191 knots (220 m.p.h.)
Maximum structural cruise	160 knots (184 m.p.h.)
Maneuvering (with 3600 lb. gross weight)	134 knots (154 m.p.h.)
Flaps extended	110 knots (127 m.p.h.)
Maximum gear extension	132 knots (152 m.p.h.)
Maximum gear retraction	110 knots (127 m.p.h.)
Maximum gear extended	132 knots (152 m.p.h.)

VIII. - Model PA-32R-301 (cont'd)

C.G. Range (gear extended)	(+91.4) to (+95.0) at 3600 lb. (+83.5) to (+95.0) at 3200 lb. (+78.0) to (+95.0) at 2400 lb. Straight line variation between points given.

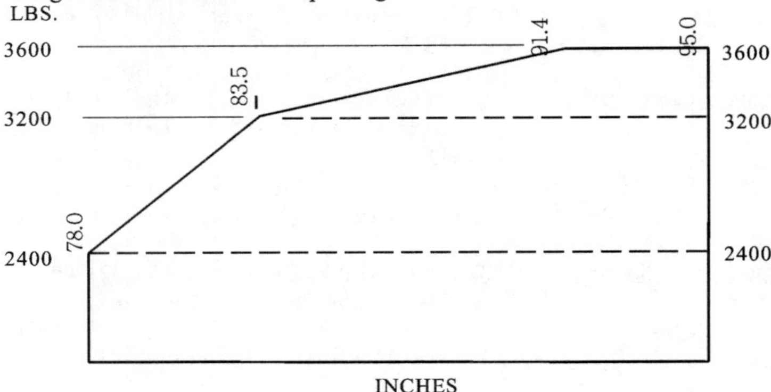

Empty Weight C.G. Range	None
Maximum Weight	Ramp: 3615 lb. Takeoff: 3600 lb. Landing: 3600 lb.
No. of Seats	6 (2 at +85.5, 2 at +119.1, 2 at +157.6)
Maximum Baggage	200 lb. (100 lb. at +42.0, 100 lb. at +178.7)
Fuel Capacity	107 gallons at +94.0 (2 wing tanks) (102 gallons usable) See NOTE 1 for data on fuel system
Oil Capacity	12 qt. at +16.6 (9-1/4 qt. usable) See NOTE 1 for data on oil system

Control Surface Movements					
	Wing Flaps	Up	0° (±1°)	Down	40° (±2°)
	Ailerons	Up	28° (±1°)	Down	22° (±1°)
	Rudder	Left	28° (±1°)	Right	28° (±1°)
	Stabilator	Up	14.5° (±.5°)	Down	5.5° (±.5°)
	Stabilator Tab	Up	5° (±1°)	Down	8° (±1°)

Manufacturer's Serial Nos.	3213029, 3213042 through 3213103 (14v), 3246001 through 3246017 (14v), and 3246018 and up (28v). The manufacturer is authorized to issue airworthiness certificates under the delegation option provisions of FAR 21.

IX. - Model PA-32R-301T (Turbo Saratoga SP), 7 PCLM (Normal Category), Approved November 7, 1979.
Same as Model PA-32R-300 except for tapered wings, turbocharged powerplant installation and other minor changes.

Engine	Lycoming TIO-540-S1AD Bendix Injector, Type RSA-10ED2 Flow Setting No. 2524791
Fuel	100 or 100LL aviation grade fuel

IX. - Model PA-32R-301T (cont'd)

Engine Limits

For airplanes equipped with standard Hartzell 2 blade propeller HC-E2YR-1( )F/F8477-4:
For 5 minute take-off, 2700 r.p.m. and 36.0" Hg MAP (300 hp) - Sea level to 16,000 ft. altitude
For maximum continuous operation, 2575 r.p.m. and 36.0" Hg MAP (294 hp) - Sea level to 16,000 ft. altitude

For airplanes equipped with optional Hartzell 3 blade propeller HC-E3YR-1( )F/F7673DR-0:
For all operations, 2700 r.p.m. and 36.0" Hg MAP (300 rated hp) - Sea level to 16,000 ft. altitude

Propeller and Propeller Limits

Hartzell constant speed Model HC-E2YR-1( )F/F8477-4 (standard 2 blade):
Pitch: High 34.0° ± 1°, Low 15.6° ± .2° at 30 in. station
Diameter: Not over 80 in., not under 78.5 in.
Governor Assembly: Hartzell F-4-11B or F-4-11B( )
Spinner: Piper P/N 98708-2 or Hartzell P/N A-2298-2

Hartzell constant speed Model HC-E3YR-1( )F/F7673DR-0 (optional 3 blade):
Pitch: High 34.5° ± 1°, Low 13.2° ± .2° at 30 in. station
Diameter: Not over 78 in., not under 76 in.
Governor Assembly: Hartzell F-4-11B or F-4-11B( )
Spinner: Piper P/N PS50077-58 or Hartzell P/N C-3575

Airspeed Limits (Indicated)

Never exceed	197 knots
Maximum structural cruise	154 knots
Maneuvering (with 3600 lb. gross weight)	134 knots
Flaps extended	112 knots
Maximum gear extension	132 knots
Maximum gear retraction	110 knots
Maximum gear extended	132 knots

C.G. Range (gear extended)

(+91.4) to (+95.0) at 3600 lb.
(+82.75) to (+95.0) at 3200 lb.
(+78.0) to (+95.0) at 2400 lb.
Straight line variation between points given.

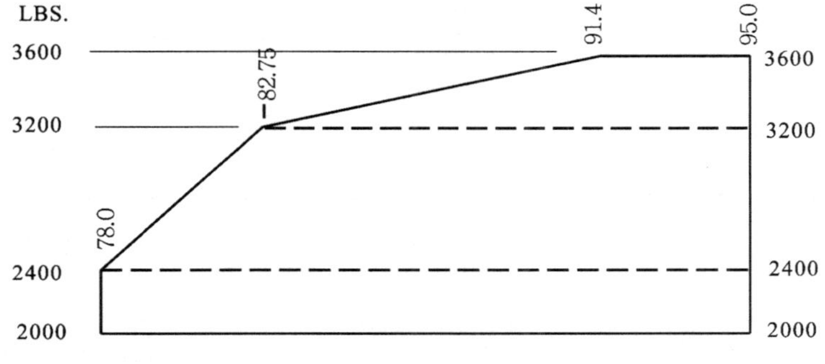

Empty Weight C.G. Range

None

Maximum Weight

Ramp: 3615 lb.
Takeoff: 3600 lb.
Landing: 3600 lb.

IX. - Model PA-32R-301T (cont'd)

No. of Seats: 7 (2 at +85.5, 3 at +118.1, 2 at +157.6)
6 (2 at +85.5, *2 at +119.1, 2 at +157.6) (See NOTE 11)
* - Optional Club Seats

Maximum Baggage: 200 lb. (100 lb. at +42.0, 100 lb. at +178.7)

Fuel Capacity: 107 gallons at +94.0 (2 wing tanks) (102 gallons usable)
See NOTE 1 for data on system fuel

Oil Capacity: 12 qt. at +16.6 (9-1/4 qt. usable)
See NOTE 1 for data on system oil

Control Surface Movements

Wing Flaps	Up	0° (±1°)	Down	40° (±2°)
Ailerons	Up	28° (±1°)	Down	22° (±1°)
Rudder	Left	28° (±1°)	Right	28° (±1°)
Stabilator	Up	14.5° (±.5°)	Down	5.5° (±0.5°)
Stabilator Tab	Up	5° (±1°)	Down	8° (±1°)

Manufacturer's Serial Nos.: 32R-8029001 through 32R-8629008, and 3229001 through 3229003. The manufacturer is authorized to issue airworthiness certificates for airplane serial numbers 32R-8029001 through 32R-8629008, and 3229001 through 3229003 under the delegation option provisions of FAR 21 (See NOTE 7).

X. - Model PA-32-301 (Saratoga), 7 PCLM (Normal Category), Approved January 9, 1980.

Same as Model PA-32-300 except for tapered wings, increased gross weight and other minor changes.

Engine: Lycoming IO-540-K1G5
Bendix Injector Type RSA-10ED1
Flow Setting No. 2524273

Fuel: 100 or 100LL aviation grade fuel

Engine Limits: For airplanes equipped with standard Hartzell 2 blade propeller HC-C2Y(K,R)-1( )F/F8475D-4:
For 5 minute takeoff, 2700 r.p.m. and full throttle (300 rated hp)
For maximum continuous operation, 2600 r.p.m. and full throttle (294 rated hp)

For airplanes equipped with optional Hartzell 3 blade propeller HC-C3YR-1( )F/F7663R-0:
For all operations, 2700 r.p.m. and full throttle (300 rated hp)

Propeller and Propeller Limits: Hartzell constant speed Model HC-C2Y(K,R)-1( )F/ F8475D-4 (standard 2 blade):
Pitch: High 34° ± 1°, Low 13.5° ± .2° at 30 in. station
Diameter: Not over 80 in., not under 78.5 in.
Governor Assembly: Hartzell F-4-11 or F-4-11( )
Spinner: P/N 67790-0 Spinner, P/N 67791-0 Bulkhead, P/N 67793-0 Bulkhead, P/N 99499-0 Plate, and two each 67794-0 Cuff (See NOTE 6)

Hartzell constant speed Model HC-C3YR-1( )F/F7663R-0 (optional 3 blade):
Pitch: High 32° ± 1°, Low 12.4° ± .2° at 30 in. station
Diameter: Not over 78 in., not under 76 in.
Governor Assembly: Hartzell F-4-11B or F-4-11B( )
Spinner: Hartzell P/N 835-47 (See NOTE 6)

X. - Model PA-32-301 (cont'd)

Airspeed Limits (Indicated)	Never exceed	197 knots
	Maximum structural cruise (with 3600 lb. gross weight)	154 knots
	Maneuvering	134 knots
	Flaps extended	112 knots

C.G. Range (gear extended)
(+90.0) to (+95.0) at 3600 lb.
(+83.5) to (+95.0) at 3200 lb.
(+78.0) to (+95.0) at 2400 lb.
Straight line variation between points given.

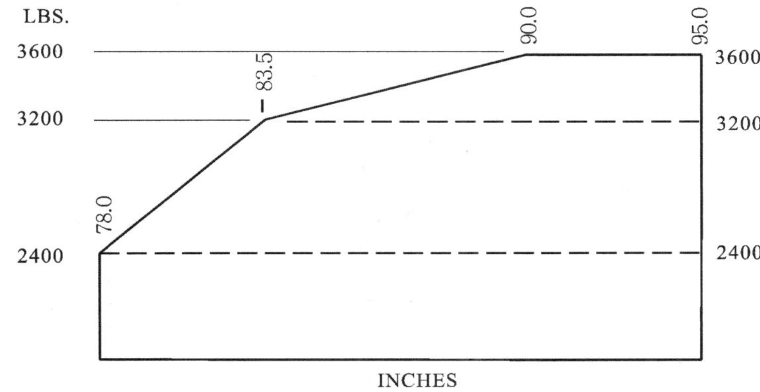

Empty Weight C.G. Range
None

Maximum Weight
Ramp: 3615 lb.
Takeoff: 3600 lb.
Landing: 3600 lb.

No. of Seats
6 (2 at +85.5, 2 at +118.1, 2 at +157.6)
7 (2 at +85.5, 3 at +118.1, 2 at +157.6)
6 (2 at +85.5, *2 at +119.1, 2 at +157.6) (See NOTE 11)
* - Optional Club Seats

Maximum Baggage
200 lb. (100 lb. at +42.0, 100 lb. at +178.7)

Fuel Capacity
107 gallons at +94.0 (2 wing tanks) (102 gallons usable)
See NOTE 1 for data on system fuel

Oil Capacity
12 qt. at +16.6 (9-1/4 qt. usable)
See NOTE 1 for data on system oil

Control Surface Movements

Wing Flaps	Up	0° (±1°)	Down	40° (±2°)
Ailerons	Up	28° (±1°)	Down	22° (±1°)
Rudder	Left	28° (±1°)	Right	28° (±1°)
Stabilator	Up	14.5° (±0.5°)	Down	5.5° (±0.5°)
Stabilator Tab	Up	5° (±1°)	Down	8° (±1°)
Nose Wheel Travel	Left	24° (±2°)	Right	24° (±2°)

Manufacturer's Serial Nos.
32-8006002 through 32-8606023, and 3206001 through 3206019, 3206042 through 3206044, 3206047, 3206050 through 3206055, and 3206060. The manufacturer is authorized to issue airworthiness certificates for airplane serial numbers 32-8006002 through 32-8606023, and 3206001 through 3206019 under the delegation option provisions of FAR 21 (See NOTE 7).

XI. - Model PA-32-301T (Turbo Saratoga), 7 PCLM (Normal Category), Approved January 9, 1980.
Same as Model PA-32-300 except for tapered wings, turbocharged powerplant, increased gross weight, and other minor changes.

Engine	Lycoming TIO-540-S1AD Bendix Injector Type RSA-10ED2 Flow Setting No. 2524791
Fuel	100 or 100LL aviation grade fuel
Engine Limits	For airplanes equipped with standard Hartzell 2 blade propeller HC-E2YR-1( )F/F8477-4: For 5 minute takeoff, 2700 r.p.m. and 36.0" Hg MAP (300 hp) - Sea level to 16,000 ft. altitude For maximum continuous operation, 2575 r.p.m. and 36.0" Hg MAP (294 rated hp) - Sea level to 16,000 ft. altitude For airplanes equipped with optional Hartzell 3 blade propeller HC-E3YR-1( )F/F7673DR-0: For all operations, 2700 r.p.m. and 36.0" Hg MAP (300 rated hp) - Sea level to 16,000 ft
Propeller and Propeller Limits	Hartzell constant speed Model HC-E2YR-1( )F/F8477-4 (standard 2 blade): Pitch: High 34° ± 1°, Low 15.6° ± .2° at 30 in. station Diameter: Not over 80 in., not under 78.5 in. Governor Assembly: Hartzell F-4-11B or F-4-11B( ) Spinner: Piper P/N 98708-2 or Hartzell P/N A-2298-2 Hartzell constant speed Model HC-E3YR-1( )F/F7673R-0 (optional 3 blade): Pitch: High 34.5° ± 1°, Low 13.2° ± .2° at 30 in. station Diameter: Not over 78 in., not under 76 in. Governor Assembly: Hartzell F-4-11B or F-4-11B( ) Spinner: Piper P/N PS50077-58 or Hartzell P/N C-3575

Airspeed Limits (Indicated)			
	Never exceed	197 knots	(226 m.p.h.)
	Maximum structural cruise (with 3600 lb. gross weight)	154 knots	(177 m.p.h.)
	Maneuvering	134 knots	(154 m.p.h.)
	Maximum flaps extended	112 knots	(129 m.p.h.)

XI. - Model PA-32-301T (cont'd)

C.G. Range (gear extended) — (+90.0) to (+95.0) at 3600 lb.
(+83.5) to (+95.0) at 3200 lb.
(+78.0) to (+95.0) at 2400 lb.
Straight line variation between points given.

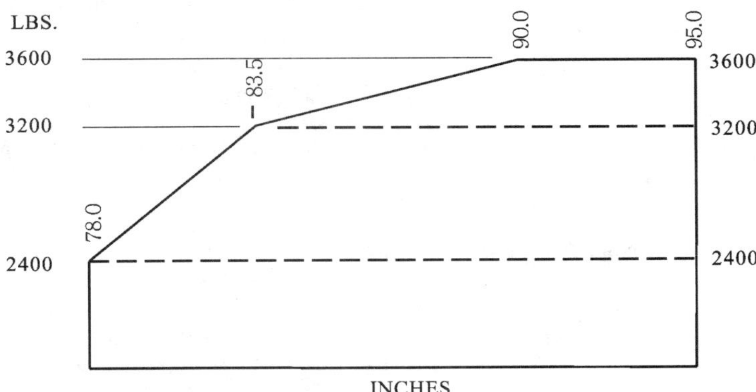

Empty Weight C.G. Range — None

Maximum Weight — Ramp: 3617 lb.
Takeoff: 3600 lb.
Landing: 3600 lb.

No. of Seats — 6 (2 at +85.5, 2 at +118.1, 2 at +157.6)
7 (2 at +85.5, 3 at +118.1, 2 at +157.6)
6 (2 at +85.5, *2 at +119.1, 2 at +157.6) (See NOTE 11)
* - Optional Club Seats

Maximum Baggage — 200 lb. (100 lb. at +42.0, 100 lb. at +178.7)

Fuel Capacity — 107 gallons at +94.0 (2 wing tanks) (102 gallons usable)
See NOTE 1 for data on system fuel

Oil Capacity — 12 qt. at +16.6 (9-1/4 qt. usable)
See NOTE 1 for data on system oil

Control Surface Movements

Wing Flaps	Up	0° (±1°)	Down	40° (±2°)
Ailerons	Up	28° (±1°)	Down	22° (±1°)
Rudder	Left	28° (±1°)	Right	28° (±1°)
Stabilator	Up	14.5° (±0.5°)	Down	5.5° (±0.5°)
Stabilator Tab	Up	5° (±1°)	Down	8° (±1°)
Nose Wheel Travel	Left	24° (±2°)	Right	24° (±2°)

Manufacturer's Serial Nos. — 32-8024001 and 32-8424002. The manufacturer is authorized to issue airworthiness certificates for airplane serial numbers 32-8024001 through 32-8424002 the delegation option provisions of FAR 21 (See NOTE 7).

XII. - Model PA-32R-301T (Saratoga II TC), 6 PCLM (Normal Category), Approved July 9, 1997.
Same as Model PA-32R-301T, Turbo Saratoga SP, except for new turbocharged powerplant, 28 Volt electrical system and other minor changes.

Engine	Lycoming TIO-540-AH1A Precision Airmotive Injector, Type RSA-10ED1 Flow Setting No. 2576554-2
Fuel	100 or 100LL aviation grade fuel
Engine Limits	For all operations, 2500 r.p.m. and 38.0" Hg MAP (300 rated hp) - Sea level to 12,000 ft. altitude Do not operate above 26.0" Hg MAP below 2100 r.p.m.
Propeller and Propeller Limits	Hartzell constant speed Model HC-I3YR-1RF/F7663DR (3 blade) Hartzell constant speed Model HC-I3YR-1RF/F7663DRB (3 blade with TKS Ice Protection System) Pitch: High 34.0° ± 0.5°, Low 15.2° ± 0.2° at 30 in. station Diameter: Not over 78 in., not under 76 in. Governor: Hartzell V-5-6 Spinner Assy: Piper P/N PS50077-90 or Hartzell P/N C-3575-1 (P) Dome: Hartzell P/N 3532-16P (with TKS Ice Protection System)
Airspeed Limits (Indicated)	Never exceed 191 knots Maximum structural cruise 167 knots Maneuvering 134 knots (with 3600 lb. gross weight) Flaps extended 110 knots Maximum gear extension 132 knots Maximum gear retraction 110 knots Maximum gear extended 132 knots
C.G. Range (gear extended)	(+91.4) to (+95.0) at 3600 lb. (+83.5) to (+95.0) at 3200 lb. (+78.0) to (+95.0) at 2400 lb. Straight line variation between points given.

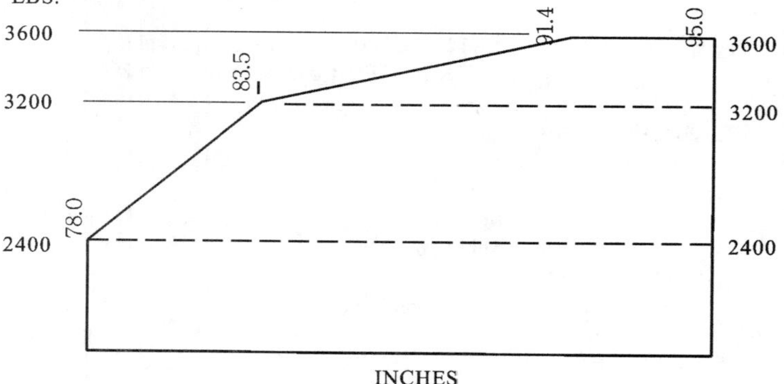

Empty Weight C.G. Range	None
Maximum Weight	Ramp: 3615 lb. Takeoff: 3600 lb. Landing: 3600 lb.
No. of Seats	6 (2 at +85.5, 2 at +119.1, 2 at +157.6) 5 (2 at +85.5, 1 at +119.1, 2 at +157.6)
Maximum Baggage	200 lb. (100 lb. at +42.0, 100 lb. at +178.7)

XII. - Model PA-32R-301T (cont'd)

Fuel Capacity	107 gallons at +94.0 (2 wing tanks) (102 gallons usable) See NOTE 1 for data on fuel system
Oil Capacity	12 qt. at +16.6 (9-1/4 qt. usable) See NOTE 1 for data on oil system

Control Surface Movements

Wing Flaps	Up	0° (±1°)	Down	40° (±2°)
Ailerons	Up	28° (±1°)	Down	22° (±1°)
Rudder	Left	28° (±1°)	Right	28° (±1°)
Stabilator	Up	14.5° (±.5°)	Down	5.5° (±.5°)
Stabilator Tab	Up	5° (±1°)	Down	8° (±1°)

Manufacturer's Serial Nos. — 3257001 and up. The manufacturer is authorized to issue airworthiness certificates for serial numbers 3257001 and up under the delegation option provisions of FAR 21.

XIII - Model PA-32-301FT (Piper 6X), 6 PCLM (Normal Category), Approved July 22, 2003.

Similar to Model PA-32R-301, Saratoga IIHP, except for fixed landing gear and other minor changes.

Engine — Lycoming IO-540-K1G5
Precision Airmotive Injector, Type RSA-10ED1
Flow Setting No. 2524273

Fuel — 100 or 100LL aviation grade fuel

Engine Limits — Equipped with Hartzell 3 blade propeller HC-I3YR-1RF/F7663DR:
For all operations, 2700 r.p.m. and full throttle (300 rated hp)

Propeller and Propeller Limits — Hartzell constant speed Model HC-I3YR-1RF/F7663DR (3 blade)
Hartzell constant speed Model HC-I3YR-1RF/F7663DRB
(3 blade with TKS Ice Protection System)
Pitch: High 32.0° ± 1°, Low 12.4° ± .2° at 30 in. station
Diameter: Not over 78 in., not under 77 in.
Governor: Hartzell V-5-4
Spinner Assy: Hartzell P/N C3575-1 (P)
Dome: Hartzell P/N C-3532-16P (with TKS Ice Protection System)
Do not exceed 23" manifold pressure below 2100 r.p.m.

Airspeed Limits (Indicated)

For serial number 3232001 and up:

Never exceed	189 knots (218 m.p.h.)
Maximum structural cruise	150 knots (173 m.p.h.)
Maneuvering (with 3600 lb. gross weight)	132 knots (152 m.p.h.)
Flaps extended	113 knots (130 m.p.h.)

XIII. - Model PA-32-301FT (cont'd)

C.G. Range

(+90.0) to (+95.0) at 3600 lb.
(+83.5) to (+95.0) at 3200 lb.
(+78.0) to (+95.0) at 2400 lb.
Straight line variation between points given.

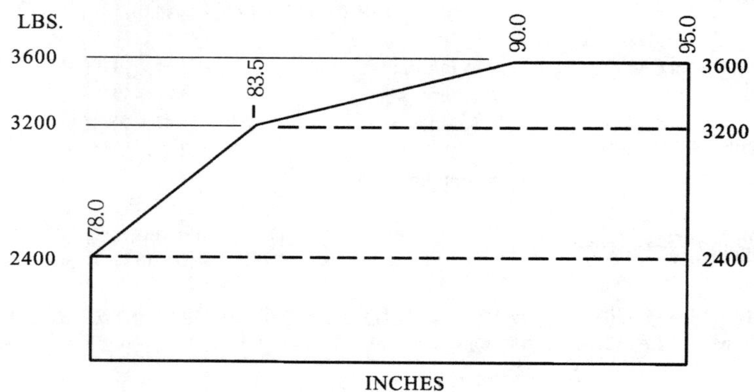

Empty Weight C.G. Range	None
Maximum Weight	Ramp: 3615 lb. Takeoff: 3600 lb. Landing: 3600 lb.
No. of Seats	6 (2 at +85.5, 2 at +119.1, 2 at +157.6)
Maximum Baggage	200 lb. (100 lb. at +42.0, 100 lb. at +178.7)
Fuel Capacity	107 gallons at +94.0 (2 wing tanks) (102 gallons usable) See NOTE 1 for data on fuel system
Oil Capacity	12 qt. at +16.6 (9-1/4 qt. usable) See NOTE 1 for data on oil system

Control Surface Movements

Wing Flaps	Up	0° (±1°)	Down	40° (±2°)
Ailerons	Up	28° (±1°)	Down	22° (±1°)
Rudder	Left	28° (±1°)	Right	28° (±1°)
Stabilator	Up	14.5° (±0.5°)	Down	5.5° (±0.5°)
Stabilator Tab	Up	5° (±1°)	Down	8° (±1°)
Nose Wheel Travel	Left	24° (±2°)	Right	24° (±2°)

Manufacturer's Serial Nos. 3232001 and up. The manufacturer is authorized to issue airworthiness certificates for serial numbers 3232001 and up under the delegation option provisions of FAR 21.

XIV. - Model PA-32-301XTC (Piper 6XT), 6 PCLM (Normal Category), Approved August 28, 2003.
Similar to Model PA-32R-301T, Saratoga IITC, except for fixed landing gear and other minor changes.

Engine
Lycoming TIO-540-AH1A
Precision Airmotive Injector, Type RSA-10ED1
Flow Setting No. 2576554-2

Fuel
100 or 100LL aviation grade fuel

Engine Limits
For all operations, 2500 r.p.m. and 38.0" Hg MAP (300 rated hp) - Sea level to 12,000 ft. altitude
Do not operate above 26.0" Hg MAP below 2100 r.p.m.

Propeller and Propeller Limits
Hartzell constant speed Model HC-I3YR-1RF/F7663DR (3 blade)
Hartzell constant speed Model HC-I3YR-1RF/F7663DRB
(3 blade with TKS Ice Protection System)
Pitch: High 34.0° ± 0.5°, Low 15.2° ± 0.2° at 30 in. station
Diameter: Not over 78 in., not under 76 in.
Governor: Hartzell V-5-6
Spinner Assy: Piper P/N PS50077-90 or Hartzell P/N C-3575-1 (P)
Dome: Hartzell P/N C-3532-16P (with TKS Ice Protection System)

Airspeed Limits (Indicated)

Never exceed	189 knots (218 m.p.h.)
Maximum structural cruise	150 knots (173 m.p.h.)
Maneuvering (with 3600 lb. gross weight)	132 knots (152 m.p.h.)
Flaps extended	113 knots (130 m.p.h.)

C.G. Range
(+90.0) to (+95.0) at 3600 lb.
(+83.5) to (+95.0) at 3200 lb.
(+78.0) to (+95.0) at 2400 lb.
Straight line variation between points given.

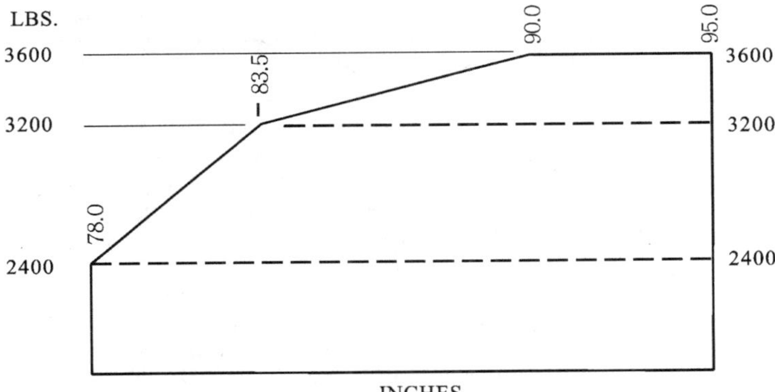

Empty Weight C.G. Range
None

Maximum Weight
Ramp: 3615 lb.
Takeoff: 3600 lb.
Landing: 3600 lb.

No. of Seats
6 (2 at +85.5, 2 at +119.1, 2 at +157.6)

Maximum Baggage
200 lb. (100 lb. at +42.0, 100 lb. at +178.7)

Fuel Capacity
107 gallons at +94.0 (2 wing tanks) (102 gallons usable)
See NOTE 1 for data on fuel system

Oil Capacity	12 qt. at +16.6 (9-1/4 qt. usable) See NOTE 1 for data on oil system				
Control Surface Movements	Wing Flaps	Up	0° (±1°)	Down	40° (±2°)
	Ailerons	Up	28° (±1°)	Down	22° (±1°)
	Rudder	Left	28° (±1°)	Right	28° (±1°)
	Stabilator	Up	14.5° (±0.5°)	Down	5.5° (±0.5°)
	Stabilator Tab	Up	5° (±1°)	Down	8° (±1°)
	Nose Wheel Travel	Left	24° (±2°)	Right	24° (±2°)

Manufacturer's Serial Nos. — 3255001 and up. The manufacturer is authorized to issue airworthiness certificates for serial numbers 3255001 and up under the delegation option provisions of FAR 21.

Data Pertinent to All Models

Datum — 78.4" forward of wing leading edge

Leveling Means — Two screws left side fuselage below window

Certification Basis — Type Certificate No. A3SO issued March 4, 1965.
Date of application for Type Certificate, February 20, 1964.
Delegation Option Authorization per FAR 21, Subpart J, granted July 17, 1968.

PA-32-260, PA-32S-300, and PA-32-300 (S/N 32-15 through 32-7840202): CAR 3, effective May 15, 1956, through Amendment 3-8, effective December 18, 1962.

PA-32-300, S/N 32-7940001 through 32-7940290: CAR 3, effective May 15, 1956, through Amendment 3-8, effective December 18, 1962. In addition, FAR 23.221 and 23.959 as amended by Amendment 23-7, effective September 14, 1969; FAR 23.1327 and 23.1547 as amended by Amendment 23-20, effective September 1, 1977; and FAR 23.1581(b)(2) as amended by Amendment 23-21, effective March 1, 1978. Equivalent Safety Finding for CAR 3.757.

PA-32R-300: CAR 3, effective May 15, 1956, through Amendment 3-8, effective December 18, 1962. In addition, FAR 23.221 and 23.959 as amended by Amendment 23-7, effective September 14, 1969; FAR 23.967(e)(2) as amended by Amendment 23-14, effective December 20, 1973; and FAR 23.1327 and 23.1547 as amended by Amendment 23-20, effective September 1, 1977.

PA-32RT-300: CAR 3, effective May 15, 1956, through Amendment 3-8, effective December 18, 1962. In addition, FAR 23.221, 23.959, and 23.1091 as amended by Amendment 23-7, effective September 14, 1969; FAR 23.427 and 23.967(e)(2) as amended by Amendment 23-14, effective December 20, 1973; FAR 23.1093 as amended by Amendment 23-15, effective October 31, 1974; FAR 23.1327 and 23.1547 as amended by Amendment 23-20, effective September 1, 1977; and FAR Part 36, through Amendment 36-7, effective October 1, 1977. Equivalent Safety Finding for CAR 3.757.

PA-32RT-300T: CAR 3, effective May 15, 1956, through Amendment 3-8, effective December 18, 1962. In addition, FAR 23.221, 23.901, 23.909, 23.959, 23.1041, 23.1043, 23.1047, 23.1091, 23.1143, and 23.1527 as amended by Amendment 23-7, effective September 14, 1969; FAR 23.427 and 23.967(e)(2) as amended by Amendment 23-14, effective December 20, 1973; FAR 23.1093 and 23.1305 as amended by Amendment 23-15, effective October 31, 1974; FAR 23.1327 and 23.1547 as amended by Amendment 23-20, effective September 1, 1977; FAR 23.1581(b)(2) as amended by Amendment 23-21 effective March 1, 1978; and FAR Part 36 through Amendment 36-7, effective October 1, 1977. Equivalent Safety Finding for CAR 3.757, 3.84 and 3.86.

PA-32R-301, S/N 32R-8013001 through 32R-8613006, 3213001 through 3213028, and 3213030 through 3213041: CAR 3, effective May 15, 1956, through Amendment 3-8, effective December 18, 1962. In addition, FAR 23.207, 23.221, 23.959, and 23.1091 as amended by Amendment 23-7, effective September 14, 1969; FAR 23.201, 23.203, and 23.967(e)(2) as amended by Amendment 23-14, effective December 20, 1973; FAR 23.1093 and 23.1557(c)(1) as amended by Amendment 23-18, effective May 2, 1977; FAR 23.1327 and 23.1547 as amended by Amendment 23-20, effective September 1, 1977; FAR 23.1581(b)(2) as amended by Amendment 23-21, effective March 1, 1978; and FAR 36 through Amendment 36-9, effective January 15, 1979. Equivalent Safety Finding for CAR 3.757 and 3.777.

PA-32R-301, S/N 3213029, 3213042 through 3213103, and 3246001 through 3246087: CAR 3, effective May 15, 1956, through Amendment 3-8, effective December 18, 1962. In addition, FAR 23.207, 23.221, 23.959, and 23.1091 as amended by Amendment 23-7, effective September 14, 1969; FAR 23.201, 23.203, and 23.967(e)(2) as amended by Amendment 23-14, effective December 20, 1973; FAR 23.1093 and 23.1557(c)(1) as amended by Amendment 23-18, effective May 2, 1977; FAR 23.1327 and 23.1547 as amended by Amendment 23-20, effective September 1, 1977; FAR 23.1581(b)(2) as amended by Amendment 23-21, effective March 1, 1978; and FAR 36, Appendix G, through Amendment 36-16, effective December 22, 1988. Equivalent Safety Finding for CAR 3.757 and 3.777.

PA-32R-301, S/N 3246088 and up: CAR 3, effective May 15, 1956, through Amendment 3-8, effective December 18, 1962. In addition, FAR 23.207, 23.221, 23.959, and 23.1091 as amended by Amendment 23-7, effective September 14, 1969; FAR 23.201, 23.203, and 23.967(e)(2) as amended by Amendment 23-14, effective December 20, 1973; FAR 23.1093 as amended by Amendment 23-18, effective May 2, 1977; FAR 23.1327 and 23.1547 as amended by Amendment 23-20, effective September 1, 1977; FAR 23.1581(b)(2) as amended by Amendment 23-21, effective March 1, 1978; FAR 23.1545 as amended by Amendment 23-23, effective December 1, 1978; FAR 23.1529 as amended by Amendment 23-26, effective October 14, 1980; FAR 23.1557(c)(1) as amended by Amendment 23-45, effective September 7, 1993; FAR 23.561(b)(3) as amended by Amendment 23-51, effective March 11, 1996; FAR 23.1305 as amended by Amendment 23-52, effective July 25, 1996; and FAR 36 through Amendment 36-16, effective December 22, 1988.

For aircraft equipped with Piper factory installed optional Avidyne Entegra system and Mid-Continent Model 4300-411 Electric Attitude Indicator, the additional certification basis for installation specific items only is: 14 CFR Part 23 regulations FAR 23.301, 23.337, 23.341, 23.561, 23.607, 23.611, as amended by Amdt. 23-48; FAR 23.303, 23.307, 23.601, 23.609, 23.1367, 23.1381 issued on 02/01/65; FAR 23.305, 23.613, 23.773, 23.1525, 23.1549 as amended by Amdt. 23-45; FAR 23.603, 23.605 as amended by Amdt. 23-23; FAR 23.777, 23.1191, 23.1337 as amended by Amdt. 23-51; FAR 23.1301, 23.1327, 23.1335 as amended by Amdt. 23-20; FAR 23.853, 23.867, 23.1303, 23.1307, 23.1309, 23.1311, 23.1321, 23.1323, 23.1329, 23.1351, 23.1353, 23.1359, 23.1361, 23.1365, 23.1431 as amended by Amdt. 23-49; FAR 23.1305 as amended by Amdt. 23-52; FAR 23.1322, 23.1331, 23.1357 as amended by Amdt. 23-43; FAR 23.1325, 23.1543, 23.1545, 23.1555, 23.1563, 23.1581, 23.1583, 23.1585 as amended by Amdt. 23-50; FAR 23. 771 as amended by Amdt. 23-14; FAR 23.1501, 23.1541 as amended by Amdt. 23-21; FAR 23.1523 as amended by Amdt. 23-34; FAR 23.1529 as amended by Amdt. 23-26; Special Condition 23-147-SC for HIRF (Docket No. CE207), dated July 16, 2004.

PA-32R-301T, S/N 32R-8029001 through 32R-8629008, and 3229001 through 3229003: CAR 3, effective May 15, 1956, through Amendment 3-8, effective December 18, 1962. In addition, FAR 23.965 of FAR 23 effective February 1, 1965; FAR 23.207, 23.221, 23.901, 23.909, 23.959, 23.1041, 23.1043, 23.1047, 23.1091, and 23.1527 as amended by Amendment 23-7, effective September 14, 1969; FAR 23.201, 23.203, and

23.967(e)(2) as amended by Amendment 23-14, effective December 20, 1973; FAR 23.1305 as amended by Amendment 23-15, effective October 31, 1974; FAR 23.1093 and 23.1557(c)(1) as amended by Amendment 23-18, effective May 2, 1977; FAR 23.1327 and 23.1547 as amended by Amendment 23-20, effective September 1, 1977; FAR 23.1581(b)(2) as amended by Amendment 23-21, effective March 1, 1978; and FAR 36 through Amendment 36-9, effective January 15, 1979. Equivalent Safety Finding for CAR 3.757 and 3.777. Compliance with FAR 23.1441 as amended by Amendment 23-9, effective June 17, 1970, will be shown with optional supplemental oxygen.

PA-32-301: CAR 3, effective May 15, 1956, through Amendment 3-8, effective December 18, 1962. In addition, FAR 23.965 of FAR 23, effective February 1, 1965; FAR 23.207, 23.221, 23.959, and 23.1091 as amended by Amendment 23-7, effective September 14, 1969; FAR 23.201, 23.203, and 23.967(e)(2) as amended by Amendment 23-14, effective December 20, 1973; FAR 23.1093 and 23.1557(c)(1) as amended by Amendment 23-18, effective May 2, 1977; FAR 23.1327 and 23.1547 as amended by Amendment 23-20, effective September 1, 1977; FAR 23.1581(b)(2) as amended by Amendment 23-21, effective March 1, 1978; FAR 23.1545 as amended by Amendment 23-23, effective December 1, 1978; and FAR 36 through Amendment 36-9, effective January 15, 1979.

PA-32-301T: CAR 3, effective May 15, 1956, through Amendment 3-8, effective December 18, 1962. In addition, FAR 23.965 of FAR 23, effective February 1, 1965; FAR 23.207, 23.221, 23.901, 23.959, 23.1041, 23.1043, 23.1047, 23.1091, 23.1143, and 23.1527 as amended by Amendment 23-7, effective September 14, 1969; FAR 23.201 and 23.967(e)(2) as amended by Amendment 23-14, effective December 20, 1973; FAR 23.1305 as amended by Amendment 23-15, effective October 31, 1974; FAR 23.1093 and 23.1557(c)(1) as amended by Amendment 23-18, effective May 2, 1977; FAR 23.1327 and 23.1547 as amended by Amendment 23-20, effective September 1, 1977; FAR 23.1581(b)(2) as amended by Amendment 23-21, effective March 1, 1978; FAR 23.1545 as amended by Amendment 23-23, effective December 1, 1978; and FAR 36 through Amendment 36-9, effective January 15, 1979. Compliance with FAR 23.1441as amended by Amendment 23-9, effective June 17, 1970, will be shown with optional supplemental oxygen.

PA-32R-301T, S/N 3257001 and up: CAR 3, effective May 15, 1956, through Amendment 3-8, effective December 18, 1962. In addition, FAR 23.965 of FAR 23, effective February 1, 1965; FAR 23.207, 23.221, 23.901, 23.909, 23.959, 23.1041, 23.1043, 23.1047, 23.1091, and 23.1527 as amended by Amendment 23-7, effective September 14, 1969; FAR 23.201, 23.203, and 23.967(e)(2) as amended by Amendment 23-14, effective December 20, 1973; FAR 23.1093 as amended by Amendment 23-18, effective May 2, 1977; FAR 23.1327 and 23.1547 as amended by Amendment 23-20, effective September 1, 1977; FAR 23.1581 as amended by Amendment 23-21, effective March 1, 1978; FAR 23.1545 as amended by Amendment 23-23, effective December 1, 1978; FAR 23.1529 as amended by Amendment 23-26, effective October 14, 1980; FAR 23.1557(c)(1) as amended by Amendment 23-45, effective September 7, 1993; FAR 23.1305 as amended by Amendment 23-52, effective July 25, 1996; and FAR 36 through Amendment 36-16, effective December 22, 1988. Compliance with FAR 23.1441 as amended by Amendment 23-9, effective June 17, 1970, has been shown with optional supplemetnal oxygen.

For aircraft equipped with Piper factory installed optional Avidyne Entegra system and Mid-Continent Model 4300-411 Electric Attitude Indicator, the additional certification basis for installation specific items only is: 14 CFR Part 23 regulations FAR 23.301, 23.337, 23.341, 23.561, 23.607, 23.611, as amended by Amdt. 23-48; FAR 23.303, 23.307, 23.601, 23.609, 23.1367, 23.1381 issued on 02/01/65; FAR 23.305, 23.613, 23.773, 23.1525, 23.1549 as amended by Amdt. 23-45; FAR 23.603, 23.605 as amended by Amdt. 23-23; FAR 23.777, 23.1191, 23.1337 as amended by Amdt. 23-51; FAR 23.1301, 23.1327, 23.1335 as amended by Amdt. 23-20; FAR 23.853, 23.867, 23.1303, 23.1307, 23.1309, 23.1311, 23.1321, 23.1323, 23.1329, 23.1351, 23.1353,

23.1359, 23.1361, 23.1365, 23.1431 as amended by Amdt. 23-49; FAR 23.1305 as amended by Amdt. 23-52; FAR 23.1322, 23.1331, 23.1357 as amended by Amdt. 23-43; FAR 23.1325, 23.1543, 23.1545, 23.1555, 23.1563, 23.1581, 23.1583, 23.1585 as amended by Amdt. 23-50; FAR 23. 771 as amended by Amdt. 23-14; FAR 23.1501, 23.1541 as amended by Amdt. 23-21; FAR 23.1523 as amended by Amdt. 23-34; FAR 23.1529 as amended by Amdt. 23-26; Special Condition 23-147-SC for HIRF (Docket No. CE207), dated July 16, 2004.

PA-32-301FT, S/N 3232001 and up and PA-32-301 XTC, S/N 3255001 and up: CAR 3, effective May 15, 1956, through Amendment 3-8, effective December 18, 1962. In addition, FAR 23.965 of FAR 23, effective February 1, 1965; FAR 23.207, 23.221, 23.901, 23.909, 23.959, 23.1091, and 23.1527 as amended by Amendment 23-7, effective September 14, 1969; FAR 23.201, 23.203, and 23.967(e)(2) as amended by Amendment 23-14, effective December 20, 1973; FAR 23.1093 as amended by Amendment 23-18, effective May 2, 1977; FAR 23.1327 and 23.1547 as amended by Amendment 23-20, effective September 1, 1977; FAR 23.1581 as amended by Amendment 23-21, effective March 1, 1978; FAR 23.1545 as amended by Amendment 23-23, effective December 1, 1978; FAR 23.1529 as amended by Amendment 23-26, effective October 14, 1980; FAR 23.853(a) and (c)(1) as amended by Amendment 23-34, effective January 15, 1987; FAR 23.1309 as amended by Amendment 23-41 for the communication and navigation LRUs only; FAR 23.1557(c)(1) as amended by Amendment 23-45, effective September 7, 1993; FAR 23.561(b)(3) as amended by Amendment 23-48, effective March 11, 1996; FAR 23.1041, 23.1043, and 23.1047 as amended by Amendment 23-51, effective March 11, 1996; FAR 23.1305 as amended by Amendment 23-52, effective July 25, 1996; and FAR 36 through the latest Amendment at the time of certification. Compliance with FAR 23.1441 as amended by Amendment 23-9, effective June 17, 1970, has been show with supplemental oxygen for the PA-32-301XTC only.

For aircraft equipped with Piper factory installed S-Tec system 55X autopilot installations, the additional certification basis for installation specific items only is: 14 CFR Part 23 regulations FAR 23.609, 23.627 issued on 02/01/65; FAR 23.611, 23.619, 23.625 as amended by Amdt. 23-7 Eff. 09/14/69; FAR 23.603 as amended by Amdt. 23-23, Eff. 12/01/78; FAR 23.1309 as amended by 23-41 Eff. 11/26/90; FAR 23.572(a)(1), 23.613(a)(b)(d) as amended by Amdt. 23-45, Eff. 09/07/93; FAR 23.561(b)(3)(e) as amended by Amdt. 23-48, Eff. 03/11/96; FAR 23.1329 as amended by Amdt. 23-49 Eff. 02/09/96.

For aircraft equipped with Piper factory installed optional Avidyne Entegra system and Mid-Continent Model 4300-411 Electric Attitude Indicator, the additional certification basis for installation specific items only is: 14 CFR Part 23 regulations FAR 23.301, 23.337, 23.341, 23.561, 23.607, 23.611, as amended by Amdt. 23-48; FAR 23.303, 23.307, 23.601, 23.609, 23.1367, 23.1381 issued on 02/01/65; FAR 23.305, 23.613, 23.773, 23.1525, 23.1549 as amended by Amdt. 23-45; FAR 23.603, 23.605 as amended by Amdt. 23-23; FAR 23.777, 23.1191, 23.1337 as amended by Amdt. 23-51; FAR 23.1301, 23.1327, 23.1335 as amended by Amdt. 23-20; FAR 23.853, 23.867, 23.1303, 23.1307, 23.1309, 23.1311, 23.1321, 23.1323, 23.1329, 23.1351, 23.1353, 23.1359, 23.1361, 23.1365, 23.1431 as amended by Amdt. 23-49; FAR 23.1305 as amended by Amdt. 23-52; FAR 23.1322, 23.1331, 23.1357 as amended by Amdt. 23-43; FAR 23.1325, 23.1543, 23.1545, 23.1555, 23.1563, 23.1581, 23.1583, 23.1585 as amended by Amdt. 23-50; FAR 23. 771 as amended by Amdt. 23-14; FAR 23.1501, 23.1541 as amended by Amdt. 23-21; FAR

23.1523 as amended by Amdt. 23-34; FAR 23.1529 as amended by Amdt. 23-26; Special Condition 23-147-SC for HIRF (Docket No. CE207), dated July 16, 2004.

Production basis

Production Certificate No. 206. The manufacturer is authorized to issue airworthiness certificates under the delegation option provisions of FAR 21.

Equipment

The basic required equipment as prescribed in the applicable airworthiness regulations (see Certification Basis) must be installed in the aircraft for certification.
In addition, the following documents are required:

MODEL	AFM/POH	REPORT NO.	APPROVED	S/N EFFECTIVITY
PA-32-260	AFM	VB-152	3- 4-65	32-1 through 32-1110
	AFM	VB-156	12-17-68	32-1111 through 32-1297, and 32-7100001 through 32-7200045
	AFM Supp.	VB-357	8-25-71	32-1 through 32-1297, and 32-7100001 through 32-7100027
	AFM	VB-478	9- 1-72	32-7300001 through 32-7300065
	AFM	VB-561	5-14-73	32-7400001 through 32-7600024
	POH	VB-820	8-18-76	32-7700001 through 32-7800008
PA-32-300	AFM	VB-154	5-27-66	32-40000 through 32-40565
	AFM	VB-158	12-17-68	32-40566 through 32-40974, and 32-7140001 through 32-7240055
	AFM Supp.	VB-357	8-25-71	32-40000 through 32-40974, and 32-7140001 through 32-7240001
	AFM	VB-393	1-20-72	32-7240056 through 32-7340191
	AFM	VB-562	5-14-73	32-7440001 through 32-7640130
	POH	VB-830	8-19-76	32-7740001 through 32-7840202
	POH	VB-830, Rev. 4	9-21-78	32-7940001 through 32-7940290
PA-32R-300	POH	VB-750	8- 1-75	32R-7680001 through 32R-7680525
	POH	VB-840	8-20-76	32R-7780001 through 32R-7880066
PA-32S-300	AFM	VB-184	2-14-67	32S-40001 through 32S-40565
	AFM	VB-186	12-17-68	32S-40566 through 32S-40974, and 32S-7140001 through 32S-7240137
	AFM Supp.	VB-357	8-25-71	32S-40001 through 32S-40974, and 32S-7140001 through 32S-7240137
PA-32RT-300	POH/AFM	VB-890	12-13-77	32R-7885002 through 32-7985106
PA-32RT-300T	POH/AFM	VB-900	5-1-78	32R-7787001, and 32R-7887002 through 32R-7987126
PA-32R-301	POH/AFM	VB-1080	11-8-79	32R-8013001 through 32R-8613006, 3213001 through 3213028, and 3213030 through 3213041
	POH/AFM	VB-1551	5-31-93	3213029, and 3213042 through 3213103
	POH/AFM	VB-1614	7-12-95	3246001 through 3246017
	POH/AFM	VB-1600	11-30-95	3246018 through 3246087
	POH/AFM	VB-1669	6-30-97	3246088 and up
PA-32R-301T	POH/AFM	VB-1090	11-8-79	32R-8029001 through 32R-8629008, and 3229001 through 3229003
	POH/AFM	VB-1647	6-30-97	3257001 and up
PA-32-301	POH/AFM	VB-1060	1-9-80	32-8006002 through 32-8606023, and 3206001 through 3206019

PA-32-301T	POH/AFM	VB-1070	1-9-80	32-8024001 through 32-8424002
PA-32-301FT	POH/AFM	VB-1850	7-22-2003	3232001 and up
PA-32-301XTC	POH/AFM	VB-1881	8-26-2003	3255001 and up

NOTE 1 Current weight and balance report, including list of equipment included in certificated empty weight, and loading instructions when necessary, must be provided for each aircraft at the time of original certification. The certificated empty weight and corresponding center of gravity locations must include undrainable system oil (not included in oil capacity) and unusable fuel as noted below:

Models PA-32-260 and PA-32-300 (S/N 32-40000 through 32-40974, and 32-7140001 through 32-7840202):
Fuel 2.3 lb. at +103.0
Models PA-32R-300, PA-32RT-300, PA-32RT-300T and PA-32-300 (S/N 32-7940001 through 32-7940290):
Fuel 24.0 lb. at +103.0
Models PA-32R-301, PA-32R-301T, PA-32-301, and PA-32-301T:
Fuel 30.0 lb. at +95.2
Model PA-32-260:
Oil 2.4 lb. at +23.0
Models PA-32-300, PA-32R-300, PA-32RT-300T, PA-32R-301, PA-32R-301T, PA-32-301 and PA-32-301T:
Oil 3.0 lb. at +23.0

NOTE 2 All placards required in the Approved Airplane Flight Manual or "Pilot's Operating Handbook and Approved Airplane Flight Manual" and Approved A.F.M. Supplements, plus the following placards, must be displayed in full view of the pilot, in the appropriate location.

(a) "THIS AIRPLANE MUST BE OPERATED AS A NORMAL CATEGORY AIRPLANE IN COMPLIANCE WITH THE OPERATING LIMITATIONS STATED IN THE FORM OF PLACARDS, MARKINGS, AND MANUALS. NO ACROBATIC MANEUVERS, INCLUDING SPINS, APPROVED."

(b) "THIS AIRCRAFT APPROVED FOR VFR, IFR, DAY AND NIGHT NON-ICING FLIGHT WHEN EQUIPPED IN ACCORDANCE WITH FAR 91 OR FAR 135."

NOTE 3 The Models PA-32-260, PA-32-300, and PA-32S-300, 6 PCLM, may be converted to the 7 place (7 PCLM) configuration by the installation of Piper Kit No. 69072-3. All weight in excess of 3112 lb. must be fuel weight only. This restriction does not apply to PA-32-300 aircraft, S/N 32-7940001 through 32-7940290.

NOTE 4 When the Model PA-32S-300 is operated in a landplane configuration, use the PA-32-300 C.G. envelope with the corresponding airplane serial number (last five digits).

NOTE 5 The Model PA-32-260, S/N 32-1 through 32-1297, and 32-7100001 through 32-7700023, and Model PA-32-300, S/N 32-40001 through 32-40974, and 32-7140001 through 32-7740113, require two nose wheel centering springs (P/N 67168) installed, if the optional nose wheel fairing or the optional nose and main wheel fairings are removed or not installed.

The Model PA-32-260, S/N 32-7800001 through 32-7800008, and Model PA-32-300, S/N 32-7840001 through 32-7940290, require rudder centering spring (P/N 37929-2) installed, if the optional nose wheel fairing or the optional nose and main wheel fairings are removed or not installed.

The Model PA-32-260, S/N 32-7800001 through 32-7800008, requires the removal of the nose gear strut fairing (P/N 37891) when the nose gear wheel fairing is removed or not installed.

NOTE 6 Models PA-32-260, PA-32-300, PA-32S-300, and PA-32R-301 (S/N 32R-8013001 through 32R-8613006, 3213001 through 3213028, and 3213030 through 3213041) may be operated with the spinner dome removed or with the spinner dome and rear bulkhead removed. Models PA-32R-300, PA-32RT-300 and PA-32-301 may be operated with spinner dome and front bulkhead removed.

NOTE 7 The following serial numbered aircraft are not eligible for import certification to the U.S.:

PA-32-300:
32-40491, 32-40503, 32-40518, 32-40532, 32-40533, 32-40544, 32-40545, 32-40965, 32-40966, 32-40968 through 32-40974, 32-7240120, 32-7240123, 32-7240126, 32-7240129, 32-7240132, 32-7340133, 32-7340155, 32-7340159, 32-7340160, 32-7340172, 32-7440144, 32-7540114, 32-7540136, 32-7640127, 32-7740100,

32-7840028, 32-7940141, and 32-7940240.
PA-32R-300:
32R-7680409, 32R-7680410, 32R-7780520, 32R-7880057, 32R-7880058, 32R-7880067, and 32R-7880068.
PA-32RT-300:
32R-7885027, 32R-7885099, 32R-7885100, 32R-7885176, 32R-7885177, 32R-7885213 through 32R-7885215, 32R-7885234 through 32R-7885237, 32R-7885259, 32R-7885260, 32R-7885285, and 32R-7985027.
PA-32RT-300T:
32R-7887036, 32R-7887081, 32R-7887222, 32R-7987050, 32R-7987085, and 32R-7987122.
PA-32R-301T:
32R-8029121, 32R-8129041, 32R-8229065, and 32R-8329017.
PA-32-301:
32-8006090, 32-8106043, and 3206005, 3206020 through 3206041, 3206045, 3206046, 3206048, 3206049, 3206056 through 3206059, 3206061 through 3206088.
PA-32-301T:
32-8024031, 32-8024032, 32-8124011, 32-8124017, 32-8124018, 32-8124035, 32-8124036, 32-8224011, 32-8224013, 32-8224014, 32-8324006, 32-8324015, and 32-8324016.

NOTE 8 The fixed pitch propeller may be used on S/N 32-1 through 32-1297, and 32-7100001 through 32-7200045.

NOTE 9 The following serial numbered aircraft are not eligible for import certification to the U.S.:

AR32-7440144, AR32-7340133, AR32-7340155, AR32-7340159, AR32-7340160, AR32-7340172.

NOTE 10 Engines with serial numbers ending with "A" require the F-4-11( ) propeller governor assembly. Other engines require the F-4-4( ) propeller governor.

NOTE 11 In the following serial numbered aircraft the rear seat location is farther aft as shown and the center seats may be removed and replaced by CLUB SEATS INSTALLATION, which has a more aft C.G. location as shown:

PA-32-260	S/N 32-7700001 through 32-7800008
PA-32-300	S/N 32-7740001 through 32-7940290
PA-32R-300	S/N 32R-7680001 through 32R-7880068
PA-32RT-300	S/N 32R-7885002 through 32R-7985106
PA-32RT-300T	S/N 32R-7787001, 32R-7887002 through 32R-7987126
PA-32R-301	S/N 32R-8013001 through 32R-8613006, 3213001 through 3213103, and 3246001 and up
PA-32R-301T	S/N 32R-8029001 through 32R-8629008, and 3229001 through 3229003
PA-32-301	S/N 32-8006002 through 32-8606023, and 3206001 through 3206019
PA-32-301T	S/N 32-8024001 through 32-8424002

NOTE 12 Lycoming engine Model IO-540-K1G5 with Hartzell propeller HC-C2YK-1(F), Blade Model 8475D-4, S/N 32-7640066 (only) and S/N 32-7640072 through 32-7940290.

NOTE 13 Lycoming engine Model IO-540-K1G5D with Hartzell propeller HC-C2YK-1(F), Blade Model 8475D-4, S/N 32R-7680141 through 32R-7880068.

NOTE 14 On Models PA-32-301, S/N 32-8006001 through 32-8606023 and 3206001 through 3206019, and PA-32-301T, S/N 32-8024001 through 32-8424002, the wheel fairings alone or the wheel fairings and landing gear strut fairings may be removed.

NOTE 15 On models PA-32-301FT, S/N 3232001 and up, and PA-32-301XTC, S/N 3255001 and up, the nose wheel centering springs must be installed when operating the aircraft with or without wheel pants.

...END...

Appendix
Answer Key

In addition to answers and Learning Statement Codes (LSC), this answer key also shows the **specific reference material** to study for each sample IA Knowledge Exam question in this *IA Test Prep*. Further study of these references provides more detailed explanations and the context in which answer alternatives were chosen.

Question	Answer	LSC	Reference

Chapter 3

Question	Answer	LSC	Reference
1	[A]	(IAR022)	14 CFR Part 1
2	[B]	(IAR022)	14 CFR Part 1
3	[C]	(IAR022)	14 CFR Part 1
3a	[C]	(IAR030)	14 CFR Part 1
4	[B]	(IAR030)	14 CFR Part 1
5	[C]	(IAR031)	14 CFR Part 1
6	[B]	(IAR031)	14 CFR Part 1
6a	[C]	(IAR031)	14 CFR Part 1
6b	[A]	(IAR031)	14 CFR Part 1
6c	[C]	(IAR031)	14 CFR Part 1
7	[B]	(IAR030)	14 CFR Part 21
8	[C]	(IAR030)	14 CFR Part 21
9	[B]	(IAR022)	14 CFR Part 21
10	[A]	(IAR030)	14 CFR Part 21
11	[B]	(IAR030)	14 CFR Part 21
12	[A]	(IAR030)	14 CFR Part 21
13	[A]	(IAR016)	14 CFR Part 21
14	[C]	(IAR016)	14 CFR Part 21
15	[B]	(IAR016)	14 CFR Part 21
16	[B]	(IAR030)	14 CFR Part 23
17	[A]	(IAR008)	14 CFR §23.2100(a)
18	[B]	(IAR008)	14 CFR §23.2100(c)
19	[C]	(IAR008)	14 CFR §23.2140(a)
20	[B]	(IAR030)	14 CFR Part 23
21	[B]	(IAR030)	14 CFR Part 23
22	[A]	(IAR030)	14 CFR Part 23
23	[A]	(IAR013)	14 CFR Part 23
24	[B]	(IAR013)	14 CFR Part 23
25	[C]	(IAR017)	14 CFR Part 23
26	[B]	(IAR017)	14 CFR Part 23
27	[C]	(IAR024)	14 CFR Part 23
28	[C]	(IAR017)	14 CFR Part 23
28a	[B]	(IAR017)	14 CFR Part 23
28b	[B]	(IAR017)	14 CFR Part 23
28c	[C]	(IAR031)	14 CFR Part 23
28d	[C]	(IAR031)	14 CFR Part 23
29	[C]	(IAR024)	14 CFR Part 23
30	[A]	(IAR031)	14 CFR Part 27

Question	Answer	LSC	Reference
Chapter 3 *(continued)*			
30a	[A]	(IAR017)	14 CFR Part 27
31	[C]	(IAR031)	14 CFR Part 27
32	[A]	(IAR031)	14 CFR Part 27
33	[B]	(IAR031)	14 CFR Part 27
34	[B]	(IAR030)	14 CFR Part 27
35	[A]	(IAR031)	14 CFR Part 27
36	[C]	(IAR017)	14 CFR Part 43
37	[B]	(IAR017)	14 CFR Part 43
38	[C]	(IAR017)	14 CFR Part 43
39	[A]	(IAR017)	14 CFR Part 43
40	[C]	(IAR021)	14 CFR Part 43
41	[B]	(IAR021)	14 CFR Part 43
42	[C]	(IAR021)	14 CFR Part 43
43	[A]	(IAR031)	14 CFR Part 43
44	[C]	(IAR031)	14 CFR Part 43
45	[A]	(IAR031)	14 CFR Part 43
46	[A]	(IAR031)	14 CFR Part 43
47	[B]	(IAR021)	14 CFR Part 43
48	[C]	(IAR021)	14 CFR Part 43
49	[A]	(IAR017)	14 CFR Part 43
50	[A]	(IAR017)	14 CFR Part 43
51	[C]	(IAR031)	14 CFR Part 43
52	[C]	(IAR021)	14 CFR Part 91
53	[A]	(IAR021)	14 CFR Part 43 and 91
54	[C]	(IAR021)	14 CFR Part 43 and 91
55	[C]	(IAR017)	14 CFR Part 91
56	[B]	(IAR031)	14 CFR Part 91
57	[C]	(IAR021)	14 CFR Part 91
58	[A]	(IAR017)	14 CFR Part 91
59	[A]	(IAR021)	14 CFR Part 91
60	[B]	(IAR010)	14 CFR Part 43
61	[A]	(IAR010)	14 CFR Part 43
62	[B]	(IAR010)	14 CFR Part 43
63	[C]	(IAR017)	14 CFR Part 43
64	[C]	(IAR017)	14 CFR Part 43
64a	[C]	(IAR021)	14 CFR Part 43
64b	[C]	(IAR021)	14 CFR Part 43
64c	[C]	(IAR032)	14 CFR Part 43, Appendix A
65	[A]	(IAR019)	14 CFR Part 43
66	[B]	(IAR019)	14 CFR Part 43
67	[A]	(IAR019)	14 CFR Part 43
68	[A]	(IAR031)	14 CFR Part 43
69	[B]	(IAR031)	14 CFR Part 43
70	[A]	(IAR031)	14 CFR Part 43
71	[C]	(IAR031)	14 CFR Part 43
72	[B]	(IAR031)	14 CFR Part 43
73	[C]	(IAR017)	14 CFR Part 43
74	[B]	(IAR019)	14 CFR Part 43

Question	Answer	LSC	Reference

Chapter 3 *(continued)*

Question	Answer	LSC	Reference
75	[A]	(IAR017)	14 CFR Part 43
76	[A]	(IAR019)	14 CFR Part 43
77	[B]	(IAR017)	14 CFR Part 43
78	[A]	(IAR019)	14 CFR Part 43
79	[A]	(IAR031)	14 CFR Part 43
80	[C]	(IAR031)	14 CFR Part 43
81	[A]	(IAR031)	14 CFR Part 43
82	[C]	(IAR031)	14 CFR Part 43
82a	[C]	(IAR031)	14 CFR Part 43
82b	[B]	(IAR031)	14 CFR Part 43
82c	[B]	(IAR031)	14 CFR Part 43
82d	[C]	(IAR031)	14 CFR Part 43
83	[C]	(IAR017)	14 CFR Part 45
83a	[B]	(IAR017)	14 CFR Part 43
83b	[A]	(IAR017)	14 CFR Part 43
84	[C]	(IAR017)	14 CFR Part 45
85	[B]	(IAR017)	14 CFR Part 45
86	[C]	(IAR017)	14 CFR Part 45
87	[C]	(IAR017)	14 CFR Part 45
88	[A]	(IAR017)	14 CFR Part 45
89	[C]	(IAR017)	14 CFR Part 45
90	[A]	(IAR017)	14 CFR Part 45
91	[B]	(IAR017)	14 CFR Part 45
91a	[A]	(IAR017)	14 CFR Part 45
91b	[B]	(IAR017)	14 CFR Part 45
91c	[A]	(IAR017)	14 CFR Part 45
91d	[C]	(IAR017)	14 CFR Part 45
91e	[B]	(IAR017)	14 CFR Part 45
92	[A]	(IAR031)	14 CFR Part 65
93	[C]	(IAR031)	14 CFR Part 65
94	[A]	(IAR031)	14 CFR Part 65
95	[A]	(IAR031)	14 CFR Part 65
96	[A]	(IAR031)	14 CFR Part 65
97	[A]	(IAR031)	14 CFR Part 65
98	[B]	(IAR031)	14 CFR Part 65
99	[A]	(IAR031)	14 CFR Part 65
100	[C]	(IAR031)	14 CFR Part 65
101	[B]	(IAR031)	14 CFR Part 65
102	[B]	(IAR031)	14 CFR Part 65
103	[C]	(IAR031)	14 CFR Part 65
104	[A]	(IAR031)	14 CFR Part 65
105	[C]	(IAR031)	14 CFR Part 65
106	[B]	(IAR031)	14 CFR Part 65
107	[A]	(IAR031)	14 CFR Part 65
108	[B]	(IAR031)	14 CFR Part 65
109	[C]	(IAR031)	14 CFR Part 65
110	[B]	(IAR031)	14 CFR Part 65
111	[C]	(IAR031)	14 CFR Part 65

Question	Answer	LSC	Reference

Chapter 3 *(continued)*

Question	Answer	LSC	Reference
112	[C]	(IAR031)	14 CFR Part 65
113	[A]	(IAR031)	14 CFR Part 65
114	[C]	(IAR031)	14 CFR Part 65
115	[C]	(IAR031)	14 CFR Part 65
116	[C]	(IAR031)	14 CFR Part 65
117	[B]	(IAR031)	14 CFR Part 65
118	[B]	(IAR030)	14 CFR Part 65
119	[A]	(IAR031)	14 CFR Part 91
120	[C]	(IAR031)	14 CFR Part 91
121	[A]	(IAR031)	14 CFR Part 91
122	[B]	(IAR017)	14 CFR Part 91
123	[C]	(IAR017)	14 CFR Part 91
123a	[B]	(IAR021)	14 CFR Part 43 and 91
124	[B]	(IAR031)	14 CFR Part 91
125	[C]	(IAR017)	14 CFR Part 91
126	[C]	(IAR017)	14 CFR Part 91
127	[C]	(IAR017)	14 CFR Part 91
128	[C]	(IAR017)	14 CFR Part 91
129	[A]	(IAR017)	14 CFR Part 91
130	[B]	(IAR017)	14 CFR Part 91
131	[C]	(IAR017)	14 CFR Part 91
132	[C]	(IAR031)	14 CFR Part 91 Part 91
132a	[C]	(IAR031)	14 CFR Part 91
133	[C]	(IAR031)	14 CFR Part 91
134	[B]	(IAR031)	14 CFR Part 91 and 135
134a	[A]	(IAR031)	14 CFR Parts 91, 125, and 135
135	[A]	(IAR031)	14 CFR Part 91
136	[C]	(IAR031)	14 CFR Part 91
137	[B]	(IAR031)	14 CFR Part 91
138	[C]	(IAR031)	14 CFR Part 91
139	[C]	(IAR031)	14 CFR Part 91
140	[C]	(IAR031)	14 CFR Part 91
141	[A]	(IAR031)	14 CFR Part 91
142	[C]	(IAR031)	14 CFR Part 125
143	[A]	(IAR031)	14 CFR Part 135
144	[A]	(IAR017)	14 CFR Part 135
145	[B]	(IAR017)	14 CFR Part 135
146	[A]	(IAR017)	14 CFR Part 135
147	[B]	(IAR017)	14 CFR Part 135
148	[C]	(IAR017)	14 CFR Part 135
149	[C]	(IAR017)	14 CFR Part 135
150	[A]	(IAR017)	14 CFR Part 135
151	[B]	(IAR017)	14 CFR Part 135
152	[B]	(IAR031)	14 CFR Part 183
152a	[C]	(IAR031)	AC 43.9-1
152b	[A]	(IAR031)	14 CFR §183.29

Question	Answer	LSC	Reference

Chapter 4

Question	Answer	LSC	Reference
153	[C]	(IAR031)	14 CFR §39.27
153a	[B]	(IAR031)	14 CFR §39.27
154	[C]	(IAR020)	AD 80-10-02
155	[B]	(IAR020)	AD 80-15-12
155a	[C]	(IAR031)	AC 39-7C
156	[C]	(IAR020)	AD 81-23-01 R1
157	[C]	(IAR020)	AD 82-06-12
158	[B]	(IAR020)	AD 82-11-05
159	[A]	(IAR020)	AD 82-11-05
160	[A]	(IAR020)	AD 90-01-06
161	[B]	(IAR020)	AD 90-08-14
162	[C]	(IAR020)	AD 90-08-14
163	[B]	(IAR020)	AD 93-24-03
163a	[C]	(IAR020)	AD 93-24-03
164	[B]	(IAR020)	AD 95-13-08
165	[C]	(IAR020)	AD 95-13-08

Chapter 5

Question	Answer	LSC	Reference
166	[C]	(IAR030)	FAA Order 8130.21
167	[A]	(IAR030)	FAA Order 8130.21
168	[C]	(IAR030)	FAA Order 8130.21
169	[C]	(IAR030)	FAA Order 8130.21
170	[B]	(IAR030)	FAA Order 8130.21
171	[C]	(IAR030)	FAA Order 8130.21
171a	[C]	(IAR030)	FAA Order 8130.21
172	[C]	(IAR030)	FAA Order 8130.21
173	[B]	(IAR030)	FAA Order 8130.21
174	[B]	(IAR030)	FAA Order 8130.21
175	[B]	(IAR030)	FAA Order 8130.21
175a	[C]	(IAR030)	FAA Order 8130.21
175b	[C]	(IAR030)	FAA Order 8130.21

Chapter 6

Question	Answer	LSC	Reference
176	[A]	(IAR031)	AC 39-7
177	[C]	(IAR031)	AC 39-7
178	[A]	(IAR031)	AC 39-7
179	[C]	(IAR031)	AC 39-7
180	[B]	(IAR031)	AC 39-7
181	[B]	(IAR031)	AC 39-7
181a	[B]	(IAR026)	AC 43-4
182	[C]	(IAR030)	AC 43-9
183	[B]	(IAR030)	AC 43-9
184	[B]	(IAR030)	AC 43-9
185	[B]	(IAR030)	AC 43.9-1
186	[A]	(IAR030)	AC 43.9-1
187	[C]	(IAR030)	AC 43.9-1
188	[A]	(IAR030)	AC 43.9-1

Question	Answer	LSC	Reference

Chapter 6 *(continued)*

Question	Answer	LSC	Reference
189	[B]	(IAR021)	AC 43.9-1
190	[A]	(IAR030)	AC 43.9-1
191	[C]	(IAR030)	AC 43.9-1
192	[B]	(IAR030)	AC 43.9-1
193	[A]	(IAR030)	AC 43.9-1
194	[B]	(IAR030)	AC 43.9-1
195	[C]	(IAR013)	AC 43.13-1
196	[B]	(IAR013)	AC 43.13-1
197	[B]	(IAR013)	AC 43.13-1
198	[B]	(IAR006)	AC 43.13-1
199	[B]	(IAR006)	AC 43.13-1
200	[B]	(IAR006)	AC 43.13-1
201	[C]	(IAR006)	AC 43.13-1
202	[C]	(IAR006)	AC 43.13-1
203	[C]	(IAR006)	AC 43.13-1
203a	[C]	(IAR014)	AC 43.13-1
204	[B]	(IAR032)	AC 43.13-1
205	[B]	(IAR006)	AC 43.13-1
206	[C]	(IAR006)	AC 43.13-1
207	[B]	(IAR032)	AC 43.13-1
208	[B]	(IAR032)	AC 43.13-1
208a	[C]	(IAR032)	AC 43.13-1
208b	[B]	(IAR032)	AC 43.13-1
208c	[C]	(IAR032)	AC 43.13-1
208d	[B]	(IAR032)	AC 43.13-1
208e	[B]	(IAR032)	AC 43.13-1
208f	[B]	(IAR032)	AC 43.13-1
208g	[B]	(IAR032)	AC 43.13-1
208h	[C]	(IAR032)	AC 43.13-1
209	[A]	(IAR032)	AC 43.13-1
210	[B]	(IAR006)	AC 43.13-1
211	[B]	(IAR032)	AC 43.13-1
212	[B]	(IAR023)	AC 43.13-1
213	[B]	(IAR018)	AC 43.13-1
214	[B]	(IAR018)	AC 43.13-1
215	[A]	(IAR027)	AC 43.13-1
215a	[B]	(IAR027)	ASA-AMT-STRUCT
215b	[A]	(IAR027)	ASA-AMT-STRUCT
215c	[C]	(IAR027)	AC 43.13-1
215d	[B]	(IAR027)	AC 43.13-1
215e	[B]	(IAR027)	AC 43.13-1
215f	[A]	(IAR006)	AC 43.13-1
216	[C]	(IAR015)	AC 43.13-1
217	[A]	(IAR015)	AC 43.13-1
218	[C]	(IAR006)	AC 43.13-1
219	[B]	(IAR006)	AC 43.13-1
220	[A]	(IAR027)	AC 43.13-1
221	[A]	(IAR027)	AC 43.13-1

Question	Answer	LSC	Reference

Chapter 6 *(continued)*

Question	Answer	LSC	Reference
222	[A]	(IAR024)	AC 43.13-1
222a	[C]	(IAR018)	AC 43.13-1
222b	[C]	(IAR018)	AC 43.13-1
223	[A]	(IAR024)	AC 43.13-1
223a	[A]	(IAR032)	ASA-AMT-STRUC
223b	[A]	(IAR024)	AC 43.13-1
223c	[C]	(IAR024)	AC 43.13-1
223d	[A]	(IAR024)	ASA-AMT-G
223e	[B]	(IAR024)	ASA-AMT-STRUCT
223f	[B]	(IAR024)	AC 43.13-1
223g	[A]	(IAR024)	AC 43.13-1
223h	[A]	(IAR024)	ASA-AMT-G
223i	[A]	(IAR032)	ASA-DAT
223j	[C]	(IAR024)	AC 43.13-1
223k	[A]	(IAR024)	AC 43.13-1
223l	[A]	(IAR024)	ASA-DAT
224	[B]	(IAR024)	AC 43.13-1
224a	[A]	(IAR024)	AC 43.13-1
224b	[A]	(IAR026)	AC 43.13-1
224c	[C]	(IAR026)	FAA-H-8083-31
224d	[B]	(IAR026)	AC 43.13-1
224e	[A]	(IAR026)	AC 43.13-1
224f	[A]	(IAR026)	AC 43.13-1
225	[B]	(IAR024)	AC 43.13-1
226	[B]	(IAR026)	AC 43.13-1
226a	[A]	(IAR026)	ASA-AMT-STRUCT
226b	[C]	(IAR026)	ASA-AMT-G
227	[B]	(IAR026)	AC 43.13-1
227a	[B]	(IAR026)	AC 43.13-1
227b	[C]	(IAR026)	AC 43.13-1
228	[A]	(IAR026)	AC 43.13-1
228a	[B]	(IAR026)	AC 43.13-1
228b	[C]	(IAR026)	AC 43.13-1
228c	[A]	(IAR026)	ASA-AMT-G
229	[A]	(IAR026)	AC 43.13-1
230	[B]	(IAR026)	AC 43.13-1
231	[B]	(IAR026)	AC 43.13-1
232	[B]	(IAR019)	AC 43.13-1
233	[A]	(IAR013)	AC 43.13-1
233a	[C]	(IAR005)	FAA-H-8083-30
234	[C]	(IAR024)	AC 43.13-1
235	[B]	(IAR024)	AC 43.13-1
236	[A]	(IAR024)	AC 43.13-1
237	[B]	(IAR015)	AC 43.13-1
238	[C]	(IAR015)	AC 43.13-1
239	[C]	(IAR014)	AC 43.13-1
240	[C]	(IAR014)	AC 43.13-1
241	[B]	(IAR015)	AC 43.13-1

Question	Answer	LSC	Reference

Chapter 6 *(continued)*

Question	Answer	LSC	Reference
242	[B]	(IAR024)	AC 43.13-1
243	[C]	(IAR005)	AC 43.13-1
244	[B]	(IAR005)	AC 43.13-1
244a	[A]	(IAR013)	AC 43.13-1
245	[C]	(IAR029)	AC 43.13-1
245a	[B]	(IAR029)	AC 43.13-1
246	[A]	(IAR018)	AC 43.13-1
247	[B]	(IAR018)	AC 43.13-1
248	[B]	(IAR024)	AC 43.13-1
249	[B]	(IAR013)	AC 43.13-1
250	[C]	(IAR028)	AC 43.13-1
251	[A]	(IAR028)	AC 43.13-1
252	[B]	(IAR028)	AC 43.13-1
253	[A]	(IAR029)	AC 43.13-1
253a	[A]	(IAR029)	AC 43.13-1
254	[B]	(IAR029)	AC 43.13-1
255	[B]	(IAR029)	AC 43.13-1
255a	[B]	(IAR008)	FAA-H-8083-1
256	[B]	(IAR029)	AC 43.13-1
257	[C]	(IAR029)	AC 43.13-1
258	[B]	(IAR008)	AC 43.13-1
259	[C]	(IAR008)	AC 43.13-1
260	[A]	(IAR003)	AC 43.13-1
260a	[B]	(IAR003)	AC 43.13-1
261	[B]	(IAR003)	AC 43.13-1
262	[C]	(IAR003)	AC 43.13-1
263	[A]	(IAR003)	AC 43.13-1
264	[A]	(IAR003)	AC 43.13-1
265	[A]	(IAR003)	AC 43.13-1
266	[A]	(IAR003)	AC 43.13-1
267	[B]	(IAR003)	AC 43.13-1
268	[A]	(IAR003)	AC 43.13-1
269	[C]	(IAR008)	AC 43.13-1
269a	[B]	(IAR001)	AC 43.13-1
269b	[B]	(IAR003)	AC 43.13-1
270	[B]	(IAR008)	AC 43.13-1
271	[A]	(IAR008)	AC 43.13-1
272	[B]	(IAR029)	AC 43.13-1
273	[A]	(IAR008)	AC 43.13-1
274	[C]	(IAR008)	AC 43.13-1
275	[A]	(IAR008)	AC 43.13-1
276	[B]	(IAR008)	AC 43.13-1
276a	[B]	(IAR008)	AC 43.13-1
277	[A]	(IAR022)	ASA-AMT-STRUC
278	[B]	(IAR032)	ASA-MHB
278a	[C]	(IAR032)	ASA-MHB
278b	[C]	(IAR032)	ASA-MHB
278c	[C]	(IAR032)	ASA-MHB

Question	Answer	LSC	Reference

Chapter 6 *(continued)*

Question	Answer	LSC	Reference
278d	[B]	(IAR005)	ASA-MHB
278e	[A]	(IAR005)	ASA-MHB
278f	[C]	(IAR005)	AC 43.13-1
278g	[B]	(IAR005)	AC 43.13-1
278h	[B]	(IAR007)	ASA-MHB
278i	[A]	(IAR007)	ASA-MHB
278j	[A]	(IAR005)	AC 43.13-1, ASA-AMT-G
279	[A]	(IAR011)	AC 43.13-2
280	[B]	(IAR011)	AC 43.13-2
281	[C]	(IAR011)	AC 43.13-2
282	[A]	(IAR011)	AC 43.13-2
283	[B]	(IAR011)	AC 43.13-2
284	[B]	(IAR011)	AC 43.13-2
285	[A]	(IAR011)	AC 43.13-2
286	[A]	(IAR011)	AC 43.13-2
287	[C]	(IAR011)	AC 43.13-2
288	[B]	(IAR011)	AC 43.13-2
289	[C]	(IAR011)	AC 43.13-2
289a	[A]	(IAR009)	AC 43.13-2
290	[C]	(IAR011)	AC 43.13-2
291	[C]	(IAR029)	FAA-H-8083-1
291a	[B]	(IAR029)	FAA-H-8083-1
291b	[C]	(IAR029)	ASA-DAT
291c	[A]	(IAR029)	FAA-H-8083-1
292	[B]	(IAR029)	FAA-H-8083-1
292a	[C]	(IAR002)	FAA-H-8083-1
293	[A]	(IAR008)	FAA-H-8083-1
293a	[B]	(IAR008)	FAA-H-8083-1
294	[A]	(IAR008)	FAA-H-8083-1
295	[B]	(IAR008)	FAA-H-8083-1
295a	[C]	(IAR008)	FAA-H-8083-1
296	[A]	(IAR029)	FAA-H-8083-1
297	[C]	(IAR008)	FAA-H-8083-1
298	[C]	(IAR029)	FAA-H-8083-1
299	[C]	(IAR011)	FAA-G-8082-11
300	[B]	(IAR011)	FAA-G-8082-11
301	[B]	(IAR011)	FAA-G-8082-11
302	[B]	(IAR010)	FAA-G-8082-11
303	[B]	(IAR011)	FAA-G-8082-11
304	[C]	(IAR010)	FAA-G-8082-11
305	[C]	(IAR011)	FAA-G-8082-11
306	[B]	(IAR011)	FAA-G-8082-11
307	[B]	(IAR010)	FAA-G-8082-11
308	[B]	(IAR017)	FAA-G-8082-11
309	[C]	(IAR025)	AC 91-67
310	[A]	(IAR025)	AC 91-67
311	[B]	(IAR025)	AC 91-67
312	[B]	(IAR025)	AC 91-67

Question	Answer	LSC	Reference

Chapter 6 *(continued)*

Question	Answer	LSC	Reference
313	[B]	(IAR025)	AC 91-67 and 14 CFR Part 91
314	[C]	(IAR025)	AC 91-67 and 14 CFR Part 91
315	[B]	(IAR025)	AC 91-67 and 14 CFR Part 91
316	[B]	(IAR025)	AC 91-67 and 14 CFR Part 91
317	[C]	(IAR025)	AC 91-67
318	[C]	(IAR025)	AC 91-67
319	[A]	(IAR025)	AC 91-67

Chapter 7

Question	Answer	LSC	Reference
320	[B]	(IAR031)	TCDS
321	[C]	(IAR011)	TCDS
321a	[C]	(IAR011)	TCDS
322	[C]	(IAR031)	TCDS
323	[C]	(IAR017)	TCDS
324	[A]	(IAR017)	TCDS
324a	[B]	(IAR031)	14 CFR Part 27
325	[C]	(IAR011)	TCDS
326	[B]	(IAR011)	TCDS
327	[C]	(IAR011)	TCDS
327a	[C]	(IAR011)	TCDS
328	[B]	(IAR011)	TCDS
329	[B]	(IAR031)	TCDS
330	[C]	(IAR011)	TCDS
331	[B]	(IAR011)	TCDS
332	[A]	(IAR011)	TCDS
333	[A]	(IAR011)	TCDS
334	[B]	(IAR017)	TCDS
334a	[B]	(IAR013)	TCDS
335	[A]	(IAR017)	TCDS
335a	[A]	(IAR017)	TCDS
336	[C]	(IAR017)	TCDS
337	[C]	(IAR013)	TCDS
337a	[B]	(IAR017)	TCDS
337b	[A]	(IAR013)	TCDS
338	[C]	(IAR013)	TCDS
338a	[B]	(IAR013)	TCDS
338b	[B]	(IAR013)	TCDS
339	[C]	(IAR013)	TCDS
340	[C]	(IAR013)	TCDS
341	[B]	(IAR013)	TCDS
342	[C]	(IAR013)	TCDS
343	[C]	(IAR013)	TCDS
344	[A]	(IAR011)	TCDS
345	[C]	(IAR011)	TCDS
346	[B]	(IAR011)	TCDS
347	[C]	(IAR011)	TCDS
348	[C]	(IAR012)	TCDS
349	[B]	(IAR012)	TCDS

Question	Answer	LSC	Reference

Chapter 7 *(continued)*

Question	Answer	LSC	Reference
350	[B]	(IAR013)	TCDS
351	[C]	(IAR020)	TCDS
352	[C]	(IAR020)	TCDS
353	[C]	(IAR020)	TCDS
354	[B]	(IAR020)	TCDS
355	[B]	(IAR020)	TCDS
356	[B]	(IAR011)	TCDS
357	[A]	(IAR013)	TCDS
358	[A]	(IAR013)	TCDS
359	[B]	(IAR013)	TCDS
360	[A]	(IAR013)	TCDS
361	[C]	(IAR020)	TCDS
362	[B]	(IAR020)	TCDS
363	[B]	(IAR020)	TCDS
364	[A]	(IAR020)	TCDS
365	[C]	(IAR020)	TCDS
366	[A]	(IAR020)	TCDS
367	[A]	(IAR020)	TCDS
368	[A]	(IAR020)	TCDS
369	[B]	(IAR020)	TCDS
370	[C]	(IAR020)	TCDS
371	[A]	(IAR020)	TCDS
372	[C]	(IAR020)	TCDS
373	[C]	(IAR020)	TCDS
373a	[B]	(IAR020)	TCDS
374	[C]	(IAR020)	TCDS
375	[C]	(IAR020)	TCDS
375a	[C]	(IAR020)	TCDS
375b	[C]	(IAR020)	TCDS
375c	[C]	(IAR020)	TCDS
375d	[C]	(IAR020)	TCDS
375e	[A]	(IAR020)	TCDS
375f	[C]	(IAR020)	TCDS
375g	[C]	(IAR020)	TCDS
375h	[B]	(IAR020)	TCDS
375i	[A]	(IAR020)	TCDS
375j	[A]	(IAR020)	TCDS
375k	[B]	(IAR020)	TCDS
375l	[B]	(IAR020)	TCDS
375m	[C]	(IAR020)	TCDS
375n	[B]	(IAR020)	TCDS
375o	[C]	(IAR029)	TCDS
375p	[B]	(IAR020)	TCDS

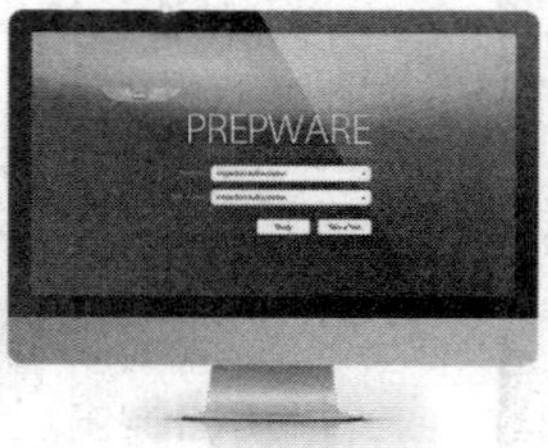

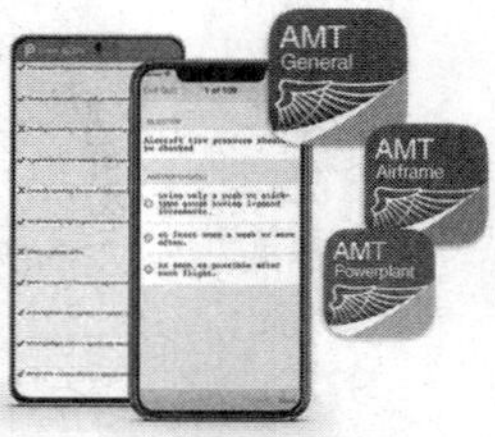